Take a break

W9-AMS-447

EXPOSED
Secret Navy Project Reveals Hidden Landscapes

HURRICANE FORECAST
Deadly Upswing Coming

Photo Contest Winners

Hunting Fossils on The Roof of the World

PLUS: Mysterious Fires in the Sky

Get your head out of the textbook and see our planet from a whole new view with *Earth* Magazine. And to show you how good life is outside of the lecture hall, we're giving you a special student rate of only $14.95. That's some serious savings off the newsstand price—over 36%!

Earth takes you to the fields and labs of real life scientists. . . recreate a horned dinosaur piece-by-piece with some of the top paleontologists. . . take a "virtual reality" trip inside a tornado. . .learn the makings of a tsunami killer wave, and what scientists are trying to do create early wave warnings.

And *Earth* will take you to remarkable places you've never been before. . . imagine traveling with scientists on dives to the bottom of the Pacific to explore underwater volcanoes. . .to glaciers atop the Andes to unravel the secrets of global warming. . . to the steamy, hot jungles of the Amazon to learn how earthquakes and rising mountains may have created an awesome number of species. . . to the beginning of life and the latest news on the dawning of human intelligence.

Order today

Photo courtesy of University of Wyoming, Geological Museum

Special Student Rate

Please enter my one-year subscription to *Earth* for only $14.95.

- ❏ Payment enclosed
- ❏ Bill me

THE REAL JURASSIC PARK LAB   BONUS FOSSIL POSTER

Earth
The Science on Our Planet

Becoming Human
A photo gallery of our ancestors

Sex and the Carbon Cycle

Tracking Stealth Quakes

SPECIAL REPORT
Mysterious Mountain-Making

visit the **Natural Hazards** web site sponsored by **Earth.**

see back for details.

Name _____

Address _____

City _____

State _____ Zip _____ Country _____

Outside the U.S.: $26.00 (GST included). Cover price $3.95. Payable in U.S. funds. Make checks payable to Kalmbach Publishing Co.

V8001E

Earth

Special Student Rate

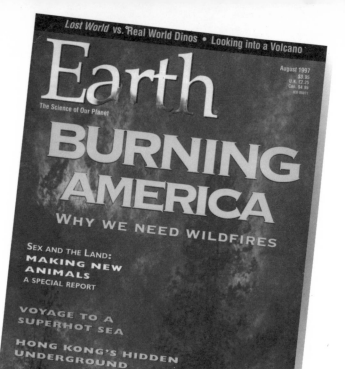

Lost World vs. Real World Dinos • Looking into a Volcano

Earth
The Science of Our Planet

August 1997
$3.95
U.K. £2.25
Can. $4.95

BURNING AMERICA
WHY WE NEED WILDFIRES

SEX AND THE LAND:
MAKING NEW ANIMALS
A SPECIAL REPORT

VOYAGE TO A SUPERHOT SEA

HONG KONG'S HIDDEN UNDERGROUND

DON'T PROCRASTINATE!

Mail back this card today and save 36%!

Natural Hazards Web Site

This exclusive web site for students is a valuable resource that will enrich your understanding and fulfill your desire to explore volcanoes, hurricanes, tornadoes and other natural phenomena.

Three easy steps to access the web site

1) Go to
   www.wiley.com/college/
   geography/hazards

2) Enter identification

3002750–62768–5546–6

3) Create User name_____
   Password_____

ORDER TODAY
AND SAVE 36%!

BUSINESS REPLY MAIL
FIRST-CLASS MAIL   PERMIT NO. 16   WAUKESHA, WI

POSTAGE WILL BE PAID BY ADDRESSEE

Earth ®

PO BOX 1612
WAUKESHA WI 53187-9950

NO POSTAGE
NECESSARY
IF MAILED
IN THE
UNITED STATES

PHYSICAL
GEOGRAPHY
OF THE GLOBAL
ENVIRONMENT

BOOK STORE
Top CASH For BOOKS Anytime

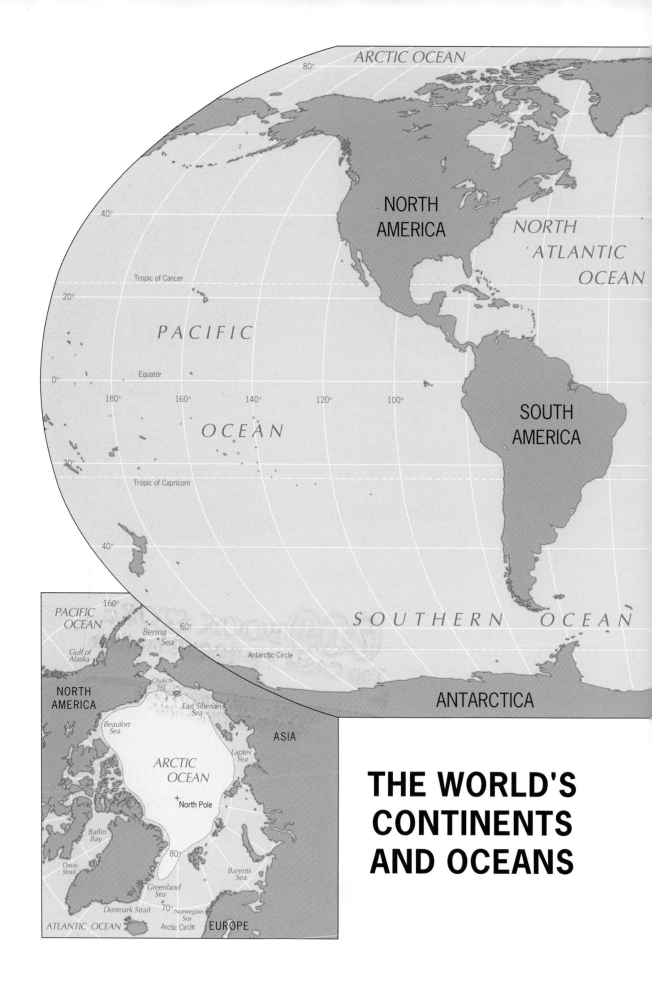

THE WORLD'S CONTINENTS AND OCEANS

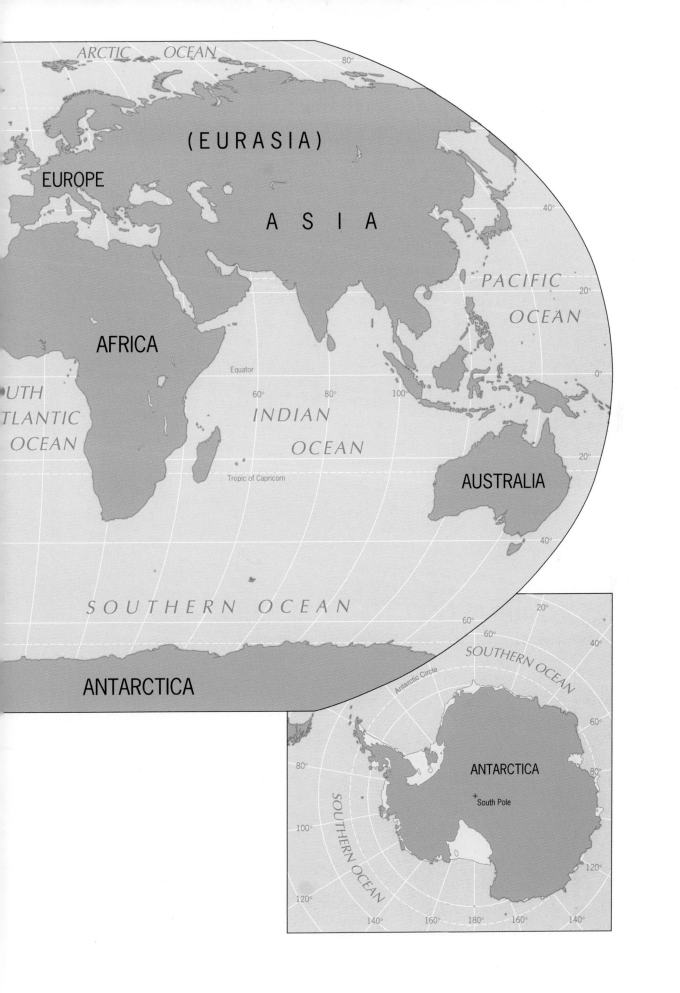

SECOND EDITION

HAZARDS UPDATE

PHYSICAL GEOGRAPHY
OF THE GLOBAL
ENVIRONMENT

H. J. de Blij
University of South Florida
St. Petersburg

Peter O. Muller
University of Miami

JOHN WILEY & SONS, INC.
New York   Chichester   Weinheim
Brisbane   Toronto   Singapore

ACQUISITIONS EDITOR   Frank Lyman
DEVELOPMENTAL EDITOR   Kathleen Dolan
SENIOR MARKETING MANAGER   Catherine Faduska
SENIOR PRODUCTION EDITOR   Jeanine Furino
SENIOR PRODUCTION MANAGER   Linda Muriello
PHOTO ASSISTANT   Kim Khatchatourian
PHOTO EDITOR   Alexandra Truitt
SENIOR DESIGNER   Dawn L. Stanley
SENIOR ILLUSTRATION COORDINATOR   Edward Starr
ASSISTANT MANUFACTURING MANAGER   Mark Cirillo
ILLUSTRATIONS   Precision Graphics/Network
                Graphics
MAPS   Mapping Specialists/Maryland
        CartoGraphics
COVER PHOTO   © Steven C. Kaufman/DRK Photo

This book was set in 10/12 ITC Garamond by University Graphics, Inc. and printed and bound by Von-Hoffmann Press. The cover was printed by Phoenix Color Inc.

Recognizing the importance of preserving what has been written, it is a policy of John Wiley & Sons, Inc. to have books of enduring value published in the United States printed on acid-free paper, and we exert our best efforts to that end.

The paper in this book was manufactured by a mill whose forest management programs include sustained yield harvesting of its timberlands. Sustained yield harvesting principles ensure that the number of trees cut each year does not exceed the amount of new growth.

Copyright © 1996, by John Wiley & Sons, Inc.

All rights reserved. Published simultaneously in Canada.

Reproduction or translation of any part of this work beyond that permitted by Sections 107 and 108 of the 1976 United States Copyright Act without the permission of the copyright owner is unlawful. Requests for permission or further information should be addressed to the Permissions Department, John Wiley & Sons, Inc.

Library of Congress Cataloging in Publication Data:
de Blij, H. J.
    Physical geography of the global environment / H. J. de Blij,
Peter O. Muller.—2nd ed.
      p.   cm.
    Includes bibliographical references.
    ISBN 0-471-03917-9 (paper : alk. paper)
    1. Physical geography.   I. Muller, Peter O.   II. Title.
  GB55.D34   1996   95-313
  910′.02—dc20   CIP

Printed in the United States of America

10 9 8 7 6 5 4 3 2 1

In the three years since the appearance of the first edition of *Physical Geography of the Global Environment*, much has happened to justify our focus on the interaction between natural environments and human societies. Hurricane Andrew ravaged much of South Florida, and its target area will bear the scars of its assault for decades to come. A devastating earthquake struck the Kobe area in Japan, killing thousands and exposing the weaknesses in the country's vaunted quake-preparedness programs. A flood-of-the-millennium devastated the American Midwest, taking a huge toll in human and animal lives and in property losses, as well as unleashing a fierce debate over the appropriateness of artificial levee-building to control great river systems. Torrential rains flooded Egyptian desert tombs not touched by moisture in thousands of years. The aftermath of the eruption of Mount Pinatubo in the Philippines was recorded in climatological as well as political events. Volcanic dust orbited the earth and contributed, researchers reported, to an interruption in a global warming trend popularly known as the enhanced greenhouse effect; more immediately, the volcano's bombardment of a nearby American air force base eased a political decision to abandon the U.S. military presence in the Philippines altogether.

Meanwhile, slower and less dramatic changes in our earthly environments were seen to have potentially far greater consequences. The enhanced greenhouse warming of the past quarter century (there was a cooling spell during the 1950s and 60s) may or may not presage a melting of glaciers and a rising of sea levels, but it does seem to portend the end of a long period of environmental stability. The loss of ozone at high altitudes may be totally or partly attributable to human pollution of the atmosphere, but in high latitudes there already are signs that serious consequences will accrue.

In this book we address such issues against a background of the environmental workings of our planet (we also touch on Earth's place in the Solar System and in the Universe). Because this is a geography book, we tend to use spatial perspectives when we address environmental questions, but our definition of physical geography is broad and encompassing.

IMPROVEMENTS IN THIS EDITION

This Second Edition brings us up to date on the ongoing debate over global climatic change, taking note of the regional character of this process. One of the most consequential findings of the past three years involves the realization that when one part of our planet records a long-term warming trend, another may experiencethe opposite. This has far-reaching implications for research ranging from global warming to human evolution.

In this edition of *Physical Geography of the Global Environment* we present the latest ideas relating to the earth's climate history, and contrast this against new information on weather extremes over the past decade. We all read about these unusual events in the press: the coldest-ever day here, the heaviest-ever rains there. Is it our imagination, or do such extremes occur more frequently these days than they used to? Our new material on environmental developments covers topics ranging from biodiversity concerns to the latest new dates on the geologic time scale, and from teleconnections (long-distance weather linkages that can relate California floods to Indonesian weather systems) to the newest satellite imagery.

Our focus on human-environmental interconnections is enhanced by many new Perspective boxes on such topics as water usage in the United States, the problem of groundwater contamination, the risk of sandstorms disrupting traffic on interstate highways, the formation of North America's most famous sinkhole, the Los Angeles earthquake of 1994, the location of the world's oldest rocks, and the rates of movement of the tectonic plates on which we live. A theme that runs throughout this book warns against careless, poorly-planned intervention in nature's design. As this Preface is being written, reports from China state that a hastily constructed series of dams in Henan Province gave way in 1975, killing as many as 230,000 people. The communist regime managed to keep this dreadful news from reaching the outside world, but now new fears arise as China, in its quest for electricity, is building a series of dams on the mighty Chang (Yangtze) River. How much better is the design for these new dams? How many thousands will be at risk downstream? As we note, mega-dam construction has become a topic of intense debate among students of the environment.

Readers of *Physical Geography of the Global Environment* will find the text updated throughout, with an expanded glossary (already the most comprehensive available in an introductory text) and a pronunciation guide found only in this book.

ORGANIZATION

We have divided the material in this text into five parts, and each part into numerous, comparatively brief *units*. These units are organized in a logical sequence well suited to the curriculum of established physical geography courses. However, by dividing this material into smaller parcels of information, we have given prominence to more than 50 themes in physical geography, ranging from ocean circulation to periglacial landforms.

PARTS

Part One—A Global Perspective

The units in this part provide an overview of the subject of physical geography and introduce four aspects of the earth: the nature and interaction of its systems, the representation of its surface features on maps, its place in the universe, and its relationship to the sun.

Part Two—Atmosphere and Hydrosphere

This part explores these two components of the earth system with units focusing on the chemical and physical characteristics of the atmosphere, the circulation of the earth's air and ocean waters, the role of atmospheric moisture in shaping weather, the geographic classification of the earth's diverse climates, and the dynamics of ever-changing climates and the human role in those changes.

Part Three—The Biosphere

The units included in this part examine the biosphere through coverage of soils and their geographic classification, biogeographic processes, and the spatial distribution of vegetation and animal life.

Part Four—The Restless Crust

The lithosphere is the focus of this part, which explores the constituents of the earth's interior and surface, movements of lithospheric plates, and tectonic processes and their effects on and beneath the surface.

Part Five—Sculpting the Surface

This part delves into the numerous and varied processes and agents—including weathering and mass movements, groundwater and surface running water, continental and mountain glaciers, wind, and coastal processes—that shape the landscapes and landforms of the earth's visible surface. It also provides an overview of the physiographic realms and regions of North America, particularly the United States.

UNIT FORMAT

Our aims in using the unit format are threefold:

1. While covering the same material that would be contained in conventional chapters, the units permit certain subjects to be covered in greater depth. Topics that usually are "hidden" in long chapters in other books here have their own heading and brief yet succinct text.
2. The flexibility of employing the units in any sequence provides a pedagogical advantage.
3. We hope that students will be attracted to more specific topics that especially interest them, which would then draw them to other areas of the book, and hence, perhaps, to further work in the field of physical geography.

FEATURES OF THIS TEXT

This text includes numerous special features that are intended to enhance and clarify a student's understanding and mastery of the material. These features include:

- *Full-color illustrations:* Over 100 computer-generated maps, over 200 diagrams, and nearly 300 photographs—many from the authors' own field experiences—are included. Captions provide additional information and link these illustrations to the text.
- *Objectives:* Objectives listed at the opening of each unit help students focus on the main points developed in that unit.
- *Perspectives on the Human Environment:* In at least one box in each unit, the theme of the unit is linked to a specific topic or case study which highlights the human-geographic dimension of that aspect of physical geography (see p. xviii).
- *Key Terms:* Major terms appear in bold-face type within each unit and are listed at the unit's end.
- *Review Questions:* Four to seven review questions, appearing at the end of each unit, tie together some of the major concepts of the unit.
- *Additional Readings:* Suggestions for additional readings are provided at the end of each unit to encourage further exploration of pertinent topics.
- *Appendix A: SI Units and Their Conversions:* Includes tables that display commonly used metric units of measurement (length, area, volume, mass, pressure, energy, and temperature) and their conversion from older units of measurement.
- *Appendix B: World Political Map:* A completely up-to-date map of the world's countries as of early 1996.

- **Pronunciation Guide:** A phonetic spelling system is employed in the end-of-text pronunciation guide that covers unfamiliar terms and locations.
- **Glossary of Terms:** Key terms and all other important terms (nearly 1,100 in total) are defined in a comprehensive glossary at the end of the text.

EARTH Magazine/World Wide Web Hazards Case Studies

Bound into the back of the text is a series of articles taken from the popular magazine *EARTH*. Focusing on some of the hazards that affect the lives of North Americans—hurricanes, tornadoes, earthquakes, volcanoes and El Niño—these articles form a series of case studies that provide an opportunity to explore these high-interest phenomena in depth. Preceding each of the articles are materials designed to relate the articles to the physical geography course and more specifically to relevant coverage within the text itself. To take the exploration of hazards a step further, we also offer a Web Site that takes you out in the "virtual field" by providing quick, direct links to a variety of current hazards resources on the world wide web. By contacting **http://www.wiley.com/college/geography/hazards**, you will find easy access to web pages such as the "Natural Hazards Center", the "National Hurricane Center", and the "Electronic Volcano" to name just a few. Of course, we also hope that students and instructors will enjoy and use these articles for any reason they choose, since we know these hazards are popular topics for discussion and exploration in the course.

KEY TO *EARTH* MAGAZINE HAZARDS CASE STUDIES

ARTICLE NUMBER	TITLE	TEXT KEY
Article 1	**El Niño–A Current Catastrophe** *When winds in the western Pacific Ocean became abnormally still, they imperiled fish and birds off the coast of South America, brought floods to Houston, Texas, and caused snowstorms in the Middle East. The culprit was El Niño.*	Unit 10, Circulation Patterns of the Atmosphere Unit 11, Hydrosphere: Circulation of the World Ocean
Article 2	**Tornado Troopers** *A small army of scientists advanced on the central United States for the largest coordinated storm hunt ever attempted. Their quarry: ten-mile high "supercell" thunderstorms and the violent tornadoes they spawn.*	Unit 13, Precipitation, Air Masses, and Fronts

ARTICLE NUMBER	TITLE	TEXT KEY
Article 3	**Hurricane Mean Season** *1995 brought a stampede of hurricanes to the Atlantic. Now there are signs that this was just a preview of a coming mean streak—a dramatic upturn in major landfalling storms. If it happens, hurricane researchers want to be ready.*	Unit 14, Weather Systems
Article 4	**Fire and Water at Krakatau** *Scientists unravel the events that wrought destruction during Krakatau's violent eruption in 1883.*	Unit 34, Volcanism and Its Landforms
Article 5	**Predicting Earthquakes** *The Los Angeles earthquake in 1994 was a rude shock not only to Angelinos. Even seismologists, who once thought they'd be able to forecast quakes as meteorologists forecast weather, failed to see it coming.*	Unit 35, Earthquakes and Landscapes

ANCILLARIES

The following ancillaries were prepared to accompany this edition of the book, and they may be obtained by contacting John Wiley & Sons:

- **Student Study Guide** (by Vernon Domingo of Bridgewater State College, Massachusetts)
  Designed to enhance student comprehension of physical geography through the use of unit summaries and objectives, key terms and concepts, self test questions, as well as special "graphicacy" and "hypothesis construction" sections for each unit.
- **Instructor's Manual and Test Bank** (by Gerald R. Webster and Roberta Haven Webster of the University of Alabama)
  Contains unit summaries, teaching objectives, and additional resources available to the instructor. Conversion tables are also included to enable users to switch easily from other texts. The Test Bank contains approximately 1200 questions.
- **Computerized Test Bank** (Delta Software)
  IBM and Macintosh versions are available with full editing features to help instructors customize tests.

- **Overhead transparencies or slide set**
  100 full-color transparencies of maps and diagrams from the text are provided in a form suitable for projection in the classroom. Instructors may choose from either transparencies or 35mm slides.
- **New Media Labs** (by David DiBiase of the Deasy GeoGraphics Laboratory, Pennsylvania State University)
  Software modules designed to give "hands-on" simulation experience on human involvement in the physical environment. Available for IBM and Macintosh.
- **Wiley's Geosciences CD-ROM** 400 images (photos and illustrations) and 8 animations organized into 58 earth science topics. CD technology allows instructors to customize lecture presentations and print their own transparencies. For IBM and MacIntosh.

We hope you will enjoy reading and using *Physical Geography of the Global Environment*, and always welcome your comments. Many suggestions made by colleagues and students helped us produce this second edition, and we look forward to hearing from you again.

H. J. de Blij
P. O. Box 608,
Boca Grande, Florida 33921

Peter O. Muller
Department of Geography,
University of Miami,
Coral Gables, Florida 33124-2060

July 3, 1997

Acknowledgments

During the preparation of this second edition of *Physical Geography of the Global Environment*, we received advice and guidance from many quarters. First of all we thank our reviewers, who supplied valuable feedback on the previous edition as well as thoughtful suggestions for improving a number of units. They were:

James Allen, Jr., Garrett Community College

David Arnold, Mississippi State University

Joseph Ashley, Montana State University

C. Mark Cowell, Indiana State University

Wade Currier, Rowan College of New Jersey

Stanford Demars, Rhode Island College

John Fox, Trenton State College

Robert Gardula, Fitchburg State College

Carol Harden, University of Tennessee

Daniel Johnson, Portland State University

Theron Josephson, Ferris State University

Donald Kiernan, Solano College

David LeBoutillier, Indiana State University

Denyse Lemaire-Ronveaux, Villanova University

James Norwine, Texas A&M-Kingsville

David May, University of Northern Iowa

James Moir, Bridgewater State College

Gary Peters, California State University-Long Beach

Virginia Ragan, University of Missouri-Kansas City

Lee Slorp, Kutztown University of Pennsylvania

Herschel Stern, Mira Costa College

We also appreciated the input of the many geographers whose comments reached us at various stages of the revision project, including all those responding to an extensive survey.

We acknowledge, too, the lasting contributions made by the following individuals during the preparation of the first edition: Edward A. Fernald, Florida State University; David E. Greenland, University of Oregon; Henry N. Michael, Temple University (emeritus); Scott E. Morris, University of Idaho; Randall J. Schaetzl, Michigan State University; and John D. Stephens, H. M. Gousha/Simon & Schuster.

At the University of Miami's Department of Geography, we are grateful for the support and/or assistance of our faculty colleagues (Thomas D. Boswell, Philip L. Keating, Bin Li, Jan Nijman, and Ira M. Sheskin), office manager Hilde Al-Mashat, work-study students Christina Barquero and Kimberly Palermo, and several students in our physical geography courses (especially Steven Shaw).

Our superb editorial and production team at John Wiley kept us on track and worked diligently to match (and we think surpass) the quality of the first edition. Developmental Editor Kathleen Dolan smoothly guided the revision process from start to finish, and constantly encouraged us to develop new ideas. Senior Production Editor Jeanine Furino deftly managed the demanding production schedule with boundless energy and sparkling humor; she also supervised the excellent copyediting job of Betty Pessagno. The illustration program was again quarterbacked marvelously by Edward Starr, who also coordinated the many improvements made to our cartographic program by Mapping Specialists, Inc. of Madison, Wisconsin (ably directed by project manager Don Larson). The complicated photography program was expertly managed by Stella Kupferberg, who was assisted by the diligent efforts of Kim Khatchatourian; Alexandra Truitt most effectively handled the day-to-day photo research tasks and merits much praise for the published results. Our designer, once again, was Dawn Stanley who deserves much of the credit for the appearance of this book.

Geography Editor Frank Lyman arrived on the scene just as this revision was gearing up; he immediately got fully involved, was a prime mover throughout, and took the leading role in improving the ancillaries that accompany this text. The revision was actually launched by Executive Editor Chris Rogers, and we appreciated his strong support throughout the process. We express our gratitude as well to Beth Brooks, who as the indispensable assistant to the Geography Editor performed several vital tasks and extended countless courtesies on our behalf. We also thank Cathy Faduska, who as marketing manager offered many useful insights as the book took shape. Finally, we acknowledge the valuable support we received from the rest of the Wiley team, particularly Linda Muriello (Production Manager), Ishaya Monokoff (Illustrations Director), Eric Stano (Supplements Editor), and Barry Harmon (former Geography Editor).

As ever, our wives, Bonnie and Nancy, deserve a huge amount of credit for so pleasantly accepting what we put them through in order to complete this project on time.

HJdB
POM

Brief Contents

Contents

Perspectives on the Human Environment

PHYSICAL GEOGRAPHY
OF THE GLOBAL
ENVIRONMENT

UNIT 1
INTRODUCING PHYSICAL GEOGRAPHY

UNIT 2
THE PLANET EARTH

PART ONE

A GLOBAL PERSPECTIVE

UNIT 3
MAPPING THE EARTH'S SURFACE

UNIT 4
THE EARTH IN THE UNIVERSE

UNIT 5
EARTH–SUN RELATIONSHIPS

Introducing Physical Geography

OBJECTIVES

◆ To introduce and discuss the contemporary focus of physical geography.

◆ To relate physical geography to the other natural and physical sciences.

◆ To introduce the systems and modeling approaches to physical geography.

This is a book about the earth's natural environments. Its title, *Physical Geography of the Global Environment*, suggests the unifying perspective. We will survey the earth's human habitat and focus on the fragile layer of life that sustains us along with millions of species of animals and plants. Features of the natural world such as erupting volcanoes, winding rivers, advancing deserts, and changing shorelines are examined not only as physical phenomena, but also in terms of their relationships with human communities and societies.

Along the way we will do more than study the physical world. We will learn just how broad a field physical geography is, and we will discover many of the topics on which physical geographers do scientific research. In order to proceed, it will be necessary to view our planet from various vantage points. We will study the earth in space and from space, from mountaintops and in underground shafts. Our survey will take us from clouds and ocean waves to rocks and minerals, from fertile soils and verdant forests to arid deserts and icy wastelands.

GEOGRAPHY

Geographers, of course, are not the only researchers studying the earth's surface. Geologists, meteorologists, biologists, oceanographers, and scientists from many other disciplines also study aspects of the planetary surface and what lies above and below it. But only one

UNIT OPENING PHOTO: Planet Earth as seen from approximately 35,000 km (22,000 mi) out in space. The African continent is at the right.

2

scholarly discipline, geography, combines and integrates and, at its best, *synthesizes* knowledge from all these other fields. Time and again in this book, you will become aware of connections among physical phenomena and between natural phenomena and human activities, of which you might not have become aware in another course.

Although geography is a modern discipline in which scientists use space-age research equipment, its roots extend to the very dawn of scholarly inquiry. When the ancient Greeks began to realize the need to organize the knowledge they were gathering, they divided it all into two areas: geography (the study of the terrestrial world) and cosmography (the study of the skies, stars, and the universe beyond). A follower of Aristotle, a scholar named Eratosthenes (ca. 273–ca. 192 B.C.), actually coined the term *geography* in the third century B.C. To him, geography was the accurate description of the earth (*geo*, meaning earth; *graphia*, meaning description), and during his lifetime volumes were written about rocks, soils, and plants. There was a magnificent library in Alexandria (in what is now Egypt) that came to contain the greatest collection of existing geographical studies.

Soon, the mass of information (data) about terrestrial geography became so large that the rubric lost its usefulness. Scientific specialization began. Some scholars concentrated on the rocks that make up the hard surface of the earth, and geology emerged. Others studied living organisms, and biology grew into a separate discipline. Eventually, even these specializations became too comprehensive. Biologists, for example, focused on plants (botany) or animals (zoology). The range of scientific disciplines expanded—and continues to do so to this day.

This, however, did not mean that geography itself lost its identity or relevance. As science became more compartmentalized, geographers realized that they could contribute in several ways, not only by conducting "basic" research, but also by maintaining that connective, integrative perspective that links knowledge from different disciplines. One aspect of this perspective relates to the "where" with which geography is popularly associated. The location or position of features on the surface of the earth (or above or below it) may well be the most significant thing about them. Thus geographers seek to learn not only about the features themselves, but also about their spatial relationships. The word **spatial** comes from the noun "space"—not the outer space surrounding our planet, but earthly, terrestrial space.

The question is not only *where* things are located, but also *why* they are positioned where they are and *how* they came to occupy that position. To use more technical language: What is the cause of the variations in the distribution of phenomena we observe to exist in geographic space? What are the dynamics that shape the spatial organization of each part of the earth's surface?

These are among the central questions that geographers have asked for centuries and continue to pose today.

In some ways, geography is similar to history: both are broad, *holistic* (all-inclusive), integrating disciplines. Historians are interested in questions concerning time and chronology, while geographers analyze problems involving space. Thus geography's scope is even broader than that of history; as we will see later, historical and chronological matters concern us in physical as well as human geography. But no body of facts or data belongs exclusively to geography. To that extent, at least, geography lost the preeminence it enjoyed in the days of the ancient Greeks.

Fields of Geography

Specialization has also developed within geography. Although all geographers share an interest in spatial arrangements, distribution, and organization, some geographers are more interested in physical features or natural phenomena, while others concentrate on people and their activities. That is why we referred in the previous paragraph to *physical* and *human* geography, the broadest possible division of the discipline. But even within these broad divisions, there are subdivisions. For example, a physical geographer may work on shorelines and beaches, on soil erosion, or on climate change. A human geographer may be interested in urban problems, in geopolitical trends, or in medical issues. As a result, geography today consists of a cluster of fields, some of which are shown in Fig. 1-1. Note that each of the geo-

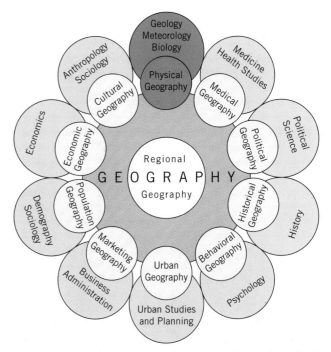

FIGURE 1-1 A schematic representation of the modern discipline of geography—physical geography highlighted—and its cognate fields.

graphic fields included (such as cultural geography, political geography, and population geography) is closely related to what is sometimes called a *cognate* (common-source) discipline. Thus cultural geography relates closely to anthropology, political geography to political science, population geography to demography, and so forth. In turn, all the fields of geography are related to each other through the spatial perspective that is geography's common bond.

Figure 1-1 is actually a simplification of the real situation in geography. Each of the fields shown consists of a combination of subfields. Cultural geography, for example, encompasses several subfields, including studies of the cultural landscape (the imprint of a culture upon the land), cultural ecology (the interrelationships between culture and nature), culture hearths (the source areas of civilizations), and cultural diffusion (the movement of innovations and ideas from place to place). Physical geography, too, is an umbrella term for an even larger number of subfields. We will encounter these subfields in the units that follow. But first let us examine physical geography as a field of geography.

PHYSICAL GEOGRAPHY

As the words imply, physical geography is the geography of the physical world. This field, like the entire discipline, is an old one. Eratosthenes was a pioneer in physical geography at a time when scholars still were unsure about the size or shape of our planet. Eratosthenes's name is permanently enshrined in physical geography's

hall of fame because of his remarkably accurate calculation of the circumference of the earth, a measurement he based on observations of the angle of the sun at various locations in Egypt (where he worked). This was the beginning of research in a field now known as *geodesy* (the study of the size and shape of the earth), but for Eratosthenes it was just one of several geographic pursuits. He realized the importance of mapping his observations, and trained himself to become a skilled *cartographer* (mapmaker and/or analyst). Having concluded that the earth was spherical (and not, as many of his colleagues believed, flat) and that parts of it were better warmed by the sun than others, Eratosthenes drew maps of the earth's environmental zones. He concluded—without ever having been anywhere near them—that the spherical earth would have a hot equatorial zone, two cold polar zones, and two temperate zones lying between these. Centuries after Eratosthenes's death, there still were scholars who did not believe that his ideas were essentially accurate. Here was a geographer far ahead of his time!

Other Greek geographers, and later Roman geographers, studied such physical features as mountains (including the Mediterranean region's volcanoes), rivers, coasts, and islands. Some of them were excellent observers and recorders, and what they wrote about the Mediterranean Basin continues to be of interest to scientists today. A scholar named Pliny the Younger (nephew of Pliny the Elder, a scientist) witnessed the eruption of Mount Vesuvius in A.D. 79 from a boat in the nearby Bay of Naples (Fig. 1-2). He described the mushroom cloud rising above the mountain and the burial of

FIGURE 1-2 An artist's rendition of the horror and chaos in the streets of Pompeii on that fateful summer day in A.D. 79 as the huge cloud of volcanic ash and toxic gases, spewed by nearby Mount Vesuvius, descended on the town and buried it in a matter of seconds.

FIGURE 1-3 A painter's depiction of the expedition of Alexander von Humboldt (standing in boat) making its way through the lowland tropics of northern South America shortly after the turn of the nineteenth century.

such towns as Pompeii and Herculaneum. What he did not know, as he chronicled the disaster from a safe distance, was that his uncle, who had rushed to the scene to help the stricken, died in Vesuvius's toxic fumes.

It is sad indeed that much of what those ancient geographers wrote has been lost over time, because what has survived is of such enormous interest. From fragments of Greek and Roman writing, we know about how wide and fast-flowing the rivers were, the activity of some now-quiet volcanoes, and the density of vegetation in places where there is none today. Worse, after the Romans carried geography so far forward, Europe descended into the Dark Ages, and geography (along with science generally) stagnated. For a thousand years, geographic learning in the Arab realm of North Africa/Southwest Asia and in China advanced far beyond that of Europe. Little of what was achieved by Arab or Chinese scholars was added to the European inventory, however. In addition, many records, maps, and books were lost as a result of wars, fires, and neglect.

Physical geography revived when the age of exploration and discovery dawned around 1500. Portuguese navigators skirted the African coast; Columbus crossed the Atlantic Ocean and returned with reports of new lands in the west. Cartographers recorded the accumulating knowledge on increasingly accurate maps. In European cities there circulated news of great rivers, snowcapped mountains, wild coasts, vast plains, forbidding escarpments, dense forests with taller trees than had ever been seen, strange and fearful animals, and alien peoples. Explorers and fortune hunters brought back hoards of gold and other valuables. Geographic knowledge could be the key to wealth.

While Europeans rushed to the new lands, scientists tried to find some order in the mass of new information that confronted them. One of the greatest of these scholars was Alexander von Humboldt (1769–1859), who traveled to the New World not for wealth but for knowledge (Fig. 1-3). He managed to travel 3000 km (1900 mi) up the uncharted Orinoco River in northern South America, did fieldwork in Ecuador and Peru, crossed what is now Mexico, and visited Cuba before reaching the United States in 1804. Later he traversed Russia, including remote Siberia. He collected thousands of rock samples and plant specimens and made hundreds of drawings of the animals he observed. After settling down in Paris, he wrote 30 books on his American travels and later produced his famous six-volume series called *Cosmos*, one of the gigantic scientific achievements of the nineteenth century.

From von Humboldt's writings we can learn about the state of physical geography in his time. It is evident that physical geography had become more than the study of the surface of the earth; now it also included studies of the soils, vegetation and animals, the oceans, and the atmosphere. Although the term *physical geography* was firmly entrenched, it might have been more appropriate to use the term *natural geography* for this wide-ranging field!

Von Humboldt demonstrated geographic research methods in many ways. While working in the Andes Mountains of western South America, for instance, he made maps of the slopes, ranges, valleys, and other terrain features of the landscape. He also mapped the vegetation and realized that altitude, temperature, and vegetation types were interrelated. This means, of course, that altitudinal zonation also influences crop cultivation, linking physical (natural) and human geography. Observing the movement of ocean water off the Pacific coast of Peru, von Humboldt identified a cold current that, he correctly concluded, began in Antarctic waters and carried a polar chill to the western shores of equatorial South America. Again, he correctly connected this cold water and prevailing onshore wind patterns with the resulting desert conditions along the narrow Peruvian coastal plain. This ocean current, in fact, was for a long time named the Humboldt Current in his honor (more recently it has become known as the Peru Current).

Subfields of Physical Geography

The stage was now set for the development of specializations within the field of physical geography, and soon these subfields began to take shape. Over the past century, physical geography has evolved into a cluster of research foci, the most important of which are diagrammed in Fig. 1-4. Remember that the entire field of physical geography is only one of those illustrated in Fig. 1-1; we now take the research sphere at the top of that diagram and reveal its internal relationships. To make things easier, we have numbered each of the eight subfields in Fig. 1-4.

The geography of landscape, **geomorphology** (1), remains one of the most productive subfields of physical geography. As the term suggests (*geo*, meaning earth; *morph*, meaning shape or form), this area of research focuses on the structuring of the earth's surface. Geo-

morphologists seek to understand the evolution of slopes, the development of plains and plateaus, and the processes shaping dunes and caves and cliffs—the elements of the (physical) landscape. Often, geomorphology has far-reaching implications. From the study of landscape, it is possible to prove the former presence of icesheets and mountain glaciers, rivers, and deserts. *Geology* is geomorphology's closest ally, but the study of geomorphology can involve far more than rocks. The work of running water, moving ice, surging waves, and restless air all contribute to landscape genesis. And while these forces shape the surface above, geologic forces modify it from below.

A field related to geomorphology, but not shown in Fig. 1-4, is **physiography**. Whereas geomorphology focuses on processes and forces, physiography looks at the regional results. Physiography literally means "landscape description," but in practice it is more complicated than that. In physiography, principles of regional geography (a field central to geography as a whole—as Fig. 1-1 demonstrates) are applied to landscape study.

Proceeding clockwise from the top of Fig. 1-4, we observe that *meteorology* (the branch of physics that deals with atmospheric phenomena) and physical geography combine to form **climatology** (2), the study of climates and their spatial distribution. Climatology involves not only the classification of climates and the analysis of their distribution, but also broader environmental questions, including climatic change, vegetation patterns, soil formation, and the relationships between human societies and climate.

As Fig. 1-4 indicates, the next three subfields relate physical geography to aspects of *biology*. Where biology and physical geography overlap is the broad subfield of **biogeography** (3–5), but there are specializations within biogeography itself. Physical geography combined with botany forms **phytogeography** (3), and combined with zoology it becomes **zoogeography** (5). Note that biogeography (4), itself linked to *ecology*, lies between these two subfields; in fact, both zoogeography and phytogeography are parts of biogeography. The next subfield of physical geography is related to soil science, or *pedology*. Pedologists' research tends to focus on the internal properties of soils and the processes that go on during soil development. In **soil geography** (6), research centers on the spatial patterns of soils, their distribution, and their relationships to climate, vegetation, and humankind.

Other subfields of physical geography are **marine geography** (7) and the study of **water resources** (8). Marine geography, which is related to the discipline of *oceanography*, actually has human as well as physical components. The human side of marine geography has to do with maritime boundaries, the competition for marine resources, and the law of the sea; therefore, this subfield is closely allied with political geography (*see* Fig. 1-1). The physical side of marine geography deals with

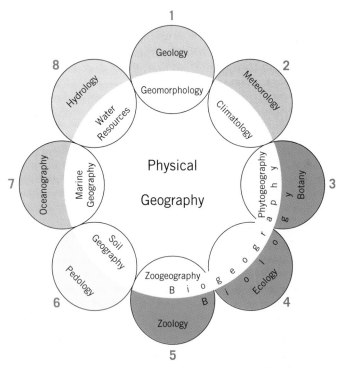

FIGURE 1-4 Schematic representation of the specialized subfields that constitute physical geography.

coastlines and shores, beaches, river mouths, and other landscape features associated with the oceanic margins of the continents. The subfield of water resources (where *hydrology* and physical geography intersect) also has human as well as physical elements. The landmasses contain fresh water at the surface (in the form of lakes and rivers) and below (as groundwater). As we will see later in this book, the study of these waters may be approached from geomorphic as well as economic standpoints.

In the 52 units that follow, these various subfields of physical geography will be examined in some detail, and the connections between physical and human geography also will be revealed. As you will see, we must go beyond the confines of Fig. 1-4 to put our work in proper perspective. To understand the basics of physical geography, we must comprehend the general properties of our planet, not only deep below its surface, but also far beyond as it orbits the sun as part of the solar system in our tiny corner of the vast universe. Comprehending general properties will be among our first tasks and constitute much of the remainder of Part One.

A CENTURY
OF PHYSICAL GEOGRAPHY

During the past century, physical geographers have made many noteworthy and interesting contributions to science. Over time, the nature of these contributions has changed. Just 100 years ago, for example, the map of the physical geography of North America was still being filled in, and physical geographers such as John Wesley Powell and Grove Karl Gilbert reported on the spectacular scenery they studied in the western United States. Through sketches and cross-sectional profiles, Powell described his journeys down the Colorado River and told the story of one of the earth's greatest natural features, the Grand Canyon (Fig. 1-5). Gilbert analyzed the origins of prominent mountains and showed how running water could remove huge amounts of rock, carrying it from hillslope to river delta. Prominent physical features were given names, and later a famous cartographer, Erwin Raisz, drew a minutely detailed map of the physiography of the entire United States—a cartographic masterpiece still in use today.

Other physical geographers, explorer–scientists all, fired the imaginations of their colleagues and students. Some of their names became permanently associated with the landscapes or physical features they studied: Louis Agassiz and glaciers, William Libbey and ocean currents, Arnold Guyot and ocean-floor topography. But perhaps the most important scholar between the 1880s and 1930s was William Morris Davis (1850–1934), who taught physical geography at Harvard and several other universities. Davis was less the explorer and more the

FIGURE 1-5 The Grand Canyon of the Colorado River, parts of which are up to 1.6 km (1 mi) deep and 29 km (18 mi) wide. This most famous natural feature of the United States extends for nearly 450 km (275 mi) across northern Arizona. John Wesley Powell led the first river expedition through the massive gorge in 1869—and gave the Grand Canyon its name.

theoretician (although he traveled worldwide in pursuit of his ideas), and in a series of significant papers he published the first comprehensive theory concerning the way rocks and geologic structures are worn down by the force of running (stream) water. He coined many terms we still use today, and he moved physical geography into the modern scientific era. In Unit 42, we look at Davis's theories and see how others built upon (or countered) them.

Following Davis, specialization in physical geography became stronger, and the subfields shown in Fig. 1-4 gained identity. Wallace Atwood took the field in a more physiographic (regional), less geomorphic direction at Harvard University, where he succeeded Davis. Atwood's book on the physiographic regions of North America was for a long time a standard work on the subject. Curtis Marbut was the leading soil geographer, and he did pioneering work on soil classification and mapping. A climatologist, C. Warren Thornthwaite, advanced his subfield through detailed studies of moisture relationships in various environments. Richard Russell put marine geography on the map through his work on coastal dynamics. A. William Küchler brought focus to the subfield of phytogeography by studying and mapping the distribution of world vegetation types.

Many other scholars, too numerous to mention here, have contributed to the recent development of physical geography in North America. You will find much of their work cited in the end-of-unit references throughout this book. There still are physical geographers working on large, theoretical questions as Davis did (now with the aid of sophisticated quantitative techniques and computers), but others work on highly specialized, very specific questions. These days, when you ask a physical geographer what his or her specialty is, you may hear such answers as "hydroclimatology," "periglacial processes," "paleogeography," or "wetland ecosystems." All this helps explain the wide range of material you will encounter in this book, which is an overview of a broad field encompassing many topics.

SYSTEMS AND MODELS IN PHYSICAL GEOGRAPHY

Physical geography is a multifaceted science that seeks to understand major elements of our complex world. In order to deal with this complexity, physical geographers employ numerous concepts and specialized methods, many of which will emerge in the units that follow. This section provides an introduction to that analysis by considering two general approaches to the subject.

Systems

In recent years, many physical geographers have found it convenient to organize their approach to the field within a systems framework of thinking. A **system** may be regarded as any set of related events or objects and their interactions. A city could be described as a large and elaborate system. Each day the system receives an inflow of energy, food, water, and vast quantities of consumer goods. Most of this energy and matter is used and changed in form by the various populations that reside in the urban center. At the same time, huge amounts of energy, manufactured goods, and services, along with sewage and other waste products, are produced in and exported from the city. Note that energy and matter freely transfer across the city's boundaries, making it an **open system** and underscoring its relationships with surrounding systems (such as the agricultural and energy-producing systems of the region of which it is a part). There are many examples of open systems throughout this book, such as a weather system or a river drainage basin. Indeed, the earth itself is an enormous open system—which comprises several interconnected lesser systems. (Although it is difficult to find one on the earth's surface, we should also know what is meant by

the term **closed system**: a self-contained system exhibiting no exchange of energy or matter across its boundaries.)

An important property of a system is its organization as an integrated whole. Accordingly, systems often contain one or more subsystems. A **subsystem** is a component of a larger system. A subsystem can act independently, but it operates within, and is linked to, the larger system. The food distribution, manufacturing, and sewage systems are subsystems of the total city system described above. Systems and subsystems have boundaries that are called *interfaces*. The transfer or exchange of energy and matter takes place at these interfaces (which may also be regarded as surfaces). Sometimes interfaces are visible: you can see where sunlight strikes the roof of a building. But often they are not visible: you cannot see the movement of groundwater, a part of the global water system, as it flows through the subterranean rocks of the geologic system. Many geographers focus their attention on these interfaces, visible or invisible, particularly when they coincide with the earth's surface. It is here that we find the greatest activity of our dynamic world.

Two other ideas commonly used in systems approaches are those of dynamic equilibrium and feedback. A system is in **dynamic equilibrium** when it is neither growing nor getting smaller but continues to be in balance and complete operation. **Feedback** occurs when a change in one part of the system causes a change in another part of the system. Let us consider two examples. If you were to look at the sand in a specific area of Miami Beach (Fig. 1-6), you would barely perceive that the currents moving along the shore are taking away some sand and bringing in a replacement supply. There is continual movement, yet over a period of weeks the beach apparently stays the same. The beach, thus, can be said to be in a state of dynamic equilibrium.

An example of feedback, or a feedback mechanism, would be the case of solar radiation being reflected from a New York City sidewalk. The sidewalk's surface receives energy from the sun, but it also reflects and radiates some of that energy back into the atmosphere as well as losing it in other ways. The more energy the sidewalk receives, the more it reflects and reradiates. Because of the reflection and reradiation, the sidewalk does not become increasingly hotter. Without the feedback mechanisms of reflection and reradiation, it would certainly be impossible to walk on that surface at midday during summer.

A feedback mechanism that operates to keep a system in its original condition, like the reflection or radiation from the sidewalk, is called a *negative feedback mechanism*. The opposite case, in which a feedback mechanism induces a progressively greater change from the original condition of a system, is called a *positive feedback mechanism*. In a later unit, we explain why the growth of a metropolitan area leads to higher average air

FIGURE 1-6 Miami Beach, Florida, looking northward along its Atlantic coastline. This beach is actually a system in dynamic equilibrium, with shore-paralleling currents constantly removing sand and depositing a replacement supply.

temperatures; a change of this kind is an example of positive feedback.

Models

Another way that physical geographers approach the study of the earth's phenomena is to make models of them. In his landmark book on geographic analysis, Peter Haggett defined a **model** as *the creation of an idealized representation of reality in order to demonstrate its most important properties*. Model building, therefore, is a complementary way of thinking about the world. It entails the controlled simplification of a complex reality, filtering out the essential forces and patterns from the myriad details with which they are embedded in a complicated world. Such abstractions, which convey not the entire truth but a valid and reasonable part of it, are highly useful because they facilitate the development of generalizations. We saw this in the preceding discussion of the city system, which underscored some universal attributes concerning the spatial interaction between cities and the surrounding regions they serve. Systems, therefore, may also be regarded as models. Models will be used frequently in this book because they allow us to quickly penetrate a complex subject and highlight its most essential aspects.

GEOGRAPHIC MAGNITUDE

In approaching the real world, we must also consider the size of the subjects and phenomena that interest us. Even speeding at 1000 km (625 mi) per hour, one may become

uncomfortable on a 15-hour flight from Los Angeles to Sydney, Australia, because, in human terms, the world is such a big place. But in studying our planet, we must not think of distance and magnitude in purely human terms.

Let us consider different sizes, or **orders of magnitude**. Figure 1-7 shows the various orders of magnitude with which we must become familiar. The scale in this figure is written in *exponential* notation. This means that 100 is written as 10^2 (10 × 10), 1000 is 10^3 (10 × 10 × 10), 0.01 is 10^{-2} (1/100), and so forth. This notation saves us from writing numerous zeros. The scales geographers use most often—which are shown on the right-hand side of Fig. 1-7—go from about 10^5 or 10^6 cm (11,000 yd), the size of Central Park in New York City (Fig. 1-8), up to about 10^{10} cm (62,500 mi), beyond the order of magnitude of the earth's circumference (*see* Perspective: Sliding Scale). Physical geographers sometimes have to expand their minds even further. Cosmic rays with wavelengths of 10^{-16} cm may affect our climate. The nearest fixed star is approximately 10^{18} cm (25 trillion mi) away; brighter stars much farther away sometimes help when navigating a path across the earth's surface. Occasionally (as in Unit 4), we have to perform mental gymnastics to conceive of such distances, but one of the beauties of physical geography is that it helps us to see the world in a different way.

Let us now embark on our detailed study of the earth. Central to this effort is the attempt to understand the environment at the earth's surface, a habitat we must all live with in a one-to-one relationship. Nobody who has become acquainted with physical geography is ever likely to forget that.

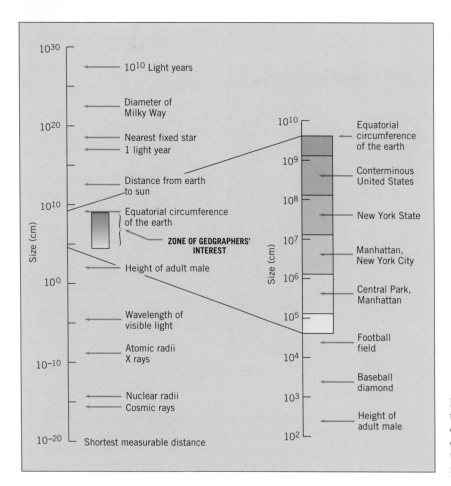

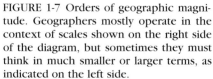

FIGURE 1-7 Orders of geographic magnitude. Geographers mostly operate in the context of scales shown on the right side of the diagram, but sometimes they must think in much smaller or larger terms, as indicated on the left side.

FIGURE 1-8 Central Park (the green rectangle left of center), located just to the north of New York City's central business district, really does occupy a central position on the island of Manhattan. Vis-à-vis our discussion here at the end of Unit 1, this park encompasses an area that is considered to rank among the smallest in size that geographers study.

Sliding Scale

Imagine a couple sunbathing on Miami Beach. We can photograph them occupying a square of sand about 1 m (3.3 ft) on a side. If we move the camera higher, a square of 10 (10^1) m reveals their companions. When we focus on a 100 (10^2) m area, we can see a crowd of people on the beach. A picture of 1000 (10^3) m includes the beach, some sea, and some land (as in Fig. 1-6). One with an edge of 10,000 (10^4) m captures most of the city of Miami Beach and parts of neighboring Miami across Biscayne Bay to the west.

Moving the camera still farther, we shoot a picture of a 100,000 (10^5) m square. It encompasses most of the South Florida region. The next step is 1,000,000 (10^6) m. This snapshot takes in the entire state of Florida, some neighboring states and Caribbean islands, and parts of the Atlantic Ocean, Gulf of Mexico, and Caribbean Sea. A photo at the next level of generalization, showing a square of 10,000,000 (10^7) m, covers most of the visible earth. And if the camera is far enough away in outer space to focus on a square of 100,000,000 (10^8) m, we see Planet Earth as a small globe. Somewhere on it is that couple lying on a square of Miami Beach's seaside sand.

KEY TERMS

biogeography (p. 6)

climatology (p. 6)

closed system (p. 8)

dynamic equilibrium (p. 8)

feedback (p. 8)

geomorphology (p. 6)

marine geography (p. 6)

model (p. 9)

open system (p. 8)

orders of magnitude (p. 9)

physiography (p. 6)

phytogeography (p. 6)

soil geography (p. 6)

spatial (p. 3)

subsystem (p. 8)

system (p. 8)

water resources (p. 6)

zoogeography (p. 6)

REVIEW QUESTIONS

1. Define the term *spatial* and show how it is central to the study of geography.
2. What contributions did the Greeks and Romans make to the early evolution of physical geography?
3. Describe the accomplishments of Alexander von Humboldt and their impact on the beginnings of modern physical geography.
4. Define the eight major subfields of physical geography.
5. Define the terms *system*, *subsystem*, *open system*, *dynamic equilibrium*, and *feedback*.
6. Define the term *model*, and describe how models can help to understand our complex physical world.

REFERENCES AND FURTHER READINGS

ATWOOD, W. W. *The Physiographic Provinces of North America* (New York: Ginn, 1940).

CHORLEY, R. J., et al. *The History of the Study of Landforms, or the Development of Geomorphology. Vol. 2: The Life and Work of William Morris Davis* (New York: Wiley, 1973).

GAILE, G. L., and WILLMOT, C. J., eds. *Geography in America* (Columbus, Ohio: Merrill, 1989), 28–94, 112–146.

GOUDIE, A. S., et al., eds. *The Encyclopedic Dictionary of Physical Geography* (Cambridge, Mass.: Blackwell, 2nd ed., 1994).

HAGGETT, P. *Locational Analysis in Human Geography* (London: Edward Arnold, 1965), 19.

JAMES, P. E., and MARTIN, G. J. *All Possible Worlds: A History of Geographical Ideas* (New York: Wiley, 3rd ed., 1993).

MARCUS, M. G. "Coming Full Circle: Physical Geography in the Twentieth Century," *Annals of the Association of American Geographers*, 69 (1979), 521–532.

PATTISON, W. D. "The Four Traditions of Geography," *Journal of Geography*, 63 (1964), 211–216.

Rediscovering Geography: New Relevance for the New Century (Washington, D.C.: National Academy of Sciences, 1995).

WOLMAN, M. G. "Contemporary Value of Geography: Applied Physical Geography and the Environmental Sciences," in A. Rogers, et al., eds., *The Student's Companion to Geography* (Cambridge, Mass.: Blackwell, 1992), 3–7.

The Planet Earth

OBJECTIVES

◆ To define and highlight the four spheres of the earth system.

◆ To highlight the general characteristics of the earth's continents.

◆ To introduce the world's ocean basins and the topographic characteristics of the seafloor.

When U.S. astronauts for the first time left Earth's orbit and reached the moon, they were able to look back at our planet and see it as no one had ever seen it before. The television cameras aboard their spacecraft beamed spectacular pictures of the earth, seen on television sets around the globe. Against the dark sky, the earth displayed a range of vivid colors, from the blue of the oceans, the green of forests, and the brown of sparsely vegetated land to the great white swirls of weather systems in the atmosphere (*see* photo, p. 2). The astronauts, however, also saw things no camera could adequately transmit. Most of all, they were struck by the smallness of our world in the vastness of the universe. It was difficult to conceive that all of humanity and its

works were confined to, and dependent upon, so tiny a planet. Every participant in those moon missions returned with a sense of awe—and a heightened concern over the fragility of our earthly life-support systems.

In the units that follow, we will examine these systems and learn, among other things, how serious the threat of irreversible damage to them may be. We should remember that all the systems we will study—weather and climate, oceanic circulation, soil formation, vegetation growth, landform development, and erosion—ulti-

UNIT OPENING PHOTO: The four spheres of the earth system in harmonious interaction. Mount Tasman, New Zealand. (Authors' photo)

mately are parts of one great earth system. Even though this total earth system is an open system with respect to energy flows, the amount of matter on and in the earth is pretty much fixed. Little new matter is being added to supplement what we use up, and so far nothing of consequence permanently leaves the earth. This means that our material resource base is finite; that is, parts of it can be used up. It also means that any hazardous products we create must remain a part of our environment.

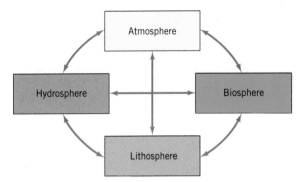

FIGURE 2-1 Relationships among the four spheres of the earth system.

SPHERES OF THE EARTH SYSTEM

In our everyday language, we recognize that our planet consists of a set of interacting shells or "spheres," some of them extending over the entire globe, others covering it partially. The **atmosphere** is the blanket of air that adheres to the earth's surface. It is our life layer, the mixture of gases we breathe. It begins a few meters within the soil or on the water's surface, and it extends to a height of about 60,000 km (37,000 mi) above the earth. Heat energy from the sun keeps the atmosphere in motion, causing weather systems to form and travel across land and sea. The atmosphere is most dense at sea level and thins out with altitude.

Below the atmosphere lies the outermost shell of the solid earth, the **lithosphere** (*lithos* means rock). The lithosphere's upper surface is sculpted into the almost endless variety of landforms and physical landscapes that form the earth's scenery. The lithosphere continues under the oceans, where the surface of the seafloor is created by forces quite different from those on land.

Constituting about 71 percent of the earth's surface, the oceans lie between the atmosphere and the lithosphere. This is the largest segment of the **hydrosphere**, which contains all the water that exists on and within the solid surface of the earth and in the atmosphere above. The oceans are the primary moisture source for the precipitation that falls on the landmasses, carried there in the constantly moving atmosphere.

The **biosphere** is the zone of life, the home of all living things. This includes the earth's vegetation, animals, and human beings. Since there are living organisms in the soil and plants are rooted in soil, part of the soil layer (which is otherwise a component of the lithosphere) may be included in the biosphere.

These four spheres—the atmosphere, lithosphere, hydrosphere, and biosphere—are the key earth layers with which we shall be concerned in our study of physical geography. But other shells of the earth also play their roles. Not only are there outer layers atop the effective atmosphere, but there also are spheres inside the earth, beneath the lithosphere. In turn, the lithosphere is affected by forces and processes from above. Thus it is important to keep in mind the interactions among the spheres—they are not separate and independent segments of our planet, but constantly interacting subsystems of the total earth system.

These systemic interrelationships are diagrammed in Fig. 2-1. The atmosphere, lithosphere, hydrosphere, and biosphere are the main components of the physical world. They are linked together in any one place and over the earth as a whole. As the diagram demonstrates, physical geographers study the phenomena within these spheres and the multiple interactions among them. Within the total earth system, the four spheres or subsystems also interconnect at any given place to form a fifth subsystem—the **regional subsystem** (Fig. 2-2). We are interested in the special ways in which the four major subsystems work together to form a regional subsystem. In a sense each regional subsystem is unique, but there are enough similarities between the ways in which our four spheres interact to produce the landscapes we live on for us to group some types of regional subsystems together. The rationale for grouping places together into regions is discussed in several units in this book.

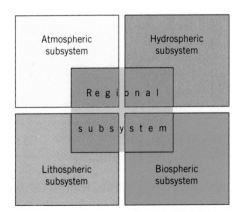

FIGURE 2-2 The earth system and its subsystems. A regional subsystem encompasses parts of the other four subsystems.

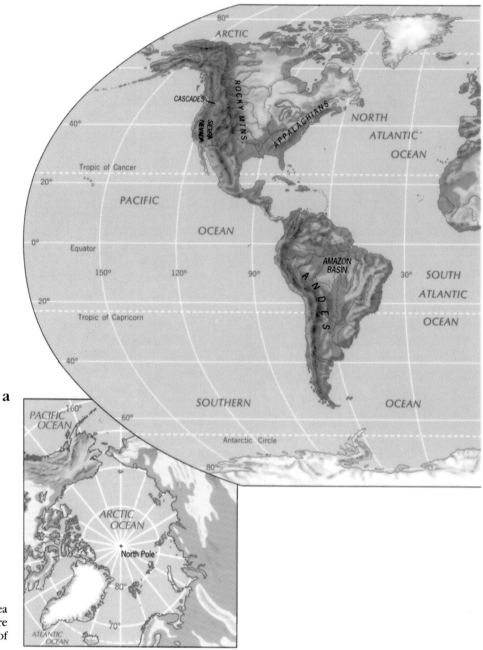

FIGURE 2-3 The distribution of land and sea on the earth's surface. The continents are shown in topographic relief; the terrain of the ocean floor is shown in Fig. 2-6.

HEMISPHERES

In addition to four layered spheres (the atmosphere, lithosphere, hydrosphere, and biosphere), the earth can also be divided into **hemispheres** (from the ancient Greek, *hemi* meaning half and *sphaira* meaning sphere). The northern half of the globe, from the equator to the North Pole, is the Northern Hemisphere; the southern half is the Southern Hemisphere. There are many differences between the two hemispheres, as we shall see in later units that treat patterns of seasonality, plan-

etary rotational effects, and the regionalization of climates. Perhaps the most obvious difference concerns the distribution of land and sea (Fig. 2-3). The earth's continental landmasses are far more heavily concentrated in the Northern Hemisphere (which contains about 70 percent of the total land area); the Southern Hemisphere has much less land and much more water than the Northern Hemisphere. Moreover, the polar areas of each hemisphere differ considerably. The Northern Hemisphere polar zone—the *Arctic*—consists of peripheral islands covered mostly by thick ice and a central mass of sea ice

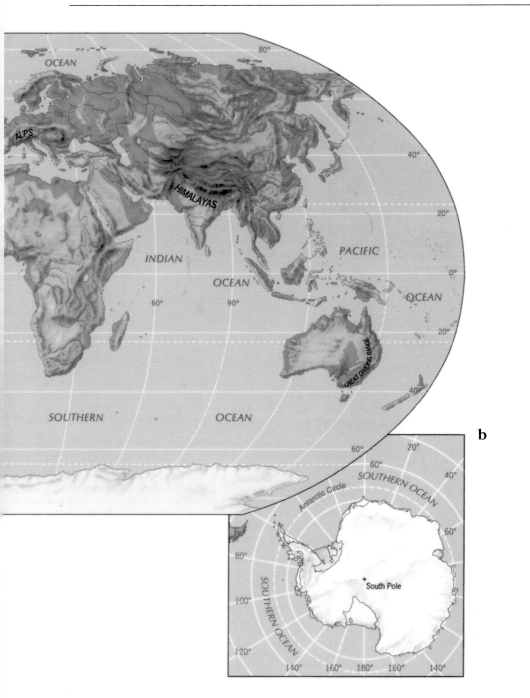

b

floating atop the Arctic Ocean (Fig. 2-3, inset map *a*). The Southern Hemisphere polar zone—the *Antarctic*—is dominated by a large continental landmass covered with the world's largest continuous icesheet (inset map *b*).

The earth can be further divided into Eastern and Western Hemispheres. Technically, the Eastern Hemisphere lies to the east of an imaginary line drawn from pole to pole through the Royal Observatory in Greenwich, England (part of the city of London), and on through the middle of the Pacific Ocean on the opposite

side of the world. In practice, however, the Western Hemisphere consists of the half of the earth centered on the Americas, and the Eastern Hemisphere contains all of Eurasia and Africa. So there is a precise use, based on a grid drawn on the globe (which is elaborated in Unit 3), and a more general use of this hemispheric division.

A look at any globe representing the earth suggests yet another pair of hemispheres: a **Land Hemisphere** and a **Water** (or Oceanic) **Hemisphere**. The landmasses are concentrated on one side of the earth to such a degree that it is appropriate to refer to that half of it as the

Land Hemisphere (Fig. 2-4). This hemisphere is centered on Africa, which lies surrounded by the other continents: the Americas to the west, Eurasia to the north and northeast, Australia to the southeast, and Antarctica to the south. The opposite hemisphere, the Water Hemisphere, is dominated by the earth's greatest ocean, the Pacific. When you look at a globe from above the center of the Pacific Ocean, only the fringes of the landmasses appear along the margins of this huge body of water.

CONTINENTS AND OCEANS

There is an old saying to the effect that "the earth has six continents and seven seas." In fact, that generalization is not too far off the mark (*see* frontispiece map and Fig. 2-3). The earth does have six continental landmasses: Africa, South America, North America, Eurasia (Europe and Asia occupy a single large landmass), Australia, and Antarctica. As for the seven seas, there are five great oceanic bodies of water and several smaller seas. The Pacific Ocean is the largest of all. The Indian Ocean lies between Africa and Australia. The North Atlantic Ocean and the South Atlantic Ocean may be regarded as two discrete oceans that are dissimilar in a number of ways. Encircling Antarctica is the Southern Ocean. And beneath the Arctic icecap lies the Arctic Ocean.

The seventh body of water often identified with these oceans is the Mediterranean Sea, which lies between Europe and Africa and is connected to the interior sea of Eurasia, the Black Sea. The Mediterranean is not of oceanic dimensions, but, unlike the Caribbean or the

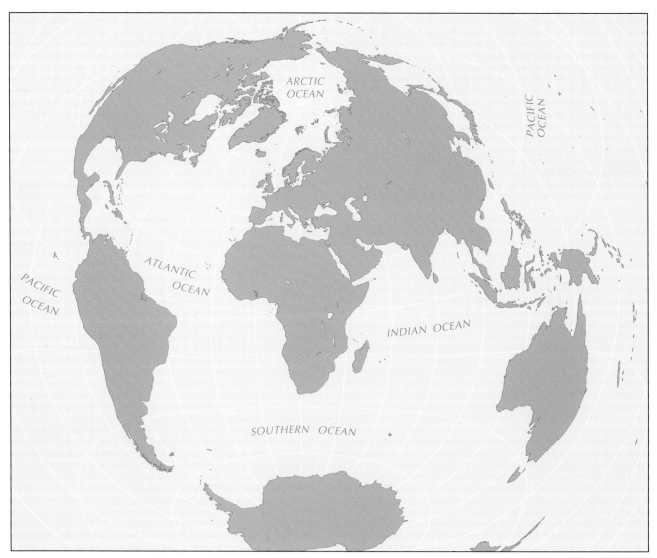

FIGURE 2-4 The Land Hemisphere. This is the half of the globe that contains most of the world's landmasses, which surround the African continent.

TABLE 2-1 DIMENSIONS OF THE LANDMASSES

| | AREA | | PERCENTAGE | | ELEVATION | |
LANDMASS	(THOUSANDS OF KM²)	(THOUSANDS OF MI²)	OF LAND SURFACE	HIGHEST MOUNTAIN	m	ft
Africa	30,300	11,700	20.2	Kilimanjaro	5,861	19,340
South America	17,870	6,900	11.9	Aconcagua	6,919	22,834
North America	24,350	9,400	16.3	McKinley	6,158	20,320
Eurasia	54,650	21,100	36.5	Everest	8,848	29,028
Australia	8,290	3,200	5.6	Kosciusko	2,217	7,316
Antarctica	13,990	5,400	9.3	Vinson Massif	5,110	16,864

Arabian Sea, it also is not merely an extension of an ocean. The Mediterranean is very nearly landlocked and has only one narrow natural outlet through the Strait of Gibraltar, between Spain and Morocco.

In our study of weather and climate, the relative location, general dimensions, and topography of the landmasses are important, because these influence the movement of moisture-carrying air.

The Landmasses

Only about 29 percent of the surface of the earth is constituted by land; 71 percent is water or ice. Thus less than one-third of our planet is habitable by human beings, but much of this area is too dry, too cold, or too rugged to allow large concentrations of settlement. Our livable world where permanent settlement is possible—known as the *ecumene*—is small indeed (*see* Perspective: World Population Distribution).

Each of the six continental landmasses possesses unique physical properties. Africa, which accounts for just over 20 percent of the total land area, is at the heart of the Land Hemisphere. Of all the landmasses, Africa alone lies astride the equator in such a way that large segments of it occupy the Northern as well as the Southern Hemisphere. Africa often is called the plateau continent, because much of its landmass lies above 1000 m (3300 ft) in elevation, and coastal plains are relatively narrow. As Fig. 2-3 reveals, a fairly steep escarpment rises near the coast in many parts of Africa, leading rapidly up to the plateau surface of the interior. African rivers that rise in the interior plunge over falls and rapids before reaching the coast, limiting their navigability. Furthermore, Africa lacks a physical feature seen on all the other landmasses: a linear mountain range comparable to South America's Andes, North America's Rocky Mountains, Eurasia's Himalayas, or Australia's Great Dividing Range. The reason for this will become clear when the geomorphic history of that continent is discussed.

South America, occupying 12 percent of the world's land, is much smaller than Africa (Table 2-1). The topography of this landmass is dominated in the west by the gigantic Andes Mountains, which exceed 6000 m (20,000 ft) in height in many places. East of the mountains, the surface becomes a plateau interrupted by the basins of major rivers, among which the Amazon is by far the largest. The Andes constitute a formidable barrier to the cross-continental movement of air, which has a major impact on the distribution of South America's climates.

North America, with one-sixth of the total land area, is substantially larger than South America. This landmass extends from Arctic to tropical environments. Western North America is mainly mountainous; the great Rocky Mountains stretch from Alaska to Mexico. West of the Rockies lie other major mountain ranges, such as the Sierra Nevada and the Cascades. East of the Rocky Mountains lie extensive plains covering a vast area from Hudson Bay south to the Gulf of Mexico and curving up along the Atlantic seaboard as far north as New York City. Another north–south-trending mountain range, the Appalachians, rises between the coastal and interior lowlands of the East. Thus the continental topography is somewhat funnel-shaped. This means that air from both polar and tropical areas can penetrate the heart of the continent, without topographic obstruction, from north and south. As a result, summer weather there can be tropical, whereas winter weather exhibits Arctic-like extremes.

Eurasia (covering 36.5 percent of the land surface) is by far the largest landmass on earth, and all of it lies in the Northern Hemisphere. The topography of Europe and Asia is dominated by a huge mountain chain that extends from west to east across the hybrid continent. It has many names in various countries, the most familiar of which are the Alps in Central Europe and the Himalayas in South Asia. In Europe, the Alps lie between the densely populated North European Lowland to the north and the subtropical Mediterranean lands to the south. In Asia, the Himalayas form but one of many great mountain ranges that emanate from central Asia northeastward into Russia's Siberia, eastward into China, and southeastward into South and Southeast Asia. Between and

World Population Distribution

The world's population is large and is currently growing at a rapid pace with little sign of slowing down. The 1995 total was approximately 5,700,000,000, with an annual net increase rate of about 1.7 percent. Thus, by the turn of the century, more than 6.2 billion people will be alive, a gain of more than 9 percent in only five years. A more revealing measure of global population growth is its *doubling time*, the number of years required to increase its size to twice its present level. That figure today is 42 years. Therefore, by 2038, 11.5 billion humans are expected on earth. If the doubling time holds constant, that 11.5 billion will become 23 billion by 2080 and 46 billion by the year 2122—an *eightfold increase* over the next 126 years!

These predictions are especially alarming because the human world is already overcrowded, and there is relatively little new living space to be opened up. The present **ecumene** (the permanently settled portion of the earth's surface) has developed over a long time period involving many centuries, and represents the totality of adjustments people have made to their environments over the course of the human experience on this planet. Thus the most productive places have been occupied for generations, and their agriculture and other technologies are being pushed to the limit to sustain large populations at close to the maximum carrying capacities of these fertile lands. And still the population keeps multiplying at a prodigious pace—at a net increase rate of over 95 million a year, nearly 8 million a month, 265,000 per day, or 11,000 every hour.

The distribution of world population is shown in Fig. 2-5. Its unevenness is immediately apparent. Fully 90 percent of humankind resides in the Northern Hemisphere. More than half of the people are concentrated on only 5 percent of the earth's land, and almost 90 percent live on less than one-fifth of the land. Because of the fertility of river valleys, plains, and deltas, lowlands contain the highest population densities, which decline markedly with increasing altitude. Well over half of humanity resides below the elevation of 200 m (660 ft), and over three-quarters live below 500 m (1650 ft). Moreover, nearly 70 percent of the population lives within 500 km (320 mi) of a seacoast.

The distributional pattern of Fig. 2-5 reveals a number of population concentrations. Three **major population clusters**, which together contain about 65 percent of humanity, are found in East Asia, South Asia, and Europe; although considerably smaller, eastern North America is often regarded as a fourth major cluster. We can also observe a number of **minor population clusters**, among them the Nile Valley and delta of Egypt in northeastern Africa, several river delta areas of Southeast Asia, the Indonesian island of Java off the Southeast Asian mainland, parts of West Africa (particularly Nigeria), and the southern Atlantic seaboard of Brazil in eastern South America.

This small number of concentrations encompasses over 80 percent of the world's population, and the remainder of the ecumene consists of much less

below these ranges lie several of the world's most densely populated river plains, including China's Huang He (Yellow) and Chang Jiang (Yangzi), and India's Ganges.

Australia is the world's smallest continent (constituting less than 6 percent of the total land area) and topographically its lowest. The Great Dividing Range lies near the continent's eastern coast, and its highest peak reaches a mere 2217 m (7316 ft). Australia's northern areas lie in the tropics, but its southern coasts are washed by the outer fringes of Antarctic waters.

Antarctica, the "frozen continent," lies almost entirely buried by the world's largest and thickest icesheet. Beneath the ice, Antarctica (constituting the remaining 9.3 percent of the world's land area) has a varied topography that includes the southernmost link in the Andean mountain chain (the backbone of the Antarctic Peninsula). Currently, very little of this underlying landscape is exposed, but Antarctica was not always a frigid polar landmass. As we will discuss in more detail later, the Antarctic ice, the air above it, and the waters around it are critically important in the global functioning of the atmosphere, hydrosphere, and biosphere.

The Ocean Basins

Until the present century, the ocean basins were unknown territory. Only the tidal fringes of the continents and the shores of deep-sea islands revealed glimpses of what might lie below the vast world ocean. Then sounding devices were developed, and some of the ocean-floor topography became apparent, at least in cross-sectional profile. Next, equipment was built that permitted the collection of rock samples from the seabed. And now marine scientists are venturing down to deep areas of the ocean floor and can watch volcanic eruptions in progress through the portholes of deep-sea submersibles. The secrets of the submerged 71 percent of the lithosphere are finally being revealed.

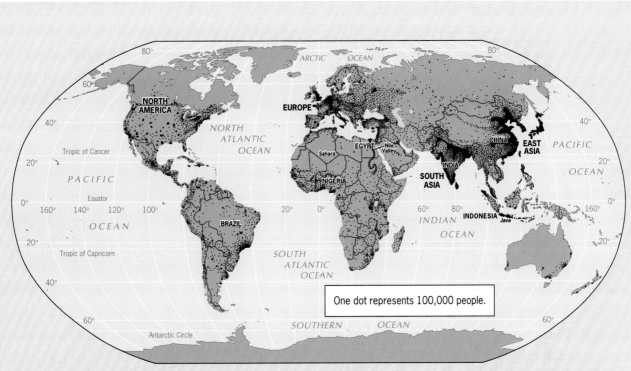

FIGURE 2-5 The spatial distribution of the world's population in 1995. The way people have arranged themselves in geographic space represents the totality of their adjustments to the environments that are capable of supporting human settlement. Individual countries are identified on the world political map in Appendix B.

densely settled areas that are less inviting in terms of topography, temperature regimes, and moisture availability. It is also worth noting that population distribution can vary within the population clusters we see on this highly generalized world map. For example, in China, the world's most populated country and heart of the East Asian cluster, the eastern half of the People's Republic ("China Proper") contains almost all the population, dominated by the enormous agglomerations of people in the Huang He and Chang Jiang river valleys. On the other hand, China's arid, mountainous west and northwest is generally as empty as North Africa's Sahara.

When the seafloors were mapped, and the composition and age of rock samples were determined, a remarkable discovery was made: the deep ocean floors are geologically different from the continental landmasses. The ocean floors have a varied topography, but this topography is not simply an extension of what we see on land. There are ridges and valleys and mountains and plains, but these are not comparable to the Appalachians or the Amazon Basin.

We return to this topic in Part Four. For the present, we should acquaint ourselves with the main features of the ocean basins. If all the water were removed from the ocean basins, they would reveal the topography shown in Fig. 2-6. The map reveals that the ocean basins can be divided into three regions: (1) the margins of the continents, (2) the abyssal zones—extensive, mound-studded plains at great depth, and (3) a system of ridges flanked by elaborate fractures and associated relief.

The continental margin consists of the continental shelf, continental slope, and continental rise. The **continental shelf** is the very gently sloping, relatively shallow, submerged plain at the edge of the continent. The map shows them to be continuations of the continental landmasses (*see* especially the areas off eastern North America, southeastern South America, and northwestern Europe). Generally, these shelves extend no deeper than 100 fathoms (180 m [600 ft]). Their average width is about 70 km (45 mi), but some are as wide as 1000 km (625 mi). Because the continental shelves are extensions of the landmasses, the continental geologic structure continues to their edges. During the most recent ice age, when much ocean water was taken up by the icesheets, sea levels dropped enough to expose most of these continental shelves. Rivers flowed across them and carved valleys that can still be seen on detailed maps. So the oceans today are fuller than in the past: they have flooded extensive plains at the margins of the continents.

At a depth of about 100 fathoms (180 m), the continental shelf ends at a break in slope that is quite marked in some places and less steep in others. At this dis-

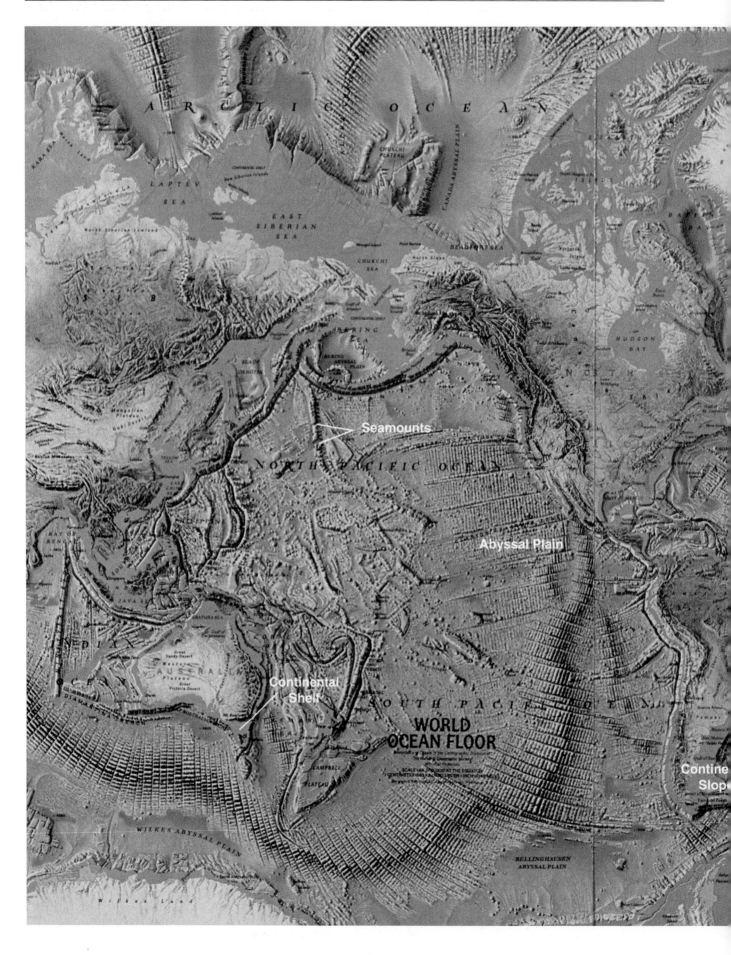

continuity the **continental slope** begins and plunges steeply downward. Often at the foot of the continental slope there is a transitional **continental rise** of gently downsloping seafloor (Fig. 2-6). The continental rise leads into the abyssal zone. This zone consists mainly of the **abyssal plains**, large expanses of lower relief ocean floor. The abyssal plains form the floors of the deepest areas of each ocean, except for even deeper trenches that occur at the foot of some continental slopes (as along the Pacific margin of Asia). The abyssal zone is not featureless, however, and the extensive plains are diversified by numerous hills and **seamounts**, all of which are of volcanic origin. (Seamounts are volcanic mountains reaching over 1000 m [3300 ft] above the seafloor.) There also are lengthy valleys, as though rivers had carved them here more than 1000 fathoms (1800 m) below the water surface. The origin of these valleys remains uncertain.

The third major ocean-floor feature is a global system of **midoceanic ridges**. These ridges are high, submarine, volcanic mountain ranges. The existence of one such ridge, the Mid-Atlantic Ridge, was long known to scientists. But its properties were not understood until quite recently, when it became clear that midoceanic ridges also extend across the Indian, Pacific, and Southern Ocean floors (Fig. 2-6). As is noted in Part Four, these midoceanic ridges are the scenes of active submarine volcanism and major movements of the earth's crust. When scientists were able to observe them for the first time, they brought back dramatic records of violent eruptions, superheated water, and exotic marine animals living along these active ridges that had never been seen before.

The active character of the midoceanic ridges, and the geologic properties of the ocean floor, become important to us when we study their implications in geomorphology. But the configuration of the ocean floors affects the movement of ocean water just as the topography on land influences the movement of air. The ocean basins are filled with water that, like the air, is in constant motion. Great permanent circulation systems have developed in the oceans, and these systems affect the weather and climate on neighboring continents. Our water-dominated planet is finally yielding secrets it has held for eons.

FIGURE 2-6 The world ocean floor. Examples of the six major features highlighted by boldface type in the text on pages 19 and 21 are labeled.

KEY TERMS

abyssal plains (p. 21)
atmosphere (p. 13)
biosphere (p. 13)
continental rise (p. 21)
continental shelf (p. 19)
continental slope (p. 21)

ecumene (p. 18)
hemisphere (p. 14)
hydrosphere (p. 13)
Land Hemisphere (p. 15)
lithosphere (p. 13)
major population clusters (p. 18)

midoceanic ridge (p. 21)
minor population clusters (p. 18)
regional subsystem (p. 13)
seamount (p. 21)
Water Hemisphere (p. 15)

REVIEW QUESTIONS

1. Name the earth's four *spheres* or interacting shells.
2. What is the *regional subsystem* and how does it relate to the earth's four spheres?
3. What is the difference between the Land Hemisphere and the Water Hemisphere?
4. Compare and contrast the distribution of continental land-masses in the Northern and Southern Hemispheres.
5. Describe the basic spatial patterns exhibited by the distribution of the world's population.
6. Describe the major topographic features of the world's ocean basins.

REFERENCES AND FURTHER READINGS

DE BLIJ, H. J., and MULLER, P. O. *Geography: Realms, Regions, and Concepts* (New York: Wiley, 7th ed., 1994).

"The Dynamic Earth," *Scientific American*, September 1983 (special issue).

GOUDIE, A. S., et al., eds. *The Encyclopedic Dictionary of Physical Geography* (Cambridge, Mass.: Blackwell, 3rd ed., 1994).

National Geographic Atlas of the World (Washington, D.C.: National Geographic Society, 6th ed., rev., 1992).

NEWMAN, J. L., and MATZKE, G. E. *Population: Patterns, Dynamics, and Prospects* (Englewood Cliffs, N.J.: Prentice-Hall, 1984).

SKINNER, B. J., and PORTER, S. C. *The Blue Planet: An Introduction to Earth System Science* (New York: Wiley, 1995).

WHITTOW, J. B. *The Penguin Dictionary of Physical Geography* (New York: Viking Penguin, 1984).

Mapping the Earth's Surface

OBJECTIVES

◆ To introduce the reference system for locations on the earth's surface.

◆ To describe the most important characteristics of maps and the features of common classes of map projections.

◆ To discuss the elements of map interpretation and contemporary cartographic techniques.

W̲hen you go to a new city, your process of learning about it begins at the hotel you stay at or in your new home. You then locate the nearest important service facilities, such as supermarkets and shopping centers. Gradually your knowledge of the city, and your activity space within it, grows. This slow pace of learning about the space you live in repeats the experience of every human society as it learned about the earth. Early in the learning process, directions concerning the location of a particular place have to be taken

from or given to someone. The most common form of conveying such information is the **map**, which Phillip Muehrcke has defined as *any geographical image of the environment*. We have all seen sketch maps directing us

UNIT OPENING PHOTO: The latest cartographic technology: satellite image of the Naples region of southern Italy, with Vesuvius volcano (right of center) lying just to the east of the city.

a

b

FIGURE 3-1 The Mesopotamian world map. The original (a) was constructed ca. 2500 B.C. A modern diagram of the original (b) more clearly indicates its principal features.

to a place for a social gathering. The earliest maps were of a similar nature, beginning with maps scratched in the dust or sand.

Although humans have been drawing maps throughout most of their history, the oldest surviving maps date only from about 2500 B.C. They were drawn on clay tablets in Mesopotamia (modern-day Iraq) and represented individual towns, the entire known country, and the early Mesopotamian view of the world (Fig. 3-1). The religious and astrological text above the map (Fig. 3-1a)

indicates that the Mesopotamian idea of space was linked to ideas about humankind's place in the universe. This is a common theme in **cartography**, the science, art, and technology of mapmaking and map use. Even today, maps of newly discovered space, such as star charts, raise questions in our minds of where we, as humans, fit into the overall scheme of things.

THE SPHERICAL EARTH

If you look out your window, there is no immediate reason for you to suppose that the earth's surface is anything but flat. It takes a considerable amount of traveling and observation to reach any other conclusion. Yet, the notion of the earth as a sphere is a longstanding one and was accepted by several Greek philosophers as far back as 350 B.C. By 200 B.C., the earth's circumference (approximately 40,000 km [25,000 mi]) had been accurately estimated by Eratosthenes to within 1 percent of its actual size. The idea of a spherical earth continued to be challenged, however, and was not universally accepted until the Magellan expedition successfully circumnavigated the globe in the early sixteenth century.

Dividing the Earth

When a sphere is cut into two parts, the edges of the cut form circles. Once the earth was assumed to be spherical, it was logical to divide it by means of circles. Sometime between the development of the wheel and the measurement of the planet's circumference, mathematicians had decided that the circle should be divided into 360 parts by means of 360 straight lines radiating from the center of the circle. The angle between each of these lines was called a **degree**. For such a large circle as the earth's circumference, further subdivisions became necessary. Each degree was divided into 60 *minutes*, and each minute was further subdivided into 60 *seconds*.

With a system for dividing the curved surface of the earth, the problem became the origin and layout of these circles. Two sets of information could be used to attack this problem. First, a sense of direction had been gained by studying the movements of the sun, moon, and stars. In particular, the sun at midday was always located in the same direction, which was designated as *south*. Knowing this, it was easy to arrive at the concepts of north, east, and west. The division of the circle could refine these directional concepts. The second piece of information was that some geographical locations in the Mediterranean region, fixed by star measurements, could be used as reference points for the division of the earth.

Using this knowledge, it was possible to imagine a series of "lines" on the earth's surface (in actuality, circles around the spherical earth), some running north–south and some running east–west, which together form a grid. The east–west lines of this grid are still called **parallels**, the name the Greeks gave them; the north–south lines are called **meridians**. As we can see in Fig. 3-2, the two sets of lines differ. Parallels never intersect one another, whereas meridians intersect at the top and bottom points (poles) of the sphere.

Latitude and Longitude

The present-day divisions of the globe stem directly from the earth grids devised by ancient Greek geographers. The parallels are called lines of latitude. The parallel running around the middle of the globe, the **equator**, is defined as zero degrees latitude. As Fig. 3-3a illustrates, **latitude** is the angular distance, measured in degrees north or south, of a point along a parallel from the equator. Lines of latitude in both the Northern and Southern Hemispheres are defined this way. Thus New York City has a latitude of 40 degrees, 40 minutes north of the equator (40°40′N), and Sydney, Australia a latitude of 33 degrees, 55 minutes south (33°55′S). The "top" of the earth, the *North Pole*, is at latitude 90°N; the "bottom," the *South Pole*, is at 90°S.

The meridians are called lines of **longitude**. In 1884, the meridian that passes through the Royal Observatory at Greenwich in London, England, was established as the global starting point for measuring longitude. This north–south line is called the **prime meridian** and is defined as having a longitude of zero degrees. Longitude is the angular distance, measured in degrees east or west, of a point along a meridian from the prime meridian. The other meridians are ascertained as if we were looking down on the earth from above the North Pole, as Fig. 3-3b indicates. We measure both east and west from the prime meridian. Therefore, New York City has a longitude of 73 degrees, 58 minutes west of the Greenwich meridian (73°58′W), and Sydney has a longitude of 151 degrees, 17 minutes east (151°17′E).

Because meridians converge at the North and South

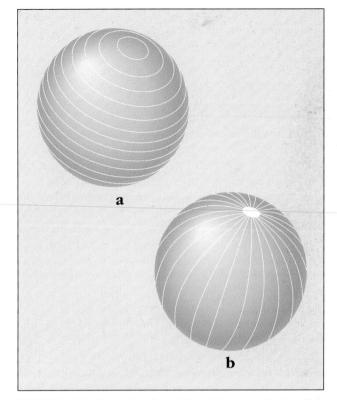

FIGURE 3-2 Parallels (a) and meridians (b) on a globe. Parallels run east–west; meridians run north–south.

Poles, the actual distance contained in one degree of longitude varies from 111 km (69 mi) at the equator to zero at the poles. In contrast, the length of a degree of latitude is about 111 km (69 mi) anywhere between 0° and 90°N or S. We say "about" because the earth is not a perfect sphere—it bulges slightly at the equator and is flattened at the poles. The Greek geographers, however, knew nothing of this. They had a more immediate and difficult

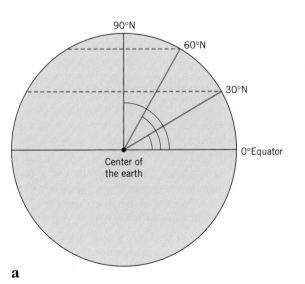

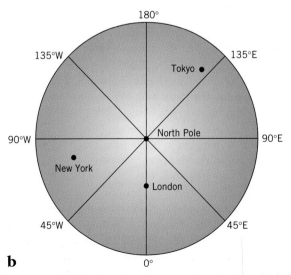

FIGURE 3-3 Latitude and longitude. Viewed from the side (a), lines of latitude (including the equator) are horizontal parallels. Lines of longitude, the meridians, appear to radiate from a center point when viewed from above the North Pole (b); they converge again at the South Pole.

problem to confront: how could they represent the three-dimensional earth on a flat chart?

MAP PROJECTIONS

If you have ever tried to cut the skin off an orange or any other spherical surface and then lay it flat on a table, you realize that this is not an easy task. At least some part of the skin must be stretched to make it completely flat. Cylinders or cones may easily be cut to be laid out flat without distortion, but not a sphere, which in geometric and cartographic terms is an *undevelopable* surface, incapable of being flattened. Once the ancient Greeks had accepted the idea that the earth was a sphere, they had to determine how best to represent the round earth on a flat surface. There is no totally satisfactory solution to this problem, but the early mapmakers soon invented many of the partial solutions that are still used commonly today.

The Greeks had noted that a light placed at the center of a globe casts shadows along the meridians and parallels. These shadows, which form lines, can be "projected" outward onto some surface that can later be cut and laid out flat. The resulting series of projected lines on the new surface is called a **map projection**, which may be defined as an orderly arrangement of meridians and parallels, produced by any systematic method, that can be used for drawing a map of the spherical earth on a flat surface.

Properties of Map Projections

Any map projection has three variable properties: scale, area, and shape. **Scale** is the ratio of the size of an object on the map to the actual size of the object it represents. For example, consider a model globe that has a diameter of 25 cm (10 in). It represents the real earth, whose diameter is 12,900 km (8000 mi). Therefore, 1 cm of the globe represents 516 km (12,900 km divided by 25 km), or 51,600,000 cm, on the real earth. So we say that the scale of the model globe is "1 to 51,600,000" or 1:51,600,000. Why must we use such large numbers instead of saying 1 (cm) to 516 (km)? Because the first rule of fractions is that the numerator and denominator must always be given in the same mathematical units.

The *area* of a section of the earth's surface is found by multiplying its east–west distance by its north–south distance. This calculation is simple for rectangular pieces of land, but tedious for territories with more complicated shapes. In many map projections, complex real areas can be well represented simply by controlled shrinking (scaling down the distances). Florida has the same area relative to other U.S. states on a model globe as in the real world. The only difference is that the scale has changed.

Now consider *shape*. When a map projection preserves the true shape of an area, it is said to be conformal. Shape can often be preserved in map projections—but

not always. In the real world, the shape of Colorado is almost a rectangle. On a map with a scale of 1:50,000 for north–south distances and a scale of 1:200,000 for east–west distances, Colorado is squeezed in the east–west direction and stretched in the north–south direction.

In creating a map projection from a globe, we can preserve one, or sometimes two, of the properties of scale, area, and shape. But it is not possible to preserve all three at once over all parts of the map. Try it for yourself by drawing a shape on a plastic ball and then cutting the ball to make a flat, two-dimensional map. You will be forced to compromise. To minimize this problem, the Greeks followed the rule cartographers still use: select a map projection best suited to the particular geographic purpose at hand.

Types of Map Projections

Since the time of the pioneering Greeks, mapmakers have devised hundreds of projections to flatten the globe so that all of it is visible at once. Indeed, mathematically, it is possible to create an infinite number of map projections. In practice, however, these cartographic transformations of the three-dimensional, spherical surface of the earth have tended to fall into a small number of categories. The four most common classes of map projections are considered here: cylindrical, conic, plane, and equal-area.

A **cylindrical projection** involves the transfer of the earth's latitude/longitude grid from the globe to a cylinder, which is then cut and laid flat. When this operation is completed, the parallels and meridians appear on the opened cylinder as straight lines intersecting at right angles. On any map projection, the least distortion occurs where the globe touches, or is *tangent* to, the geometric object it is projected onto; the greatest distortion occurs farthest from this place of contact. In Fig. 3-4a, we observe in the left-hand diagram that the globe and cylinder are tangent along the parallel of the equator. The parallel of tangency between a globe and the surface onto which it is projected is called the **standard parallel**. The right-hand diagram of Fig. 3-4a shows a globe larger than the cylinder, and two standard parallels result. This projection reduces distortion throughout the map, and is particularly useful for representing the low-latitude zone straddling the equator between the pair of standard parallels.

Mathematical modifications have increased the utility of cylindrical projections, the best known of which was devised in 1569 by the Flemish cartographer Gerhardus Mercator. In a **Mercator projection** (Fig. 3-5), the spacing of parallels increases toward the poles. This increase is in direct proportion to the false widening between normally convergent meridians that is necessary to draw those meridians as parallel lines. Although this produced extreme distortion in the area of the polar latitudes, it provided a tremendously important service for

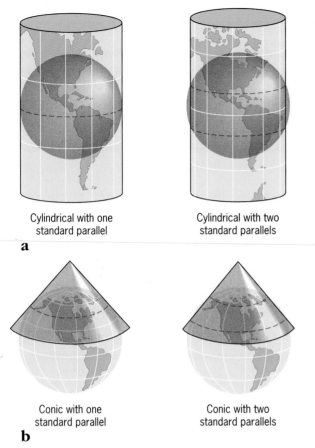

Cylindrical with one
standard parallel

Cylindrical with two
standard parallels

a

Conic with one
standard parallel

Conic with two
standard parallels

b

FIGURE 3-4 The construction of cylindrical (a) and conic (b) projections with one and two standard parallels.

FIGURE 3-5 The Mercator projection, which produces straight parallels and meridians. Any straight line drawn on this map is

navigators using the newly perfected magnetic compass. Unlike any other map projection, a straight line drawn on this one is a line of true and constant compass bearing. Such lines are called *rhumb lines*. Once a navigator has determined from the Mercator map of the world the compass direction to be traveled, the ship can easily be locked onto this course.

Cones can be cut and laid out flat as easily as cylinders, and the **conic projection** has been in use almost as long as the cylindrical. A conic projection involves the transfer of the earth's latitude/longitude grid from a globe to a cone, which is then cut and laid flat. Figure 3-4b shows the derivation of the two most common conic projections, the one- and two-standard-parallel cases. On a conic projection, meridians are shown as straight lines that converge toward the (North) Pole. Parallels appear as arcs of concentric circles with the same center point that shorten as latitude increases. This projection is best suited for the middle latitudes, such as the United States and Europe, where distortion is minimal if the apex of the imaginary cone is positioned directly above the North Pole.

Planar projections, in which an imaginary plane touches the globe at a single point, exhibit a wheel-like symmetry around the point of tangency between the plane and the sphere. Planar projections were the first map projections developed by the ancient Greeks. Today

a (rhumb) line of constant compass bearing, a tremendous advantage for long-distance navigation.

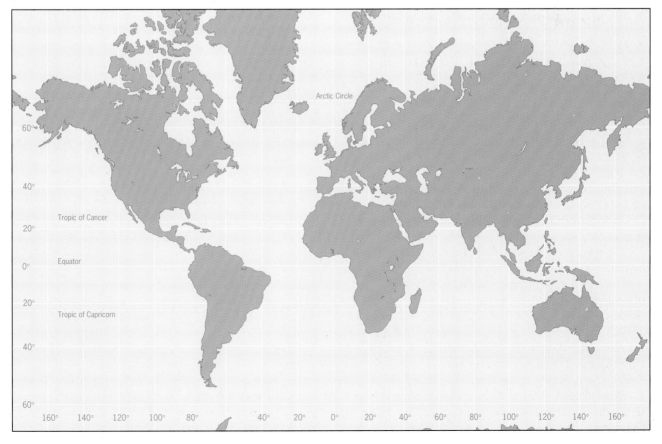

they are most frequently used to represent the polar regions (Fig. 3-6). One type of planar projection, the gnomonic, possesses an especially useful property: a straight line on this projection is the shortest route between two

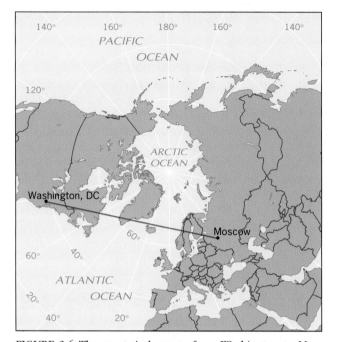

FIGURE 3-6 The great-circle route from Washington to Moscow, which becomes a straight line on this polar gnomonic projection.

points on the earth's surface, which has vital implications in this age of intercontinental jet travel. Long international flights seek to follow the shortest routes, and these are found on the spherical earth by imagining the globe to be cut exactly in half along a straight line running through the origin and destination cities. When a sphere is cut in half, the circle formed along the edge of the cut is called a *great circle*. (*Small circles* are the edges of all other cuts when a sphere is divided into two unequal portions.) Long-distance air traffic usually follows great-circle routes, such as the one shown between Moscow and Washington in Fig. 3-6.

We have noted that, through the mathematical manipulation of projective geometry, it is possible to derive an unlimited number of map projections that go beyond the convenience and simplicity of the cylinder, cone, and plane. Among these other types of map projections, **equal-area projections** rank among the most important. An equal-area projection is one in which all the areas mapped are represented in correct proportion to one another. Thus, on a world map of this type, the relative areal sizes of the continents are preserved. True shape, however, must be sacrificed. Nonetheless, good equal-area projections attempt to limit the distortion of shape so that the continental landmasses are still easily recognizable. This should be carefully noted in Fig. 3-7, which displays the flat polar quartic projection. Because they maintain the areal relationships of every part of the

FIGURE 3-7 An equal-area projection: all areas mapped are represented in their correct relative size. This is the flat polar quartic equal-area projection—in interrupted form—which was developed in 1949 for the U.S. Coast and Geodetic Survey by F. W. McBryde and P. D. Thomas.

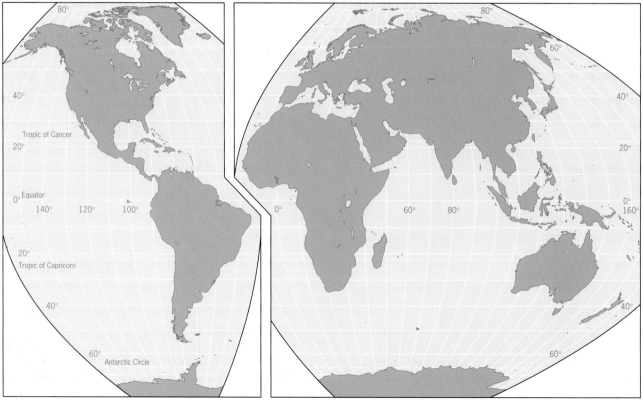

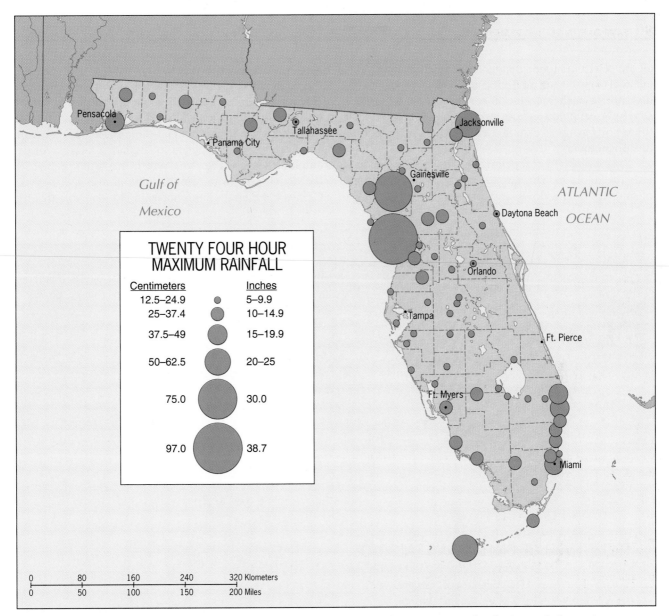

TWENTY FOUR HOUR
MAXIMUM RAINFALL

Centimeters		Inches
12.5–24.9		5–9.9
25–37.4		10–14.9
37.5–49		15–19.9
50–62.5		20–25
75.0		30.0
97.0		38.7

FIGURE 3-8 Florida's daily rainfall maxima in excess of 12.5 cm (5 in), as of 1980, an example of point-symbol mapping.

globe, equal-area projections are particularly useful for mapping the worldwide spatial distributions of land-based phenomena. The map in Fig. 3-7 also possesses another feature you have undoubtedly observed by now: it is not a continuous representation but is *interrupted*. This device helps the cartographer minimize distortion, devote most of the projection to the parts that project best, and deemphasize areas of the globe that are not essential to the distribution at hand (such as omitting large parts of the oceans in mappings of land-based phenomena).

MAP INTERPRETATION

One of the most important functions of maps is to communicate their content effectively and efficiently. Because so much spatial information exists in the real world, cartographers must first carefully choose the in-

formation to be included and deleted in order to avoid cluttering the map with less relevant data. Thus maps are models: their compilers simplify the complexity of the real world, filtering out all but the most essential information. Even so, a considerable amount of information remains, and all of it must be compressed into the small confines of the final map. To facilitate that task, cartographers have learned to encode their spatial messages through the use of symbolization. Decoding this cartographic shorthand is not a difficult task, and the place to begin is the map's key or **legend** (or sometimes its written caption), in which symbols and colors are identified. These symbols usually correspond to the different categories of geographic data: dimensionless *points*, one-dimensional *lines*, two-dimensional *areas*, and three-dimensional *volumes* or *surfaces*.

Point symbols tell us the location of each occurrence of the phenomenon being mapped and, frequently, its quantity. This is illustrated in Fig. 3-8, which is a map of all the places in Florida that have recorded 24-hour rainfall totals in excess of 12.5 cm (5 in). A dot or circle marks the spot of each such location, whose rainfall amount can be ascertained in the legend. Taken together, all these point symbols exhibit the statewide distribution of this phenomenon.

Line symbols represent linkages and/or flows that exist between places. The map in Fig. 3-9 shows the pattern of intercity freight flows on Florida's railroad network.

Here we observe both the rail system's connections and the volume of freight that flows among these connected cities (represented by varying line thickness that can be measured in the legend).

Area symbols portray two-dimensional spaces, with colors representing specific quantitative ranges. An example is seen in Fig. 3-10, which maps the number of days each year with temperatures above 32.2 degrees Celsius (90 degrees Fahrenheit) in Florida. Note that the

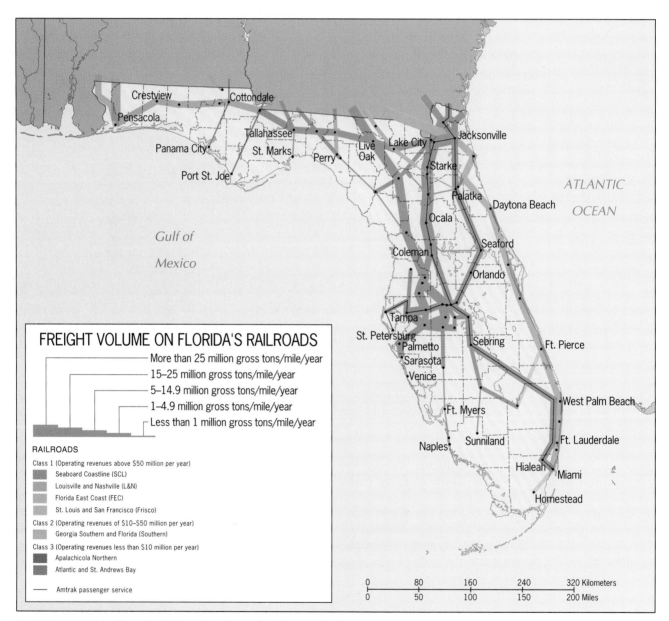

FIGURE 3-9 Freight flows in 1981 on Florida's railroad network, demonstrating the line-symbol technique.

highest quantity (125-plus days) is represented by the darkest symbol and that the area symbols become progressively lighter as number of days decreases. This map also illustrates how color printing can enhance the precision and versatility of cartographic communication. The color scheme of these temperature zones is designed to allow a second set of areas to show through in the background—Florida's 67 counties, which are very help-

ful for accurately determining the placement of the boundaries of each zone.

Volume symbols describe *surfaces*, which can be generalizations of real surfaces (such as the world topographic relief map in Fig. 2-3) or representations of conceptual surfaces. An example of the latter is provided in Fig. 3-11, which shows Florida's yearly rainfall as a "surface" that possesses "ridges" representing higher rainfall

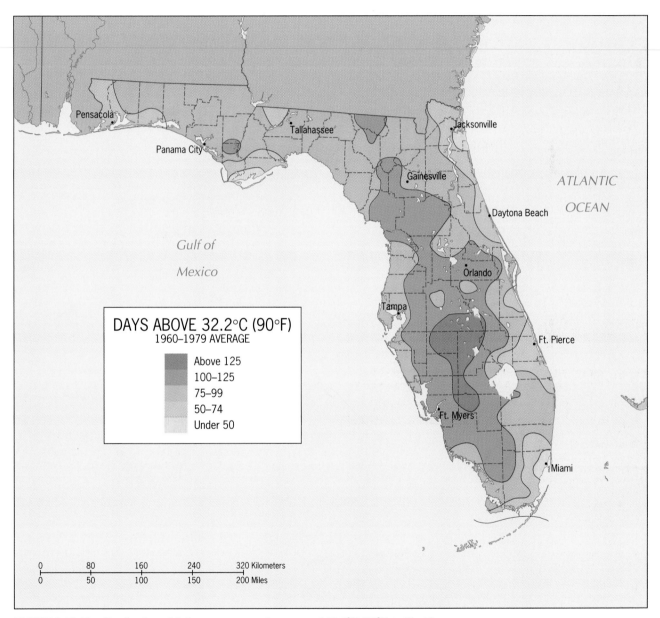

FIGURE 3-10 The distribution of daily temperatures in excess of 32.2°C (90°F) in Florida. This map illustrates area-symbol mapping, which is significantly enhanced when color printing is possible.

amounts along the northwestern and southeastern coasts, and a "trough" representing lower rainfall totals extending down through the interior of the peninsula. The mapping technique employed in this figure is commonly used by cartographers to solve the problem of portraying three-dimensional volumetric data on a two-dimensional map. It is known as **isarithmic (isoline) mapping**, and consists of numerous **isolines** that con-

nect all places possessing the same value of a given phenomenon or "height" above the flat base of the surface. In Fig. 3-11, the boundary lines between each color zone, as the legend indicates, connect all points reporting that particular rainfall amount. For example, the cigar-shaped, dark green ridge just west of Miami is enclosed by the 152-cm (60-in) isoline (or *isohyet*—the specific name given to an isoline connecting points of identical

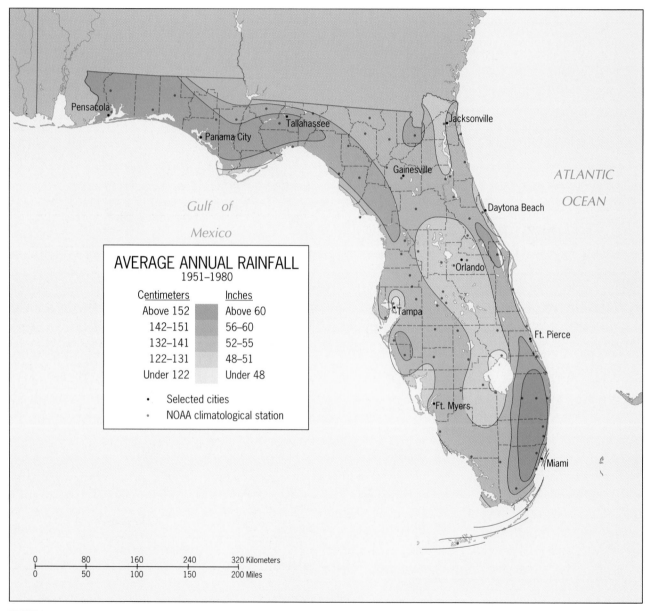

FIGURE 3-11 Average annual rainfall distribution in Florida, displaying the technique of isarithmic mapping, which allows three-dimensional volumetric data to be shown on a two-dimensional map.

rainfall total). Similarly, the 132-cm (52-in) isohyet encloses the long north–south trough running south from Orlando through the center of the peninsula.

Perhaps the best known use of isolines in physical geography is the representation of surface relief by **contouring**. As Fig. 3-12 demonstrates, each contour line represents a specific and constant elevation, and all the contours together provide a useful generalization of the surface being mapped. A more practical example of contouring is shown in Fig. 3-13, where the portion of landscape in Fig. 3-13a corresponds to the *topographic map* of that terrain in Fig. 3-13b. By comparing the two, it is

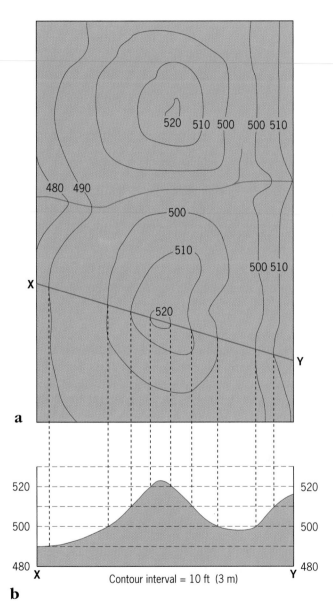

a

b

FIGURE 3-12 In topographic contouring, each contour line's points have the identical height above sea level. The surface relief described by the contour map (a) can be linked to a cross-sectional profile of the terrain. (In diagram b, this is done for line **X-Y** on map a.) Note how contour spacing corresponds to slope patterns: the wider the spacing, the gentler the slope and vice versa.

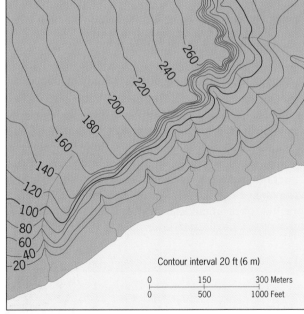

a

b

FIGURE 3-13 A perspective sketch of a coastal landscape (a) and its corresponding topographic map (b), taken from U.S. Geological Survey sources. Note that this map is scaled in feet and that the contour interval—most appropriate to this map—is 20 ft (6 m).

easy to read the contour map and understand how this cartographic technique accurately portrays the configuration of the earth's surface relief.

NEW CARTOGRAPHIC FRONTIERS

The past three decades have witnessed rapid advances in the field of computing. As computer power and speed increased, graphic performance capabilities greatly improved and increasingly allowed this technology to be used for mapmaking. Thus today it is ever more likely that any given map is an automated product, "drawn" by a computer that has manipulated spatial data compiled by a geographic information system. A **geographic information system—GIS** for short—is an assemblage of computer hardware and software that enable spatial data to be collected, recorded, stored, retrieved, manipulated, analyzed, and displayed to the user. Increasingly, the data acquired for GIS processing are obtained by **remote sensing** of the earth's environment from high-altitude observation platforms (*see* Perspective: Remote Sensing of the Environment). This approach allows for the simultaneous collection of several layers of data pertaining to the same study area. These layers can then be integrated by multiple map overlays in order to assemble the components of the complex real-world pattern (Fig. 3-14).

Perhaps the most revolutionary aspect of GIS cartography is its break with the static map of the past. The use of GIS methodology involves a constant dialogue, via computer commands and feedback to queries, between the map and the map user. This instantaneous, two-way communication is known as **interactive mapping**, and it is expected to become the cornerstone of cartography in the future. Just one of the many possibilities of this technique is the use of video-disk maps displayed on automobile dashboards, allowing the driver to ask questions that elicit immediate map directions showing the best route to the desired destination. Cartographers have

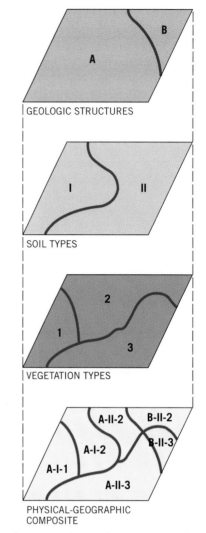

FIGURE 3-14 GIS processing of overlays to produce map composite.

always been quick to adopt new technologies, but the pace of change today is so swift that the venerable science of mapmaking is being completely transformed in the final years of this century.

KEY TERMS

cartography (p. 24)

conic projection (p. 27)

contouring (p. 33)

cylindrical projection (p. 26)

degree (p. 24)

equal-area projection (p. 28)

equator (p. 25)

geographic information system (p. 34)

interactive mapping (p. 34)

isarithmic (isoline) mapping (p. 32)

isoline (p. 32)

latitude (p. 25)

legend (p. 29)

longitude (p. 25)

map (p. 23)

map projection (p. 26)

Mercator projection (p. 26)

meridians (p. 24)

parallels (p. 24)

planar projection (p. 27)

prime meridian (p. 25)

remote sensing (pp. 34, 35)

scale (p. 26)

standard parallel (p. 26)

Remote Sensing of the Environment

Maps have been used for centuries to graphically communicate information about the earth's surface. The Greek legend of Icarus, who flew too close to the sun and melted his wax-and-feather wings, shows how badly the scientific ancestors of modern geographers wanted to view the earth from the sky. Today this is possible, and the work of mapmakers and spatial analysts is greatly facilitated and enhanced by powerful new tools and techniques. The most important of these is remote-sensing technology, the ability to scan the earth from distant observation platforms.

Remote sensing has been defined by Benjamin Richason as *any technique of imaging objects without the sensor being in direct contact with the object or scene itself.* He goes on to point out that geographers who use this method normally *collect* data via an appropriate imaging system, *interpret* that spatial information (which is stored in the system, usually on film or computer tape), and *display* and *communicate* the results on a map.

Aerial photography is a remote-sensing technique that has been used since the advent of cameras early in the nineteenth century. Even though the Wright brothers did not take off until 1903, hot-air balloons and even trained birds were able to carry cameras aloft before 1850. With the rapid proliferation of aircraft in this century, several methods were developed (many by the military to improve its advantages in ground warfare) that permitted maps to be produced directly from series of photographs taken from planes.

At the same time, photographic technology was being perfected to expand this capability. Along with improved camera systems came ever more sensitive black-and-white and then color films. Moreover, by World War II, ultrasensitive film breakthroughs extended the use of photography into the *infrared radiation (IR)* range beyond the visible capacities of the human eye. By directly "seeing" reflected and radiated solar energy, aerial infrared photography could, for the first time, penetrate clouds, haze, and smoke—and even obtain clear images of the ground at night. The U.S. Air Force had further pioneered the use of IR color imagery, although the "colors" obtained—known as *false-color images*—bore no resemblance to the natural colors of the objects photographed (but could readily be decoded by analysts).

Over the past half-century, nonphotographic remote sensing has developed rapidly as new techniques and instruments have opened up a much wider portion of the *electromagnetic spectrum*. This spectrum is diagrammed in Fig. 3-15, and consists of a continuum of electric and magnetic energy as measured by wavelength, from the high-energy *shortwave radiation* of cosmic rays (whose waves are calibrated in billionths of a meter) to the low-energy *longwave radiation* of radio and electric power (with waves measured in units as large as kilometers).

Figure 3-15 also shows the *spectral bands*—or the portions of the electromagnetic spectrum—that can be picked up by radio, radar, IR sensors, and other detectors. As the new science of remote sensing matured, its practitioners learned more and more about those parts of the spectrum and about which types of equipment are best suited to studying various categories of environmental phe-

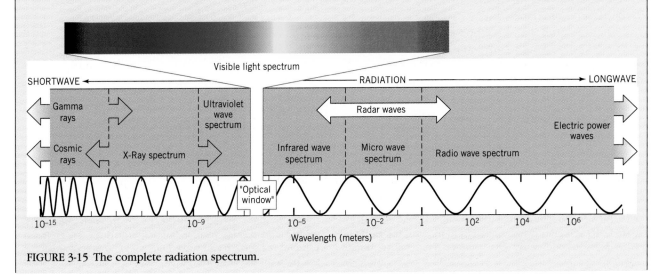

FIGURE 3-15 The complete radiation spectrum.

nomena. This is important because each surface feature or object emits and reflects a unique pattern of electromagnetic energy—its *spectral signature*—that can be used to identify it, much like a fingerprint can identify any human individual.

Access to the nonvisible portions of the electromagnetic spectrum is a great technological triumph that is now paying rich dividends. Before this breakthrough, our perception was narrowly limited to the visible portion of the spectrum (the "optical window" shown in Fig. 3-15), which one scientist has likened to the width of a pencil in comparison to the earth's circumference (40,000 km [25,000 mi]).

Two kinds of remote-sensing systems have been devised to collect and record electromagnetic pulses. *Passive systems* measure energy radiated and/or reflected by an object, such as the IR photography method described above. More common

today are *active systems* that transmit their own pulsations of energy, thereby "illuminating" target objects that "backscatter," or reflect, some of that energy to receiving sensors that "see" the image. *Radar* is a good example of such a system, and Fig. 3-16 shows how high-altitude radar can be used to plot the ocean surface and measure wave heights. The same diagram also demonstrates how natural IR emissions can be sensed to provide data on land, sea, and cloud-top temperatures. Thus many remote-sensing platforms employ both active and passive systems. In fact, they increasingly employ *multispectral systems*, arrays of scanners attuned simultaneously to several different spectral bands, which greatly enhances the quality of observations and their interpretations.

Although remote sensors can be ground-based (e.g., for use in profiling the atmosphere above), most systems that facilitate our understanding of phys-

ical geography need to collect data at high altitudes. Aircraft have been and continue to be useful, but they are limited by how high they can fly (approximately 20 km [12 mi]) and the weather conditions in which they can operate. With planes unable to reach the very high altitudes required to obtain the small-scale imagery for seeing large areas of the surface in a single view, the opening of the space age in 1957 soon provided the needed alternative in artificial orbiting satellites.

By the 1980s, dozens of special-purpose satellites were circling the earth at appropriate altitudes and providing remotely sensed data to aid in assembling the "big picture." Among the most important of these today is the NOAA (National Oceanic and Atmospheric Administration) GOES-8 weather satellite, launched in 1994, which is fixed in geosynchronous orbit so that it remains stationary above the same point on earth.

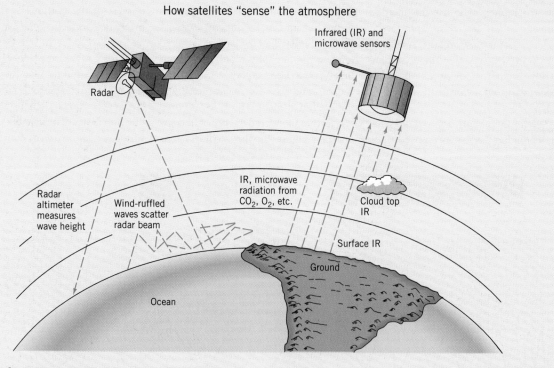

FIGURE 3-16 Satellites can use radar and naturally emitted infrared (IR) and microwave emissions to monitor the earth. IR emitted by ground or sea gives surface temperature. IR and microwave emissions from such air constituents as carbon dioxide or oxygen give average temperature for several layers (typically nine) of the atmosphere up through the stratosphere, and average relative humidity for three layers. Radar can plot sea surface and wave height and, indirectly, map the sea bottom. Radar scattered by waves reveals surface wind speed.

GOES-8 keeps watch on the oceans bordering the United States in order to detect approaching storms and increase the accuracy of daily weather forecasts.

Also vital are the four *Landsat* satellites that were launched between 1972 and 1982 to provide a steady stream of data about the earth and its myriad resources. Using a battery of state-of-the-art multispectral scanners and special television cameras, *Landsat's* sensors have provided new insights into geologic structures, the expansion of deserts, and the growth and contraction of algae and other organisms that are crucial to food chains in the oceans. At the same time, these satellites carefully monitor world agriculture, forestry, and countless other environment-impacting human activities (Fig. 3-17).

FIGURE 3-17 A *Landsat* image of a large portion of Southern California, exhibiting an array of natural and artificial landscape features. The most prominent natural features are the western corner of the Mojave Desert (upper-right quadrant) and its framing mountain ranges. Dominating the cultural landscape is metropolitan Los Angeles, the bluish, red-speckled coastal lowland in the bottom right-center of the image.

REVIEW QUESTIONS

1. What is the difference between latitude and longitude? What are the reference lines and/or points for each measurement system?

2. Describe how the properties of *scale, area*, and *shape* relate to a map projection.

3. What are the differences between cylindrical, conical, and planar map projections?

4. What is an equal-area projection, and what mapping task(s) is it well suited to?

5. Define and give examples of point, line, area, and volume map symbols.

6. What are the distinguishing features of the geographic information system and remote-sensing techniques?

REFERENCES AND FURTHER READINGS

BRACKEN, I., and WEBSTER, C. *Information Technology in Geography and Planning: Principles of Geographical Information Systems* (London/New York: Routledge, 1991).

BYLINSY, G. "Managing With Electronic [GIS-generated] Maps," *Fortune*, April 24, 1989, 237–254.

CAMPBELL, J. *Map Use and Analysis* (Dubuque, Iowa: Wm. C. Brown, 2nd ed., 1994).

CAMPBELL, J. B. *Introduction to Remote Sensing* (New York: Guilford Press, 1987).

CARTER, J. R. *Computer Mapping: Progress in the '80s* (Washington, D.C.: Association of American Geographers, Resource Publications in Geography, 1984).

DENT, B. D. *Cartography: Thematic Map Design* (Dubuque, Iowa: Wm. C. Brown, 3rd ed., 1993).

GREENHOOD, D. *Mapping* (Chicago: University of Chicago Press, 1964).

MUEHRCKE, P.C. *Map Use: Reading–Analysis–Interpretation* (Madison, Wis.: JP Publications, 3rd ed., 1992).

RICHASON, B. F., JR. "Remote Sensing: An Overview," in B. F. Richason, Jr., ed., *Introduction to Remote Sensing of the Environment* (Dubuque, Iowa: Kendall/Hunt, 2nd ed., 1983), 3–15 (definition of remote sensing on p. 5).

ROBINSON, A. H., et al. *Elements of Cartography* (New York: Wiley, 6th ed., 1995).

SNYDER, J. P. *Flattening the Earth: Two Thousand Years of Map Projections* (Chicago: University of Chicago Press, 1993).

STAR, J., and ESTES, J. *Geographic Information Systems: An Introduction* (Englewood Cliffs, N.J.: Prentice-Hall, 1990).

The Earth in the Universe

OBJECTIVES

◆ To introduce the basic structure of the universe, speculations about its origin, and the position of our home galaxy and star within it.

◆ To describe the functions of the sun as the dominant body of the solar system.

◆ To briefly survey each of the sun's nine planets and the lesser orbiting bodies that constitute the remainder of the solar system.

It was pointed out in Unit 2 that we live on a small, fragile planet. Even from the vantage point of the moon—our nearest neighbor in outer space, only 390,000 km (240,000 mi) away—the earth appears greatly shrunken in size (*see* photo above). If we were to view the earth from the vicinity of the sun, from a distance of 150 million km (93 million mi), it would only be a tiny speck. But these perspectives do not even begin to suggest the unimaginable vastness of the universe.

How big is the universe? What is its structure? How do our sun and its family of planets (including Earth) fit into the overall scheme of things? These are some of the key questions to be considered in this unit.

UNIT OPENING PHOTO: Earthrise above the moon's barren surface, as photographed by the Apollo astronauts from the lunar orbiter in 1969.

THE UNIVERSE

We may define the universe as the entity that contains all of the matter and energy that exists anywhere in space and time. As for its size, to understand the enormity of the universe we must consider space and time together. The fastest thing that moves in the universe is light, a form of radiant energy that travels at a speed of 300,000 km (186,000 mi) per second. Thus, in a single second, a ray of light travels a distance equal to 7½ times the earth's circumference. Even at this speed, however, it takes about 8 minutes for light to travel from the sun to the earth. Light from the nearest star takes *more than four years* to get here. If that star exploded today, we would not know it for another four-plus years; our telescopes, therefore, show us history.

Given the untold billions of stars that populate the universe (Fig. 4-1) and the fact that years are required for the light to reach us even from the nearest star beyond our own sun, astronomers calibrate interstellar distances in light-years. A **light-year** is the distance traveled by a pulse of light in one year: 9.46 trillion (9.46×10^{12}) kilometers or 5.88 trillion (5.88×10^{12}) miles. To travel that distance aboard an airplane at 800 km (500 mi) per hour would require a journey of almost 1,350,000 years!

Now let us train our sights on really deep space. Our planetary system belongs to a **galaxy** (an organized, disk-like assemblage of billions of stars) called the *Milky Way*, which is about 120,000 light-years in diameter. As recently as the 1920s, this was believed to be the entire universe, but newer astronomical discoveries have drastically transformed that perception. We now know that the Milky Way Galaxy itself is merely one of about 30 loosely bound galaxies that have clustered to form what astronomers call the *Local Group* of galaxies, which measures nearly 3 million light-years across its longest dimension. This Local Group, in turn, is but a small component of the *Local Supercluster* (a supercluster is a conglomeration of galaxies, comprising the largest of all celestial formations), measuring approximately 100 million light-years in diameter.

To this point, we have progressed through three time–space levels—galaxy, galaxy cluster, galaxy supercluster—but only now are we ready to tackle the full dimensions of the known universe. That ultimate level is now under intense investigation by astrophysicists around the world, and research frontiers continue to expand. Today, astronomers record images from their telescopes that may have traveled 15 billion light-years from what are believed to be the outer edges of the universe. Since our Local Supercluster lies near the universe's center, the *radius* of the universe probably exceeds 15 billion light-years and the *diameter* should be at least 30 billion light-years.

The hierarchical organization of the universe's various time–space levels is seen in the "cones of resolution" diagrammed in Fig. 4-2. Each level is highly complex in its internal structure. For instance, the Milky Way Galaxy alone consists of more than 100 billion stars, of which our sun is only one—and a middle-sized and most ordinary star at that. In all, there are billions of galaxies: their sheer numbers are beyond our comprehension, and most lie beyond the view of our most powerful telescopes. Then how many of those huge balls of glowing gas that we call stars does the total universe actually contain? The latest estimate is more than *200 billion-billion* (200×10^{18}), about 50 billion stars for every human now alive. Where did all these celestial bodies and the other diverse matter and energy of the universe come from, and how was everything scattered across such an immense space?

The answers to these questions rest with theories concerning the origin of the universe, which is widely believed to be the result of the so-called *Big Bang*. The Big Bang was a massive explosion of truly cosmic proportions, in which all the primordial matter and energy that existed before the formation of the universe was compressed together at almost infinite density, heated to trillions of degrees, and blown apart. This stupendous blast propelled matter and energy outward in a rapidly expanding fireball, which has been cooling and slowing ever since. In the wake of this violent advancing wave lay an amorphous cloud of debris, from which the contents of the universe gradually formed. Galaxies and larger star clusters slowly took shape from the conden-

FIGURE 4-1 A slice of the night sky, magnified by a powerful telescope, showing a nearby galaxy surrounded by a veritable blizzard of individual stars.

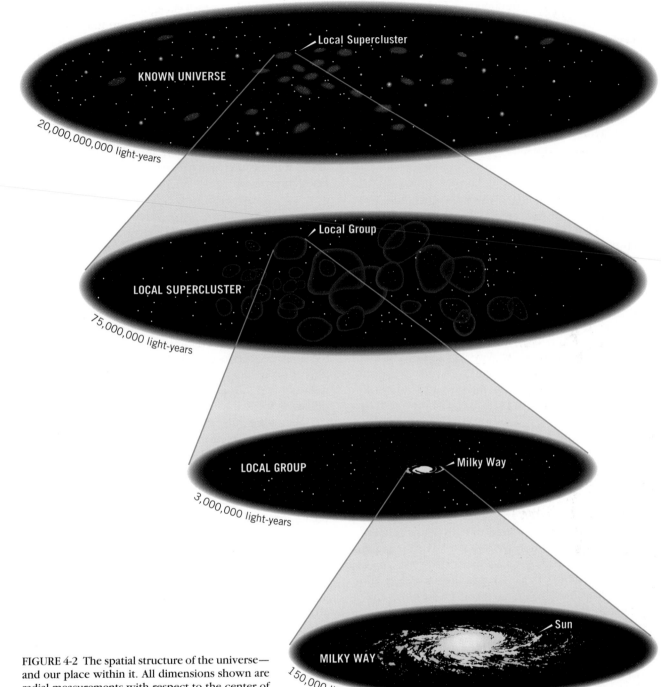

FIGURE 4-2 The spatial structure of the universe—and our place within it. All dimensions shown are radial measurements with respect to the center of each disk.

sation and consolidation of cooling gas and dust clouds. A key force in binding them together was **gravity**, the force of attraction acting among all physical objects due to their *mass*, the quantity of material they are composed of. Indeed, gravitational forces shape the structure of every one of the universe's time–space levels, bonding enormous interstellar clusters as well as the superheated gases of individual stars.

The Big Bang, which is based on Albert Einstein's general theory of relativity, took place approximately 15 billion years ago. As was noted above, the farthest objects in the universe appear to be located at a distance of almost 15 billion light-years from Earth. If the universe is still expanding in the wake of the Big Bang, then the matter and energy at the outermost extremities of the universe represent the leading edge of that advance. The matter that exists at these extremities has been formed into mysterious, star-like objects; these brightly glowing masses are called *quasars*, shorthand for "quasi-stellar objects." Interestingly, quasars are embedded in a uniform glow of radio-wave-frequency radiation, an energy environment consistent with the hypothesis that this cosmic radiation (confirmed to exist everywhere in space) is the faint "echo" of the Big Bang that occurred so long ago.

Many scientists are also studying the consequences of these recent revelations for the future evolution of the universe. If the universe should turn out to be a finite or closed system, then gravity will inevitably reverse the expanding edge, force an implosion leading to another Big Bang, and spawn infinite expansion–contraction cycles beyond it. However, should the universe prove to be an open system, then it may end with a whimper rather than a bang as galaxies inexorably overcome the pull of gravity and eventually drift away from one another. In the search for answers, some researchers now subscribe to even more complex outcomes.

THE SOLAR SYSTEM

Our home galaxy, the Milky Way, began to form more than 12 billion years ago. The star we know as the sun, however, was a relatively late addition and did not appear until about 4.6 billion years ago. The processes that formed the sun mirrored the forces at work throughout the universe. One particular rotating cloud of gas and dust began to cool, and soon its center condensed to form a star. Simultaneously, the remaining materials in the swirling cloud around this new star formed a disk and began to sort themselves out as the consolidating mass of the sun exerted an ever stronger gravitational pull. Millions of eddies within this disk now began to condense as well and formed sizeable conglomerations of solid matter called *planetesimals*. As these objects grew in mass, gravity began to draw them together. Soon these planetesimals were traveling in swarms, and it wasn't long before they were compressed together to form nine planets that began to circle the sun in regular orbits. (**Planets** are dark solid bodies, much smaller in size than stars, whose movements are controlled by the gravitational effects of nearby stars.)

Most of the larger residual planetesimals were captured by the gravitational fields of the evolving planets and began to orbit them as satellites or *moons*. A large belt of smaller planetesimal-like materials (known as *asteroids*) congregated between the fourth and fifth planets, and remain in orbit today. The sun's gravitational field also contains small bodies of frozen gases and related materials called *comets*, tiny clusters of rock known as *meteoroids*, and vast quantities of *dust* that may be remnants of the system's formation.

The sun, its planets, and related residual materials were born together some 4.6 billion years ago and collectively constitute the **solar system** (Fig. 4-3a). The sun is located at the center of the solar system, the source of

light, heat, and the overall gravitational field that sustains the planets. The nine planets may be grouped as follows. Mercury, Venus, Earth, and Mars comprise the four terrestrial or *inner planets*, which are rather small in size (*see* Fig. 4-3b). The next four—Jupiter, Saturn, Uranus, and Neptune—constitute the major or *outer planets*, and are much larger in size. The outermost ninth planet, Pluto, was only discovered in 1930 (though its presence had long been suspected). Although it seems to be more like the inner planets, astronomers still know so little about it that Pluto is not classified within either group. Let us now examine the major components of the solar system.

The Sun

The sun is the dominant body of the solar system. Its size relative to the planets and lesser orbiting materials is so great that the sun accounts for 99.8 percent of the mass of the entire solar system, more than 750 times the mass of all the planets combined (the sun's diameter alone is 109 times larger than the earth's). This, of course, enables the sun to extend its gravitational field far out into space. The effect of that gravity on the planets does weaken with distance but at a rather slow rate. Earth, the third planet, is located at an average distance of 150 million km (93 million mi). The orbit of outermost Pluto, however, is about 40 times that distance from the center of the solar system, demonstrating that the sun's gravitational pull is powerful enough to control the movements of a planet nearly 6 billion km (3.7 billion mi) away. Remember, too, that the largest planets are the fifth through eighth, located in a zone roughly 1 to 5 billion km (0.5 to 3.1 billion mi) distant from the sun.

As with most stars, the sun is a churning thermonuclear furnace, composed mainly of superheated hydrogen and helium gases mixed in a ratio of approximately 3 to 1. Surface temperatures average 5500°C (ca. 10,000°F).* The enormous quantities of light and heat given off by the sun, in the form of a stream of rapidly moving atomic particles, come from its surface and atmosphere (Fig. 4-4). Most of that gaseous flow of energy, which radiates outward in every direction and is known as the *solar wind*, is lost in space. (The earth receives less than one-billionth of the light and heat that are expelled by the sun.) Although the solar wind "blows" at a fairly steady speed, disturbances originating deep inside the sun occasionally rise to the surface and modify the outflow of solar energy. Solar scientists are particularly familiar with an 11-year cycle of magnetic storm-like activity that is associated with large, dark "spots" on the sun's surface (*see* photo, p. 224). At their peak, *sunspots* can affect the earth by triggering magnetic storms

---

*The Celsius (C) and Fahrenheit (F) temperature scales are defined and explained on p. 84.

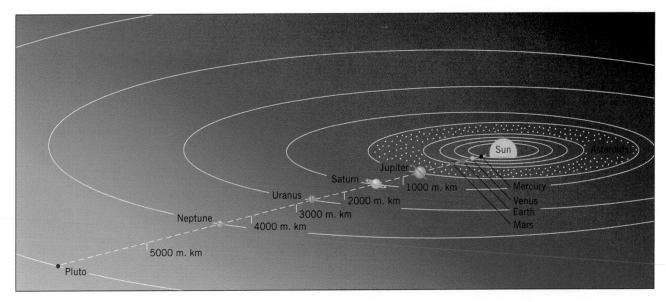

a

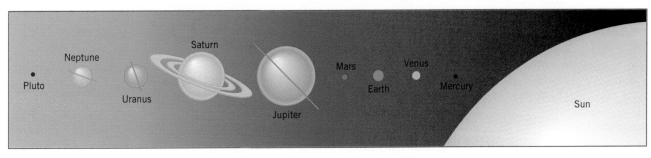

b

FIGURE 4-3 The solar system (a) and the relative sizes of the nine planets with respect to one another and the sun (b). In diagram a, the planets are aligned only for demonstration purposes; distances are given in millions of kilometers.

here that interfere with radio communications, cause powerline surges, and produce especially brilliant auroras in the night skies of the polar and subpolar regions. The last sunspot activity cycle peaked in 1989; the mid-cycle minimum was observed on schedule in 1995, and the next maximum should occur just around the turn of the century.

The Planets

All the planets except Neptune and Pluto are bright enough to be seen in the night skies without a telescope, and they have been observed by humans for thousands of years. The ancient Greeks were particularly fascinated by these moving celestial bodies and coined the word *planet* (which meant wanderer in their language). The arrangement of the nine planets in the solar system is shown in Fig. 4-3, which highlights their concentric orbits (only Pluto deviates from this pattern, as we shall see). The farther a planet is located from the sun, the greater the length of its orbit. One complete circling of the sun within such an orbital path is called a **revolution**. The closest planet to the sun, Mercury, needs only 88 earth-days to complete one revolution. The earth, of course,

FIGURE 4-4 This stunning view of a total solar eclipse reveals the *corona*, the sun's outer atmosphere and source of the solar wind. Also visible are a pair of solar prominences, bright arches of gas that are shaped by the local magnetic fields of the sun's most active regions.

TABLE 4-1   CHARACTERISTICS OF THE PLANETS OF THE SOLAR SYSTEM

PLANET	MEAN ORBITAL DISTANCE FROM THE SUN (MILLIONS KM/MI)	PERIOD OF ONE REVOLUTION (DAYS/YEARS)	PERIOD OF ROTATION ON AXIS	DIAMETER AT EQUATOR	MASS (EARTH=1)	MAIN ATMOSPHERIC COMPONENTS	SURFACE TEMPERATURE	NUMBER OF MOONS
Mercury	57.9 (36.0)	88 (0.24 yrs)	58.7 days	4,878 km (3,032 mi)	0.06	Sodium Potassium Helium	−173°C to 427°C (−279°F to 801°F)	0
Venus	108.2 (67.2)	224.7 (0.62 yrs)	243 days	12,104 km (7,523 mi)	0.82	Carbon dioxide Carbon monoxide Hydrogen chloride	470°C (878°F)	0
Earth	149.6 (93.0)	365.3 (1.0 yrs)	23 hrs, 56 mins	12,756 km (7,928 mi)	1.00	Nitrogen Oxygen Water vapor	−88°C to 58°C (−125°F to 136°F)	1
Mars	227.9 (141.6)	687 (1.9 yrs)	24 hrs, 37 mins	6,794 km (4,222 mi)	0.11	Carbon dioxide Carbon monoxide Water vapor	−63°C to 27°C (−81°F to 81°F)	2
Jupiter	778.4 (483.8)	4,332 (11.9 yrs)	9 hrs, 51 mins	142,984 km (88,865 mi)	317.9	Hydrogen Helium Methane	−163°C to −123°C (−261°F to −189°F)	16
Saturn	1,424.0 (885.0)	10,761 (29.5 yrs)	10 hrs, 14 mins	120,536 km (74,914 mi)	95.1	Hydrogen Helium Methane	−178°C (−288°F)	18
Uranus	2,872.0 (1,785.0)	30,865 (84.0 yrs)	17 hrs, 14 mins	51,118 km (31,770 mi)	14.56	Hydrogen Helium Methane	−215°C (−355°F)	15
Neptune	4,499.0 (2,796.1)	60,195 (164.8 yrs)	16 hrs, 3 mins	50,538 km (31,410 mi)	17.24	Hydrogen Helium Methane	−217°C (−359°F)	8
Pluto	5,943.0 (3,693.6)	90,471 (247.7 yrs)	6.4 days	2,280 km (1,417 mi)	0.002	Nitrogen Carbon dioxide Methane	−233°C (−387°F)	1

requires exactly one year (365¼ days). The outer planets take far longer to revolve around the sun. (Saturn requires almost 30 years, whereas outermost Pluto requires more than 247 years.) Information on revolution times is presented in column 3 of Table 4-1, which also displays seven other categories of vital planetary data.

The Inner Planets

The four planets nearest the sun—Mercury, Venus, Earth, and Mars—are classified as *terrestrial* or earth-like. All are much smaller than the four outer planets (*see* columns 5 and 6 in Table 4-1). Each is a solid sphere, composed largely of iron and rock, built around a dense metallic core. The surface layer of each inner planet received vast quantities of gases that were exhaled by volcanoes as the planet gradually cooled. In each case, an atmosphere formed. Only Mercury no longer possesses one; its atmospheric envelope quickly boiled away by proximity to the searing heat of the sun. All of these planets experienced considerable volcanic and seismic

(earthquake) activity. Again, only Mercury no longer exhibits those geologic disturbances, while the other three planets remain quite active. Another characteristic of the terrestrial planets is the paucity of moons. While the four outer planets claim a total of 57 moons, only 3 are found in the inner solar system. Earth's moon is the largest of these, with the remaining two tiny moons in orbit around Mars. Let us now take a closer look at each planet, proceeding in order away from the sun.

Mercury, as is clear from the discussion above, is a dead planet. It could not hold onto its atmosphere, and volcanism and seismic activity appear to have ceased more than 3 billion years ago. Every aspect of Mercury is overshadowed by its innermost position in the solar system, which brings the planet's orbit to within 46 million km (29 million mi) of the sun at its closest point. Surprisingly, Mercury is not the hottest of the planetary surfaces: Venus, because of its very dense atmosphere, averages about 300°C (540°F) higher (*see* Table 4-1, column 8). Moreover, Mercury's surface temperature on the side facing away from the sun dips as low as −173°C

FIGURE 4-5 A closeup view of Venus, taken from the Pioneer Orbiter in 1979 at a distance of 59,000 km (36,000 mi). Whereas thick clouds unrelentingly shroud this mysterious planet, more recent space probes have successfully penetrated that barrier to map the hellish Venusian surface.

(−279°F). It should be noted, in column 4 of the table, that this planet spins or **rotates** on its axis approximately once every 59 days (a complete rotation of the earth occurs once every [24-hour] day). Astronomers have assembled a fairly detailed picture of Mercury, thanks to the closeup imagery obtained during the 1974–1975 fly-by of NASA's Mariner 10 space probe. The surface is like that of our moon—lifeless and heavily cratered from severe meteorite impacts that could not be cushioned by the frictional effects of an overlying atmosphere.

Venus is often called the earth's twin because the two planets are so close in size and mass, but they really have very little else in common. The extremely high temperatures of the Venusian surface (averaging 470°C [878°F]), the hottest of any planet, have already been noted. These are undoubtedly heightened by the weight of an oppressive atmosphere that is 90 times as dense as the earth's. Thick layers of yellowish-white, sulfuric acid clouds constantly obscure Venus (Fig. 4-5). But the space explorations of the past quarter-century, particularly the U.S. Magellan mission that mapped Venus from an orbiting satellite in 1990–1991, have provided detailed images of its forbidding surface. Violent volcanism is widespread, and there are highly varied landforms all across the planet's relatively flat face. To earthlings, Venus has always been the third brightest object in the sky (after the sun and moon). It is often observed as the "evening star" or "morning star" as its position alternates while both planets revolve around the sun. At its closest point, Venus comes within 40 million km (25 million mi)

of the earth, a distance about 100 times greater than that between the earth and the moon. Even though it is the second closest planet to the sun, Venus rotates on its axis much more slowly than any of the other eight planets. This oddity results from its aberrant rotational behavior, with Venus being the only planet to rotate in the direction opposite to that of its orbital revolution.

Earth, the third planet, possesses a physical environment that is significantly shaped by its rotational and orbital behavior (earth–sun relationships are the subject of Unit 5). At this point, the earth should be compared to its planetary cousins by referring to the data displayed in Table 4-1. Perusal of this table immediately reveals that the earth and the conditions necessary to support terrestrial life are unique. They simply do not exist anywhere else in the solar system (though, conceivably, they could occur on planets elsewhere in the universe). Columns 7 and 8 in Table 4-1 are especially important because they illustrate the range of atmospheric compositions and surface temperatures, and underscore how different our planet is from all the others in these environmental variables that are so critical to the survival of Earth's plants and animals.

Mars has fascinated skywatchers and scientists for generations because it is here that conditions appear to come closest to supporting at least the hardier known forms of life. Speculations concerning life on Mars, however, were dashed following the analysis of environmental data transmitted back to Earth by the two Viking landers that the United States successfully put on this planet in 1976. The surface itself is quite reminiscent of the earth's most arid landscapes; the Viking images reveal a reddish, rocky, wind-blown—and decidedly frigid—desert. Higher-altitude photographs obtained by Viking and other space probes were just as interesting, revealing gigantic canyons that were sculpted by running water and the solar system's highest known mountain—volcanic Olympus Mons, towering 27 km (16.8 mi or nearly 90,000 ft) above its surrounding plain (Fig. 4-6). Although earthly life is not possible at this time, the Martian surface almost certainly possessed substantial supplies of water in the past. (Water still exists on Mars today, but low atmospheric pressure [force per unit area] and temperature prevent it from flowing freely across the surface.) The polar regions are particularly intriguing because they contain "icecaps" that wax and wane with the seasons as the Martian axis of rotation alternatively tilts toward and away from the sun during the planet's 687-day revolution. These polar caps, which consist mainly of water ice, would obviously be attractive targets in future explorations of the planet.

The Outer Planets

The outer planets—Jupiter, Saturn, Uranus, Neptune—differ radically from their inner solar system counterparts. They are known as the *major* planets (or the *Jovian* plan-

FIGURE 4-6 Olympus Mons soaring skyward from the surface of Mars. This most prominent feature of the red planet is also the highest volcano ever observed by earthbound scientists.

ets—they all resemble giant, prototypical Jupiter), because they account for over 99 percent of all the matter of the solar system exclusive of the sun itself. Each of these planets is a huge sphere composed largely of gases (hydrogen and helium dominate, as column 7 of Table 4-1 indicates), and each possesses a deep atmosphere with clouds. All except Uranus give off more heat than they absorb from the distant sun. This heat is a residue from the formation of these planets, a physical property that can be likened to still-warm pieces of coal long after their fires have gone out. Perhaps the single most striking feature in this outer part of the solar system is the planet Saturn surrounded by its spectacular rings. In recent years, space scientists have discovered that Jupiter, Uranus, and Neptune also possess rings, but these are far less prominent and dramatic than Saturn's.

Jupiter, the biggest planet in the sun's family, is one-tenth the size of the sun and so large that more than a thousand Earths would be needed to fill its volume. Jupiter is also the solar system's most rapidly spinning planet, requiring less than 10 hours to complete one full rotation. This rapid rotation produces a decided bulge at the equator and the corresponding flattening of the polar regions. It also shapes the formation of latitudinal belts across the gaseous face of Jupiter. These horizontal bands, together with the Great Red Spot (Fig. 4-7), are the planet's most prominent features. They suggest an incredibly violent atmosphere and "surface" (no sharp boundary marks the contact between the atmosphere and the liquid interior). In fact, the Red Spot is a slow-moving, raging, hurricane-like storm that has been observed by astronomers since its discovery in 1630. Of

Jupiter's 16 moons, 4 are large enough to be of planetary dimensions. Jupiter also possesses a system of thin, essentially transparent rings that consist, quite literally, of millions of additional tiny moons.

Saturn, the second largest planet, also is the second fastest in rotation (about 10¼ hours) and, like Jupiter, also exhibits equatorial bulging and polar flattening. Saturn's rings have fascinated skywatchers for centuries, but it was not until the 1980s that NASA's Voyager 1 and 2 space probes flew through them and sent back detailed information (Fig. 4-8). Now we know that there are seven major rings, which consist of literally hundreds of tiny ringlets. The rings themselves (which do not touch Saturn) are composed of icy particles of water or rock that orbit above Saturn's equator like a pulverized moon. Although the rings extend outward into space for at least 400,000 km (250,000 mi), they are quite flat, with an average thickness of less than 15 km (9 mi). Beyond its ring system, Saturn has 18 moons, the most of any planet. The largest, Titan, is the only moon in the solar system with an atmosphere, composed mainly of nitrogen.

Uranus and **Neptune** have frequently been called twins because, at least from a distance, they were seen to share many characteristics (*see* Table 4-1). But the closeup data transmitted back to Earth by the Voyager 2 fly-by (Uranus in 1986; Neptune in 1989) revealed major structural and atmospheric differences that have forced planetary scientists to revise their perceptions. Among the new discoveries: Uranus spins on its side and possesses a system of at least 11 rings, whereas Neptune gives off internal heat that helps drive a much more active atmosphere and was found to have spawned a four-ring system.

FIGURE 4-7 Jupiter photographed from a distance of 28 million km (18 million mi) by Voyager 1 in 1979. Clearly visible are the banded surface, Great Red Spot (lower left), and two inner moons (Io and Europa). The remnants of Comet Shoemaker-Levy, which slammed into Jupiter with great fanfare in mid-1994, produced a series of scars in the white band near the bottom of this image.

FIGURE 4-8 One of the solar system's most spectacular sights: the intricate patterns and dazzling colors of Saturn's rings, as photographed by Voyager 2 in 1981.

The Lesser Bodies of the Solar System

Pluto, the ninth and outermost planet, is treated separately from the other planets because of its small size and the fact that so little is known about it (no space probe has yet ventured to its vicinity). Perhaps its most notable characteristic is its highly eccentric orbit, which passes inside Neptune's orbital path for 20 of the 247½ years of the Plutonic revolution, and is also inclined at an angle of more than 17° to the geometric plane in which the earth and other planets circle the sun (Fig. 4-3). Pluto has a single large moon (Charon) about half its size, and many astronomers now consider the pair to constitute a double-planet system.

Moons are satellites that orbit every planet except Mercury and Venus. Moons probably originated as planetesimals that were subsequently captured by the gravitational pull of the emerging planets. The best known satellite to us on Earth is our own moon, on whose surface American astronauts first landed in 1969. The moon revolves around the earth once every 27.3 days and is located at an average distance of 385,000 km (240,000

mi)—a rocket journey of about six days. The comparative sizes of the two bodies can best be visualized by reference to an often used analogy: if the earth is imagined to be a basketball, then the relative size of the moon would equal that of a tennis ball. Following a violent beginning from the time of the birth of the solar system (ca. 4.6 billion years ago) until about 3 billion years ago, the moon has been a lifeless body during the most recent two-thirds of its existence. The lunar surface can be broadly divided into three physiographic categories: plains, highlands, and craters. The plains and highlands were created by volcanic activity during the moon's early active stage, the lowlands originating as sheet-like lava flows and the hilly uplands shaped by eruptions. Superimposed across most of the moon's face are billions of impact craters (Fig. 4-9), with more than 300,000 measuring at least 1 km (0.62 mi) across.

We have already identified the remaining smaller objects of the solar system—asteroids, comets, meteoroids, and dust. However, as we saw in the case of the cratering

FIGURE 4-9 The crater-studded surface of the moon in the vicinity of the lunar equator's western segment (as seen from Earth). This photo was taken during the final manned (Apollo 17) mission to the moon in 1972. Most appropriate for a physical geography textbook, the large crater left of center is named Eratosthenes (*see* Unit 1). On the horizon at the right is the rim of Copernicus, the lunar surface's largest crater.

Perspectives on THE HUMAN ENVIRONMENT

When Worlds Collide

A movie carrying this title in the 1950s intrigued millions with its doomsday plot about a runaway planet that smashed into the earth. In the 1990s, we know that such a scenario is no longer purely science fiction because there is mounting evidence that extraterrestrial objects have repeatedly struck our planet over the course of its existence. Even a comet or meteorite of sufficient size could cause tremendous environmental damage: on June 30, 1908 a tiny asteroid perhaps 80 m (265 ft) in diameter exploded above the remote Tunguska region of northeastern Russia, leveling trees and triggering fires over a 2500-sq-km (965-sq-mi) zone. Millions of such objects litter the solar system, and hundreds, maybe thousands, intersect the earth's orbit. Inevitably, some will hit, with the bigger objects capable of producing much wider devastation than the 1908 impact (Fig. 4-10).

Scientific concern about such collisions intensified during the 1980s in conjunction with research into the mass extinctions—involving perhaps two-thirds of all living species—that marked the end of the Cretaceous period some 65 million years ago (*see* geologic time scale on p. 399). New geologic and other evidence increasingly pointed toward a 10-km-wide (6-mi-wide) asteroid crashing into Earth, which pulverized enough surface debris to throw up a globe-girdling dust cloud that shut out sunlight for at least a year—thereby destroying the food supply of countless plants and animals. Proponents of this theory still encounter considerable opposition, but their argument was

FIGURE 4-10 A NASA artist's impression of the cataclysmic forces that would be unleashed if a very large asteroid (800 km [500 mi] in diameter) were to strike the earth. In this doomsday scenario, the energy of such an impact would equal 5 trillion nuclear bombs, sufficient to vaporize the world ocean, melt the earth's crust, and "sterilize" our planet of all existing life forms.

strengthened in 1991 by the discovery of a buried impact crater of exactly the right age and size on Mexico's Yucatán Peninsula. Today the debate still rages about the biological effects of such an event, but more scientists than ever are convinced that the threat of such a catastrophe, every 100,000 years or so, is real.

If that is the case, can we take measures to protect ourselves? A number of scientists think we can and should, and they have begun to make their case to military planners and national governments. A first step would to be to discover exactly what is out there, and plans are underway to create a global telescope network to monitor every sector of the skies. Such a system could be operational by 2020 at a relatively modest annual cost. Some astronomers have already started searching on their own, and many new asteroids have been found and their routes plotted. If such an object were discovered to be on a collision course with the earth, there would hopefully be sufficient time to develop the technology to destroy or at least deflect the interloper, perhaps by employing nuclear explosives.

of the moon (a process that also occurs on many planets and every other moon), these objects make up in quantity what they lack in size, and their presence is continuously and ubiquitously felt. Moreover, astronomers are increasingly concerned that asteroids and *meteorites* (meteoroids that penetrate the atmosphere and reach the surface) pose a significant threat to Earth (*see* Perspective: When Worlds Collide).

This unit has discussed the overall context of the earth, focusing on its place in the tiny corner of the universe it travels through and making comparisons with the other major bodies of the solar system. The next unit elaborates the special relationships between Planet Earth and the life-giving sun, and forms the foundation for understanding the seasonal rhythms that are so important to its physical geography.

KEY TERMS

galaxy (p. 40)	revolution (p. 43)	Earth (p. 45)	Uranus (p. 46)
gravity (p. 41)	rotation (p. 45)	Mars (p. 45)	Neptune (p. 46)
light-year (p. 40)	solar system (p. 42)	Jupiter (p. 46)	Pluto (p. 47)
moons (p. 47)	Mercury (p. 44)	Saturn (p. 46)	
planets (p. 42)	Venus (p. 45)		

REVIEW QUESTIONS

1. Describe the hierarchical organization of the universe's time–space levels.

2. Define the term *planet* and describe how the solar system's planets formed.

3. Define the term *revolution* and discuss its application to the orbital patterns of the sun's planets.

4. Name the *inner planets* and list the major characteristics they share.

5. Name the *outer planets* and list the major characteristics they share.

6. What are the differences between *moons, comets, asteroids,* and *meteoroids?*

REFERENCES AND FURTHER READINGS

ABELL, G. O. *Exploration of the Universe* (New York: Holt, Rinehart, & Winston, 3rd ed., 1975).

AUDOUZE, J., and ISRAEL, G., eds. *The Cambridge Atlas of Astronomy* (New York: Cambridge University Press, 2nd ed., 1988).

BEATTY, J. K., et al., eds. *The New Solar System* (Cambridge, Mass.: Sky Publishing Corp., 2nd ed., 1982).

CLOUD, P. *Cosmos, Earth, and Man* (New Haven, Conn.: Yale University Press, 1978).

HAWKING, S. W. *A Brief History of Time: From the Big Bang to Black Holes* (New York: Bantam Books, 1988).

LIGHTMAN, A. *Ancient Light: Our Changing View of the Universe* (Cambridge, Mass.: Harvard University Press, 1993).

SAGAN, C. *Cosmos* (New York: Random House, 1980).

WHIPPLE, F. L. *Orbiting the Sun: Planets and Satellites of the Solar System* (Cambridge, Mass.: Harvard University Press, 1981).

ZEILIK, M. *Astronomy: The Evolving Universe* (New York: Wiley, 7th ed., 1994).

Earth–Sun Relationships

OBJECTIVES

◆ To examine the earth's motions relative to the sun.

◆ To demonstrate the consequences of the earth's axis tilt for the annual march of the seasons.

◆ To introduce the time and spatial variations in solar radiation received at surface locations.

arth is a small, fragile planet. As the third planet of the solar system it orbits the sun—the source of light, heat, and the gravitational field that sustains all nine planets (*see* Fig. 4-3). Among the planets, however, only the earth exhibits the unique physical conditions that are essential for the support of life as we know it. Conceivably, these conditions could exist on planets elsewhere in the vast universe. The earth's natural environment, particularly its vital heating, is significantly shaped by the movements of our planet relative to the sun. In this unit, we explore basic earth–sun relation-

ships and their profound consequences for the temperature patterns that occur on the earth's surface.

EARTH'S PLANETARY MOTIONS

Unit 4 introduced two basic concepts of planetary motion: revolution and rotation. A **revolution** is one com-

UNIT OPENING PHOTO: An astronaut's eye view of the rising sun above the Indian Ocean off Australia.

plete circling of the sun by a planet within its orbital path; the earth requires precisely one year to revolve around the sun. As it revolves, each planet also exhibits a second simultaneous motion—**rotation** or spinning on its axis. It takes the earth almost one calendar day to complete one full rotation on its **axis**, the imaginary line that extends from the North Pole to the South Pole through the center of the earth.

As the earth revolves around the sun and rotates on its axis, the sun's most intense rays constantly strike a different patch of its surface. Thus, at any given moment, the sun's heating (or solar energy) is unevenly distributed, always varying in geographic space and time. By time, we mean not only the hour of the day, but also the time of the year—the major rhythms of Earth time for all living things as measured in the annual march of the seasons. Before examining seasonality, we need to know more about the concepts of revolution and rotation.

Revolution

The earth revolves around the sun in an orbit that is almost circular. Its annual revolution around the sun takes 365¼ days, which determines the length of our year. Rather than starting the New Year at a time other than midnight, one full day is added to the calendar every fourth year, when February has 29 days instead of 28. Such a year (occurring in 1996, 2000, and 2004, for example) is called a *leap year*.

Like the earth itself, which is *nearly* a sphere, the earth's orbital path is *nearly* circular around the sun. In fact, the earth is slightly closer to the sun in early January than it is in early July. This makes its orbital trajectory slightly elliptical. The average distance from the earth to the sun is approximately 150 million km (93 million mi). But on January 3, when the earth is closest to the sun, the distance is about 147.3 million km (91.5 million mi). This position is called the moment of **perihelion** (from the ancient Greek *peri* meaning near and *helios* meaning sun). From that time onward, the earth–sun distance increases slowly until July 4, half a year later, when it reaches about 152.1 million km (94.5 million mi). This position is called **aphelion** (*ap* means away from). Thus the earth is farthest from the sun during the Northern Hemisphere summer and closest during the northern winter (or Southern Hemisphere summer). But the total difference is only about 5 million km (3.1 million mi), not enough to produce a significant variation in the amount of solar energy received by our planet.

Rotation

Our planet ranks among the solar system's fast-spinning bodies, which produces equatorial bulging and polar-area flattening. Accordingly, geophysicists have discovered that the earth's diameter when measured pole to pole (12,715 km [7,900 mi]) is slightly less than it is at the equator (12,760 km [7,927 mi]). Thus the earth is not a perfect sphere; it is an *oblate spheroid*, the technical term used to describe the departure from a sphere that is induced by the bulging/flattening phenomenon just described. The earth's deviation from a true sphere, however, is a minor one. In fact, the difference between the equatorial and polar diameters is so small (just 45 km [27 mi] or 0.35 percent) that it matters only to earth scientists and other specialists involved in activities that demand exactness—for example, in space flight, detailed cartography, or geodesy (precise planetary measurement).

As the earth rotates on its axis—which occurs in a west-to-east direction—this motion creates the alternations of day and night, as one half of the planet is always turned toward the sun while the other half always faces away. One complete rotation takes roughly 24 hours (23 hours, 56 minutes, to be exact), or one calendar day. During one full revolution around the sun, the earth makes 365¼ rotations. Consider this: the earth's circumference at the equator is slightly less than 40,000 km (25,000 mi). Thus a place on or near the equator, say the city of Quito, Ecuador, rotates at a speed of 1666 km (1040 mi) per hour—continuously! But the distance traveled during a complete rotation diminishes northward and southward from the equator, until it becomes zero at the poles. A person standing on the North or South Pole would merely make one full turn in place every 24 hours. This contrast between the force of rotation at or near the poles on one hand, and at or near the equator on the other, causes the earth to develop that slight bulge at the equator.

Actually, the person standing at the pole does not feel any effect different from someone standing on the equator. We do not notice the effect of the earth's rotation, because everything on the planet—land, water, air—moves along at the same rate of speed. And the rate of rotation does not vary, so no slowing down or speeding up is sensed. But look up into the sky and watch the stars or moonrise, and the reality of the earth's rotation soon presents itself. This is especially true concerning the stationary sun. To us, it appears to "rise" in the east (the direction of rotation) as the leading edge of the unlit half of the earth turns back toward the sun. Similarly, the sun appears to "set" in the west as the trailing edge of that sunlit half of the earth moves off toward the east.

Note that we just referred to a person *standing* at certain places on the earth's surface. The fact is that a person or object not in motion does not experience any effect from our planet's rotation. But *moving* people and objects do. Moving currents of water and streams of air are affected by a force that tends to deflect them away from their original direction of movement. This force was not known until it was identified by the French scientist Gustave Gaspard de Coriolis in the 1830s. Everything that moves under the influence of our rotating earth is affected by this force, which is appropriately named the **Coriolis force** after its discoverer. The Coriolis force is

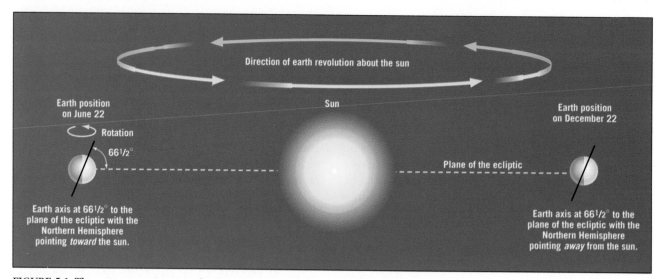

FIGURE 5-1 The extreme summer and winter positions of the earth with regard to the sun. The earth's axis is tilted at the same angle to the plane of the ecliptic throughout the year.

an important factor in the earth's climate and weather, ocean currents, and related parts of other environmental systems (*see* Unit 9).

The earth rotates eastward, so that sunrise is always observed on the eastern horizon. The sun then traverses the sky to "set" in the west; but, of course, it is not the sun but the *earth* whose movement causes this illusion. Looking down on a model globe, viewing it from directly above the North Pole (as in Fig. 3-6), we see that rotation occurs in a counterclockwise direction. This might seem to be a rather simple exercise. But not many years ago a leading television network opened its nightly national

news program with a large model globe turning . . . the wrong way!

SEASONALITY

If you draw onto a flat piece of paper the orbital path of the earth around the sun, the paper could be described as a geometric plane. The actual plane in space, which contains the line traced by the earth's slightly elliptical orbit and the stationary sun, is called the **plane of the**

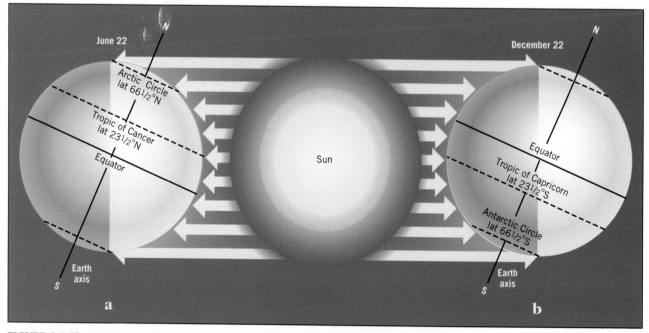

FIGURE 5-2 The relative positions of the earth and the sun on June 22 and December 22. Points on Earth receive the sun's rays at different angles throughout the year.

FIGURE 5-3 A time-lapse photograph of the post–midnight sky of Lapland in late June. At this time of the year, here in northernmost Finland close to 70°N, the sun never sets. Photo courtesy of John D. Stephens.

ecliptic. The seasons occur because the earth is tilted with respect to the plane of the ecliptic.

Axis Tilt

The earth's axis is always tilted at an angle of 66½ degrees to the plane of the ecliptic and is always tilted in the same direction no matter where the earth is in its orbit. The constant tilt of the axis is the key to these seasonal changes. Sometimes the term *parallelism* is used to describe this axial phenomenon, meaning that the earth's axis remains parallel to itself at every position in its orbital revolution. Thus at one point in its revolution, around June 22, the northern half of the earth, the Northern Hemisphere, is maximally tilted toward the sun. At this time, the Northern Hemisphere receives a much greater amount of solar energy than the Southern Hemisphere does. When the earth has moved to the opposite point in its orbit six months later, around December 22, the Northern Hemisphere is maximally tilted away from the sun and receives the least energy. This accounts for the seasons of heat and cold, summer and winter. Figure 5-1 summarizes these earth–sun relationships and shows how these seasons occur at opposite times of the year in the Northern and Southern Hemispheres.

It is worthwhile for us to examine in greater detail the particular positions of the orbiting earth illustrated in Fig. 5-1. Now consider Fig. 5-2a, which shows that on or about June 22, parallel rays from the sun fall vertically at noon on the earth at latitude 23½°N. This latitude, where the sun's rays strike the surface at an angle of 90 degrees, is given the name **Tropic of Cancer**—the most northerly latitude where the sun's noontime rays strike vertically. We can also see that all areas north of latitude 66½°N, which is called the **Arctic Circle**, remain totally in sunlight during the earth's 24-hour rotation (Fig. 5-3). If a vertical pole were placed at the equator at noon on

this day of the year, the sun should appear to be northward of the pole, making an angle of 23½ degrees with the pole and an angle of 66½ degrees with the ground (Fig. 5-4). Note that the summation of these two angles equals 90 degrees.

Precisely six months later, on December 22, the position of the earth relative to the sun causes the sun's rays to strike vertically at noon at 23½°S, the latitude called the **Tropic of Capricorn** (the southernmost latitude where the sun's noon rays can strike the surface at 90 degrees). The other relationships between the earth and the sun for June 22 described above are exactly reversed (*see* Fig. 5-2b). Accordingly, the entire area south of the **Antarctic Circle**, located at latitude 66½°S, receives 24 hours of sunlight. Simultaneously, the area north of the Arctic Circle is in complete darkness (note that in Fig. 5-2a, the area south of the Antarctic Circle was similarly darkened on June 22).

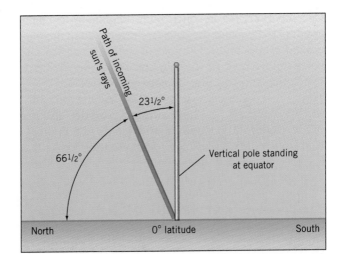

FIGURE 5-4 Angular relationships of incoming solar rays with the ground and a vertical pole standing at the equator at noon on June 22.

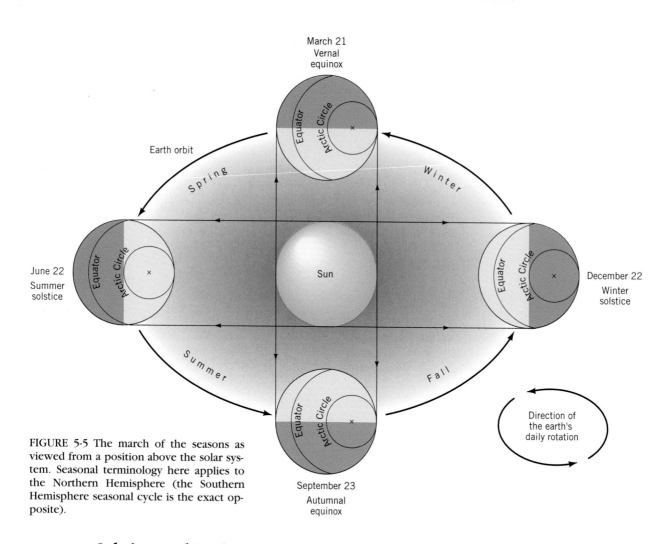

FIGURE 5-5 The march of the seasons as viewed from a position above the solar system. Seasonal terminology here applies to the Northern Hemisphere (the Southern Hemisphere seasonal cycle is the exact opposite).

Solstices and Equinoxes

To us on earth, it appears that the highest daily position of the sun at noontime gets lower in the sky as the seasons progress from summer to fall to winter. If you were to plot the position of the noontime sun throughout the year at a location in the middle latitudes of the Northern Hemisphere, it would seem to climb higher and higher until June 22, when it would appear to stop. Then it would move lower and lower, until it stopped again at December 22, before once more beginning to climb. South of the equator the dates are reversed, but the phenomenon is identical. The ancient Greeks plotted the apparent movement of the sun and called the points at which the stops occurred **solstices** ("sun stands still"). Today we continue to use their word, calling the earth–sun position of June 22 the **summer solstice** and that of December 22 the **winter solstice**. In the Southern Hemisphere, of course, these dates are reversed.

Exactly halfway between the two solstice dates, there are two positions where the rotating globe receives 12 hours of sunlight and 12 of darkness at all latitudes. These positions occur on or about March 21 and September 23. Because of the equal lengths of night at every latitude, these special positions are called **equinoxes**, which in Latin means equal nights. On these two occasions the sun's rays fall vertically over the surface at the equator, and the sun rises and sets due east and west. The equinox of March 21 is known as the **spring** or **vernal equinox**, and that of September 23 as the **fall** or **autumnal equinox**.

The Four Seasons

You can achieve a clear idea of the causes of the seasons if you imagine you are looking down on the earth's orbit around the sun (the plane of the ecliptic) from a point high above the solar system, a perspective diagrammed in Fig. 5-5. The North Pole always points to your right. At the summer solstice, the Arctic Circle receives sunlight during the entire daily rotation of the earth, and all parts of the Northern Hemisphere have more than 12 hours of daylight. These areas receive a large amount of solar energy in the summer season. At the winter solstice, the area inside the Arctic Circle receives no sunlight at all, and every part of the Northern Hemisphere receives less than 12 hours of sunlight. Thus winter is a time of cooling, when solar energy levels are at a minimum. However, at both the spring and fall equinoxes, the Arctic Circle and the equator are equally divided into day and night. Both hemispheres receive an equal amount of sunlight and darkness, and energy from the sun is equally distributed.

Measuring Time on Our Rotating Earth

The measurement of time on the earth's surface is important in the study of our planet. One of the most obvious ways to start dealing with time is to use the periods of light and darkness resulting from the daily rotation of the earth. One rotation of the earth, one cycle of daylight and nighttime hours, constitutes one full day. The idea of dividing the day into 24 equal hours dates from the fourteenth century.

With each place keeping track of its own time by the sun, this system worked well as long as human movements were confined to local areas. But by the sixteenth century, when sailing ships be-

gan to undertake transoceanic voyages, problems arose because the sun is always rising in one part of the world as it sets in another. On a sea voyage, such as that of Columbus in the *Santa Maria*, it was always relatively simple to establish the latitude of the ship. Columbus's navigator had only to find the angle of the sun at its highest point during the day. Then, by knowing what day of the year it was, he could calculate his latitude from a set of previously prepared tables giving the angle of the sun at any latitude on a particular day.

It was impossible, however, for him to estimate longitude. In order to do that,

he would have to know accurately the difference between the time at some agreed meridian, such as the prime meridian (zero degrees longitude), and the time at the meridian where his ship was located. Until about 1750, no portable mechanical clock or chronometer was accurate enough to keep track of that time difference.

With the perfection of the chronometer in the late eighteenth century, the problem of timekeeping came under control. But by the 1870s, as long-distance railroads began to cross the United States, a new problem surfaced. Orderly train schedules could not be devised if

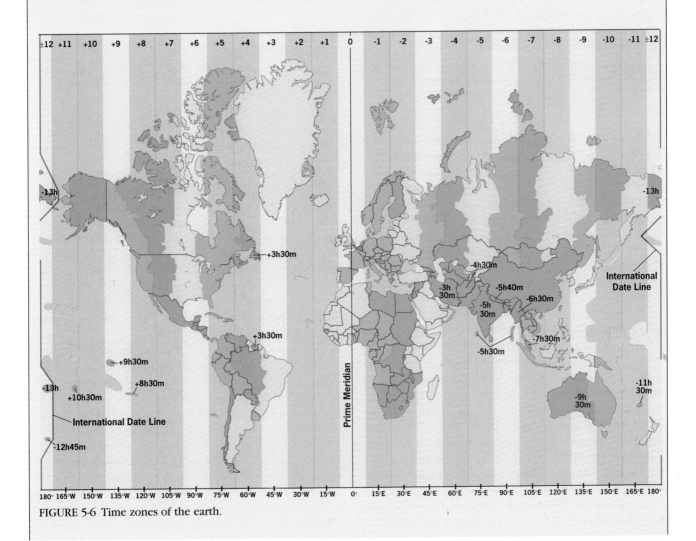

FIGURE 5-6 Time zones of the earth.

each town and city operated according to local sun time, and the need for a system of time organization among different regions became essential.

The problem of having different times at different longitudes was finally resolved at an international conference held in 1884 in Washington, D.C. There it was decided that all earth time around the globe would be standardized against the time at the prime meridian (a concept introduced on p. 25), which passed through Britain's Royal Observatory at Greenwich (London). The earth was divided into the 24 time zones shown in Fig. 5-6, each using the time at standard meridians located at intervals of 15 degrees of longitude with respect to the prime meridian ($24 \times 15° = 360°$). Each time zone differs by one hour from the next, and the time within each zone can be related in one-hour units to the time at Greenwich. When the sun rises at Greenwich, it has already risen in places east of the observatory. Thus the time zones to the east are designated as *fast*; time zones west of Greenwich are called *slow*.

This solution led to a peculiar problem. At noon at Greenwich on January 2, 1996, it will be midnight on January 2 at 180°E longitude (12 time zones ahead) and midnight on January 1 at 180°W (12 time zones behind). However, 180°E and 180°W are the *same* line. This meridian was named the **international date line** by the Washington conference. It was agreed that travelers crossing the date line in an eastward direction, toward the Americas, should repeat a calendar day; those traveling west across it, toward Asia and Australia, should skip a day. The international date line did not pass through many land areas (it lies mainly in the middle of the Pacific Ocean), thereby avoiding severe date problems for people living near it. Where the 180th meridian did cross land, the date line was arbitrarily shifted to pass only over ocean areas.

Similarly, some flexibility is allowed in the boundaries of other time zones to allow for international borders and even for state borders in such countries as Australia and the United States (*see* Fig. 5-6). Some countries, such as India, choose to have standard times differing by half or a quarter of an hour from the major time zones. Others, such as China, insist that the *entire country* adhere to a single time zone!

A further arbitrary modification of time zones is the adoption in some areas of **daylight-saving time**, whereby all clocks in a time zone are set forward by one hour from standard time for at least part of the year. The reason for this practice is that many human activities start well after sunrise and continue long after sunset, using considerable energy for lighting and heating. Energy can be conserved by setting the clocks ahead of the standard time. In the United States today, most states begin daylight-saving time during the first weekend in April and end it on the last weekend in October.

The annual revolution of the earth around the sun and the constant tilt of its axis give our planet its different seasons of relative warmth and coldness. The yearly cycle of the four seasons may be traced using Fig. 5-5. *Spring* begins at the vernal equinox on March 21 and ends at the summer solstice on June 22; *summer* runs from that date through the autumnal equinox on September 23; *autumn* occurs from then until the arrival of the winter solstice on December 22; *winter* then follows and lasts until the vernal equinox is again reached on March 21. This cycle, of course, applies only to the Northern Hemisphere; the Southern Hemisphere's seasonal march is the mirror image, with spring commencing on the date of the northern autumnal equinox (September 23).

Throughout human history, the passage of the seasons has been used as a basis for establishing secure reference points for measuring time (*see* Perspective: Measuring Time on Our Rotating Earth). The changing spatial relationships between the earth and sun, produced by the planetary motions of revolution and rotation on a constantly tilted axis, cause important variations in the amount of solar energy received at the earth's surface. Those patterns are explored at some length in Unit 7, but certain basic ideas are introduced here because they follow directly from the preceding discussion.

INSOLATION AND ITS VARIATION

At any given moment, exactly one-half of the rotating earth is in sunlight and the other half is in darkness. The boundary between the two halves is called the **circle of illumination**, an ever shifting line of sunrise in the east and sunset in the west. The sunlit half of the earth is exposed to the sun's radiant energy, which is transformed into heat at the planetary surface, and, to a lesser extent, in the atmospheric envelope above it. There is, however, considerable variation in the surface receipt of **insolation** (a contraction of the term **in**coming **sol**ar radi**ation**).

Let us imagine for a moment that the earth's axis had no tilt, that it was always perpendicular to the plane of the ecliptic. If that were the case, our planet would maintain its equinox position throughout the year. In such a situation, insolation would strictly be dependent on latitude—the amount of solar energy received at a point would depend upon its distance from the equator. The equator would receive the greatest solar radiation because the sun's rays strike it most directly.

This can be demonstrated in Fig. 5-7, which shows how the parallel rays of the sun fall on various parts of the spherical earth. Note that three equal columns of so-

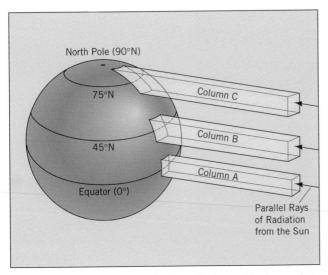

FIGURE 5-7 The reception of solar radiation, with the earth in the equinox position. A surface at a higher latitude receives less radiation than a surface of equal area at a lower latitude.

lar radiation strike the curved surface differently, with the lower latitudes receiving more insolation per unit area than the higher latitudes. At the equator, all the solar rays in column A are concentrated on a small square box; in the midlatitudes, at 45°N, an equal number of rays in column B are diffused across a surface area about twice as large; and in the polar zone, at 75°N latitude, that same number of rays in column C are scattered across an area more than three times the size of the box illuminated by column A.

In this particular instance, the equator receives the most intense insolation because the midday sun's rays strike it vertically, from the point directly overhead (known as the **zenith**), 90 degrees above the horizon. Thus, if one travels north or south from the equator, the solar radiation received decreases with progressively higher latitude as the angle of the sun's noontime rays declines from the zenith point in the sky. However, that angle of **solar elevation** above the horizon, also known as the *angle of incidence*, is not the only determinant of annual insolation received at a point on the earth's surface. The duration of daily sunlight is an equally important factor.

The axis of the real earth, as we already know, is tilted at an angle of 23½ degrees from a straight line perpendicular to the plane of the ecliptic (or 66½ degrees with reference to the plane itself). This causes considerable variation in the length of a day at most latitudes during the course of a year. If you look at the weather page in today's newspaper or the weather segment of this evening's local television news program, you will quickly realize that today's sunrise and sunset times are slightly different from yesterday's or tomorrow's. This reflects the constant change of the latitude where the mid-

day sun shines on the earth from the zenith point. From our earlier discussion of the seasons, you should be aware that the variation in this latitude occurs between 23½°N (on the day of the Northern Hemisphere summer solstice) and 23½°S (the winter solstice). The equatorial position illustrated in Fig. 5-7 occurs only on the days of the spring and fall equinoxes.

The combined effects of solar elevation and daily sunlight duration are graphed in Fig. 5-8. It should be stressed that this graph is a model of a much more complicated real world. Whereas its main purpose is to show how insolation varies on our planet, the patterns in Fig. 5-8 depict solar radiation received at the top of the atmosphere (or at the surface if we assumed the earth had no atmosphere at all).

In Fig. 5-8, the vertical axis represents the complete range of latitudes while the horizontal axis represents the months of the calendar year, with the solstices and equinoxes specially drawn in. The units of solar radiation measurement are not important for understanding this graph (insolation here is calibrated in megajoules per square meter per day). However, variations of insolation can readily be interpreted by looking at the isoline pattern (isolines are explained on p. 32): the higher the value, the greater the amount of radiation received.

Thus, if we wanted to trace the global latitudinal profile of insolation for the summer solstice, we would simply follow the vertical line labeled June 22 from pole to pole. Starting at the top at the North Pole, we begin with some of the highest recorded values on the graph (greater than 44), which are equaled or slightly surpassed only in the high latitudes of the Southern Hemisphere around the time of the winter solstice. (Remember, that date is only 12 days before perihelion, when the earth's orbit makes its closest approach to the sun; hence the higher values at the start of the southern summer.) The polar-area values on June 22 are high because the length of a day north of the Arctic Circle is 24 hours. Even though the sun is at a fairly low angle there (23½ degrees or less above the horizon), the many extra hours of sunlight are sufficient to raise the polar radiation-receipt level beyond the highest value ever recorded for the equator (above 44 versus an equatorial range from about 33 to 38).

As we descend in latitude from the Arctic Circle, the June 22 line does not fall but rather levels off across the middle latitudes all the way to the Tropic of Cancer, near where the value of 40 is finally reached. This high, stable radiation value is maintained because, despite the decrease in daylight length as we move south, insolation is increasingly reinforced by the rising angle of the sun in the sky as we approach zenith at 23½°N latitude (note that the zenithal position of the sun across the year is denoted by the bell-shaped curve drawn in red). Once we proceed south of the Tropic of Cancer on the June 22 line, insolation values exhibit a very different trend:

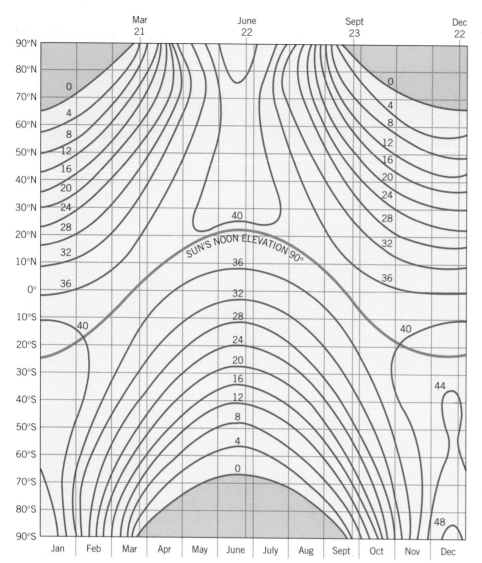

FIGURE 5-8 The spatial distribution of solar radiation falling on the top of the atmosphere (in megajoules per sq m per day). The annual variations you see are the result of the combined effects of solar elevation and the duration of daily sunlight.

they begin to decline swiftly, and south of the equator they also fall off regularly (about seven to eight units every 10 degrees of latitude) until zero is reached at the Antarctic Circle. From here to the South Pole, of course, we enter the zone of seasonal darkness, which is illustrated in Fig. 5-2a.

A great deal more could be said about the latitudinal distribution of incoming solar radiation shown in Fig. 5-8, which in many ways summarizes the earth–sun relationships we have covered here in Unit 5. You are encouraged to study this graph, and by tracing patterns along various latitudes or for specific times of the year, you should find the exercise a useful application of the concepts associated with revolution, rotation, axis tilt, seasonality, solstices, equinoxes, and insolation. Even though this graph is a simplification of reality, it is a useful beginning to the study of the atmosphere (which commences in Unit 6) because it marks the point of transition from astronomic controls to the terrestrial forces that shape weather and climate across the face of the earth.

KEY TERMS

Antarctic Circle (p. 53)

aphelion (p. 51)

Arctic Circle (p. 53)

axis (p. 51)

circle of illumination (p. 56)

Coriolis force (p. 51)

daylight-saving time (p. 56)

equinox (p. 54)

fall (autumnal) equinox (p. 54)

insolation (p. 56)

international date line (p. 56)

perihelion (p. 51)

plane of the ecliptic (p. 52)

revolution (p. 50)

rotation (p. 51)

solar elevation (p. 57)

solstice (p. 54)

spring (vernal) equinox (p. 54)

summer solstice (p. 54)

Tropic of Cancer (p. 53)

Tropic of Capricorn (p. 53)

winter solstice (p. 54)

zenith (p. 57)

REVIEW QUESTIONS

1. Describe the motions of the earth in its revolution around the sun and its rotation on its axis.

2. Describe the seasonal variation in the latitude of the vertical, noontime sun during the course of the year.

3. Differentiate between the spring and autumnal equinoxes and between the summer and winter solstices.

4. What is an *oblate spheroid*, and why is the earth an example of this phenomenon?

5. What is the *international date line*, and why is it a necessary part of the earth's meridional system?

6. Why is insolation at the North Pole on the day of the summer solstice greater than that received at the equator on the equinoxes?

REFERENCES AND FURTHER READINGS

BARTKY, I. R., and HARRISON, E. "Standard and Daylight-Saving Time," *Scientific American*, May 1979, 46–53.

GEDZELMAN, S. D. *The Science and Wonders of the Atmosphere* (New York: Wiley, 1980), Chapters 4 and 6.

HARRISON, L. C. *Sun, Earth, Time and Man* (Chicago: Rand McNally, 1960).

JOHNSON, W. E. *Mathematical Geography* (New York: American Book Co., 1907).

NEIBURGER, M., et al. *Understanding Our Atmospheric Environment* (San Francisco: Freeman, 2nd ed., 1982), Chapter 3.

U.S. NAVAL OBSERVATORY. *The Air Almanac* (Washington, D.C.: U.S. Government Printing Office, annual).

ZEILIK, M. *Astronomy: The Evolving Universe* (New York: Wiley, 7th ed., 1994).

PART TWO

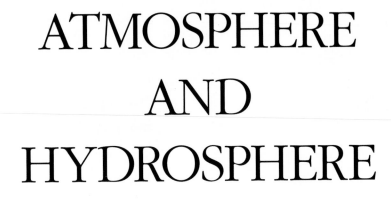

ATMOSPHERE AND HYDROSPHERE

Composition and Structure of the Atmosphere

OBJECTIVES

◆ To describe the constituents of the atmosphere and their relative concentrations.

◆ To survey the four layers of the atmosphere together with their major properties.

◆ To discuss the problem of ozone depletion and its consequences.

Our atmosphere, one of our most precious natural resources, constitutes a vital component in the systematic study of our planet. This thin, shell-like envelope of life-sustaining air that surrounds the earth (*see* photo above) is a place of incredible activity—as the units of Part Two demonstrate. It has been called the working fluid of our planetary heat engine, and its constant motions shape the course of environmental conditions at every moment in every locality on the surface. The short-term conditions of the restless atmospheric system that impinge on daily human activities are called **weather**; the long-term conditions of aggregate weather over a region, summarized by averages and measures of variability, constitute its **climate**.

The atmosphere extends from a few meters below the ground on land, or at the water's surface in oceanic areas, to a height of about 60,000 km (37,000 mi). Most of the mass of the atmosphere is concentrated near the planetary surface. Physical geographers are especially in-

UNIT OPENING PHOTO: Cross-sectional view of the atmosphere, which occupies the narrow zone between the blue of the surface and the blackness of outer space.

terested in the lower parts of the atmosphere, those below 50 km (31 mi) and particularly below 10 km (6 mi). Important flows of energy and matter occur within these lower layers, which constitute the effective atmosphere for all life forms at the surface. Here, too, great currents of air redistribute heat across the earth. These currents are part of the systems that produce our daily weather. Over time, weather and the flows of heat and water across the earth are eventually translated into our surface patterns of climate. We are affected by both the local climate and the larger atmosphere, and we have the power to alter them to some degree. In the past, climatic changes have occurred without human intervention; but in the future, and even today, humankind may be playing a more active role.

CONTENTS OF THE ATMOSPHERE

The atmosphere may be broadly divided into two vertical regions (Fig. 6-1). The lower region, called the **homosphere**, extends from the surface to 80 to 100 km (50 to 63 mi) above the earth and has a more or less uniform chemical composition. Beyond this level, the chemical composition of the atmosphere changes in the upper region known as the **heterosphere**. The homosphere is the more important of the two atmospheric regions for human beings because we live in it. If you experimented by collecting numerous air samples of the homosphere, you would find that it contains three major groups of components—**constant gases**, **variable gases**, and **impurities**. The constant gases are always found in the same proportions, but the variable gases are present in differing quantities at different times and places. Impurities are solid particles floating in the atmosphere, whose quantities also vary in time and space.

Constant Gases

Two major constant gases make up 99 percent of the air, and both are crucial to sustaining human and other forms of terrestrial life. They are nitrogen, which constitutes 78 percent of the air, and oxygen, which accounts for another 21 percent. Thus the bulk of the atmosphere that we breathe consists of nitrogen. As far as we are directly concerned, nitrogen is important only because it is not poisonous. Indirectly, it is vital: atmospheric nitrogen is converted by bacteria into other nitrogen compounds essential for plant growth.

Immediately necessary to our survival, of course, is oxygen. We absorb oxygen into our bodies through our lungs and into our blood; one of its vital functions there is to "burn" our food so that its energy can be released. Such burning actually involves the chemical combination of oxygen and other materials to create new products. The biological name for this process is *respiration*, and the chemical name is *oxidation*. An example of rapid oxidation is the burning of *fossil fuels* (coal, oil, and natural gas): without oxygen, this convenient way of releasing the energy stored in these fuels would be lost to us. Slow oxidation can also occur, as in the rusting of iron. Therefore, oxygen is essential not only for respiration but also for its role in many other chemical processes.

In 1894, when scientists first removed oxygen and nitrogen from a sample of air, they noticed that a gas remained that seemed chemically inactive: it would not combine with other compounds, thereby making it an *inert gas*. The discoverers named the gas argon and found that it makes up almost 1 percent of dry air (air not combined with moisture). Although this inert gas has some commercial uses (Fig. 6-2), it does not play a major role in the workings of environmental systems.

Variable Gases

Although they collectively constitute only a tiny proportion of the air, we must also recognize the importance of certain atmospheric gases that are present in varying quantities. Three of these variable gases are essential to human well-being: carbon dioxide, water vapor, and ozone.

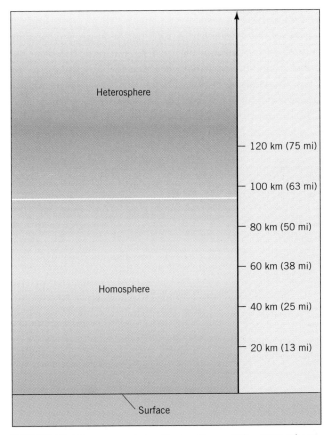

Heterosphere

— 120 km (75 mi)

— 100 km (63 mi)

— 80 km (50 mi)

— 60 km (38 mi)

Homosphere

— 40 km (25 mi)

— 20 km (13 mi)

Surface

FIGURE 6-1 The two main vertical regions of the atmosphere.

FIGURE 6-2 Argon is frequently used inside "neon" lights. This is part of the famous Strip in the heart of Las Vegas, Nevada.

Carbon Dioxide

Carbon dioxide (CO_2), which on average comprises only 0.04 percent of dry air, is a significant constituent of the atmosphere in terms of its climatic influence. Despite the comparatively small amounts present, carbon dioxide fulfills two vital functions for the earth. The first is in the process of *photosynthesis*, in which plants use carbon dioxide and other substances to form carbohydrates, which are an essential part of the food and tissue of both plants and animals. The second function of carbon dioxide is to absorb some of the energy transferred to the atmosphere from the earth's surface (a process discussed in Unit 7). Because most of the other atmospheric constituents are such poor absorbers of this energy, carbon dioxide helps to keep the atmosphere at temperatures that permit life (which globally now average just over 15°C [59°F]).

Carbon dioxide plays still other environmental roles. It helps dissolve limestone, which leads to the intriguing features of certain limestone-based landscapes (*see* Unit 44). Futhermore, a number of scientists believe that carbon dioxide plays a role in both major and minor climatic change. It has been estimated that over the past two centuries the total quantity of this gas in the atmosphere has risen by as much as 25 percent. The primary cause is believed to be increasing industrialization and the associated burning of fossil fuels. The rise in the atmospheric carbon dioxide level since 1960 has been unusually swift (Fig. 6-3), accounting for no less than half the total increase since the onset of the Industrial Revolution more than 200 years ago. Surprisingly, that upward trend, for unknown reasons, slowed down after 1991. Nonetheless, a resumption is expected, and because carbon dioxide is a factor in the warming of the atmosphere, some re-

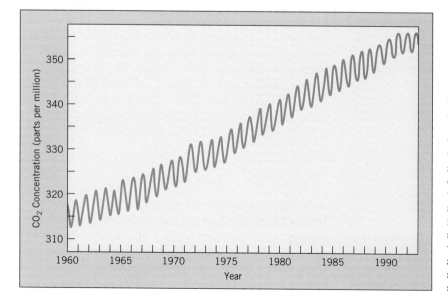

FIGURE 6-3 Changes in the carbon dioxide content of the earth's atmosphere, 1960–1993. These data were collected at Hawaii's Mauna Loa Observatory, far from the world's industrial and urban areas. Yearly atmospheric CO_2 levels reach a minimum during the Northern Hemisphere summer, when global vegetation and photosynthesis are at their annual maxima. During the northern winter, with carbon-dioxide-absorbing vegetation at a minimum worldwide, atmospheric levels reach their annual peak. The steady rise in CO_2 observed since the late 1950s unexpectedly slackened during the early 1990s.

searchers worry that its continued prodigious production might significantly affect the future climate of the earth.

Water Vapor

The ability to absorb energy from the earth's surface and atmosphere is also found in the most widely distributed variable atmospheric gas—**water vapor**, the invisible gaseous form of water (H_2O). Water vapor is more efficient than carbon dioxide in capturing radiant energy because it not only can absorb energy but can store it as well. When water vapor is moved around by currents of air, stored energy is transported along with it. This is part of an essential process by which the surface temperatures of most parts of the earth are kept moderate enough for human habitation. In deserts or cold regions, water vapor makes up only a minute fraction of 1 percent of the air. But over warm oceans or moist tropical land areas it may make up as much as 3 or 4 percent.

These variations demonstrate an important relationship: *the warmer the air the more moisture or water vapor it can hold.* Because the parts of the atmosphere near the earth's surface have relatively high temperatures, that is where most of the water vapor occurs. Without water vapor, there would be no clouds or rainfall. Thus most portions of the land surface would be too dry to permit agriculture. Without the great cycle in which water moves from the surface into the atmosphere and back again, little life of any kind would be found on our planet.

Ozone

The other variable gases in the lower parts of the atmosphere are found in much smaller quantities than water vapor. The most important is ozone, the rarer type of oxygen molecule composed of three atoms (O_3) instead of two (O_2). Ozone is confined mainly to the so-called **ozone layer** (sometimes referred to as the *ozonosphere*), which extends between 15 and 50 km (9 and 31 mi) above the earth. The greatest concentrations of ozone, however, are found between about 20 and 25 km (13 and 16 mi), although this gas is usually formed at higher levels and transported downward. Even where it is most highly concentrated, ozone often constitutes less than six parts per 100,000 of the atmosphere. But like carbon dioxide it is very important. It, too, has the ability to absorb radiant energy, particularly the *ultraviolet radiation* associated with incoming solar energy. Ultraviolet radiation can give us a suntan, but large doses cause severe sunburn, blindness, and skin cancers. The ozone layer shields us from excessive quantities of this high-energy radiation.

Other Variable Gases

Minute quantities of many other variable gases are present in the atmosphere. The most noteworthy are hydro-

gen, helium, sulfur dioxide, oxides of nitrogen, ammonia, methane, and carbon monoxide. Some of these are air pollutants (foreign matter injected into the atmosphere) derived from cities and industry. They can have harmful effects even when the concentrations are one part per million or less. Other pollutants are found in the form of solid particles, which can be classified as the impurities of the atmosphere.

Impurities

If you were to collect air samples, particularly near a city, they would likely contain a great number of impurities in the form of *aerosols* (tiny floating particles suspended in the atmosphere). Typical rural air might contain about four particles of dust per cubic millimeter, whereas city parks often have four times that density. A business district in a metropolitan area might have 200 particles per cubic millimeter, and an industrial zone over 4000. Both smoke and dust particles are common in urban air, but dust particles are the most prevalent type in rural air. Bacteria and plant spores are found in all parts of the lower atmosphere. Salt crystals are another major impurity, with large quantities usually formed by evaporation above breaking ocean waves.

Collectively, the impurities play an active role in the atmosphere. Many of them help in the development of raindrops (*see* Unit 12). Moreover, the small particles can affect the color of the sky. Air and the smallest impurities scatter more blue light from the sun than any other color. This is why the fair-weather sky looks blue. But when sunlight travels a longer distance through the atmosphere to the surface, as at sunrise or sundown, most of the blue light has been scattered. We see only the remaining yellow and red light which, of course, produces colorful sunrises and sunsets. Occasionally, when there is an abnormally large amount of impurities in the atmosphere, such as after the eruption of a major volcano, this process is carried to some spectacular extremes (Fig. 6-4).

Atmospheric Cycles

As the new planet Earth cooled following its birth about 4.6 billion years ago, the atmosphere was formed from gases expelled by volcanoes and the hot surface itself. During this formation, the atmospheric constituents achieved a state of dynamic equilibrium (a systems concept discussed in Unit 1), an efficient balance that is still maintained today—if the air is not significantly altered by pollutants. The prevailing composition of the atmosphere we have just described is not static but is the result of constant gains and losses of its major and minor components. A critical part of this component flow takes place because the boundary or surface layer of the atmosphere adjoins the lithosphere, hydrosphere, and biosphere. Four vital cycles have developed at this interface,

FIGURE 6-4 Volcanic dust high in the atmosphere is often responsible for breathtaking sunsets dominated by fiery reds and yellows.

involving the transfer of water, oxygen, nitrogen, and carbon dioxide.

The **hydrologic cycle** is a complex system of exchange involving water as it circulates among the atmosphere, lithosphere, hydrosphere, and biosphere; this cycle is so important that much of Unit 12 is devoted to it. In the **oxygen cycle**, oxygen is put back into the atmosphere as a byproduct of photosynthesis and is lost when it is inhaled by animals or chemically combined with other materials during oxidation.

The **nitrogen cycle** is maintained by plants, whose roots contain bacteria that can extract nitrogen from the air or soil. These *nitrogen-fixing bacteria* convert atmospheric nitrogen into the organic compounds of the plants, especially organic protein. Some of this organic material is transferred to animals, including human beings, when the plants are eaten. When the plants and animals die, the nitrogen is transformed by other bacteria and microorganisms first into ammonia, urea, and nitrates, and then eventually back into the gaseous form of nitrogen, which returns to the atmosphere.

The **carbon dioxide cycle** is dominated by exchanges that occur between the air and the oceans. This atmospheric gas enters the sea by direct absorption from the air, by plant and animal respiration, and by the oxidation of organic matter. Alternatively, carbon dioxide is released from the ocean following the decomposition of countless millions of small organisms known as *plankton*. Another major carbon dioxide exchange takes place between the atmosphere and land plants of the biosphere, with the gas taken from the air by plants during photosynthesis and released by them during respiration and decay. In addition, carbon dioxide is released in the burning of fossil fuels, with possible consequences for climatic change.

THE LAYERED STRUCTURE OF THE ATMOSPHERE

Earlier we noted that the atmosphere consists of two broad regions: a lower homosphere and an upper heterosphere (Fig. 6-1). A more detailed picture of the structure of the atmosphere emerges if we subdivide it into a number of vertical layers according to temperature characteristics. Altitude has a major influence on temperature, and the overall variation of atmospheric temperature with height above the surface is shown in Fig. 6-5.

The bottom layer of the atmosphere, where temperature usually decreases with altitude, is called the **troposphere**. The rate of a decline in temperature is known as the **lapse rate**, and in the troposphere the average lapse rate is 6.5°C/1000 m (3.5°F/1000 ft). The upper boundary of the troposphere, along which temperatures stop decreasing with height, is called the **tropopause**.

Beyond this discontinuity, in a layer called the **stratosphere**, temperatures either stay the same or start increasing with altitude. Layers in which temperature increases with altitude exhibit positive lapse rates. These are called **inversions** because they invert or reverse what we on the surface believe to be the normal state of

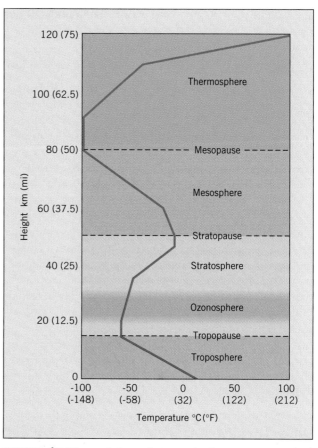

FIGURE 6-5 The variation of atmospheric temperature (red line) with height.

temperature change with elevation—a decrease with height.

As the top of the stratosphere is approached, beyond about 52 km (32 mi) above the earth, temperatures remain constant with increasing altitude. This boundary zone is called the **stratopause**, and is topped by a layer known as the mesosphere. In the **mesosphere**, temperatures again fall with height, as they did in the troposphere. Eventually, the decline in temperature stops, at a boundary you might rightly guess to be called the **mesopause**. This occurs at about 80 km (50 mi) above the earth's surface. Not far beyond the mesopause, temperatures once more increase with height in a layer called the **thermosphere**.

The Troposphere

Because the troposphere is the atmospheric zone in which we live, and the layer where almost all weather happens, we need to know quite a bit about it. That survey of its processes is undertaken in Units 7 through 10, with Unit 8 focusing on temperature relationships. Before that detailed treatment, let us summarize the most significant interactions across the tropopause and the nature of the layers that lie above it.

The tropopause, as Fig. 6-5 indicates, is positioned at an average height of about 12 km (7.5 mi or 40,000 ft). Actually, this altitude varies with latitude: it is lowest over the poles (about 8 km [5 mi]) and highest above the equator (about 16 km [10 mi]). There are usually two distinct breaks in the tropopause, which are characterized by areas of variable lapse rates. These breaks are generally found at latitudes of about 25° and 50°N and S. The breaks, associated with fast-flowing winds in the upper atmosphere, are important because, through them, the troposphere and the stratosphere exchange materials and energy. Small amounts of water vapor may find their way up into the stratosphere at these breaks, while ozone-rich air may be carried downward into the troposphere through them.

The Stratosphere

Above the tropopause is the calmer, thinner, clear air of the stratosphere. Jet aircraft mainly fly through the lower stratosphere because it provides the easiest flying conditions (Fig. 6-6). The nearly total absence of water vapor in this layer prevents the formation of clouds, thus providing pilots with fine visibility. And temperature inversion prohibits vertical winds, so the horizontal winds in the stratosphere are almost always parallel to the earth's surface, ensuring smoother flights than in the troposphere.

The *ozone layer* (ozonosphere) lies within the stratosphere (Fig. 6-5). Here ozone is naturally produced by the action of ultraviolet sunlight on oxygen, and naturally destroyed when it is turned back into oxygen. Ozone is

also transported by natural processes from one part of the stratosphere to another. These chemical and transport processes create a constant balance of stratospheric ozone. The critical importance of the ozone layer in shielding the surface of the earth from ultraviolet radiation has already been noted. At the same time, the absorption of ultraviolet radiation heats the stratosphere, giving it the positive temperature lapse rate we noted earlier. Thus it is vitally important to maintain the proper ozone balance; and increasing reports of *ozone holes* in the atmosphere have raised concerns among environmental scientists (*see* Perspective: Ozone Holes in the Stratosphere).

The Mesosphere

Above the stratosphere, in the altitudinal zone between about 50 and 80 km (31 and 50 mi), lies the layer of decreasing temperatures called the mesosphere. Over high latitudes in summer, the mesosphere at night sometimes displays high, wispy clouds that are presumed to

FIGURE 6-6 A typical daytime view from a window aboard a commercial flight. Jet planes usually cruise at stratospheric altitudes in order to avoid the bumpier air and obscuring cloud decks of the troposphere below.

Ozone Holes in the Stratosphere

In 1982 a British environmental research team in Antarctica made a startling discovery: its instruments could not detect the ozone layer in the stratosphere overhead. Atmospheric scientists had never before encountered this phenomenon, but artificial satellites and high-flying aircraft by 1985 confirmed the readings of ground-based spectrophotometers and established that a large "ozone hole" existed over most of the southern polar continent. Concerned investigators soon learned that this was a seasonal occurrence that peaked in the spring (Fig. 6-7); but it was also clear that the overall level of ozone was declining.

By the end of the 1980s, scientists had reached a consensus as to the causes of stratospheric ozone destruction. During the southern winter (late June through late September), Antarctic air is isolated from the rest of the Southern Hemisphere by a strong circumpolar windflow above the surrounding Southern Ocean. With warmer air walled off and daylight reduced to a minimum, the intense cold of Antarctica's surface gradually penetrates the overlying atmosphere. This supercold air even affects the stratosphere, where icy cloud layers form.

The surfaces of the ice particles that constitute these clouds are sites of chemical reactions involving chlorine, which are triggered by ultraviolet radiation as soon as sunlight returns near the end of the long Antarctic winter. The chlorine atoms released by these reactions swiftly destroy ozone molecules by breaking them down into other forms of oxygen. Moreover, each single freed atom of chlorine can trigger hundreds of destructive ozone reactions. Thus the ozone depletion process spreads rapidly, slowing only as rising temperatures evaporate the

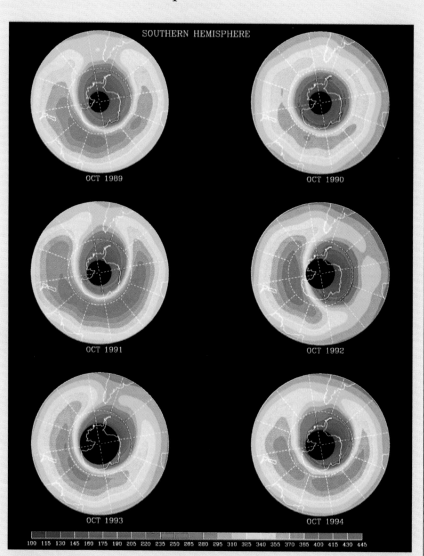

FIGURE 6-7 The ozone hole overlying Antarctica has been closely monitored since its detection in 1982. This sequence of remotely sensed satellite imagery, centered on the South Pole, shows the average distribution of ozone above the entire Southern Hemisphere for the same early spring month for the 1989–1994 period. Additional information on the technicalities of these observations is provided in the entry for Fig. 6-7 in Photo Credits section at the back of the book.

stratospheric clouds. By midsummer, ozone levels are again on the rise; they will peak in winter, but with the return of spring another cycle of destruction will be spawned.

This scenario should not necessarily suggest that the summer-through-winter buildup replenishes all the ozone lost in the spring. As Fig. 6-7 indicates, large holes in the ozone layer—shown by the blue-to-black color sequence— appeared above Antarctica in every year. The 1992 ozone hole was the largest ever observed, at one point (shortly after this image was taken) covering 24.3 million sq km (9.4 million sq mi), an area almost twice the size of the Antarctic continent itself. Events such as these make it quite likely that a net overall depletion of ozone is taking place.

The villain, of course, is the chlorine, most of which is not of natural origin but a byproduct of modern technology that adversely impacts our fragile atmosphere. The evidence today is overwhelming that artificial compounds called **chlorofluorocarbons (CFCs)** are at fault. Before the 1990s, these chemicals were widely used in everyday life for a variety of purposes, such as coolants in refrigerators and air-conditioning systems, propellants in aerosol sprays, cleaning solvents for computer components, and plastic foam in hundreds of products. Thus vast quantities of CFCs entered the atmosphere for many years. Only recently, however, did scientists learn that these compounds can rise easily into the stratosphere and disperse themselves across the entire globe.

Even though initial research on the ozone hole in the mid-1980s tentatively concluded that the phenomenon was limited to Antarctica, the world scientific community was concerned enough to call for action. That resulted in a 1987 conference in Montreal, Canada, where more than 30 countries took the first steps to limit their CFC production. The Montreal Protocol established a scientific assessment panel, whose recommendations would be followed as part of an ongoing process to counteract the depletion of atmospheric ozone. A second conference took place in 1990, at which nearly 100 countries agreed to phase out the manufacturing of CFCs by 2000. In 1991 that deadline was advanced to 1996 (developing countries have received dispensations to delay implementation) when the panel reported an intensifica-

tion in global ozone loss; in 1992 additional CFC-type chemicals were added to the list of products to be phased out (by 2030). By the mid-1990s, these measures appeared to be paying off because the annual rise in atmospheric CFC concentration was half of what it had been at the end of the 1980s. And, most encouragingly, the number of treaty signatories had risen to 132 countries, which contain almost 85 percent of the world's population.

Nonetheless, new research findings underscore the environmental damage that has already taken place. We now know that fully 10 percent of the earth's ozone layer has disappeared since 1970. A great deal also has been learned about the spatial variation of ozone depletion, which now extends far beyond the southern polar latitudes. Antarctica continues to be the leading region, and the 1993 spring ozone level observed there was the lowest ever recorded anywhere (and may represent an ominous upward extension of the ozone hole). Elsewhere in the Southern Hemisphere, particularly in southern South America, there is growing evidence of ozone depletion. But the main concern today is with the Northern Hemisphere.

Ozone destruction in recent years has reached serious levels in the higher latitudes of the Northern Hemisphere. Readings in the spring of 1993 were the lowest on record, showing an astounding 10 percent plunge from the previous year in Canada, Russia, and northern Europe. Accompanying this drop were unusually high levels of stratospheric chlorine, probably enhanced by the massive 1991 eruption of Pinatubo volcano in the Philippines. Fortunately, no ozone hole has yet opened above the northern polar region, whose winters are warmer and shorter than those of Antarctica (with its isolating windflows, higher lying terrain, and thick icesheet). But an Arctic ozone hole remains a distinct possibility—it nearly occurred in 1993—and would undoubtedly have more immedi-

ate consequences for humans because almost 90 percent of the world's population resides in the Northern Hemisphere.

Perhaps most sobering of all is the realization that humans have set off a sequence of atmospheric processes that cannot quickly be reversed. It takes about 10 years for rising CFC gases to reach the stratosphere; thus we have witnessed the effects of CFC usage only through the mid-1980s. And when CFCs reach the ozone layer, they do not dissipate but remain in place as active chemicals for perhaps as long as another 140 years. Given these interpretations, we should not be surprised at the rapidity of ozone depletion because a new view of human-induced environmental modification is now emerging. Instead of slow change, pollutants may build up for years without noticeable effect; quietly, a critical mass is approached and surpassed, and then sudden, convulsive change occurs with far-reaching consequences.

The global depletion of atmospheric ozone must be ranked among the most serious problems of potential environmental change. In view of the current situation, predictions have been made that the quantity of ultraviolet radiation reaching the earth's surface will increase 5 to 20 percent over the next three decades. Even if the lower estimate is correct, at least 1 million new cases of skin cancer can be expected to materialize annually; other medical problems that would intensify include cataracts and the weakening of the immune system. Animal life of all kinds would be gravely threatened. One study has reported that a 10 percent increase in ultraviolet radiation could eliminate most forms of plankton, the biological cornerstone of food chains in the oceans. Land plants would undoubtedly be adversely affected too, and crop yields could drop by as much as 25 percent. This would present a disaster of unparalleled magnitude for a rapidly growing human population that can barely feed itself today.

FIGURE 6-8 *Aurora borealis*, popularly called Northern Lights, illuminates the midwinter Alaskan landscape.

be sunlight reflected from meteoric dust particles that become coated with ice crystals. Another common phenomenon in this layer occurs when sunlight reduces molecules to individual electrically charged particles called ions in a process known as *ionization*. Ionized particles concentrate in a zone called the **D**-layer, which reflects radio waves sent from the earth's surface; "blackouts" in communications between the ground and astronauts occur as the **D**-layer is crossed by space vehicles during reentry.

The Thermosphere

The thermosphere is found above 80 km (50 mi) and continues to the edge of space, about 60,000 km (37,000 mi) above the surface. Temperature rises spectacularly in this layer and likely reaches 900°C (1650°F) at 350 km (220 mi). However, because the air molecules are so far apart at this altitude, these temperatures really apply only to individual molecules and do not have the same kind of environmental significance they would on the earth's surface.

Ionization also takes place in the thermosphere, pro-

ducing two more belts (known as the **E**- and **F**-layers) that reflect radio waves. Intermittently, ionized particles penetrate the thermosphere, creating vivid sheet-like displays of light called the *aurora borealis* in the Northern Hemisphere and the *aurora australis* in the Southern Hemisphere (Fig. 6-8). In the upper thermosphere, there are further concentrations of ions that comprise the Van Allen radiation belts. This outermost layer is sometimes referred to as the *magnetosphere* because here the earth's magnetic field is frequently more influential in the movement of particles than its gravitational field is. The thermosphere has no definable outer boundary and gradually blends into interplanetary space.

RESEARCH FRONTIERS

Scientists still understand relatively little about the layers above the effective atmosphere. These outer regions beyond the troposphere, which consist of concentrated ozone, electrically charged particles, bitter cold and extreme heat, meteoric dust, and weirdly illuminated clouds, lie at the frontiers of our knowledge. The ozone-depletion crisis, however, is now unleashing an unprecedented scientific effort to learn much more about these higher layers, because it is increasingly evident that what happens along the fragile outer fringes of our planetary domain is of significance to atmospheric and related processes in the surface layer. With human technology demonstrating an ever greater capacity to alter the chemistry of the air, much research is focusing on the nature of such change, its rates in various parts of the world, and the long-term consequences of its intensification.

As this work proceeds, a heightening sense of urgency prevails in certain quarters because climate and other environmental changes tend not to occur gradually and incrementally. Rather, they often seem to exhibit sharp jumps in response to the subtle but steady reorganization of the earth's atmospheric system. The challenge lies not only in identifying the problems, but also in reversing the sequence of events that produce them (as in the attempt to halt human-induced ozone destruction). Such concerns—which also apply to the biosphere, the hydrosphere, and even the lithosphere—are becoming an integral part of physical geography in the 1990s.

KEY TERMS

carbon dioxide cycle (p. 66)

chlorofluorocarbons (CFCs) (p. 69)

climate (p. 62)

constant gases (p. 63)

heterosphere (p. 63)

homosphere (p. 63)

hydrologic cycle (p. 66)

impurities (p. 63)

inversion (p. 66)

lapse rate (p. 66)

mesopause (p. 67)

mesosphere (p. 67)

nitrogen cycle (p. 66)

oxygen cycle (p. 66)

ozone layer (p. 65)

stratopause (p. 67)

stratosphere (p. 66)

thermosphere (p. 67)

tropopause (p. 66)

troposphere (p. 66)

variable gases (p. 63)

water vapor (p. 65)

weather (p. 62)

REVIEW QUESTIONS

1. What are the constituents of dry air in the atmosphere?

2. Discuss the role of ozone in absorbing incoming solar radiation. Where does this absorption take place?

3. Give the approximate altitudinal extents of each of the atmosphere's layers, and describe the temperature structure of each.

4. Define and describe the basic function of the oxygen, nitrogen, carbon dioxide, and hydrologic cycles.

5. What is the extent of the world's *ozone hole* problem, and what are its likely causes?

REFERENCES AND FURTHER READINGS

BRIMBLECOMBE, P. *Air: Composition and Chemistry* (New York: Cambridge University Press, 1986).

CRAIG, R. A. *The Edge of Space* (Garden City, N.Y.: Anchor/Doubleday, 1968).

GRAEDEL, T. E., and CRUTZEN, P. J. "The Changing Atmosphere," *Scientific American*, September 1989, 58–68.

GRIBBIN, J. "The Ozone Layer," *New Scientist*, May 5, 1988, "Inside-Science" Supplement No. 9.

INGERSOLL, A. P. "The Atmosphere," *Scientific American*, September 1983, 162–174.

MCELROY, M. B., and SALAWITCH, J. B. "Changing Composition of the Global Stratosphere," *Science*, Vol. 243, February 10, 1989, 763–770.

MESZAROS, E. *Global and Regional Changes in Atmospheric Composition* (Boca Raton, Fla.: CRC Press, 1993).

MINNAERT, M. The *Nature of Light and Color in the Open Air* (New York: Dover, 1954).

PARKS, N. "Build-Up of Carbon Dioxide Slows Down," *Earth*, May 1994, 20–23.

SCHAEFER, V. J., and DAY, J. *A Field Guide to the Atmosphere* (Boston: Houghton Mifflin, 1981).

STOLARSKI, R. S. "The Antarctic Ozone Hole," *Scientific American*, January 1988, 30–37.

TOON, O. B., and TURCO, R. P. "Polar Stratosphere Clouds and Ozone Depletion," *Scientific American*, June 1991, 68–74.

"Vanishing Ozone: The Danger Moves Closer to Home" (cover story), *Time*, February 17, 1992, 60–68.

YOUNG, L. B. *Earth's Aura* (New York: Avon, 1979).

YOUNG, L. B. *Sowing the Wind: Reflections on the Earth's Atmosphere* (Englewood Cliffs, N.J.: Prentice-Hall, 1990).

Radiation and the Heat Balance of the Atmosphere

OBJECTIVES

◆ To understand the sun-generated flows of energy that affect the earth and its atmosphere.

◆ To discuss the greenhouse effect and its purported linkage to climate change.

◆ To introduce the earth's heat flows and their spatial patterns.

To understand the workings of weather and climate, we need to become familiar with the atmospheric processes that shape them. In Unit 6 we describe the atmosphere as a dynamic, constantly churning component of a gigantic heat engine. In this unit we focus on the functioning of that engine, which is fueled by incoming solar radiation (*insolation*). Its main operations coordinate and distribute this radiant heat energy between the earth's surface and the envelope of air that surrounds it. As the earth is heated by the sun's rays, the air in contact with the surface becomes warmer. That air begins to rise, cooler air descends to replace it, and the atmosphere has been set into motion. On a global scale, as insolation constantly changes, there is always considerable variation in heat energy across the planetary surface. To maintain equilibrium, large amounts of that energy must be moved from place to place to balance heat surpluses and deficits.

UNIT OPENING PHOTO: Mirrors focus the sun's rays to power an electric generator at this experimental solar energy facility in California's Mojave Desert.

THE RADIATION BALANCE

The sun provides 99.97 percent of the energy required for all the physical processes that take place on the earth and in its atmosphere. As a result of absorbed insolation, different types of radiant heat or radiation flow throughout the earth–atmosphere system, and inputs and outputs of radiation are balanced at the planetary surface.

We may regard **radiation** as a transmission of energy in the form of electromagnetic waves. The wavelength of the radiation is the distance between two successive wave crests. This wavelength varies in different types of radiation and is inversely proportional to the temperature of the body that sent it out: the higher the temperature at which the radiation is emitted, the shorter the wavelength of the radiation. The sun has a surface temperature of about 5500°C (10,000°F), whereas the average surface temperature of the earth is approximately 15°C (59°F). Thus radiation coming from the sun is **shortwave radiation**, and that emitted from the earth is **longwave radiation**. There is, in fact, a wide spectrum of radiation of different wavelengths, which is depicted in Fig. 3-15 (p. 35). This *electromagnetic spectrum* ranges from very short waves, such as cosmic rays and gamma rays, to very long waves, such as radio and electric-power waves.

Radiation from the Sun

Measurements indicate that, on average, 1.95 calories* of energy per sq cm (0.16 per sq in) are received every minute at the top of the earth's atmosphere. This value, called the *solar constant*, would equal in one day all the world's industrial and domestic energy requirements for the next 100 years based on current rates of consumption.

When radiation travels through the atmosphere, several things can happen to it (Fig. 7-1). According to estimates based on available data and global averages, of all the incoming solar energy only 31 percent travels directly to the earth's surface; this energy flow is called **direct radiation**. An almost equal amount, 30 percent, is reflected and scattered back into space by clouds (25 percent) and dust particles (5 percent) in the atmosphere. Another 17 percent of the incoming solar rays is absorbed by clouds (3 percent) and dust and other components of the atmosphere (14 percent). Some of the scattered rays, 22 percent in all, eventually find their way down to the earth's surface and are collectively known

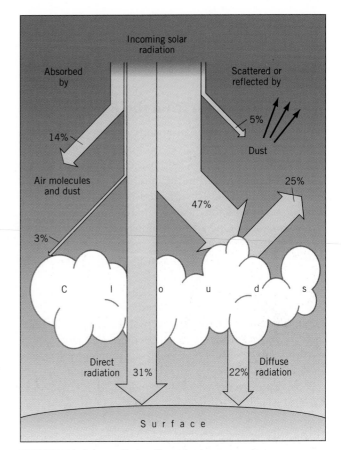

FIGURE 7-1 Solar radiation flows in the atmosphere.

as **diffuse radiation**. Altogether, just over half (53 percent) of the solar energy arriving at the outer edge of the atmosphere reaches the surface as either direct or diffuse radiation. The rest is either absorbed by the atmosphere (17 percent) or scattered and/or reflected back into space (30 percent).

No matter where radiation strikes the earth, one of two things can happen to it. It can either be *absorbed* by—and thereby heat—the earth's surface, or it can be *reflected* by the surface, in which case there is no heating effect. The amount of radiation reflected by the surface depends mainly on the color, composition, and slope of the surface. A ray of solar energy falling on the equator on the day of the equinox, because it strikes perpendicularly, is less likely to be reflected than one falling on the same day at 45°N (*see* Fig. 5-7). And if the surface is a dark color, such as black soil or asphalt, the energy is more likely to be absorbed than if the object has a light color, such as a white building.

The proportion of incoming radiation that is reflected by a surface is called its **albedo**, a term derived from the Latin word *albus*, meaning white. The albedo of a snowy surface, which reflects most of the incoming radiation, might be 80 percent, whereas the albedo of a dark-green-colored rainforest, which reflects very little radiation, might be as low as 10 percent. Not surprisingly,

---

*One *calorie* is the amount of heat energy required to raise the temperature of 1 gram of water 1°C. (This should not be confused with the calories associated with the energy value of food, which are 1000 times larger than the calories mentioned above.) Another metric unit used to measure energy is the *joule* (one calorie equals 4.184 joules); power, or energy per unit time, is often measured in *watts* (one watt equals one joule per second).

albedo varies markedly from place to place. Of all the solar radiation entering the atmosphere, only about half is absorbed by the earth's surface. We now turn to the other side of the coin—the radiation from the earth itself.

Radiation from the Earth

The earth does more than absorb or reflect shortwave insolation: it constantly gives off longwave radiation on its own. When the earth's landmasses and oceans absorb shortwave radiation from the sun, it is transformed into longwave radiation. This process is triggered by rising temperature, and the heated surface now emits longwave radiation. One of two things can happen to this radiation leaving the planetary surface: it is either absorbed by the atmosphere or it escapes into space (Fig. 7-2).

The major atmospheric constituents that absorb the earth's longwave radiation are carbon dioxide, water vapor, and ozone (*see* Unit 6). Each of these variable gases absorbs radiation at certain wavelengths but allows other wavelengths to escape through an atmospheric "window." Up to 9 percent of all terrestrial radiation is thereby lost to space, except when the window is shut by clouds. Clouds absorb or reflect back to earth almost all the outgoing longwave radiation. Therefore, a cloudy winter night is likely to be warmer than a clear one.

The atmosphere is heated by the longwave radiation it absorbs. Most of this radiation is absorbed at the lower,

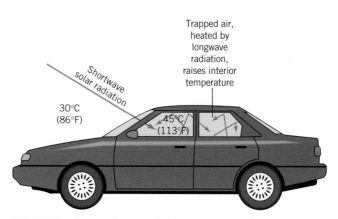

FIGURE 7-3 A parked automobile demonstrates the greenhouse effect. Shortwave radiation from the sun enters through the glass windows and strikes interior surfaces. Now, transformed into longwave radiation, that energy heats the interior air. But this air cannot pass through the glass and is trapped inside the car—at a temperature 15°C (27°F) higher than the air outside.

denser levels of the atmosphere, a fact that helps account for the air's higher temperatures near the earth's surface. *Thus our atmosphere is actually heated from below, not directly by the sun above.* The atmosphere itself, being warm, can also emit longwave radiation. Some goes off into space, but some, known as **counter-radiation**, is reradiated back to the earth (Fig. 7-2). Without this counter-radiation from the atmosphere, the earth's mean surface temperature would be about −20°C (−4°F), 35°C (63°F) colder than its current average of approximately 15°C (59°F). The atmosphere, therefore, acts as a blanket.

The blanket effect of the atmosphere is similar to the action of radiation and heat in a garden greenhouse. Shortwave radiation from the sun is absorbed and transmitted through the greenhouse's glass windows, strikes the interior surface, and is converted to heat energy. The longwave radiation generated by the surface heats the inside of the greenhouse. But the same glass that let the shortwave radiation in now acts as a trap to prevent that heat from being transmitted to the outside environment, thereby raising the temperature of the air inside the greenhouse. Another example of this same principle is the heating of a closed automobile parked in direct sunlight (Fig. 7-3).

A similar process takes place on the earth, with the atmosphere replacing the glass. Not surprisingly, we call this basic natural process of atmospheric heating the **greenhouse effect**. As explained in the discussion of ozone depletion in the stratosphere in Unit 6, human beings may be influencing the atmosphere's delicate natural processes. The greenhouse effect is now under intensive scrutiny because a number of scientists have recently voiced concern that human activities are triggering a sequence of events that could produce a *global warming* trend, with possibly dire consequences for near-future environmental change (*see* Perspective: The Greenhouse Effect and Climate Change).

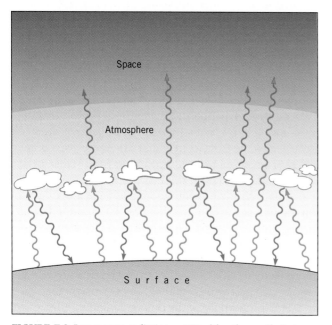

FIGURE 7-2 Longwave radiation emitted by the earth (orange arrows) and atmosphere (green arrows). Most of the terrestrial radiation is reflected back to the surface or absorbed by the atmosphere. The latter process simultaneously warms the atmosphere, which can now emit its own longwave radiation both downward toward the earth (*counter-radiation* that is critically important in heating the planetary surface) and upward into space.

The Greenhouse Effect and Climate Change

The greenhouse effect makes our planet habitable. Just as warmth is trapped beneath the glass of a greenhouse, the atmosphere retains heat emitted by the earth's sun-radiated surface. This long-wave radiation is absorbed by various constituents of the atmosphere, chief among them carbon dioxide (CO_2), water vapor, and ozone. Without this absorption, the earth's surface heat would escape into space and our planet would be frigid.

Carbon dioxide, the key "greenhouse gas," is one of the atmosphere's variable gases. If there is more of it in the atmosphere, the earth should warm up; when there is less, temperatures should cool down. The amount of CO_2 in the atmosphere is not constant. Carbon dioxide enters and leaves the atmosphere through several complex, interrelated cycles. Land plants remove CO_2 from the atmosphere during photosynthesis, but when they die and decay, the gas is returned to the air. Carbon dioxide also is absorbed directly from the atmosphere by ocean water, to be used by plankton floating on the ocean surface. When plankton die, they sink to the ocean floor and release CO_2. That CO_2 eventually comes back to the surface and is released into the atmosphere. All this makes it difficult to assess long-term trends in the CO_2 content of the atmosphere.

Enter now the human factor. During the more than 200 years since the onset of the Industrial Revolution, the burning of coal, oil, and natural gas—the fossil fuels—has produced enormous quantities of carbon dioxide. As a result, the CO_2 content of the atmosphere has increased substantially. No reliable data exist to tell us what the atmosphere's CO_2 content was two centuries ago, but scientists report that its concentration has increased from about 300 parts per million (ppm) to more than 350 over the past 50 years, an increase of more than 15 percent (*see* Fig. 6-3).

Simultaneously, scientists reported an increase in average global temperatures. By the mid-1980s, there were warnings that the continued pollution of the atmosphere by human industrial activity would lead to an enhanced greenhouse effect that would melt glaciers, raise sea levels by as much as 10 feet, and inundate coastal cities and lands. As if to confirm these predictions, the 1980s produced four of the warmest years ever recorded in North America and in Western Europe. As people sweltered in New York, London, and Paris, the prospect of an overheated world seemed real.

For some time, the evidence appeared overwhelming, and a majority of scientists concurred that human-made greenhouse gases (not only CO_2, but also trace gases such as methane produced by crop farming and livestock herding) were responsible for observed temperature increases. But not all climatologists were convinced. One unresolved issue, for example, had to do with the relative amounts of the CO_2 increase contributed by nature and by human activity. We have no long-term baseline from which to measure natural fluctuations in the atmosphere's CO_2 content; how can we therefore be sure that part of the observed increase does not represent a natural cycle?

Doubt also was cast on the twentieth-century temperature record (*see* Unit 20). According to some interpretations of available data, the earth actually underwent a *cooling* phase from about 1940 to 1970. But there was no corresponding reduction in the measured amount of CO_2 in the atmosphere, which appears to contradict the axiom that increased CO_2 means enhanced greenhouse warming.

Another point of debate involved the conclusion that the extraordinarily warm years of the late 1980s in the Northern Hemisphere represented worldwide conditions. Some geographic research suggested that these warm years were not matched by similar conditions south of the equator. Indeed, the decade of the 1980s did not, in the Southern Hemisphere, produce similar consistent warming.

Still another argument centered on nature's capacity to sustain or recover its equilibrium. Time and again, throughout the earth's history, the atmosphere has cleansed itself of massive pollution; why would this not be happening now, when excess CO_2 enters it? Those who took this position seemed to have a point when, in 1991, the Philippine volcano Pinatubo erupted, spewing ash and dust high into the atmosphere. That comparatively minor event was enough to interrupt the global warming trend, and when the mid-1990s brought unusually cold weather to many areas of the world, support for the enhanced-greenhouse theory began to wane.

In fact, the evidence suggests we are dangerously polluting the atmosphere at the very time that nature, too, seems to have changes in store for us. This may mean that human-engendered greenhouse gases may not cause the kind of consistent warming some predict, but will have a "trigger" effect that could worsen the changes of nature's design. Thus, rather than anticipating global warming, we should anticipate global change involving extremes of heat *and* cold, flood and drought, storm and calm. Now ponder Fig. 7-4, which indicates that may already be happening: the past decade has witnessed some startling weather extremes that, in terms of frequency of occurrence and worldwide distribution, surpass those of any previous ten-year period on record.

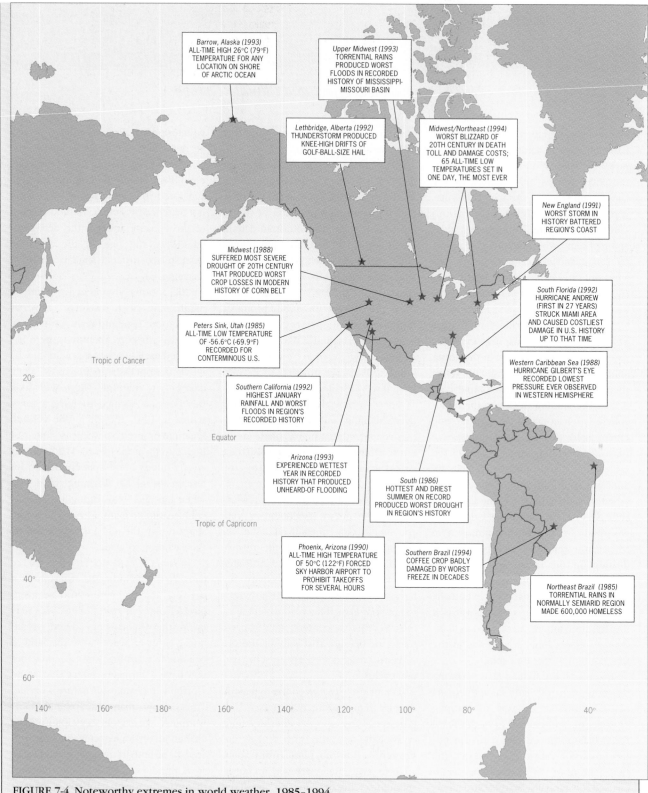

Barrow, Alaska (1993)
ALL-TIME HIGH 26°C (79°F)
TEMPERATURE FOR ANY
LOCATION ON SHORE
OF ARCTIC OCEAN

Upper Midwest (1993)
TORRENTIAL RAINS
PRODUCED WORST
FLOODS IN RECORDED
HISTORY OF MISSISSIPPI-
MISSOURI BASIN

Lethbridge, Alberta (1992)
THUNDERSTORM PRODUCED
KNEE-HIGH DRIFTS OF
GOLF-BALL-SIZE HAIL

Midwest/Northeast (1994)
WORST BLIZZARD OF
20TH CENTURY IN DEATH
TOLL AND DAMAGE COSTS;
65 ALL-TIME LOW
TEMPERATURES SET IN
ONE DAY, THE MOST EVER

New England (1991)
WORST STORM IN
HISTORY BATTERED
REGION'S COAST

Midwest (1988)
SUFFERED MOST SEVERE
DROUGHT OF 20TH CENTURY
THAT PRODUCED WORST
CROP LOSSES IN MODERN
HISTORY OF CORN BELT

South Florida (1992)
HURRICANE ANDREW
(FIRST IN 27 YEARS)
STRUCK MIAMI AREA
AND CAUSED COSTLIEST
DAMAGE IN U.S. HISTORY
UP TO THAT TIME

Peters Sink, Utah (1985)
ALL-TIME LOW TEMPERATURE
OF -56.6°C (-69.9°F)
RECORDED FOR
CONTERMINOUS U.S.

Western Caribbean Sea (1988)
HURRICANE GILBERT'S EYE
RECORDED LOWEST
PRESSURE EVER OBSERVED
IN WESTERN HEMISPHERE

Southern California (1992)
HIGHEST JANUARY
RAINFALL AND WORST
FLOODS IN REGION'S
RECORDED HISTORY

Arizona (1993)
EXPERIENCED WETTEST
YEAR IN RECORDED
HISTORY THAT PRODUCED
UNHEARD-OF FLOODING

South (1986)
HOTTEST AND DRIEST
SUMMER ON RECORD
PRODUCED WORST DROUGHT
IN REGION'S HISTORY

Phoenix, Arizona (1990)
ALL-TIME HIGH TEMPERATURE
OF 50°C (122°F) FORCED
SKY HARBOR AIRPORT TO
PROHIBIT TAKEOFFS
FOR SEVERAL HOURS

Southern Brazil (1994)
COFFEE CROP BADLY
DAMAGED BY WORST
FREEZE IN DECADES

Northeast Brazil (1985)
TORRENTIAL RAINS IN
NORMALLY SEMIARID REGION
MADE 600,000 HOMELESS

Tropic of Cancer

20°

Equator

Tropic of Capricorn

40°

60°

140°   160°   180°   160°   140°   120°   100°   80°   40°

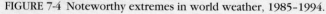

FIGURE 7-4 Noteworthy extremes in world weather, 1985–1994.

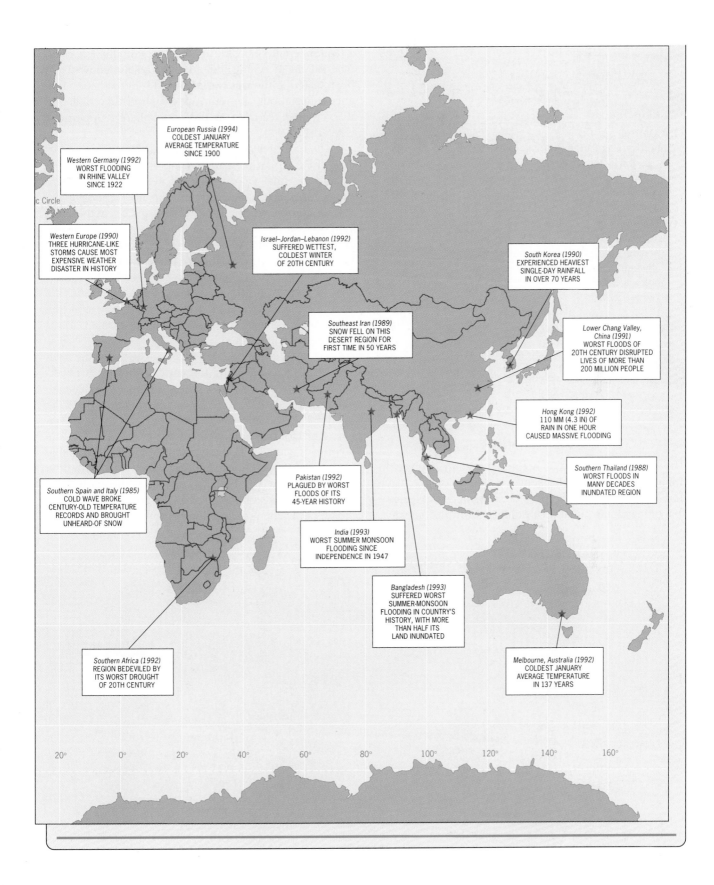

European Russia (1994)
COLDEST JANUARY
AVERAGE TEMPERATURE
SINCE 1900

Western Germany (1992)
WORST FLOODING
IN RHINE VALLEY
SINCE 1922

c Circle

Western Europe (1990)
THREE HURRICANE-LIKE
STORMS CAUSE MOST
EXPENSIVE WEATHER
DISASTER IN HISTORY

Israel–Jordan–Lebanon (1992)
SUFFERED WETTEST,
COLDEST WINTER
OF 20TH CENTURY

South Korea (1990)
EXPERIENCED HEAVIEST
SINGLE-DAY RAINFALL
IN OVER 70 YEARS

Southeast Iran (1989)
SNOW FELL ON THIS
DESERT REGION FOR
FIRST TIME IN 50 YEARS

Lower Chang Valley,
China (1991)
WORST FLOODS OF
20TH CENTURY DISRUPTED
LIVES OF MORE THAN
200 MILLION PEOPLE

Hong Kong (1992)
110 MM (4.3 IN) OF
RAIN IN ONE HOUR
CAUSED MASSIVE FLOODING

Southern Spain and Italy (1985)
COLD WAVE BROKE
CENTURY-OLD TEMPERATURE
RECORDS AND BROUGHT
UNHEARD-OF SNOW

Pakistan (1992)
PLAGUED BY WORST
FLOODS OF ITS
45-YEAR HISTORY

Southern Thailand (1988)
WORST FLOODS IN
MANY DECADES
INUNDATED REGION

India (1993)
WORST SUMMER MONSOON
FLOODING SINCE
INDEPENDENCE IN 1947

Bangladesh (1993)
SUFFERED WORST
SUMMER-MONSOON
FLOODING IN COUNTRY'S
HISTORY, WITH MORE
THAN HALF ITS
LAND INUNDATED

Southern Africa (1992)
REGION BEDEVILED BY
ITS WORST DROUGHT
OF 20TH CENTURY

Melbourne, Australia (1992)
COLDEST JANUARY
AVERAGE TEMPERATURE
IN 137 YEARS

20°   0°   20°   40°   60°   80°   100°   120°   140°   160°

TABLE 7-1   ESTIMATED ANNUAL RADIATION BALANCE (NET RADIATION) OF THE EARTH'S SURFACE IN THOUSANDS OF CALORIES PER SQUARE CENTIMETER

INCOMING

Shortwave radiation (insolation) reaching the top of the atmosphere	263
Longwave counter-radiation from the atmosphere absorbed at the earth's surface	206
	469

OUTGOING

Longwave radiation emitted by the earth	258
Shortwave radiation reflected into space by the atmosphere and the earth's surface	94
Shortwave radiation absorbed by the atmosphere	45
	397
***NET RADIATION BALANCE** (incoming minus outgoing)	**72**

Source: Adapted from W. D. Sellers, *Physical Climatology* (Chicago: University of Chicago Press, 1965), 32, 47.

Net Radiation

The annual radiation balance for the earth is given in Table 7-1. We can see that similar quantities of shortwave and longwave radiation arrive at our planet's surface, but that the outgoing radiation is dominated by the longwave radiation emitted by the earth. The amount left over, when all the incoming and outgoing radiation flows have been tallied, is the **net radiation**. This net radiation balance totals about one-fourth of the shortwave radiation that originally arrives at the atmosphere's uppermost layer.

The reflectivity (albedo) of the earth's surface and its temperature play particularly important roles in determining the final value of global net radiation. For instance, there is usually a difference in albedo and surface temperature between an area of land and sea at the same latitude (a topic treated in Unit 8). As is shown in Fig. 7-5, this variation results in a difference in net radiation values over the land and the ocean (which is represented by the red and blue isolines, respectively). That difference is greatest in the low latitudes and diminishes toward the poles. Overall, net radiation is greatest at low latitudes and smallest, or even negative (especially above ice-covered surfaces), at high latitudes.

Net radiation, moreover, may well be the single most important factor affecting the earth's climates. It is certainly basic to the majority of physical processes that take place on the earth because it provides their initial driving energy. For example, net radiation is by far the most significant factor determining the evaporation of water. The amount of water evaporated and the quantity of available net radiation together can largely explain the distribution of vegetation across the land surfaces of the earth, from the dense forests of the equatorial tropics to the sparse mosses and lichens of the subarctic environmental zones.

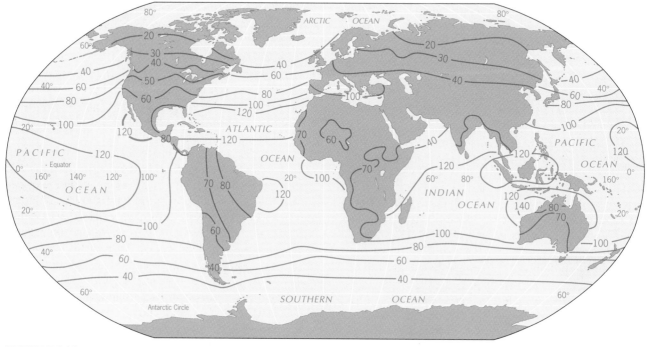

FIGURE 7-5 The annual distribution of net radiation at the surface of the earth. Values are in thousands of calories per square centimeter. Red isolines show the pattern over land, and blue isolines over the oceans.

Furthermore, net radiation is vital in shaping the *heat energy balance* of the earth.

THE HEAT BALANCE

Climate is often considered to be something derived from the atmosphere, and it is true that the climate of a place is essentially the result of the redistribution of heat energy across the face of the earth. However, the events of the atmosphere are greatly affected by the processes that operate on the earth's surface itself. Flows of heat energy to and from the surface are as much a part of the climate of an area as the winter snow or summer thunderstorm—more so, in fact, because these heat flows operate continuously.

The heat energy balance of the earth's surface is composed, in its simplest form, of four different kinds of flows. One of these—the composite flows of *radiant heat* that make up net radiation—is already familiar to us. The second—*latent heat* (which causes evaporating liquids to change into gases)—is treated in Unit 12. The remaining two—*sensible heat flow* and *ground heat flow*—are introduced here.

All air molecules contain heat energy, the heat that we feel on our skins, and this sensed heat is given the name **sensible heat flow**. Usually, during the day, the ground warms the air above it. Warmed air rises, and parcels of air move upward in a vertical heat-transfer process known as **convection**—thereby causing a sensible heat flow. We can occasionally see the results of this process, as in the case of shimmering air above a parking lot on a very hot day. Sometimes, when the ground surface is colder than the overlying air, sensible heat flows downward. This often happens at night or during the severe winters of cold climates.

Whereas sensible heat flow depends on convection, the heat that flows into and out of the ground depends on **conduction**, the transport of heat energy from one molecule to the next. The heat that is conducted into and out of the earth's surface is collectively called **ground heat flow** or *soil heat flow*. These terms are used for convenience, even though this heat sometimes travels into plants, buildings, or the ocean. Ground heat flow is the smallest of the four heat balance components. Generally, the heat that passes into the ground during the day is approximately equal to that flowing out at night. Thus, over a 24-hour period, the balance of ground heat flow often is so small that it can be disregarded.

Except for the usually small amount of energy used by plants in photosynthesis, the total heat balance of any part of the earth, say the part just outside your window, is made up of the flows of radiant heat (comprising net radiation), latent heat, sensible heat, and ground (soil)

heat. We could examine the heat balance of a single leaf or football stadium or continent, but we are more concerned here with the initial explanation of climates through the heat balance approach.

Climates and the Heat Balance

At any location, the temperature of the atmosphere depends on how much heat is involved in local radiant, latent, and sensible (as well as ground) heat flows. Net radiation is usually a source of heat for the earth, and the heat gained in this way is used mainly for evaporation (in which case it is termed latent heat) or in a sensible heat flow into the air. But there are significant variations on this theme across the earth's surface, and these lead to significant variations in climate. With that in mind, let us examine and compare the heat balance characteristics of four locations at widely separated latitudes.

Deep in the equatorial rainforest of South America at latitude 3°S, 1100 km (680 mi) inland from the mouth of the Amazon River, lies the northern Brazilian city of Manáos. Its hot humid climate is explained by the high amount of net radiation it receives, which in turn evaporates much of its large annual quantity (1800 mm [71 in]) of rainfall. If we examine the heat balance diagram for Manáos, shown in Fig. 7-6a, we can see that most of the heat received in net radiation (**NR**) is lost through the latent (evaporative) heat flow (**LH**). A rather small amount is left over for the passage of sensible heat (**SH**) into the air. These conditions are almost constant throughout the year.

In contrast, at the subtropical latitude of Aswan, Egypt (located astride the Tropic of Cancer [23½°N]), net radiation varies with the season of the year, being highest in summer (Fig. 7-6b). There is little surface water to be evaporated, so the loss by latent heat is virtually nil (**LH** values are too small to appear on the graph). But Aswan's scorching temperatures—it lies in the heart of the North African desert zone—would be even higher if most of the heat gained by net radiation did not pass, via sensible flow, higher into the atmosphere.

Paris, the capital of France, lies within the middle latitudes near 50°N and exhibits another type of heat balance, as Fig. 7-6c indicates. The seasonal variation of net radiation is again a factor, but in Paris the loss of latent heat is only somewhat greater than the loss of sensible heat. However, a rather curious pattern occurs in Paris during the winter months. The net radiation becomes negative—more radiant heat is lost than is gained. Net radiation is no longer a heat source. Fortunately, this loss is offset: air that has been warmed in its journey across the North Atlantic Ocean can now provide heat to warm the earth. Accordingly, during the winter months, the sensible heat flow is directed toward the earth's surface, as is shown by its negative values in Fig. 7-6c. The sen-

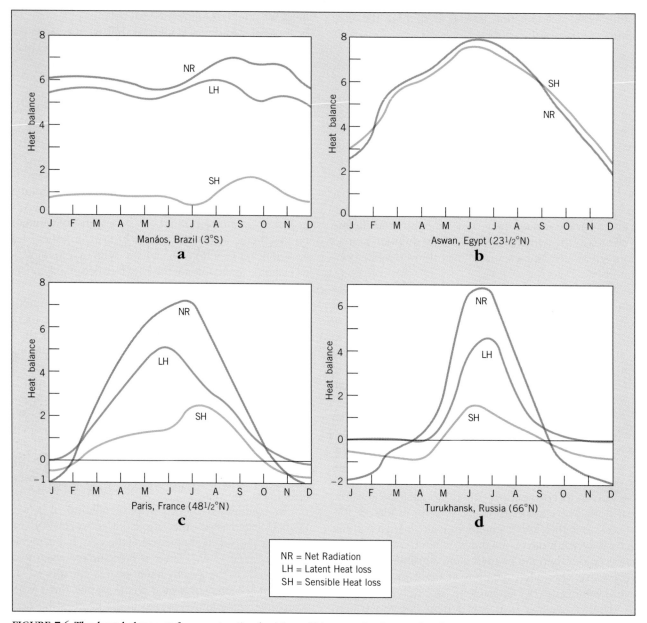

FIGURE 7-6 The heat balance at four contrasting locations. Values are in thousands of calories per square centimeter per month.

sible heat flow, therefore, is responsible for keeping Parisian winter air temperatures fairly mild.

In central Siberia, deep inside the northern Asian component of Russia, this does not happen. Turukhansk, located at latitude 66°N, is typical, and air coming to this town in the winter has not traveled over a warm ocean but over a cold continent. Although the air passes some sensible heat toward the ground, it does not pass enough to offset the large net radiation deficit experienced in winter near the Arctic Circle (Fig. 7-6d). The result is bone-chilling temperatures. Yet here the seasonal change of climate is extreme. Paradoxically, the balance of heat in the summer months is rather like that in trop-

ical Brazil! There are many such variations of heat balance across the ever changing face of the earth.

GLOBAL DISTRIBUTION OF HEAT FLOWS

We have already examined the geographic variation of net radiation (Fig. 7-5). Now we will consider the disposal of net radiation through latent and sensible heat, losses that are necessary to keep the totality of radiation in balance for the earth's surface as a whole. To find the

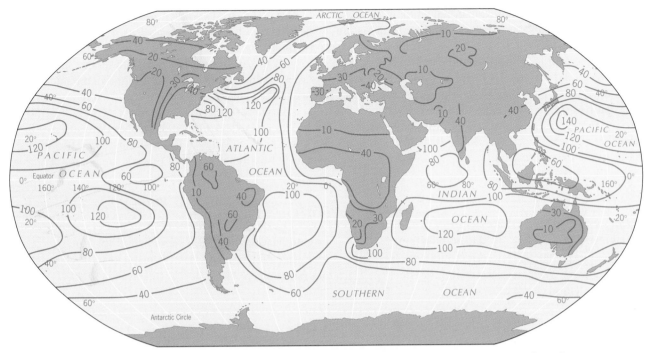

FIGURE 7-7 The global distribution of latent heat loss. The heat used in evaporation is expressed in thousands of calories per square centimeter per year. Red isolines show the pattern over land, and blue isolines over the oceans.

amount of heat lost as latent heat, we multiply the amount of water evaporated by the value of the *latent heat of vaporization* (the amount of energy required to evaporate water). The global distribution of latent heat loss is mapped in Fig. 7-7. Over land surfaces (red isolines), the largest amount of latent heat loss occurs in the tropics on both sides of the equator. Latent heat loss gen-

erally declines across subtropical latitudes, increases in the middle latitudes, and then further declines in the higher latitudes. Over ocean surfaces (blue isolines), where water is always available for evaporation, latent heat loss is greatest in the subtropics. Here there are fewer clouds, on average, to reduce radiant heat input. Because of the effect of cloud cover, latent heat loss over

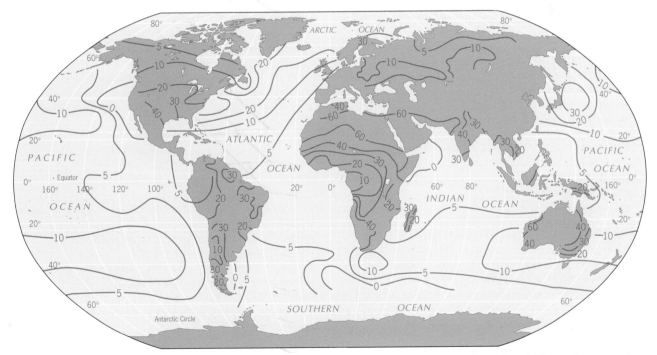

FIGURE 7-8 The global distribution of sensible heat loss. Values are in thousands of calories per square centimeter per year. Red isolines show the pattern over land, and blue isolines over the oceans.

oceans is not as great in the equatorial latitudes as in the subtropical latitudes. As over the land surfaces, latent heat loss is least above oceans at high latitudes.

Sensible heat loss over the land surface is greatest in the subtropics; from there it decreases toward the poles and the equator. This is shown by the red isolines in Fig. 7-8, mapped in the same units as the world latent-heat-loss map (Fig. 7-7). Above the oceans (blue isolines), however, the amount of sensible heat loss generally tends to increase with latitude.

Over the earth's surface as a whole, the net radiation heat gain is itself balanced by 70 percent of that heat being lost in the latent form and 30 percent being used to heat the air as sensible heat. But at any single point on our planet's surface, there is a unique interaction between the values of net radiation and latent and sensible heat flow. Temperature is the result of these heat flows, and its properties and variations are the subject of Unit 8. In Unit 9, global heat flows are discussed further, especially their linkage to atmospheric circulation patterns.

KEY TERMS

albedo (p. 73)

conduction (p. 79)

convection (p. 79)

counter-radiation (p. 74)

diffuse radiation (p. 73)

direct radiation (p. 73)

greenhouse effect (p. 74)

ground heat flow (p. 79)

longwave radiation (p. 73)

net radiation (p. 78)

radiation (p. 73)

sensible heat flow (p. 79)

shortwave radiation (p. 73)

REVIEW QUESTIONS

1. What are the differences between solar and terrestrial radiation?

2. How much of the solar energy entering the atmosphere is absorbed by the atmosphere and how much by the earth's surface? How much is reflected by the atmosphere and by the surface?

3. Describe in your own words the meaning of the term *greenhouse effect*.

4. Differentiate among the flows of radiant heat, latent heat, sensible heat, and ground heat.

5. What is meant by the term *albedo*? Give some examples of its application in your daily life.

6. What is meant by the popular term *global warming*? Can a strong case be made for it based on current evidence?

REFERENCES AND FURTHER READINGS

ABRAHAMSON, D. E., ed. *The Challenge of Global Warming* (Washington, D.C.: Island Press, 1989).

BALLING, R. C., JR. "Global Temperature Data," *Research and Exploration (National Geographic Society)*, 9 (Spring 1993), 201–207.

BENARDE, M. A. *Global Warning . . . Global Warming* (New York: Wiley, 1992).

COWEN, R. C. "Scientists Question Global Warming Theory," *Christian Science Monitor*, September 14, 1994, 8.

DE BLIJ H. J., ed. *Nature on the Rampage* (Washington, D.C.: Smithsonian Institution Press, 1994).

FRÖHLICH, C., and LONDON, J. *Radiation Manual* (Geneva: World Meteorological Organization, 1985).

Global Warming Debate: Special Issue. *Research and Exploration (National Geographic Society)*, 9 (Spring 1993), 142–249.

HERMAN, J. R., and GOLDBERG, R. A. *Sun, Weather, and Climate* (New York: Dover, 1985).

KERR, R. A. "Is the Greenhouse Here?" *Science*, February 5, 1988, 559–561.

KONDRATYEV, K. *Radiation in the Atmosphere* (New York: Academic Press, 1969).

LIOU, K.-N. *An Introduction to Atmospheric Radiation* (New York: Academic Press, 1980).

MILLER, D. H. *A Survey Course: The Energy and Mass Budget at the Surface of the Earth* (Washington, D.C.: Association of American Geographers, Commission on College Geography, Publication No. 7, 1968).

MINTZER, I. *A Matter of Degrees: The Potential for Controlling the Greenhouse Effect* (Washington, D.C.: World Resource Institute, 1987).

REVELLE, R. "Carbon Dioxide and World Climate," *Scientific American*, August 1982, 35–43.

SCHNEIDER, S. H. "The Changing Climate," *Scientific American*, September 1989, 70–79.

STEVENS, W. K. "Searching for Signs of Global Warming," *New York Times*, January 29, 1991, B1, B9.

WYMAN, R. L., ed. *Global Climate Change and Life on Earth* (New York: Chapman & Hall, 1990).

YOUNG, L. B. *Sowing the Wind: Reflections on the Earth's Atmosphere* (Englewood Cliffs, N.J.: Prentice-Hall, 1990).

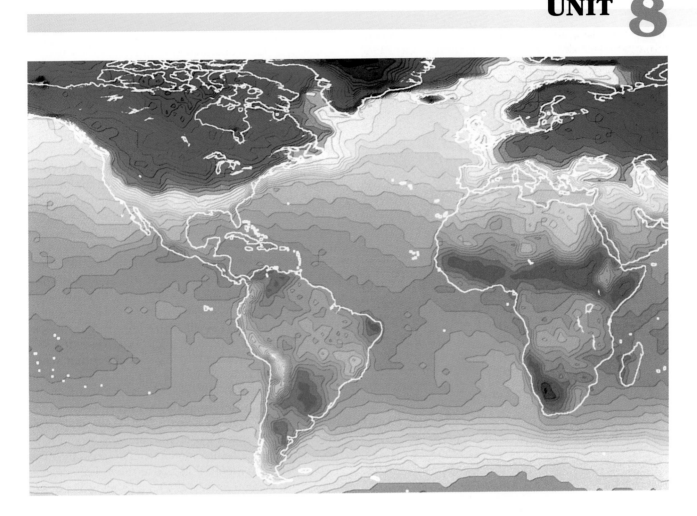

Atmospheric and Surface Temperature

OBJECTIVES

◆ To discuss the measurement and characteristics of temperature and heat.

◆ To explain the adiabatic process whereby vertically moving air heats and cools.

◆ To discuss the global distribution of temperatures and their variation in time and space.

On the rather cloudy day of June 26, 1863, British scientist James Glaisher and his assistant climbed into the basket of a balloon in Wolverton, England. This flight, one of their 28 flights between 1862 and 1866 (Fig. 8-1), lasted an hour and a half. They ascended to 7050 m (23,500 ft) and traveled 80 km (50 mi) before descending at Ely. En route they encountered rain, snow, and fog. One of the main purposes of this flight was to note the temperatures along the way. In Glaisher's own words, these varied from the "extreme heat of summer" to the "cold of winter." In fact, however, the temperatures varied from 19°C (66°F) at the ground to −8.3°C (17°F) at 7050 m. The pair had good reason to be wary of the hazards of high altitudes: on a flight the previous year, Glaisher had fainted at 8700 m (28,700 ft)

from lack of oxygen. His assistant, arms paralyzed with cold, climbed the rigging of the balloon to release the gas control with his teeth. Those flights firmly established that temperature decreases with height, at least as far as 8700 m above the surface.

These findings of vertical temperature changes were an important addition to existing knowledge. Together with what was already known about heating patterns of the planetary surface, they reinforced the notion that temperatures could be highly variable in any direction.

UNIT OPENING PHOTO: A satellite image of January surface temperatures in the hemisphere centered on the Atlantic Ocean. Temperatures range from 40°C (104°F) in southwestern Africa (brown) to −38°C (−36°F) in northernmost Canada (dark purple).

FIGURE 8-1 James Glaisher and his assistant, Coxwell, during a balloon flight on September 5, 1862.

Before we further examine those vertical and horizontal temperature relationships, we must briefly digress to inquire more explicitly about what is meant by the concept of temperature.

WHAT IS TEMPERATURE?

Imagine an enclosed box containing only molecules of air. They are likely to be moving constantly in all directions, in what is called *random motion*. The molecules move because they possess the energy of movement, known as **kinetic energy**. The more kinetic energy molecules possess, the faster they move. The index we use to measure their kinetic energy is called **temperature**. Thus temperature is an abstract term describing the energy, and therefore the speed of movement, of molecules. In a gas such as air, the molecules actually change their location when they move. But in a solid, like ice,

they only vibrate in place. Nonetheless, the speed of this vibration is described by their temperature.

It is almost impossible to examine individual molecules, so we usually use an indirect method to measure temperature. We know that changes of temperature make gases, liquids, and solids expand and contract. Therefore, temperature is most commonly measured by observing the expansion and contraction of mercury in a glass tube. Such an instrument is called a **thermometer**, and you are probably familiar with the medical and weather varieties. The mercury thermometer is placed in the mouth, air, or some other place where it can come into thermal equilibrium with the medium whose temperature it is measuring.

A thermometer is calibrated according to one of several internationally accepted scales. The scale most commonly used throughout the world is the **Celsius scale** (formerly Centigrade scale), the metric measurement of temperature we have routinely been using in this book. On this scale, the *boiling point* of water is set at 100°C and its *freezing point* at 0°C. On the **Fahrenheit scale**, presently used only in the United States, water boils at 212°F and freezes at 32°F. Scientists also employ an *absolute scale*—the **Kelvin scale**—which is based on the temperature of *absolute zero* (−273°C [−459.4°F]). (Scientists theorize that a gas at absolute zero would have no volume, no molecular motion, and no pressure.) The Kelvin degree is identical in size to the Celsius degree, except that water freezes at 273°K and boils at 373°K. We will not mention Kelvin degrees again in this book, but will continue to use Celsius degrees accompanied by their Fahrenheit equivalents in parentheses.

It is important to distinguish temperature from heat. Temperature merely measures the kinetic energy of molecules: it does not measure the number of molecules in a substance or its *density* (the amount of mass per unit of volume). But the heat of a substance depends on its volume, its temperature, and its capacity to hold heat. Thus a bowl of soup, with a high heat capacity, might burn your tongue at the same temperature at which you could comfortably drink a glass of hot water. Because it contains many more molecules, a large lake with a water temperature of 10°C (50°F) contains much more heat than a cup of hot coffee at 70°C (158°F). Now that we know something about temperature, let us return to discussing the temperatures of the atmosphere.

THE VERTICAL DISTRIBUTION OF TEMPERATURE

Glaisher's observations for the lower part of the atmosphere were correct—temperature does indeed decrease with height. However, subsequent unmanned balloon observations showed that, above about 12 km (40,000 ft), temperature stops decreasing with height and begins to increase. Nobody believed this at first, and only after several hundred balloon ascents was it finally accepted. Later ascents in this century to still higher altitudes revealed an even more complex temperature pattern. These upper atmospheric layers are discussed in Unit 6, and their temperature characteristics are graphed in Fig. 6-5. Our focus here is on the lowest layer, between the ground and the tropopause.

Tropospheric Temperature and Air Stability

The troposphere is the layer of the atmosphere we live in, and it is here that the weather events and climates affecting humans occur. As shown in Fig. 6-5 (p. 66), its temperature decreases with height until the tropopause is reached. Another distinctive feature of the troposphere is the possibility and frequency of vertical, as well as horizontal, movement of air. Any long continuation of vertical movement depends on the rate of change of temperature with height—the *lapse rate*. As we note in Unit 6, the average tropospheric lapse rate is 0.65°C/100 m (3.5°F/1000 ft) of elevation.

The lapse rate determines the **stability** of the air, a concept illustrated by the wedge of wood in Fig. 8-2. When it is resting on its side (a), a small push at the top may move it horizontally, but its vertical position remains the same. It is therefore *stable*. When the wood rests on its curved base (b), a similar push might rock it, but it will still return to its original position; it is still stable. But if we balance the wedge of wood on its pointed edge (c), a small push at the top knocks it over. It does not return to its original position; hence, it is *unstable*.

We use the same terminology to refer to the vertical movement of a small parcel of air. If it returns to its original position after receiving some upward force we say it is stable. But if it keeps moving upward after receiving

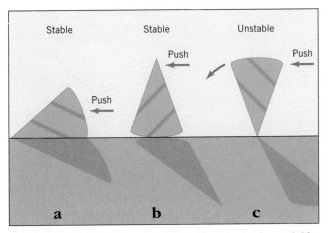

FIGURE 8-2 The concept of stability. The block of wood, like parcels of air, is considered stable as long as it returns to its original position after a small push.

the force, then we say it is unstable. In order to understand air stability, we must first consider that air is a poor conductor of heat. Without any air movement, it takes a long time for heat to pass from one air molecule to the next. Therefore, a parcel of air of one temperature that is surrounded by a mass of air at another temperature will neither gain nor lose heat energy in a short period of time. When heat is neither gained from nor lost to the surrounding air, the process is called **adiabatic**.

Adiabatic Lapse Rates

If you have ever used a bicycle pump, you know that when the air is compressed at the bottom of the pump the air's temperature rises and makes the bottom of the pump hot. The opposite occurs when the volume of a given mass of air is forced to expand: the temperature of the air decreases. A similar thing happens in the atmosphere. If a parcel of air rises to a higher altitude, it expands and cools (as if it were contained inside an expanding balloon). This is an adiabatic process because that air parcel neither gains heat from nor loses heat to its surroundings. Hence, such air-parcel lapse rates in the troposphere are called **adiabatic lapse rates**.

Dry Adiabatic Lapse Rate (DALR)

When an air parcel is not saturated with water vapor, it cools with height at a constant rate of 1°C/100 m (5.5°F/1000 ft). This is the **dry adiabatic lapse rate (DALR)**. But because air sometimes is saturated, and because mechanisms (especially radiant energy exchange) are involved in heating and cooling the atmosphere, any particular atmospheric lapse rate may not be the same as the DALR. The lapse rate at any particular time or place is called the **environmental lapse rate (ELR)**. In the context of our example, the ELR is the temperature decline with height in the stationary mass of air in the atmosphere that surrounds our cooling parcel of air. The troposphere's (nonadiabatic) normal lapse rate—which we have now more specifically defined as the ELR—averages 0.65°C/100 m (3.5°F/1000 ft).

Now let us go back to our parcel of air and see what happens to it in two different environments with two different ELRs. In Fig. 8-3a, a parcel of air rises from the ground (either because it is warmed by the surface or forced upward mechanically). Because it neither gains heat from nor loses it to the surrounding air mass, it cools at the DALR, decreasing its temperature 1°C for each 100 m of ascent. But in this case the ELR is 0.5°C/100 m, so after rising 100 m, the air parcel has a temperature 0.5°C lower than its surroundings. The colder the air, the more dense it is. Therefore, the air parcel is now denser and heavier than the surrounding air and tends to fall back to earth. This would happen even if the parcel rose to 300 m, where it would be 1.5°C colder than its surroundings. The cooled parcel thus returns to its original posi-

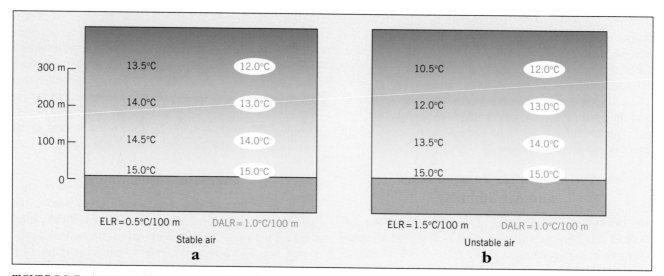

FIGURE 8-3 Environmental lapse rate conditions for a mass of stationary air (shown in blue) that surrounds an individual parcel of air (shown in white) rising through it. When the ELR is less than the DALR, an air parcel is stable (a). When the ELR is greater than the DALR, an air parcel is unstable (b).

tion on the surface. We would say the whole of the air in that environment is stable, meaning *it resists vertical displacement.*

In contrast, Fig. 8-3b shows the ELR to be 1.5°C/100 m. Under these conditions, an air parcel rising and cooling at the DALR would be warmer than its surroundings. The warmer the air, the less dense it is, so the air parcel that is lighter and less dense than the surrounding air continues to rise. We would call the air in this environment unstable, because it does not return to its original position.

We can often tell whether or not a portion of the atmosphere is stable by looking at it. A stable atmosphere is marked by clear skies or by flat, layer-like clouds. An unstable atmosphere is typified by puffy, vertical clouds, which sometimes develop to great heights. A photograph of Florida taken from a spacecraft, shown in Fig. 8-4, illustrates the two conditions. Over the Atlantic Ocean to the east and the Gulf of Mexico to the west, the ELR is less than the DALR, so parcels of air remain near the sea surface. But the higher temperatures of the land surface in daytime make the ELR higher than the DALR. The resulting instability allows parcels of hot air to rise in the atmosphere, forming those puffy clouds as they cool.

Saturated Adiabatic Lapse Rate (SALR)

The situation is somewhat different when the air contains water vapor that is changing to water droplets as it cools. Heat is given off when the state of water changes from a gas to a liquid, a process called *condensation*. The resultant lapse rate when condensation is occurring is less than the DALR. This lapse rate is called the *wet* or **saturated adiabatic lapse rate (SALR)**.

Unlike the DALR, the value of the SALR is variable, depending on the amount of water condensed and latent heat released. A typical value for the SALR at 20°C (68°F)

is 0.44°C/100 m (2.4°F/1000 ft)—compared with 1°C/100 m for the DALR. As a rule, we may assume that the atmosphere will be stable if the ELR is less than the SALR, and unstable if the ELR is greater than the DALR. If the

FIGURE 8-4 Looking northward at the Florida peninsula and surrounding waters from a spacecraft situated approximately above Havana, Cuba. In this classic summertime view, the land area is almost perfectly defined by the distribution of the myriad puffy clouds, the signature of atmospheric instability associated with rising hot air.

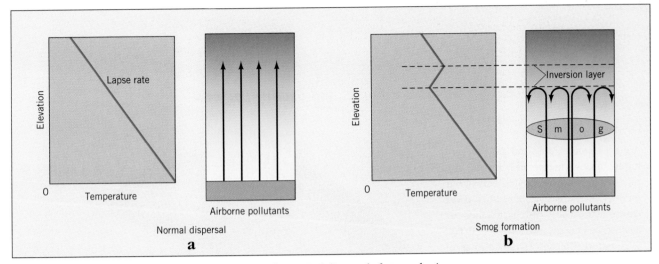

FIGURE 8-5 The effect of a temperature inversion on the vertical dispersal of atmospheric pollutants. A normal dispersal (a); the formation of smog (b).

ELR lies between the SALR and DALR, the atmosphere is said to be *conditionally unstable*. The conditions depend on whether the gaseous water vapor in the air changes into liquid water and thereby adds heat to the air.

Temperature Inversions and Air Pollution

In dealing with the stability of the troposphere, we are really considering the possibility—and vigor—of the vertical mixing of air within it. This has practical implications, the most important of which is how well pollutants will disperse when released into the atmosphere. The initial vertical (and horizontal) distribution of pollutants depends on the location of their sources. Any further spread of air pollution is associated with two main factors: (1) the stability of the air and its propensity to allow vertical mixing, and (2) how well air stability combines with the flushing effect of horizontal winds. Both are related to the temperature structure of the lower troposphere, particularly the influence of temperature inversions.

Under usual conditions, tropospheric temperature decreases with height as ground-warmed air rises, expands, and cools (*see* Fig. 6-5). Thus any pollutants contained in that surface layer of air would disperse along with it. At times, however, this vertical cleansing mechanism does not operate because the usual negative lapse rate is replaced by a positive one. Such an increase in temperature with height is defined in Unit 6 as a **temperature inversion** because it inverts what we, on the surface, believe to be the "normal" behavior of temperature change with altitude.

Because of the nightly cooling of the earth's surface and the atmosphere near the ground, it is common for a temperature inversion—warm air lying above cold air—to develop in early morning over both city and countryside. These inversions, which form an atmospheric "lid," can be broken down by rapid heating of the surface or by windy conditions. Without these conditions, air pollution is trapped and intensifies beneath the inversion layer (Fig. 8-5). This is especially true in certain urban areas, where *dust domes* frequently build up (*see* Perspective: Urban Dust Domes and Heating Patterns).

Cold-air drainage at night and the subsidence (vertical downflow) of air from higher in the troposphere can also contribute to the trapping of pollutants, as residents of metropolitan Los Angeles know only too well (Fig. 8-7). Cold air drains into the Los Angeles Basin from the mountains that form its inland perimeter, and high-level inversions at 1000 m (3300 ft) are common given the subsiding motion of air over the Pacific Ocean that marks the lowland's coastal rim. These airflows reinforce the urban region's inversion layer, resulting in poor-quality, surface-level air that is popularly known as *smog* (the contraction of "smoke" and "fog").

Horizontal flushing by winds can help to relieve air pollution. Air pollution potential may be estimated by calculating the vertical range of well-mixed pollutants and the average wind speed through the mixing layer. Figure 8-8 shows the average number of days per year in the United States when inversions and light wind conditions create a high air pollution potential. On average, it appears that ventilation conditions are best in the Northeast, the Midwest, and the Great Plains, and poorest in the Appalachian corridor and most of the Far West. Air pollution potential alone might be regarded as a matter of climatological luck, but the frequency of pollution sources also determines the level of air pollution. Important, too, is local topography, which does not show up at the scale of the map in Fig. 8-8.

Perspectives on THE HUMAN ENVIRONMENT

Urban Dust Domes and Heating Patterns

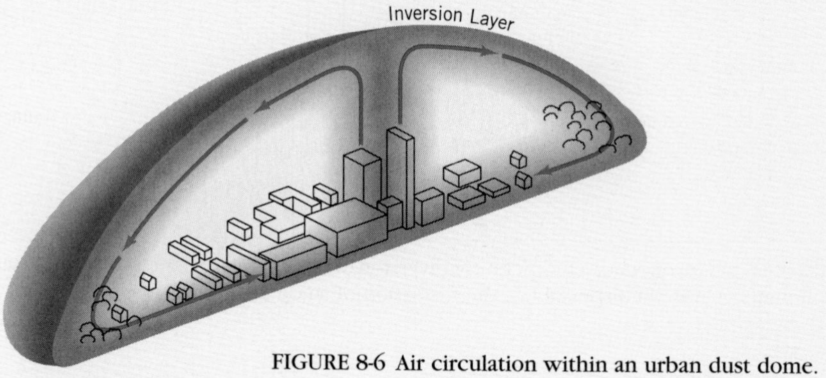

A giant dome of dust and gaseous pollution buries Chicagoland! It's not the plot of a new horror movie. Many of our metropolitan areas lie beneath a **dust dome**, whose brownish haze stands out against the blue sky (*see* top photo in Fig. 8-7). Before Chicago spawned rings of automobile suburbs, its dust dome was sharply defined. A half-century ago, the wall of dust and other pollutants began near Midway Airport, about 12 km (7.5 mi) from the city's downtown Loop. In the nearby countryside visibility might be 25 km (15 mi), but within the dome it was only 0.5 to 0.75 km (0.3 to 0.5 mi). Los Angeles and many other large cities are also often enshrouded by a thick layer of polluted air. Why don't these effluents simply just blow away?

Studies of the movement of dust and gaseous pollutants over cities show that heat generated in urban areas forms a local circulation cell. Air currents capture the dust and mold it into a dome (Fig. 8-6). Dust and pollution particles rise in air currents around the center of the city where the temperature is warmest. As they move upward, the air cools and diverges. The particles gradually drift toward the edges of the city and settle downward. Near the ground they are drawn into the center of the city to complete the circular motion. Temperature inversions above the city prevent escape upward, and the particles tend to remain trapped in this continuous cycle of air movement.

All parts of the radiation balance discussed in Unit 7 (*see* pp. 73–79) are altered in the urban environment. The interception of incoming shortwave radiation is most apparent during the winter, when the pollution is worse and the sun's rays are striking at a lower angle. In London, 8.5 percent of the direct solar radiation is lost when the sun's elevation is 30 degrees above the horizon, but 12.8 percent is lost with a solar elevation of 14 degrees. On very cloudy days, 90 percent of this radiation can be lost. Consequently, the number of hours with bright sunshine is reduced. Researchers have discovered that outside London it is sunny an average of 4.3 hours per day; in the center of the city, however, their findings reveal it is sunny only 3.6 hours per day.

The amount of insolation absorbed by a city's surface depends on the albedo of that surface. This, in turn, depends on the actual materials used in construction, and varies from city to city. British cities are usually made of dark materials or materials that have been blackened by smoke. They have a lower albedo (17 percent) than agricultural land does (approximately 23 percent), and therefore absorb more radiation. In contrast, the central areas of Los Angeles have lighter colors and therefore exhibit a higher albedo than the more vegetated residential zones.

Scientists know less about the behavior of longwave radiation in cities, but many studies show that both upward and downward longwave radiation increase. This increased radiation can offset the decreased shortwave radiation. The result is a net radiation that is not too different from that in surrounding rural areas. Of the net radiation arriving at the city surface, some studies suggest that about 80 percent of it is lost as sensible heat warming the city air and that the rest acts mainly as ground heat flow to warm the materials constituting the urban landscape. Very little heat appears to be used in evaporative cooling.

FIGURE 8-6 Air circulation within an urban dust dome.

Nearby mountains can decidedly counteract the flushing effect of horizontal winds. Los Angeles and Denver—two U.S. metropolitan areas that rank among the worst in air pollution—are classic examples. And when cities are situated in valleys near mountain ranges, wind flushing may be even further diminished. Mexico City, probably the world's most polluted (and soon-to-be most populated) metropolis, not only is an example of this phenomenon, but also suffers additionally because it is located at a high elevation (2240 m [7350 ft]) where the air already contains 30 percent less oxygen than at sea level.

FIGURE 8-7 Downtown Los Angeles, with and without its frequently heavy shroud of smog. The Los Angeles Basin is particularly susceptible to the development of temperature inversions—and high levels of surface air pollution.

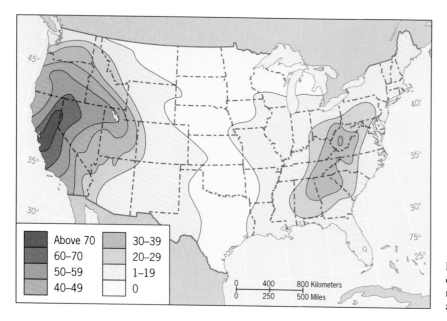

FIGURE 8-8 Air pollution potential in the contiguous United States, in terms of the number of days per year with light winds and inversion conditions.

THE HORIZONTAL DISTRIBUTION OF TEMPERATURE

The spatial distribution of temperatures across the landmasses, oceans, and icecaps that constitute the earth's surface represents the response to a number of factors. Certainly, insolation is one such factor, and the amount of that solar radiation received depends on the length of daylight and the angle of the sun's rays (both a function of latitudinal position). A second major factor is the nature of the surface. Land heats and cools much more rapidly than water, and this strongly affects not only air temperatures directly above each type of surface, but also adjacent areas influenced by them. A number of lesser factors can be locally important as well. Since tropospheric temperatures usually decline with height, places located at higher altitudes tend to experience temperatures lower than those recorded at places closer to sea level. Another moderator of surface temperature is cloudiness, and places with more extensive cloud cover generally experience lower daytime high temperatures than similar places with clearer skies (this may also be true in areas where pollution haze plays a role identical to that of clouds). And for coastal locations and islands, air temperatures can be influenced to a surprising extent by the warmth or coolness of local ocean currents.

Daily and Yearly Cycles

The pattern of temperature change during a day is called the **diurnal cycle**. Shortly after dawn, radiation from the sun begins to exceed the radiant loss from the earth's surface. The earth begins to heat the air, so air temperature rises. It continues to rise as the net radiation rises. But the heating of the ground and the flow of sensible heat take some time to develop fully. Thus maximum air temperatures usually do not occur simultaneously with the maximum net radiation peaks at solar noon, but an hour or more later. In late afternoon, net radiation and sensible heat flow decline markedly, and temperatures begin to fall. After the sun has set, more radiation leaves the earth than arrives at the surface, which produces a negative net radiation. The surface and the air above it enter a cooling period that lasts all through the night. Temperatures are lowest near dawn; with sunrise, the diurnal cycle starts again.

The pattern of temperature change during a year is called the **annual cycle** of temperature, which in the middle and high latitudes is rather similar to the diurnal cycle. In the spring, net radiation becomes positive and air temperatures begin to rise. The highest temperatures do not occur at the time of the greatest net radiation, the summer solstice, but usually about a month later. In autumn, decreasing net radiation leads to progressively lower temperatures. The lowest winter temperatures occur toward the end of the period of lowest (and often negative) net radiation and when the ground has lost most of the heat it gained during summer. Then it is spring again, and net radiation once more begins its cyclical increase.

Land/Water Heating Differences

The time it takes to heat the surface at any particular location determines when the highest air temperatures will occur. The difference between land and ocean offers the most clear-cut example. Dry land heats and cools relatively rapidly because radiation cannot penetrate the solid surface to any meaningful extent.

Unlike solid surfaces, water requires far more time to heat up and cool down. For one thing, compared to land, radiation can penetrate the surface layer of water to a relatively greater depth. There is also considerable vertical mixing—driven by waves, currents, and other water movements—that constantly takes place between newly warmed (or cooled) surface water and cooler (or warmer) layers below. Moreover, the energy required to raise the land temperature by a given number of degrees would have to be tripled in order to increase the surface temperature of a body of water by an equivalent amount.

Not surprisingly, the ocean surface exhibits a decidedly smaller annual temperature range: from −2°C (28.4°F) to about 32°C (90°F), as opposed to the land-surface extremes of −88°C (−127°F) and 58°C (136°F). In addition, seasonal ocean temperature change is particularly moderate: in the tropics this variation averages 1 to 4°C (2 to 7°F), and even the upper-middle latitudes record only a modest 5 to 8°C (9 to 15°F) swing between seasonal extremes. On a diurnal basis, the ocean-surface range is almost always less than 1°C (1.8°F).

As a consequence of this heating differential, the air above an ocean remains cooler in summer and warmer in winter than does the air over a land surface at the same latitude. This can be seen in Table 8-1, which displays data on the annual range of temperatures, at 15-degree latitudinal intervals, for each hemisphere. Note that the Southern Hemisphere, which is only about 20 percent land, consistently exhibits smaller yearly temperature ranges than the Northern Hemisphere, whose surface is approximately 40 percent land.

In places where the oceanic air is transported onto the continents (as in our discussion of Paris in Unit 7), air temperatures are ameliorated accordingly—that is, they do not become extremely hot or cold. As the distance from the coast increases, however, this moderating effect diminishes (it terminates more abruptly if high mountain ranges parallel to the shore block the inland movement of oceanic air). This is illustrated in Fig. 8-9,

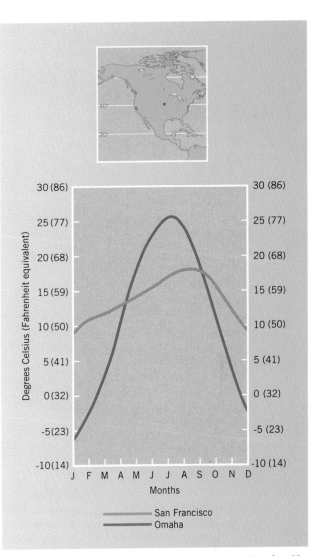

FIGURE 8-9 The annual temperature regimes at Omaha, Nebraska (41°N, 96°W), and San Francisco, California (38°N, 122°W). Note the contrast between interior continental and coastal locations.

which graphs the annual temperature regimes for San Francisco, California, and Omaha, Nebraska, both located at approximately the same latitude (40°N). Interior Omaha, located in the heart of the North American continent, experiences both a hotter summer and a much colder winter, while San Francisco, located on the Pacific coast, enjoys a temperature regime that is free of extremes in both summer and winter.

This moderating influence of the ocean on air temperature is called the **maritime effect** on climate. In the opposite case, where the ocean has a minimal ameliorating influence on air temperatures well inland, there is a **continental effect**. This property of **continentality** is strongly suggested in the case of Omaha (Fig. 8-9), but the most dramatic examples are found deep inside Eurasia, in the heart of the world's largest landmass. The Russian town of Verkhoyansk, located in far northeastern Siberia, is well known to climatologists in this regard, and its annual temperature regime is plotted in Fig. 8-10.

	NORTHERN	SOUTHERN
LATITUDE	HEMISPHERE	HEMISPHERE
0	0 (0.0)	0 (0.0)
15	3 (5.4)	4 (7.2)
30	13 (23.4)	7 (12.6)
45	23 (41.4)	6 (10.8)
60	30 (54.0)	11 (19.8)
75	32 (57.6)	26 (46.8)
90	40 (72.0)	31 (55.8)

TABLE 8-1 VARIATION IN AVERAGE ANNUAL TEMPERATURE RANGE BY LATITUDE, °C (°F)

Source: Adapted from Frederick K. Lutgens and Edward J. Tarbuck, *The Atmosphere: An Introduction to Meteorology* (Englewood Cliffs, N.J.: Prentice-Hall, 6th ed., 1995), 57.

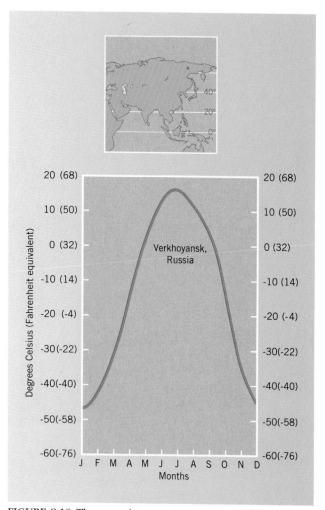

FIGURE 8-10 The annual temperature regime at Verkhoyansk (68°N, 133°E) in Russia's far northeast, demonstrating the extremes of continentality.

Sometimes air from outside an area has more influence on air temperatures than do local radiation and sensible and latent heat flows. For instance, there might be quite a large amount of net radiation at midday during a Nebraska winter, but the air temperatures may still be very low. This is because the overlying air may have come from the Arctic, thousands of kilometers to the north, where a completely different heat balance prevailed. Thus, although air temperature is a function of the amount of heat that makes up the local heat balance, it can also be affected by the **advection** (horizontal transport through the atmosphere via wind) of air from a region exhibiting a different heat energy balance. The results of different heat balances and the large-scale advection of air can be seen in the worldwide distribution of surface air temperatures.

Global Temperature Variations

The global distribution of air temperatures is mapped in Figs. 8-11a and 8-11b. In order to avoid the distorting effects of altitude, all temperatures have been converted to their averages at sea level. Insolation, which is deter-

mined by the angle of the sun's rays striking the earth and the length of daylight, can change significantly in most places over a period of weeks. Thus physical geographers have always faced a problem in trying to capture the dynamic patterns of global temperatures on a map. For our purposes, the worldwide shifting of air temperatures, or their *seasonal march*, is best visualized by comparing the patterns of the two extreme months of the year—January and July—which immediately follow the solstices. The cartographic technique of isarithmic mapping (defined in Unit 3) is used in Fig. 8-11, whose January and July distributions employ **isotherms**—lines connecting all points having the same temperature.

Both maps reveal a series of latitudinal temperature belts that are shifted toward the "high-sun" hemisphere (Northern Hemisphere in July, Southern Hemisphere in January). The tropical zone on both sides of the equator experiences the least change between January and July, because higher total amounts of net radiation at low latitudes (*see* Fig. 7-5) lead to higher air temperatures. The middle and upper latitudes, particularly in the hemisphere experiencing winter, exhibit quite a different pattern. Here the horizontal rate of temperature change over distance—or **temperature gradient**—is much more pronounced, as shown by the "packing" or bunching of the isotherms (note that the isotherms are much farther apart in the tropical latitudes).

Another major feature of Figs. 8-11a and 8-11b is the contrast between temperatures overlying land and sea. There is no mistaking the effects of continentality on either map: the Northern Hemisphere landmasses vividly display their substantial interior annual temperature ranges. Clearly visible, too, are the moderating influence of the oceans and the maritime effect on the air temperatures over land surfaces near them. Where warm ocean currents flow, as in the North Atlantic just west and northwest of Europe, the onshore movement of air across them can decidedly ameliorate winter temperatures; note that northern Britain, close to 60°N, lies on the 5°C (41°F) January isotherm, the same isotherm that passes through North Carolina at about 35°N, the northern edge of the U.S. Sunbelt! In general, we observe a poleward bending of the isotherms over all the oceans, an indicator of their relative warmth with respect to land at the same latitude. The only notable exceptions occur in conjunction with cold ocean currents, such as off the western coasts of Africa and South America south of the equator, or off the northwestern coast of Africa north of the equator.

The flow patterns associated with ocean currents and the movement of air above them remind us that the atmosphere and hydrosphere are highly dynamic entities. Indeed, both contain global-scale circulation systems that are vital to understanding weather and climate. Now that we are familiar with the temperature structure of the atmosphere, we are ready to examine the forces that shape these regularly recurring currents of air and water.

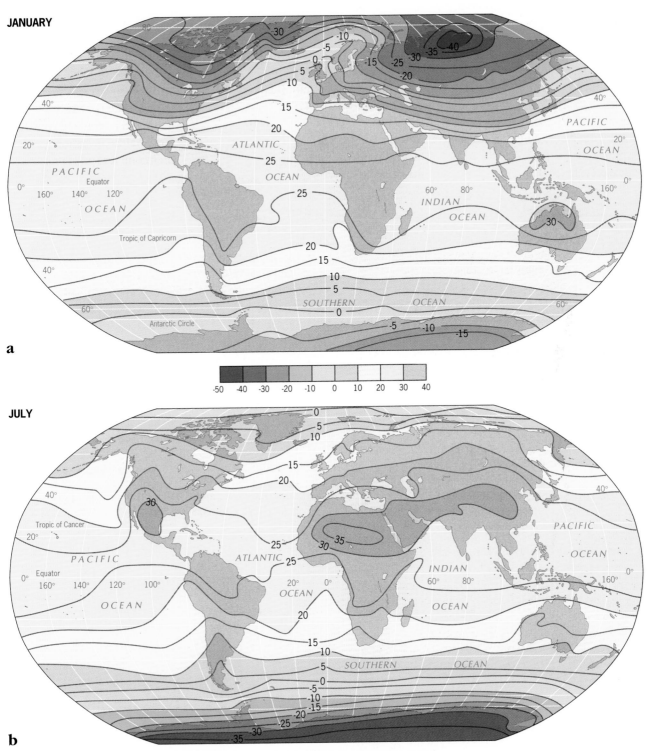

FIGURE 8-11 Mean sea-level air temperatures, in degrees Celsius, for January (a) and July (b).

KEY TERMS

adiabatic (p. 85)

adiabatic lapse rate (p. 85)

advection (p. 92)

annual cycle (p. 90)

Celsius scale (p. 84)

continentality (continental effect) (p. 91)

diurnal cycle (p. 90)

dry adiabatic lapse rate (DALR) (p. 85)

dust dome (p. 88)

environmental lapse rate (ELR) (p. 85)

Fahrenheit scale (p. 84)

isotherms (p. 92)

Kelvin scale (p. 84)

kinetic energy (p. 84)

maritime effect (p. 91)

saturated adiabatic lapse rate (SALR) (p. 86)

stability (p. 85)

temperature (p. 84)

temperature gradient (p. 92)

temperature inversion (p. 87)

thermometer (p. 84)

REVIEW QUESTIONS

1. Discuss the major differences between the Celsius, Fahrenheit, and Kelvin temperature scales.

2. What is the difference between stable and unstable air?

3. What is the *adiabatic* process? Define *DALR* and *SALR*.

4. What is a *temperature inversion*, and what is its relationship to an urban dust dome?

5. What are the factors that shape the spatial distribution of temperature across the earth's surface?

6. What are the main differences in annual temperature regimes between maritime and interior continental locations?

REFERENCES AND FURTHER READINGS

ELSOM, D. *Atmospheric Pollution: A Global Problem* (Cambridge, Mass.: Blackwell, 2nd ed., 1992).

GEIGER, R. *The Climate Near the Ground* (Cambridge, Mass.: Harvard University Press, 1965).

HENDERSON-SELLERS, A., and ROBINSON, P. J. *Contemporary Climatology* (London/New York: Longman, 1986), Chapter 2.

KONDRATYEV, K. *Radiation in the Atmosphere* (New York: Academic Press, 1969).

MATHER, J. R. *Climatology: Fundamentals and Applications* (New York: McGraw-Hill, 1974), Chapter 2.

MIDDLETON, W. *A History of the Thermometer and Its Use in Meteorology* (Baltimore: Johns Hopkins University Press, 1966).

OKE, T. R. *Boundary Layer Climates* (New York: Methuen, 2nd ed., 1987).

TREWARTHA, G. T., and HORN, L. H. *An Introduction to Climate* (New York: McGraw-Hill, 5th ed., 1980), Chapters 2, 8–12.

Air Pressure and Winds

OBJECTIVES

◆ To explain atmospheric pressure and its altitudinal variation.

◆ To relate atmospheric pressure to windflow at the surface and aloft.

◆ To apply these relationships to the operation of local wind systems.

In our previous discussions of the atmosphere, we have likened it to a blanket and a protective shield. We will now add yet another analogy: an ocean of air surrounding the earth. Such imagery is often used in physical geography because the atmosphere resembles the world ocean in its circulation patterns. This unit is about atmospheric pressure and winds, the dynamic forces that shape the regularly recurring global movements of air (and water). Units 10 and 11 focus on the patterns that result—the atmospheric and oceanic currents that constitute the general circulation systems affecting our planetary surface and form the framework for weather and climate.

The leading function of the general circulation of the atmosphere is to redistribute heat and moisture across the earth's surface. Were it not for the transport of heat

from the equator to the poles, most of the earth's surface would be uninhabitable because it would be either too hot, too cold, or too dry. The atmospheric circulation accounts for about 87 percent of this heat redistribution, and oceanic circulation accounts for the remainder. The atmosphere moves the way it does primarily because of the variation in the amount of net radiation received at the surface of the earth. The resulting temperature differences produce the global wind system. We will begin to examine this system by detailing the relationships between air pressure, heat imbalances, winds, and the rotational effects of the earth.

UNIT OPENING PHOTO: Moving air shapes many features on the earth's surface, including this wind-pruned Douglas fir on the northern California coastline.

ATMOSPHERIC PRESSURE

Wind, the movement of air relative to the earth's surface, is a response to an imbalance of forces acting on air molecules. This is true whether the air is moving horizontally or vertically, and indeed these two movement dimensions are related via the concept of atmospheric pressure. First we consider atmospheric pressure, its fundamental cause, and the resultant vertical distribution produced. Atmospheric pressure is then linked to windflow by considering both the additional forces that come into play once motion begins and the patterns of air circulation within the atmosphere that result.

The Concept of Pressure

The primary force exerting an influence on air molecules is gravity. The atmosphere is "held" against the earth by gravitational attraction. The combined weight of all the air molecules in a column of atmosphere exerts a force on the surface of the earth. Over a given area of the surface, say a square centimeter (0.39 sq in), this force produces a **pressure**. Although several different units are used to measure atmospheric pressure, the standard unit of pressure in atmospheric studies is the *millibar (mb)*. The average weight of the atmospheric column pressing down on the earth's surface (or *standard sea-level air pressure*) is 1013.25 mb, equivalent to 14.7 lb per sq in.

Atmospheric pressure is commonly measured as the length of a column of liquid it will support. In 1643 the Italian scientist Evangelista Torricelli performed an experiment in which he filled a glass tube with mercury and then placed the tube upside down in a dish of mercury. Figure 9-1 depicts his experiment. Instead of the mercury in the tube rushing out into the dish, the atmospheric pressure pushing down on the mercury in the surrounding dish supported the liquid still in the tube. The height of the column in the tube was directly proportional to the atmospheric pressure: the greater the pressure, the higher the column in the tube. (Note that standard sea-level air pressure produces a reading of 760 mm [29.92 in] in the height of the mercury column.) Torricelli had invented the world's first pressure-measuring instrument, known as a **barometer**.

Atmospheric Pressure and Altitude

Once scientists found they could measure atmospheric pressure, they set about investigating its properties. They soon discovered that atmospheric pressure does not vary all that much horizontally but does decrease very rapidly with increasing altitude. Measurements of atmospheric pressure from both higher land elevations and balloons showed dramatic results. The standard pressure at sea level (1013 mb) decreases to about 840 mb at Denver,

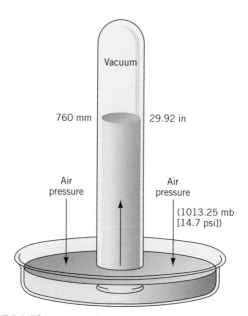

FIGURE 9-1 The mercury barometer invented by Torricelli. The greater the atmospheric pressure, the higher the column of mercury. These data exhibit the reading when standard sea-level pressure exists.

the "Mile-High City" whose elevation is 1584 m (5280 ft). On top of Mount Whitney in California's Sierra Nevada Mountains, at 4418 m (14,495 ft), the air pressure is approximately 600 mb. On top of the world's tallest peak—Mount Everest in South Asia's Himalayas, at 8848 m (29,028 ft)—the pressure is only 320 mb.

Because air pressure depends on the number of molecules in motion and is highest in the lower atmospheric layers, we may deduce that most of the molecules are concentrated near the earth's surface. This is confirmed in Fig. 9-2, which graphs the percentage of the total mass of the atmosphere below certain elevations. For example, 50 percent of the air of the atmosphere is found below 5 km (3.1 mi), and 85 percent lies within 16 km (10 mi) of the surface.

AIR MOVEMENT IN THE ATMOSPHERE

Since the days when it became common for sailing ships to make transoceanic voyages, people have known that the large-scale winds of the planet flow in certain generalized patterns. This information was vital in planning the routes of voyages that might take two or three years. However, it was often of little assistance in guiding the ships through the more localized, smaller-scale winds that fluctuate from day to day and place to place. It is therefore useful to separate large-scale air movement from smaller-scale movement, even though the two are

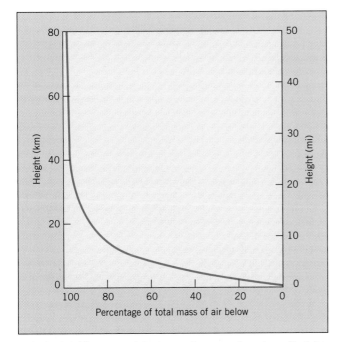

FIGURE 9-2 The mass of the atmosphere as a function of height. A greater proportion of the atmosphere is concentrated near the earth's surface. Atmospheric pressure depends on mass, so it also decreases with altitude.

related to the same phenomenon, atmospheric pressure. We begin by considering the causes, and resultant patterns, of the large-scale movement of air that is in contact with the surface of the earth.

Causes of Atmospheric Circulation

Two basic factors explain the circulation of air in the atmosphere: (1) the earth receives an unequal amount of heat energy at different latitudes, and (2) it rotates on its axis. If we examine the amount of incoming and outgoing radiation by latitude, as shown in Fig. 9-3 for the Northern Hemisphere (the Southern Hemisphere's general pattern is identical), we find that there is a marked surplus of net radiation between the equator and the 35th parallel. At latitudes poleward of 35°N, outgoing radiation exceeds incoming radiation. The main reason for this is that rays of energy from the sun strike the earth's surface at higher angles, and therefore at greater intensity, in the lower latitudes than in the higher latitudes (*see* diagram, p. 57). As a result, the equator receives about two and one-half times as much annual solar radiation as the poles do.

If this latitudinal imbalance of energy were not somehow balanced, the low-latitude regions would be continually heating up and the polar regions cooling down. Energy, in the form of heat, is transferred toward the poles, and the amount of heat transferred by atmospheric circulation (and to a much lesser extent, oceanic circulation) is also indicated in Fig. 9-3. We can see from the heat-transfer curve that the maximum transfer occurs in the middle latitudes. Thus the weather at these latitudes is characterized by frequent north–south movements of air masses.

Imagine for a moment that the earth is stationary and that there are no thermal differences between the land-

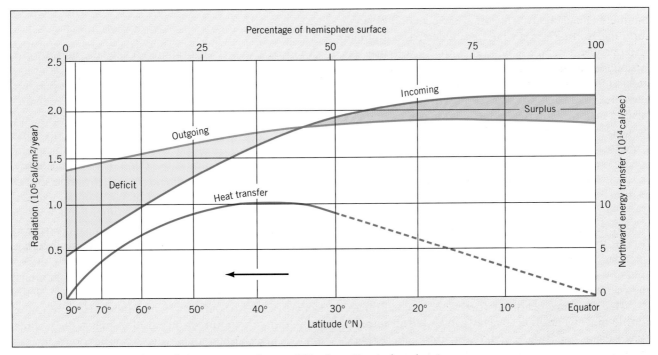

FIGURE 9-3 Latitudinal radiation balance averaged over all Northern Hemisphere longitudes, and the consequent poleward transfer of heat.

masses and oceans. Under these circumstances, heat transfer could occur by a simple cellular movement: warm air would rise at low latitudes, travel toward the poles at a high altitude, descend as it cools, and then return to the low latitudes as a surface wind. This type of circulation, however, is prohibited by the rotation of the earth and the differing energy-absorbing characteristics of land and water.

Strange as it may seem, the simple rotation of the earth complicates the operation of the general atmospheric circulation. The most important effect is expressed as an apparent deflective force. An understanding of the nature of this force is best approached by an analogy. Suppose you are sitting on a horse on a moving merry-go-round and try to throw a ball to a friend sitting on a horse ahead of you; but when you throw the ball straight at your friend, it travels toward the outside of the merry-go-round. This happens because between the time you release the ball and the time it would have reached your friend, your friend is no longer where he or she was when you released the ball. To you, it appeared that the ball was deflected to the right of its intended path (assuming that the merry-go-round was spinning counterclockwise when viewed from above). But to an observer standing next to the merry-go-round, the ball traveled in a straight path.

This deflective force affecting movement on a rotating body is called the **Coriolis force**. As noted in Unit 5, anything that moves over the surface of our spinning planet—from stream currents to missiles to air particles—is subjected to the Coriolis force. In the absence of any other forces, moving objects are deflected to their right in the Northern Hemisphere and to their left in the Southern Hemisphere. Thus, if we have a wind blowing from the North Pole, it would be deflected to the right and become an easterly wind (Fig. 9-4). (Note that this easterly wind blows toward the west: *winds are always named according to the direction from which they come.*) A little later, we will see how the Coriolis force plays a major role in determining the general pattern of atmospheric circulation. Before we do that, however, we need to consider the actual forces involved in windflow.

Forces on an Air Molecule

The earth's energy imbalance and its rotation may be regarded as the foremost causes of the general circulation of the atmosphere. However, specific wind speed and direction are determined by three forces: (1) the pressure-gradient force, (2) the Coriolis force, and (3) the frictional force.

The **pressure-gradient force** is the actual trigger for the movement of air. Gravity causes the air to press down against the surface of the earth; this is atmospheric pressure. The pressure, however, may be different at two locations. The difference in surface pressure over a given

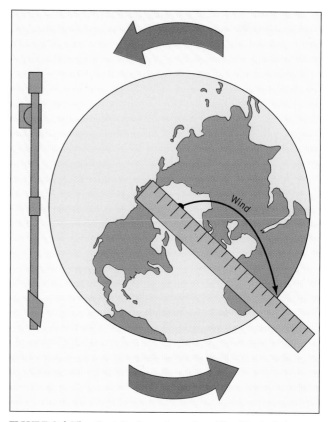

FIGURE 9-4 The Coriolis force in action. The North Pole on a map of the Northern Hemisphere is placed over the spindle of a record turntable, which is spun in a counterclockwise direction to simulate the earth's rotation. A line starting from the Pole and drawn along the edge of a stationary ruler (on the earth rotating beneath the ruler) describes an arc and ends up traveling toward the west.

distance between two locations is called the *pressure gradient*. When there is a pressure gradient, it acts as a force that causes air to move from the place of higher pressure to that of lower pressure. This force, called the pressure-gradient force, increases as the difference in air pressure across a specified distance increases. A common cause of the differences in air pressure is differences in air temperature, which, in turn, cause differences in air density. Warm air is less dense and tends to rise, lowering surface pressure as the outflow of air at higher altitudes exceeds the inflow near the surface. On the other hand, cold air tends to sink, reinforcing and raising surface pressure as the outflow near the surface exceeds the inflow aloft (Fig. 9-5).

We are already familiar with the *Coriolis force*, which acts to deflect moving air to the right (Northern Hemisphere) or left (Southern Hemisphere). Here we need only to note two further observations. First, the Coriolis force is not constant, but is greatest at the poles and decreases as one approaches the equator (Fig. 9-6). Second, we can deduce that it will be significant only over

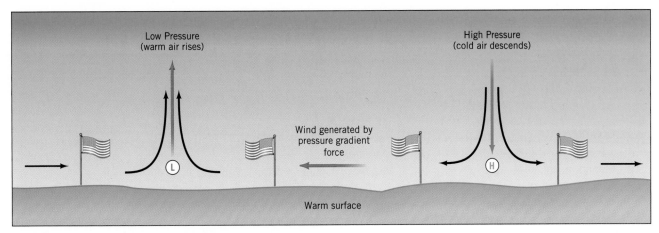

FIGURE 9-5 Air movement at the surface is always from areas of higher pressure (**H**) toward areas of lower pressure (**L**). The greater the pressure difference between **H** and **L**, the higher the pressure gradient and the stronger the wind.

fairly large distances, since it is dependent on the rotation of the earth.

Finally, some of the motion of the air in the atmosphere takes place very near the earth's surface. Thus individual air molecules close to the surface are slowed by drag, or a **frictional force**. The magnitude of the frictional force depends primarily on the "roughness" of the surface. There is less friction with movement across a smooth snow or water surface than across the ragged skyline of a metropolitan area.

We are now ready to examine how these three forces act together in the atmosphere. Collectively, they determine the pattern of windflow within any area—something that impinges upon a wide range of human activities (*see* Perspective: Air Pressure and Wind in Our Daily Lives).

LARGE- AND SMALLER-SCALE WIND SYSTEMS

Except for local winds (which affect only a relatively small area) and those near the equator, the wind never blows in a straight path from an area of higher pressure to an area of lower pressure. Once motion begins, the pressure-gradient, Coriolis, and frictional forces come into play and heavily influence the direction of windflow.

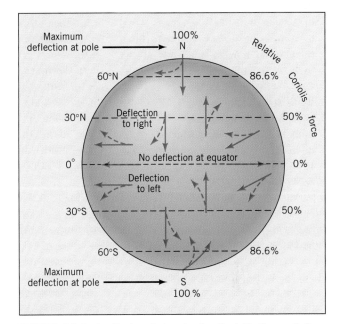

FIGURE 9-6 Latitudinal variation in the Coriolis force: deflection is zero at the equator, increases with latitude, and is most pronounced at each pole (note the percentages along the right side of the globe).

Geostrophic Winds

Once a molecule of air starts moving under the influence of a pressure-gradient force, the Coriolis force deflects it to the right if it is in the Northern Hemisphere. Figure 9-7 diagrams the path that results: eventually, the pressure-

FIGURE 9-7 Formation of a geostrophic wind in the Northern Hemisphere, looking directly down toward the surface.

Pressure gradient force

Coriolis force

1005 mb
1010 mb
1015 mb
Geostrophic wind
1020 mb

(Dashed lines are isobars)

Air Pressure and Wind in Our Daily Lives

Even a 98-pound weakling can support 10 tons! It's not an advertisement for a super body-building program. We all bear the weight of the atmosphere pressing down on us. At sea level, the air pressure can be from 9 to 18 metric tons depending on our size. Like the deep-sea creatures who live their entire lives with the weight of hundreds of meters of water above them, we have adapted to functioning and moving efficiently in our particular environmental pressure zone. Only a sharp change—such as a Florida sea-level flatlander taking a vacation trip high in Colorado's Rocky Mountains—reminds us of our adjustment to, and dependence on, a specific atmospheric environment. Two factors influence this sensitivity to altitude.

The first is the density of air molecules, particularly oxygen, at any altitude. At higher elevations the air is "thinner"; that is, there is more space between the oxygen molecules, and consequently fewer of them in any given air space. We have to do more breathing to get the oxygen necessary to maintain our activity levels. When the Olympic Games were held in Mexico City in 1968, the low density of oxygen at that elevation (2240 m [7350 ft]) was a decisive factor in the unimpressive competition times recorded by most of the participating athletes.

The second factor is the response of our internal organs to changes in atmospheric pressure. Our ears may react first as we climb to higher elevations. The "pop" we hear is actually the clearing of a tiny tube that allows pressure between the inner and middle ear to equalize, thereby preventing our eardrums from rupturing. At very high altitudes, we travel in pressurized aircraft. Astronauts also use pressurized cabins, and when they leave their vehicles for walks in atmosphere-less space, they require spacesuits to maintain a safe pressurized (and breathing) environment.

Winds play a constant role in our lives as well because they are a major element of local weather and climate. In coastal areas or on islands, atmospheric conditions can vary considerably over short distances. Oceanfront zones facing the direction of oncoming wind experience more air movement, cloudiness, and moisture than nearby locations protected from this airflow by hills or mountains. Places exposed to wind are called **windward** locations; areas in the "shadow" of protecting topographic barriers are known as **leeward** locations.

The city of San Francisco, which sits at the tip of a narrow, longitudinally ridged peninsula between the Pacific Ocean to the west and San Francisco Bay to the east, provides an excellent ex-

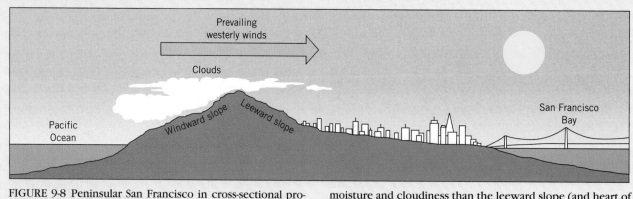

FIGURE 9-8 Peninsular San Francisco in cross-sectional profile: the windward slope facing the Pacific experiences more moisture and cloudiness than the leeward slope (and heart of the city) facing the bay.

gradient force and the Coriolis force acting on the wind balance each other out. The resultant wind, called a **geostrophic wind**, follows a relatively straight path that minimizes deflection. Geostrophic windflow is common in the "free" atmosphere, that is, above the contact layer where friction with the surface occurs. Because of the balancing of forces, a geostrophic wind always flows parallel to the **isobars**—lines that join points of equal atmospheric pressure (Fig. 9-7). If we have a map showing the distribution of atmospheric pressure well above the surface, we can get a rather good idea of where these winds are blowing (as we shall see in Unit 15). We can, moreover, predict the wind speed if we also know the pressure-gradient force, the air density, and the latitude (which determines the strength of the Coriolis force; *see* Fig. 9-6).

Frictional Surface Winds

Near the surface of the earth, below an elevation of about 1000 m (3300 ft), frictional force comes into play and disrupts the balance represented by the geostrophic wind.

ample of these influences (Fig. 9-8). Prevailing westerly winds blow across the cool nearshore waters of the Pacific, which produces much cloudiness, ground fog, and general dampness along the city's windward slopes. But these conditions are less likely to occur on the leeward slopes facing the bay to the east, and it is here that the historic heart of the city is situated.

At another place and time, a similar windward/leeward relationship helped shape the locational pattern of the textile industry in nineteenth-century Britain. West of England's "backbone" of the Pennine Mountains, in Lancashire, the humid air transported in from the Atlantic Ocean by westerly winds was ideal for cotton-textile manufacturing. Woolen-textile manufacturing, however, required a drier environment, which was readily available to the east in Yorkshire

on the leeward side of the moisture-screening Pennines.

There are countless other examples of associations between wind and human activities. As we note in Unit 8, the horizontal flushing effects of wind are vital to maintaining acceptable air quality in urban areas, where large quantities of pollutants are dumped into the local atmosphere. Many outdoor sporting events are affected by wind conditions during games. For example, certain baseball stadiums are famous—or infamous—for their unpredictable wind currents. San Francisco's Candlestick Park (located in the peninsula's lee alongside the bay) is notorious in this regard. And in Chicago, the "Windy City," hundreds of Cubs games have been influenced by the wind at Wrigley Field.

In America's continental heartland, the reinforcing effects of wind on cold

winter temperatures continue to be an unpleasant fact of life. To give us a precise idea of how cold we would feel under given conditions of wind speed and air temperature, scientists have developed the *wind-chill index*, which is displayed in Table 9-1 (presented in the more familiar Fahrenheit degrees). Although this index does not take into account evaporative heat loss and the amount of protective clothing we wear, the wind-chill index is closely related to the occurrence of frostbite. The index, therefore, applies mainly to sensible heat loss. You can see from the table that a comparable feeling of cold can be expected at different combinations of wind speeds and air temperatures. You would begin to feel very cold if the temperature were $-15°F$ ($-26°C$) in calm conditions or if the temperature were 35°F (2°C) in a wind of 15 mph (7 m/s).

TABLE 9-1 THE WIND-CHILL INDEX, A MEASURE OF COOLING POWER ON EXPOSED FLESH. WIND SPEEDS OVER 40 MILES PER HOUR HAVE LITTLE ADDITIONAL CHILLING EFFECT.

MPH	DRY-BULB TEMPER-ATURE						WIND CHILL INDEX (°F)										
	35	30	25	20	15	10	5	0	−5	−10	−15	−20	−25	−30	−35	−40	−45
Calm	35	30	25	20	15	10	5	0	−5	−10	−15	−20	−25	−30	−35	−40	−45
5	33	27	21	16	12	7	1	−6	−11	−15	−20	−26	−31	−35	−41	−47	−54
10	21	16	9	2	−2	−9	−15	−22	−27	−31	−38	−45	−52	−58	−64	−70	−77
15	16	11	1	−6	−11	−18	−25	−33	−40	−45	−51	−60	−65	−70	−78	−85	−90
20	12	3	−4	−9	−17	−24	−32	−40	−46	−52	−60	−68	−76	−81	−88	−96	−103
25	7	0	−7	−15	−22	−29	−37	−45	−52	−58	−67	−75	−83	−89	−96	−104	−112
30	5	−2	−11	−18	−26	−33	−41	−49	−56	−63	−70	−78	−87	−94	−101	−109	−117
35	3	−4	−13	−20	−27	−35	−43	−52	−60	−67	−72	−83	−90	−98	−105	−113	−123
40	1	−4	−15	−22	−29	−36	−45	−54	−62	−69	−76	−87	−94	−101	−108	−118	−128
45	1	−6	−17	−24	−31	−38	−46	−54	−63	−70	−78	−87	−94	−101	−108	−118	−128
50	0	−7	−17	−24	−31	−38	−47	−56	−63	−70	−79	−88	−96	−103	−110	−120	−128

Very cold

Bitterly cold

Extremely cold

Source: U.S. Army Quartermaster Research and Engineering Center, Report EP-143, 1961.

Friction both reduces the speed and alters the direction of a geostrophic wind. The frictional force acts in such a way as to cause the pressure-gradient force to overpower the Coriolis force, so that the (no-longer-geostrophic) wind at the surface blows across the isobars instead of parallel to them. This produces a flow of air out of high-pressure areas and into low-pressure areas, but at an angle to the pressure gradient rather than straight across.

Since surface pressure systems are often roughly cir-

cular when viewed from above, we can deduce the general circulation around cells of low and high pressure (Fig. 9-9). Surface winds converge toward a **cyclone** (a low-pressure cell—**L** in Fig. 9-9a); this converging air has to go somewhere, so it rises vertically in the center of the low-pressure cell. The reverse is true in the center of an **anticyclone** (a high-pressure cell—**H** in Fig. 9-9b); diverging air moves outward and draws air down in the center of the high-pressure cell (*see also* Fig. 9-5). Thus cyclones are associated with *rising air* at their centers,

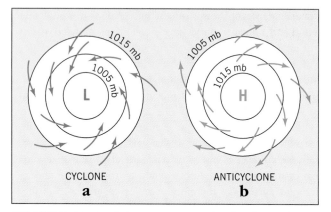

FIGURE 9-9 Air circulation patterns associated with a cyclonic low-pressure cell (a) and an anticyclonic high-pressure cell (b) in the Northern Hemisphere. In the Southern Hemisphere, the windflows move in the opposite direction (clockwise toward cyclones and counterclockwise away from anticyclones).

and anticyclones are associated with *subsiding air* at their centers. This simple vertical motion produces very different weather associated with each type of pressure system.

LOCAL WIND SYSTEMS

Although we cannot overlook the forces outlined above when we focus our attention on small-scale airflows, local wind systems are often more significant in day-to-day weather because they respond to much more subtle variations in atmospheric pressure than are depicted in Fig. 9-9. Moreover, because smaller distances are involved, the effect of the Coriolis force can usually be disregarded. A number of common local wind systems serve to illustrate how topography and surface type can influence the pressure gradient and its resultant windflow.

Sea/Land Breeze Systems

In coastal zones and on islands, two different surface types are in close proximity: land and water. As we have noted before, land surfaces and water bodies display sharply contrasting thermal responses to energy input. Land surfaces heat and cool rapidly, whereas water bodies exhibit a more moderate temperature regime.

During the day, a land surface heats up quickly, and the air layer in contact with it rises in response to the increased air temperature. This rising air produces a low-pressure cell over the coastal land or island. Since the air over the adjacent water is cooler, it subsides to produce a surface high-pressure cell. A pressure gradient is thereby produced, and air in contact with the surface now moves from high pressure to low pressure. Thus, during the day, shore–zone areas generally experience air moving from the water to land, a **sea breeze** (Fig.

9-10a). At night, the circulation reverses because the warmer air (and lower pressure) is now over the water. This results in air moving from land to the water, a **land breeze** (Fig. 9-10b).

Note that when generated, sea and land breezes produce a circulation cell composed of the surface breeze, rising and subsiding air associated with the lower and higher pressure areas, respectively, and an airflow aloft in the direction opposite to that of the surface (Fig. 9-10). Although it modifies wind and temperature conditions at the coast, the effect of this circulation diminishes rapidly as one moves inland. Note also that we use the word *breeze*. This accurately depicts a rather gentle circulation in response to a fairly weak pressure gradient. The sea breeze/land breeze phenomenon can easily be overpowered if stronger pressure systems are nearby.

Mountain/Valley Breeze Systems

Mountain slopes, too, are subject to the reversal of day and night local circulation systems. This wind circulation is also thermal, meaning that it is driven by temperature differences between adjacent topographic features. During the day, mountain terrain facing the sun tends to heat up more rapidly than shadowed, surrounding slopes. This causes low pressure to develop, spawning an upsloping *valley breeze*. At night, greater radiative loss from the mountain slopes cools them more sharply, high pressure develops, and a downsloping *mountain breeze* results. Figure 9-11 shows the operation of this type of oscillating, diurnal wind system in a gently rising highland valley.

Other Local Wind Systems

Another category of local wind systems involves **cold-air drainage**, the steady downward oozing of heavy, dense, cold air along steep slopes under the influence of gravity. The winds that result are known as **katabatic winds** and are especially prominent under calm, clear conditions where the edges of highlands plunge sharply toward lower lying terrain. These winds are fed by large pools of very cold air that collect above highland zones. They are also common around major icesheets, such as the huge, continental-scale glaciers that cover most of Antarctica and Greenland.

Katabatic winds can attain destructive intensities when the regional windflow steers the cold air over the steep edges of uplands, producing a cascade of air much like water in a waterfall. The most damaging winds of this type occur where local topography channels the downward surge of cold air into narrow, steep-sided valleys. The Rhône Valley of southeastern France is a notable example: each winter it experiences icy, high-velocity winds (known locally as the *mistral* winds) that drain the massive pool of cold air that develops atop the snowy French and Swiss Alps to the valley's northeast.

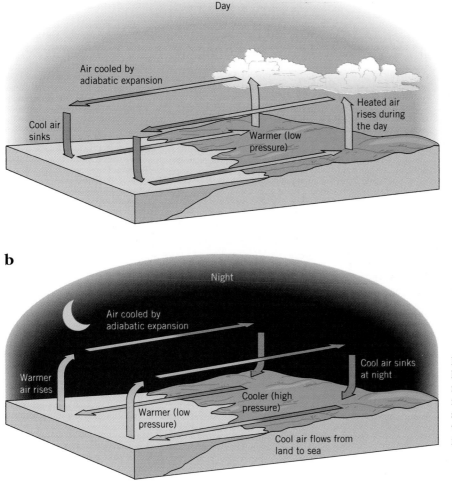

a

Day

Air cooled by adiabatic expansion

Cool air sinks

Warmer (low pressure)

Heated air rises during the day

b

Night

Air cooled by adiabatic expansion

Warmer air rises

Warmer (low pressure)

Cooler (high pressure)

Cool air sinks at night

Cool air flows from land to sea

FIGURE 9-10 Sea breeze/land breeze local air circulation systems. These reversing, cell-like airflows develop in response to pressure differentials associated with day/night temperature variations at the land and water surfaces.

Yet another type of local wind system is associated with the forced passage of air across mountainous terrain (which is discussed in detail in Unit 13; *see* pp. 141–144). Briefly, this transmontane movement wrings out most of the moisture contained in the original mass of air, and it also warms the air adiabatically as it plunges downward after its passage across the upland. Thus the area that extends away from the base of the mountain's backslope experiences dry and relatively warm winds, which taper off with increasing distance from the highland. Occasionally, such winds can exceed hurricane-force intensity (greater than 120 kph [75 mph]). This happens when they are reinforced by an anticyclone upwind from the mountains that feeds air into a cyclone located on the downwind side of the upland.

Surprisingly, atmospheric scientists have yet to provide a generic name for such wind systems, which still go only by their local names. The best known U.S. example is the **chinook wind**, which occurs on the (eastern) downwind side of the Rocky Mountains along the western edge of the central and northern Great Plains. Another well-known example is the **Santa Ana wind** of coastal Southern California, an occasional hot, dry airflow whose unpleasantness is heightened by the downward funneling of this wind from the high inland desert (where it is generated by an anticyclone) through narrow passes in the mountains that line the coast.

In many locations, as we have just seen, local winds can at times become more prominent than larger-scale airflows. But the regional expressions of the global system of wind currents still play a far more important role overall. That is the subject of Unit 10, which investigates the general circulation of the atmosphere.

KEY TERMS

anticyclone (p. 101)
barometer (p. 96)
chinook wind (p. 103)
cold-air drainage (p. 102)
Coriolis force (p. 98)

cyclone (p. 101)
frictional force (p. 99)
geostrophic wind (p. 100)
isobar (p. 100)
katabatic wind (p. 102)

land breeze (p. 102)
leeward (p. 100)
pressure (p. 96)
pressure-gradient force (p. 98)

Santa Ana wind (p. 103)
sea breeze (p. 102)
wind (p. 96)
windward (p. 100)

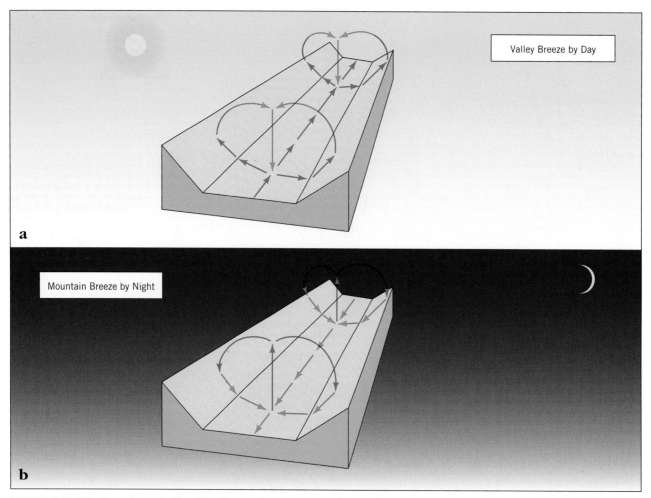

FIGURE 9-11 The formation of valley (daytime) and mountain (nighttime) breezes. Red arrows represent warm winds; blue arrows indicate colder winds.

REVIEW QUESTIONS

1. Define atmospheric pressure, and describe its vertical structuring within the atmospheric column.

2. What are the forces that determine wind speed and direction?

3. Define the Coriolis force, and describe its operations in both the Northern and Southern Hemispheres.

4. What is a *geostrophic* wind? Where does it occur and why?

5. Describe the air circulation patterns associated with cyclones and anticyclones.

6. Describe the operation of the sea breeze/land breeze local wind system.

REFERENCES AND FURTHER READINGS

ATKINSON, B. W.   *Meso-Scale Atmospheric Circulations* (New York: Academic Press, 1981).

DUTTON, J. A.   *The Ceaseless Wind: An Introduction to the Theory of Atmospheric Motion* (New York: McGraw-Hill, 1976).

EDINGER, J. G.   *Watching for the Wind* (Garden City, N.Y.: Doubleday, 1967).

GEDZELMAN, S. D.   *The Science and Wonders of the Atmosphere* (New York: Wiley, 1980), Chapter 15.

GROSS, J.   "When the Fog Rolls In, the Bay Area Hears Music," *New York Times*, June 22, 1988, 10.

HIDY, G. M.   *The Winds* (New York: Van Nostrand-Reinhold, 1967).

MIDDLETON, W. K.   *The History of the Barometer* (Baltimore: Johns Hopkins University Press, 1964).

PALMÉN, E., and NEWTON, C. W.   *Atmospheric Circulation Systems* (New York: Academic Press, 1969).

"The Santa Ana Winds," *New York Times*, October 29, 1993, A-10.

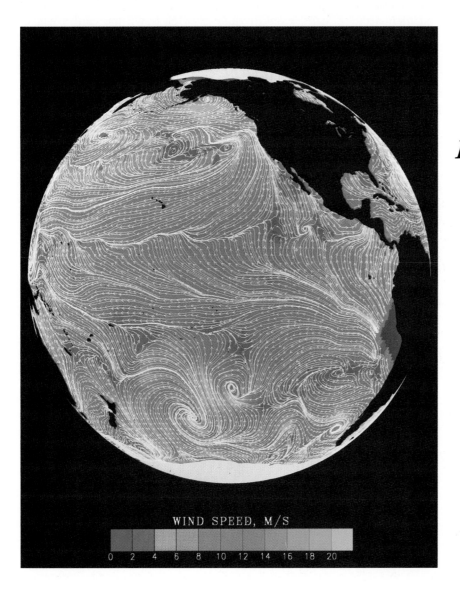

WIND SPEED, M/S

0 2 4 6 8 10 12 14 16 18 20

Circulation Patterns of the Atmosphere

OBJECTIVES

- To develop a simple model of the global atmospheric circulation.
- To discuss the pressure systems and wind belts that constitute that model circulation, and the complications that arise when the model is compared to the actual atmospheric circulation.
- To introduce the basic workings of the upper atmosphere's circulation.

Unit 9 (Air Pressure and Winds) introduced the basic causes of air movements in the atmosphere. In this unit, we focus on the global air currents that constitute the general atmospheric circulation. In the short run these air currents carry along, and to a certain extent cause, the weather systems that affect us daily. The longer-term operation of this general circulation, in conjunction with atmospheric energy flows, produces the climates of the earth.

To be sure, the workings of the general atmospheric circulation are very complex, and we still do not understand the exact nature of some of the circulation features we are about to examine. We can, however, deduce many of these features using our knowledge of the basic

UNIT OPENING PHOTO: Satellite image of airflows above the Pacific Ocean, with colors representing wind speed. Note how local storms are embedded in the global-scale wind belts.

causes of air movement. Let us begin by considering the atmosphere's near-surface circulation, and then make some observations regarding windflow in the upper atmosphere.

A MODEL OF THE SURFACE CIRCULATION

We note in Unit 9 the probable arrangement of the global atmospheric circulation on a uniform, nonrotating earth: a single, girdling cell of low pressure around the equator where air would rise, and a cell of high pressure at each pole where air would subside. The surface winds on such a planet would be northerly in the Northern Hemisphere and southerly in the Southern Hemisphere, moving directly from high pressure to low pressure across the pressure gradient (Fig. 10-1).

We might further speculate that adding the rotation of the earth to this simple model would produce surface northeasterly winds in the Northern Hemisphere (as the air moving toward the equator was deflected to the right) and southeasterly winds in the Southern Hemisphere. In fact, such a model would be "unstable," breaking down

because of one simple problem: achievement of this scenario would require slowing down the earth's rate of spin, since everywhere on the planet the atmosphere would be moving *against* the direction of earth rotation. As it turns out, we do have areas of low pressure at the equator and high pressure at the poles, but the situation in the midlatitudes is more complex.

Let us now introduce the actual effects of rotation on our hypothetical planet but, for the time being, ignore seasonal heating differences and the land/water contrast at the surface. The model that results is an idealized but reasonable generalization of the surface circulation pattern, and its following elaboration is keyed to Fig. 10-2.

The Equatorial Low and Subtropical High

Year-round heating in the equatorial region produces a thermal low-pressure belt in this latitudinal zone. That belt of rising air is called the **Equatorial Low** or **Inter-Tropical Convergence Zone (ITCZ)**. (The reason for this latter terminology will become evident shortly.) The air rises from the surface to the tropopause and flows poleward in both the Northern and Southern Hemisphere.

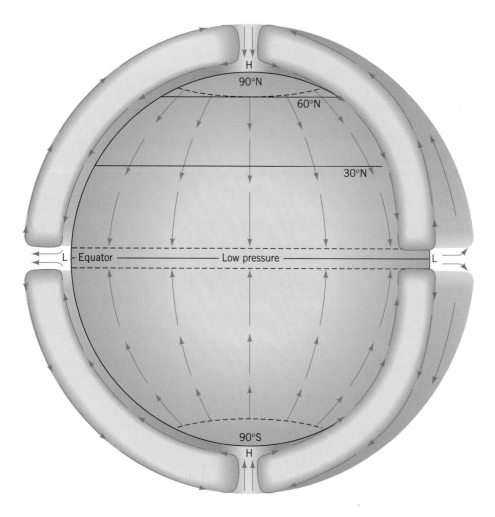

FIGURE 10-1 Hypothetical atmospheric circulation on a featureless, nonrotating earth. Polar high pressure and equatorial low pressure would result in northerly surface winds in the Northern Hemisphere and southerly surface winds in the Southern Hemisphere. The rotation of the earth, the variation in the latitude of the vertical sun position, and land/water heating contrasts at the surface prevent this simple general circulation from developing.

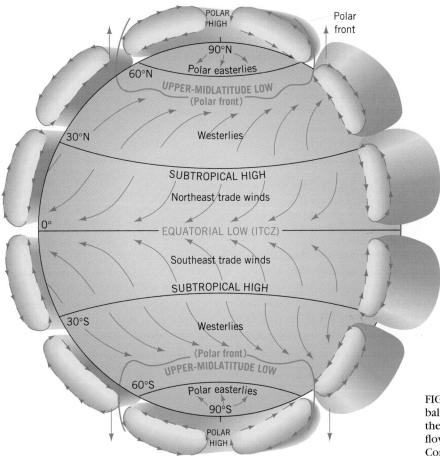

FIGURE 10-2 An idealized version of the global atmospheric circulation pattern showing the major pressure belts, the cell-like airflows that develop between them, and the Coriolis deflection of surface winds.

Much of this now-cooled, poleward-moving air descends at around latitudes 30°N and 30°S. (Subsidence here is most likely associated with the "piling up" of air aloft because of its increased westerly flow, a directional curvature caused by the strengthening Coriolis force; *see* Fig. 9-6.) The descending air produces a belt of high pressure at the surface near both these latitudes; the two high-pressure belts, understandably, are termed the **Subtropical Highs**.

The Trade Winds and the Westerlies

Remembering that surface airflows diverge out of areas of high pressure, we can easily deduce the nature of the wind movement, both equatorward and poleward, of this high-pressure belt. Air returning toward the equator in the Northern Hemisphere is deflected to the right and forms a belt of northeasterly winds, called the **Northeast Trades** (*see* Perspective: The Sailor's Legacy—Naming the Winds). Air returning toward the equator from the Subtropical High in the Southern Hemisphere is deflected to its left to form the **Southeast Trades**. As you can see in both Fig. 10-2 and the unit-opening photo, these two wind belts converge—hence the rationale for calling this low-latitude area the Inter-Tropical Convergence Zone.

Air moving poleward from the two Subtropical Highs acquires the appropriate Coriolis deflection and forms two belts of generally west-to-east-flowing winds (one in the Northern Hemisphere, one in the Southern Hemisphere) known as the **Westerlies**. These prevailing winds form broad midlatitude belts from about 30°N to 60°N and 30°S to 60°S.

The Polar High, Polar Easterlies, and Polar Front

Let us leave the Westerlies for a moment and consider the windflows emanating from the **Polar Highs**, large cells of high pressure centered over each pole. Here, air moving toward the equator is sharply deflected to become the **Polar Easterlies**. You can see in Fig. 10-2 that the Polar Easterlies flowing out of the Polar Highs will meet the Westerlies flowing out of the Subtropical Highs.

The atmospheric boundary along which these wind systems converge is called the **Polar Front**. Along the Polar Front in each hemisphere (located equatorward of 60°N and 60°S, respectively), the warmer air from the Subtropical High is forced to rise over the colder, and thus denser, polar air. This rising air produces a belt of low pressure at the surface called the **Upper-Midlatitude Low**.

The Sailor's Legacy—Naming the Winds

Spanish sea captains headed to the Caribbean and the Philippines in search of gold, spices, and new colonial territory for the crown. They depended on a band of steady winds to fill the sails of their galleons as they journeyed westward in the tropics. Those winds were named the trade winds, or the *trades*. These were the winds that first blew Christopher Columbus and his flotilla to North America in 1492.

In the vicinity of the equator, the Northern and Southern Hemisphere trades converge in a zone of unpredictable breezes and calm seas. Sailors dreaded being caught in these so-called *doldrums*. A ship stranded here might drift aimlessly for days. That was the fate of the ship described in these famous lines from Samuel Taylor Coleridge's *The Rime of the Ancient Mariner*:

> Day after day, day after day,
> We stuck, nor breath nor motion;
> As idle as a painted ship
> Upon a painted ocean.

Ships also were becalmed by the light and variable winds in the subtropics at about latitudes 30°N and 30°S. Spanish explorers who ran afoul of the breezes in these hot regions threw their horses overboard to lighten their loads and save water for the crew. The trail of floating corpses caused navigators of the seventeenth century to label this zone the *horse latitudes*.

In the middle latitudes of the Southern Hemisphere, ships heading eastward followed the strong westerly winds between 40° and 60°S. These winds were powerful but stormier than the trades to the north; so, depending on their approximate latitude, they became known as the *Roaring Forties*, the *Furious Fifties*, and the *Screaming Sixties*. Thus some important terminology still applied to wind belts dates from the early days of transoceanic sailing.

Overall, our now-completed model of the surface circulation for our uniform, rotating planet has seven pressure features (Equatorial Low, two Subtropical Highs, two Upper-Midlatitude Lows, and two Polar Highs) and six intervening wind belts (the Northeast and Southeast Trades plus the Westerlies and Polar Easterlies of each hemisphere). Note that the circulation patterns of the Northern and Southern Hemispheres are identical except for the opposite Coriolis deflection (Fig. 10-2).

THE ACTUAL SURFACE CIRCULATION PATTERN

In contrast to the idealized pattern of the surface circulation model, the actual pattern (Fig. 10-3) is considerably more complex because it incorporates two influences ignored in the model. Remember that the location of maximum solar heating shifts throughout the year as the latitude of the (noonday) vertical sun changes from 23½°N at the Northern Hemisphere's summer solstice to 23½°S at its winter solstice. Of great significance is that the continents respond more dramatically to this latitudinal variation in heating than do the oceans (as we note in Unit 8). This has the effect of producing individual pressure cells (which we can call *semipermanent highs and lows*) rather than uniform, globe-girdling belts of low and high pressure. Moreover, because the Northern Hemisphere contains two large landmasses while the Southern Hemisphere is mostly water, the two hemispheres exhibit somewhat different atmospheric circulations. Nonetheless, they are similar enough to consider together.

The Equatorial Low (ITCZ)

Careful inspection of Fig. 10-3 reveals the following modifications to the simplified picture seen in Fig. 10-2. The Equatorial Low, or ITCZ, migrates into the "summer" hemisphere (the Northern Hemisphere during July, the Southern Hemisphere during January), a shift most prominent over landmasses. Note that in July (Fig. 10-3b), the Equatorial Low is located nearly 25 degrees north of the equator in the vicinity of southern Asia. But ITCZ migration is subdued over oceanic areas because bodies of water are much slower to respond to seasonal changes in solar energy input received at the surface. Note, too, that the migration of the Equatorial Low into the Southern Hemisphere in January (Fig. 10-3a) is far less pronounced. This makes sense because there are fewer large landmasses in the tropics south of the equator; the Equatorial Low is, however, considerably displaced poleward over Africa and Australia, and to a lesser extent above South America.

The Bermuda and Pacific Highs

In contrast to our model's Subtropical High pressure belts (straddling 30°N and 30°S), the actual semipermanent highs at these latitudes are more cellular. There are five

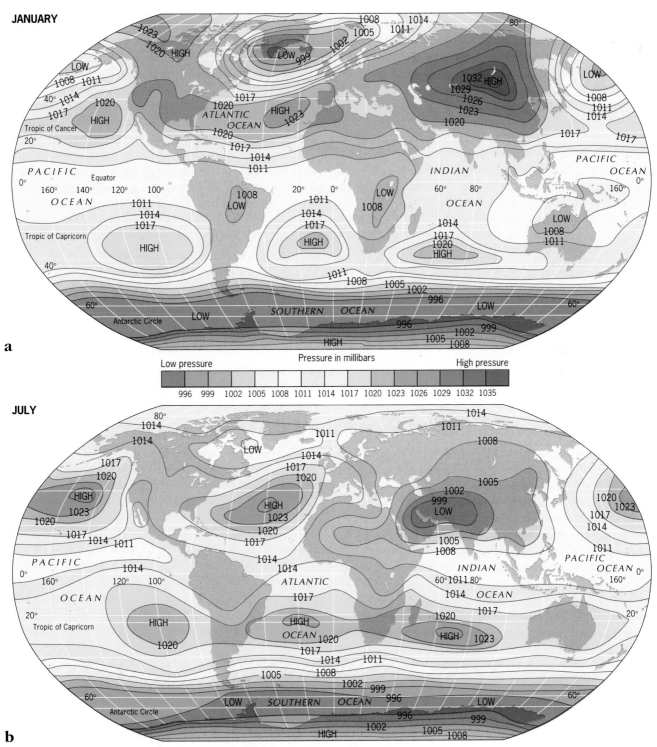

FIGURE 10-3 Global mean surface pressure patterns in January (a) and July (b). Maps adapted from *Goode's World Atlas*, 19th edition. © 1995 by Rand McNally R.L. 92-S-82-renewal 95.

such cells on the map, one above each subtropical ocean (Fig. 10-3). These cells are most evident in the "summer" hemisphere. In the Northern Hemisphere, the North Atlantic's high-pressure cell is called the *Bermuda* (or *Azores*) *High*, and the North Pacific cell is referred to as the *Pacific* (or *Hawaiian*) *High*. These cells also shift north and south with the sun, but to a much lesser extent, latitudinally, than the Equatorial Low. This again is due

to the more subdued response of water to seasonal changes in solar energy receipt.

The Canadian and Siberian Highs

Now directing attention to the polar regions, we see that our simplified picture of a Polar High centered over each pole needs considerable revision, particularly in the

109

Northern Hemisphere where the large landmasses of Eurasia and North America markedly protrude into the high latitudes. It is here that the seasonal cooling will be most extreme, rather than over the more northerly Arctic Ocean with its floating polar icecap. As a result, the Polar High in the Northern Hemisphere is actually two separate cells, a weaker cell centered above northwestern Canada (the *Canadian High*) and a much more powerful cell covering all of northern Asia (the *Siberian High*). Note, however, that these features are evident only during the winter (Fig. 10-3). In the Southern Hemisphere, because of the dominance of the Antarctic landmass in the high latitudes, the model's single cell of high pressure over the pole is reasonably accurate.

The Aleutian, Icelandic, and Southern Hemisphere Upper-Midlatitude Lows

Finally, the actual configuration of the Upper-Midlatitude Low between the Polar and Subtropical Highs needs to be reexamined. In the Northern Hemisphere in January (Fig. 10-3a), when both the Polar and Subtropical Highs are apparent, the Upper-Midlatitude Low is well defined as two cells of low pressure, one over each subarctic ocean. These are the *Aleutian Low* in the northeastern Pacific off Alaska and the *Icelandic Low* in the North Atlantic centered just west of Iceland. The convergence and ascent of air within these cells are more complex than this still rather generalized map suggests; thus these features are covered in greater depth in the discussion of air masses and storm systems of the midlatitudes in Unit 14. For the time being we also note that these cells, too, weaken to the point of disintegration during the summer (Fig. 10-3b).

As for the Southern Hemisphere, note again on the map that the Upper-Midlatitude Low is evident in both winter and summer. The persistence of the Polar High over Antarctica makes this possible. Moreover, the absence of landmasses in this subpolar latitudinal zone causes the Upper-Midlatitude Low to remain belt-like rather than forming distinct cells over each ocean.

SECONDARY SURFACE CIRCULATION: MONSOONAL WINDFLOWS

The global scheme of wind belts and semipermanent pressure cells we have just described constitutes the **general circulation** (or *primary circulation*) system of the atmosphere. At a more localized scale, there are countless instances of "shifting" surface wind belts that create pronounced winter/summer contrasts in weather patterns. Here, to illustrate such regional, or *secondary*, circulation systems, we will confine our attention to one of the most spectacular examples: the Asian monsoon.

A **monsoon** (derived from *mawsim*, the Arabic word for season) is a regional wind that blows onto and off of certain landmasses on a seasonal basis. Monsoonal circulation occurs most prominently across much of southern and eastern Asia, where seasonal wind reversals produced by the shifting systems cause alternating wet and dry seasons. Specifically, the moist onshore winds of summer bring the *wet monsoon*, whereas the offshore winds of winter are associated with the *dry monsoon*. These reversing wind systems override the expected pattern of the general atmospheric circulation—and yet, as we are about to see, are still a part of it.

In January, high pressure over the interior of southern Asia (particularly the Indian subcontinent) produces northeasterly surface winds for much of the region (Fig. 10-4a). This cool continental air contains very little mois-

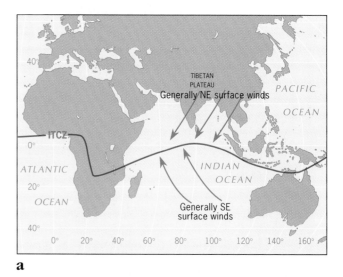

a

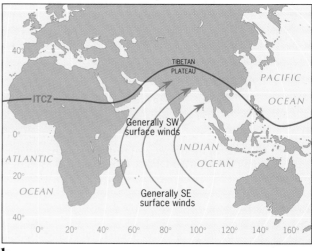

b

FIGURE 10-4 The wind reversals associated with the dry winter monsoon (a) and wet summer monsoon (b) in southern Asia. This phenomenon is usually most pronounced on the Indian subcontinent, one of humankind's greatest concentrations (*see* also Fig. 2-5).

ture, so precipitation during winter is at a minimum. But as spring gives way to summer the high-pressure cell dissipates, and the ITCZ (Equatorial Low) shifts far northward to a position over the Tibetan Plateau. As a result, the airflow from the Southeast Trades now crosses the equator and is recurved—by the opposite Coriolis deflection of the Northern Hemisphere—into a southwesterly flow (Fig. 10-4b).

This air has passed above most of the warm tropical Indian Ocean and therefore now possesses a very high moisture content. The arrival of this saturated air over the Indian subcontinent marks the onset of the wet summer monsoon, and precipitation is frequent and heavy. As a matter of fact, the world record one-month precipitation total is held by the town of Cherrapunji in the hills of northeastern India (*see* p. 203), where in July of 1861 930 cm (30 *feet*!) of rain fell. During the winter months of the dry monsoon, however, average precipitation values at Cherrapunji are normally on the order of 1 to 2 cm (0.5 to 1.0 in).

The southwestern wet monsoon consists of two main branches, as Fig. 10-5 shows. One branch penetrates the Bay of Bengal to Bangladesh and northeastern India, where it is pushed westward into the densely populated Ganges Plain of northern India by the Himalayan mountain wall (Fig. 10-6). A second branch to the west of the subcontinent, with a tendency to split into two

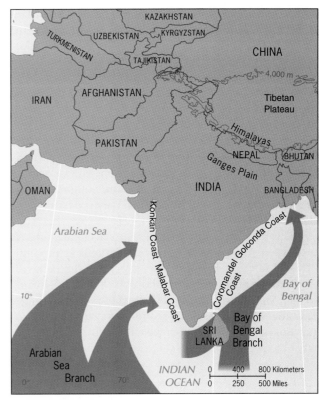

FIGURE 10-5 The main branches of the southwestern, wet-monsoon windflow over South Asia.

FIGURE 10-6 The topographic gradient from the low-lying Ganges Plain (bottom) to the main ranges of the Himalayas (top) is one of the earth's most dramatic. This gigantic mountain barrier blocks the onshore windflow of the summer monsoon, steering it westward onto the Gangetic lowland where its torrential rains nourish one of humanity's largest and most heavily populated agricultural regions.

airflows, arrives from the Arabian Sea arm of the Indian Ocean. The rains from both branches gradually spread across much of the subcontinent, soak the farm fields, and replenish the wells.

These wet-monsoon rains, however, are not continuous, even during the wettest of years. They depend on the recurrence of smaller-scale low-pressure cells within the prevailing southwesterly airflow. These depressions reinforce the lifting of the moist air, and enhance the formation and continuation of rain throughout the summer months. When they fail to materialize, disastrous drought can result; their dramatic failure in the summer of 1987 triggered one of India's worst dry spells and crop losses of the past century.

This peculiar South Asian monsoonal circulation also owes its identity to subtle seasonal variations in the windflows of the upper atmosphere, especially the behavior of the tropical, subtropical, and Polar Front jet streams (which are introduced in the following section). In fact, the same is true everywhere: airflow patterns in the upper atmospheric circulation exert a decisive influence on what happens at the surface.

CIRCULATION OF THE UPPER ATMOSPHERE

Windflow in the upper atmosphere is geostrophic (perpendicular to the pressure gradient, as shown in Fig. 9-7), with the higher pressure on the right looking downwind in the Northern Hemisphere and on the left in the Southern Hemisphere. Furthermore, the general circulation aloft is much simpler, since we lose the effects of the land/water contrast that made the surface pattern decidedly cellular. However, the specific nature of the upper atmospheric circulation is complex, and we will not attempt to explain how it is maintained, but rather will note some relevant generalizations.

First and foremost, the upper atmospheric circulation is dominated by **zonal flow**, meaning *westerly* in its configuration. Thus the winds of the upper atmosphere generally blow from west to east throughout a broad latitudinal band in both hemispheres; essentially, windflow is westerly poleward of 15°N and 15°S. But the pressure gradient in the upper atmosphere is not uniform, and two zones of concentrated westerly flow occur: one in the subtropics and one along the Polar Front. These concentrated, high-altitude, tube-like "rivers" of air are called **jet streams**.

A cross-sectional profile of the atmosphere between the equator and the North Pole (Fig. 10-7) reveals that these two jet streams—appropriately called the *subtropical jet stream* and the *Polar Front jet stream*—are located near the altitude of the tropopause (12 to 17 km [7.5 to 10.5 mi]). The diagram also shows the existence of a third jet stream, the *tropical easterly jet stream*, which

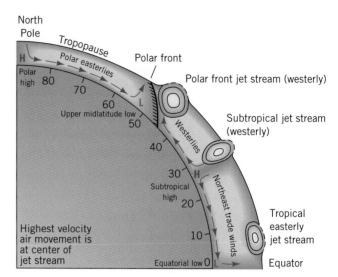

FIGURE 10-7 The atmosphere of the Northern Hemisphere in cross-section, with its three jet streams highlighted.

is a major feature of the opposite, east-to-west flow in the upper atmosphere of the equatorial zone south of 15°N. Interestingly, the tropical easterly jet stream occurs in the Northern Hemisphere only, whereas the subtropical and Polar Front jet streams exist in both hemispheres. The two latter jet streams are instrumental in moving large quantities of heated air from the equatorial to higher latitudes. They usually flow at extremely high rates of speed (in excess of 350 kph [220 mph]), and can thus achieve the heat and volume transfers that could not be accomplished at the far more moderate velocities associated with the cell circulations depicted in Fig. 10-2. It should also be pointed out that the subtropical and tropical jet streams are evident throughout the year, but the Polar Front jet stream is strong only during the half-year centered on winter. You would be correct in presuming that this jet stream must be related to the Polar High semipermanent pressure cell, but the details need not concern us here.

Another important generalization concerns the frequency of deviations from zonal windflow in the midlatitudes, particularly above the heart of North America. For a variety of rather complicated reasons, waves develop in the upper atmospheric pressure pattern. These alternating sequences of *troughs* (areas of low pressure) and *ridges* (areas of high pressure) cause the geostrophic wind to flow northwesterly and southwesterly around them (Fig. 10-8). Those deviations from westerly airflow are important because they reflect substantial *meridional* (north–south) air exchange. As you are aware, the fundamental cause of atmospheric circulation is a heat imbalance between the polar and tropical regions; these periods of meridional or **azonal flow** help correct that heat imbalance (Fig. 10-8). It is also worth noting that episodes of pronounced azonal flow produce unusual weather for the surface areas they affect. The next time

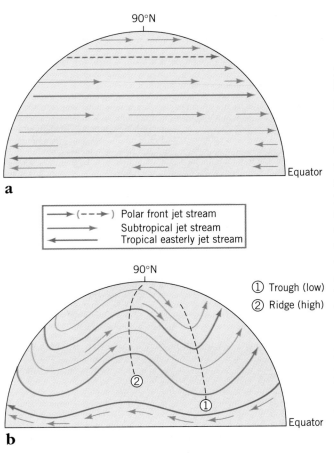

a

→ (---→) Polar front jet stream
→ Subtropical jet stream
← Tropical easterly jet stream

① Trough (low)
② Ridge (high)

b

you notice an unseasonable weather event, it is likely to be a result of waves developing in the upper atmospheric pressure pattern.

In a sense, it is misleading to separately treat the surface and upper atmospheric pressure patterns and their resultant windflows because they must always be interrelated. Indeed, strong surface pressure gradients are invariably reinforced by strong upper atmospheric pressure gradients. In truth, they are both the cause and effect of each other. But keep in mind that there is another component of this cause-effect relationship: the circulation of the world ocean. Although the effects of the ocean surface on the general circulation of the atmosphere have been discussed, the enormous influence of the circulating atmosphere on the ocean surface has yet to be examined. That is our next task.

FIGURE 10-8 Zonal (westerly) flow dominates in the upper atmosphere of the Northern Hemisphere (a). Azonal (meridional) flow (b) is characterized by pronounced troughs and ridges, and a generally stronger Polar Front jet stream. Azonal flow promotes warm-air movement to the north beneath the ridges and cold-air movement to the south beneath the troughs.

KEY TERMS

azonal flow (p. 112)
Equatorial Low (p. 106)
general circulation (p. 110)
Inter-Tropical Convergence Zone (ITCZ) (p. 106)

jet stream (p. 112)
monsoon (p. 110)
Northeast Trades (p. 107)
Polar Easterlies (p. 107)

Polar Front (p. 107)
Polar High (p. 107)
Southeast Trades (p. 107)
Subtropical High (p. 107)

Upper-Midlatitude Low (p. 107)
Westerlies (p. 107)
zonal flow (p. 112)

REVIEW QUESTIONS

1. List the seven semipermanent pressure belts of the surface atmospheric circulation and give their approximate locations.

2. List the six wind belts that connect these semipermanent highs and lows.

3. Discuss the shifting of these wind and pressure systems with the seasons of the year.

4. What are some of the main differences between the ideal

model and the actual pattern of surface atmospheric circulation?

5. Describe the mechanisms of the monsoonal circulation of South Asia.

6. Describe the zonal circulation pattern of the upper atmosphere, its relation to the jet streams, and why azonal flow occurs.

REFERENCES AND FURTHER READINGS

ATKINSON, B. W. *Meso-Scale Atmospheric Circulations* (Orlando, Fla.: Academic Press, 1981).

CHANG, J. *Atmospheric Circulation Systems and Climates* (Honolulu: Oriental Publishing Co., 1972).

FEIN, J. S., and STEPHENS, P. L., eds. *Monsoons* (New York: Wiley, 1987).

GEDZELMAN, S. D. *The Science and Wonders of the Atmosphere* (New York: Wiley, 1980), Chapter 16.

LORENZ, E. *The Nature and Theory of the General Circulation of the Atmosphere* (Geneva: World Meteorological Organization, 1967).

PALMÉN, E., and NEWTON, C. W. *Atmospheric Circulation Systems: Their Structure and Physical Interpretation* (New

York: Academic Press, 1969).

PERRY, A. H., and WALKER, J. M. *The Ocean–Atmosphere System* (London/New York: Longman, 1977).

RAMAGE, C. S. *Monsoon Meteorology* (New York: Academic Press, 1971).

REITER, E. R. *Jet Streams* (Garden City, N.Y.: Anchor/Doubleday, 1967).

RIEHL, H. *Introduction to the Atmosphere* (New York: McGraw-Hill, 3rd ed., 1978), Chapter 6.

WEBSTER, P. J. "Monsoons," *Scientific American*, August 1981, 109–118.

WEISMAN, S. R. "Worst Drought in Decades Hits Vast Area of India," *New York Times*, August 16, 1987, 8.

MULTI-CHANNEL SEA SURFACE TEMPERATURE °C (30-MAR-89 TO 5-APR-89)

ICE   -4 -2 0 2 4 6 8 10 12 14 16 18 20 22 24 26 28 30 32 34 36 38   LAND

Hydrosphere: Circulation of the World Ocean

OBJECTIVES

◆ To relate the surface oceanic circulation to the general circulation of the atmosphere.

◆ To describe the major currents that constitute the oceanic circulation.

◆ To demonstrate the role of oceanic circulation in the transport of heat at the earth's surface.

I n this unit, we focus on the large-scale movements of water, known as **ocean currents**, that form the oceanic counterpart to the atmospheric system of wind belts and semipermanent pressure cells treated in Unit 10. The two systems are closely integrated, and we will be examining that relationship in some detail. Ocean currents, by the way, affect not only the 71 percent of the face of the earth covered by the world ocean, but the continental landmasses as well. Therefore, we must become familiar with the oceanic circulation on a global basis, because it is vital to our understanding of the weather and climate of each part of the planet's surface.

SURFACE CURRENTS

Like the global atmospheric circulation above it, the world ocean is a significant transporter of heat from equatorial to polar regions. The oceans account for 13 percent of the total movement of heat from low to high latitudes; the atmosphere is responsible for the other 87

UNIT OPENING PHOTO: Global distribution of sea-surface temperatures (°C), ranging from very warm to near freezing, at their approximate equinox position.

percent. But in the broad zone between the tropics and the upper midlatitudes, both north and south of the equator, the oceans are estimated to account for up to 25 percent of the poleward heat movement (Fig. 11-1).

Specifically, it is through the circulation of water masses in large-scale currents that the world ocean plays its vital role in constantly adjusting the earth's surface heat imbalance. Although the sea contains numerous horizontal, vertical, and even diagonal currents at various depths, almost all of the oceanic heat-transfer activity takes place via the operation of horizontal currents in the uppermost 100 m (330 ft) of water. Thus most of our attention in this unit is directed toward the 10 percent of the total volume of the world ocean that constitutes this surface layer.

Although they transport massive volumes of water, most global-scale ocean currents differ only slightly from the surface waters through which they flow. So-called "warm" currents, which travel from the tropics toward the poles, and "cold" currents, which move toward the equator, usually exhibit temperatures that deviate by only a few degrees from those of the surrounding sea. Yet these temperature differences are often sufficient to markedly affect atmospheric conditions over a wide area. As demonstrated in our discussion of the maritime effect (p. 91), onshore winds blowing across warm currents pick up substantial moisture from the heightened evaporation of seawater, whose latent heat (p. 79) can generate rising currents of air.

In their rates of movement, too, most currents are barely distinguishable from their marine surroundings. Currents tend to move slowly and steadily, averaging only about 8 km (5 mi) per hour. They are often called **drifts** because they lag far behind the average speeds of surface winds blowing in the same general direction. Faster-moving currents are usually found only where narrow straits squeeze the flow of water, such as between Florida and Cuba or in the Bering Sea between Alaska and northeasternmost Russia. Slower-moving currents exist as well but are mainly confined to the deeper oceanic layers below 100 m (330 ft), where the friction caused by the high pressure of overlying water is much greater.

GENERATION OF OCEAN CURRENTS

Ocean currents can be generated in several ways. Sometimes water piles up along a coastline, yielding a slightly higher sea level than in the surrounding ocean. A good example is the tropical South Atlantic Ocean just south of the equator, where the landmass of northeastern Brazil protrudes well out to sea. When westward-flowing water piles up against this shore, gravity forces it back, forming an eastward-moving current along the equator (Fig. 11-2). The eastward rotation of the earth reinforces this piling-up phenomenon, which occurs to some degree along the western edge of every ocean basin. Alternatively, surface waters do not experience such squeezing along the eastern margins of an ocean, and water movement there is more diffuse.

Another source of oceanic circulation, which largely affects deeper zones below the surface layer, is variation in the density of seawater. Density differences can arise from temperature differences, as when the chilled surface water of the high latitudes sinks and spreads toward the equator, or from salinity differences. The ocean beneath a dry subtropical high-pressure zone is more saline than under an equatorial rain belt. The saltier water, being denser than the less saline water, tends to sink and give way to a surface current of lower salinity.

These influences notwithstanding, the leading generator of ocean currents is the frictional drag on the water surface set up by prevailing winds. Frictional drag transfers kinetic energy, the energy of movement, from the air to the water. Once set in motion, as is the case with

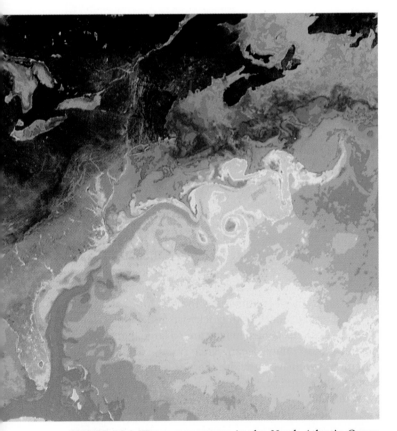

FIGURE 11-1 Water movements in the North Atlantic Ocean east of the United States as shown by variations in sea-surface temperatures. Dominating the circulation pattern is the northeastward flow of warm tropical water (indicated by reds, oranges, and yellows) from the near-tropical latitudes south of Florida toward the upper-middle latitudes. This warm-water flow carries across the entire North Atlantic and is a major influence on the atmospheric environment of Western Europe.

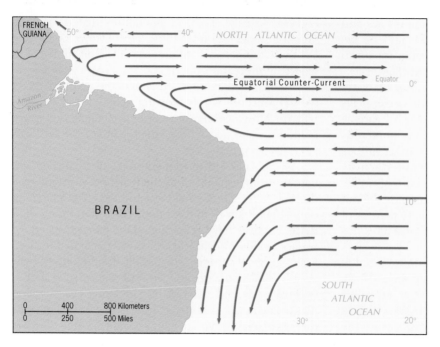

FIGURE 11-2 The pileup of ocean water against the northeastern coast of Brazil forces a return flow, generating the Equatorial Counter-Current, which moves eastward against the surrounding dominant westward current.

moving air, the water is subjected to the deflective Coriolis force. As a rule, when the prevailing wind blows over the ocean surface of the rotating earth, the Coriolis force steers the surface current to flow at an angle of about 45° to the right of the wind in the Northern Hemisphere (and at approximately 45° to the left in the Southern Hemisphere). This surface motion also influences the waters below to a depth of around 100 m (330 ft). Within this column, the motion in each underlying water layer is increasingly to the right (or left in the Southern Hemisphere) as depth increases, and exhibits a decreasing speed of flow.

FLOW BEHAVIOR OF OCEAN CURRENTS

In our discussion of oceanic circulation so far, we have for the most part been dealing with models that describe ideal situations. By this point in the book, of course, we are well aware that in nature things are more complex, and the currents of the world ocean are no exception. It would, therefore, be erroneous to presume that a large-scale ocean current is an unswerving river of water that follows the same exact path and exhibits constant movement characteristics. Deviations from the "norm" occur all the time. With the rapid expansion of oceanographic research based on satellite data since the 1960s, much has been learned about the detailed dynamics of surface currents. In many ways, their flow patterns (if not their speeds) resemble those of the Polar Front jet stream discussed in Unit 10.

Most ocean currents develop river-like *meanders*, or curves, which can become so pronounced (especially after the passage of storms) that many detach and form localized *eddies*, or loops, that move along with the general flow of water. These phenomena are most common along the boundaries of currents, where opposing water movements heighten the opportunities for developing whorl-like local circulation cells. Figure 11-3 diagrams such a situation involving the western edge of the warm Gulf Stream current off the Middle Atlantic coast of the United States.

Gyre Circulations

Prevailing winds, the Coriolis force, and sometimes the configuration of bordering landmasses frequently combine to channel ocean currents into cell-like circulations that resemble large cyclones and anticyclones. In the ocean basins these continuously moving loops are called **gyres**, a term used for both clockwise and counterclockwise circulations. Gyres, in fact, are so large that they can encompass an entire ocean. Since ocean basins are usually more extensive in width than in length, most gyres assume the shape of elliptical cells elongated in an east–west direction.

The ideal model of gyre circulation in the world ocean, shown in Fig. 11-4, displays a general uniformity in both the Northern and Southern Hemisphere. As with the model of the general circulation of the atmosphere, these hemispheric flow patterns are essentially mirror images of one another. Each hemisphere contains tropical and subtropical gyres; differences in high-latitude land/water configurations give rise to a fifth (subpolar) gyre in the Northern Hemisphere that is not matched in the Southern.

Subtropical Gyres

The **subtropical gyre** dominates the oceanic circulation of both hemispheres. In each case, the subtropical gyre circulates around the Subtropical High that is stationed above the center of the ocean basin (*see* Fig. 10-3). The two clockwise-circulating gyres of the Northern Hemisphere are found beneath two such high-pressure cells: the Pacific (Hawaiian) High over the North Pacific Ocean and the Bermuda (Azores) High centered above the North Atlantic Ocean. In the Southern Hemisphere, there are three subtropical gyres that each exhibit a counter-clockwise flow trajectory; these are located beneath the three semipermanent zones of subtropical high pressure,

respectively centered over the South Pacific Ocean, the South Atlantic Ocean, and the southern portion of the Indian Ocean.

The broad centers of each of these five gyres are associated with subsiding air and generally calm wind conditions, and are therefore devoid of large-scale ocean currents. The currents are decidedly concentrated along the peripheries of the major ocean basins, where they constitute the various segments or limbs of the subtropical gyre. Along their equatorward margins, the subtropical gyres in each hemisphere carry warm water toward the west. These currents diverge as they approach land. Some of the water is reversed and transported eastward along the equator as the Equatorial Counter-Current, but most of the flow splits and is propelled poleward as warm currents along the western edges of each ocean basin (Fig. 11-2).

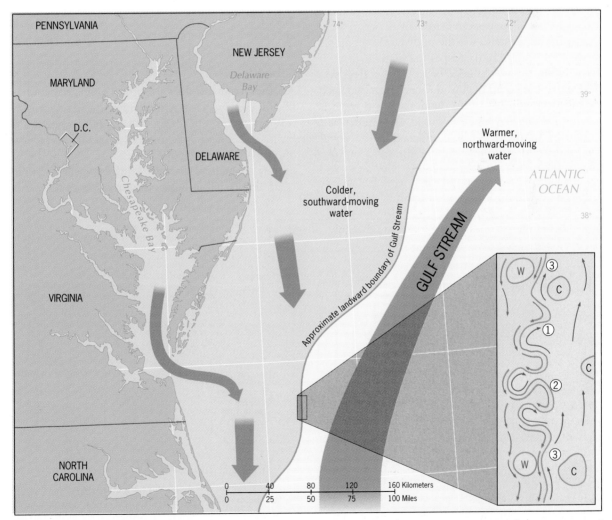

FIGURE 11-3 When ocean currents consisting of waters with contrasting temperatures make contact as they move in opposite directions, their boundaries become masses of swirling eddies. As the northward-moving Gulf Stream begins to move off the North American east coast, colder waters moving southward create such a situation. In this sketch of a part of the Middle Atlantic coastline, a section of the contact zone is enlarged (inset). At ①, the contact surface between the contrasting waters becomes indented. At ②, the swirling eddies create balloon-like protrusions of cold and warm water. At ③, these warm and cold masses are separated from the main bodies on the opposite sides of the boundary.

As Fig. 11-4 indicates, when polar waters are encountered in the upper midlatitudes, the currents swing eastward across the ocean. In the Northern Hemisphere, these eastward drifts remain relatively warm currents because the colder waters of the subpolar gyre to their north are largely blocked by landmasses from mixing with this flow. When they reach the eastern edge of the ocean, the now somewhat cooled waters of the subtropical gyre turn toward the equator and move southward along the continental coasts. These cool currents parallel the eastern margins of the ocean basins and finally converge with the equatorial currents to complete the circuit and once again form the westward-moving (and now rapidly warming) equatorial stream.

Gyres and Windflow

The side panels in Fig. 11-4 remind us that a subtropical gyre circulation is continuously maintained by the operation of the wind belts above it. The trade winds in the tropical segments of the eastern margin of an ocean work together with the Westerlies in the midlatitude segments of the western periphery to propel the circular flow. If you blow lightly across the edge of a cup of coffee, you will notice that the liquid begins to rotate; with two such reinforcing wind sources to maintain circular flow, it is no wonder that these gyres never cease. The circulations of the narrow **tropical gyres**, composed of the equatorial currents and returning counter-currents, are also reinforced by winds—the converging Northeast and Southeast Trades.

The circulatory interaction between the sea and overlying windflows is more complicated in the case of the Northern Hemisphere **subpolar gyres**, where landmasses and sea ice interrupt surface ocean flows. Along the southern limbs of these gyres, westerly winds drive a warm current across the entire ocean basin. This flow remains relatively warm because, as indicated above, the southward penetration of cold Arctic waters is largely blocked by continents, squeezing through only via the narrow Bering Strait in the northernmost Pacific and the slender channel separating Canada and Greenland in the northwestern North Atlantic. In the more open northeastern Atlantic between Greenland and Europe, a branch of the warm eastward-moving drift enters the subpolar gyre (Fig. 11-4), frequently driven across the Arctic Circle by a reinforcing southwesterly tail wind associated with storm activity along the Polar Front. The subpolar gyres are nonexistent in the Southern Hemisphere; instead, an eastward-moving cold current is propelled by the upper-midlatitude Westerlies that, in the absence of continental landmasses, girdle the globe.

Upwelling

Another feature of the subtropical gyre circulation is represented in Fig. 11-4 by the wider spacing of arrows representing currents along the eastern sides of ocean basins. This wide spacing indicates that surface waters are not squeezed against the eastern edge of the basins, as they are against the west. It also suggests the presence of an additional influence that reinforces the actions of

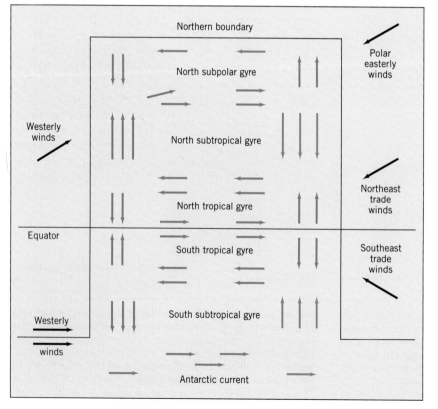

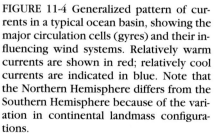

FIGURE 11-4 Generalized pattern of currents in a typical ocean basin, showing the major circulation cells (gyres) and their influencing wind systems. Relatively warm currents are shown in red; relatively cool currents are indicated in blue. Note that the Northern Hemisphere differs from the Southern Hemisphere because of the variation in continental landmass configurations.

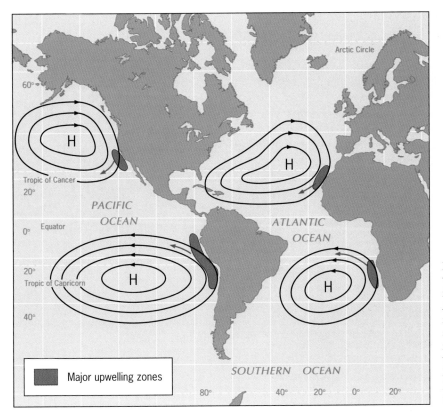

FIGURE 11-5 Four major cold-water upwelling zones, all adjacent to west coasts in the subtropical latitudes of the eastern Atlantic and Pacific ocean basins. The Coriolis force and high-pressure atmospheric systems drive waters offshore (red arrows), to be replaced by nutrient-rich, upwelling waters near the coast (indicated in light green). In each such location coastlines are arid, but offshore fishing industries are very productive.

prevailing winds—upwelling. **Upwelling** involves the rising of cold water from the ocean depths to the surface where the Coriolis force prompts ocean currents to diverge from continental coastlines. As warmer surface waters are transported out to sea they are replaced by this cold water, which lowers surface air temperatures and the local rate of evaporation. Not surprisingly, some of the driest coastal areas on earth are associated with upwelling, particularly in latitudes under the influence of the semipermanent subtropical high-pressure cells.

Figure 11-5 maps four such upwelling zones in the Pacific and Atlantic Oceans. Each subtropical west coast on the continent adjacent to the shaded upwelling zone experiences aridity. Much of coastal northwestern Mexico as well as northern Chile and Peru bordering the Pacific exhibit desert conditions. Moreover, in the areas of northwestern and southwestern Africa bordering the Atlantic upwelling zones, we find two of the driest deserts in the world—the Sahara and the Namib, respectively. Although it can result in desiccation on nearby coasts, upwelling does produce one important benefit for humans: it carries to the surface nutrients that support some of the most productive fishing grounds in the world ocean.

The Geography of Ocean Currents

The basic principles of oceanic circulation are now familiar to us, and they can be applied at this point to

the actual distribution of global-scale surface currents mapped in Fig. 11-6. As we know, these currents do respond to seasonal shifts in the wind belts and semipermanent highs and lows. However, those responses are minimal because seawater motion changes quite slowly and usually lags weeks or even months behind shifts in the atmospheric circulation above. The geographic pattern of ocean currents shown in Fig. 11-6, therefore, is based on the average annual position of these flows. But most currents deviate only slightly from these positions. On the world map, the only noteworthy departures involve the reversal of smaller-scale currents under the influence of monsoonal air circulations near the coasts of southern and southeastern Asia (*see* Fig. 10-4). Now let us briefly survey the currents of each major ocean basin.

Pacific Ocean Currents

The Pacific Ocean's currents closely match the model of gyre circulations displayed in Fig. 11-4. In both the Northern and Southern Hemisphere components of this immense ocean basin, surface flows are dominated by the subtropical gyres. In the North Pacific, the limbs of this gyre are constituted by the clockwise flow of the North Equatorial, Japan (Kuroshio), North Pacific, and California Currents. Because the Bering Strait to the north admits only a tiny flow of Arctic seawater to the circulation, all of this gyre's currents are warm except for the California Current. That current is relatively cold as a re-

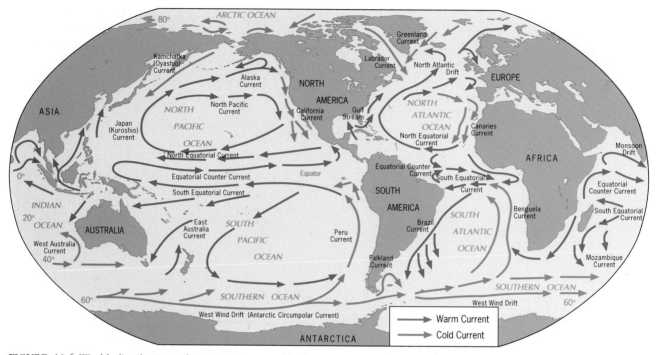

FIGURE 11-6 World distribution of ocean currents, showing average positions and relative temperatures in each of the ocean basins.

sult of upwelling and its distance from the tropical source of warm water. The lesser circulations of the tropical and subpolar gyres are also evident in the North Pacific. The tropical gyre comprises the low-latitude Equatorial Counter-Current and the North Equatorial Current. The subpolar gyre consists of the upper-midlatitude loop of the North Pacific, Alaska, and Kamchatka (Oyashio) Currents.

The South Pacific Ocean is a mirror image of the Northern Hemisphere flow pattern, except for the fully expected replacement of the subpolar gyre by the globe-encircling movement of the West Wind Drift (Antarctic Circumpolar Current), where the Pacific gives way to the Southern Ocean at approximately 45°S. The strong subtropical gyre that dominates the circulation of the South Pacific includes the South Equatorial, East Australian, West Wind Drift, and Peru Currents. The South Pacific's tropical gyre is constituted by the Equatorial Counter-Current and the South Equatorial Current.

Atlantic Ocean Currents

The Atlantic really consists of two ocean basins because of the hourglass-like narrowing between South America and Africa in the vicinity of the equator. Nevertheless, the overall pattern of Atlantic currents is quite similar to that of the Pacific. The subtropical gyre—composed of the North Equatorial Current, Gulf Stream, North Atlantic Drift, and Canaries Current—dominates circulation in the North Atlantic Ocean. The subpolar gyre, as noted above, is modified by sea ice and high-latitude land bodies, but

the rudiments of circular flow are apparent on the map. In the equatorial latitudes, despite the east–west proximity of continents, both tropical gyres have enough space to develop their expected circulations.

Proceeding into the South Atlantic Ocean, the subtropical gyre is again dominant, with the warm waters of the Brazil Current bathing the eastern shore of South America and the cold waters of the Benguela Current, reinforced by upwelling, paralleling the dry southwestern African coast. South of 35°S the transition to the Southern Ocean and its West Wind Drift is identical to the pattern of the South Pacific.

Indian Ocean Currents

The Indian Ocean's circulatory system is complicated by the configuration of surrounding continents, with an eastward opening to the Pacific in the southerly low latitudes and the closure of the northern part of the ocean by the South Asian coast. Still, we can observe some definite signs of subtropical gyre circulation in the Indian Ocean's Southern Hemisphere component. Moreover, a fully developed pair of tropical gyres in the latitudinal zone straddling the equator operates for a good part of the year. During the remaining months, as is explained in the discussion of the wet monsoon in Unit 10, the Inter-Tropical Convergence Zone (ITCZ) is pulled far northward onto the Asian mainland. This temporarily disrupts the air and even the oceanic circulations that prevail north of the equator between October and June (*see* Fig. 10-4).

DEEP-SEA CURRENTS

As we hinted earlier in this unit, significant water movements occur below the surface layer of the sea, which extends to a depth of about 100 m (330 ft). In fact, nothing less than another complete global system of water flows exists below that level, where the bulk (approximately 90 percent) of the world ocean's water lies. This deep-sea system of oceanic circulation operates in sharp contrast to the surface system, which interfaces with the atmosphere and is largely driven by prevailing winds in tandem with the Coriolis force.

The deep-sea system of oceanic movement can be categorized as a **thermohaline circulation**, because it is controlled by differences in the temperature and/or salinity of water masses. Thermohaline circulation involves the flow of currents driven by differences in water density. Because of the greater pressure of overlying water below 100 m (330 ft), increased frictional resistance acts to substantially slow the speed of currents. Another factor that makes deep-sea currents much slower than their surface-layer counterparts is that the latter are strictly horizontal whereas the former are much likelier to exhibit vertical motion.

The temperature and salinity differences that trigger thermohaline circulation are generated at the ocean surface in the high-latitude wind belts. Water density gradients high enough to spawn deep-sea currents are developed by the actions of two related processes. One process entails the sinking of surface water, which gets colder, and therefore denser, when it is in contact with polar-area air temperatures. The other is the freezing of surface seawater, which increases the salinity—and density—of the water just under the ice (which consists mainly of nonsaline, fresh water).

The generalized world map of deep-sea circulation is shown in Fig. 11-7. The dominant flow pattern is one of water masses moving from high to lower latitudes. Note that each ocean basin has its own deep-water circulation, with the only interoceanic exchanges occurring in the depths of the Southern Ocean. Note, too, that each polar oceanic zone "feeds" different oceans: the deep-sea currents of the North and South Atlantic emanate from the Arctic Ocean, while those of the Pacific and Indian Oceans are generated in the waters surrounding Antarctica.

THE OCEAN–ATMOSPHERE SYSTEM

This unit has discussed numerous interactions between the world ocean and the atmosphere, and by now it should be evident that a two-way relationship exists. The atmosphere creates surface currents and, indirectly, deep-sea circulation. At the same time, the ocean affects the atmosphere in several ways. Two of the most important of those influences are the transport of heat from the equator toward the poles, and the capacity of the world ocean to store huge amounts of heat.

We began this unit by emphasizing that the oceans reinforce the processes of the general atmospheric circulation by transporting significant quantities of heat

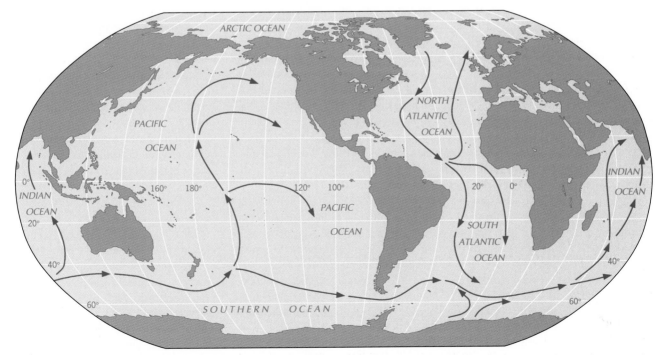

FIGURE 11-7 The global deep-sea circulation pattern, at a depth of about 4000 m (13,000 ft). The general flow of water masses is from higher toward lower latitudes. Note that the interchange of water between ocean basins occurs only in the Antarctic-girdling Southern Ocean.

El Niño–Southern Oscillation

A vital component in the general circulation of the atmosphere is the rising motion of air in the warm tropics near the equator. Since nearly 80 percent of the surface that the equator traverses is ocean, significant fluctuations in the temperature of that sea surface could temporarily modify the pattern of semipermanent low-pressure cells that mark the ITCZ in these latitudes. This in turn could produce short-term, global-scale changes in weather patterns capable of disrupting human activities.

Such linkages within the ocean–atmosphere system are no longer the stuff of scientific speculation: recent climatological research has confirmed both their existence and their importance in the functioning of the global environment. Much of this work has focused on sea-surface warming in the largest stretch of equatorial water—the vast central Pacific that sprawls across 160 degrees of longitude between Indonesia and South America.

Intermittent anomalies in seawater temperature off the coasts of Peru and Ecuador in northwestern South America have long been known, but their larger significance became apparent only during the 1980s. The warming of coastal Pacific waters there by about 2°C (4°F) is, in fact, a yearly occurrence that temporarily reduces the fish catch when a local southward-drifting warm current suppresses the usually present upwelling (*see* Fig. 11-5) that transports nutrients crucial to the surface-layer food chain. Because this three-month-long phenomenon usually arrives around Christmas time, it is called **El Niño** (''the child'') in honor of the Christ Child. But once every two to seven years (most recently in 1982–1983, 1986–1987, and 1992–1994), a much more pronounced El Niño develops.

This abnormal El Niño—which can linger up to three years—is accompanied by an expanded zone of warm coastal

waters. The upwelling of cold water associated with the Peru Current ceases, and the absence of nutrients results in massive fish kills and the decimation of the bird population (which must feed on the fish). In addition, Ecuador and Peru experience increases in rainfall that can result in crop losses as well as flooding in the heavily populated valleys of the nearby Andes Mountains. These localized El Niño effects, in fact, are but one symptom of a geographically much wider anomaly in the relationship between the equatorial ocean and the atmosphere. The cause of this abnormal El Niño is rooted in the temporary reversal of surface sea currents and airflows throughout the Pacific's equatorial zone.

The eastern portion of that oceanic zone is normally an area of high atmospheric pressure, because the upwelling Peru and California Currents converge (Figs. 11-5 and 11-6), and the cool surface water creates a condition of stability that inhibits the air from rising to form the equatorial low-pressure trough. These cool waters then move westward into the Pacific as the Equatorial Current, propelled not only by the converging trade winds but also by a strong surface airflow from the eastern Pacific high toward the semipermanent low in the western equatorial Pacific, which is positioned over Indonesia and northern Australia. As Fig. 11-8a shows, a cell of air circulation forms above the equator, with air rising above the western low, flowing eastward at high altitude, and subsiding over the eastern Pacific.

For reasons that are still not fully understood, during pronounced El Niño episodes there is a collapsing of both this pressure difference (between the eastern Pacific high and the western Pacific low) and the resultant westward surface windflow. (These corresponding atmospheric events are called the **Southern Oscillation**.) There then occurs a swift reversal in the flow of equatorial water

and wind—known in combination as **ENSO** (El Niño–Southern Oscillation)—as the midoceanic circulation cell now operates in the opposite direction (Fig. 11-8b). Most importantly, the piled-up warm water in the western Pacific surges back to the east as the greatly enhanced Equatorial Counter-Current, and the eastern Pacific equatorial zone is now overwhelmed by water whose temperatures can be as much as 8°C (14°F) higher

than normal. Moreover, as Fig. 11-8b shows, these events are accompanied by a subsurface invasion of warm water, which makes the Peruvian upwelling flow warm and reinforces the anomalous heating of the ocean surface.

The effects of ENSO are now known to spread so far beyond the equatorial Pacific that climatologists today rank the phenomenon as a leading cause of disturbance in global weather patterns. The El Niño of 1982–1983 was the strongest ever observed, with sea-surface warming between Indonesia and Peru double the expected ENSO temperature anomaly. Reports of severe weather abnormalities soon flowed in from around the world and fell into two categories. Heavy rains and disastrous flooding occurred in Ecuador and Peru, the islands of the central Pacific, and the westernmost portion of the United States. The other abnormality caused by El Niño was drought, which occurred in northern India (where the wet monsoon rains failed), interior Australia, Central America, Indonesia, and southern Africa. For more information on El Niño and its effects, see the *Earth* Magazine article on pp. *Earth* 2–7.

ENSO research efforts have multiplied in the 1990s, and much is being learned. Among the most important new findings is that extratropical Pacific sea-surface temperatures can remain elevated years after an El Niño event has ended. Satellite imagery has demonstrated that the 1982–1983 El Niño produced so much eastward-moving warm water that a sizeable mass of it in the northern tropics ricocheted off the North American landmass. Subsequently, this huge pool of water migrated slowly northwestward and was still evident a decade later, thousands of miles away in the midlatitudes east of Japan. Since that part of the northern Pacific is a spawning ground for North American weather systems, there may be a connection between this pool and the extreme weather events that have plagued the United States since the mid-1980s.

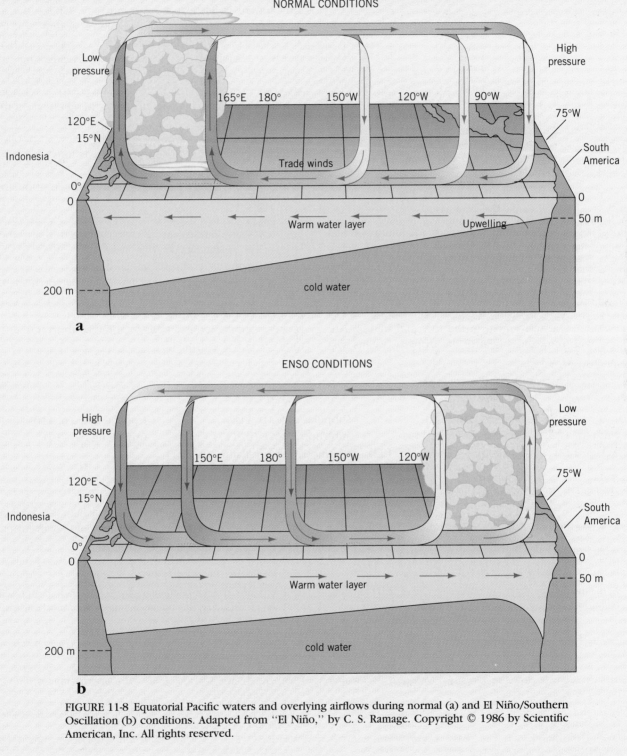

FIGURE 11-8 Equatorial Pacific waters and overlying airflows during normal (a) and El Niño/Southern Oscillation (b) conditions. Adapted from ''El Niño,'' by C. S. Ramage. Copyright © 1986 by Scientific American, Inc. All rights reserved.

TABLE 11-1 AVERAGE ANNUAL HEAT FLOW AT THE EARTH'S SURFACE IN UNITS OF 10^{14} CALORIES PER SECOND

LATITUDE	TOTAL	OCEAN (%)	ATMOSPHERE (%)
60°	7.6	0.7 (9.2)	6.9 (90.8)
50°	8.2	1.3 (15.9)	6.9 (84.1)
40°	12.0	1.8 (15.0)	10.2 (85.0)
30°	11.0	2.1 (19.1)	8.9 (80.9)
20°	8.4	1.3 (15.5)	7.1 (84.5)
10°	4.6	−0.3 (—)	4.9 (106.5)
Totals	51.8	6.9 (13.3)	44.9 (86.7)

Source: Adapted from M. I. Budyko, *The Heat Balance at the Earth's Surface* (Washington, D.C.: U.S. Department of Commerce, Office of Technological Services, 1958).

from low to higher latitudes. Table 11-1 indicates that almost seven-eighths of this poleward movement of heat is accounted for by the atmosphere and the remaining one-eighth by the sea. However, the breakdown shown in the table reveals that the oceans transport considerably more heat at certain latitudes, and it has been estimated that the sea is responsible for as much as 25 percent of the global heat movement in the low and middle latitudes. With the excess heat of the tropical oceans sys-

tematically dissipated by its poleward transport, the world ocean exhibits a general pattern of temperature decline as latitude increases (*see* Fig. 8-11).

A number of climatologists today assert that the single most important influence of the world ocean on the atmosphere is its heat storage capacity. Unlike the atmosphere, which is far less capable of storing heat for any length of time, the sea is known to act as a vast heat reservoir. Scientists are now closely investigating this component of the ocean–atmosphere system, and some are concerned that if this excess heat were to be released rapidly it could have a significant warming effect on the global climate.

We have already learned that the periodic pooling of warmer water can cause some far-reaching temporary changes in atmospheric circulation. The cyclical equatorial Pacific temperature abnormality, known popularly as *El Niño*, has received particular attention since its disastrous episode of 1982–1983 (*see* Perspective: El Niño–Southern Oscillation). It is also believed that other large-scale anomalies of this type may affect global weather patterns, and research continues.

On a normal day-to-day basis, the latent heat released in the evaporation of seawater helps to power the general circulation of the atmosphere. That evaporative process simultaneously adds vital moisture to the air, a topic explored in Unit 12.

KEY TERMS

drift (p. 115)

El Niño (p. 122)

ENSO (p. 122)

gyre (p. 116)

ocean current (p. 114)

Southern Oscillation (p. 122)

subpolar gyre (p. 118)

subtropical gyre (p. 117)

thermohaline circulation (p. 121)

tropical gyre (p. 118)

upwelling (p. 119)

REVIEW QUESTIONS

1. Describe the ways in which ocean currents develop.

2. What is the relationship between the subtropical gyres and the overlying atmospheric circulation?

3. Name the major currents of the Pacific Ocean and their general flow patterns.

4. Name the major currents of the North and South Atlantic Oceans and their general flow patterns.

5. Describe the general pattern of deep-sea currents. What factors influence those currents?

6. Briefly describe what is meant by El Niño/Southern Oscillation, and discuss this phenomenon's major mechanisms.

REFERENCES AND FURTHER READINGS

"Changing Climate and the Oceans," *Oceanus*, 29 (Winter 1986–1987), special issue.

COUPER, A. D., ed. *Atlas and Encyclopedia of the Sea* (New York: Harper & Row, 1989).

GROVES, D. G., and HUNT, L. M., eds. *Ocean World Encyclopedia* (New York: McGraw-Hill, 1980).

HARVEY, J. *Atmosphere and Ocean: Our Fluid Environments* (New York: Crane Russak, 1978).

KING, C.A.M. *Oceanography for Geographers* (London: Edward Arnold, 1962).

PERRY, A. H., and WALKER, J. M. *The Ocean–Atmosphere System* (London/New York: Longman, 1977).

PHILANDER, G. *El Niño, La Niña and the Southern Oscillation* (Orlando, Fla.: Academic Press, 1989).

RAMAGE, C. S. "El Niño," *Scientific American*, June 1986, 76–85.

STEVENS, W. K. "Effects of El Niño Reach Across Ocean and Linger a Decade," *New York Times*, August 9, 1994, B–7.

STEWART, R. W. "The Atmosphere and the Ocean," *Scientific American*, September 1969, 76–86.

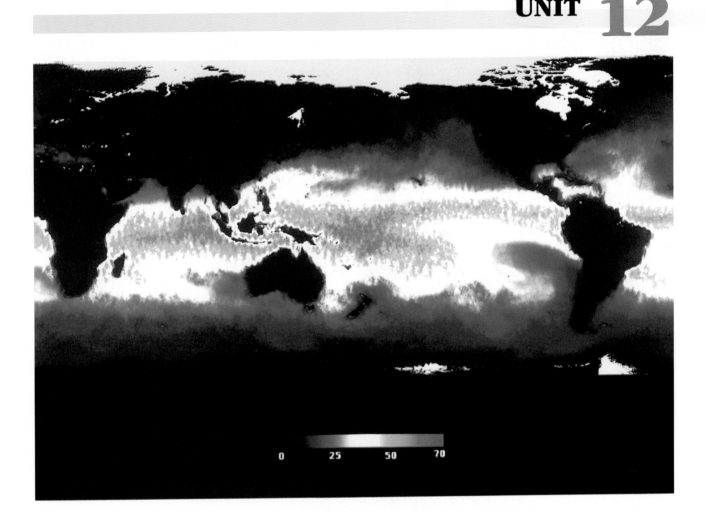

| 0 | 25 | 50 | 70 |

Atmospheric Moisture and the Water Balance

OBJECTIVES

◆ To discuss the various forms of water and to understand the important heat transfers that accompany changes of these physical states.

◆ To explain the various measures of atmospheric humidity, how they are related, and the processes responsible for condensation.

◆ To outline the hydrologic cycle and the relative amounts of water that flow within this cycle.

◆ To introduce the concept of precipitation.

◆ To describe the earth's surface water balance and its variations.

O ur physical world, as we are discovering, is characterized by energy flows and mass transfers. Matter and energy are never destroyed; they continually pass from one place to another. One of the best examples of such cyclical motion is the flow of water on the earth. The *hydrosphere* encompasses the global water system, whose flows occur in the world ocean, on and within the land surface, and in the atmosphere. Unit 11 treats the oceans, and in Part Five we examine water movements on and beneath the ground.

In this unit, we focus on water in the atmosphere and its relationships with the earth's surface. Our survey considers the continual movement of water among var-

UNIT OPENING PHOTO: A global view of water vapor in the atmosphere at the height of the Northern Hemisphere winter, mapped directly by satellite-based microwave sensors.

ious earth spheres that leads to a balance of water on the earth's surface. This circulation of water is powered by radiant energy from the sun, a form of heat whose inflows and outflows also balance at the planetary surface.

PHYSICAL PROPERTIES OF WATER

Human bodies are 70 percent water. Each of us requires 1.4 liters (1.5 quarts) of water a day in order to survive, and our food could not grow without it. Water is everywhere. It constitutes 71 percent of our planet's surface. We breathe it, drink it, bathe in it, travel on it, and enjoy the beauty of it. We use it as a raw material, a source of electrical power, a coolant in industrial processes, and a medium for waste disposal.

The single greatest factor underlying water's widespread importance is its ability to exist in three physical states (shown in Fig. 12-1) within the temperature ranges encountered near the earth's surface. The solid form of water, ice, is composed of molecules linked together in a uniform manner. The bonds that link molecules of ice can be broken by heat energy. When enough heat is applied, ice changes its state and becomes the liquid form we know as water. The molecules in the liquid are not arranged in an evenly spaced pattern but exist together in a random form. In the liquid state, the individual molecules are freer to move around. The introduction of additional heat completely frees individual molecules from their liquid state, and they move into the air. These airborne molecules now become a gas called **water vapor**.

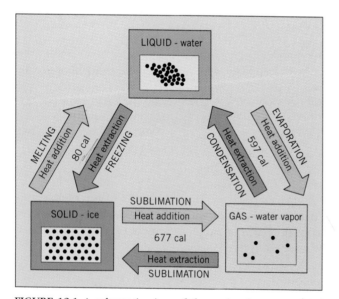

FIGURE 12-1 A schematic view of the molecular structure of water in its three physical states and the heat-energy exchanges among those states. The latent heat-exchange numbers between the arrows are explained in the text.

Water is so common and consistent in its physical behavior that we use it as a reliable measure of heat. We say that 1 *calorie* (cal) is the amount of heat energy required to raise the temperature of 1 gram (g) (0.04 oz) of water by 1°C (1.8°F). It takes about 80 cal to change 1 g of water from the solid state to the liquid state, a process we call **melting**. When it was first discovered that 80 cal of heat was needed to break the molecular bonds in solid water, the required heat appeared to be hidden, or *latent*. Thus the heat involved in melting is called the **latent heat of fusion**. Similarly, it takes 597 cal to change the state of 1 g of water at 0°C (32°F) from a liquid to a gas. This change is called **evaporation** or *vaporization*, and the heat associated with it is known as the **latent heat of vaporization**. Sometimes ice can change directly into water vapor. In this process, called **sublimation**, the heat required (677 cal) is the sum of the latent heats of fusion and vaporization.

One of the beauties of these physical processes is that they are completely reversible. Water vapor can change back into water in the **condensation** process; water can change into ice through **freezing**; and water vapor can change directly into ice as well, a process that is also called sublimation (or *deposition*). In reversing these processes, the identical quantities of latent heat are given off.

MEASURING WATER VAPOR

Lord Kelvin, the inventor of the absolute temperature scale, once said that we do not know anything about anything until we can measure it. How, then, do we measure water in its three physical states? The measurement of solid ice and liquid water is quite straightforward—we simply weigh them. We also employ this method indirectly when we measure the vapor pressure of water vapor in a column of air. (*Vapor pressure* is the pressure exerted by the molecules of water vapor in air.) Other measurements of water vapor are also of value, the most useful of which are relative humidity, specific humidity, and the mixing ratio.

To use any of these methods of measurement, water vapor must first be condensed into liquid water. One way of condensing water vapor is to cool it together with the surrounding air. *Saturated air* is air that is holding all the water vapor molecules it can possibly contain at a given temperature. Condensation takes place when a parcel of saturated air is cooled further.

Condensation often happens around the surface of a cold soft-drink bottle or can. The air in contact with the container is cooled. Because cooler air holds less water vapor than warmer air, the saturation level is soon reached, and the excess vapor in the cooled air condenses into liquid water droplets on the container. The same process occurs when the earth cools at night, subsequently cooling and saturating the air next to it. The excess water vapor beyond the saturation level contained in such a layer of air condenses into fine water droplets, on surfaces at and near the ground, that we call **dew**. Accordingly, the temperature at which air becomes saturated, and below which condensation occurs, is called the **dew point**.

Relative Humidity

Relative humidity tells us how close a given parcel of air is to its dew point, or saturation level. Consequently, we can define **relative humidity** as the proportion of water vapor present in a parcel of air relative to the maximum amount of water vapor that air could hold at the same temperature. Relative humidity is expressed as a percentage, so that air of 100 percent relative humidity is saturated and air of 0 percent relative humidity is completely dry. Relative humidity, which is dependent on the air temperature, often varies in opposition to that temperature. Relative humidity is usually lower in early and midafternoon when the diurnal temperature reaches its high, because the warmer the air the more water vapor it can hold. At night, when the temperature falls, the colder air holds less water vapor, and so the air approaches its dew point and relative humidity is higher.

Specific Humidity and the Mixing Ratio

The other measurements used to assess the amount of water vapor in the air need less explanation. *Specific humidity* is the ratio of the weight (mass) of water vapor in the air to the combined weight (mass) of the water vapor plus the air itself. The *mixing ratio* is the ratio of the mass of water vapor to the total mass of the dry air containing the water vapor.

The relative humidity, the mixing ratio, and the specific humidity may be found by using a *psychrometer*, an instrument with two thermometers. The bulb of one thermometer is surrounded by a wet cloth; water from the cloth evaporates into the air until the air surrounding the bulb is saturated. The evaporation process results in cooling, and that thermometer reaches a temperature called the *wet-bulb temperature*. The other thermometer, not swaddled in cloth, indicates the *dry-bulb temperature*. The difference in temperature between the wet-bulb and dry-bulb thermometers is then computed, and the relative humidity and/or the mixing ratio is then determined by referring to the appropriate set of published humidity tables.

Now that we are armed with some basic terminology and an understanding of the ways in which water changes from one physical state to another, we can proceed to find out how it circulates through the earth system. The natural cycle describing this circulation is called the hydrologic cycle.

THE HYDROLOGIC CYCLE

Early scientists believed that the wind blew water from the sea through underground channels and caverns and into the atmosphere, removing the salt from the water in the process. Today's scientists talk in terms of the **hydrologic cycle**, whereby water continuously moves from the atmosphere to the land, plants, oceans, and freshwater bodies and then back into the atmosphere. The hydrologic cycle model consists of a number of stages, with Fig. 12-2 showing the relative amounts of water involved in each:

1. The largest amounts of water transferred in any segment of the total cycle are those involved in the direct evaporation from the sea to the atmosphere and in precipitation back to the sea. As noted previously, evaporation is the process by which water changes from the liquid to the gaseous (water vapor) form. **Precipitation** includes any liquid water or ice that falls to the surface through the atmosphere.

2. The passage of water to the atmosphere through leaf pores is called *transpiration*, and the term *evapotranspiration* encompasses the processes by which water evaporates from the land surface and plants. Evapotranspiration combines with the precipitation of water onto the land surface to play a quantitatively smaller, but possibly more important, part in the hydrologic cycle.

3. If surplus precipitation at the land surface does not evaporate, it is removed via the surface network of streams and rivers, a phenomenon called **runoff**. In Fig. 12-2, the runoff value includes some water that *infiltrates* (penetrates) the soil and flows beneath the surface, eventually finding its way to rivers and the ocean.

The hydrologic cycle can readily be viewed as a closed system, in which water is continuously moved among the component spheres of the earth system. For example, the values in Fig. 12-2 indicate that the amount of water transported in the atmosphere over the continents equals the amount transported by surface runoff back to the ocean. Water circulates between the lower atmosphere, the upper lithosphere, the plants of the biosphere, and the oceans and freshwater bodies of the hydrosphere. The system can also be split into two subsystems, one consisting of the precipitation and evaporation over the oceans, and the other involving evapotranspiration and precipitation over land areas. The two subsystems are linked by horizontal movement in the atmosphere (known as *advection*) and by surface runoff flows.

The time required for water to traverse the full hydrologic cycle can be quite brief. A molecule of water can pass from the ocean to the atmosphere and back

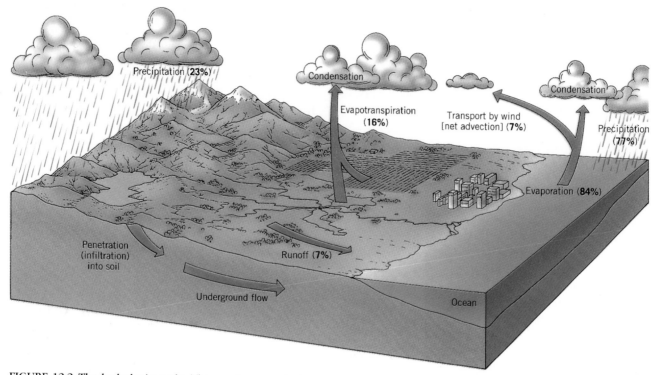

FIGURE 12-2 The hydrologic cycle. The numbers attached to each stage of the cycle show the percentage of the total water annually circulating in the system that is involved in any single stage. Surface flow and underground flow are considered as land runoff.

again within a matter of days. Over land, the cycle is less rapid. Groundwater goes into the soil or subsurface and can remain there for weeks, months, or years. The circulation is even slower where water in the solid form of ice is concerned. Some water has been locked up in the major icesheets and glaciers of the world for many thousands of years.

Water is quite unequally distributed within the hydrosphere, as Fig. 12-3 reminds us. The world ocean contains 97 percent of all terrestrial water, but the high salt content makes it of little direct use to people. Of the remaining 3 percent that is constituted by the world's freshwater supply, three-quarters is locked up in icesheets and glaciers. The next largest proportion of fresh water, about one-seventh, is accessible only with difficulty because it is groundwater located below 750 m (2500 ft).

Therefore, the fresh water needed most urgently for our domestic, agricultural, and industrial uses must be taken from the relatively small quantities found at or near the surface—in the rivers, lakes, soil layer, and atmosphere. These storage areas contain less than one two-hundredth of the 3 percent of fresh water in the hydrosphere, and that finite supply is increasingly taxed by growing human consumption and abuse (*see* Perspective: Water Usage in the United States). Moreover, this very modest amount of fresh water must keep circulating within the hydrologic cycle through evaporation, condensation, and precipitation.

EVAPORATION

We cannot see evaporation occurring, but its results are sometimes visible. When you see mist rising from a lake that is warm in comparison to the cold air above it, you are seeing liquid water droplets that have already condensed. It is not too hard to imagine molecules of invisible water vapor rising upward in the same way.

Conditions of Evaporation

Evaporation occurs when two conditions are met. First, heat energy must be available at the water surface to change the liquid water to a vapor. This heat is sometimes provided by the moving water molecules, but most often the radiant heat of the sun, or both sources together, provides the necessary heat energy.

The second condition is that the air must not be saturated—it must be able to absorb the evaporated water molecules. Air near a water surface normally contains a large number of vapor molecules. Thus the vapor pressure, caused by the density and movement of the vapor molecules, is high. But air at some distance from the water surface has fewer molecules of vapor and, consequently, a lower vapor pressure. In this case we say there is a *vapor-pressure gradient* between the two locations. A vapor-pressure gradient exists above a water surface as long as the air has not reached its saturation level (dew point). Just as people tend to move from a very crowded

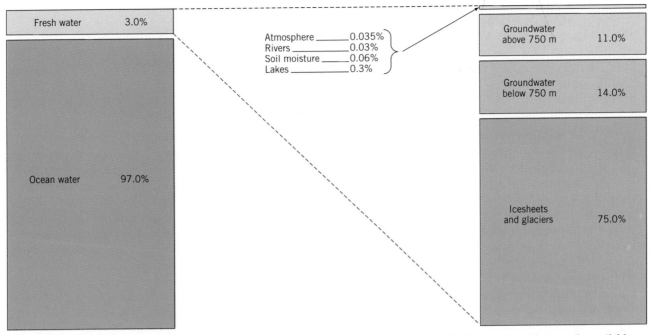

FIGURE 12-3 The distribution of water in the hydrosphere. The column on the right shows the percentage distribution of the 3 percent of total hydrospheric water that is fresh. Of that fresh- water component, only about one-tenth is easily available to humans.

room to a less crowded one, molecules of water vapor tend to move along the vapor-pressure gradient to areas of less pressure, which usually means moving higher in the atmosphere.

Knowing the requirements for evaporation—a heat source and a vapor-pressure gradient—we can infer the kinds of situations that would yield maximum evapora- tion. Because radiation from the sun is a leading heat source, large amounts of evaporation can be expected where there is a great deal of sunlight. This is particularly true of the tropical oceans. Water molecules also move most easily into dry air because of the sizeable vapor- pressure gradient. So the drier the air the more evapo- ration will occur, as can be seen by the high values associated with the subsiding air of the subtropical high- pressure zones on the world map of latent heat loss (Fig. 7-7, p. 81). Evaporation is even faster in windy condi- tions, when the air containing new vapor molecules can be continually replaced.

Evapotranspiration

We have seen that most of the evaporation into the at- mosphere occurs over the ocean. Over land, water evap- orates from lakes, rivers, damp soil, and other moist surfaces; the water that plants lose to the air during photosynthesis (transpiration) is another major source. As we know, **evapotranspiration** is the passage of mois- ture from the land surface to the atmosphere through the combined processes of evaporation and transpiration.

Physical geographers draw a distinction between potential evapotranspiration and actual evapotranspira- tion. **Potential evapotranspiration (PE)** is the maxi- mum amount of water that can be lost to the atmosphere from a land surface with abundant available water. **Actual evapotranspiration (AE)** is the amount of wa- ter that can be lost to the atmosphere from a land sur- face with any particular soil-moisture conditions. AE can equal PE when the land surface is saturated, but when the soil moisture is less than its maximum value, AE is usually less than PE.

Now that water has entered the atmosphere, let us see how it makes its way through this component of the hydrologic cycle and moves back to the surface. We be- gin with the formation of clouds and then trace the de- velopment of precipitation.

CONDENSATION AND CLOUDS

Clouds are visible masses of suspended, minute water droplets or ice crystals. Two conditions are necessary for the formation of clouds:

1. The air must be saturated, either by cooling below the dew point (causing water vapor to condense) or by evaporating enough water to fill the air to its maxi- mum water-holding capacity. Parcels of air may cool enough to produce condensation when they rise to the higher, cooler parts of the atmosphere, or when they come into contact with colder air or a colder sur- face.

Perspectives on THE HUMAN ENVIRONMENT

*Water Usage in the United States**

In the United States today, water is consumed in prodigious quantities at a daily rate exceeding 4900 l (1300 gal) per person. That figure represents a reduction of about 10 percent over the past two decades, thanks to conservation efforts and slowed economic growth. But the American level of water usage is still three times higher than in the countries of Western Europe and hundreds of times higher than in most developing nations. Given this massive rate of consumption, we—and the rest of humankind—are facing a critical shortage of usable fresh water in the foreseeable future. A geographic and sectoral breakdown of the problem sheds light on where further changes will be necessary.

The spatial distribution of surface and underground water withdrawal is shown in Fig. 12-4. Surprisingly, the heavily populated (and industrialized) Northeast and Midwest are comparatively moderate water consumers. Much greater regional water use—and overuse—occurs in the drier western United States, most notably in the croplands of California and Idaho where there is very heavy reliance on irrigation systems.

Among the major sectors of the economy, *agriculture* accounts for about 40 percent of national water withdrawal; these supplies create the unusually high productivity of the 5 percent of U.S. farmlands that are irrigated, especially in California where they claim almost 80 percent of the state's daily water consumption. Another 20 percent of the nation's withdrawals are about evenly divided between the *domestic* and *industrial* sectors; both have recorded declines since 1985, led by heavy industries whose manufacturing processes were altered to cut back on water intake. The remaining 40 percent is withdrawn by the energy sector to produce *electricity*, mainly by converting heated water to steam for use in generators; since most of that electrical power is subsequently consumed by the domestic and industrial sectors, their final totals are obviously greater than the modest 20 percent indicated above.

*The source for most of this box is "Water: The Power, Promise, and Turmoil of North America's Fresh Water," *National Geographic*, special issue, November 1993.

FIGURE 12-4 Water withdrawal from surface and underground sources in the conterminous United States. White represents the highest levels, and dark blue the lowest. This computer map was prepared by the U.S. Geological Survey in 1993.

2. There must be a substantial quantity of small airborne particles called **condensation nuclei**, around which liquid droplets can form when water vapor condenses. Condensation nuclei are almost always present in the atmosphere in the form of dust or salt particles.

The greater the moisture content of cooling air, the greater the condensation and the development of the cloud mass. Although most clouds form and remain at some elevation above the earth's surface, they can come into direct contact with the surface; we call such clouds **fog** (Fig. 12-5).

FIGURE 12-5 The morning sun "burning off" the ground fog that shrouds this portion of the rural landscape in western New York State.

Besides being the source of all precipitation, clouds play a key role in the atmosphere's heat balance. Clouds both reflect some incoming shortwave radiation back to space at their tops and scatter another part of this incoming solar radiation before it can strike the surface directly (*see* Fig. 7-1, p. 73). At the same time, clouds absorb part of the earth's longwave radiation and reradiate it back toward the land and the sea.

Cloud Classification

Cloud-type classification, a common practice in meteorology and climatology, is based on the criteria of general structure, appearance, and altitude (Fig. 12-6).

One major cloud-type grouping encompasses *stratus* clouds. As this term implies, stratus clouds are layer-like in appearance; they also are fairly thin, and normally cover a wide geographic area. Stratus clouds are classified according to their altitude. Below 3 km (1.9 mi or 10,000 ft), they are simply called stratus clouds—or *nimbostratus* if precipitation is occurring. Between 3 and 6 km (3.8 mi or 20,000 ft), they are designated *altostratus* clouds, and above 6 km, *cirrostratus* clouds.

A second major cloud-type category involves *cumulus* clouds, which are thick, puffy, billowing masses that often develop to great heights (Fig. 12-6). These clouds are also subclassified into the same lower, middle, and upper altitudinal levels, proceeding in ascending order through *cumulus* or *stratocumulus*, *altocumulus*, and *cirrocumulus*. Very tall cumulus clouds, extending from 500 m (1600 ft) at the base to about 12 km (7.6 mi or 40,000 ft) at their anvil-shaped heads, are called *cumulonimbus*. These are often associated with violent weather, including heavy rain, high winds, lightning, and thunder.

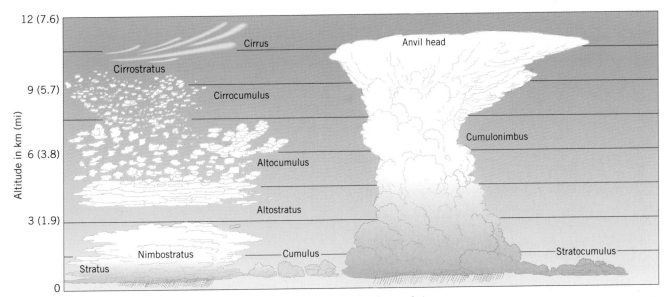

FIGURE 12-6 Schematic diagram of the different cloud types: stratus, cumulus, and cirrus.

A third cloud-type category is constituted by *cirrus* clouds. These are thin, wispy, streak-like clouds that consist of ice particles rather than water droplets. They invariably occur at altitudes higher than 6 km (3.8 mi or 20,000 ft) and are signaled by the prefix *cirro* (Fig. 12-6, upper left).

PRECIPITATION

Clouds are a necessary prerequisite for precipitation to fall to earth. However, not all clouds produce precipitation because water droplets and ice crystals first must undergo some intermediate transformations before they are ready to fall to the surface. When water droplets within clouds first form, they are so small that the slightest upward air current keeps them airborne for long periods. We now examine the processes by which these droplets grow large enough to fall out of a cloud as precipitation.

The Ice-Crystal Process

Probably the most common process, called the *ice-crystal process*, was first identified in the 1930s by meteorologists Tor Bergeron and Von Findeisen. This process requires both liquid droplets and ice particles in a cloud. Ice particles are normally present if the temperature is below 0°C (32°F) and if there are small particles called *freezing nuclei*. Freezing nuclei perform the same function for ice particles that condensation nuclei perform for water droplets.

When a cloud contains both ice particles and water droplets, the water droplets tend to evaporate and then sublimate (change from a vapor to a solid) directly onto the ice crystals. The ice crystal attracts more of the water vapor because the vapor pressure over the ice crystal is lower than the vapor pressure over the water droplet. Thus the ice crystal grows at the expense of the liquid droplet.

The ice crystals become larger and often join together to form a snowflake. When the snowflake is heavy enough, it drops out of the cloud. On its way down it usually encounters higher temperatures and melts, eventually reaching the surface as a liquid raindrop. Most rainfall and snowfall in the midlatitudes are formed by the ice-crystal process, but in the tropics the temperature of many clouds does not necessarily drop below the freezing point. Therefore, a second process—coalescence—is thought to make raindrops large enough to fall from clouds.

The Coalescence Process

The *coalescence process* requires some liquid droplets to be larger than others, which happens when there are giant condensation nuclei. As they fall, the larger droplets collide and join with the smaller ones. But narrowly missed smaller droplets may still be caught up in the wake of the larger ones and drawn to them. In either case, the larger droplets grow at the expense of the smaller ones and soon become heavy enough to fall to earth.

Forms of Precipitation

Precipitation reaches the earth's surface in several forms, as the photographic display in Fig. 12-7 indicates. Large liquid water droplets form **rain**. If the ice crystals in the ice-crystal process do not have time to melt before reaching the earth's surface, the result is **snow**. **Sleet** refers to pellets of ice produced by the freezing of rain before it hits the surface; if it freezes after reaching the ground, it is called *freezing rain* (or *glaze*). Soft **hail** pellets (sometimes called snow pellets) can form in a cloud that has more ice crystals than water droplets, and eventually fall to the surface. True *hailstones* result when falling ice crystals are blown upward from the lower, warmer part of a cloud, where they gain a water surface, to the higher, freezing part, where the outer water turns to ice. This process, which often occurs in the vertical air circulation of thunderstorms, may be repeated over and over to form ever larger hailstones (*see* Fig. 12-7d).

THE SURFACE WATER BALANCE

As far as human beings are concerned, the most crucial segment of the hydrologic cycle occurs at the planetary surface. Here, at the interface between earth and atmosphere, evaporation and transpiration help plants grow, and precipitation provides the water needed for that evapotranspiration. And it is here at the surface that we may measure the *water balance*. An accountant keeps a record of financial income and expenditures and ends up with a bottom-line balance. The balance is positive when profits have been earned and negative when excess debts have been incurred. We can describe the balance of water at the earth's surface in similar terms, using methods devised by climatologist C. Warren Thornthwaite and his associates.

Water can be gained at the surface by precipitation or, more rarely, by horizontal transport in rivers, soil, or groundwater. Water may be lost by evapotranspiration or through runoff along or beneath the ground. The water balance at a location is calculated by matching the gains from precipitation with the losses through runoff and evapotranspiration. When actual evapotranspiration is used for the computation, the balance (in the absence of such human intervention as importing irrigation water) is always zero because no more water can run off or evaporate than is gained from precipitation. However, when potential evapotranspiration is taken into account, the balance may range from a constant surplus of water

a

b

c

d

FIGURE 12-7 The major forms of precipitation. A rare cloud-burst drives torrents of rain onto the desert surface outside Tucson, Arizona (a). Falling snow accumulating rapidly at a Vermont farm (b). The icy after-effects of a Connecticut storm dominated by freezing rain (c). A North Texas field moments after being pounded by golf-ball-sized hail (d).

at the earth's surface to a continual deficit. Figure 12-8 illustrates this range.

The Range of Water Balance Conditions

Bellary, located in the center of southern India, exemplifies the water balance at a deficit (Fig. 12-8a). Throughout the year, the potential evapotranspiration exceeds the water gained in precipitation. On average,

even during the time of the late-summer rains, the soil contains less water than it could hold. Because plants depend on water, the vegetation in this region is sparse, except where irrigation is possible.

At Bogor, on the main Indonesian island of Java, the situation is reversed (Fig. 12-8b). During every month of the year, rainfall, sometimes as much as 45 cm (18 in) in a single month, exceeds the amount of water that can be lost through evapotranspiration. The surplus water pro-

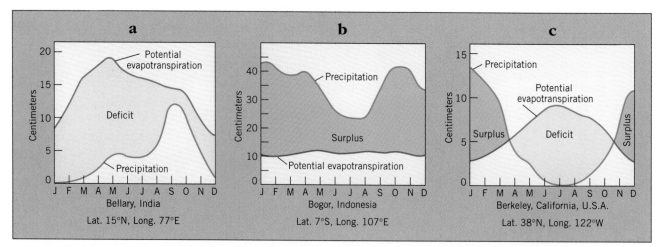

FIGURE 12-8 The range of water balance conditions. Bellary, India (a) experiences a constant deficit because potential evapotranspiration always exceeds precipitation. At Bogor in Indonesia (b) the situation is reversed, and a constant water surplus is recorded. At Berkeley, California (c), across the bay from San Francisco, the intermediate situation occurs, with a combination of surplus and deficit at different times of the year.

vides all that is needed for luxuriant vegetation, and still leaves copious quantities to run off the land surface.

An intermediate situation exists in Berkeley, California, adjacent to San Francisco (Fig. 12-8c). From November to March, precipitation exceeds the potential evapotranspiration, but from April through October there is a water deficit. Starting in April, when potential evapotranspiration surpasses precipitation, soil moisture from below the ground is used in evaporation. Most of it is drawn up through the roots of plants and evaporates from their leaves. This process continues until the end of October, when rainfall once more exceeds potential evapotranspiration and the stock of available soil water is recharged. During this time runoff is more plentiful from California's winter storms.

The amount of runoff in any location cannot exceed the amount of precipitation, and usually there is much less runoff than precipitation. This is because some water almost always evaporates and/or infiltrates the soil. In the United States, runoff approaches the amount of precipitation only in parts of western Oregon and Washington in the Pacific Northwest. In the much drier southwestern states, runoff is quite small in comparison to precipitation.

Water Balance Variations and Latitude

Across the globe as a whole, the values of precipitation, evaporation, and runoff vary greatly with latitude. As Fig. 12-9 shows, annual precipitation and runoff are highest near the equator. Evaporation is also high in this low-latitude zone, but it is greatest in the subtropical latitudes (20 to 35 degrees). The upper midlatitudes (45 to 60 degrees) exhibit water surpluses. Precipitation, runoff, and evaporation are lowest in the highest latitudes. All the forces underlying this distribution of the three variables graphed in Fig. 12-9 are discussed in Units 7 to 11.

On this graph we are mainly looking at the results of differences in radiant energy that reaches different lat-

itudinal zones, and global-scale currents of atmospheric and oceanic circulation. Near the equator, stronger radiant energy from the sun leads to high evaporation rates. It also causes air to rise, cool, and thereby yield large quantities of precipitation. In subtropical areas descending and warming air, in association with semiper-

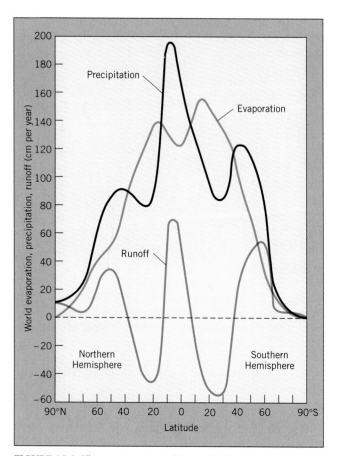

FIGURE 12-9 The average annual latitudinal distribution of precipitation, evaporation, and runoff for the entire surface of the earth.

134

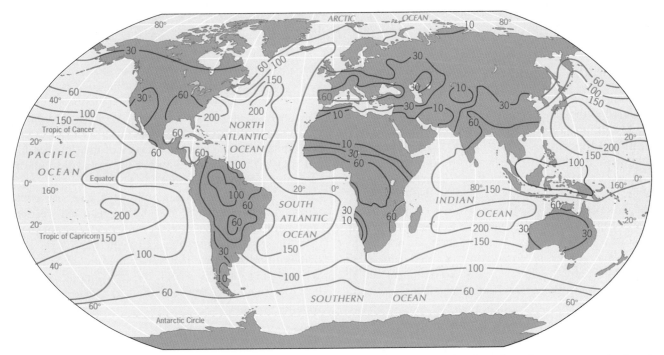

FIGURE 12-10 The global distribution of annual evaporation and evapotranspiration in centimeters, with land elevations adjusted to sea level. Red isolines show the pattern over land, and blue isolines over the oceans.

manent high-pressure cells, produces clear weather. Here, there are high rates of evaporation over the oceans, but little evaporation over the land surfaces because of the scarcity of moisture to be evaporated.

In the higher midlatitudes, eastward-moving storms (driven by the Westerlies and the Polar Front jet stream) provide moderate amounts of precipitation in most areas, but smaller quantities of radiant energy evaporate less of that water than would be the case in the low latitudes. In the high latitudes, the cold air can hold little water vapor. Consequently, there is little precipitation, and, given the low amounts of radiant energy received here, rates of evaporation are minimal. These relationships should be kept in mind as you now consider and compare the world distributions of evapotranspiration (Fig. 12-10) and precipitation (Fig. 12-11).

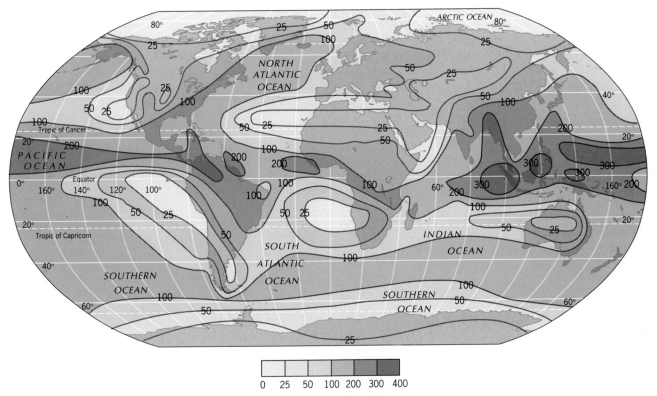

FIGURE 12-11 The global distribution of annual precipitation in centimeters.

Population and the Water Balance

As a final exercise in this unit, let us reconsider the distribution of the earth's population (Fig. 2-5, p. 19) in the context of the global patterns of evapotranspiration (Fig. 12-10) and precipitation (Fig. 12-11). This comparison will reveal that most people live in areas where there are neither great surpluses nor great deficits in the water balance.

For example, the great fertile zones of North America, Europe, and much of China are concentrated in midlatitude areas, which, year in and year out, do not usually experience excessive conditions of precipitation or evapotranspiration. Thus 75 cm (30 in) of annual rainfall in the U.S. Corn Belt or the North China Plain may be far more effective for crop raising than the 200 cm (80 in)

received at a tropical location, where much of the moisture is removed by evapotranspiration.

The variability of precipitation is a phenomenon that concerns farmers around the world. In general, variability increases as the yearly precipitation total decreases. We pursue this matter in our survey of the dry climates in Unit 17, which features a map of the global distribution of annual precipitation variability on p. 192.

Before we study the geography of climate, however, we need to become familiar with the organized weather systems that continually parade across the earth's surface. These systems play a major role in shaping the temperature and moisture regimes of many climate types. We begin in Unit 13 by relating the atmospheric moisture flows we have just learned about to the formation of weather systems.

KEY TERMS

actual evapotranspiration (p. 129)
cloud (p. 129)
condensation (p. 126)
condensation nuclei (p. 131)
dew (p. 127)
dew point (p. 127)

evaporation (p. 126)
evapotranspiration (p. 129)
fog (p. 131)
freezing (p. 126)
hail (p. 132)
hydrologic cycle (p. 127)

latent heat of fusion (p. 126)
latent heat of vaporization (p. 126)
melting (p. 126)
potential evapotranspiration (p. 129)
precipitation (p. 127)

rain (p. 132)
relative humidity (p. 127)
runoff (p. 127)
sleet (p. 132)
snow (p. 132)
sublimation (p. 126)
water vapor (p. 126)

REVIEW QUESTIONS

1. Describe the energy requirements for the melting of ice and the evaporation of water.

2. Describe the various measures of atmospheric humidity and their relation to one another.

3. How does evaporation differ from evapotranspiration?

4. How does potential evapotranspiration differ from actual evapotranspiration?

5. Describe the two processes of raindrop formation.

6. Describe the necessary conditions for surface runoff within the context of the water balance.

REFERENCES AND FURTHER READINGS

BAUMGARTNER, A., and REICHEL, E.   *The World Water Balance: Mean Annual Global, Continental, and Maritime Precipitation, Evaporation, and Runoff* (Amsterdam: Elsevier, 1975).

COTTON, W. R., and ANTHES, R. A.   *Storm and Cloud Dynamics* (San Diego: Academic Press, 1989).

HOUZE, R. A., JR.   *Cloud Dynamics* (San Diego: Academic Press, 1993).

LEGATES, D. R., and MATHER, J. R.   "An Evaluation of the Average Annual Global Water Balance," *Geographical Review*, 82 (1992), 253–267.

LEOPOLD, L. B.   *Water: A Primer* (San Francisco: Freeman, 1974).

LUDLAM, F. H.   *Clouds and Storms: The Behavior and Effect of Water in the Atmosphere* (University Park, Penn.: Pennsylvania State University Press, 1980).

MASON, B. J.   *Clouds, Rain and Rainmaking* (New York: Cambridge University Press, 2nd ed., 1975).

MATHER, J. R.   *The Climate Water Budget in Environmental Analysis* (Lexington, Mass.: Heath, 1978).

MILLER, D. H.   *Water at the Surface of the Earth* (New York: Academic Press, 1977).

SCHAEFER, V. J., and DAY, J. A.   *A Field Guide to the Atmosphere* (Boston: Houghton Mifflin, 1981).

SCORER, R. S.   *Clouds of the World* (North Pomfret, Vt.: David & Charles, 1972).

SUMNER, G.   *Precipitation: Process and Analysis* (New York: Wiley, 1988).

THORNTHWAITE, C. W., and MATHER, J. R.   *The Water Balance* (Centerton, N.J.: Drexel Institute of Technology, Laboratory of Climatology, Publications in Climatology, Vol. 8, 1955).

"Water: The Power, Promise, and Turmoil of North America's Fresh Water," *National Geographic*, special issue, November 1993.

Precipitation, Air Masses, and Fronts

OBJECTIVES

◆ To discuss the four basic mechanisms for producing precipitation.

◆ To develop the concept of air masses—their character, origin, movement patterns, and influence on precipitation.

◆ To distinguish between cold fronts and warm fronts, and to describe their structure and behavior as they advance.

On a day-to-day basis, the atmosphere is organized into numerous weather systems that blanket the earth. In this unit, we make the connection between the atmospheric moisture flows covered in Unit 12 and the formation of those weather systems. Much of our attention is directed at the forces that cause the atmospheric lifting of moist air, thereby producing precipitation. We also introduce the concept of air masses—large uniform bodies of air that move across the surface as an organized whole—and the weather contrasts that occur along their advancing edges.

PRECIPITATION-PRODUCING PROCESSES

All precipitation originates from parcels of moist air that have been adiabatically cooled below their condensation level (dew-point temperature). This is accomplished through the lifting of air from the vicinity of the surface

UNIT OPENING PHOTO: Awesome lightning bolts herald an approaching late afternoon thunderstorm on the outskirts of Tucson, Arizona.

to higher levels in the atmosphere. The occurrence of at least one of four processes is necessary to induce the rising of moist air that will result in significant precipitation. On many occasions more than one process occurs, which heightens the production of precipitation. These four precipitation-producing mechanisms involve (1) the forced lifting of air where low-level windflows converge; (2) the spontaneous rise of air, or convection; (3) the forced uplift of moving air that encounters mountains; and (4) the forced uplift of air at the edges of colliding air masses associated with cyclonic storms.

CONVERGENT-LIFTING PRECIPITATION

Where warm, moist airflows converge at or very near the surface, particularly in the tropical latitudes, their molecules are forced to crowd together. This heightens molecular kinetic energy, warms the combining windstreams, and induces the air to rise. Because the equatorial zone of convergence is already an area of relatively low atmospheric pressure (having pulled those winds toward it in the first place), the lifting of air here becomes more pronounced. The cooling of large quantities of water vapor in the uplifted air causes the rainfall that is so common in the wet tropics. As we would expect, the most prominent and durable tropical weather systems marked by this process of **convergent-lifting precipitation** lie where the trade winds from the Northern and Southern Hemispheres come together at the ITCZ.

The Inter-Tropical Convergence Zone (ITCZ)

In our discussion of global wind belts and semipermanent pressure zones in Unit 10, we noted that the Northeast and Southeast Trade Winds converge in the equatorial trough of low pressure. The rising air that re-

sults is responsible for the cloudiness and precipitation that mark the Inter-Tropical Convergence Zone. The ITCZ occurs at low latitudes all around the earth, most notably above the oceans, particularly the equatorial Pacific. But the rainfall that its clouds deliver over the intervening continents and islands is vital to millions of inhabitants of tropical lowlands and hillsides, especially in the Indonesian archipelago of Southeast Asia and the Zaïre (Congo) Basin in western equatorial Africa.

The Inter-Tropical Convergence Zone changes its location throughout the year, generally following the latitudinal corridor of maximum solar heating. Accordingly, as Fig. 13-1 demonstrates, the average July position of the ITCZ lies at about 10°N, whereas in January, at the opposite seasonal extreme, it is found hundreds of kilometers to the south. Although the ITCZ owes its origin to the position of the overhead sun, this is not the only factor that determines its location. The distribution of land and sea as well as the flows of the tropical atmosphere are also important. Therefore, at any given moment, the ITCZ may not be where generalized theory tell us it ought to be.

Closer inspection of Fig. 13-1 shows that in July the ITCZ is entirely absent in the Southern Hemisphere, when it is drawn northward to a latitudinal position beyond 20°N over southern Asia—as was pointed out in our discussion of the wet monsoon in Unit 10. Yet in January the ITCZ ranges from a southernmost extreme of 20°S, above northern Australia, to a northernmost position within the Northern Hemisphere tropics in the oceans adjoining South America. Not unexpectedly, the moisture regimes of areas lying inside the broad latitudinal band mapped in Fig. 13-1 are decidedly boosted as the ITCZ continually shifts across this region.

CONVECTIONAL PRECIPITATION

Convection is the name given to spontaneous vertical air movement in the atmosphere, and **convectional precipitation** occurs after condensation of the upward-

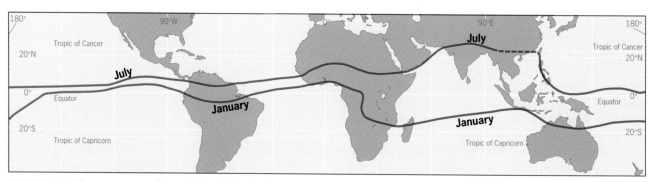

FIGURE 13-1 The average positions of the ITCZ in January and July. The wide annual swing over the northern Indian Ocean is associated with the regional-scale atmospheric circulations that produce the dry and wet monsoons in South Asia, a topic treated in Unit 10.

moving air. The process of convection is generally localized, usually covering only a few square kilometers of the surface. Convection begins when this relatively small area is steadily and intensely heated by insolation, and the parcel of overlying air is rapidly warmed through its contact with the now hot surface.

A rising column of air, known as a *convection cell*, develops quickly. The temperature of this upward-flowing air cools at the dry adiabatic lapse rate (DALR) of 1°C per 100 m (5.5°F per 1000 ft) until the dew point is reached and condensation commences. (As noted in Unit 8, the DALR is the cooling rate of rising unsaturated air.) A small cumulus cloud forms, which soon mushrooms as the added energy of the latent heat released by the condensing water vapor converts the original convection cell into an ever more powerful updraft.

Very large cumulus clouds of the towering cumulonimbus type can now develop if three conditions are met. First, there must be sufficient water vapor in the updraft to sustain the formation of the cloud. Second, the immediate atmosphere must be unstable, so that the upward airflow originally triggered in the surface-layer convection cell is able to persist to a very high altitude. And third, there must be relatively weak winds aloft. With these conditions satisfied, we are likely to encounter the type of weather associated with *thunderstorms*.

Thunderstorms

Thunderstorms are common in both the low and middle latitudes, especially during afternoon hours in the warmest-weather months. Like the larger-scale storm systems discussed in Unit 14, thunderstorms have a distinct life cycle. Although that cycle rarely lasts longer than a few hours, these smaller-scale weather systems can attain sizeable dimensions. For example, the massive thunderclouds shown in Fig. 13-2 were easily discerned by astronauts as they orbited above the Amazon region of equatorial South America.

The life cycle of a thunderstorm begins with the onset of the convection process in an unstable atmosphere as moist heated air rises, undergoes condensation, and releases large quantities of latent heat. This heat energy makes the air much warmer than its surroundings, and the resulting updrafts of warm air attain speeds of about 10 m per second (22 mph). They can sometimes move as fast as 30 m per second (67 mph) in this *developing stage*, as Fig. 13-3a indicates. Raindrops and ice crystals may form at this stage, but they do not reach the ground because of the updrafts.

When the middle or *mature stage* is reached the updrafts continue, producing towering cumulonimbus clouds. More significantly, however, enough raindrops fall to also cause downdrafts of cold air. Evaporation from the falling drops accentuates the cooling. Heavy rain now begins to fall from the bottom of the cloud (Fig. 13-3b), and cold air at that level spreads out in a wedge

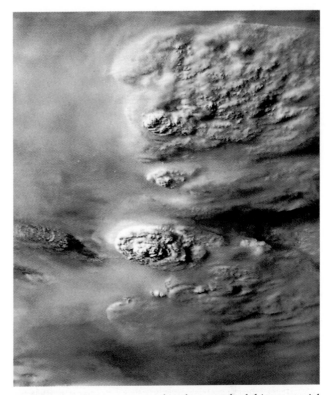

FIGURE 13-2 The astronauts who photographed this equatorial cloud deck erupting with numerous thunderstorms likened the scene to that of a thick, creamy soup coming to a boil. The concentric cloud rings of the cell just below the center of the photograph, visible only from above, reveal the widening rims of successive central updrafts in the storm cell, which failed to break through the tropopause and spread out in all directions.

formation. At the same time, the top of the cloud is often drawn out by upper-air winds to form an *anvil top*, so named because it resembles the shape of a blacksmith's anvil (Fig. 13-4).

Within a few hours, the moisture in the storm is used up. The final, *dissipating stage* occurs when the latent heat source starts to fail. Downdrafts gradually predominate over updrafts, and this situation continues until the storm dies away (Fig. 13-3c).

Thunderstorm-Related Phenomena

A phenomenon associated with thunderstorms in the mid-latitudes is the formation of *hail*. Figure 13-3b shows that, in the mature stage of the storm, an ice crystal might be caught in a circulation that continually moves it above and below the freezing level. This circulation pattern can create concentric shells of ice in the hailstone. In some storms, such as a *squall-line storm* (see unit-opening photo), these circulations can be exaggerated. Figure 13-5 demonstrates how falling hailstones can be scooped back into the main cloud by the intense circulation around such a storm system. Thus hailstones may experience several journeys through the freezing level before falling to earth;

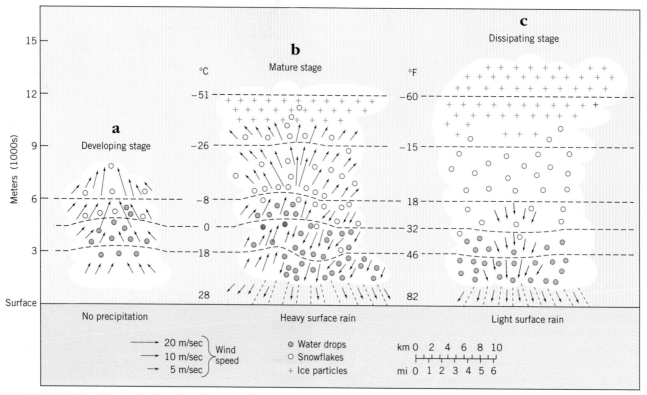

FIGURE 13-3 The life cycle of a thunderstorm. The *developing stage* (a) involves increasingly rapid updrafts; the *mature stage* (b) is associated with both updrafts and downdrafts; and the *dissipating stage* (c) is largely one of cool downdrafts. The arrows indicate the direction and speed of the vertical air currents.

many a farmer's crop has been destroyed because of these conditions, when hailstones as large as softballs can bombard the ground (*see* Fig. 12-7d). Figure 13-5 also indicates that squall-line storms can generate even more severe weather in the form of *tornadoes* (*see* Perspective: Tornadoes and Their Consequences).

Lightning and *thunder* are related phenomena, and both are a result of the thunderstorm's powerful vertical air currents. Although the exact mechanism is not fully understood, cloud droplets and ice crystals acquire electrical charges. This electrical energy, in the form of lightning strokes, makes intermittent contact with the ground. These lightning strokes also rapidly heat the surrounding air, which expands explosively—producing thunder.

Thunderstorms do not usually exist as a single cell such as in the simplified storm in Fig. 13-3. Radar studies have shown that thunderstorms actually consist of several cells organized into clusters, ranging from 2 to 8 km

FIGURE 13-4 Above Singapore's harbor, a classic anvil top crowns a massive cumulonimbus cloud in an approaching thunderstorm. (Authors' photo)

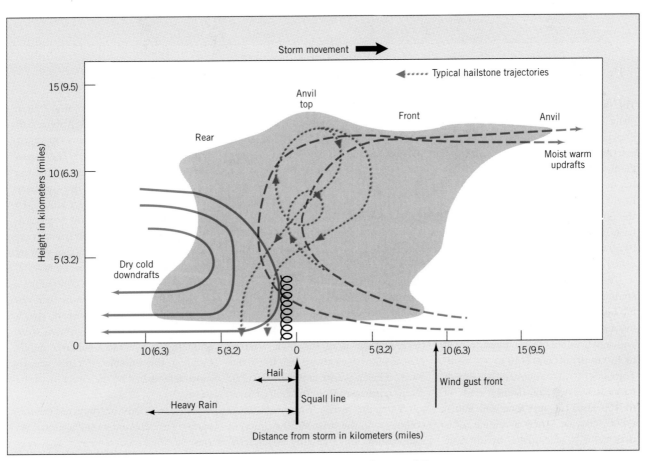

FIGURE 13-5 Model of a thunderstorm producing sizeable hailstones. This squall-line storm is advancing from left to right.

(1.2 to 5 mi) in diameter. Occasionally, these clusters expand over a much wider area and assume their own rotation, thereby forming *supercells* and even larger *mesoscale convective complexes*. As the storms advance, new cells develop to replace old ones that dissipate. If the downdrafts of two cells meet at the surface, a new updraft and a new cell can be initiated.

Convectional thunderstorms are the dominant type of precipitation in the equatorial zone and are especially active in the vicinity of the ITCZ (where the convergence of the trade winds accentuates the uplifting of air). Heat is a constant feature of the surface environment in the low latitudes. That heat, coupled with the generally high water vapor content of tropical equatorial air, produces impressive convectional storms of the type seen in Fig. 13-2.

OROGRAPHIC PRECIPITATION

Mountains, and highlands in general, strongly influence air and moisture flows in the atmosphere. During *orographic uplift* (*oros* is the Greek word for mountain), a moving mass of air encounters a mountain range or other upland zone it is forced to rise over, propelled by surface

winds and the push of air piling up behind it. When a mountain range is close to and parallels the coast, and when prevailing onshore winds carry maritime air laden with moisture, not much cooling is needed to trigger condensation during the forced ascent of this airflow. The rain (and often snow) that results from this lifting process is called **orographic precipitation**.

This precipitation on the range's windward slopes, however, is not matched on the inland-facing leeward slopes. Quite to the contrary, the lee side of the range is marked by dryness because much of the water vapor gained over the ocean has been precipitated out during the windward-slope ascent. Once the air mass crests the range, it usually descends to lower altitudes, thereby rapidly warming and reducing its relative humidity.

Stages of Orographic Uplift

A classic example of this orographic effect exists along most of the west coast of the conterminous United States, where the Sierra Nevada and Cascade Mountains form a north–south wall extending from Southern California well into Canada. Figure 13-7 illustrates the eastward-moving passage of moist Pacific air across the Cascades of central Oregon, specifically between Eugene in the

Tornadoes and Their Consequences

The interiors of large continents, especially in the spring and early summer, experience severe thunderstorms that can produce tornadoes, nature's most vicious weather. A **tornado** is a small vortex of air, averaging 100 to 500 m (330 to 1650 ft) in diameter, that descends to the ground from rotating clouds at the base of a violent thunderstorm. If the twister forms over a water surface, it is called a *waterspout*.

The pressure in the center of such a spinning funnel cloud may be 100 to 200 mb below the pressure of the surrounding air. This triggers highly destructive winds, whose speeds can range from 50 to 130 m per second (110 to 300 mph). Tornadoes tend to follow rather straight paths that can leave damage trails up to 160 km (100 mi) long and 900 m (3000 ft) wide. Their awesome, often dark color is accentuated by the vegetation, loose soil, and other objects sucked into the center tube of this gigantic vacuum cleaner (Fig. 13-6).

The United States experiences more severe thunderstorms than any other land area on earth. Each year, these 10,000-plus storms spawn more than 1000 tornadoes; about 20 percent of the latter are classified as "strong," and even those tornadoes categorized as "weak" can exhibit winds in excess of 45 m per second (100 mph). Tornadoes most frequently develop from squall-line thunderstorms over the Great Plains and Midwest, particularly in the north–south corridor known as "Tornado Alley" that extends through central Texas, Oklahoma, Kansas, and eastern Nebraska. Here, dry air from the high western plateaus moves eastward into the Plains where it meets warm, moist, low-level air swept northward from the Gulf of Mexico. When this maritime tropical Gulf air collides with the drier air, the different temperature and humidity characteristics of these air masses create a volatile mixture that can unleash especially violent thunderstorms.

Many residents of the central United States have tragic tales to tell about tornadoes because these storms kill an average of 90 people each year. The largest outbreak of tornadoes ever recorded in a 24-hour period occurred one April day in 1974, when 148 twisters killed 315 people in the Midwest and Southeast. One of the worst single disasters in recent years was the destruction of the tiny West Texas town of Saragosa on May 22, 1987. Of this poor farming community's 350 residents, 29 died and more than 120 were injured as a massive tornado struck just before dark. Even though many were attending an event in a concrete public building, this durable structure was almost demolished by the direct hit of the twister.

The conditions that triggered this tornado were detected by the National Severe Storms Forecast Center in Kansas City, and a warning was issued, virtually at the last moment. Although the alert came too late for Saragosa, forecasters

near-coastal Willamette Valley and Bend on the interior Columbia Plateau. Between these two cities lie the Three Sisters Peaks, a complex of high summit ridges atop the Cascades reaching an elevation of approximately 3000 m (10,000 ft).

Let us assume that a parcel of air with a temperature of 21°C (70°F) arrives at the near-sea-level western base of the Cascades in Eugene. This parcel of air begins its forced ascent of the mountain range and starts to cool at the DALR (1°C/100 m [5.5°F/1000 ft]). At this rate, the air parcel cools to 15°C (59°F) by the time it reaches 600 m (2000 ft). Now assume that the air reaches its dew point at this temperature, and condensation begins. Clouds quickly form, and continued cooling soon leads to heavy rainfall as the air parcel continues to make its way up the windward slope.

Because saturation has occurred and condensation has begun, however, the air above 600 m cools more slowly at the saturated adiabatic lapse rate (SALR)—here assumed to average 0.65°C/100 m (3.5°F/1000 ft). By the time the uplifted air reaches the mountaintop, at an

altitude of 3000 m (10,000 ft), its temperature is some 15.5°C (28°F) lower, or −0.5°C (31°F). Since this temperature is below the freezing point, precipitation near the summit will fall as snow. (In winter, of course, freezing will commence at a lower elevation on the mountainside.)

Now having crested the range, the air parcel seeks to return to its original level and descends to the warmer layer of the lower atmosphere. The air is now no longer saturated, so the air parcel warms at the DALR. When it reaches the bottom of the 2400-m (8000-ft) leeward slope, its temperature will have risen by 24.5°C (44°F) to 24°C (75°F). Thus our air parcel is noticeably warmer on the lee side of the range, even though the elevation around Bend is 600 m (2000 ft) higher than at Eugene. It is also much drier, as explained earlier, and the generally dry conditions caused by this orographic process is termed the **rain shadow effect**.

In the absence of new moisture sources, rain shadows can extend for hundreds of kilometers. This is the case throughout most of the far western United States,

FIGURE 13-6 Nature's most violent weather is associated with tornadoes. This unusually close view shows a ferocious twister striking Hesston, Kansas on March 13, 1990.

continue to improve their predictions with Doppler radars and other advanced weather-tracking equipment. In fact, meteorologists today increasingly talk about a revolution that is linking radar, satellite, and computer technology to make tornado detection and warning more precise than ever. For more information on tornadoes and the scientists who monitor them, see the *Earth* Magazine article on pp. *Earth* 7-16.

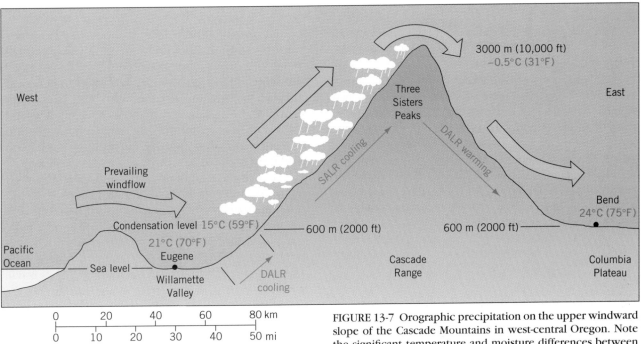

West

3000 m (10,000 ft)
−0.5°C (31°F)

Three
Sisters
Peaks

East

SALR cooling

DALR warming

Prevailing
windflow

Condensation level 15°C (59°F)

21°C (70°F)

600 m (2000 ft)

600 m (2000 ft)

Bend
24°C (75°F)

DALR
cooling

Pacific
Ocean

Sea level

Eugene

Willamette
Valley

Cascade
Range

Columbia
Plateau

| 0 | 20 | 40 | 60 | 80 km |
| 0 | 10 | 20 | 30 | 40 | 50 mi |

DALR = ± 1°C per 100 m/±5.5°F per 1000 ft
SALR = −0.65°C per 100 m/−3.5°F per 1000 ft

FIGURE 13-7 Orographic precipitation on the upper windward slope of the Cascade Mountains in west-central Oregon. Note the significant temperature and moisture differences between the windward and leeward sides of this major mountain barrier. (Horizontal scale is not exact.)

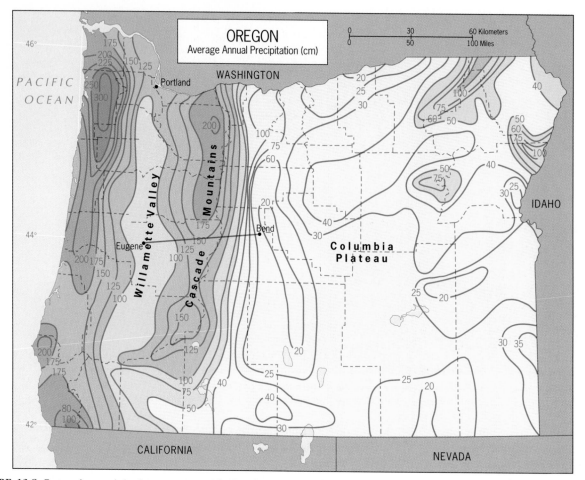

FIGURE 13-8 Oregon's precipitation pattern, with the distribution of isohyets exhibiting the results of the orographic effect as westerly winds off the Pacific are forced across the north-south-trending Cascade Mountains. The transect across the Cascades between Eugene and Bend, diagrammed in Fig. 13-7, is marked by the red line.

because once Pacific moisture is removed from the atmosphere it cannot be locally replenished. Figure 13-8 vividly illustrates the rain shadow covering the eastern half of Oregon, which is interrupted only by secondary orographic precipitation effects associated with smaller mountain ranges in the state's northeastern corner. Leeward areas also frequently experience the rapid movement of warm, dry air known as a *foehn* (pronounced *fern* with the "r" silent) wind. Above the western plateaus of the United States and Canada, these airflows are called *chinook* winds and can sometimes reach sustained speeds approaching 50 m per second (110 mph).

FRONTAL (CYCLONIC) PRECIPITATION

A fourth process that generates precipitation is associated with the coming together of air masses of significantly different temperatures. Such activity is a common occurrence in the middle and upper-middle latitudes, where the general atmospheric circulation causes poleward-flowing tropical air to collide with polar air moving toward the equator. Because converging warm and cold air masses possess different densities, they do not readily mix. Rather, the denser cold air will inject itself beneath and push the lighter warm air upward. And if the warmer air approaches a stationary mass of cooler air, it will ride up over the cooler air. In both situations we have the lifting of warm, often moist air, the cooling of which soon produces condensation, clouds, and precipitation.

Air masses are bounded by surfaces along which contact occurs with neighboring air masses possessing different characteristics. Therefore, such narrow boundary zones mark sharp transitions in density, humidity, and especially temperature. Whenever warm and cool air are in contact like this, the boundary zone is called a **front**. Although fronts can be stationary, they usually advance. Thus a moving front is the leading edge of the air mass built up behind it. When warm air is lifted, cooled, and its water vapor condensed as a result of frontal movement, **frontal precipitation** is produced.

We already know that when a warm air mass infringes upon a cooler one, the lighter warmer air overrides the cooler air. This produces a boundary called a

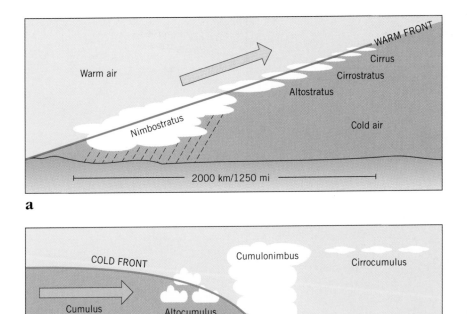

a

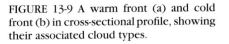

b

FIGURE 13-9 A warm front (a) and cold front (b) in cross-sectional profile, showing their associated cloud types.

warm front (Fig. 13-9a). Warm fronts, because of their gentle upward slope, are associated with wide areas of light to moderate precipitation that extend out well ahead of the actual surface passage of the front. As indicated in the diagram, warm fronts involve the entire sequence of stratus clouds, ranging from upper-level cirrostratus down through altostratus and nimbostratus layers (*see* Fig. 12-6).

The behavior of a **cold front** is far more dramatic. A cold front is produced when the cold air, often acting like a bulldozer, hugs the surface and pushes all other air upward as it wedges itself beneath the preexisting warmer air (Fig. 13-9b). A cold front also assumes a steeper slope than a warm front, causing more abrupt cooling and condensation in the warm air that is uplifted just ahead of it. This produces a smaller but much more intense zone of precipitation that usually exhibits the kind of stormy weather associated with the convectional precipitation process discussed earlier. Not surprisingly, cold fronts exhibit a variety of cumulus-type clouds, particularly the clusters of cumulonimbus clouds that are the signature of squall-line thunderstorms. But once the swiftly passing violent weather ends, the cool air mass behind the front produces generally fair and dry weather.

Frontal precipitation is frequently called *cyclonic precipitation* because it is closely identified with the passage of the warm and cold fronts that are essential components of the cyclones that shape the weather patterns of the midlatitudes. In Unit 14, which treats weather systems, we examine the life cycle of these storm cells and

their fronts, extending the discussion developed here. The rest of this unit takes a closer look at air masses (whose movement triggers these weather systems) and introduces a geographic classification based on source regions of air masses.

AIR MASSES IN THE ATMOSPHERE

In large, relatively uniform expanses of the world, such as the snow-covered Arctic wastes or the warm tropical oceans, masses of air in contact with the surface may remain stationary for several days. As the air hovers over these areas, it takes on the properties of the underlying surface, such as the icy coldness of the polar zones or the warmth and high humidity of the maritime tropics. Extensive geographic areas with relatively uniform characteristics of temperature and moisture form the **source regions** where large air masses can be produced.

The air above a source region will reach an equilibrium with the temperature and moisture conditions of the surface. An **air mass** can thus be defined as a very large parcel of air in the boundary layer of the troposphere that possesses relatively uniform qualities of temperature, density, and humidity in the horizontal dimension. In addition to its large size, which is regarded by meteorologists to be at least 1600 km (1000 mi) across, an air mass must be bound together as a cohesive unit. This is necessary because air masses travel as distinct

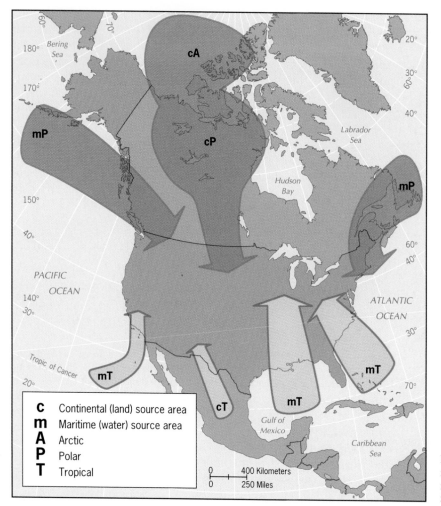

FIGURE 13-10 Source regions and the most likely paths of the principal air masses that affect the continental United States.

entities, covering hundreds of kilometers as they are steered away from their source regions by the airflows of the general atmospheric circulation.

Classifying Air Masses

To keep track of their identities as they migrate, air masses are given letter codes that reflect the type of surface and general location of their source region. These codes are as follows: maritime (**m**) or continental (**c**) and tropical (**T**) or polar (**P**). In a warm source region, such as the Caribbean Sea, the equilibrium of temperature and moisture in an overlying air mass is established in two or three days. Unstable air in either a maritime tropical (**mT**) air mass or a continental tropical (**cT**) air mass is warmed to a height of about 3000 m (10,000 ft).

In contrast, within cold source regions air masses can take a week or more to achieve equilibrium with their underlying surfaces. Only a relatively shallow layer of air, up to about 900 m (3000 ft), cools in continental polar (**cP**) or maritime polar (**mP**) air masses. Of course, the air above that level is cold as well, but for reasons other than contact with the surface. It takes such a relatively long time for a cold air mass to become established because cooling at its base stabilizes lapse rates, thereby preventing vertical mixing and prohibiting efficient heat exchange.

In addition to the four leading types of air masses (**mT**, **cT**, **mP**, and **cP**) already discussed, three more can be recognized. The first is found in the highest latitudes and is called continental Arctic (**cA**) (or continental Antarctic [**cAA**] in the Southern Hemisphere); paradoxically, these form at higher latitudes than the polar air masses do. The final two are found in the lowest latitudes astride the equator and are labeled continental equatorial (**cE**) and maritime equatorial (**mE**). The letter pairs used to label air masses are sometimes supplemented by letters that indicate whether the mass is colder (**k**) or warmer (**w**) than the underlying surface, and whether the air is stable (**s**) or unstable (**u**). Thus a maritime tropical air mass that is warmer than the ocean below it and unstable is designated as **mTwu**.

Movements of Air Masses

If you live in the midwestern United States, you can testify to the bitterness of the polar air that streams southward from Canada in the winter and the oppressiveness of the warm humid air that flows northward from the Gulf of Mexico in the summer. Air masses affect not only their source areas, but also the regions they migrate across. As an air mass moves, it may become somewhat modified by contact with the surface or by changes within the air, but many of its original characteristics remain identifiable far from its source region. This is a boon to weather forecasters, because they can predict weather conditions more accurately if they know the persisting characteristics of a moving air mass as well as its rate and direction of movement.

The principal air masses affecting North America are shown in Fig. 13-10. The arrows represent the most common paths of movement away from the various source regions indicated on the map. If the contents of this map look like a military battle plan, that may be because opposing air masses fight for supremacy throughout the year. Losses and gains of territory in this battle are controlled by one important factor: the Polar Front and its associated jet stream. The main current of a river does not let eddies pass from one bank to another, and so it is with the Polar Front jet stream, which provides an unseen barrier to advancing air masses. This is a key to understanding weather in most of the United States, and our discussion of the Polar Front jet stream, begun in Unit 10, is developed further in Unit 14.

KEY TERMS

air mass (p. 145)
cold front (p. 145)
convection (p. 138)
convectional precipitation (p. 138)
convergent-lifting precipitation (p. 138)

front (p. 144)
frontal (cyclonic) precipitation (p. 144)
orographic precipitation (p. 141)
rain shadow effect (p. 142)

source region (p. 145)
tornado (p. 142)
warm front (p. 145)

REVIEW QUESTIONS

1. What are the necessary conditions for convectional precipitation? What are some of the more severe types of weather associated with this type of precipitation?

2. Describe the geographic pattern of precipitation on the windward and leeward flanks of a mountain range.

3. Describe the general precipitation characteristics associated with a warm front.

4. Describe the general precipitation characteristics associated with a cold front.

5. Describe the characteristics of the four major air mass types: **mT**, **cT**, **mP**, and **cP**.

6. Describe the internal structure of a tornado and explain why these twisters are capable of so much destruction.

REFERENCES AND FURTHER READINGS

APPLEBOME, P. "After a Twister [in Saragosa, Texas]: Coping in Town That Isn't There," *New York Times*, May 24, 1987, 1, 18.

BATTAN, L. J. *Fundamentals of Meteorology* (Englewood Cliffs, N.J.: Prentice-Hall, 2nd ed., 1984), Chapters 6 and 7.

BYERS, H. R., and BRAHAM, R. R. *The Thunderstorm* (Washington: U.S. Weather Bureau, 1949).

COTTON, W. R., and ANTHES, R. A. *Storm and Cloud Dynamics* (San Diego: Academic Press, 1989).

COWEN, R. C. "New Spin on Tracking Destructive Tornadoes," *Christian Science Monitor*, April 20, 1994, 16.

DE BLIJ, H. J., ed. *Nature on the Rampage* (Washington, D.C.: Smithsonian Institution Press, 1994).

EAGLETON, J. R., et al. *Thunderstorms, Tornadoes, and Building Damage* (Lexington, Mass.: Heath, 1975).

FEW, A. A. "Thunder," *Scientific American*, July 1975, 80–90.

LUDLAM, F. H. *Clouds and Storms: The Behavior and Effect of Water in the Atmosphere* (University Park, Penn.: Pennsylvania State University Press, 1980).

MIDDLETON, W. *History of Theories of Rain and Other Forms of Precipitation* (Chicago: University of Chicago Press, 1968).

ROBINSON, A. *Earth Shock: Hurricanes, Volcanoes, Earthquakes, Tornadoes and Other Forces of Nature* (New York: Thames & Hudson, 1993).

SCHONLAND, B.F.J. *The Flight of Thunderbolts* (Oxford, U.K.: Oxford University Press [Clarendon], 1964).

Significant Tornadoes: 1871–1991 (St. Johnsbury, Vt.: Tornado Project, 1993).

SNOW, J. T. "The Tornado," *Scientific American*, April 1984, 86–96.

WEEMS, J. E. *The Tornado* (Garden City, N.Y.: Doubleday, 1977).

WILLIAMS, E. R. "The Electrification of Thunderstorms," *Scientific American*, November 1988, 88–99.

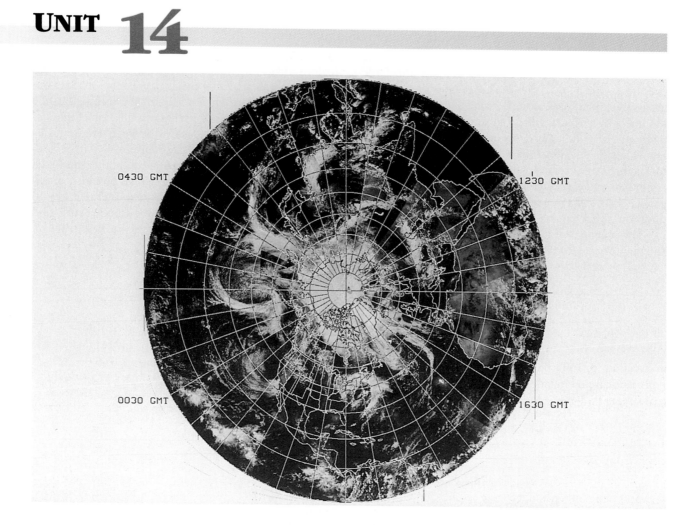

0430 GMT 1230 GMT

0030 GMT 1630 GMT

Weather Systems

OBJECTIVES

◆ To demonstrate the importance of migrating weather systems in the global weather picture.

◆ To discuss the significant tropical weather systems, particularly hurricanes.

◆ To explain how midlatitude cyclones are formed, and to describe the weather patterns associated with them.

Weather systems are organized phenomena of the atmosphere—with inputs and outputs and changes of energy and moisture. Unlike the semipermanent pressure cells and windflows of the general circulation, these transient weather systems are secondary features of the atmosphere that are far more limited in their magnitude and duration. Such recurring weather systems, together with the constant flows of moisture, radiation, and heat energy, make up our daily weather. This unit focuses on the migrating atmospheric disturbances we call **storms**. We begin by looking at the storm systems of the tropics, which are typically associated with heavy precipitation. We then shift our focus from the low latitudes to the storm systems that punctuate the weather patterns of the middle and higher latitudes.

LOW-LATITUDE WEATHER SYSTEMS

The equatorial and tropical latitudes are marked by a surplus of heat, which provides the energy to propel winds and to evaporate the large quantities of seawater that are often carried by them. The abundance of wa-

UNIT OPENING PHOTO: Satellite image mosaic of the Northern Hemisphere for April 25, 1984, with numerous weather systems visible.

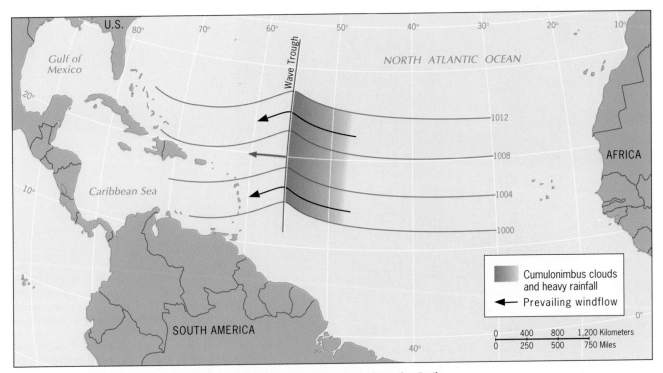

FIGURE 14-1 Isobars showing a westward-moving easterly wave approaching the Caribbean Sea and the development of towering rain clouds along its trailing limb.

ter vapor, convergence of the trade winds, and frequent convection produce the heavy rains that are observed in the tropics. Two of the major low-latitude weather systems are covered in other units. In Unit 13 we discuss the largest system, the Inter-Tropical Convergence Zone (ITCZ), and in Unit 10 we examine the related monsoonal circulation of coastal Asia. Here we investigate the smaller-scale moving weather systems that recur in tropical regions: easterly waves and hurricanes.

Easterly Waves

For centuries, people have known about the trade winds, the constant easterly surface flow of tropical air between 30°N and 30°S latitude. But only within the past four decades has it been discovered that this flow is frequently modified by wave-like phenomena that give rise to distinctive weather systems. Unit 9 explains that geostrophic winds parallel isobars. If we apply this relationship to Fig. 14-1, we see that the isobars indicate the easterly trade winds blowing across the tropical North Atlantic Ocean. The **easterly wave** depicted represents a vertical perturbation within the general flow of these winds.

Through fairly complex mechanisms, this easterly wave produces both uplift and descent of air. Westward-moving air is forced to rise on the upwind side of the wave and descend on the downwind side. Thus, on the eastern side of such a wave, we can expect to find towering cumulus clouds and often heavy rainfall. But clear weather occurs on the leading western side of the low-pressure wave trough because the air column decreases in depth, and atmospheric subsidence prevails.

Weather systems associated with easterly waves (which also go by their more popular name, *tropical waves*) are most frequently observed in the Caribbean Basin. Similar phenomena are also common in the west-central Pacific and in the seas off China's central east coast. These systems travel toward the west at about 5 to 7 m per second (11 to 16 mph) and are relatively predictable in their movement. But for reasons that are not well understood, each year certain easterly waves increase their intensity. As these weak low-pressure troughs deepen, they begin to assume a rotating, cyclonic organization. Such a disturbance is called a *tropical depression*. If the low intensifies and sustained wind speeds surpass 15 m per second (34 mph), the weather system becomes a *tropical storm*. Further development is possible, and if the now fully formed tropical cyclone exhibits sustained winds in excess of 33 m per second (74 mph), a hurricane is born.

Hurricanes

An intensely developed tropical cyclone is one of the most fascinating and potentially destructive features of the atmosphere. These severe storms go by different names in different regions of the world. In the western Atlantic and eastern Pacific Oceans they are called **hur-**

FIGURE 14-2 A view from space of Hurricane Gilbert, centered over the western Caribbean Sea south of Cuba on September 13, 1988. Gilbert, which ravaged Jamaica and then parts of Mexico, was one of the most powerful hurricanes ever observed, and at one point registered the lowest surface pressure reading ever recorded in the Western Hemisphere.

ricanes, the term we employ here. Elsewhere they are known as *typhoons* (western North Pacific), *cyclones* (Indian Ocean), or *willy-willies* (Australia).

The tropical cyclone is a tightly organized, moving low-pressure system, normally originating at sea in the warm moist air of the low-latitude atmosphere. As with all cyclonic storms, it has distinctly circular wind and pressure fields (*see* Fig. 9-9a). In Fig. 14-2, a satellite image of Hurricane Gilbert which ravaged parts of the Caribbean and Mexico in 1988, the circular windflow extends vertically, and the pinwheel-like cloud pattern associated with the center of the system reminds us of its cyclonic origin.

The World Meteorological Organization describes a hurricane as having wind speeds greater than 33 m per second (74 mph) and a central surface pressure below 900 mb. However, in the most severe storms, such as Hurricane Gilbert (which exhibited the lowest pressure ever recorded in the Western Hemisphere), wind speeds can exceed 90 m per second (200 mph) and cause massive destruction.

The hurricane's structure is diagrammed in Fig. 14-3. A striking feature is the open **eye** that dominates the middle of the cyclonic system, the "hole in the doughnut" from which spiral bands of cloud extend outward, often reaching an altitude of 16 km (10 mi). The strongest winds and heaviest rainfall (up to 50 cm [20 in] per day) are found in the **eye wall**, the towering cloud tube that marks the rim of the eye. Within the eye itself, which extends vertically through the full height of the storm, winds are light and there is little rain as dry cool air de-

scends down the entire length of the central column. Temperatures vary little throughout the area of a tropical cyclone, which has an average diameter of about 500 km (310 mi). But the diameter can vary from 80 km (50 mi) to about 2500 km (1550 mi). In general, the greater the amounts of energy and moisture involved in the system, the more extensive and powerful the hurricane will be.

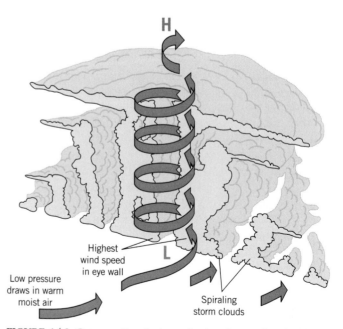

H

Highest wind speed in eye wall

L

Low pressure draws in warm moist air

Spiraling storm clouds

FIGURE 14-3 Cross-sectional view of a hurricane showing its mechanics and component parts.

Predicting Atlantic Hurricanes

The annual North Atlantic hurricane season lasts from June 1 through November 30, the period when tropical sea-surface temperatures are at their warmest. The average number of hurricanes in this region each year is six, with four additional tropical storms that never reach hurricane-strength wind speeds. But, as we well know, departures from climatological averages are common, and since 1900 the yearly number of hurricanes has ranged from 1 to 21.

Weather predictions, even for a one- or two-day period, can be a very chancy business, and longer-range forecasting seems many years away from entering the realm of reliability. Yet astonishingly, William Gray, a meteorologist at Colorado State University, began to successfully predict the annual level of Atlantic hurricane activity in the mid-1980s. From 1984 to 1994, Gray respectively predicted 7, 8, 4, 5, 7, 4, 6, 3, 4, 7, and 4 hurricanes; the actual totals were 5, 7, 4, 3, 5, 6, 8, 4, 4, 4, and 3.

Gray's impressively on-target long-range forecasts are based on an analysis of several atmospheric factors, which he

believes are the determining influences in the region centered on the tropical North Atlantic. (These factors do not apply to tropical cyclones in the Pacific and Indian Oceans.) The first has to do with expanded areas of unusually warm sea-surface temperatures in the eastern equatorial Pacific (the El Niño phenomenon discussed in Unit 11): the greater the El Niño effect, the fewer the number of Atlantic hurricanes. The second factor involves stratospheric winds that girdle the globe at a height of 20 km (13 mi) above the equator, an airflow that reverses direction approximately every two years. When these winds blow from the west or weakly from the east, hurricane activity increases; a stronger easterly flow decidedly suppresses tropical cyclone formation.

The third factor is the upper-level tropospheric windflow that operates at a height of 12 km (7.5 mi) above the Caribbean Sea; the stronger the westerly flow, the smaller the number of hurricanes. The fourth factor involves surface-level air pressures over the western Caribbean and Gulf of Mexico to the

northwest; lower-than-normal barometric readings in May and June presage increased hurricane activity. A possible fifth factor has also come under consideration: the degree of drought in the Sahel region (just south of the arid Sahara) of western Africa. When the wet-monsoon rains in the Sahel are abundant, they seem to correlate with heightened hurricane activity in the nearby tropical Atlantic.

Gray's studies have also focused on the historical pattern of Atlantic hurricane occurrence in recent decades. His findings have shown that the atmospheric factors listed above combined to sharply weaken hurricane activity throughout the 1970s and 1980s. Miami, whose vulnerable position near the tip of Florida exposed it to more than a dozen serious tropical cyclones between 1925 and 1960, did not experience even a mild storm between Hurricane Betsy in 1965 and Hurricane Andrew in 1992. For more information on hurricane prediction and an overview of the unusually active 1995 hurricane season, see the *Earth* Magazine article on pp. *Earth* 17–24.

Hurricane Development

Hurricanes are efficient machines for drawing large quantities of excess heat away from the ocean surface and transporting them into the upper atmosphere. Accordingly, they are most likely to form in late summer and autumn when tropical sea surfaces reach their peak annual temperatures (i.e., greater than 27°C [80°F]). The origin of these tropical storms appears to be related to easterly waves and the equatorial trough of low pressure. However, many aspects of hurricane formation are still poorly understood, particularly the exact triggering mechanism that transforms fewer than one of every ten easterly waves into a tropical cyclone. Nonetheless, meteorologists are making progress in their attempts to forecast seasonal hurricane activity (*see* Perspective: Predicting Atlantic Hurricanes).

From studies of past storms, it is known that hurricanes form when latent heat warms the center of a preexisting storm and helps to intensify an anticyclone

present in the upper troposphere. Once the tropical cyclone has developed, it becomes self-sustaining. Vast quantities of heat energy are siphoned from the warm ocean below and transported aloft as latent heat. This energy is released as sensible heat when clouds form. The sensible heat provides the system with potential energy, which is partially converted into kinetic energy, thereby causing the hurricane's violent winds.

Hurricanes originate between 5 and 25 degrees latitude in all tropical oceans except the South Atlantic and the southeastern Pacific, where the ITCZ seldom occurs. These latitudinal limits are determined by the general conditions necessary for hurricane formation: below 5 degrees the equatorial Coriolis force is too weak to generate rotary air motion, while poleward of 25 degrees sea-surface temperatures are too cool. Once formed, the movement of hurricanes is erratic (they have been likened to balloons floating in rivers with complex currents).

151

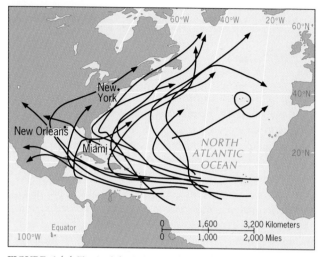

FIGURE 14-4 Typical hurricane tracks in the North Atlantic Ocean. The path of every storm since 1871 is mapped in the National Hurricane Center's "track book," the tenth item cited in the reference list at the end of this unit.

In the Northern Hemisphere, hurricanes usually travel first westward and then to the northwest before curving around to the north and east, where they come under the influence of the westerly winds of the middle latitudes. The paths of typical Atlantic hurricanes are mapped in Fig. 14-4, and they underscore the vulnerability of the U.S. southeastern coast, from Chesapeake Bay south to the Florida Keys, as well as the entire rim of the Gulf of Mexico. Hurricanes usually advance with forward speeds of 16 to 24 kph (10 to 15 mph). If they pass over a large body of land or cooler waters, their energy source—the warm ocean—is cut off, and they gradually weaken and die.

Hurricane Destruction

Before these tropical cyclones die, normally within a week of their formation, they can cause widespread damage on land. Much of this damage is done by high winds and torrential rains. However, near coastlines, waves and tides rise to destructive levels, particularly when wind-driven water known as the *storm surge* (which can surpass normal high-tide levels by more than 5 m [16 ft]) is hurled ashore. The destructive potential can be respected when you consider that the amount of energy unleashed by a hurricane in one hour equals the total electric power generated in the United States in an entire year. With the exception of tornadoes—which sometimes are spawned in the inner spiral rain bands of the strongest hurricanes—tropical cyclones are the most dangerous of all atmospheric weather systems.

Not surprisingly, the most powerful hurricanes rank among the landmark weather events of this century. Before World War II (and the development of sophisticated weather tracking), hurricanes in 1900 and 1935 caused enormous destruction on the coasts of Texas and South Florida, respectively; in 1938, much of southern New England was devastated. After 1945, hurricanes were named in alphabetical order of their occurrences each year. Among the most destructive of these have been Hazel (1954), Camille (1969), Agnes (1972), Gilbert (1988), and Hugo (1989).

Most devastating of all, however, was Hurricane Andrew, which slammed into the Bahamas, southern Florida, and Louisiana on August 23–25, 1992 (Fig. 14-5). Because it was the first major hurricane to strike one of the largest U.S. metropolitan areas—Miami—with its full force (average wind speed: 235 kph [145 mph]), it pro-

FIGURE 14-5 Some of the worst devastation caused by Hurricane Andrew in the suburbs south of Miami on August 24, 1992. This is part of Cutler Ridge, located just inland from where the eye wall passed over the Florida coast.

duced the nation's costliest natural disaster up to that time. The numbers on the destruction caused by Andrew in the Miami area are truly staggering, yet they only begin to suggest the upheaval that followed in a community still not fully recovered more than three years after the storm. The hurricane damaged or destroyed 107,800 private homes (95,400 had to be completely rebuilt), resulting in a total of $16.04 billion paid to settle 795,912 insurance claims. At least 125,000 people were left homeless, and 86,000 were left jobless. And the monumental cleanup effort saw the removal of *35 million* tons of debris (at a cost of $600 million), traffic light replacements in 2300 intersections, and the installation of more than 50,000 new street signs.

Since 1900, hurricanes have cost more than 1500 American lives and untold billions of dollars in damage. In other countries, however, it is not unusual for the death toll of a single storm to be in the thousands. One of the twentieth century's greatest natural disasters was the tropical cyclone that smashed into Bangladesh in 1970, killing upward of 300,000 people; perhaps as many as half that number perished when another cyclone struck the country in 1991.

Although the tropics are generally marked by monotonous daily weather regimes, violence and drama accompany the occasional storms that are spawned in the energy-laden air of the low latitudes. Weather systems in the middle and higher latitudes are frequently less dramatic, but are usually more frequent and affect wider areas.

WEATHER SYSTEMS OF THE MIDDLE AND HIGHER LATITUDES

The general atmospheric circulation of the middle and higher latitudes exhibits some basic differences from that of the tropical latitudes. The principal difference is the frequent interaction of dissimilar air masses. In the higher latitudes, high-pressure and low-pressure systems moving eastward are carried along by the westerly winds of the upper air. In the middle latitudes, air of different origins—cold, warm, moist, dry—constantly comes together, and fast-flowing jet streams are associated with sharp differences in temperature. Therefore, the weather is much more changeable than in the tropics.

The Polar Front Jet Stream

Unit 10 explains how the fast-flowing, upper-air currents of the jet streams are related to the process of moving warm air from the tropics and cold air from the high latitudes. Unit 13 points out that jet streams may also be thought of as boundaries between large areas of cold and warm air. One of them, the Polar Front jet stream, although not in itself a weather system, controls the most important midlatitude weather systems.

Figure 14-6 shows the upper-air isotherms (red dashed lines of constant temperature) sloping away from the tropics toward the polar regions. In the middle latitudes these temperatures drop abruptly, and warm and cold air face each other horizontally. As we know, such

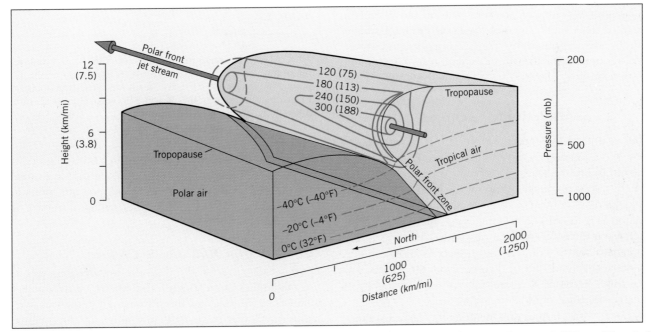

FIGURE 14-6 A cross-sectional model perspective of the Polar Front zone and (high-velocity) jet stream over the central United States. The view is from the northwest toward the southeast. The north–south extent of this diagram is roughly the distance from Lake Superior to the Louisiana Gulf Coast. The wind speeds surrounding the jet stream are calibrated in kph with mph equivalents shown.

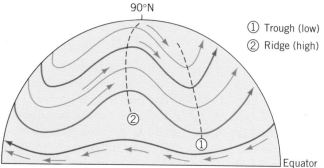

FIGURE 14-7 The spatial context of the Polar Front jet stream, represented by the solid red line, in this generalization of the upper atmosphere of the Northern Hemisphere.

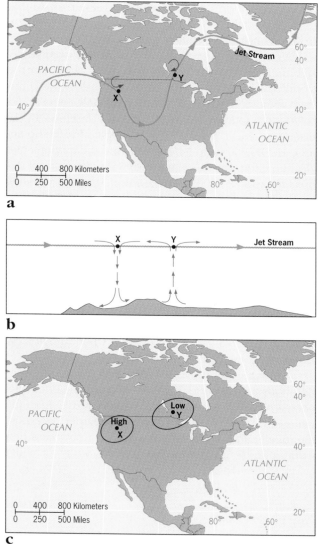

FIGURE 14-8 The relationship between the Polar Front jet stream and surface pressure patterns. (a) Location of the jet stream 9000 m (30,000 ft) above North America. (b) Associated air movement around points **X** and **Y** in vertical cross-section. (c) Resulting surface pressure conditions.

a narrow zone of contact constitutes a front. The *polar front*, therefore, separates relatively cold polar air from relatively warm tropical air, and the jet stream associated with this temperature divide is called the **Polar Front jet stream** (Fig. 14-6). As noted in Unit 10, at the core of the wave-like westerly winds of the upper troposphere, the Polar Front jet stream snakes its way around the globe in large meanders (Fig. 14-7).

These meanders develop in response to fairly complex physical laws that govern the movement of high-velocity air currents in the upper atmosphere. The important thing to remember is that they cause airflows to converge and diverge strongly in the vicinity of the meander bends. This upper atmospheric convergence and divergence causes the air to subside (under the area of convergence) or ascend (under the area of divergence). As a result, these zones have important implications for the surface pressure pattern and attendant weather. This significance of the Polar Front jet stream can be demonstrated by considering the process of **cyclogenesis**—the formation, evolution, and movement of midlatitude cyclones.

Midlatitude Cyclones

The weather at the earth's surface in the midlatitudes is largely determined by the position of the Polar Front jet stream. In Fig. 14-8a, the jet stream passes across North America 9000 m (30,000 ft) above the ground. The air in the jet stream converges at point **X**, as shown in cross-section (Fig. 14-8b). The converging air undergoes an increase in density, and so it descends. The descending air causes high pressure at the surface, as shown in Fig. 14-8c, and spreads out or diverges from the surface anticyclone. At point **X**, therefore, we could expect fair weather because of the descending air.

Meanwhile, farther downwind in the eastward-flowing jet stream, at point **Y**, the air in the jet stream diverges. The diverging air aloft must be replaced, and air is drawn up from below. As air is drawn up the column, a converging cyclonic circulation develops at the surface, exhibiting its characteristic low pressure. The rising air

cools, forming clouds that soon produce precipitation. In this manner, the jet stream becomes the primary cause of fair weather when it generates upper atmospheric convergence, and stormy weather when it produces divergence.

Life Cycle of a Midlatitude Cyclone

On most days, spiral cloud bands can be observed in the midlatitude atmosphere. The unit-opening satellite image (p. 148) indicates the presence of numerous midlatitude cyclones; particularly prominent are the cyclonic spirals in the North Pacific and North Atlantic Oceans. These are the most common large-scale weather systems found outside the tropics. The midlatitude cyclone, like its low-latitude counterpart, goes by several names. It is often called a *depression* or an *extratropical* (meaning outside the tropics) *cyclone*. A midlatitude cyclone is characterized by its circular windflow and low-pressure field as well as the interaction of air of different properties. As with many atmospheric weather systems, it possesses a definite life cycle.

Figure 14-9a shows the early stage in the development of a midlatitude cyclone. A mass of cold air and a mass of warm air lie side by side, with the boundary between them called a **stationary front**. Divergent upper airflow causes a slight cyclonic motion at the surface, and a small kink appears in the stationary front. As the surface cyclonic motion develops, the kink grows larger and becomes an **open wave**, as indicated in Fig. 14-9b. In this open-wave stage, the warm and cold air interact in distinct ways. On the eastern side of the cyclone, warm air now glides up over the colder air mass along a surface that has become a warm front. To the west, the advancing cold air—now a cold front—injects itself beneath the warm air in an action resembling that of a snowplow.

The open-wave stage represents maturity in the life cycle of a midlatitude cyclone, but generally the energy of the circulation is spent within a few days. Figure 14-9c shows how the better defined cold front travels faster than the warm front, overtaking it first at the center of the cyclone. When this happens, the snub-nosed cold front lifts the warm air entirely off the ground, causing an **occluded front**—the surface boundary between the cold and cool air—to form. In time, the entire wedge of surface warm air is lifted to the colder altitudes aloft, and the cyclone dissipates (Fig. 14-9d). This weather system, however, does not die where it was born: the Polar Front jet stream has continually steered the cyclone eastward during its life span.

This wave model of the midlatitude cyclone was first proposed by Norwegian weather forecasters more than 75 years ago. It has turned out to be a good forecasting tool, and we can see why if we look more closely at the open-wave stage.

Weather Forecasting and the Open-Wave Stage

The most obvious features of Fig. 14-10a, which shows a wave cyclone above the southeastern United States, are the roughly circular configuration of the isobars and the positions of the fronts. Although the isobars are generally circular, they tend to be straight within the area enclosed by the cold and warm fronts. This wedge of warm air is called the *warm sector*. Also noteworthy is the kinking or sharp angle the isobars form at the fronts themselves.

The wind arrows tend to cross the isobars at a slight angle, just as they should when friction at the earth's surface partially upsets the geostrophic balance. The wind arrows indicate a large whirl of air gradually accumulating at the center of the low-pressure cell. This converging air rises and cools, especially when it lifts at the fronts, and its water vapor condenses to produce a

significant amount of precipitation. Maritime tropical air (**mTw**), warmer than the surface below it, forms the warm sector, and cold continental polar air (**cPk**) advances from the northwest.

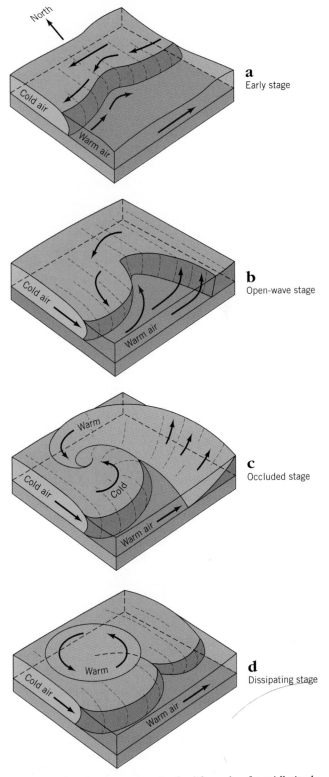

a
Early stage

b
Open-wave stage

c
Occluded stage

d
Dissipating stage

FIGURE 14-9 The four stages in the life cycle of a midlatitude cyclone: (a) early; (b) open-wave; (c) occlusion; and (d) dissipation. Copyright © Arthur N. Strahler. Reproduced by permission of Arthur N. Strahler.

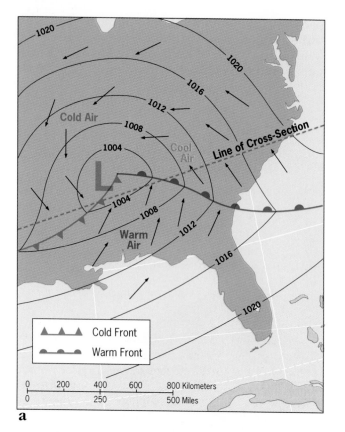

a

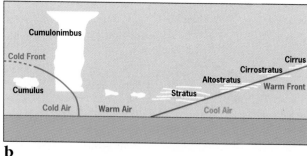

b

Post Cold Front	Warm Sector	Pre Warm Front	Sector / Weather Condtion
Heavy Rain, Then Clearing	Showers	Rain and Fog	**Precipitation**
↘ ↘	↗ ↗	↖ ↗	**Surface Wind Direction**
Rising	Low Steady	Falling	**Pressure**
Lower	Higher	Lower	**Temperature**

c

FIGURE 14-10 The open-wave stage of a midlatitude cyclone over the southeastern United States. Pressure fields, wind-flows, and fronts are shown in diagram (a). Diagram (b) provides a cross-sectional view along the dashed line mapped in (a). Surface weather conditions along the cross-sectional transect are summarized in diagram (c).

In the cross-sectional view in Fig. 14-10b, the warm front has the expected gentle slope. As the warm air moves up this slope, it cools and condenses to produce the characteristic sequence of stratus clouds. The cold front, however, is much steeper than the warm front. This forces the warm air to rise much more rapidly, creating towering cumulus clouds that can yield thunderstorms and heavy rains.

Although no two wave cyclones are identical, there are enough similarities for an observer on the ground to predict the pattern and sequence of the atmospheric events outlined above. These events are summarized in Fig. 14-10c and are as follows:

1. As the midlatitude cyclone approaches along the line of cross-section shown in Fig. 14-10a, clouds thicken, steady rain falls, and the pressure drops.
2. As the warm front passes, the temperature rises, wind direction shifts, pressure remains steady, and the rain lets up or turns to occasional showers with the arrival of the warm sector.
3. Then comes the frequently turbulent cold front, with its high-intensity but short-duration rain, another wind shift, and an abrupt drop in temperature.

The whole system usually moves eastward, roughly parallel to the steering upper-air jet stream. Thus a weather forecaster must predict how fast the system will move and when it will occlude. But those two predictions are often difficult and account for most of the forecasts that go awry.

Because it is so influential and prevalent, the midlatitude cyclone is a dominant climatic control outside the tropics. These weather systems follow paths that move toward the poles in the summer and toward the equator in the winter, always under the influence of the Polar Front and its jet stream. Their general eastward movement varies in speed from 0 to 60 kph (0 to 35 mph) in winter; in summer, the velocities range from 0 to 40 kph (0 to 25 mph). These wave cyclones may be anywhere from 300 to 3000 km (200 to 2000 mi) in diameter and range from 8 to 11 km (5 to 7 mi) in height. They provide the greatest source of rain in the midlatitudes, and, as we are about to see, they generate kinetic energy that helps power the primary circulation of the atmosphere.

ENERGY AND MOISTURE WITHIN WEATHER SYSTEMS

As can be seen in satellite images and as your own experiences probably confirm, every day a multitude of weather systems parade across the planetary surface. These systems entail the large-scale atmospheric disturbances of the tropical, middle, and higher latitudes

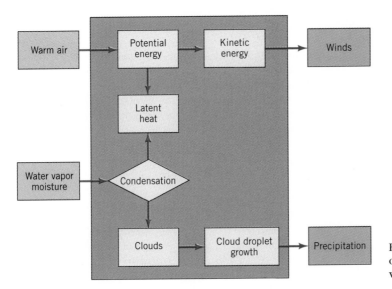

FIGURE 14-11 A schematic representation of the energy and moisture transformations within a weather system.

discussed in this unit, which are fundamental to an understanding of the climates of those zones. They also involve the meso- (medium-) and small-scale systems that make up the daily texture of the atmosphere, such as monsoonal circulations (covered in Unit 10) and thunderstorms (Unit 13). Each weather system, though distinctive in detail, has a characteristic internal organization. In each, warm air is transformed into faster-moving air, and moisture becomes precipitation.

The processes of such transformations form a unique system, which is diagrammed in Fig. 14-11. There are two inputs into this system: (1) warm air carrying the energy of heat, and (2) moisture in the form of water vapor. The heat energy of the rising warm air adds potential energy to the system, and the system's center of gravity lifts higher off the ground. At the same time, the moisture condenses, which has two effects. First, the latent heat that is released helps to swell the store of potential energy; and second, the condensed moisture appears as clouds. There are, of course, outputs from the system as well. The growing store of potential energy changes to kinetic energy, and air movement increases. This kinetic energy leaves the system in the form of winds. Meanwhile, the droplets in the clouds grow to precipitable size. They, too, are exported from the system, via the various forms of precipitation that fall to earth.

These events occur in clouds of the ITCZ, in tropical easterly waves, and in the wet monsoons. You may be able to envision them best in the more dramatic weather produced by tropical cyclones and thunderstorms. Yet, whether in a tornado or in a midlatitude cyclone, these events take place in a similar way. Because of the inputs and outputs of energy and moisture to many of these phenomena, we have considerable justification for calling them weather *systems*. Unit 15 looks at how we track and forecast weather systems such as these, and Units 16 to 19 demonstrate how these systems combine together with the general atmospheric circulation to form the climates of the earth.

KEY TERMS

cyclogenesis (p. 154)

easterly wave (p. 149)

eye (p. 150)

eye wall (p. 150)

hurricane (p. 149)

occluded front (p. 155)

open wave (p. 155)

Polar Front jet stream (p. 154)

stationary front (p. 155)

storm (p. 148)

weather system (p. 148)

REVIEW QUESTIONS

1. Describe the general conditions necessary for the formation of a hurricane.

2. Describe the internal structure of a fully developed hurricane.

3. How does the Polar Front jet stream contribute to the formation of a midlatitude cyclone?

4. Describe the internal structure of a midlatitude cyclone.

5. What is an occluded front, and what sort of weather is associated with it?

6. Describe the general weather changes at a location experiencing the complete passage of a midlatitude cyclone.

REFERENCES AND FURTHER READINGS

CARLSON, T. *Mid-Latitude Weather Systems* (London/New York: Routledge, 1991).

COWEN, R. C. "Global Weather Factors Set Hurricane Factory Spinning," *Christian Science Monitor,* July 12, 1994, 7.

DE BLIJ, H. J., ed. *Nature on the Rampage* (Washington, D.C.: Smithsonian Institution Press, 1994).

EAGLEMAN, J. R. *Severe and Unusual Weather* (Lenexa, Kan.: Trimedia, 2nd ed., 1990).

HARMAN, J. R. *Synoptic Climatology of the Westerlies: Process and Patterns* (Washington, D.C.: Association of American Geographers, Publications for College Geography, 1991).

HARMAN, J. R. *Tropospheric Waves, Jet Streams, and United States Weather Patterns* (Washington, D.C.: Association of American Geographers, Commission on College Geography, Resource Paper No. 11, 1971).

HASTENRATH, S. *Climate Dynamics of the Tropics* (Dordrecht, Netherlands: Kluwer, 1991).

LEITH, C. E., and LITTLE, C. G. *Severe Storms: Prediction, Detection, and Warning* (Washington, D.C.: National Academy of Sciences, 1977).

MUSK, L. F., and BAKER, S. *Weather Systems* (New York: Cambridge University Press, 1988).

National Oceanic and Atmospheric Administration. *Tropical Cyclones of the North Atlantic Ocean, 1871–1986* (Asheville, N.C.: NOAA, National Climatic Data Center, Historical Climatology Series 6-2, 3rd ed., September 1987).

NIEUWOLT, S. *Tropical Climatology: An Introduction to the Climates of the Low Latitudes* (New York: Wiley, 1977).

PIELKE, R. A. *The Hurricane* (London/New York: Routledge, 1990).

REITER, E. R. *Jet Streams* (Garden City, N.Y.: Anchor/Doubleday, 1966).

RIEHL, H. *Climate and Weather in the Tropics* (New York: Academic Press, 1979).

RILEY, D., and SPALTON, L. *World Weather and Climate* (New York: Cambridge University Press, 2nd ed., 1981).

ROBINSON, A. *Earth Shock: Hurricanes, Volcanoes, Earthquakes, Tornadoes and Other Forces of Nature* (New York: Thames & Hudson, 1993).

SIMPSON, R. H., and RIEHL, H. *The Hurricane and Its Impact* (Baton Rouge, La.: Louisiana State University Press, 1981).

Weather Tracking and Forecasting

OBJECTIVES

◆ To discuss the general network of weather stations and the types of data collected from each.

◆ To illustrate typical weather maps compiled from weather data and to provide some elementary interpretations of them.

◆ To outline weather forecasting methods and comment on their formulation and reliability.

By late afternoon of August 8, 1988, the aging northside Chicago neighborhood surrounding Wrigley Field was already jammed with cars and thousands of people on foot. Everyone was streaming toward the old ballpark in anticipation of a historic event—the arrival of night baseball. The luckiest 39,008 of the crowd managed to squeeze into the stadium; thousands more were content to mill around outside Wrigley Field and participate vicariously as the Cubs took on the Phillies. About an hour before the 7:05 game time,

a 91-year-old Cubs fan pushed a button as players from the Chicago Symphony Orchestra struck up the awesome first chords of Richard Strauss's *Also Sprach Zarathustra* (better known as the opening theme from *2001: A Space Odyssey*). The 540 new high-intensity lights began to flicker and soon reached full strength.

UNIT OPENING PHOTO: Ever more sophisticated high-technology instruments monitor today's weather, as we see here atop the control tower at Miami International Airport.

The game began on time, and by the fourth inning the Cubs had even built a 3–1 lead. Then, suddenly, a wall of dark clouds bore in from the northwest, a long downpour ensued, and the umpires were forced to cancel the game at 10:30 P.M. before it had gone far enough for the score to become official. Thus, once again, the best laid plans of people were all for naught. The weather prediction for that Monday evening had mentioned the chance of thundershowers in the metropolitan area, but forecasters could not say with any degree of accuracy which city neighborhoods and suburbs would receive rain and which would not.

This experience is a reminder that, even with today's sophisticated knowledge of the atmosphere and high-technology equipment to monitor its every murmur, weather forecasting remains an art as well as a science. In this unit, we explore the forecasting process by discussing the collection of weather data, the manipulation and mapping of these data, and the critical interpretations made by analysts that lead to weather predictions.

The American public today is probably more aware of weather conditions than ever before (witness the stampede to supermarkets and hardware stores when a snowstorm or hurricane is forecast). This is due, in no small part, to the improvement in the news media's weather coverage in recent years. On television, the inarticulate little bow-tied man, with his broken car antenna pointing at a messy, postage-stamp-sized map, is long gone. Today we are far more likely to see weather reporters with meteorology degrees who use animated maps, color radar, and frequent references to jet stream behavior; and with millions of U.S. homes now able to receive *The Weather Channel* via cable TV, viewers are better informed than ever. The daily press was slower to follow, but when *USA Today* introduced its full-color weather page in the mid-1980s, every major newspaper in the nation swiftly upgraded its weather reporting and graphic presentation.

WEATHER DATA ACQUISITION

The World Meteorological Organization, an agency of the United Nations, supervises the World Weather Watch, a global network of more than 10,000 weather stations and moving vehicles. About 150 nations participate in this monitoring system, which coordinates and distributes weather information from its processing centers in Washington, D.C., Moscow, and Melbourne, Australia. Every three hours, beginning at midnight Greenwich Mean Time (G.M.T.), a set of standard observations on the local state of the atmosphere are taken around the globe and reported to those centers from about 4000 land-based stations, more than 1000 upper-air observation stations, and at least 5000 ships, aircraft, and satellites in transit. High-capacity computers then

rapidly record, manipulate, and assemble these data to produce *synoptic weather charts*, which map meteorological conditions at that moment in time across wide geographic areas.

Although reliable weather instruments have been available since about 1700, comprehensive simultaneous observation could not be attempted before the invention of long-distance communications. The arrival of the telegraph in the 1840s provided the initial breakthrough, but it wasn't until a half-century later that the U.S. Congress, responding to a number of weather-related disasters (such as the great blizzard of 1888), funded the establishment of a monitoring network by the Army's Signal Corps. By the turn of the twentieth century, these efforts intensified with the founding of the Weather Bureau in the U.S. Department of Agriculture, and forecasting improved.

The next impetus was supplied by the fledgling aviation industry, whose successful lobbying campaign led to an enlargement of the activities of the Weather Bureau with its switch from the Agriculture to the Commerce Department just before the United States entered World War II. After the nation entered the war in 1941, weather forecasters were required to expand technologies and hone their prediction skills for the military services, and these were applied to civilian forecasting after 1945. As postwar technology advanced ever more rapidly, including the use of orbiting satellites after 1958, the Weather Bureau kept pace. In 1965 a major reorganization of the environmental agencies of the federal government resulted in the creation of the National Oceanic and Atmospheric Administration (NOAA), and the Bureau was incorporated as the National Weather Service.

Weather Stations

Within the United States, more than 1000 manned observation stations (augmented by many automatic stations at inaccessible locations) report surface weather conditions at least once every three hours. Each station deploys a number of instruments to obtain their readings. These include thermometers, barometers, rain gauges, hygrometers (to measure the moisture of the air), weather vanes (to indicate wind direction), and anemometers (to measure wind speed). The following data are then reported to the synoptic network: air temperature, dew-point temperature, air pressure and direction of change, precipitation (if any), wind speed and direction, visibility, relative amount of cloudiness, cloud types present, and estimated height of the cloud base.

This data cluster for each station is then processed by the central facility for placement in shorthand, symbolic form on the synoptic weather chart, as shown in Fig. 15-1. The importance of radar, especially for storm detection and tracking, has long been recognized by the National Weather Service (Fig. 15-2). But the high cost

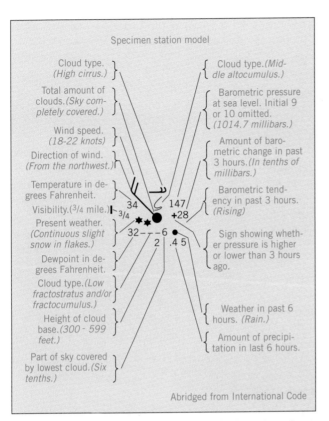

Specimen station model

Cloud type. (High cirrus.)

Total amount of clouds. (Sky completely covered.)

Wind speed. (18-22 knots)

Direction of wind. (From the northwest.)

Temperature in degrees Fahrenheit.

Visibility. (3/4 mile.)

Present weather. (Continuous slight snow in flakes.)

Dewpoint in degrees Fahrenheit.

Cloud type. (Low fractostratus and/or fractocumulus.)

Height of cloud base. (300 - 599 feet.)

Part of sky covered by lowest cloud. (Six tenths.)

Cloud type. (Middle altocumulus.)

Barometric pressure at sea level. Initial 9 or 10 omitted. (1014.7 millibars.)

Amount of barometric change in past 3 hours. (In tenths of millibars.)

Barometric tendency in past 3 hours. (Rising)

Sign showing whether pressure is higher or lower than 3 hours ago.

Weather in past 6 hours. (Rain.)

Amount of precipitation in last 6 hours.

Abridged from International Code

FIGURE 15-1 The data cluster for each weather station that is entered in this shorthand form on the synoptic weather chart. This diagram, which represents a specimen station, is printed on the front of the *Daily Weather Map* that is published by the National Weather Service.

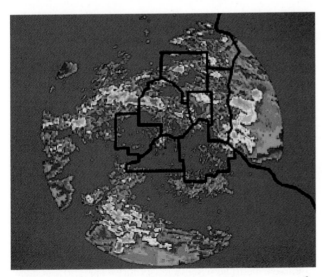

FIGURE 15-2 Radar displays of regional weather are now a familiar part of newscasts on local television stations. This image shows a band of thunderstorms (oranges and reds) crossing the heart of the Minneapolis-St. Paul metropolitan area.

of such equipment has limited the installation of radar to less than 300 weather stations through the mid-1990s (plans do call for adding radar to many more observation facilities by the turn of the century).

Surface conditions, however, represent only a single dimension of the overall weather picture. To gain a fuller understanding, meteorologists must go beyond this bottom slice of the atmosphere and acquire a more complete profile of its current vertical structuring. One way this is accomplished is through the use of **radiosondes**, radio-equipped instrument packages that are carried aloft by balloon. The United States has about 130 radiosonde stations, part of a worldwide network that transmits readings to receiving facilities on the ground every 12 hours (noon and midnight G.M.T.). Whereas most of these upper-air observations of temperature, humidity, and pressure are obtained for altitudes up to 32 km (20 mi) above the surface, a number of radiosonde stations now release more powerful balloons that can reach heights well beyond 32 km. High-altitude radar is also a useful tool for acquiring information about the upper atmosphere, and equipped stations regularly make *rawinsonde* observations (radar trackings of radiosonde balloons) that provide information about wind speed and directions at various vertical levels.

Weather Satellites

Perhaps the most important source of three-dimensional atmospheric data today are the dozens of orbiting weather satellites that constantly monitor the earth. Many of these satellites operate within longitudinal orbits about 1100 km (700 mi) high that pass close to the poles, so that they survey a different meridional segment of the surface during each revolution; in this manner, a complete picture of the globe can be assembled every few hours. More and more satellites are now being placed in much higher orbits (around 35,000 km [22,000 mi]) above the equator, called **geosynchronous orbits**—i.e., they revolve at the same speed as the planet rotates and are therefore stationary above a given surface location. A "fixed" satellite at this altitude can continually monitor the same one-third or so of the earth, and with infrared (IR) capability it can perform that task in darkness as well as daylight.

One of the best known orbiters of this kind is the GOES-8 satellite. It transmits images to ground facilities twice an hour, and many of them appear in daily television and newspaper weather reports (Fig. 15-3). GOES-8 (GOES stands for Geostationary Orbiting Environmental Satellite) was launched in 1994 to replace older satellites that monitor the oceans bordering the United States. It will be joined in late 1995 by GOES-9, allowing GOES-8 to be shifted to a fixed position (above 0°, 75°W) that scans the tropical Atlantic while GOES-9 hovers over the eastern equatorial Pacific Ocean about 3700 km (2300 mi) southeast of Hawaii. By 2005, NOAA expects to have three additional GOES satellites in orbit to further enhance its coverage with the most advanced weather observation systems available.

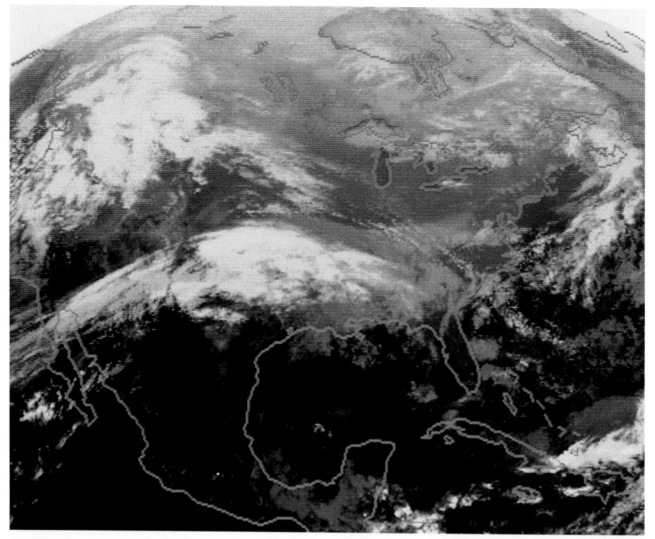

FIGURE 15-3 A GOES-8 infrared image of central North America, with darker shades representing warm areas of the surface and atmosphere, and lighter (whiter) colors the colder clouds, water bodies, and land areas. This view was taken on November 25, 1994 at 5:45 P.M., E.S.T., and shows storm systems in the U.S. Northeast and Northwest.

MAPPING WEATHER DATA

The massive quantity of surface and upper-air weather data reported by thousands of stations around the globe are collected and organized by various international and national processing centers so that forecasters may begin their never-ending work. Synoptic weather charts are crucial to this task, and the meteorological services of most of the world's countries prepare and publish these maps at frequent intervals. In the United States, the National Weather Service's National Meteorological Center publishes its *Daily Weather Map*, which includes a large-surface weather map and smaller maps showing 500-mb height contours, highest and lowest temperatures of the previous 24 hours, and precipitation areas and amounts of the previous 24 hours. These maps for any given day can be purchased from NOAA's Climate Analysis Center (World Weather Building, Room 808, Washington, DC

20233). Because the surface and 500-mb maps are fundamental tools in learning about the geography of American weather, we will review each in detail. The maps of Sunday, October 12, 1986 were selected for presentation here.

The Surface Weather Map

The map of surface weather conditions for 7:00 A.M. Eastern Standard Time on October 12, 1986 is shown in Fig. 15-4. But before we proceed to interpret its contents, it is necessary to become familiar with the map's point, line, and area symbols.

Each weather station is represented by a point symbol that shows up on the map as a data cluster arranged around a central circle—an application of the specimen station model displayed in Fig. 15-1. In the central circle, the percentage of blackening indicates the proportion of

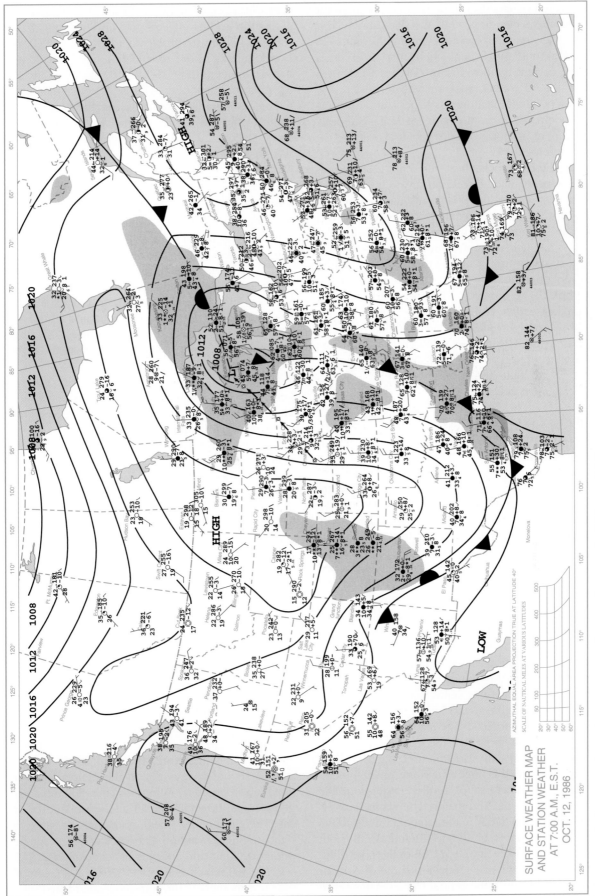

SURFACE WEATHER MAP
AND STATION WEATHER
AT 7:00 A.M., E.S.T.
OCT. 12, 1986

FIGURE 15-4 The National Weather Service's surface weather map for 7:00 A.M., E.S.T., Sunday, October 12, 1986.

the sky covered by clouds; on the October 12 map, for instance, Seattle is cloudless while the skies above New Orleans are seven-tenths covered. Among the numbers distributed around the central circle, the most noteworthy are current temperature in degrees Fahrenheit (shown at the 10 o'clock position with respect to the circle), dew-point temperature (8 o'clock), and air pressure (1 o'clock). Thus, on the map, Los Angeles reports a temperature of 64°F (18°C), a dew point of 56°F (13°C), and a pressure of 1015.6 mb.

Among these symbols, most important are wind direction and speed. Wind direction (always indicated by where the wind comes *from*) is represented as the line extending outward from the central circle; wind speed is indicated by the number of hash marks attached to the end of that line. Thus Miami's wind is from the north-northwest at about 4 to 13 kph (3 to 8 mph), while the wind in Albuquerque, New Mexico, is from the east at 33 to 40 kph (21 to 25 mph). The line symbols in Fig. 15-4 represent either isobars or fronts. The grayish area symbols represent ongoing precipitation, which is largely rain in the Great Lakes region but snow in the shaded zone centered on Colorado; also represented as area symbols are high- and low-pressure cells.

Open-Wave Cyclone

The main weather pattern exhibited in Fig. 15-4 should not be unfamiliar to you, and probably removes any remaining mystery as to why we chose this particular synoptic chart. Although no two storm systems are exactly alike, the wave cyclone shown on this map conforms quite closely to the model and discussion in Unit 14. If you compare the map's air mass and frontal characteristics to the diagram shown in Fig. 14-9 (p. 155), you will recognize that the October 12, 1986 cyclone has reached a stage of development that is about three-quarters of the way toward reaching full open-wave status. The eastward-sweeping cold front has just reached the Mississippi Valley, and it can be expected both to begin occluding and to continue moving rapidly to the east and north toward the stationary warm front. The latter extends northeastward, through central Ontario and Quebec, from the low-pressure center located astride the border between Wisconsin and Michigan's Upper Peninsula. Comparison should also be made with Fig. 14-10a (p. 156), which will show that the October 12 system unmistakably displays the isobaric and windflow patterns associated with this stage of cyclonic development.

Temperature relationships further underscore the linkage between model and reality. The warm sector is especially well delineated by the temperature contrasts behind (Moline, Illinois, at 46°F [8°C]) and ahead (both Chicago and Milwaukee at 62°F [17°C]) of the cold front. Note, too, that the massive southward surge of cold air behind the cold front in the central United States is undoubtedly an early-autumn outburst of fast-moving polar

(**cPk**) air. Even in subtropical South Texas the difference across the leading edge of cold air is striking: while San Antonio behind the front recorded 55°F (13°C), nearby Laredo, ahead of the advancing tongue of cold air, still had a temperature of 76°F (24°C). Yet, modification of the polar outburst of cold air is also quite evident in the temperature gradient that rises from the Canadian border south to the Rio Grande Valley: 18°F (−8°C) at Minot, North Dakota; 29°F (−2°C) at Pierre, South Dakota; 33°F (0.5°C) at Dodge City, Kansas; 41°F (5°C) at Abilene, Texas; and 55°F (13°C) at San Antonio.

Storm Track

Another notable feature of Fig. 15-4, which is represented by a series of point symbols (white **X**'s inside tiny black squares), is the path or *track* of the moving storm system. The three squared **X**'s on this map—located in southeastern Nebraska, north-central Iowa, and north-western Wisconsin—indicate the positions of the low-pressure center, respectively, 18, 12, and 6 hours before map time. The arrows that connect these **X**'s, together with the low's presently mapped position, define the storm track.

Midlatitude cyclones tend to form in certain areas and often follow similar tracks in their eastward journeys. (We noted the same tendency for poleward-curving North Atlantic hurricanes; *see* Fig. 14-4.) Figure 15-5 maps the most common cyclone tracks in the United States, with each average storm path named after its place of origin. Note that these disturbances are generated at various latitudes and that most tracks converge toward the northeastern seaboard. When the October 12 cyclone is compared to Fig. 15-5, it appears to have formed in the Central source region but is pursuing a more northerly course.

The Upper-Air Weather Map

The significance of vertical atmospheric data for weather forecasting has already been established. Accordingly, the National Weather Service, based on radiosonde and rawinsonde observations, issues a 500-Millibar Height Contours map as part of its *Daily Weather Map* synoptic-chart package. This is an important tool in the analysis of the upper atmosphere. In reading a 500-mb chart, it should be remembered that the higher the altitude of the 500-mb level, the warmer the surface temperature of the air beneath it; therefore, the gradient of the 500-mb surface should always dip in the general direction of the higher latitudes. Winds on the 500-mb chart are geostrophic, flowing parallel to the isobars (or, in this case, height contour lines), with high pressure on the right looking downwind. Extreme curvature signifies upper-air cyclonic and anticyclonic circulations.

The 500-mb chart for October 12, 1986, compiled at the same moment as the surface weather map (Fig. 15-4), is shown in Fig. 15-6. Symbolization here should be quite

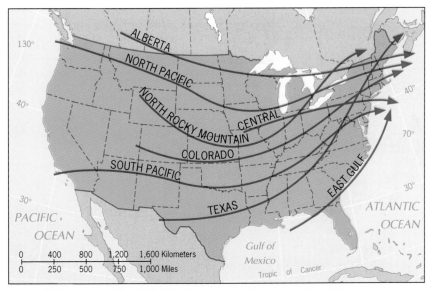

FIGURE 15-5 Common midlatitude cyclone tracks in the contiguous United States. These weather systems are named after their region of origin and generally move in a west-to-east direction.

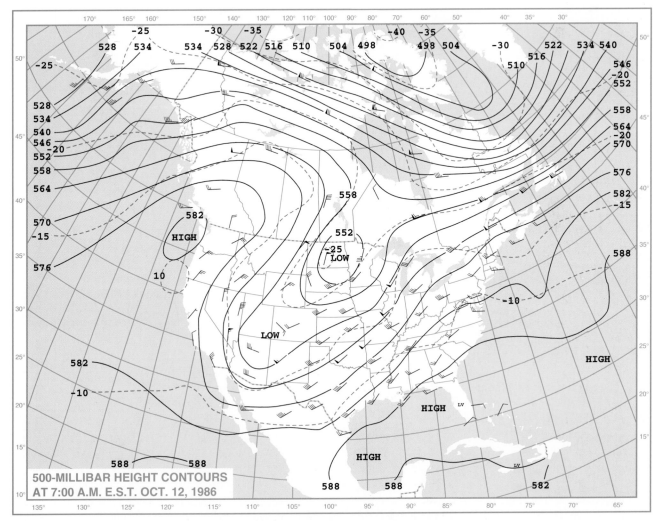

FIGURE 15-6 The 500-mb chart for October 12, 1986, compiled at the same moment as the surface weather map in Fig. 15-4. Heights are given in 10-m units known as dekameters.

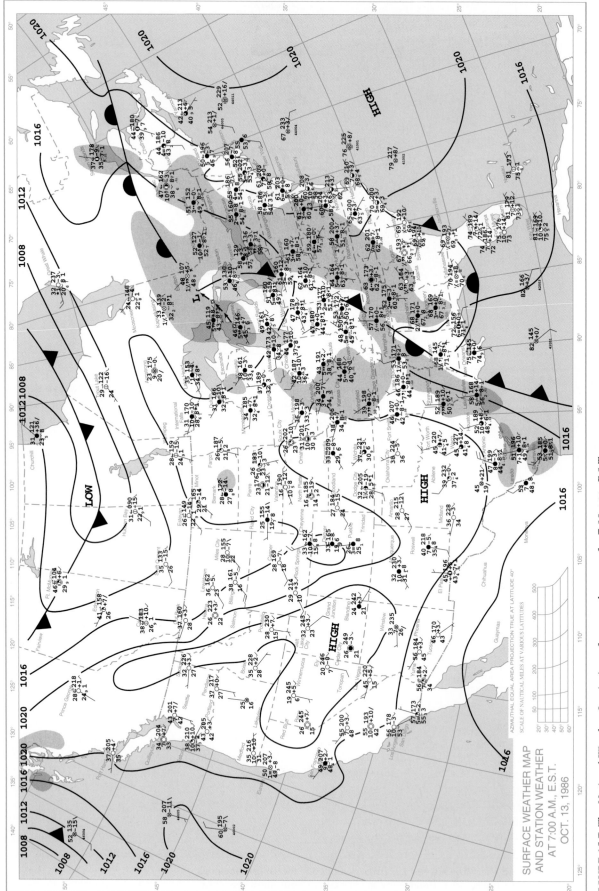

SURFACE WEATHER MAP
AND STATION WEATHER
AT 7:00 A.M., E.S.T.
OCT. 13, 1986

FIGURE 15-7 The National Weather Service's surface weather map for 7:00 A.M., E.S.T.,
Monday, October 13, 1986.

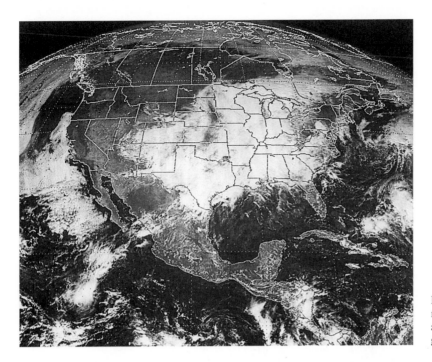

FIGURE 15-8 Satellite image of the conterminous United States for October 12, 1986. Usually, the brightest white clouds contain the greatest precipitation.

easy to interpret. The height contours on this map are calibrated in 10-m units called dekameters (one is equal to 32.8 ft), and, in a sudden embrace of the metric system, the National Weather Service employs Celsius instead of Fahrenheit degrees in the mapping of key upper-air isotherms. Wind speeds and directions are represented by the same symbols used for surface weather stations. A departure in wind-speed depiction is the triangular pennant (which equals five hash marks in the style of Fig. 15-1), indicating velocities in excess of 88 kph (55 mph).

The configuration of the 500-mb surface resembles the ground-level pressure pattern (Fig. 15-4) in a number of ways. Most prominent is the upper-air trough that dips out of Canada to cover much of the interior United States. At its heart is the fast-flowing Polar Front jet stream, which can be picked out by following the highest speed winds represented by pennant symbols. Thus we may trace this jet stream—which adheres closely to the 564-dekameter height contour—from Montana south to Arizona and then northeastward through the central Great Plains across Lake Superior into southeastern Canada. As we would expect, bad weather in the form of frontal precipitation occurs where the jet stream flows toward the North Pole. Conversely, fair weather is observed to the west of the Rocky Mountains where the jet moves toward the equator.

Next we can recognize correspondence between the surface cyclone and the upper-air trough located over the northern Plains, but note the latter is centered above eastern North Dakota somewhat to the west of the ground-level low. This westward sloping of the low with increasing altitude occurs because of the concentration of cold air to the west of the storm system, which brings the 500-mb level significantly closer to the ground than it would in the wave cyclone's warmer sectors. We can also say something about the track that the storm is likely to follow in the immediate future, because the upper-air

steering winds are so strikingly parallel to the cyclone's current path.

Atmospheric conditions, of course, are subject to change in a matter of hours, particularly with the imminent occlusion of our weather system. Nevertheless, if you had to issue the forecast at 7:00 on that Sunday morning with only these maps to go by, you would not be going out on a limb to predict that the cyclone was likely to continue moving straight across Lake Superior and then turn in a more easterly direction.

WEATHER FORECASTING

Unlike weather forecasters who must make projections into an unknown future, we have the luxury of being able to check the outcome of our "prediction" by simply consulting the next daily weather map (October 13, 1986). That surface map is shown in Fig. 15-7 and immediately reminds us how fickle these storms can be. Note that the cyclone has indeed moved to the northeast, but instead of crossing Lake Superior the track shifted eastward across northern Wisconsin. From there this weather system turned to the northeast, proceeded across the northern portions of Lakes Michigan and Huron, and was over Ontario north of Lake Huron at 7:00 A.M. on October 13. (The cyclone subsequently deepened and moved down the St. Lawrence Valley, entering the Atlantic Ocean late in the day on October 14.)

As for the fronts, the anticipated occlusion never occurred. This particular wave cyclone left the continent before reaching that stage of development, underscoring that models of cyclogenesis, by themselves, are not always reliable predictors of weather behavior. Returning to the morning of October 12, it is also worth noting the satellite image of the United States shown in Fig. 15-8, which was taken only a few hours after the map in Fig.

Weather Extremes

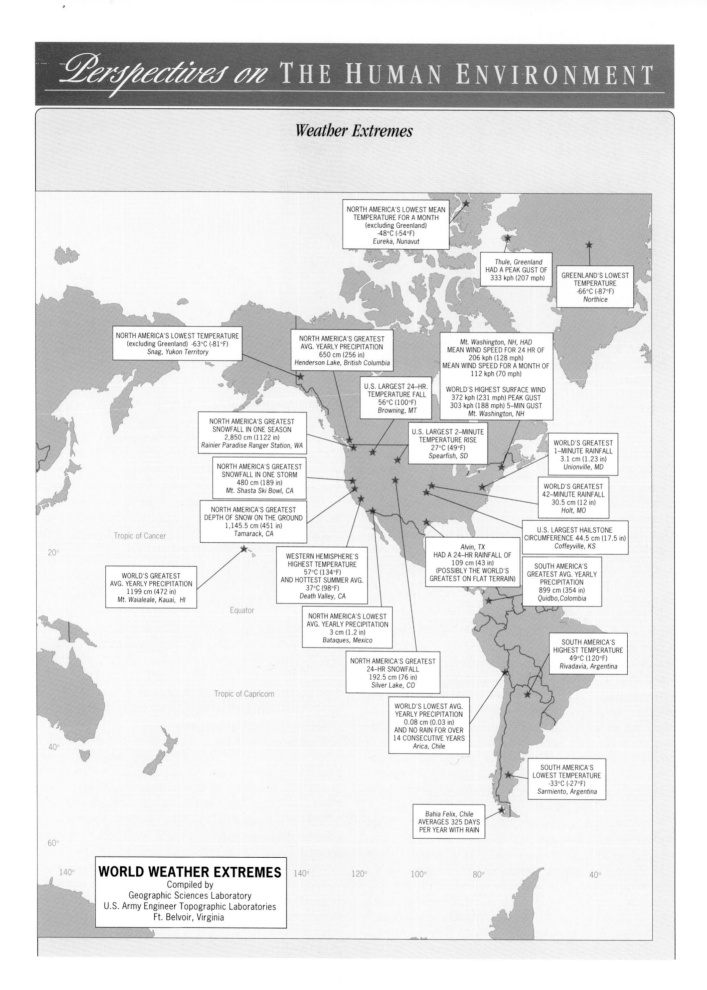

NORTH AMERICA'S LOWEST MEAN
TEMPERATURE FOR A MONTH
(excluding Greenland)
-48°C (-54°F)
Eureka, Nunavut

Thule, Greenland
HAD A PEAK GUST OF
333 kph (207 mph)

GREENLAND'S LOWEST
TEMPERATURE
-66°C (-87°F)
Northice

NORTH AMERICA'S LOWEST TEMPERATURE
(excluding Greenland) -63°C (-81°F)
Snag, Yukon Territory

NORTH AMERICA'S GREATEST
AVG. YEARLY PRECIPITATION
650 cm (256 in)
Henderson Lake, British Columbia

Mt. Washington, NH, HAD
MEAN WIND SPEED FOR 24 HR OF
206 kph (128 mph)
MEAN WIND SPEED FOR A MONTH OF
112 kph (70 mph)
WORLD'S HIGHEST SURFACE WIND
372 kph (231 mph) PEAK GUST
303 kph (188 mph) 5-MIN GUST
Mt. Washington, NH

U.S. LARGEST 24-HR.
TEMPERATURE FALL
56°C (100°F)
Browning, MT

NORTH AMERICA'S GREATEST
SNOWFALL IN ONE SEASON
2,850 cm (1122 in)
Rainier Paradise Ranger Station, WA

U.S. LARGEST 2-MINUTE
TEMPERATURE RISE
27°C (49°F)
Spearfish, SD

WORLD'S GREATEST
1-MINUTE RAINFALL
3.1 cm (1.23 in)
Unionville, MD

NORTH AMERICA'S GREATEST
SNOWFALL IN ONE STORM
480 cm (189 in)
Mt. Shasta Ski Bowl, CA

WORLD'S GREATEST
42-MINUTE RAINFALL
30.5 cm (12 in)
Holt, MO

NORTH AMERICA'S GREATEST
DEPTH OF SNOW ON THE GROUND
1,145.5 cm (451 in)
Tamarack, CA

U.S. LARGEST HAILSTONE
CIRCUMFERENCE 44.5 cm (17.5 in)
Coffeyville, KS

Tropic of Cancer

20°

WORLD'S GREATEST
AVG. YEARLY PRECIPITATION
1199 cm (472 in)
Mt. Waialeale, Kauai, HI

WESTERN HEMISPHERE'S
HIGHEST TEMPERATURE
57°C (134°F)
AND HOTTEST SUMMER AVG.
37°C (98°F)
Death Valley, CA

Alvin, TX
HAD A 24-HR RAINFALL OF
109 cm (43 in)
(POSSIBLY THE WORLD'S
GREATEST ON FLAT TERRAIN)

SOUTH AMERICA'S
GREATEST AVG. YEARLY
PRECIPITATION
899 cm (354 in)
Quidbo,Colombia

Equator

NORTH AMERICA'S LOWEST
AVG. YEARLY PRECIPITATION
3 cm (1.2 in)
Bataques, Mexico

SOUTH AMERICA'S
HIGHEST TEMPERATURE
49°C (120°F)
Rivadavia, Argentina

NORTH AMERICA'S GREATEST
24-HR SNOWFALL
192.5 cm (76 in)
Silver Lake, CO

Tropic of Capricorn

WORLD'S LOWEST AVG.
YEARLY PRECIPITATION
0.08 cm (0.03 in)
AND NO RAIN FOR OVER
14 CONSECUTIVE YEARS
Arica, Chile

40°

SOUTH AMERICA'S
LOWEST TEMPERATURE
-33°C (-27°F)
Sarmiento, Argentina

Bahia Felix, Chile
AVERAGES 325 DAYS
PER YEAR WITH RAIN

60°

WORLD WEATHER EXTREMES
Compiled by
Geographic Sciences Laboratory
U.S. Army Engineer Topographic Laboratories
Ft. Belvoir, Virginia

140°   140°   120°   100°   80°   40°

Since climatologists no longer concentrate their work on average figures, a brief consideration of the extreme weather conditions that increasingly concern them is a worthwhile exercise. (We first encountered this topic in the Perspective box in Unit 7, where we noted the unusual number of weather extremes recorded during the past decade [*see* Fig. 7-4 on pp. 76–77]; the aim here is to take a longer view of world-wide patterns, with emphasis on the United States.)

In early 1987, Athens, Greece experienced its first recorded snowfall and

FIGURE 15-9 Some noteworthy global weather extremes.

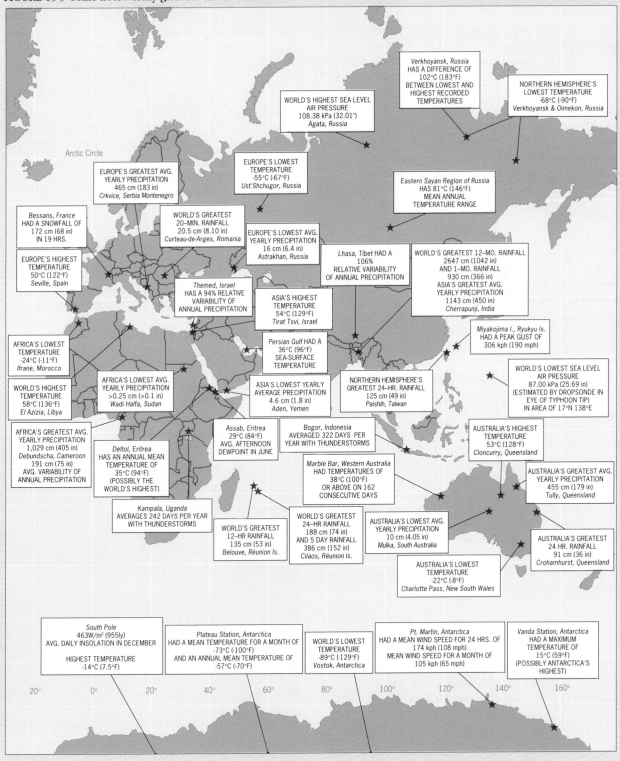

then, a few months later, its hottest summer ever—only to learn the following December that the average yearly temperature came out almost exactly "normal." On August 11, 1984, Miami residents experienced both the highest (35.5°C [96°F]) and lowest (21°C [70°F]) temperatures ever recorded on that day—only to see it noted that the mean temperature for the day came very close to the normal daily average of 28°C (82°F). These are only two isolated cases, but they underscore how averages can misrepresent actual weather conditions and the accompanying human adjustments to the physical environment of a particular place.

The most changeable weather occurs in the midlatitudes, and the United States certainly makes an impact on the global map of extreme weather (Fig. 15-9). Among the world weather records held by places in the United States are greatest yearly average precipitation, highest surface wind speed, and the greatest rainfalls of 1 minute, 42 minutes, and 24 hours on flat terrain. Moreover, the United States holds the hemispheric record for highest temperature and hottest average summer, as well as six other North American records for overall precipitation and snowfall.

Continentality has something to do with the startling swings of U.S. seasonal temperature patterns, which so often seem to produce such "record" weather episodes as the terrible Northeast/Midwest winter of 1993–1994 or the memorable heat and drought of mid-1988 almost everywhere east of the Rocky Mountains. But we also know that the great, unremitting tug of war between polar and tropical air masses, heightened by the volatile twistings of the Polar Front jet stream, produce many more weather extremes in the United States than in the continental heart of much larger Eurasia. (Note how, in Fig. 15-9, Siberia records only a handful of global weather extremes in temperature and air pressure readings.) And as we learn more about global temperature trends (Unit 7) and phenomena of the El Niño type (Unit 11), atmospheric scientists may well discover additional explanations for the many still-puzzling occurrences of extreme weather that mark the world map.

15-4 was compiled. Our cyclone dominates the national weather pattern, and precipitation zones are likeliest where the white clouds are brightest (which usually indicates the highest cloud tops).

Although it is safe to say that those who issued the actual forecast on October 12, 1986 had more data and experience on which to base their predictions, much of the salient information they needed was indeed summarized in the surface and 500-mb synoptic charts we have discussed. Whereas the 7:00 A.M. E.S.T. maps represent "freeze frames," forecasters had access to the "motion picture" of the dynamic atmosphere as satellites, radar, and other instruments constantly monitored its every change. Yet even these data, when run through highly sophisticated computers, could provide only small additional clues as to what was about to happen. Thus even in the 1990s weather prediction, especially in the United States, is still an art as well as a science, because American forecasters oversee a geographic domain that experiences some of the most variable weather on earth that frequently swings from one extreme to another (*see* Perspective: Weather Extremes).

The Forecasting Industry

Weather forecasts in the United States are prepared at the National Meteorological Center at Camp Springs, Maryland, adjacent to Andrews Air Force Base in the eastern suburbs of Washington, D.C. From the broad national forecast, weather predictions are issued for regions and localities that are then communicated to the public via local National Weather Service facilities around the country. The only departure from this procedure is the forecasting of hurricanes and severe local disturbances that have the potential for generating tornadoes. Special-purpose forecasting centers in Miami and Honolulu are in charge of Atlantic and Eastern/Central Pacific hurricanes, respectively; the National Severe Storms Forecast Center in Kansas City monitors conditions that may allow for the development of dangerous local convective storm cells and squall lines (Fig. 15-10).

National Weather Service forecasts since the 1950s have been produced by computers programmed with mathematical models of the atmosphere, which translate

FIGURE 15-10 NOAA's special-purpose forecasting facilities utilize state-of-the-art computer and imaging systems to study and track storms. This installation is part of the new National Lightning Detection Network.

current temperature, wind, air pressure, and moisture data into projections and maps of likely weather conditions 12, 24, 36, and 48 hours into the future. This computer forecasting method is called *numerical weather prediction* and is based on projections by small increments of time. For example, a forecast is prepared for the weather 15 minutes from now. Once those conditions are determined, the computer repeats the process by using them to make predictions for 15 minutes after that and so forth until the desired future time of the forecast is reached. Obviously, unavoidable errors begin to creep in immediately, and they are magnified as the length of the forecast period increases. The American Meteorological Society has pronounced that the accuracy of forecasts up to 48 hours ahead is now "considerable." While that may be true nationally, your local experiences may not always fit that designation!

It has been pointed out that the northeastern United States, home to over 25 percent of the U.S. population, is a region of particularly changeable weather. Thus, over the years, more forecasts have gone wrong there than in any other part of the nation. Winter storms have proven especially difficult to predict even a few hours ahead, and the need for better snow and ice warnings in this heavily urbanized region has prompted the intensification of research efforts. Meteorologists are most interested in learning more about the sudden genesis of fierce snowstorms they call "bombs."

> [A "bomb" is a] relatively mild winter weather disturbance that drifts northward in the Atlantic from the Cape Hatteras area off North Carolina and then, when it is off New York or Cape Cod, the "bombing range," suddenly and inexplicably explodes within six hours or so into a surprising, freezing fury that threatens unsuspecting shipping and paralyzes unprepared coastal cities and towns with heavy sheets of snow and ice (Ayres, 1988).

Over the past quarter-century, more than a hundred of these storms have caught the Northeast off guard, and forecasters are still unable to do anything more than closely watch the area when conditions are ripe and report a storm as it swiftly materializes. Meteorologists suspect that "bombs" are the product of the collision of Arctic and moist tropical air. This contrast is heightened along the Mid-Atlantic seaboard by the trapping of large quantities of cold air in the nearby Appalachian Mountains and the infusion of warm-surface seawater by the Gulf Stream (*see* Fig. 11-3). A new data-gathering network of sea buoys and monitoring instruments is now in place offshore, and researchers are hopeful that this technology will enable them to better understand and predict these dangerous storms.

Because the predictions of the National Weather Service are for broad areas and are occasionally unreliable,

businesses whose activities are closely related to weather conditions have given rise to a private forecasting industry. About 50 such companies now operate in the United States, and they concentrate on tailoring forecasts to the special needs of their corporate clients. Agricultural concerns comprise the biggest single user category. Private forecasters can prepare surprisingly specific predictions, particularly as to moisture conditions, and the recent growth of their clientele demonstrates that an important need is being met. Many subscribers are electrical utilities and fuel-oil distributors that require hour-by-hour information in order to be ready to meet customer demands as temperatures rise or fall. And a whole host of firms engaged in outdoor activities are also frequent consumers of private forecasting services, ranging from airline companies to motion-picture producers to construction contractors. The general public is probably best acquainted with the news media's use of private forecasts: Accu-Weather of State College, Pennsylvania, for instance, issues daily customized forecasts to hundreds of newspapers and radio and television stations.

Long-Range Forecasting

Long-range forecasting constitutes a lively new frontier for weather scientists and poses some of the toughest research challenges they face today. To be sure, in response to the need for longer-term outlooks, the National Weather Service has for years issued monthly and seasonal forecasts (Fig. 15-11), but the American Meteorological Society understandably regards the accuracy of such prognostications as "minimal." Nonetheless, interest in these 30- and 90-day forecasts remains high, and they are widely reported by the news media.

The monthly projections are based on analyses of upper-air pressure patterns, which are evaluated against atmospheric behavior in previous years and then translated into surface-level temperature and precipitation zones. Seasonal (90-day) forecasts are derived from the experiences of the two previous years and can often go awry as soon as they are issued. Not surprisingly, atmospheric scientists rate the 30-day forecasts as only 14 percent better than random guesswork for temperature, and just 6 percent better for precipitation; the 90-day forecasts score a dismal 5 and 3 percent, respectively.

Units 13, 14, and 15 have focused on the moving weather systems that blanket the earth at any given moment. These recurring systems, together with the constant flows of radiation, heat energy, and moisture, make up our daily weather. Over time, a region's weather conditions exhibit regular rhythms and patterns that are characteristic of a certain type of climate. Unit 16 provides a basis for the classification and regionalization of climate types, which is subsequently applied (in Units 17–19) to all of the earth's land areas.

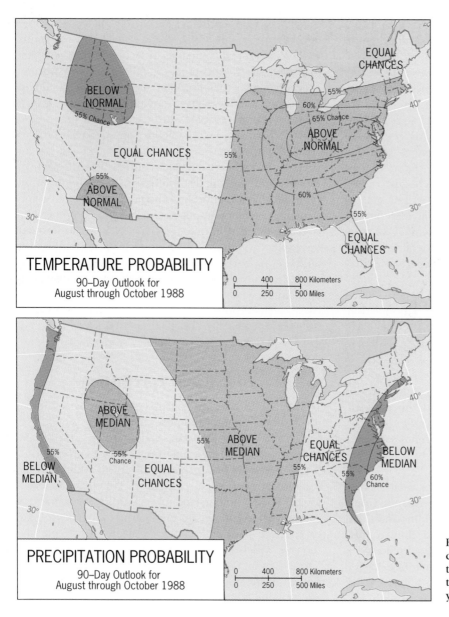

FIGURE 15-11 Long-range weather forecasting by the National Weather Service: the 90-day outlook for August through October of 1988, issued on July 29 of that year.

KEY TERMS

geosynchronous orbit (p. 161)   radiosonde (p. 161)

REVIEW QUESTIONS

1. Describe the basic weather elements whose variations are observed and recorded at weather stations.

2. What is a radiosonde station and what kinds of atmospheric data does it provide?

3. How does the 500-mb chart aid in the forecasting of surface weather?

4. Why is long-range weather forecasting still such a tenuous science?

REFERENCES AND FURTHER READINGS

AYRES, B. D., JR. "Volatile Weather 'Bomb' Is Tracked," *New York Times*, March 14, 1988, 8.

BURROUGHS, W. *Watching the World's Weather* (New York: Cambridge University Press, 1991).

CARLETON, A. M. *Satellite Remote Sensing in Climatology* (London: Belhaven, 1991).

DOVIAK, R. J. and ZRNIC, D. S. *Doppler Radar and Weather Observations* (San Diego: Academic Press, 2nd ed., 1993).

EAGLEMAN, J. R. *Severe and Unusual Weather* (Lenexa, Kan.: Trimedia, 2nd ed., 1990).

FISHMAN, J., and KALISH, R. *The Weather Revolution: Innovations and Imminent Breakthroughs in Accurate Forecasting* (New York: Plenum, 1994).

GODSKE, C. L., et al. *Dynamic Meteorology and Weather Forecasting* (Boston: American Meteorological Society, 1975).

HOUSTON, P. "[Private] Weather Forecasters Enjoy Boom," *New York Times*, July 18, 1988, 27.

HUGHES, P. J. *American Weather Stories* (Washington, D.C.: U.S. Environmental Data Service, 1976).

National Oceanic and Atmospheric Administration. *Operations of the National Weather Service* (Washington, D.C.: NOAA, 1981).

PEARCE, E. A., and SMITH, C. G. *World Weather Guide* (New York: Random House, 1990).

PETTERSSEN, S. *Weather Analysis and Forecasting* (New York: McGraw-Hill, 2 vols., 1956).

RILEY, D. and SPALTON, L. *World Weather and Climate* (New York: Cambridge University Press, 2nd ed., 1981).

RUFFNER, J. A., and BLAIR, F. E. *Weather Almanac* (Detroit: Gale Research Co., 2nd ed., 1977).

TAKAHASHI, K., ed. *World-Wide Weather* (Rotterdam: A. A. Balkema, 1985).

WAGNER, R. L., and ADLER, B., JR. *The Weather Sourcebook: Your One-Stop Resource for Everything You Need to Feed Your Weather Habit* (Old Saybrook, Conn.: Globe Pequot, 1994).

Climate Classification and Regionalization

OBJECTIVES

◆ To define climate and discuss the general problems of climate classification based on dynamic phenomena.

◆ To outline a useful climate classification scheme devised by Köppen, based on temperatures and precipitation amounts and timing.

◆ To apply the modified Köppen classification system to the earth, and briefly describe appropriate climate regions as they appear on a hypothetical continent and the world map.

C limatology is the study of the earth's regional climates, whereas meteorology is the study of short-term atmospheric phenomena that constitute weather. To understand this distinction, we may use the rule that weather happens now but climate goes on all the time. Thus climate involves the day-to-day, aggregate weather conditions each of us expects to experience in a particular place. If we live in Florida or Hawaii, we expect to wear light clothes for most of the year; if we live in Alaska, we expect to wear heavy jackets. Climate, therefore, is a synthesis of the succession of weather events we have learned to expect in any particular location.

Because there is a degree of regularity in the heat and water exchanges at the earth's surface and in the general circulation of the atmosphere, there is a broad and predictable pattern of climates across the globe. But this pattern is altered in detail by such additional *climatic controls* as the location of land and water bodies, ocean currents, and mountain ranges and other highlands. More specifically, the **climate** of a place may be defined as the average values of weather elements, such as tem-

UNIT OPENING PHOTO: At the Sahara's southern margin, a zone of climatic transition, the arid desert gives way to the semi-arid grassland.

perature and precipitation, over at least a 30-year period, and the important variations from those average values. In Units 17 to 19, we examine the ways in which broad patterns and fine details create distinctly different climates across the earth. This unit provides a framework for that investigation by discussing the classification of climate types and their global spatial distribution.

CLASSIFYING CLIMATES

Because of the great variety and complexity of recurring weather patterns across the earth's surface, it is necessary for atmospheric scientists to reduce countless local climates to a relative few that possess important unifying characteristics. Such classification, of course, is the organizational foundation of all the modern sciences. Where would chemistry be without its periodic table of the elements, geology without its time scale of past eras and epochs, and biology without its Linnaean system of naming plant and animal species?

The ideal climate classification system would achieve five objectives:

1. It should clearly differentiate among all the major types of climates that occur on earth.
2. It should show the relationships among these climate types.
3. It should apply to the whole world.
4. It should provide a framework for further subdivision to cover specific locales.
5. It should demonstrate the controls that cause any particular climate.

Unfortunately, in the same way that no map projection can simultaneously satisfy all of our requirements (*see* Unit 3), no climate classification system can simultaneously achieve all five of these objectives. There are two reasons. First, so many factors contribute to climate that we must compromise between simplicity and complexity. We have values for radiation, temperature, precipitation, evapotranspiration, wind direction, and so forth. We could use one variable—as the Greeks used latitude to differentiate "torrid," "temperate," and "frigid" zones—and have a classification system that would be too simple to be useful. Or, on the other hand, we could use all the values to gain infinite detail—but overwhelming complexity.

The second reason is that the earth's climates form a spatial continuum. We almost never find sharp areal breaks between the major types of climate, because both daily and generalized weather patterns change from place to place gradually, not abruptly. Yet, any classificatory map forces climatologists to draw lines separating a given climate type from its neighbors, thereby giving

the impression that such a boundary is a sharp dividing line rather than the middle of a broad transition zone.

When confronted with these problems, climatologists focus their compromises on a single rule: the development or choice of a climatic classification must be determined by the particular use for which the scheme is intended. With so many potential uses, it is not surprising that many different classification systems have been devised. Whereas qualitative and/or subjective criteria could be used to distinguish climates (*see* Perspective: Climate in Daily Human Terms), the practicalities of contemporary physical geography demand an objective, quantitative approach. Moreover, for our purposes, the classification system must be reasonably simple and yet reflect the full diversity of global climatic variation. The **Köppen climate classification system** provides us with that balanced approach, offering a descriptive classification of world climates that brilliantly negotiates the tightrope between simplicity and complexity. To Wladimir P. Köppen (1846–1940), the key to classifying climate was plant life.

THE KÖPPEN CLIMATE CLASSIFICATION SYSTEM

In 1874, the Swiss botanist Alphonse de Candolle produced the first comprehensive classification and regionalization of world vegetation based on the internal functions of plant organs. It is to Wladimir Köppen's everlasting credit that he recognized that a plant, or assemblage of plants, at a particular place represents a synthesis of the many variations of the weather experienced there. He therefore looked to de Candolle's classification of vegetation to solve the puzzle of the global spatial organization of climates.

Köppen (pronounced "KER-pin" with the "r" silent) compared the global distribution of vegetation mapped by de Candolle with his own maps of the world distribution of temperature and precipitation. He soon identified several correlations between the atmosphere and biosphere and used them to distinguish one climate from another. For example, he observed that in the high latitudes the boundary marking the presence or absence of trees closely coincided with the presence or absence of at least one month in the year with an average temperature of 10°C (50°F). He noted many other such correlations, and in 1900 he published the first version of his classification system. This regionalization scheme was subsequently modified by Köppen and other researchers (notably Rudolf Geiger) and was to become the most widely used climatic classification system.

Figure 16-1 presents the simplified version of the Köppen system that will be used in this book. Just as code letters are used to describe air masses, the Köppen

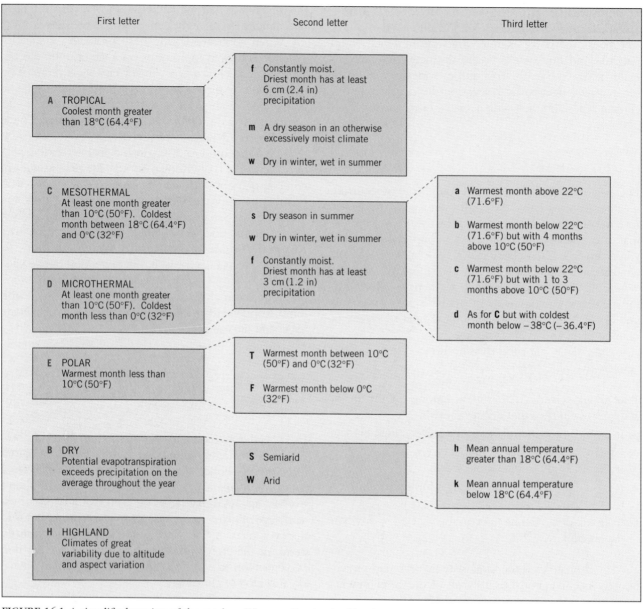

FIGURE 16-1 A simplified version of the modern Köppen climate classification system.

classification uses a shorthand notation of letter symbols to distinguish different characteristics of the major climates. The six major climate groups (column 1) are labeled **A**, **B**, **C**, **D**, **E**, and **H**. The major tropical (**A**), mesothermal (**C**), microthermal (**D**), and polar (**E**) climates are differentiated according to temperature. (*Mesothermal* implies a moderate amount of heat, and *microthermal* implies a small amount—as the temperature criteria in their respective boxes in Fig. 16-1 indicate.) The major dry (**B**) climates are distinguished by potential evapotranspiration exceeding precipitation. The variable highland climates (**H**) are grouped into a separate major category.

As Fig. 16-1 shows, these major climate groups are further subdivided in terms of heat or moisture by the use of a second letter (column 2) and in some cases a third letter (column 3). In the second column, the lowercase letters **f**, **m**, **w**, and **s** tell us when precipitation occurs during the year, and are applicable to **A**, **C**, and **D** climates. Capital letters **S** and **W** indicate the degree of aridity in dry (**B**) climates, with **S** designating semiarid conditions and **W** full aridity. Letters referring to moisture conditions are shown in the green-colored boxes in Fig. 16-1; letters referring to heat are shown in the peach-colored boxes. The heat modifiers of the third column—lower-case letters **a**, **b**, **c**, and **d**—provide details on the temperatures of **C** and **D** climates. The third letters **h** and **k** do the same for the **B** climates, and the second letters **T** and **F** subdivide the temperatures in polar (**E**) climates.

We can define many different climates with this shorthand code, including all that can be represented and mapped at the world regional scale. As we proceed

Perspectives on THE HUMAN ENVIRONMENT

Climate in Daily Human Terms

1. "The butter when stabbed with a knife flew like very brittle toffee. The lower skirts of the inner tent are solid with ice. All our [sleeping] bags were so saturated with water that they froze too stiff to bend with safety, so we packed them one on the other full length, like coffins, on the sledge."

2. "The rain poured steadily down, turning the little patch of reclaimed ground on which his house stood back into swamp again. The window of this room blew to and fro: at some time during the night, the catch had been broken by a squall of wind. Now the rain had blown in, his dressing table was soaking wet, and there was a pool of water on the floor."

3. "It was a breathless wind, with the furnace taste sometimes known in Egypt when a *khamsin* came, and, as the day went on and the sun rose in the sky it grew stronger, more filled with the dust of Nefudh, the great sand desert of Northern Arabia, close by us over there, but invisible through the haze."

4. "The spring came richly, and the hills lay asleep in grass—emerald green, the rank thick grass; the slopes were sleek and fat with it. The stock, sens-

ing a great quantity of food shooting up on the sidehills, increased the bearing of the young. When April came, and warm grass-scented days, the flowers burdened the hills with color, the poppies gold and the lupines blue, in spreads and comforters."

5. "The east wind that had blown coldly across the [English] Channel that morning had brought a dusting of snow to Picardy. Snow in April! It lay in a thin covering on hillsides, like long, torn bed sheets, the earth showing through in black streaks. It made the ordinary-looking landscape seem dramatic, the way New Jersey looks in bad weather, made houses and fences emphatic, and brought a sort of cubism to villages that would otherwise have been unmemorable. Each place became a little frozen portrait in black and white."

What do these rather poetic excerpted selections have in common? They attempt to convey a sense of climate, the impact a particular climate—especially a demanding and harsh climate—has on people who attempt to explore and live in it. Description of cli-

mate is one of the cornerstones of the literature of exploration and travel. Perhaps nothing conveys a sense of an environment as efficiently and dramatically as notation of its winds, precipitation, and temperatures. And, as these skillful writers show, climate tells you a great deal about the "feel" of a place.

Now to relieve your curiosity. Selection 1 is from Robert Falcon Scott's memoirs of his explorations in the Arctic, published posthumously in 1914. Selection 2 is by Graham Greene, who, after being stationed in West Africa during World War II, wrote *The Heart of the Matter*. Selection 3 is by Thomas Edward Lawrence ("Lawrence of Arabia"), the British soldier who organized the Arab nations against the Turks during World War I. Selection 4 is by John Steinbeck, writing in *To a God Unknown* during the 1930s about the Mediterranean climate of central California. Selection 5 is from travel writer Paul Theroux, who, in *Riding the Iron Rooster*, recorded his mid-1980s, train-window impression of northern France en route from London to a year of journeying on China's railroads.

through each major climate type in the next three units, you will be able to decode any combination of Köppen letter symbols by referring back to Fig. 16-1. For instance, an **Af** climate is a tropical climate in which the average temperature of every month exceeds 18°C (64.4°F) and total monthly precipitation always exceeds 6 cm (2.4 in). Before considering each of the six major climate groups, we need to establish their spatial dimensions, both in general terms and on the world map.

THE REGIONAL DISTRIBUTION OF CLIMATE TYPES

The easiest way to picture the distribution of world climates is first to imagine how they would be spatially

arranged on a **hypothetical continent** of uniform low elevation. Such a model is shown in Fig. 16-2 and represents a generalization of the world's landmasses. Comparison with the actual global map reveals that this hypothetical continent tapers from north to south in close correspondence with the overall narrowing of land areas between 50°N and 40°S. Note that on the world map the bulk of both Eurasia and North America is greatest in the latitudinal zone lying between 50°N and 70°N. Conversely, at 40°S the only land interrupting the world ocean is the slender cone of southern South America and small parts of Australia and New Zealand.

Five of the six major climate groups—**A**, **B**, **C**, **D**, and **E**—are mapped on the hypothetical continent in Fig. 16-2. (**H** climates are not shown beyond a token presence because this model deemphasizes upland areas.)

THE HYPOTHETICAL CONTINENT MODEL

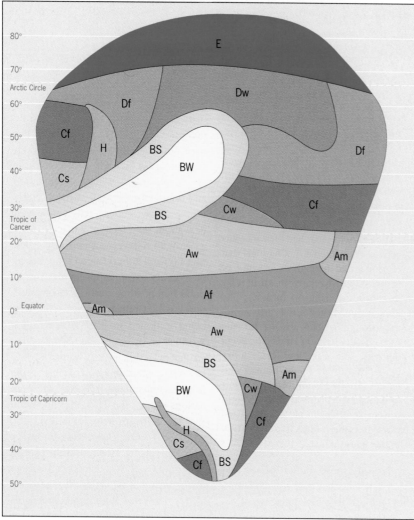

FIGURE 16-2 The distribution of climate types across a hypothetical continent of low, uniform elevation. Climate types on this world-island are identified in the legend of the map in Fig. 16-3.

The hypothetical continent model, of course, represents an abstraction of the more complex *empirical* or actual distribution of climate regions (shown in Fig. 16-3)—which, like all regional schemes, is itself a simplification of the still greater complexity that exists across the earth's surface.

The following discussion of world climate regions is based upon a side-by-side consideration of Figs. 16-2 and 16-3. The hypothetical continent is useful for gaining a general understanding of the latitudinal extent and continental position of each climate type; the empirical Köppen climate map makes the connections to the specific regions that blanket the continental landmasses. The emphasis in the remainder of this unit is on broad patterns. The following three units consider each major climate type in greater detail.

The Distribution of A Climates

The tropical **A** climates straddle the equator, extending to approximately 25 degrees latitude in both the North-

ern and Southern Hemispheres (Fig. 16-2). The heart of the **A** climate region is constituted by its wet subtype, the *tropical rainforest* climate (**Af**), named for the vegetation it nurtures. On the poleward margins of the tropical rainforest, forming transitional belts between the **A** and **B** climates, lies the **Aw** climate, which exhibits a distinct winter dry season. This is called the *savanna* climate because of the tall grasses that grow there, dominating the vegetational spaces between clumps of trees and/or thorny bushes. A third **A** climate, the **Am**, or *monsoon*, variety of the rainforest climate, is often found at coastal corners of the **Af** region and is subject to even more pronounced seasonal fluctuations in rainfall.

Closer inspection of the tropical climate zone on the hypothetical continent will also reveal that areas of **A** climate are widest on the eastern side of the world-island. This occurs because the trade winds blow onshore from the northeast and southeast, and the difference in overall latitudinal extent of the **A** region is about 20 degrees greater on the east coast of a continent than on the west coast.

The world Köppen map (Fig. 16-3) displays a more intricate pattern but is consistent with the generalizations we have just made with respect to the distribution of **A** climates. In the Americas, the **A** climates are centered around the equator, reaching to about 25°N and S on the eastern coasts and islands, and to about 20°N and S in western Central and South America, respectively. The African pattern is similar south of the equator, but to the north the **A** climates are compressed because the moist northeast trades are blocked by the landmass of southern Asia. For reasons identified in Unit 10, the Asian tropical climates are dominated by the monsoon effect. The prevalence of oceans and seas between Southeast Asia and northern Australia has sharply reduced the areal extent of tropical climates over land, but many of the islands of the lengthy Indonesian archipelago still exhibit rainforest climates over most of their surfaces.

The Distribution of **B** Climates

The dry **B** climates are located poleward of the **A** climates on the western sides of continents. These climates are associated with the subsiding air of the subtropical high-pressure zones, whose influence is greatest on the eastern side of oceans. This leads to a major intrusion of arid climates onto neighboring continents in both the Northern and Southern Hemispheres. As the hypothetical continent model shows (Fig. 16-2), this occurs at the average latitudes of the Subtropical High between 20° and 30°N and S. Once inland from the west coast of the continent, however, the **B** climate region curves poleward, following the airflows of the Westerlies in each hemisphere and reaching to about 55°N and 45°S, respectively. As shown in Unit 17, continentality also plays a significant role in the occurrence of **B** climates, particularly on the Eurasian landmass.

The heart of the **B** climate region contains the driest climatic variety, true *desert* (**BW**—with **W** standing for *wüste*, the German word for desert). It is surrounded by the semiarid short-grass prairie, or *steppe* (**BS**), which is a moister transitional climate that lies between the **BW** core and the more humid **A**, **C**, and **D** climates bordering the dry-climate zone.

On the world map (Fig. 16-3), we again observe the consistencies between model and reality. The most striking similarity occurs with respect to the vast bulk of subtropical-latitude land that stretches east–west across northern Africa and then curves northeastward through southwestern and central Asia. The desert climate (**BW**) dominates here and forms a more or less continuous belt between northwestern Africa and Mongolia, interrupted only by narrow seas, mountain corridors, and a handful of well-watered river valleys. Included among these vast arid basins are North Africa's Sahara, the deserts of the Arabian Peninsula and nearby Iran, and China's Taklimakan and Gobi Deserts. Note that these **BW** climates become colder north of 35°N (as indicated by the third

Köppen letter **k** replacing **h**) and that all the deserts are framed by narrow semiarid or steppe (**BS**) zones that also change from **h** to **k** in the vicinity of the 35th parallel.

In North America the arid climates occur on a smaller scale, but in about the same proportion and relative areal extent vis-à-vis this smaller continental landmass. The rain shadow effect (*see* Unit 13) plays an important role in the dryness of the western United States and Canada, where high mountains parallel the Pacific coast and block much of the moisture brought onshore by prevailing westerly winds. In the Southern Hemisphere, only Australia is reminiscent of the **B** climate distributions north of the equator because this island continent comprises the only large body of land within the subtropical latitudes. On the tapering landmasses of South America and southern Africa, the arid climate zones are compressed but still correspond roughly to the regional patterns of the hypothetical continent.

The Distribution of **C** Climates

As Fig. 16-2 reveals, the mesothermal (moderate-temperature) **C** climates are situated in the middle latitudes. The land/water differential between the Northern and Southern Hemispheres, however, becomes more pronounced poleward of the subtropics, and the hypothetical continent no longer exhibits regional patterns that mirror each other. North of the equator, as the world-island's greatest bulk is approached, the absence of oceanic-derived moisture in the continental interior produces an arid climate zone that interrupts the east–west belt of **C** climates lying approximately between the latitudes of 25° and 45°N. In the Southern Hemisphere, where the mesothermal climatic belt is severely pinched by the tapering model continent, **C** climates reach nearly across the remaining landmass south of 40°S.

The subtypes of the **C** climate group reflect the hemispheric differences just discussed, but some noteworthy similarities exist as well. On the western coast, adjacent to the **B** climate region, small but important zones of *Mediterranean* climate (**Cs**) occur where the subtropical high-pressure belt has a drying influence during the summer. Poleward of these zones, *humid climates with moist winters* (**Cf**) are encountered and owe their existence, especially on the western side of the continent, to the storms carried by the midlatitude westerly winds. **Cf** climates are also found on the eastern side of the continent, where they obtain moisture from humid air on the western side of the adjoining ocean's subtropical high-pressure zone. Another mesothermal subtype, sometimes called the *subtropical monsoon* (**Cw**) climate, is found in the interior of the **C** climate region and represents a cooler version of the savanna climate where the **Aw** regime spills across the poleward margin of the tropics.

The world Köppen map (Fig. 16-3) again demonstrates the applicability of the distributional generalizations of the model continent. Both Eurasia and North

WORLD CLIMATES
After Köppen–Geiger

A HUMID EQUATORIAL CLIMATE

Af	No dry season
Am	Short dry season
Aw	Dry winter

B DRY CLIMATE

| BS | Semiarid |
| BW | Arid |

h=hot
k=cold

C HUMID TEMPERATE CLIMATE

Cf	No dry season
Cw	Dry winter
Cs	Dry summer

a=hot summer
b=cool summer
c=short, cool summer
d=very cold winter

D HUMID COLD CLIMATE

| Df | No dry season |
| Dw | Dry winter |

E COLD POLAR CLIMATE

| E | Tundra and ice |

H HIGHLAND CLIMATE

| H | Unclassified highlands |

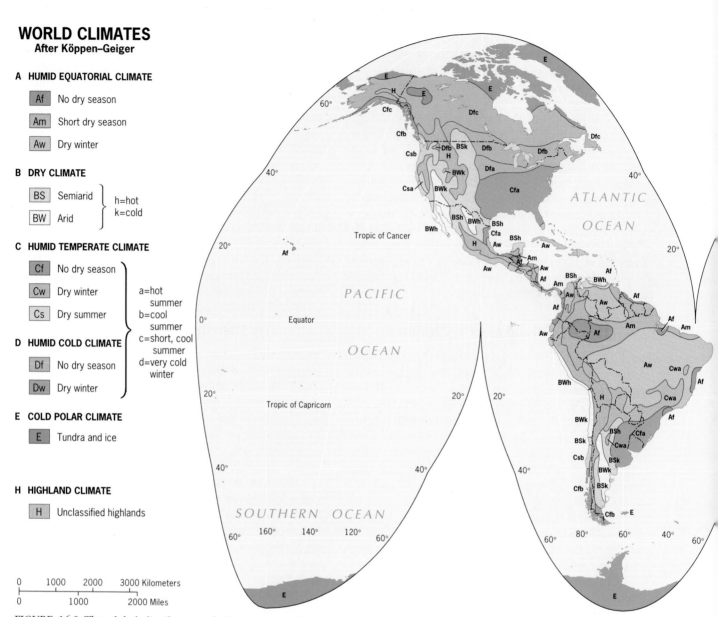

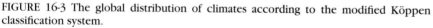

FIGURE 16-3 The global distribution of climates according to the modified Köppen classification system.

America exhibit the central **B** climate interruption of the mesothermal climatic belt, with the **C** climates less prominent on the Eurasian landmass where distances from maritime moisture sources are much greater. Southern Hemisphere patterns are also generally consistent with the model, although the **C** climates are often confined to narrow coastal strips that result from rain-shadow effects produced by South America's Andes Mountains, South Africa's Great Escarpment, and Australia's east-coast Great Dividing Range.

The individual **C** climates further underscore the linkages between the hypothetical continent and the real world. Mediterranean (**Cs**) climates are located on every west coast in the expected latitudinal position (around 35 degrees), and even penetrate to Asia's "west coast,"

where the eastern Mediterranean Sea reaches the Middle East.

The west-coast marine variety of the **Cf** climate is generally observed on the poleward flank of the **Cs** region, but mountains paralleling the ocean often restrict its inland penetration; the obvious exception is Europe, where blocking highlands are absent and onshore Westerlies flow across warm ocean waters to bring mild weather conditions hundreds of kilometers inland. The **Cf** climates on the eastern sides of continents are particularly apparent in the United States (thanks to the moderating effects of the Gulf of Mexico), South America, and the Pacific rim of Australia. In East Asia, this **Cf** subtype is limited by the more powerful regional effects of monsoonal reversals, but where it does occur in eastern China

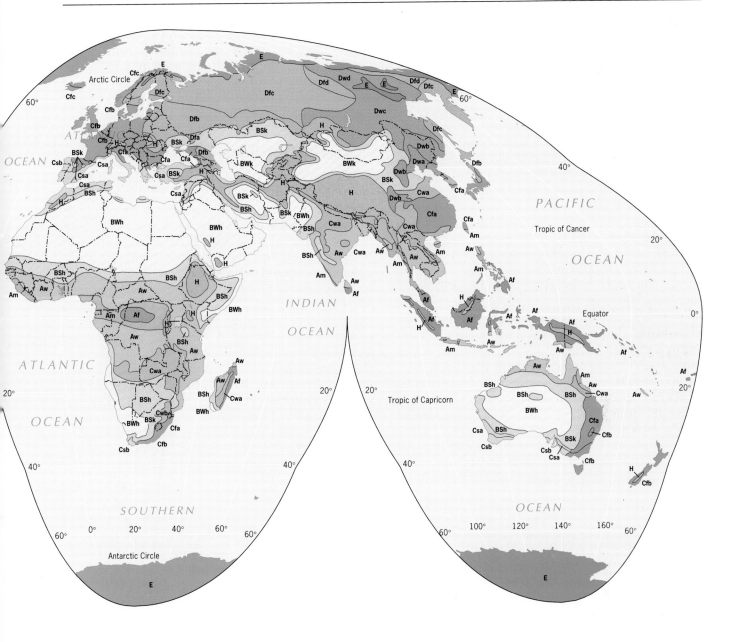

and Japan it is associated with a massive population concentration that contains over one-eighth of humankind.

The **Cw** climates of the Northern Hemisphere are confined mainly to the subtropical portions of southern and eastern Asia, and, of course, are closely related to the monsoonal weather regime. The Southern Hemisphere **Cw** climate regions function as transitional zones, in both Africa and South America, between the equatorial-area savannas to the north and the subtropical arid and humid climates to the south.

The Distribution of D Climates

The microthermal **D** climates, which receive relatively small amounts of heat, are found exclusively in the Northern Hemisphere. There are no large landmasses in the Southern Hemisphere between 50° and 70°S, so here latitude and the effect of continentality do not support the cold winters of the **D** climates. On the world-island (Fig. 16-2) the microthermal climates are associated with the large landmasses in the upper-middle and subpolar latitudes, and occupy an east–west belt between approximately 45° and 65°N. This climatic band of severe winters is at its widest in the interior of the hypothetical continent, moderating only near the coasts (particularly the west coast), where the ameliorating influences of maritime air penetrate inland.

Two subtypes of the **D** climate group can be observed. The **Df** variety experiences precipitation throughout the year; it can exhibit a fairly warm summer near the **C** climate boundary and certain coastal zones, but in the interior and toward the higher latitudes the summers are shorter and much cooler. The other sub-type—**Dw**—is encountered where the effects of continentality are most pronounced. Annual moisture totals there are lower, and the harsh winters are characterized by dryness. The extreme cooling of the ground in mid-winter is associated with a large anticyclone that persists throughout the cold season and blocks all surface winds that might bring in moisture from other areas.

The actual distribution of **D** climates (Fig. 16-3) accords quite well with the model continent. The greatest east–west extents of both Eurasia and North America are indeed the heart of the **D** climate region, which dominates the latitudes lying between 50° and 65°N. The only exceptions occur along the western coasts where the warm North Atlantic Drift and Alaskan Current bathe the subpolar coastal zone with moderating temperatures. In the Eurasian interior, extreme continentality pushes the microthermal climates to a higher average latitude, a pattern not matched in eastern Canada, where the colder polar climates to the north penetrate farther south.

As for the **D** subtypes, they correspond closely to the model. However, note that North America, wide as it is, does not contain sufficient bulk to support the **Dw** climate or even the coldest variant of the **Df** climate (**Dfd**). Only Eurasia exhibits these extreme microthermal climates, which are concentrated on the eastern side of that enormous landmass. Two reasons underlie the eastward

deflection: (1) the overall altitude of the land surface is much higher in the mountainous Russian Far East and northeastern China, and (2) northeastern Asia lies farthest from the warm waters off western Eurasia that pump moisture into the prevailing westerly winds that sweep around the globe within this latitudinal zone.

The Distribution of **E** Climates

In both hemispheres, land areas poleward of the Arctic and Antarctic Circles (66½°N and S, respectively), with deficits of net radiation, exhibit polar **E** climates. The hypothetical continent and world Köppen map both display similar patterns, with **E** climates dominant in Antarctica, Greenland, and the poleward fringes of northernmost Eurasia and North America. As for subtypes, the *tundra* climate (**ET**), where the warmest month records an average temperature between 0°C (32°F) and 10°C (50°F), borders the warmer climates on the equatorward margins of each polar region. The highest latitudes in the vicinity of the poles themselves—as well as most of interior Greenland and high-lying Antarctica—experience the *icecap* or *frost* climate (**EF**), in which the average temperature of the warmest month fails to reach 0°C (32°F). As explained in Unit 19, **E** climates are also often associated with the high-altitude areas of lower latitudes (where they are labeled **H** or *highland* climates), particularly in the upper reaches of mountainous zones that are too cold to support vegetation.

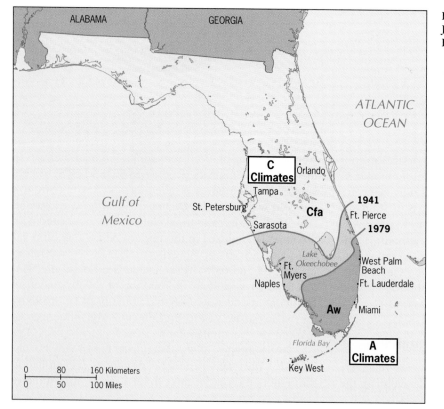

FIGURE 16-4 The changing location of the January 18°C (64.4°F) isotherm in southern Florida, 1941–1979.

BOUNDARIES OF CLIMATE REGIONS

The classification system developed by Köppen and his associates has been criticized because it does not consider the causes of climate and because some of its climate–vegetation links are not very strong. Yet it remains the most often used classification scheme, and it has obviously proven itself both useful and appropriate for the purpose of introducing students to the complexity of global climatic patterns. Earlier in this unit, we discussed the shortcomings of any geographical classification of climate. We conclude by again reminding you that what look like sharp regional boundaries on the hypothetical continent and world Köppen map are really zones of transition from one climate type to another.

The Köppen regional scheme, therefore, which was designed for the broad global or regional scale, is not well suited for microclimatological studies. Thus it is not possible to accurately place a climate–region boundary in a local area, because only a few places in any city or county have compiled the necessary weather data for climatic analysis. Think of your local weather report. The "official" temperature and other weather statistics are ob-tained at a single site for the whole area, such as at the airport or a downtown park, or at best for a handful of communities scattered across your area.

Another problem in applying Köppen boundaries at the microscale is that these boundaries change over time. Climates, as we will see, are highly dynamic, and substantial change can occur over long periods. And even over shorter time spans, that change can be seen on the map. A good example is shown in Fig. 16-4, which maps the poleward boundary of the **Aw** climate in tropical South Florida. Note that in 1941 the January 18°C (64.4°F) isotherm was located between 40 km (25 mi) and 120 km (75 mi) to the northwest of its 1979 position. Thus, over at least those four decades, the tropical zone contracted, and cities such as Ft. Myers, Naples, and Ft. Pierce "shifted" into the mesothermal (**Cfa**) climatic zone. These "changes," of course, are very subtle, and even a resident of Ft. Myers who has lived there since 1940 would almost certainly not be aware of any meaningful change. But this little exercise about boundaries is instructive and should interject a little healthy skepticism in preparation for the more detailed overview of climate regions covered in Units 17 to 19.

KEY TERMS

climate (p. 174)   hypothetical continent (p. 177)   Köppen climate classification system (p. 175)

REVIEW QUESTIONS

1. How is a location's climate determined from weather data?

2. What are the thermal criteria for tropical, mesothermal, microthermal, and polar climates in the Köppen system?

3. Outline the interpretation of the second and third letters of the codes used in the Köppen system. For example, how does an **Af** climate differ from an **Aw** climate?

4. Describe two advantages of the Köppen climate classification system.

5. Describe two problems or shortcomings of the Köppen climate classification system.

6. Why should Köppen regional-scale boundaries not be used to ascertain climate-zone boundaries at the local scale?

REFERENCES AND FURTHER READINGS

CRITCHFIELD, H. J. *General Climatology* (Englewood Cliffs, N.J.: Prentice-Hall, 4th ed., 1983).

EAGLEMAN, J. R. *The Visualization of Climate* (Lexington, Mass.: Heath, 1976).

HAURWITZ, B., and AUSTIN, J. M. *Climatology* (New York: McGraw-Hill, 1944).

HOUGHTON, J. T., ed. *The Global Climate* (New York: Cambridge University Press, 1984).

KENDREW, W. G. *The Climates of the Continents* (New York: Oxford University Press, 5th ed., 1961).

LINACRE, E. *Climate Data and Resources: A Reference and Guide* (London/New York: Routledge, 1992).

LOCKWOOD, J. *World Climatic Systems* (London: Edward Arnold, 1985).

LYDOLPH, P. E. *The Climate of the Earth* (Totowa, N.J.: Rowman & Allanheld, 1985).

OKE, T. R. *Boundary Layer Climates* (London/New York: Methuen, 2nd ed., 1987).

RILEY, D., and SPALTON, L. *World Weather and Climate* (New York: Cambridge University Press, 2nd ed., 1981).

TREWARTHA, G. T. *The Earth's Problem Climates* (Madison, Wis.: University of Wisconsin Press, 2nd ed., 1981).

TREWARTHA, G. T., and HORN, L. H. *An Introduction to Climate* (New York: McGraw-Hill, 5th ed., 1980).

WILCOCK, A. A. "Köppen After Fifty Years," *Annals of the Association of American Geographers*, 58 (1968), 12–28.

UNIT 17

Tropical (A) and Arid (B) Climates

OBJECTIVES

◆ To expand the discussion of tropical (A) and arid (B) climates using climographs developed for actual weather stations.

◆ To highlight climate-related environmental problems within tropical and arid climate zones.

◆ To examine the causes and consequences of tropical deforestation and desertification.

U
nit 16 set the stage for a global survey of the principal climate types based on the Köppen classification and regionalization system. Our overview begins in this unit with the **A** (tropical) and **B** (dry or arid) climates, and you should use Figs. 16-2 (p. 178) and 16-3 (pp. 180–181) as a guide to the spatial distribution of all the world's major climate types. In addition to providing greater detail about each of the principal subdivisions of the tropical and arid climates, we focus on the leading environmental problem of each major climate group. For the **A** climates, *deforestation* is an intensifying crisis that could have far-reaching environmental consequences in the foreseeable future. For the **B** climates, our concern is with *desertification*, the spread

of desert conditions engendered by human abuse of dryland environments. Based on a similar approach, the rest of the major climate types are treated in Units 18 and 19.

THE MAJOR TROPICAL (A) CLIMATES

Contained within a continuous east–west belt astride the equator, varying latitudinally from 30 to 50 degrees wide,

UNIT OPENING PHOTO: The landscapes of **A** and **B** climates in local juxtaposition. The thin fertile strip bordering the Nile holds back the vast surrounding desert in upper Egypt.

Deforestation of the Tropics

Only a generation ago, lush rainforests clothed most of the land surfaces of the equatorial tropics and accounted for 10 percent of the earth's vegetation cover. Since 1980, however, industries and millions of residents of low-latitude countries have been destroying billions of those trees through rapid **tropical deforestation**—the clearing and destruction of rainforests to make way for expanding settlement frontiers and the exploitation of new economic opportunities. In Brazil's Amazon Basin in 1988 alone, an area almost the size of Colorado was burned, half of it virgin forest and the other half jungly regrowth, to briefly replenish the infertile soil. If this prodigious removal rate endures, the planet's tropical rainforests will be reduced to two large patches early in the twenty-first century (in west equatorial Africa and the central Amazon Basin), and these will disappear by 2050.

As rainforests are eradicated, they do not just leave behind the ugly wastelands seen in Fig. 17-1. The smoke from their burning releases particles and gases that rise thousands of meters into the atmosphere and are then transported thousands of kilometers around the world. A number of atmospheric scientists believe that the effects of such fires, together with the removal of massive stands of carbon-dioxide-absorbing trees, could have an increasingly harmful effect on global climate. Let us quickly review the causes and climatic consequences of this deepening environmental crisis.

The causes of deforestation in tropical countries lie in the swift reversal of perceptions about tropical woodlands in recent years. Long regarded as useless resources and obstacles to settlement, rainforested areas suddenly assumed new importance as rapidly growing populations ran out of living space in the 1970s and 1980s. Governments began to build roads and subsidize huge colonization schemes, instantaneously attract-

FIGURE 17-1 The charred landscape in the aftermath of a massive rainforest fire. This scene is in Brazil's Rondônia State on the Amazon Basin's southwestern rim, one of the most heavily burned rainforest zones in the equatorial tropics.

ing millions of peasants who were now able to buy land cheaply for the first time in their lives. Even though the deforested soil soon proved infertile, enough affordable land was available locally to allow these farmers to relocate to newly cleared plots every few years and begin the cycle over again.

In addition to this shifting-cultivation subsistence economy, many commercial opportunities arose as well. Cattle ranchers were only too happy to buy up the abandoned farmlands and replace them with pasture grasses, and a large industry mushroomed during the 1980s in Amazonia, Costa Rica, and elsewhere. At the same time, lumbering became a leading activity as developed countries, including Japan and many in Europe, turned from their exhausted midlatitude forests to the less expensive and more desirable trees of the tropics as sources for paper, building materials, and furniture. By 1990, perhaps 300 million Third World workers depended on the rainforests for their livelihoods, with the total steadily mounting each year. But with two-thirds of the earth's population expected to re-

side in the tropical countries within the next half-century, deforestation portends a major economic crisis in the near future.

There are many serious threats to the global environment as well, not the least of which are mass extinctions of plant and animal life (at the rate of about 100 species a day for the next quarter-century), because the tropical rainforest is home to as many as 80 percent of the planet's species (*see* box p. 243). As mentioned above, rainforest destruction may be linked to the earth's climatic patterns, and a research effort is underway to explore possible connections and the longer-term consequences of their operation. Two major climatic problems associated with the wholesale burning of tropical woodlands have already been identified, and both are likely to reach global proportions as deforestation continues to intensify.

The first of these problems involves the fires themselves, which not only pump large quantities of carbon and other particles into the atmosphere but also release massive quantities of gases.

These gases are now being studied for the first time, and there is evidence that they affect the chemical balance of the atmosphere by entering into the global airflows that rise and diverge above the tropics. Particular concern is focusing on a possible link with the depletion of the ozone layer (*see* Unit 6), because these gases released from rainforest fires have been observed over Antarctic latitudes and contain methane and nitrogen oxides that are known to react with (and alter) stratospheric ozone.

The second climatic problem is that the trees of the tropical rainforest have always absorbed large quantities of atmospheric carbon dioxide, thereby helping to maintain a stable global balance of this gas. With the destruction of this principal agent for removing CO_2 from the air, climatologists are unsure of what to expect. Some predict a steady rise in atmospheric concentrations of CO_2. Others see a worldwide intensification of pollution, because rainforests have been known to function as large-scale cleansing mechanisms for effluents that entered its ecosystems via air- and water flows.

The deforestation crisis came to the world's attention in the late 1980s, but it is unlikely to be reversed by any last-minute crusade to save the remaining trees.

is the warmth and moisture of the **A** or **tropical climates**. Warmth is derived from proximity to the equator (**A** climate temperatures must average higher than 18°C [64.4°F] in the coolest month); moisture comes from the rains of the ITCZ (Inter-Tropical Convergence Zone), wet-monsoon systems, easterly waves, and hurricanes. The tropical climates can be subdivided into three major types: the *tropical rainforest*, the *monsoon rainforest*, and the *savanna*.

The Tropical Rainforest (Af) Climate

The **Af** or tropical rainforest zone exhibits the greatest effects of heat and moisture. In areas undisturbed by human intervention, tall evergreen trees completely cover the land surface (Fig. 17-2). This virgin forest has very little undergrowth, however, because the trees let through only a small amount of sunlight. But any serious discussion of this environment in the 1990s must take into consideration the appalling rush toward the destruction of rainforests throughout the tropical latitudes (*see* Perspective: Deforestation of the Tropics).

The climatic characteristics of the tropical rainforest are illustrated in Fig. 17-3, which uses three graphic displays for a typical weather station in the **Af** zone. The place chosen for such presentation is São Gabriel de Cachoeira, a Brazilian village situated almost exactly on the equator deep within the heart of the Amazon Basin. Of the three graphs, the most important for our purposes is the first—the climograph, a device employed repeatedly in Units 17, 18, and 19. A **climograph** for a given location simultaneously displays its key climatic variables of average temperature and precipitation, showing how they change month by month through the year.

The climograph in Fig. 17-3 reveals that average monthly temperatures in São Gabriel are remarkably constant, hovering consistently around 26°C (79°F) throughout the year. In fact, the diurnal (daily) temperature may vary as much as 6°C (11°F), an amount greater than the annual range of 1.6°C (3°F). Rainfall also tends to occur in a diurnal rather than seasonal rhythm. A normal daily pattern consists of relatively clear skies in the morning, followed by a steady buildup of convectional clouds from the vertical movement of air with the increasing heat of the day. By early afternoon, thunderstorms burst forth (*see* Fig. 13-2) with torrential rains. After several hours, these stationary storms play themselves out as temperatures decline a few degrees with the approach of night. This pattern monotonously repeats itself day after day.

In São Gabriel, no month receives less than 130 mm (5.1 in) of rain, and in May about 10 mm (0.4 in) of rain falls per day. The average total rainfall for the year is 2800 mm (109 in). Such large amounts of rain keep the water balance in perpetual surplus, as the second graph in Fig. 17-3 indicates. Evapotranspiration in this area constantly occurs at its potential rate; much of the surplus water

FIGURE 17-2 The dense tropical rainforest of the western Amazon Basin near Iquitos, Peru.

186

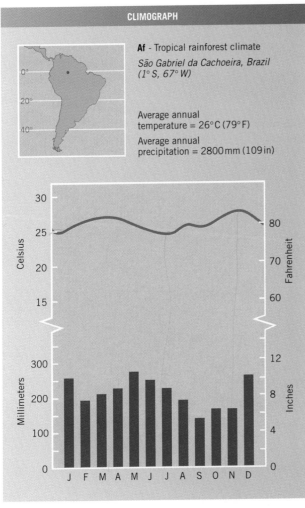

a

FIGURE 17-3 Climograph and related graphic displays for a representative weather station in the tropical rainforest (**Af**) climate zone.

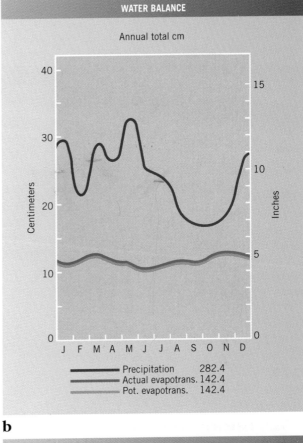

b

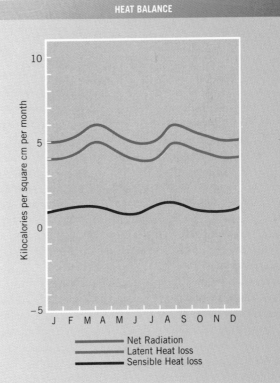

c

runs off into the Rio Negro (on which the village is located), and from this tributary directly into the Amazon itself. As a result, most of the net radiation is used in latent heat (third graph in Fig. 17-3), and so the moisture in the highly humid air is continuously replaced. When the moisture condenses, latent heat adds to the monotonous warmth of the atmosphere. When the sun passes directly overhead at the equinoxes, the rainfall values are not noticeably affected because they are always high. But the heat balance is markedly affected, as demonstrated by the relatively higher values of net radiation and latent heat loss in the graph.

The Monsoon Rainforest (Am) Climate

The monsoon rainforest (**Am**) climate is restricted to tropical coasts that are often backed by highlands. It has a distinct dry season in the part of the year when the sun is lower, but almost always a short one; a compensa-

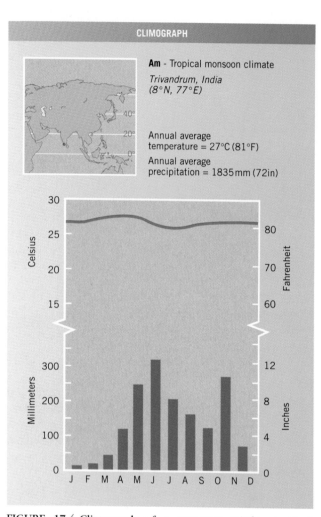

FIGURE 17-4 Climograph of a monsoon rainforest (**Am**) weather station.

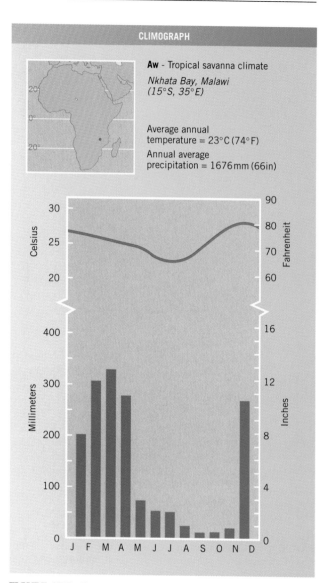

FIGURE 17-5 Climograph of a tropical savanna (**Aw**) weather station.

ting longer season of heavy rainfall generally prevents any soil-moisture deficits. These characteristics are seen in the representative climograph for this climate type (Fig. 17-4), the city of Trivandrum located near India's southern tip on the narrow Malabar Coast at the base of the Cardamom Uplands.

In some places, such as the hills leading up to the Himalayas in northeastern India, the highlands add an orographic effect to the wet-monsoon rains. The station of Cherrapunji (*see* Fig. 18-13) in this hilly region averages an annual total of 11,450 mm (451 in) of rain—most of it occurring during the three summer months! (Until the 1950s this was believed to be the rainiest place on earth, but that distinction now belongs to Mount Waialeale on the Hawaiian island of Kauai, where 11,990 mm [472 in] of rain falls in an average year.) The vegetation of the monsoon rainforest zone consists mainly of evergreen trees, with occasional grasslands interspersed. The

trees, however, are not as dense as those found in the remaining areas of true tropical rainforest.

The Savanna (Aw) Climate

The savanna (**Aw**) climates are found in the transitional, still-tropical latitudes between the subtropical high-pressure and equatorial low-pressure belts. Although these areas often receive between 750 and 1750 mm (30 and 70 in) of rain per year, there is an extended dry season in the months when the angle of the sun is at its lowest. Moisture deficits in the soil can occur during the dry season, whose length and severity are proportional to a given savanna area's distance from the equator. The climograph for Nkhata Bay, Malawi in southeastern Africa (Fig. 17-5) shows the typical march of temperatures and precipitation totals in a normal year. Do not be misled, however, by the decline of both variables in the mid-

clumps of trees or individual trees and thorn bushes. There are extensive areas of savanna climate in the world, most prominently covering parts of tropical South America, South Asia, and Africa (*see* Fig. 16-3). Along the poleward margins of this climate zone, however, rainfall is so unreliable that these peripheral areas are considered too risky for agricultural development in the absence of irrigation systems.

THE MAJOR ARID (B) CLIMATES

The **B** or **arid climates** provide a striking visual contrast to the **A** climates (Fig. 17-7) because in these areas potential evapotranspiration exceeds the moisture supplied by precipitation. Dry climates are found in two general locations (Fig. 16-3). The first group consists of two interrupted bands near 30°N and 30°S, where the semipermanent subtropical high-pressure cells are dominant. The largest areas of this type are the African-Eurasian desert belt (stretching from the vast Sahara northeastward to Mongolia) and the arid interior of Australia. The driest of these climates are often located near the western coasts of continents. Weather stations in northern Chile's Atacama Desert, for instance, can go 10 years or more without recording precipitation.

The second group of dry climates can be attributed to two different factors. First, a continental interior remote from any moisture source, experiencing cold high-pressure air masses in winter, can have an arid climate (the central Asian countries east of the Caspian Sea are a good case in point). And second, rain-shadow zones in the lee of coast-paralleling mountain ranges often give rise to cold deserts, as in the interior Far West of North America and southern Argentina's Patagonia.

Average annual temperatures in arid climates are usually typical of those that might be expected at any given latitude—high in the low latitudes and low in the higher latitudes. However, the annual *range* of temperature is often greater than might be expected. In North Africa's central Sahara, for example, a range of 17°C (30°F) at 25°N is partly the result of a continental effect. The diurnal range of temperature can also be quite large, averaging between 14° and 25°C (25° and 45°F). On one unforgettable day in the Libyan capital city of Tripoli, the temperature went from below freezing (−0.5°C [31°F]) at dawn to 37°C (99°F) in midafternoon! Rocks expanding through such extreme changes of heat sometimes break with a sharp crack; North African soldiers in World War II occasionally mistook these breaking rocks for rifle shots.

The **B** climates are not necessarily dry all year round. Their precipitation can range from near 0 to about 625 mm (25 in) per year, with averages as high as 750 mm (30 in) in the more tropical latitudes. In summer, these equatorward margins of the **B** climates are sometimes

FIGURE 17-6 The pastoral Maasai constantly move their animals in search of seasonal grass spawned by the often elusive rains of the savanna climate. Here, in Tanzania's Ngorongoro Crater, the latest rains have been bountiful, and the cattle herders will remain at this spot for a while.

dle months of the year, because on this occasion we deliberately selected a Southern Hemisphere station to remind you of the opposite seasonal patterns that exist south of the equator.

In the region of East Africa to the north of Malawi, the pastoral Maasai people (Fig. 17-6) traditionally move their herds of cattle north and south in search of new grass sprouted in the wake of the shifting rains that correspond to the high sun. The vegetation of this region, besides its tall, coarse seasonal grasses, also includes

FIGURE 17-7 One of the world's driest landscapes: the heart of the Sahara in southern Algeria.

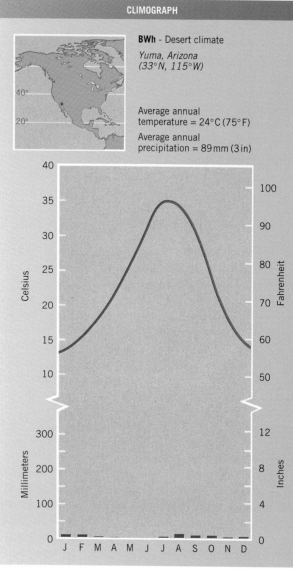

CLIMOGRAPH

BWh - Desert climate

*Yuma, Arizona
(33°N, 115°W)*

Average annual
temperature = 24°C (75°F)

Average annual
precipitation = 89 mm (3 in)

a

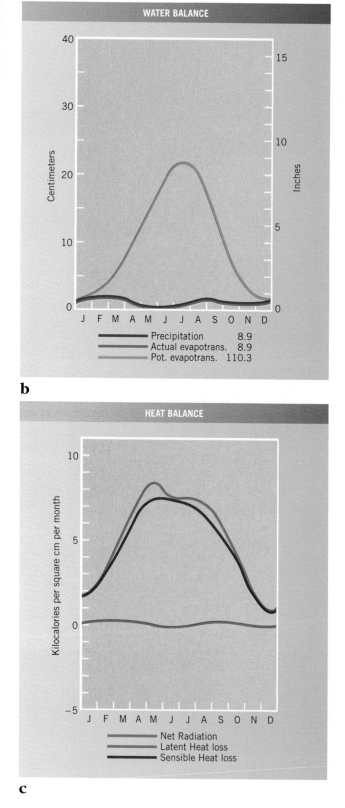

WATER BALANCE

Precipitation	8.9
Actual evapotrans.	8.9
Pot. evapotrans.	110.3

b

HEAT BALANCE

Net Radiation
Latent Heat loss
Sensible Heat loss

c

FIGURE 17-8 Climograph and related graphic displays for a representative weather station in the hot dry desert (**BWh**) climate zone.

affected by the rains of the ITCZ as they reach their poleward extremes. In winter, the margins of the arid areas nearest the poles may be subject to midlatitude-cyclone rains. There also are rare thunderstorms, which occasionally lead to flash floods. At Hulwan, just south of Cairo, Egypt, seven of these storms in 20 years yielded a total of 780 mm (31 in) of rain. Moreover, along coasts bathed by cold ocean currents, rather frequent fogs provide some moisture; Swakopmund, in southwestern Africa's Namibia, adjacent to the Benguela Current (*see* Fig. 11-6), experiences about 150 days of such fog each year.

The Desert (BW) Climate

For the most part, however, considerable dryness is the rule in **B** areas. This aridity is exemplified by Yuma, Arizona, where the average annual temperature is 23.5°C

(74.3°F), with an annual range of some 23°C (41°F). The climate here is **BWh**—a hot, dry desert. The small amount of rainfall shown in Fig. 17-8, 89 mm (3.4 in), comes either from occasional winter storms or from the convectional clouds of summer. The water balance diagram below the climograph shows a marked deficit throughout the year: potential evapotranspiration greatly exceeds precipitation. Most water that falls quickly evaporates back into the dry atmosphere. But as the heat balance diagram for Yuma indicates, there is little water for evaporation, so latent heat loss is negligible. Consequently, most of the heat input from net radiation, which is markedly seasonal, is expended as sensible heat—adding even more warmth to the already hot air (daily highs between May and October almost always exceed 38°C [100°F]).

The Steppe (BS) Climate

As the world climate map (Fig. 16-3) shows, the most arid areas—deserts (**BW**)—are always framed by semiarid zones. This is the domain of the *steppe* (also known in western North America as the *short-grass prairie*) or **BS** climate, which can be regarded as a transitional type between fully developed desert conditions and the subhumid margins of the **A**, **C**, and **D** climates that border **B** climate zones. The maximum and minimum amounts of annual rainfall that characterize semiarid climates tend to vary somewhat by latitude. In general, lower-latitude **BS** climates range from 375 to 750 mm (15 to 30 in) yearly; in the middle latitudes, a lower range (from 250 to 625 mm [10 to 25 in]) prevails. Akmola in the central Asian republic of Kazakhstan, which lies at the heart of the Eurasian landmass, is a classic example of the cold subtype of the steppe climate (**BSk**). The climograph for this manufacturing center (Fig. 17-9) shows its proximity to the Turkestan Desert that lies just to the south: rainfall totals only 279 mm (11 in) per year. As for the monthly temperature curve, an enormous range of 39°C (70°F) is recorded, reflecting one of the most pronounced areas of continentality on the earth's surface.

Human Activities in B Climates

During the late 1950s and early 1960s, Akmola (then called Tselinograd) was the headquarters for the so-called Virgin and Idle Lands Program, one of the most ambitious agricultural development schemes in the history of what until 1991 was the Soviet Union. In an attempt to emulate the highly mechanized, large-scale wheat farming of a similar climatic zone in the northern Great Plains of the United States (Fig. 17-10), Moscow's planners invested heavily in the Kazakh Soviet Socialist Republic, hoping to substantially boost food production which had been a troublesome problem since the Communist Revolution of 1917. For a few years the gamble of expanding agriculture into this risky region appeared to be paying off. But those turned out to be wetter-than-normal years, and by the mid-1960s it became evident that the plan was doomed to failure because extreme continentality worked against the accumulation of sufficient moisture to support the massive wheat-raising that the Soviet leaders had in mind.

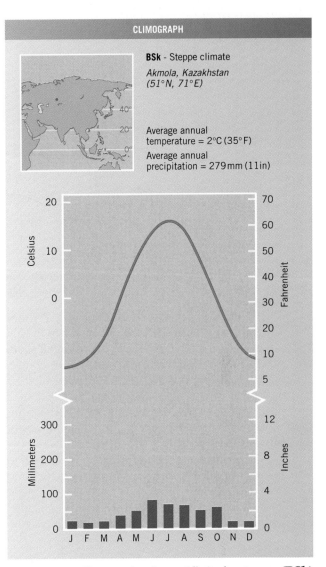

CLIMOGRAPH

BSk - Steppe climate
Akmola, Kazakhstan
(51°N, 71°E)

Average annual
temperature = 2°C (35°F)

Average annual
precipitation = 279mm (11in)

FIGURE 17-9 Climograph of a midlatitude steppe (**BSk**) weather station.

FIGURE 17-10 A pair of combines (left) harvesting ripe wheat in the heart of the Great Plains region in western Nebraska. This is one of the world's most mechanized farmscapes, a model of large-scale agricultural productivity.

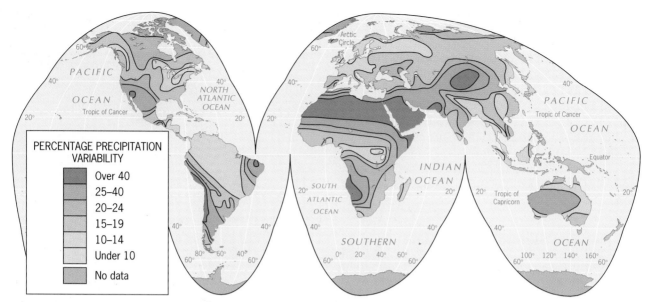

FIGURE 17-11 The global distribution of annual precipitation variability, shown as percentage departures from normal. Map from *Goode's World Atlas*, 19th Edition. © 1995 by Rand McNally R.L. 92-S-82-renewal 95.

PERCENTAGE PRECIPITATION VARIABILITY

- Over 40
- 25–40
- 20–24
- 15–19
- 10–14
- Under 10
- No data

The problem of drought, which is examined in Unit 18, is an ever-present environmental hazard in all semi-arid areas, and the Great Plains region has, of course, suffered many severe dry spells since it was settled by farmers over a century ago. The global distribution of precipitation variability is mapped in Fig. 17-11 and underscores this dilemma. Perversely, for human purposes, rain falls most reliably where it is abundant, while the arid areas experience the greatest variations from year to year (as even a cursory comparison of Figs. 16-3 and 17-11 reveals).

Throughout the human experience on this planet, dry climates have proven inhospitable, restricting settlement to the rare water sources of oases or to the valleys of rivers that originate outside a given zone of aridity. In certain cases, such as Egypt's Nile Valley (*see* this unit's opening photo), irrigation can maximize these scarce water resources, but the practice often tends to wash important minerals from the soil. Environmental abuse has been a particular problem at the margins of arid zones in recent years, and *desertification* (*see* Perspective: Desertification) now constitutes a serious crisis in many such places. This, however, is only the latest reminder that the balance among the earth's natural systems in dryland areas is extremely fragile.

KEY TERMS

arid climate (p. 189)   desertification (p. 193)   tropical deforestation (p. 185)

climograph (p. 186)   tropical climate (p. 186)

REVIEW QUESTIONS

1. Describe the general latitudinal extent of **A** climates.
2. Describe the general latitudinal extent of **B** climates.
3. Contrast the rainfall regimes of **Af**, **Aw**, and **Am** climates.
4. What is the primary difference between **BS** and **BW** climates?
5. Why is ongoing tropical deforestation regarded as a catastrophe to so many natural scientists?
6. Describe the circumstances leading to desertification.

REFERENCES AND FURTHER READINGS

ALLAN, T., and WARREN, A. *Deserts: The Encroaching Wilderness* (New York: Oxford University Press, 1993).

AUERBACH, J. D. "Turning Sand into Land: Desert Farms in Israel Grow Lush Crops from Sand and Salty Water," *Christian Science Monitor*, May 19, 1987, 21, 22.

BATES, M. *Where Winter Never Comes: A Study of Man and Nature in the Tropics* (New York: Scribner's, 1952).

BEAUMONT, P. *Drylands: Environmental Management and Development* (London/New York: Routledge, 1993).

CAUFIELD, C. *In the Rainforest* (New York: Knopf, 1985).

COLE, M. M. *The Savannas: Biogeography and Geobotany* (Orlando, Fla.: Academic Press, 1986).

COUPLAND, R. T., ed. *Grassland Ecosystems of the World* (New York: Cambridge University Press, 1979).

(*References continue.*)

Desertification

Desertification may be an awkward word, but it is on target in what it defines: the process of desert expansion into steppelands, largely as a result of human degradation of fragile semiarid environments (Fig. 17-12). This term was added to the physical geographer's lexicon in the 1970s following the disastrous famine, caused by rapid and unforeseen desiccation, that struck the north-central African region known as the *Sahel*. This name means "shore" in Arabic—specifically the southern shore of the Sahara, an east–west semiarid (**BSh**) belt straddling latitude 15°N, ranging in width from 320 to 1120 km (200 to 700 mi), that stretches across the entire continent of Africa (Fig. 16-3). During the height of the tragic starvation episode from 1968 to 1973, as paralyzing drought destroyed both croplands and pasturelands, the Sahel suffered a loss of at least 250,000 human lives and 3.5 million head of cattle.

This catastrophe prompted an international research effort, guided by the United Nations Environmental Program (UNEP). One of its first products was a world map of spreading deserts (Fig. 17-13), whose most hazardous zones coincided closely with the semiarid regions mapped in Fig. 16-3. Another accomplishment was the creation of a global desertification monitoring network, which has provided UNEP with some alarming data.

By 1990, some 15 million km² (5.8 million mi²) of the world's arid and semiarid drylands—an area almost the size of South America—had become severely desertified, thereby losing over 50 percent of its potential productivity. Another 34.8 million km² (13.4 million mi²)—an area greater than the size of Africa—had become at least moderately desertified, losing in excess of 25 percent of its potential productivity. Moreover, an area half the size of New York State (64,800 km² [25,000 mi²]) was being added to the total of severely desertified land each year. Perhaps most worrisome is that today more than a billion people (over one-sixth of humankind) are affected by desertification—and the world's arid-land population is growing at a rate 50 percent faster than in adequately watered areas.

FIGURE 17-12 The advancing desert bears down on the stricken Sahel near Mauritania's capital, Nouakchott. In a desperate attempt to anchor the ever shifting dunes, grids of branches have been pressed into the sand.

When desertification was first studied, it was believed that climate change played a significant role. Today, however, we have learned enough to know that the process is more heavily controlled by human actions, whose consequences are accentuated in years of lower than normal precipitation. The British geographer/ecologist Andrew Goudie has noted the particular suscep-

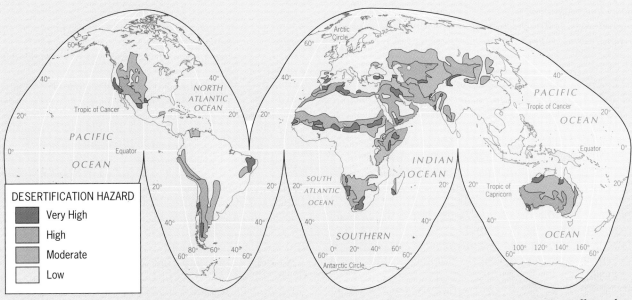

DESERTIFICATION HAZARD
- Very High
- High
- Moderate
- Low

FIGURE 17-13 The degree of the desertification hazard according to the scientists of the United Nations Environmental Program. The red-brick-colored areas indicate a critical situation; the orange-colored zones experience serious effects; the tan-colored areas represent a significant problem.

tibility of semiarid areas, because rainfall there is sufficient to swiftly erode carelessly used soils and because farmers often misinterpret short-term success during wetter than normal years as a sign of long-term crop-raising stability (as the Soviets did before they canceled Kazakhstan's Virgin and Idle Lands Program).

Desperate for more food-producing, income-earning space, Third World agriculturalists readily expand their cash cropping and cattle herding into these marginal lands during wet years, ploughing furiously and introducing more grazing animals than the limited grass/shrub vegetation can support in dry years. Once these bad land-use practices are established,

> . . . a depletion of vegetation occurs which sets in [motion] such insidious processes as [soil erosion and the blowing away of loosened topsoil by the wind]. The vegetation is removed by clearance for cultivation, by the cutting and uprooting of woody species for fuel, by overgrazing [cattle], and by the burning of vegetation for pasture and charcoal (Goudie, 1986, p. 49).

As suggested in Fig. 17-12, the final stage of the desertification process is nothing less than the total denudation of once-productive land. In the more extreme cases, the cultural landscape may even be obliterated by the encroaching desert sands. In the western Sahel, for instance, encroachment is now so far advanced that some predict that the entire upper West African country of Mali will soon become uninhabitable (most of its permanently settled adult residents already shovel sand to keep daily life from grinding to a halt).

With desertification heavily tied to artificial causes, planners and policymakers are focusing on human activities in their search for solutions to the crisis. Although no scientist has yet demonstrated that the process cannot be reversed, there is no question that desertification has already caused much irreparable environmental damage through widespread soil erosion and vegetation loss. Passive solutions, such as teaching good conservation practices to tens of millions of impoverished farmers in the remote areas of the Third World, do not offer much promise given the enormity and immediacy of the problem. Bolder solutions, therefore, are receiving considerable attention by United Nations agencies, international banking organizations, private foundations, and affected governments.

One such possibility is the new technology of *fertigation*, perfected by the Israelis in their recent large-scale development of the Negev Desert. Here, plants have been genetically engineered to be irrigated by the brackish underground water that is available throughout the region (most semiarid areas are underlain by similar water resources). To avoid evaporation, the water is pumped through plastic pipes directly to the roots of plants, where it combines with specially mixed fertilizers to provide exactly the right blend of moisture and nutrients. This system, which can be completely controlled by computers, also protects the plants from salt burn (common in drylands), pests, and diseases.

By the early 1990s, more than 500,000 people were thriving in the Negev's agricultural settlements—and producing enormous quantities of high-quality fruits and vegetables for domestic and foreign markets. Clearly, this kind of high-technology agriculture portends revolutionary change in the farming potential and carrying capacity of desert-area land all over the world. Can the forces now marshalling to confront desertification turn the tide before it is too late?

(*References continued.*)

GLANTZ, M. H., ed. *Desertification: Environmental Degradation In and Around Arid Lands* (Boulder Colo.: Westview Press, 1977).

GOODALL, D. W., et al. *Arid Land Ecosystems* (New York: Cambridge University Press, 1979).

GOUDIE, A. S. *The Human Impact on the Natural Environment* (Cambridge, Mass.: MIT Press, 2nd ed., 1986 [Blackwell, 4th ed., 1993]).

GOUDIE, A. S., and WILKINSON, J. C. *The Warm Desert Environment* (New York: Cambridge University Press, 1977).

GOUROU, P. *The Tropical World: Its Social and Economic Conditions and Its Future Status* (London/New York: Longman, trans. S. H. Beaver, 5th ed., 1980).

GRAINGER, A. *The Threatening Desert: Controlling Desertification* (London: Earthscan, 1990).

HARE, F. K. *Climate and Desertification: A Revised Analysis* (Geneva: World Meteorological Organization/UNEP, World Climate Program, Vol. 44, 1983).

HEATHCOTE, R. L. *The Arid Lands: Their Use and Abuse* (London/New York: Longman, 1983).

MIDDLETON, N. J. *World Atlas of Desertification* (Sevenoaks, U.K.: Edward Arnold, 1992).

NIEUWOLT, S. *Tropical Climatology: An Introduction to the Climates of the Low Latitudes* (New York: Wiley, 1977).

PARK, C. *Tropical Rainforests* (London/New York: Routledge, 1994).

RAVEN, P. H. "The Cause and Impact of Deforestation," in H. J. de Blij, ed., *Earth '88: Changing Geographic Perspectives* (Washington, D.C.: National Geographic Society, 1988), pp. 212–229.

READING, A. J., et al. *Humid Tropical Environments* (Cambridge, Mass.: Blackwell, 1994).

REPETTO, R. "Deforestation in the Tropics," *Scientific American*, April 1990, 36–42.

RICHARDS, J. F., and TUCKER, R. P., eds. *World Deforestation in the Twentieth Century* (Durham, N.C.: Duke University Press, 1988).

THOMAS, D.S.G., and MIDDLETON, N. J. *Desertification: Exploding the Myth* (New York: Wiley, 1994).

United Nations Conference on Desertification, Secretariat. *Desertification: Its Causes and Consequences* (Elmsford, N.Y.: Pergamon, 1977).

Humid Mesothermal (C) Climates

OBJECTIVES

- ◆ To expand our understanding of the various **C** climates.
- ◆ To interpret representative climographs depicting actual conditions in these **C** climate areas.
- ◆ To highlight a major environmental-climatic problem of many **C** climate areas—drought.

The moderately heated **C** or **mesothermal climates** are dominant on the equatorward sides of the middle latitudes, where they are generally aligned as interrupted east–west belts (Figs. 16-2 and 16-3). On a global scale, mesothermal climates may be viewed as transitional between those of the tropics and those of the upper-midlatitude zone, where polar influences begin to produce climates marked by harsh winters. The specific limiting criteria are (1) an average temperature below 18°C (64.4°F) but above 0°C (32°F) in the coolest month, and (2) an average temperature of not less than 10°C (50°F) for at least one month of the year. Such limits tell us that temperature is now a more important climatic indicator than in the lower latitudes, and that the annual rhythms of the **C** climates are more likely

UNIT OPENING PHOTO: Landscape typical of the dry-summer Mediterranean climate. Mediterranean coast of Sicily, the island off the southern tip of Italy.

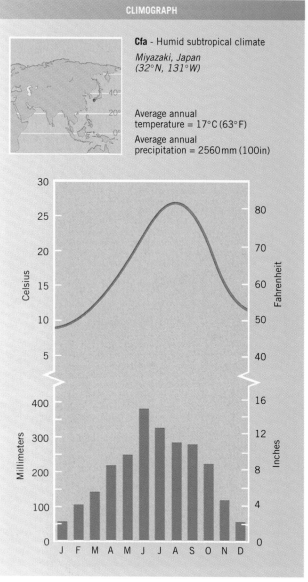

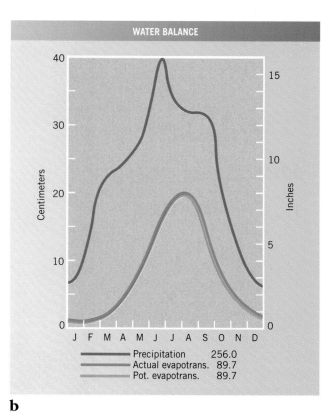

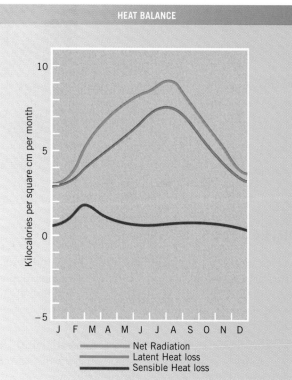

a

FIGURE 18-1 Climograph and related graphic displays for a representative weather station in the humid subtropical (**Cfa**) climate zone.

to involve cyclical shifts between warm and cool seasons rather than rainy and dry seasons.

Humid mesothermal climates are also associated with that great lower-atmosphere battleground, the latitudinal belt where polar and tropical air masses collide and mix. As a result, changeable weather patterns are the rule as meandering jet streams periodically steer a parade of anticyclones and cyclonic storms across the midlatitudes from west to east. Because so many of the world's large population concentrations lie in the **C** climate zone (*see* Fig. 2-5), these swings are of considerable importance. In daily human terms, the episodes of extreme weather frequently experienced there are of particular concern. These departures from average conditions can also occur over longer time periods, as we see in our focus on drought as a major environmental problem of the mesothermal climates.

We can distinguish three major subtypes of humid mesothermal climate according to their pattern of precipitation occurrence. The **Cf** climates are perpetually moist; the **Cs** climate has a dry season in the summer; and the **Cw** climate is dry in the winter.

THE PERPETUALLY MOIST (Cf) CLIMATES

The perpetually moist mesothermal climates (**Cf**) are found in two major regional groupings, both of which are near a major source of water.

The Humid Subtropical (Cfa) Climate

The first such group is the warmer, perpetually moist mesothermal climate, usually called the *humid subtropical* climate (**Cfa**), which is situated in the southeastern portions of the five major continents. It owes its existence to the effects of the warm moist air traveling northwestward (in the Northern Hemisphere) around the western margins of the oceanic subtropical high-pressure zones. Additional moisture comes from the movement of midlatitude cyclones toward the equator in winter and from tropical cyclones in summer and autumn.

Miyazaki, located on the east coast of the southern Japanese island of Kyushu, exhibits the strong seasonal effects of the humid subtropical climate (Fig. 18-1). Average monthly temperatures there range from 6.8°C (44°F) in mildly cold January to 26.7°C (80°F) in hot humid August; the mean annual temperature is 16.7°C (63°F). Rainfall for the year almost totals that of the tropical rainforest—2560 mm (100 in), peaking in early summer. The water balance at Miyazaki is never at a deficit, not even during the relatively dry winter. Throughout the year, actual evapotranspiration always equals the potential evapotranspiration. The heat balance diagram also demonstrates the humid nature of this climate. Most of the net radiation is used to evaporate water, and a relatively small proportion passes into the air as sensible heat.

The similarities of the humid subtropical climate in diverse parts of the world are underscored in the climographs for Charleston, South Carolina (Fig. 18-2), and Shanghai, China (Fig. 18-3). Both cities lie on the southeastern seaboard of a large landmass (North America and Asia, respectively) near latitude 32°N. Compared to Miyazaki (Fig. 18-1), Charleston and Shanghai exhibit significantly lower annual precipitation totals—undoubtedly a function of stronger wind flows off the continent lying to their west—while offshore Kyushu is more con-

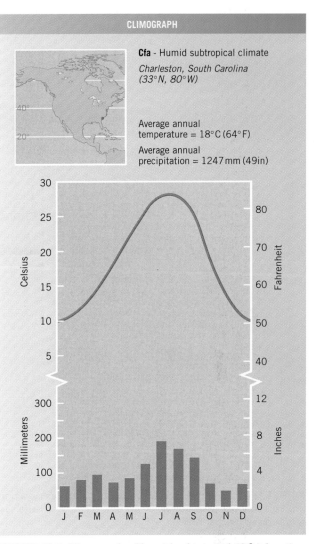

FIGURE 18-2 Climograph of humid subtropical (**Cfa**) location with a particularly mild winter.

stantly bathed by the moist southeasterly winds generated by the North Pacific's Subtropical High.

Slight differences can also be detected in a comparison of all three **Cfa** stations. For example, whereas summer temperatures are strikingly alike, Charleston has a noticeably milder winter because of the smaller bulk of North America (which does not develop as cold and durable a winter high-pressure cell as interior Asia) and the proximity of the warm Gulf of Mexico to the southwest. The overall congruence of climatic characteristics between Charleston and Shanghai, however, is not observed in the cultural landscapes of the southeastern United States and east-central China. In fact, these are about as different as any two on earth (Fig. 18-4), and demonstrate the completely different uses that human societies can make of nearly identical natural environments.

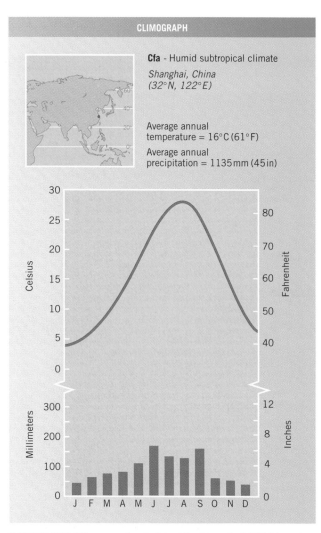

CLIMOGRAPH

Cfa - Humid subtropical climate

*Shanghai, China
(32°N, 122°E)*

Average annual
temperature = 16°C (61°F)

Average annual
precipitation = 1135mm (45in)

FIGURE 18-3 Climograph of a humid subtropical (**Cfa**) location with a pronounced winter, enhanced in this case by the dry monsoon of interior Asia.

a

b

FIGURE 18-4 A pair of farming areas in similar **Cfa** climatic zones, but these agricultural landscapes could not be more different. In the southeastern United States (a), large fields, independent farmsteads, and highway transport dominate. In east-central China (b), small rectangular plots, clustered villages, and water transport dominate.

FIGURE 18-5 The verdant landscape of southern England's Sussex Downs in the marine west coast climate zone of Atlantic-facing Western Europe.

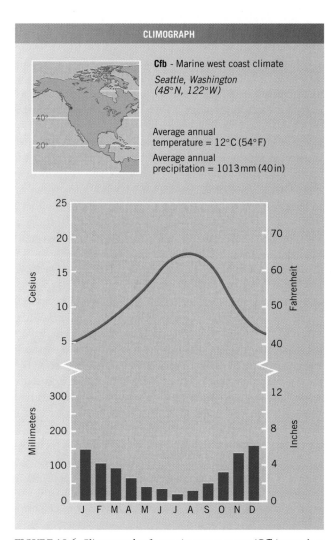

FIGURE 18-6 Climograph of a marine west coast (**Cfb**) weather station in the U.S. Pacific Northwest.

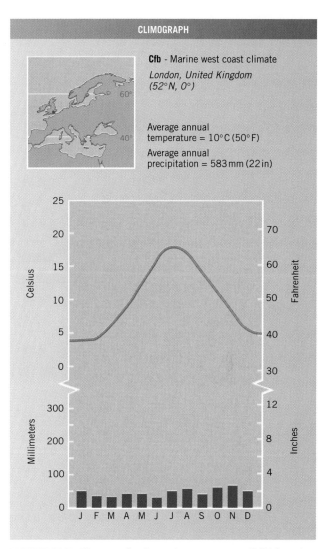

FIGURE 18-7 Climograph of a marine west coast (**Cfb**) location in maritime Europe. Despite its image as a rainy city, note that London annually receives only about half the precipitation of Seattle (Fig. 18-6).

The Marine West Coast (Cfb, Cfc) Climate

The second group of perpetually moist mesothermal climates, with coasts caressed by the prevailing Westerlies year-round, is usually called the *marine west coast climate* (**Cfb**, **Cfc**). Figure 18-5 shows one such area in Western Europe; others are found in the Pacific Northwest of the United States, the adjacent west coast of Canada, southern Chile, southeastern Australia, and New Zealand. In all these places, storms generated by midlatitude cyclones bring a steady flow of moist, temperate, maritime air from the ocean onto nearby land surfaces. The extent of inland penetration depends on topography. Where few orographic barriers exist, such as in Europe north of the Alps, the **Cfb** climate reaches several hundred kilometers eastward from the North Atlantic; but where mountains block the moist onshore winds, such as in the far western United States, Chile, and Australia,

these climates are confined to coastal and near-coastal areas.

There are rarely any extremes of temperature in marine west coast climates. Average monthly temperatures never exceed 22°C (71.6°F), and at least four months record mean temperatures above 10°C (50°F) in the **Cfb** zones (the cooler **Cfc** subtype experiences less than four months of mean temperatures above 10°C). The **Cfb** patterns are seen in the climographs for Seattle, Washington and London, England (Figs. 18-6 and 18-7). In your comparisons you will note that, despite its popular image as a rainy city, London receives on average only 583 mm (22 in) of precipitation annually, equivalent to the yearly total in the moistest semiarid climate. But London does indeed experience a great number of cloudy and/or rainy days each year (the hallmark of many a marine west coast

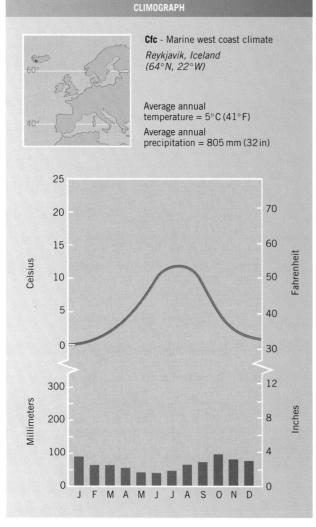

CLIMOGRAPH

Cfc - Marine west coast climate

Reykjavik, Iceland
(64°N, 22°W)

Average annual
temperature = 5°C (41°F)

Average annual
precipitation = 805 mm (32 in)

FIGURE 18-8 Climograph of a weather station that experiences the cooler variety of the marine west coast climate (**Cfc**).

climate location), with much of its precipitation occurring as light drizzle. The cooler **Cfc** pattern is shown in the climograph for Reykjavik, Iceland (Fig. 18-8), which records only three months with mean temperatures at or above 10°C (50°F).

The **Cfb** climates are situated precisely in the middle of the midlatitude, prevailing westerly wind belts. The cooler **Cfc** climates are found only in the Northern Hemisphere at higher latitudes, where strong warm ocean currents sweep poleward along western continental shores (such as off Iceland, Norway, and Alaska). Western civilization moved to the **Cfb** climates of northwestern Europe after leaving its Mediterranean hearth. The adequate year-round precipitation naturally produces a forest of evergreen conifers and broadleaf trees that shed their leaves in winter. Although dry spells are not unknown in these areas, the soil rarely has a moisture deficit for a long period of time.

THE DRY-SUMMER (Cs) CLIMATES

The second major mesothermal subtype is the dry subtropical or *Mediterranean* climate (**Cs**). It is frequently described, particularly by people of European descent, as the most desirable climate on earth. This is the climate that attracted filmmakers to Southern California, so famous for its clear light and dependable sunshine. The warmer variety of Mediterranean climate, **Csa**, is found in the Mediterranean Basin itself as well as in other interior locations. The cooler variety, **Csb**, is found on coasts near cool offshore ocean currents, most notably in coastal areas of California, central Chile, southern and southwestern Australia, South Africa, and Europe's Iberian Peninsula (northern Portugal and northwestern Spain).

In the **Cs** climate, rainfall arrives in the cool season, largely as a result of the winter storms produced by midlatitude cyclones. Annual precipitation totals are moderate, ranging from approximately 400 mm (16 in) to 650 mm (25 in). The long dry summers are associated with the temporary poleward shift of the wind belts, specifically the warm-season dominance of subsiding air on the eastern side of the oceanic subtropical high-pressure zone. Thus extended rainless periods are quite common in Mediterranean climates. But when they occur in the more widely distributed (and more heavily populated) **Cf** climates, they can lead to serious *drought*, which disrupts human-environmental relationships (*see* Perspective: The Drought of '88).

Some aspects of the "perfect" climate can be observed for San Francisco in Fig. 18-9. The cool offshore California Current, with its frequent fogs (*see* Fig. 9-8), keeps average monthly temperatures almost constant throughout the year, with a range of only 6.6°C (12°F) around an annual mean of 13.6°C (56.5°F). The water balance values for San Francisco indicate an average annual rainfall total of 551 mm (22 in), with the rains mostly emanating from winter storms steered southward by the Polar Front jet stream. Although in winter actual evapotranspiration reaches the potential amount, in summer it falls well short.

The effect of the dry summer is reflected in the heat balance diagram for nearby Sacramento, located about 150 km (95 mi) northeast of San Francisco. During winter much of the net radiation is expended as latent heat loss. But in the summer soil-moisture deficits require that a large proportion of the net radiation be used in the form of sensible heat to warm the air; with intensive fruit and vegetable agriculture dominating California's Central Valley south of Sacramento, widespread irrigation is an absolute necessity (Fig. 18-11). For purposes of comparison, Fig. 18-12 shows the climograph for Athens, Greece, a classic example of the warmer **Csa** Mediterranean climate variety. Note the similarities to the San Francisco temperature and precipitation pattern (Fig.

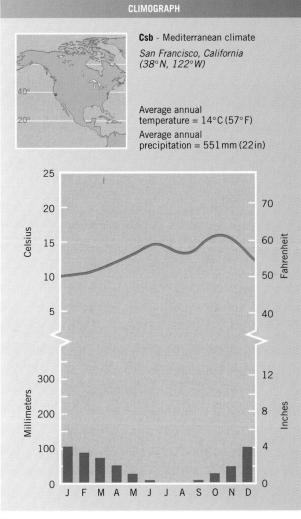

CLIMOGRAPH

Csb - Mediterranean climate

San Francisco, California
(38°N, 122°W)

Average annual
temperature = 14°C (57°F)

Average annual
precipitation = 551 mm (22 in)

a

FIGURE 18-9 Climograph and related graphic displays for a representative weather station in the Mediterranean (**Csb**) climate zone. The heat balance graph is based on data from nearby Sacramento.

18-9); the only noteworthy difference is the greater annual range of temperatures produced by the more continental location of Athens.

THE DRY-WINTER (Cw) CLIMATES

The **Cw** climate, the third major subtype of humid mesothermal climate, is little different from the tropical savanna (**Aw**) climate discussed in Unit 17. The only departures are the amounts of rainfall and a distinct cool season, with the average temperature of at least one month falling below 18°C (64.4°F). The **Cw** climate, therefore, is the only subdivision of the **C** climates found extensively in the tropics. For this reason it often shares with the **Aw** climate the designation *tropical wet-and-dry* climate.

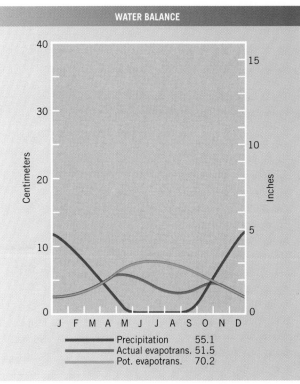

WATER BALANCE

Precipitation	55.1
Actual evapotrans.	51.5
Pot. evapotrans.	70.2

b

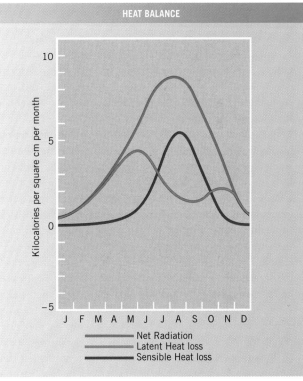

HEAT BALANCE

Net Radiation
Latent Heat loss
Sensible Heat loss

c

In many cases, the winter dry season of **Cw** climates is a result of offshore-flowing winds that accompany a winter monsoon (indeed, the **Cw** climate is sometimes called the *subtropical monsoon* climate). In South and Southeast Asia as well as in northeastern Australia, rain-

The Drought of '88

The drought that plagued so much of the United States in 1988, the worst since the mid-1920s, was a classic example of this major environmental problem. At its peak in late summer, extreme and severe drought conditions covered more than 40 percent of the nation (Fig. 18-10). For the **B** climate areas, which occupy most of the conterminous United States west of 100°W longitude, these dry spells have always been part of the gamble of living and farming there. Even in the Mediterranean climate of the Pacific coastal zone, rainless summers are a constant reminder that water resources must be managed wisely because recurrent shortages have been part of California's modern settlement history. But what made the 1988 drought one of this century's most memorable environmental events was its grip on the normally humid **Cfa** climate zone that blankets the southern two-thirds of the country's eastern half.

Unlike desertification, which is largely a process of human degradation that affects many **B** climate environments (*see* Unit 17), drought is a natural hazard that recurs in seemingly irregular cycles. Specifically, a **drought** involves the below-average availability of water in a given area over a period of at least several months. The conditions that constitute a drought are clear enough: a decrease in precipitation accompanied by warmer-than-normal temperatures and the shrinkage of surface and soil water supplies. If the drought reaches an extreme stage of development, all of these conditions intensify further and may result in spreading grass and/or forest fires as well as the blowing away of significant quantities of topsoil by hot dry winds.

However dramatic it may become at its height, a drought has no clear beginning or end. It develops slowly until it becomes recognized as a crisis, and it tends to fade away as more normal moisture patterns return. Because droughts encompass no spectacular meteorological phenomena, they were not intens-

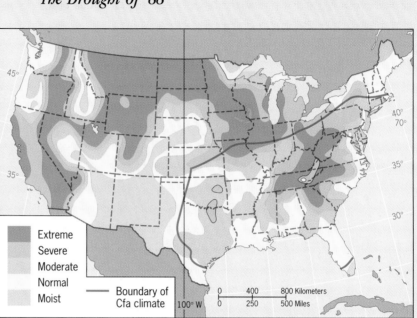

FIGURE 18-10 The distribution of drought conditions in the conterminous United States, August 20, 1988.

ively studied before the 1970s. But the Sahel disaster of two decades ago in northern Africa (*see* Unit 17) finally aroused the interest of climatologists, whose subsequent research is providing a clearer understanding of droughts and their often far-reaching consequences.

As these studies proceeded, it quickly became evident that a drought was not simply a meteorological/hydrological aberration but a complex phenomenon that also encompassed economic, sociological, and even political dimensions. During 1988, for example, the worsening water shortage in the Midwest became a national news story that daily dominated newscasts and headlines from June through September. As the heavily publicized plight of farmers and the disruptions of their lives reached unprecedented levels, the election-year Congress responded by passing the $4 billion Federal Disaster Relief Act of 1988 (hurriedly signed into law in mid-August) to ameliorate the financial losses suffered by agricultural producers. Any definition of drought, therefore, must now go well beyond the physical-geographic aspects

of a prolonged water shortage. As the 1988 experience demonstrated, drought is no longer mainly an agricultural disaster.

Today new links are constantly being forged between the natural environment and the economic activities of advanced, postindustrial societies. The high-technology computer industry, for example, is increasingly dependent on uninterrupted supplies of pure water; when water-use cutbacks were ordered for the San Francisco Bay Area in mid-1988, the semiconductor industry in Silicon Valley was forced to slow down and operate at a temporary disadvantage. In other manufacturing spheres, the proper disposal of chemical and other toxic wastes depends upon sufficient water supplies to dilute and flush the pollutants—a process that quickly slows to the point of danger when a drought takes hold.

The growing strains on U.S. water resources, however, are most noticeable where new urban development mushrooms, particularly in the outer reaches of Sunbelt and reviving northern metropolitan areas. Pressures are at their greatest in places like Arizona's Phoenix and Tucson in the already arid Southwest. But the 1988 drought revealed the vulnerability of the Southeast as well; in rapidly expanding metropolitan Atlanta, for instance, Lake Lanier, a chief source of that region's drinking water, experienced so substantial a decline in water level that steep cutbacks in consumption were mandated—among other things virtually paralyzing the lake's thriving resort and recreational functions.

Northern Georgia, of course, was but one of many large areas in the U.S. humid subtropical climate zone that was severely affected by the Drought of '88 (Fig. 18-10). In the northwestern corner of the **Cfa** zone, as already noted, the Corn Belt of the Midwest suffered one of its worst crop losses (of both corn and soybeans) in modern history. Moreover,

while the eastern part of the Corn Belt received significant precipitation during the following winter of 1988–1989, the parched western portion of the region did not.

As the map shows, other hard-hit areas were the upper Ohio Valley, the central Appalachians (including most of Tennessee and western North Carolina), and wide segments of the Atlantic Coastal Plain from central Georgia northeast to Virginia. In Georgia and the Carolinas, conditions were especially difficult because the 1988 drought aggravated a pre-existing water shortage dating from 1985, itself following hard on the heels of a disastrous drought in 1980–1981. Also negatively affected were areas that did not directly experience the drought. The lower Mississippi Valley, for example, suffered major economic disloca-

tions because so little water came down from the upper basin that river levels fell to record lows, often exposing sandbars that halted barge traffic—and local industry—for weeks on end.

The cause of all this misery was a northward shifting of the subtropical jet stream. Instead of following its usual route eastward from California to the Gulf of Mexico and then northeastward between the Midwest and Southeast, the mid-1988 path of this high-speed, upper-air windflow mainly followed the U.S.-Canada border from the Pacific to the upper Great Lakes and then swung even further north to Hudson Bay. Although a weaker branch of the jet stream snaked south across northern Mexico, most of the U.S. interior east of the Rocky Mountains sweltered under the influence of a persistent, anomalous high-pressure cell.

fall is provided by the wet summer monsoon. In other cases, such as in south-central Africa, the winter dry season occurs because the rains of the ITCZ migrate into the Northern Hemisphere with the sun.

Cw climates are also associated with higher elevations in the tropical latitudes, which produce winter temperatures too cool to classify these areas as **A**. Cherrapunji, in the Khasi Hills of northeastern India, is a good example (Fig. 18-13). Because of its location near the subtropics at latitude 25°N, combined with its altitude

of 1313 m (4300 ft), this highland town exhibits the characteristics of the cooler variety of this climatic subtype (**Cwb**). Cherrapunji's enormous annual rainfall (which averages 11,437 mm [450 in]), as explained in the discussion of the **Am** climate in Unit 17, is the result of wet-monsoon winds that encounter particularly abrupt orographic lifting on their way inland from the Bay of Bengal to the blocking Himalayan mountain wall to the north.

FIGURE 18-11 California's agriculturally rich San Joaquin Valley (the southern component of the Central Valley) is highly dependent on irrigation for its productivity. These lush cotton fields border the dry foothills that lead up to the Sierra Nevada on the valley's eastern margin.

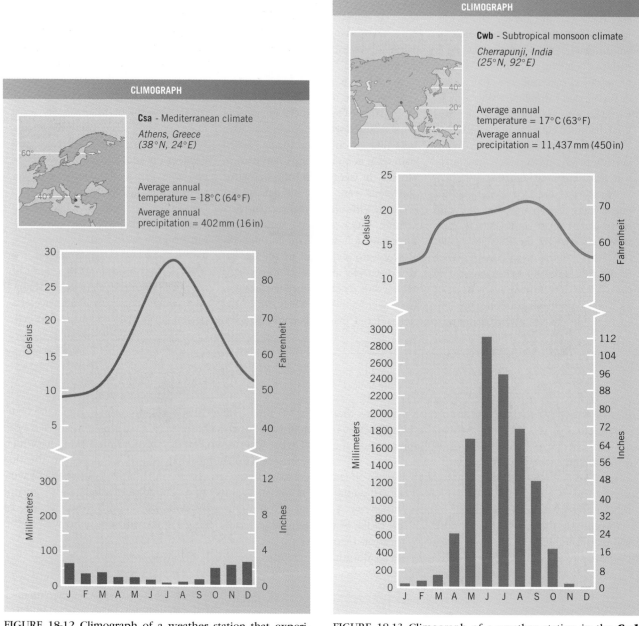

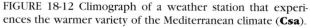

FIGURE 18-12 Climograph of a weather station that experiences the warmer variety of the Mediterranean climate (**Csa**).

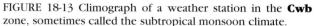

FIGURE 18-13 Climograph of a weather station in the **Cwb** zone, sometimes called the subtropical monsoon climate.

KEY TERMS

drought (p. 202)   mesothermal climate (p. 195)

REVIEW QUESTIONS

1. What distinguishes **Cfa** from **Cfb** and **Cfc** climates?

2. How do **Cw** climates differ from **Aw** climates?

3. What are the major precipitation patterns associated with Mediterranean (**Cs**) climates?

4. What is meant by the term *drought*?

REFERENCES AND FURTHER READINGS

ARAKAWA, H., ed. *Climates of Northern and Eastern Asia* (Amsterdam: Elsevier, World Survey of Climatology, Vol. 8, 1969).

BRYSON, R. A., and HARE, F. K., eds. *Climates of North America* (Amsterdam: Elsevier, World Survey of Climatology, Vol. 11, 1974).

DI CASTRI, F., and MOONEY, H. A., eds. *Mediterranean-Type Ecosystems: Origin and Structure* (New York: Springer-Verlag, 1973).

HOUSTON, J. M. *The Western Mediterranean World: An Introduction to Its Regional Landscapes* (New York: Praeger, 1964).

"The Long, Hot Summer of '88," *Natural History*, January 1989 (special issue).

MATHER, J. R. *Water Resources: Distribution, Use, and Management* (New York: Wiley/V. H. Winston, 1984), 362–382.

PALMER, W. C. *Meteorological Drought: Its Measurement and Classification* (Washington, D.C.: U.S. Weather Bureau, Research Paper No. 45, 1965).

STEWART, G. R. *Storm* (New York: Modern Library, 1947).

WALLEN, C. C., ed. *Climates of Central and Southern Europe* (Amsterdam: Elsevier, World Survey of Climatology, Vol. 6, 1977).

WALLEN, C. C., ed. *Climates of Northern and Western Europe* (Amsterdam: Elsevier, World Survey of Climatology, Vol. 5, 1970).

Higher Latitude (D, E) and High-Altitude (H) Climates

OBJECTIVES

◆ To expand the discussion of typical **D**, **E**, and **H** climates and to interpret representative climographs for these zones.

◆ To highlight a major environmental-climatic problem of many D climate regions—acid precipitation.

◆ To characterize the general influence of altitude on climatic conditions.

I n this unit, we survey the remaining climates in the Köppen global classification and regionalization system, which range from the upper-middle to the polar latitudes and into the highest elevations capable of supporting permanent human settlement. We begin with the microthermal (**D**) climates that dominate the Northern Hemisphere continents poleward of the mesothermal (**C**) climates. Some large population clusters are located within the milder portions of the microthermal climate zone. The major environmental problem we consider here, one now expanding in every **D** climate region, is associated with the effluents of urban-industrial concen-

trations—*acid precipitation.* Our attention next turns to the polar (**E**) climates that blanket the highest latitudes. Finally, we treat the highland (**H**) climates, which are often reminiscent of **E**-type temperature and precipitation regimes because they lie at altitudes high enough to produce Arctic-like conditions regardless of latitude.

UNIT OPENING PHOTO: Densely packed coniferous trees of the vast Canadian snowforest at the Athabasca River, Jasper National Park, Alberta.

THE MAJOR HUMID MICROTHERMAL (D) CLIMATES

Northern Hemisphere continental landmasses extend in an east–west direction for thousands of kilometers. In northern Eurasia, for instance, the distance along 60°N between Bergen on Norway's Atlantic coast and Okhotsk on Russia's Pacific shore is almost 9000 km (5600 mi); indeed, Russia alone is so broad that the summer sun rises above the eastern Pacific shoreline before it has set over the Baltic Sea in the west—ten time zones away! Consequently, vast areas of land in the middle and upper latitudes are far away from the moderating influence of the oceans. The resulting continentality shapes a climate in which the seasonal rhythms of the higher latitudes are carried to extremes, one of distinctly warm summers balanced by harsh frigid winters. These are the **D** or **humid microthermal climates**, distinguished by a warm month—or months—when the mean temperature is above 10°C (50°F), and a period averaging longer than a month when the mean temperature is below freezing (0°C [32°F]). The people who live in **D** climates have acquired lifestyles as varied as their wardrobes to cope with such conditions.

The cold-winter **D** climates are found mainly in the middle- and upper-latitude continental expanses of Asia and North America (Fig. 16-3). As the world map shows, these areas are sometimes punctuated by regions of the entirely summerless **E** climates, especially in northeastern Asia and the Canadian Arctic. The **D** climates are also found in certain Northern Hemisphere midlatitude uplands and on the eastern sides of continental landmasses. These microthermal climates represent the epitome of the continentality effect: although they cover only 7 percent of the earth's surface overall, they cover no less than 21 percent of its land area.

The Humid Continental (Dfa/Dwa, Dfb/Dwb) and Taiga (Dfc/Dwc, Dfd/Dwd) Climates

The humid microthermal climates are usually subdivided into two groups. In the upper-middle latitudes, where more heat is available, we observe the *humid continental* climates (**Dfa/Dwa** and **Dfb/Dwb**). Farther north toward the Arctic, where the summer net radiation cannot raise average monthly temperatures above 22°C (71.6°F), lie the subarctic *taiga* climates (**Dfc/Dwc** and **Dfd/Dwd**). (*Taiga* is the Russian word for "snowforest.") Both groups of climates usually receive enough precipitation to be classified as moist all year round (**Df**). But in eastern Asia, the cold air of winter cannot hold enough moisture, so a dry-winter season (**Dw**) results.

All **D** climates show singular extremes of temperature throughout the year. In Unit 7 we observe the heat

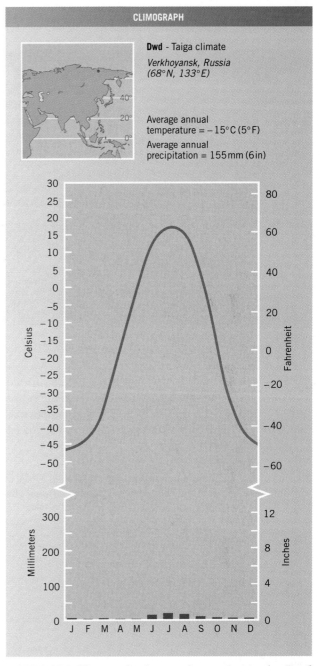

CLIMOGRAPH

Dwd - Taiga climate

Verkhoyansk, Russia (68°N, 133°E)

Average annual temperature = −15°C (5°F)

Average annual precipitation = 155mm (6in)

FIGURE 19-1 Climograph of a weather station in the **Dwd** zone, the harshest extreme of the humid microthermal climate regions.

balance for the northern Siberian town of Turukhansk (Fig. 7-6d, p. 80). In July the temperature there averages a pleasant 14.7°C (58.5°F), but the January mean plunges to −31°C (−24°F). However, brutal as the latter reading might be, Turukhansk is only an example of the **Dfc** climatic subtype, which is not the most extreme category. The harshest subtype, **Dwd**, is exemplified by Verkhoyansk, a remote far northeastern Russian village 2100 km (1300 mi) northeast of Turukhansk, whose particularly extreme temperatures are discussed in Unit 8 (*see* Fig. 8-10, p. 92). As the climograph in Fig. 19-1 shows, the

FIGURE 19-2 These giant cabbages are among the many varieties of summer vegetables raised during Alaska's short but intense growing season.

average January temperature in Verkhoyansk is an astonishing −46.8°C (−52°F). Also fascinating are the *ranges* of temperature, the difference between the highest and lowest monthly averages: 45.7°C (82°F) for Turukhansk and 62.5°C (112.5°F) for Verkhoyansk. These are among

FIGURE 19-3 A once-smooth road after several annual cycles of freezing and thawing in the underlying surface layer of soil. This valley is located near Mount Drum in south-central Alaska.

the greatest annual temperature ranges found anywhere on our planet's surface.

Such temperatures have a far-reaching impact. There is a short, intense growing season, because the long summer days at these high latitudes permit large quantities of heat and light to fall on the earth. Special quick-growing strains of vegetables and wheat have been developed to make full use of this brief period of warmth and moisture (Fig. 19-2). On the Chukotskiy Peninsula, the northeasternmost extension of Asia where Russia faces Alaska's western extremity across the narrow Bering Strait, one variety of cucumbers usually grows to full size within 40 days. But agriculture is severely hampered by the thin soil (most was removed by the passing of extensive icesheets thousands of years ago) and permafrost. **Permafrost**, discussed at length in Unit 48, is a permanently frozen layer of the subsoil that sometimes exceeds 300 m (1000 ft) in depth.

Although the top layer of soil thaws in summer, the ice beneath presents a barrier that water cannot permeate. Thus the surface is often poorly drained. Moreover, the yearly freeze–thaw cycle expands and contracts the soil, making construction of any kind difficult, as Fig. 19-3 demonstrates. Because of the soil's instability, the Trans-Alaska oil pipeline built during the 1970s must be elevated above ground on pedestals for much of its 1300-km (800-mi) length. This prevents the heated oil in the pipeline from melting the permafrost and causing land instability that might damage the pipe.

Long-lasting snow cover has other effects. It reflects most of the small amount of radiation that reaches it, so that little is absorbed. It cools the air and contributes to the production of areas of high atmospheric pressure. It presents difficulties for human transport and other activities, but, paradoxically, it does keep the soil and dormant plants warm because snow is a poor conductor of heat. Measurements in St. Petersburg, Russia's second largest city, have shown that temperatures below a snow cover can be as high as −2.8°C (27°F) when the air overlying the cover is a bitter −40°C (−40°F).

In the humid microthermal climates, most of the precipitation comes in the warmer months. This is not just because the warmer air can hold more moisture. In winter, large anticyclones develop in the lower layers of the atmosphere. Within the anticyclones the air is stable, and these high-pressure cells tend to block midlatitude cyclones. In summer, convection in the unstable warmer air creates storms, so midlatitude cyclones can pass through **D** climate areas more frequently during that time of year. In some places there is also a summer monsoon season. China's capital city of Beijing (**Dwa**) experiences a pronounced wet monsoon (Fig. 19-4), which is typical of upper-midlatitude coastal areas on the East Asian mainland.

The more detailed features of the microthermal climate type are seen in the temperature, precipitation, water balance, and heat balance data displayed in Fig. 19-5. These graphs are for the southern Siberian city of Barnaul, located in a **Dfb** region within the heart of Eurasia at 53°N near the intersection of Russia and China with the western tip of Mongolia. The extreme seasonal change of Barnaul's climate is apparent everywhere in Fig. 19-5. Average monthly temperatures range from 20°C (68°F) in July to −17.7°C (0°F) in January; the annual mean temperature is 1.4°C (34.4°F). Most of the precipitation falls during summer, but it is not enough to satisfy the potential evapotranspiration rate.

The heat balance diagram shows the large amount of net radiation that supports Barnaul's intense summer growing season. Lack of naturally available surface water means that only a little of this energy is used in evaporation in summer; thus most of it warms the air as sensible heat. As in Turukhansk and Verkhoyansk, the low winter temperatures are the result of net radiation deficits that are not balanced by a significant flow of sensible heat to the surface. Barnaul, a major manufacturing center in its own right, is also located near the Kuznetsk Basin, one of the largest heavy-industrial complexes in Russia. Such regions are important features of the economic geography of the middle latitudes, and they involve physical geography as well because industrial areas are the sources of pollutants that produce **acid precipitation**—abnormally acidic rain, snow, or fog resulting from high levels of the oxides of sulfur and nitrogen in the air. This is a critical environmental problem that is worsening throughout the **D** climate zone (*see* Perspective: Acid Precipitation).

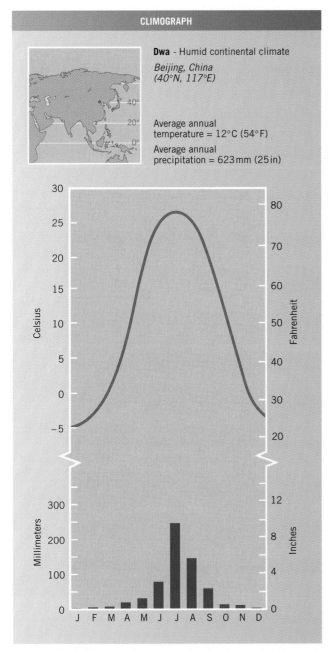

CLIMOGRAPH

Dwa - Humid continental climate

Beijing, China
(40°N, 117°E)

Average annual
temperature = 12°C (54°F)

Average annual
precipitation = 623 mm (25 in)

FIGURE 19-4 Climograph of a weather station in the **Dwa** climate zone that also experiences a marked summer monsoon.

THE POLAR (E) CLIMATES

Beyond the Arctic and Antarctic Circles (66½°N and S, respectively), summer and winter become synonymous with day and night. Near the poles there are six months of daylight in summer, when the monthly average temperature might "soar" to −22°C (−9°F). In winter, six months of darkness and continual outgoing radiation lead to the lowest temperatures and most extensive icefields on the planetary surface. The lowest temperature ever recorded on earth—−89°C (−129°F)—was for a memorably chilly day at one of the highest altitude stations in central Antarctica, the Russian research facility, Vostok.

The Tundra (ET) Climate

E or **polar climates** are defined as those climates in which the mean temperature of the warmest month is less than 10°C (50°F). There are two major subtypes of polar climate. If the warmest average monthly tempera- ture is between 0°C (32°F) and 10°C (50°F), the climate is called *tundra* (**ET**), after its associated vegetation of mosses, lichens, and stunted trees (Fig. 19-6). Practically all of the **ET** climates are found in the Northern Hemisphere, where the continentality effect yields particularly

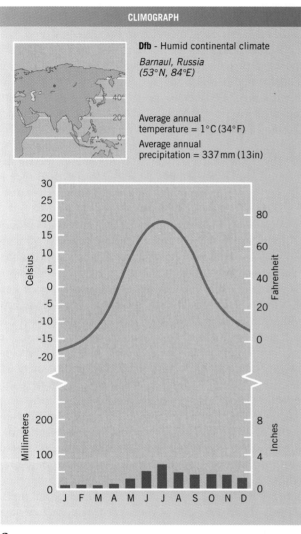

a

FIGURE 19-5 Climograph and related graphic displays for a representative weather station in the **Dfb** climate zone.

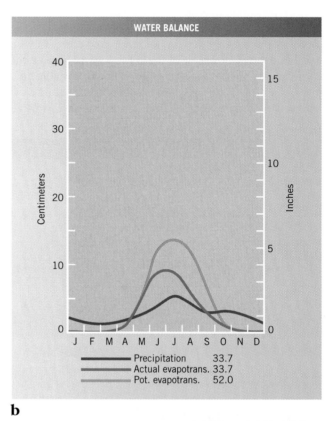

b

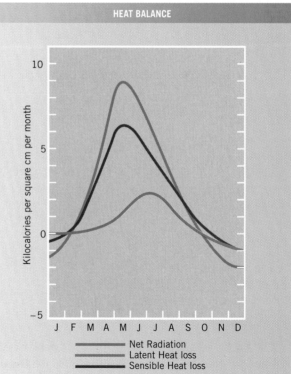

c

FIGURE 19-6 The pre-winter carpet of tundra that blankets the foothills of the Alaska Range near Mount McKinley in Denali National Park.

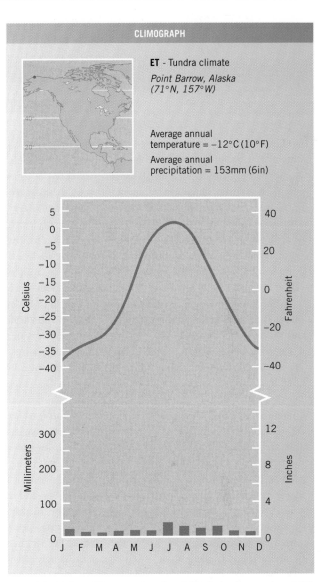

CLIMOGRAPH

ET - Tundra climate
Point Barrow, Alaska
(71°N, 157°W)

Average annual
temperature = −12°C (10°F)

Average annual
precipitation = 153mm (6in)

FIGURE 19-7 Climograph of a high-latitude tundra (**ET**) weather station.

long and bitter winters. Furthermore, many of the **ET** areas border the Arctic Ocean, which provides a moisture source for frequent fogs when it is not frozen. In summer, poor drainage leads to stagnant water, the breeding grounds for enormous swarms of flies and mosquitoes. Precipitation, mainly from frontal midlatitude depressions in the warmer months, seldom exceeds 300 mm (12 in) for the year. The climograph for Point Barrow at Alaska's northern tip (Fig. 19-7) displays the typical **ET** regimes of temperature and moisture.

The Icecap (EF) Climate

In *icecap* (**EF**) climates, the second major polar subtype, we find the lowest annual temperatures on earth. In a given year about 90 mm (3.5 in) of precipitation, usually in the form of snow, falls onto the barren icy surface. The inhospitable climate has made it difficult to collect data from these areas, which lie mainly in Antarctica and interior Greenland as well as atop the Arctic Ocean's floating icecap. But in recent years international cooperation has made more information available. The Russian station, Mirnyy, is located in Antarctica just inside the Antarctic Circle. Its relatively low latitude hosts temperatures that are rather moderate for an icecap climate, but the mean annual temperature of −11°C (12°F) is not high, and the warmest monthly average does not rise above the freezing point—the hallmark of an **EF** climate (Fig. 19-9).

Year-round snow makes it almost impossible to obtain accurate water balance data, but the Russians have measured the heat balance components, which vividly characterize the frigid **EF** climate. There is significant positive net radiation at Mirnyy for only about four months of the year; more radiation leaves the earth than

enters during the other months. Sensible heat flow throughout the year is directed from the air toward the ground, the final result of the general circulation of the atmosphere moving heat toward the poles. Sensible heat and net radiation can provide energy for some evaporation in the warmer months, but in the winter condensation of moisture onto the surface provides only a minor source of heat. These extreme conditions notwithstanding, a number of scientists now live and work in icecap climates, studying the environment or searching for oil and other secrets of the earth. But every one of them depends for survival on artificial heating and food and water supplies flown in from the outside world, without which human life could not exist in the coldest climate on the planet.

Perspectives on THE HUMAN ENVIRONMENT

Acid Precipitation

Acid precipitation, which receives considerable attention in the news media, would be near the top of any physical geographer's list of global environmental problems. Acid rain is the most common form of this phenomenon, but the effects of acid precipitation are also associated with snow, fog, clouds, and even dust particles contained in the boundary layer of the atmosphere.

When fossil fuels (oil, coal, and natural gas) are burned, they release sizeable quantities of sulfur dioxide (SO_2) and nitrogen oxides (NO_x) into the surrounding air. These effluent gases then react chemically with water vapor in the atmosphere and are transformed into precipitable solutions of both sulfuric and nitric acid. When these gaseous pollutants are present in sufficient quantity, they can produce acid concentrations in precipitation that are capable of causing major damage to vegetation as well as to animal and aquatic life.

Acidity is measured by the *pH scale*, which ranges from 0 to 14. Above the neutral level of 7, alkalinity is observed (which strengthens as the pH rises toward 14); below 7, a solution becomes increasingly acidic as its pH approaches zero. In most humid environments, the slightly acidic pH of 6.5 is considered normal in standing bodies of fresh water (by way of comparison, the salty seawater of the oceans exhibits an average pH of 7.8). Recent research has shown that significant environmental damage occurs with increasing acidification. For instance, even a relatively small drop in pH from 6.5 to 5.9 changes a lake's phytoplankton composition so that certain fish species are eliminated from the food chain and disappear. When a pH of 5.6 is reached, large patches of slimy algae appear on the surface and begin to choke

off the lake's oxygen supply and block sunlight from filtering down into lower water layers. These effects would climax should the pH drop to 5.0 (about 30 times the normal acidity level), a point at which no fish species would be able to survive.

Besides freshwater lakes, there is now ample evidence that acid precipitation is also devastating oceanic life near certain coastal zones: a recent study by the Environmental Defense Fund demonstrated that acid rain linked to nitrogen oxide air pollution was destroying fish and other marine creatures in numerous estuaries along the heavily industrialized northeastern U.S. seaboard. Forests are another major casualty of waterborne acid pollution. Vast areas of the upper-middle latitudes have experienced damage to once-healthy woodlands over the past two decades—a shocking 25 percent of Europe's forests are now affected (*see* Fig. 22-7)—and the problem is continuing to spread, particularly in the southern and central portions of the **D** climate zone. Even human endeavors are being increasingly affected; in rural areas acid-sensitive crops are constantly threatened, while in the cities the deterioration of older buildings and concrete is noticeably hastened.

The sources of sulfur dioxide and nitrogen oxides are most closely linked to major urban-industrial areas containing the highest densities of fossil-fuel combustion. Electric power plants, serving both manufacturers and the general public, are the leading polluters in the United States, producing almost 70 percent of the sulfur dioxide and one-third of the nitrogen oxide emissions. Transportation is the next single biggest offender, accounting for approximately 40 percent of all nitrogen oxide pollution (mainly via the exhaust systems of motor vehicles).

The sources of the remaining emissions that result in acid rain are fairly evenly dispersed among the industrial, commercial, and household/office sectors.

The geography of acid rain in North America, however, decreasingly corresponds to the spatial distribution of sulfur dioxide and nitrogen oxide sources because of measures taken to improve atmospheric pollution since the 1960s. In an effort to improve the air quality of industrial areas, much higher smokestacks (often in excess of 300 m [1000 ft]) were constructed to disperse the effluents released by fossil-fuel burning. These measures did achieve their local goals, but instead of dispersing, the pollutants entered higher-level, longer-distance windflows that tended to channel and transport them in still-high concentrations. Thus, in effect, certain distant areas thereby became the new dumping grounds for these emissions. In the case of the U.S. Midwest, the largest North American regional source of sulfur and nitrogen oxides, prevailing winds steered these acid-precipitation-producing wastes hundreds of kilometers toward the northeast (and eventually south as well).

As the maps in Fig. 19-8 reveal, a strong relationship exists between those pollutant-laden winds and the spatial distribution of acid rain. The crisis is particularly acute in the heavily wooded **Dfb** zone of southeastern Canada, where fish kills and other serious environmental damage have been widespread. North America, however, is not alone: similar effects of acid precipitation are now spreading across Eurasia too, most notably in Northern and Eastern Europe as well as in many of the vast wilderness areas that blanket Siberia and the Russian Far East.

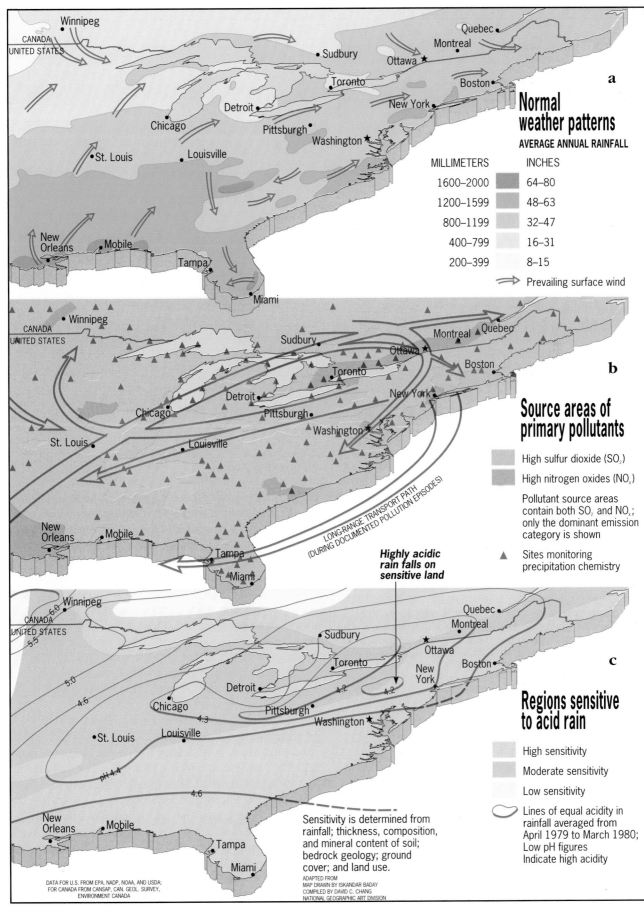

FIGURE 19-8 Eastern North America weather patterns (a), pollution sources and associated windflows (b), and land sensitivity related to acid rain occurrence (c). Copyright © National Geographic Society.

213

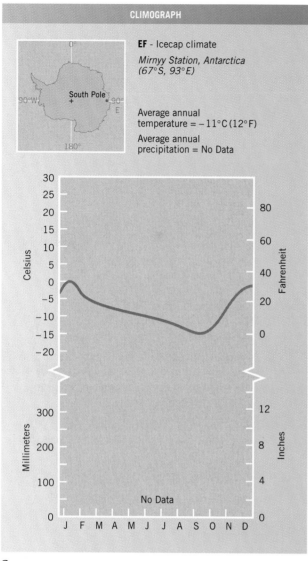

a

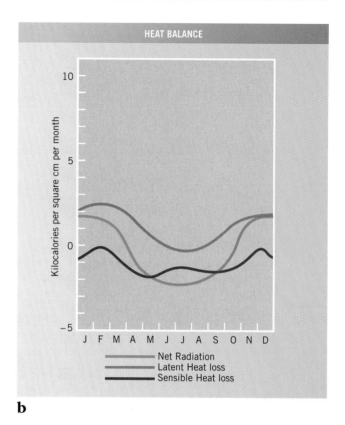

b

FIGURE 19-9 Climograph and heat balance diagram for a representative weather station in the icecap (**EF**) climate zone. There are no precipitation data available for this remote Antarctic location.

HIGH-ALTITUDE (H) CLIMATES

Our overview of the mosaic of climates covering Earth's land surface would not be complete without some mention of the climates of highland regions. The mountains of the highest uplands reach into the lower temperatures and pressures of the troposphere. Thus, as one moves steadily upward into a highland zone, the corresponding changes in climate with increasing elevation mimic those observed in a horizontal passage from equatorial to progressively higher latitudes.

One of the outstanding features of **H** or **highland climates** is their distinct *vertical zonation* according to altitude. Nowhere has this been better demonstrated than in the tropical Andes mountain ranges of north-

western South America. Figure 19-10 combines the characteristics of many of these highlands into a single model. The foothills of the Andes lie in a tropical rainforest climate in the Amazon Basin in the east and in arid climates tempered by a cool Pacific Ocean current in the west. Above 1200 m (4000 ft), tropical climates give way to the subtropical zone. At 2400 m (8000 ft) the mesothermal climates appear, with vegetation reminiscent of that found in Mediterranean climatic regions. These in turn give way to microthermal climates at 3600 m (12,000 ft), and above 4800 m (16,000 ft) permanent ice and snow create a climate like that of polar areas (Fig. 19-11). It is sometimes said that if one misses a bend at the top of an Andean highway, the car and driver will plunge through four different climates before hitting bottom!

These altitudinal zones mainly reflect the decrease of temperatures with rising elevation. But wind speeds tend to increase with height, as can rainfall (and snowfall), fog, and cloud cover in many instances. Moreover, the radiation balance is markedly altered by altitude. Because less shortwave radiation is absorbed by the atmosphere at higher elevations, greater values are recorded at the surface. This is especially true of the ultraviolet radiation responsible for snow blindness as well as the suntans of mountaineers and skiers. Where there are snow-covered surfaces, much of the incoming radiation is reflected and not absorbed, which further acts to keep the temperatures low.

Mountainous areas are often characterized by steep slopes, and these slopes have different orientations or

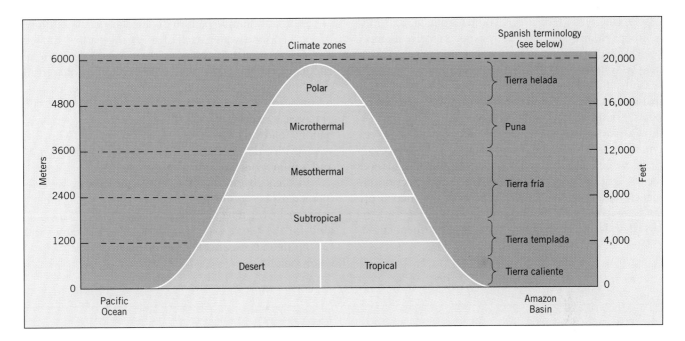

Vertical zone	Elevation range	Average annual temperature range
Tierra helada	4800+ m (16,000+ ft)	< −7°C (< 20°F)
Puna	3600 - 4800 m (12,000 - 16,000 ft)	−7° - 13°C (20° - 55°F)
Tierra fría	1800 - 3600 m (6000 - 12,000 ft)	13° - 18°C (55° - 65°F)
Tierra templada	750 - 1800 m (2500 - 6000 ft)	18° - 24°C (65° - 75°F)
Tierra caliente	0 - 750 m (0 - 2500 ft)	24° - 27°C (75° - 80°F)

FIGURE 19-10 A highly generalized west-to-east cross-section of the Andes in equatorial South America. The vertical climatic zonation shown on the mountain above corresponds to the Spanish terminology at its right, which is interpreted in the table at the left.

FIGURE 19-11 The spectacular Kibo cone that crowns East Africa's Mount Kilimanjaro (elevation 5861 m [19,340 ft]). Even though this mountain lies within sight of the equator, its upper 20 percent exhibits a polar-like climate.

aspects (they may face in any compass direction). In the Northern Hemisphere, southerly aspects receive far more solar radiation than do north-facing slopes, which may receive no direct solar radiation for much of the winter season. These differences in radiation receipt are often manifested as vegetation contrasts on different slope aspects. South-facing slopes are commonly drier and exhibit a sparser vegetation cover. North-facing slopes are typically more lush, since temperatures, and hence evaporation rates, are lower. Rugged terrain also influences local windflow patterns that may have an effect on climatic conditions. Finally, the climate of a particular upland area depends on its location with respect to the global-scale factors of climate, such as the general circulation of the atmosphere.

A broadly simple pattern of climates occurs because of the heat and water exchanges at the earth's surface and because of the spatial organization of the general circulation of the atmosphere. This pattern is altered in detail by the specific location of land and water bodies, ocean currents, and upland regions. The resulting mosaic of climates may be classified in different ways. We have mainly followed the system first devised by Köppen, who divided the climates of the earth into six major types and a number of additional subtypes.

In Units 20 and 21, respectively, we explore two additional aspects of the geography of climate: (1) climatic variations over time and what these changes may portend for the future distribution of climate regions, and (2) the interactions between humans and their climatic environment.

KEY TERMS

acid precipitation (p. 209)   humid microthermal climate (p. 207)   polar climate (p. 210)
highland climate (p. 214)   permafrost (p. 208)

REVIEW QUESTIONS

1. How do humid continental climates differ from taiga climates?

2. What is permafrost and how does it form?

3. What is acid precipitation, and why is it often found so far from its source areas?

4. Why do polar climates invariably exhibit low precipitation values?

5. Describe the effects of increasing altitude on temperature, precipitation (types and amounts), and wind speeds.

REFERENCES AND FURTHER READINGS

ALLAN, N.J.R., et al., eds.   *Human Impact on Mountains* (Totowa, N. J.: Rowman & Littlefield, 1988).

BARRY, R. G.   *Mountain Weather and Climate* (London/New York: Routledge, 2nd ed., 1992).

BENISTON, M., ed.   *Mountain Environments in Changing Climates* (London/New York: Routledge, 1994).

BLISS, R. G.   *Tundra Ecosystems: A Comparative Analysis* (New York: Cambridge University Press, 1981).

BRYSON, R. A., and HARE, F. K., eds.   *Climates of North America* (Amsterdam: Elsevier, World Survey of Climatology, Vol. 11, 1974).

FRENCH, H. M., and SLAYMAKER, O., eds.   *Canada's Cold Environments* (Montreal/Kingston: McGill-Queen's University Press, 1993).

GERRARD, A. J.   *Mountain Environments: An Examination of the Physical Geography of Mountains* (Cambridge, Mass.: MIT Press, 1990).

HARE, F. K., and THOMAS, M. K.   *Climate Canada* (New York: Wiley, 2nd ed., 1980).

HARRIS, S. A.   *The Permafrost Environment* (Totowa, N.J.: Rowman & Littlefield, 1986).

IVES, J. D., and BARRY, R. G., eds.   *Arctic and Alpine Environments* (London: Methuen, 1974).

LUOMA, J. R.   "Bold Experiment in Lakes Tracks the Relentless Toll of Acid Rain," *New York Times*, September 13, 1988, 21, 24.

LYDOLPH, P. E., ed.   *Climates of the Soviet Union* (Amsterdam: Elsevier, World Survey of Climatology, Vol. 7, 1977).

PARK, C. C.   *Acid Rain: Rhetoric and Reality* (London/New York: Routledge, 1989).

PIELOU, E. C.   *A Naturalist's Guide to the Arctic* (Chicago: University of Chicago Press, 1994).

PRICE, L. W.   *Mountains and Man: A Study of Process and Environment* (Berkeley, Calif.: University of California Press, 1981).

SCHWARTZ, S. E.   "Acid Deposition: Unraveling a Regional Phenomenon," *Science*, Vol. 243, February 10, 1989, 753–763.

STONEHOUSE, B.   *Polar Ecology* (London/New York: Chapman & Hall, 1989).

SUGDEN, D. E.   *Arctic and Antarctic: A Modern Geographical Synthesis* (Totowa, N.J.: Barnes & Noble, 1982).

UPADHYAY, D. S.   *Cold Climate Hydrometeorology* (New York: Wiley, 1993).

Dynamics of Climate Change

OBJECTIVES

◆ To examine various lines of evidence for climate change.

◆ To give a brief history of climatic change over the past 1.5 million years.

◆ To discuss mechanisms that cause variations in climatic conditions.

A courageous Viking named Eric the Red discovered the coast of Greenland around A.D. 982 and a few years later founded the Norse colony of Osterbygd there. This colony and others flourished at first, yet less than 500 years later the last colonist died. The disappearance of these settlements was the only recorded instance of a well-developed European outpost being completely extinguished. Was it because of inbreeding? Was it because of plague brought to Greenland by pirates? Or was it because the climate became so cold that even the hardy Norsemen were pushed beyond the margin of survival? The exact causes will never be ascertained, but we do know that the demise of the colony coincided with the coldest temperatures Greenland has experienced in the past 1400 years.

The disappearance of this Viking community is not an isolated incident stranded in the distant past. An average cooling of only about 0.4°C (0.7°F) in the Northern Hemisphere between 1940 and 1975 reduced the growing season in England by over a week. We do not know if this trend will continue—indeed, there is much more talk today about global *warming*! But we can be sure that people, including scientists, will be asking such questions for years to come. In this unit, we try to discover just what is—and what is not—known about changing climates. Climatic change is an urgent issue,

UNIT OPENING PHOTO: Bristlecone pines in the White Mountains of eastern California. These trees' growth rings yield a climatic record of the past 8000-plus years.

217

with implications for our future well-being; yet it is normal for the earth's climate to undergo change. The significance of ongoing changes can be fully assessed, however, only when viewed against the background of past changes.

DOES THE CLIMATE CHANGE?

In our definition of climate, we noted a statistical focus on the average values of the various weather elements taken over at least a 30-year period. A widely accepted recent 30-year period is that between 1931 and 1960. Mean values of many weather parameters during this period, called *climatic normals*, have been published by the World Meteorological Organization for hundreds of locations around the globe.

An even more specific definition is used for studies of possible changes in climate. A **climatic state** is defined as the average (together with the variability and other statistics) of the complete set of atmospheric, hydrospheric, and cryospheric (ice) variables over a specified period of time in a specific domain of the earth-atmosphere system. It is therefore possible to have monthly, seasonal, annual, or decadal climatic states. From here on in this unit, the word "climate" is used as an abbreviation for climatic state.

There are three kinds of evidence that might show a change in climate:

1. There is *direct evidence*, as in the daily weather records gathered all over the world. These are extremely valuable, but have been available for only about the past two centuries or less.
2. There are *historical climatic data*, consisting of such sources as written records and observations on crop yields and drought. In some cases, as in Egypt's Nile Valley, these data have been kept for thousands of years.
3. There is evidence to be found all over the natural world, from the sediments deposited on the seafloor to the concentric growth rings of tree trunks. This kind of evidence has been called *proxy climatic data* because it is indirect. The methods of gathering these data are much more complicated than those used in assembling direct data. Let us consider some examples.

Evidence from Oceanic and Lake Sediments

Sediments accumulated on the floors of the oceans yield data on sea-surface temperature, the amount of ice on the globe, and other factors. Samples of these deposits are tested for accumulations of ash and sand, the fossils of tiny sea creatures, and mineral composition. Nuclear

FIGURE 20-1 The cataloguing of core samples of ocean-floor sediments showing their careful positional calibration. This is the laboratory aboard the famous drilling vessel, *Glomar Challenger*.

particles within the atoms of these substances present especially valuable information. The nucleus of an atom is made up of positively charged protons and uncharged neutrons. Two atoms of one element, such as oxygen, may have different numbers of neutrons, although they will always have the same number of protons. The related forms of the element, differentiated by the number of neutrons each has, are called *isotopes*. The concentration and ratio of different isotopes in samples or *cores* of ocean sediment (and other related factors) yield information on past temperatures and the volume of polar ice (Fig. 20-1).

Sediments at the bottom of lakes are often useful in researching past temperature and precipitation patterns. This is particularly true of lakes that at one time existed at the edges of continental icesheets or highland glaciers. Each year spring and summer runoff carries both fine and coarse sediment into the lake. The coarse sediment, because of its weight, is deposited on the lake bed first. Then, in the still water conditions under the ice of winter, the fine sediment is gradually deposited. Thus the passing of each year is accompanied by a layer composed of coarse sediment below and fine sediment on top. These alternating layers of sediments are called *varves*. Their testimony to melting and runoff conditions, indicative of past temperature and precipitation regimes, may sometimes be traced as far back as 5000 years.

Evidence on the Land Surface

Ice on the earth was the first phenomenon to set scientists' minds thinking about climatic change. Between 1800 and 1830, investigators in Switzerland and Norway put forward the idea that the landscapes in their moun-

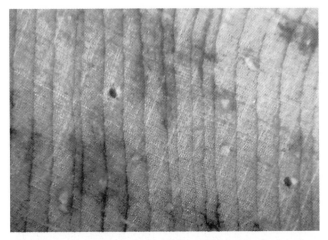

FIGURE 20-2 A close-up view of the annual growth rings in a tree trunk, showing their variation from year to year. The central portion of this photograph covers the decade from 1530 to 1540. (Photo courtesy of Henry N. Michael.)

tainous areas could be explained only by a great extension of glaciers in the past. Since that time, landscapes carved out by moving ice have been studied to determine the previous extent of glaciers and icesheets (*see* Units 45 through 48). Recently, the ice itself has yielded some amazing secrets. Cores of ice taken from the Greenland Icesheet, when examined isotopically, have provided a detailed temperature record for about the past 200,000 years.

Biological features on the land surface can also provide investigators with significant details. The types of fossil pollen in bogs give clues to past climates and their variations. Trees are sometimes good indicators as well. *Dendrochronology,* the study of the width of their annual growth rings (Fig. 20-2), may be used to determine details of the hydrologic and thermal conditions prevailing prior to and during the growth of these rings. Studies of bristlecone pines in the White Mountains of eastern California (*see* unit opening photo) have yielded a climatic record of the past that extends back for more than 8700 years. Most tree-ring records are much shorter than this but are still valuable.

Evidence Below the Earth's Surface

Scientists also look for clues within the lithosphere, the earth's crust. Climate helps determine the type of soil that develops in an area; ancient soils that have been buried and later exposed can, therefore, indicate the climates that prevailed when they were formed. The rocks of the earth also contain the story of past climates. Sometimes, the rock type evinces the climatic conditions under which it was formed. On other occasions, certain features of the rock tell a story. Fossilized sand dunes, for instance, reveal not only desert conditions but also the prevailing wind direction during their formation. The most

detailed rock record of past climates comes from the fossilized plants and animals found embedded in rock. These fossils, when interpreted, can provide much information on past life forms and their associated climatic environments.

These proxy data and other evidence are the pieces of a gigantic jigsaw puzzle. When put together, they give us information about past climates. But each form of evidence tells about a limited span of time. Furthermore, the geographic extent of any particular set of evidence is often small. Therefore, not all of the pieces of the jigsaw are available. Even so, there is no doubt that the climate does change, and the next item on our agenda is to find out how it has changed.

THE CLIMATIC HISTORY OF THE EARTH

Most history books start with the oldest times and work their way forward. In contrast, most descriptions of past climates start with the most recent times and work backward. To begin with, much more information is available on recent times, which allows statements to be made with greater certainty. Another advantage of such an approach is that it allows investigators to use certain climatic events as landmarks as they delve farther into the increasingly obscure past. The following account also begins with the more recent time periods, and we use certain events for guides as we shift time scales within our examination of the climatic past.

The Past 150 Years

Although there is much evidence of past climate change and weather variability, accurate recordkeeping based on reliable instruments began only during the last 150 years. Weather records spanning a century or more still are rare, and come from weather stations mostly in Western Europe and North America. Many crucial areas of the world have climate records less than a half-century old. Political upheavals have interrupted recordkeeping in other areas; in West Africa, for example, good records from weather stations established by the French and the British ended, or were interrupted, during the instabilities of the 1960s. Moreover, satellite observations of global weather conditions are just a few decades old, and to this day surface verification is inadequate over large parts of the earth's landmasses and oceans. Undeniably, the absence of dependable long-term data is a major obstacle in our efforts to gauge climate change.

Nevertheless, the available record indicates considerable temperature variation over the past 130 years (Fig. 20-3). The most notable features of the average annual temperature of the world's landmasses over the past century and a quarter are a warming trend beginning in the

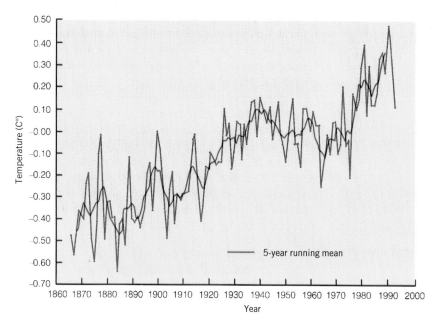

FIGURE 20-3 *The variation of annual mean surface temperatures for the world's land areas, 1866–1992. Because this graph incorporates data from many different places, annual mean temperatures are expressed as a deviation from the average annual mean temperature for the 1951–1980 period.*

1880s, a cooling trend from 1940 into the 1960s, and another apparent warming trend since then.

Different parts of the world have experienced these changes to a greater or lesser degree. For instance, the 1940–1975 cooling trend was most strongly felt in the Atlantic sector of the Arctic region, where average winter temperatures in some locations dropped almost 3°C (5.6°F). On the other hand, no cooling at all was detected during that 35-year period in parts of the Southern Hemisphere (such as New Zealand), where the warming trend launched in the 1880s continued.

Recent research indicates that this regional variation in warming and cooling is a hallmark of climate change. Computer models that project global warming during the twenty-first century, for example, are now indicating that any such warming will be felt most strongly in the higher latitudes—not evenly around the world. *Global* warming (or cooling) may, therefore, turn out to be a misnomer: while certain regions experience one trend, others may record the opposite.

The Past 1500 Years

To gauge climate and weather over the past 1500 years, we must depend on less exact information. Tree rings, lake sediments, ice cores, cave deposits, harvest records, diaries, and other sources produce a picture of considerable variation and even quite sudden change.

Apparently, the earth was warm and becoming warmer during the centuries prior to A.D. 1000—a trend not unlike the one we are experiencing today. Sea level was high, low-lying coastal countries coped with rising water (in Holland the Dutch learned to build dikes and polders during this period), and high-latitude environs, even Greenland and Iceland, proved amenable to per-

manent settlement. The Romans had planted grapevines in Britain, and the wine industry thrived there under the mild conditions of what climatologists have come to call the *Medieval Optimum*.

In her 1988 book entitled *The Little Ice Age*, Jean Grove describes conditions thus:

> For several hundred years climatic conditions in Europe had been kind; there were few poor harvests and famines were infrequent. The pack ice in the Arctic lay to the north and long sea voyages could be made in the small craft then in use . . . Icelanders made their first trip to Greenland about A.D. 982 and later they reached the Canadian Arctic and may even have penetrated the North West Passage. Grain was grown in Iceland and even in Greenland; the northern fisheries flourished and in mainland Europe vineyards were in production [500 km (320 mi)] north of their present limits.

Were these conditions representative only of the higher latitudes of the Northern Hemisphere or of the world as a whole? We get a hint from the Southern Hemisphere that the Medieval Optimum may have been a global phenomenon. Even as the Scandinavians traversed the Atlantic and, led by Leif Eriksson (the son of Eric the Red), reached northern North America, Polynesians in the Southern Hemisphere for the first time managed to land in (upper-midlatitude) New Zealand. Was it a coincidence? Perhaps. On the other hand, this group of Polynesians (the Maori) had been sailing the South Pacific for centuries, never having reached the largest islands in their realm. The warmth and tranquility of the Medieval Optimum may have given them the same opportunity the Scandinavians seized in the far north.

The Medieval Optimum seems to bear some resemblance to the current warming trend. Many glaciers were in retreat, winters were generally mild, crops were plentiful. The frontiers of human settlement pushed into higher latitudes. But then, as Grove tells us, things changed quite suddenly:

> The beneficent times came to an end. Sea ice and stormier seas made the passages between Norway, Iceland, and Greenland more difficult after A.D. 1200; the last report of a voyage to Vinland [North America] was made in 1347. Life in Greenland became harder; the people were cut off from Iceland and eventually disappeared from history toward the end of the fifteenth century. Grain would no longer ripen in Iceland, first in the north and later in the south and east . . . life became tougher for fishermen as well as for farmers. In mainland Europe, disastrous harvests were experienced in the latter part of the thirteenth and in the early fourteenth century . . . extremes of weather were greater, with severe winters and unusually hot or wet summers.

From the late thirteenth century onward, Europe was in the grip of climatic extremes as the Medieval Optimum gave way to the *Little Ice Age*. Britain's wine industry was extinguished in a few years; the limit of agriculture was driven southward by hundreds of kilometers. Not until the fifteenth century did the Little Ice Age moderate somewhat (it may be no coincidence that this marked the revival of Atlantic navigation, including Columbus's crossings). But in the mid-sixteenth century the cold returned, not to yield again until the mid-nineteenth—marking the start of the warming trend to which we may now be contributing through our industrial pollution of the atmosphere.

Is the Little Ice Age over? At the moment, we have no more evidence to suggest that it is than we have to conclude that it is not. The current warming phase may be just a prelude to the return of the conditions of the thirteenth century, the time of weather extremes that threw Europe into disorder (for some tantalizing, post–1985 evidence, *see* Fig. 7-4).

The Past 15,000 Years

By going back 15,000 years, we must rely on even less dependable evidence for our reconstruction of climate. There is no written human record of the weather, although archeologists draw useful conclusions from the seeds, tools, and other artifacts recovered from ancient inhabited sites. And now another line of evidence becomes critical: the geologic record. Just 15,000 years ago, the world was a very different place, and its geology bears witness.

Fifteen thousand years ago, the earth still was in the grip of a major *glaciation* that had lasted about 80,000 years. Great icesheets covered most of northern North America; ice stood as far south as the Ohio River and New York City. Virtually all of what is today Canada was buried under thousands of meters of ice. Much of the United States resembled Siberia; in Europe, the ice covered Scandinavia and nearly all of Britain and Ireland; most of northern Asia was under glaciers. And the higher mountains of the earth (the Rockies, Andes, Alps, and Himalayas among them) lay under icecaps.

Yet there were signs, 15,000 years ago, that this glaciation was coming to an end. The ice had reached its maximum extent just 3000 years earlier, and the margins were melting. Vast amounts of meltwater poured into the U.S. Midwest and flowed into the oceans. Pulverized rock, pebbles, and boulders carried by the ice were deposited by these meltwaters. Landmarks such as the Great Lakes and Cape Cod were in the making.

By about 10,000 years ago, it was clear that this was no temporary warming (such as this glaciation had witnessed several times before), but that a new epoch was in the making. Physical geographers call this epoch the *Holocene interglaciation* (an interglaciation being a warm period between glaciations). By about 7000 years ago, the great icesheets had melted away approximately to their present dimensions, and for about 2000 years the earth was even warmer than it is today. Since then, the global climate appears to have been not only warm but also relatively stable. Even the fluctuations of the Little Ice Age are merely mild aberrations against the background of full-scale glaciation.

One lesson learned from research into the Holocene (the epoch that has witnessed the entire drama of the emergence of human civilization) is that reversals of temperature are sudden and seem to be preceded by opposite trends. For example, the warmest period of the Holocene was preceded by a temporary but severe cooling called the Younger Dryas. As in the case of the Little Ice Age, trends in one direction seem to presage sudden reversals to the other. This is one concern arising from the current "greenhouse warming" phenomenon—that this global warming, far from foreshadowing an overheated world, portends a precipitous return to full-scale glaciation.

The Past 150,000 Years

One hundred fifty thousand years ago, the earth was locked in another great glaciation, which preceded the one just discussed. We have the geologic evidence to confirm that this glaciation, too, was experienced worldwide and spread its icesheets deep into the heart of the present-day United States. Of particular interest is the warm spell that separated this glaciation from its successor, because this interglaciation is similar to the Holocene epoch of today.

This warm spell, called the Eemian interglaciation, began about 130,000 years ago and lasted about 10,000 years. Evidence from ice cores taken from the Greenland Icesheet, which has survived the interglaciations and thus carries valuable information, indicates that the Eemian resembled the Holocene in some ways but differed in others. One similarity lies in the timing of maximum warmth: like the Holocene, the Eemian began with nearly 3000 years of high temperatures, but even higher than those of the early Holocene. Then, however, the Eemian's equable climate was frequently interrupted by cold episodes much more severe than those we have known during the Holocene.

The well-defined length of the Eemian—just 10,000 years—gives rise to concern. The Holocene already has lasted longer, and the question is how long our interglaciation will continue. When the Eemian ended, one last surge of rising temperatures was summarily terminated by a massive onset of severe cold. That cold marked the beginning of 110,000 years of global glaciation, interrupted only by a comparatively slight warming (about halfway through) that was not intense enough to be called a true interglaciation.

We should also take note of a cataclysmic event that occurred during the period under discussion. About 73,000 years ago, a huge volcanic eruption took place on the Indonesian island of Sumbawa, where the volcano named Toba exploded and spewed vast quantities of ash and dust high into the atmosphere. The earth already was in the grip of a glaciation; now the crucial tropical latitudes were deprived of sunlight as these airborne ejected materials were carried around the globe by high-altitude winds. Anthropologists and archeologists report that humankind came very close to extinction during this twin assault by nature. When the skies finally cleared, there was little to suggest that the next interglaciation would witness the rise of modern civilization.

The Past 1,500,000 Years

The sequence just described—of long-lasting glaciations separated by relatively brief interglaciations—has marked the last 1.5 million years and beyond. The past 1.5 million years constitute the second half of an intensified ice age that began between 2.5 and 3.0 million years ago. The word "intensified" is appropriate, because the cooling that foreshadowed the current ice age may have begun as long as 20 million years ago; the evidence from Antarctica points to the early formation of permanent ice there. But the full-scale impact of this *Late Cenozoic Ice Age* did not come until a significant cooling occurred between 2.5 and 3 million years ago, and the oscillations producing lengthy glaciations and short interglaciations began.

This is why the Late Cenozoic Ice Age is sometimes called the *Pleistocene Ice Age*, the Pleistocene epoch

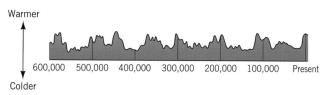

FIGURE 20-4 This graph of surface temperature changes over the past 600,000 years, based on a composite of seven independent estimates, demonstrates the repeated alternation between cold and warm conditions.

marking the past two-plus million years of severe cold and brief warmth. During this Pleistocene phase of the Late Cenozoic Ice Age, the icesheets and icecaps have formed as many as 30 times, advancing deep into the middle latitudes and lower altitudes, lingering as long as 90,000 to 100,000 years, and then retreating for about 10,000 years.

As noted in detail in Unit 46, the geologic record of the glaciations marking the past 1.5 million years is quite strong both in North America and Europe. So is the accumulated evidence concerning climatic fluctuations: protracted cold and fleeting warm conditions alternate constantly (Fig. 20-4). Earlier glaciations extending beyond 1.5 million years, however, are more difficult to determine because subsequent icesheet advances have destroyed most of the evidence.

From the broader perspective of all geologic time, the great surges of ice and frigid temperatures of the Late Cenozoic Ice Age appear unusual in the climatic history of the earth. Most evidence points to an overall warm earth climate, with temperatures about 5°C (9°F) higher than in the present. It has been suggested that such balmy conditions have existed for more than 90 percent of the past 500 million years, at least at latitudes north of 40°N.

It is known for certain that the "normal" warmth was punctuated at least twice before the Late Cenozoic Ice Age. One global ice age occurred about 300 million years ago, and another one took place some 600 million years ago. Evidence for these events comes from geologic deposits left by icesheets moving over the earth's surface. Some investigators have suggested that as many as five major ice ages occurred prior to 600 million years ago; they still lack sufficient proof to confirm these hypotheses, but theirs is an active research frontier.

MECHANISMS OF CLIMATE CHANGE

The picture of past climates and their variations that we have just sketched is still a hazy one. It does become clearer as we focus on more recent times, but there is still a need for more factual information. One fact, however, will be apparent to you by now—climates certainly

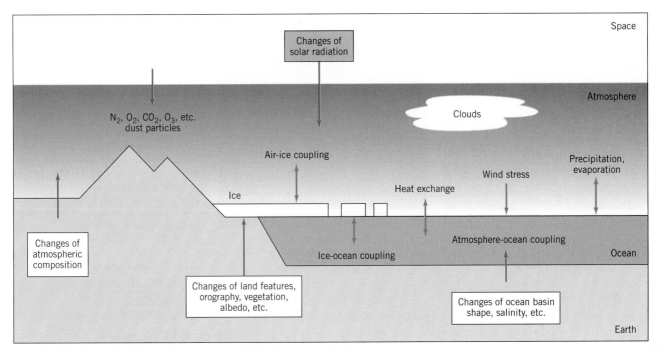

FIGURE 20-5 The atmosphere–ocean–ice–earth climatic system. Red arrows denote external processes; green arrows indicate internal processes.

do change. Perhaps you have already anticipated the next question we must address: how?

Earlier, we considered the evidence for climatic change as pieces of a gigantic jigsaw puzzle, many of which are missing. It is not possible, therefore, to be sure that the picture of climate change in need of explanation is the real picture. This is only the first difficulty in searching for the causes of climatic change. Another problem is that there are many probable causes, all acting at different scales in time and geographic space. A further complicating factor is that many forces shaping climate are linked and interact with one another, so that changes in one trigger changes in others. As if this were not enough, there are still gaps in our knowledge of the exact way the atmosphere and oceans operate, both separately and in tandem. And there may even be additional major variables that have not yet been considered.

Given these difficulties, we nonetheless build a framework for ideas and facts concerning the causes of climatic change. But first we must specify the parameters of the system within which climate changes take place. This system cannot deal with the atmosphere alone: it must also account for the ice of the world (known as the *cryosphere*) as well as the oceanic component of the hydrosphere. Such an interlinked atmosphere–ocean–ice–earth system is shown in Fig. 20-5. Within this system, both the heat-energy system and the hydrologic cycle are at work; moreover, the diagram models the interactions of wind, ice, and ocean characteristics (including surface currents). Outside the climatic system, a number of boxes list forces, such as a change in radiation coming from the

sun, that have the power to alter the system externally. The important thing to remember is that a change in any one or more of these processes creates a climatic change.

External Processes

At night, any particular earth location receives far less radiation from the sun than during the day. In a similar way, solar radiation changing over a longer period of time affects the earth's climate. There are both short-term and long-term variations in the behavior of the sun. Short-term changes take place when storm areas occur on the surface of the sun; these are called *sunspots*, and occur in cycles that peak about every 11 years (Fig. 20-6 shows sunspots at their maximum and minimum development). Sunspots may also affect terrestrial weather, but research findings are still inconclusive as to whether they significantly influence the amount of solar radiation received at the earth's surface.

Longer-term changes also occur because of three cyclical peculiarities in the orientation of the earth's orbit and axis. One is a periodic variation in the shape of the earth's orbit around the sun (Fig. 20-7a): during cycles lasting about 100,000 years, the earth's orbit "stretches" from nearly circular to markedly elliptical and back to nearly circular. The resulting fluctuation in the distance between the earth and sun is as much as 17.5 million km (11 million mi). Another variation in the earth–sun relationship, which follows an approximately 41,000-year cycle, is in the obliqueness of the earth's axis. In other words, the earth "rolls" like a ship (Fig. 20-7b), so that

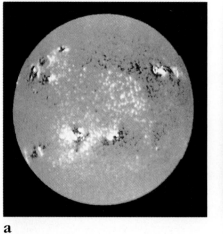

a

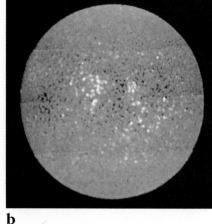

b

FIGURE 20-6 Sunspots are stormy areas on the solar surface which develop in cycles that peak approximately every 11 years (a). In between these maxima, sunspot activity wanes to a low level (b). These images are called magnetograms, the state-of-the-art technique for portraying sunspot development. Sunspots are shown in two colors (blue and yellow-green), differentiated according to their magnetic polarities.

the angle between our planet's axis and the plane of the ecliptic (*see* Unit 5) changes from 65.6 to 68.2 degrees (the present angle of 66.5 degrees is therefore not constant). The major effect of this axial shifting is to alter the annual distribution of solar radiation received at the surface. The third variation might best be described as a "wobble" because, like a spinning top, the earth's axis swivels once every 21,000 years or so (Fig. 20-7c). This affects the distance between the earth and the sun during any given season, which gradually changes as the cycle proceeds.

These variations individually produce episodes of some cooling and warming. But when the cooling periods of all three cycles coincide, the variation in solar radiation estimated to reach the earth is strikingly parallel to the waxing and waning of the global ice cover over the past 1.5 million years. However, these orbital and axial variations presumably also took place during the 90 percent of earth history when ice ages did not occur.

Other external processes that might lead to climatic change are considered elsewhere in this book (start with the Perspective box in Unit 45). They include volcanism (the subject of Unit 34), the uplifting and wearing away of the land surface (Units 32, 37, and 38), the shifting

distribution of landmasses and oceans caused by plate tectonics (Unit 33), and the hypothesized intensification of the greenhouse effect (Unit 7).

Internal Processes

Units 7 through 14 describe the heat and water exchanges within the atmosphere, the way the general circulation distributes heat and moisture across wide areas of the earth, and the weather systems forming the fine grain of atmospheric movement and operation. All these are internal processes, and many of them function as systems by themselves. Although changes in any one of them could lead to climatic variation, they are linked by feedback mechanisms (*see* Unit 1).

An increase in unusual variations in ocean temperature provides an example of *positive feedback*. A change in the temperature of the sea surface may modify the amount of sensible heat transferred to the overlying air, thereby altering atmospheric circulation and cloudiness. Variations in radiation, wind-driven mixing of ocean water, and other factors may, in turn, affect the temperature of the original ocean surface. In the tropical Pacific Ocean, the sea-surface temperature has increased for

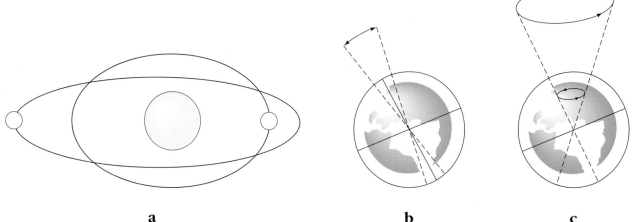

a   **b**   **c**

FIGURE 20-7 Long-term changes in the earth's orbit and axis: (a) stretch, (b) roll, and (c) wobble.

Teleconnections

In the quest for a better understanding of the forces that shape climate change, one promising new avenue of research is focused on teleconnections. **Teleconnections** are long-distance linkages between weather patterns that occur in widely separated parts of the world. The most studied of these relationships is ENSO, the El Niño-Southern Oscillation (discussed in the Perspective box in Unit 11), which links midlatitude weather extremes to cyclical abnormalities in the warming of the tropical Pacific Ocean surface.

New research has also established teleconnections between ENSO and weather cycles in the lower latitudes, particularly variations in the wet monsoon of South Asia (a topic covered in Unit 10). The one-billion-plus residents of India and neighboring Bangladesh are well aware of fluctuations in these torrential summer-monsoon rains. In some years this activity is mild and intermittent, but in other years it is much more active and often destructive. Even though the causal mechanism has yet to be identified, there appears to be a relationship between warm-water El Niño peaks in the Pacific and mild wet monsoons on the Indian subcontinent. Conversely, when Pacific surface temperatures reach a cool ebb between El Niños (such a lull is now called *La Niña*), South Asia is likely to experience its most violent summer rains (Fig. 20-8). Very recently, researchers made an important discovery that strengthens these linkages: sea-surface temperatures warm and cool along the equatorial Indian Ocean in lockstep with the El Niño and La Niña events in the Pacific Ocean to the east.

Teleconnection research is rapidly expanding today and concentrates on two related areas. The first is the search for mechanisms that control the evolution of meteorological conditions which, for months or even years at a time, spawn high- and low-pressure cells whose

FIGURE 20-8 The country of Bangladesh—one of the world's poorest and most overcrowded—is formed by the double delta of two of South Asia's largest rivers, the Ganges and the Brahmaputra. Thus residents are used to major floods in association with the wet summer monsoon. But the late-summer deluge of 1993 produced the worst flood in memory; at one point, more than half of this Iowa-sized nation (population: 120 million) was under water.

movements result in atypical weather patterns. The second concern is the specification and interpretation of additional ENSO-like systems. Three such large-scale relationships are now under investigation:

1. Linkages between the northern Pacific and North America involving weather cycles in Alaska's Aleutian Islands, western Canada, the U.S. Southeast, and the Caribbean.

2. The so-called North Atlantic Oscillation, shaped by varying pressure gradients between the subpolar Icelandic Low and the subtropical Bermuda (Azores) High.

3. The influence of the seasonal snow cover in northern Eurasia on Asia's monsoonal airflows and even the El Niño phenomenon itself.

Two goals propel teleconnection research in contemporary climatology: (1) the long-range prediction of weather patterns for the coming season (and perhaps even the coming year); and (2) the development of longer-term explanations that will help to unlock the remaining secrets of how and why the global climate changes.

several years at a time because of positive feedback mechanisms such as those just described.

An example of *negative feedback* occurs when a snow-covered surface reduces atmospheric temperatures. The cooler atmosphere holds less water vapor, and thus less snow falls; Antarctica's low precipitation levels are partly a result of such negative feedback. Many other examples of both positive and negative feedback exist. Researchers today are particularly interested in *teleconnections*, newly discovered relationships between weather phenomena that involve distant parts of the globe (*see* Perspective: Teleconnections).

THE CLIMATIC FUTURE

If we understand the mechanisms of climate, is it possible to predict the climatic future of the earth? Unfortunately, despite the availability of better information than ever before and the advances in knowledge that have recently taken place, forecasting climates continues to be a supremely difficult challenge. These constraints notwithstanding, it is still possible to conjecture about what lies ahead. One of the more thoughtful current statements was offered by a leading student of global climate change, Reid Bryson, Professor Emeritus of Meteorology, Geography, and Environmental Studies at the University of Wisconsin:

> From the beginning of the atmosphere to the present the climate has been changing, on all time-scales and for many reasons. Paralleling the changes of climate, reacting to and interacting with it, has been the ever-changing pattern of life. . . . It

can be said with some certainty that the climate will continue to change, and with great probability that there will be more periods of continental glaciation, and not in the far distant geological future. In a brief century the widespread use of meteorological instrumentation has replaced human memory and qualitative chronicles with a storehouse of quantitative information about the climate. At the same time, research has shown us how to read the clock and climatic record provided by nature. A belief in a constant climate has been replaced by a knowledge of how climate has changed, and how the changes of the past have affected the biota and cultures that embellish the earth. . . . The problem now is to understand fully the meaning of this rich natural history in order to anticipate the coming climatic events and prepare for them. This is a very large order for humanity if it is going to last to and through the next ice age without losing the hard-won fruits of civilization. (Bryson, 1988, 245-246)

As we draw conclusions about the tendencies of climate to undergo change, it is important also to keep in mind another force—human interference (a topic discussed in Unit 21). Modification of the ozone layer, increased atmospheric carbon dioxide, and other human-induced factors could alter the unknown natural pattern of climatic change. Only one thing is certain, as stated recently by the U.S. National Academy of Sciences: "The clear need is for greatly increased research on both the nature and the causes of climatic variation." Undoubtedly, the spirit of Eric the Red would echo that statement in light of the fate of the colony that was founded on the temporarily hospitable shore of Greenland more than 1000 years ago.

KEY TERMS

climatic state (p. 218)   teleconnections (p. 225)

REVIEW QUESTIONS

1. What are proxy climatic data, and why are they important in the reconstruction of past climates?
2. Describe the general climatic history of the past 1.5 million years.
3. What is the Late Cenozoic Ice Age?

4. Describe the primary external processes that might contribute to climatic fluctuations.
5. What are teleconnections? Give two examples of this phenomenon.

REFERENCES AND FURTHER READINGS

BALLING, R. C., JR. *The Heated Debate: Greenhouse Predictions Versus Climate Reality* (San Francisco: Pacific Research Institute for Public Policy, 1992).

BRADLEY, R. S., and JONES, P., eds. *Climate Since A.D. 1500* (London/New York: Routledge, 1992).

BRYSON, R. A. "What the Climatic Past Tells Us About the Environmental Future," in H. J. de Blij, ed., *Earth '88: Changing Geographic Perspectives* (Washington, D.C.: National Geographic Society, 1988), 230–246. Quotation taken from pp. 245–246.

BUDYKO, M. I. *The Earth's Climate: Past and Future* (Orlando, Fla.: Academic Press, 1982).

GLANTZ, M., et al., eds. *Teleconnections Linking Worldwide Climate Anomalies* (New York: Cambridge University Press, 1990).

"Global Warming Debate," *Research & Exploration* (National Geographic Society), theme issue, 9 (Spring 1993), 142–249.

GOODESS, C., et al. *The Nature and Causes of Climate Change: Assessing the Long-Term Future* (Boca Raton, Fla.: CRC Press, 1992).

GROVE, J. M. *The Little Ice Age* (London/New York: Methuen, 1988). Quotations taken from pp. 1–2.

HOUGHTON, J. T., et al., eds. *Climate Change: The IPCC Scientific Assessment* (New York: Cambridge University Press, 1991).

HOUGHTON, R. A., and WOODWELL, G. M. "Global Climatic Change," *Scientific American*, April 1989, 36–44.

"How Climate Changes," *The Economist*, April 7, 1990, 13; 95–100.

JAGER, J., and FERGUSON, H. L., eds. *Climate Change: Science, Impacts, Policy* (London/New York: Cambridge University Press, 1992).

LADURIE, E.L.R. *Times of Feast, Times of Famine: A History of Climate Since the Year 1000* (Garden City, N.Y.: Doubleday, 1971).

LAMB, H. H. *Climate, History, and the Modern World* (London: Methuen, 1982).

MATHER, J. R., and SDASYUK, G. V., eds. *Global Change: Geographical Approaches* (Tucson: University of Arizona Press, 1992).

MAUNDER, W. J. *Dictionary of Global Climate Change* (New York: Routledge, Chapman & Hall, 2nd ed., 1994).

ROBERTS, N. *The Holocene* (Boston: Blackwell, 1989).

ROLAND, F. S., and ISAKSEN, I.S.A., eds. *The Changing Atmosphere* (New York: Wiley, 1988).

ROTBERG, R. I., and RABB, T. K. *Climate and History* (Princeton, N.J.: Princeton University Press, 1981).

SCHNEIDER, S. H. "The Changing Climate," *Scientific American*, September 1989, 70–79.

SILVER, C. S., and DE FRIES, R. S. *One Earth, One Future: Our Changing Global Environment* (Washington, D.C.: National Academy Press, 1990).

WRIGHT, H. E., et al. *Global Climates Since the Last Glacial Maximum* (Minneapolis: University of Minnesota Press, 1994).

WYMAN, R. L., ed. *Global Climate Change and Life on Earth* (London/New York: Routledge, 1991).

UNIT 21

Human-Climate Interactions and Impacts

OBJECTIVES

♦ To relate our understanding of atmospheric processes to the human environment.

♦ To illustrate the utility of using energy balance concepts to characterize systems of the human environment.

♦ To focus on several of the impacts humans have had, and may come to have, on our climatic environment.

This closing unit on the physical geography of the atmosphere treats some key relationships between humans and their climatic environment, and highlights the impact of human activities on the functioning of climate at various scales. The unit begins by specifying the interaction of the human body with its immediate surroundings and shows how our comfort depends on access to adequate shelter. Most modern dwellings, particularly in the United States, tend to be clustered in urban areas that exhibit **microclimates** (climate regions on a localized scale) of their own. The cities that anchor this metropolitan landscape usually contain concentrations of the wrong things in the wrong place at the wrong time—a phenomenon known as pollution.

We examine those aspects of pollution that occur in the atmosphere and how urban air affects the surrounding countryside. Our inquiry then expands to the macroscale as we briefly discuss humankind as a possible agent of world climatic change. We therefore proceed from the level of the individual to that of the entire globe. A good way to start that progression is to consider the human body as a heating and cooling system.

UNIT OPENING PHOTO: The bleak landscape of Bedzin, Poland, testimony to the appalling environmental abuses that accompanied industrialization in the world of Soviet-led communism that collapsed in the early 1990s.

THE HEAT BALANCE OF THE HUMAN BODY

Unit 7 describes the concept of the heat balance with respect to the surface of the earth. The idea of examining the flows of heat energy to and from an object can usefully be applied to the human body as well. It is essential that the internal temperature of our bodies remain at about 37°C (98.6°F). Depending on the person, temperature fluctuations exceeding 3° to 6°C (5.5° to 11°F) result in death. The body, therefore, has to be kept within a very limited range of temperature by regulating the flows of heat to and from it. Four kinds of heat flows can be altered—radiant, metabolic, evaporative, and convectional—and they are diagrammed in Fig. 21-1.

To begin with, we all live in a radiation environment. We receive shortwave radiation from the sun and longwave radiation from our surroundings—clothing, walls, the planetary surface, and the like. We humans also emit longwave radiation. Our radiation balance, the sum of incoming and outgoing radiation, can be positive or negative, depending on the environment we are in. More often than not, the balance is a positive one. So net radiation is usually a heat gain for us, especially during daylight hours.

Another heat gain is the heat that our bodies produce, called *metabolic heat*. The body produces metabolic heat by converting the chemical energy in the food we assimilate into heat energy. The amount of metabolic

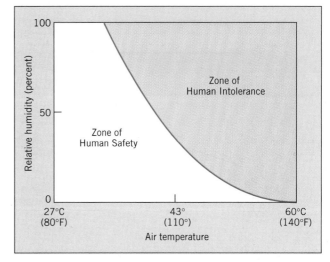

FIGURE 21-2 The limitation imposed on our evaporative cooling system by relative humidity.

heat we produce depends on, among other things, our age, activity level, and environmental temperature. An older person at rest produces metabolic heat equal to that used to power a 75-watt lightbulb. A five-year-old child produces the equivalent of 120 watts, and an active adult about 260 watts (this rate could double if the adult were playing tennis). We produce even more metabolic heat when exposed to colder temperatures. At 33°C (92°F) an adult creates 3100 calories (cal) per day, whereas at 0°C (32°F) an adult's metabolism creates 3930 cal per day.

Metabolic heat is always a heat source to the human body. But the evaporation of water from the skin through perspiration is always a heat loss. Perspiring is a vital function, because some of the heat used in evaporating the perspired water is taken from the body and thereby cools it down. One of the factors determining how much evaporation can take place is the amount of water vapor already present in the surrounding air, as measured by relative humidity. As we all know, we can feel fairly comfortable on a hot day if the relative humidity is low, but we feel distinctly uncomfortable, even at a lower temperature, if the humidity is high. You can see in Fig. 21-2 that if other heat losses and gains are kept constant, relative humidity acts as an index of how efficient our evaporative cooling system is. It could even become a deciding factor between life and death.

Finally, a flow of sensible heat can act as either a cooling or a heating mechanism, depending on the relative temperatures of the body and the air, by means of a *convectional* heat flow. Hot air blowing onto our bodies makes us gain heat, whereas a cold wind leads to a rapid heat loss. When we breathe we pass air into and out of our lungs; this air may be warmer or cooler than our lungs. For convenience, we can regard the loss or gain of heat through breathing as a convective heat flow.

Our bodies consciously and unconsciously regulate these four types of heat flows so that body temperature

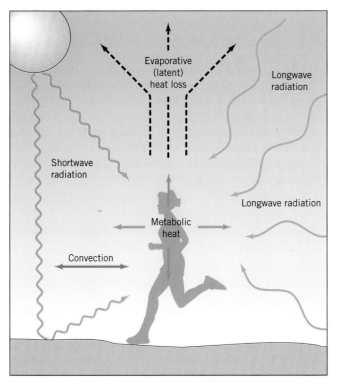

FIGURE 21-1 Heat-energy flows to and from the human body.

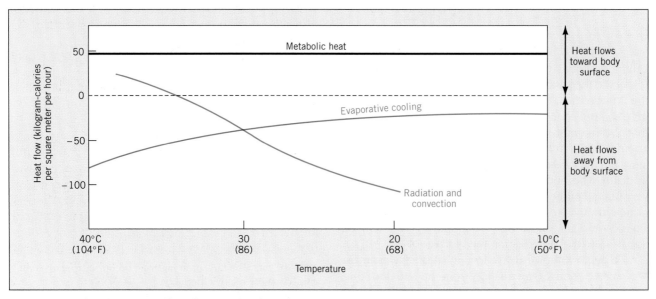

FIGURE 21-3 The adjustment of heat flows under changing environmental temperatures to maintain a constant internal body temperature.

stays within the narrow vital range. For example, if our environmental temperature changes over a short period of time from 40°C (104°F) to 10°C (50°F), the metabolic heat production might stay the same, as Fig. 21-3 shows. But radiant and convectional heat flows change from being a slight heat gain to being a marked heat loss. Evaporative cooling of the body moves from a high level to a relatively low level in the cooler environment. Many of our feelings of comfort or discomfort and almost all of our artificial adaptations to the environment, such as shelter and clothing, are related to the balance of these energy flows that maintain our constant internal temperature.

SHELTER, HOUSES, AND CLIMATE

Ever since people first felt the effects of sensible heat loss by winds or an increased evaporative cooling when they were rained upon, they have sought some form of shelter. The form of shelter depended on the most prevalent features of the climate and on the available building materials. Climate still plays a large role in some of the design features of our dwellings. In hot climates, for instance, where there is a need to promote convectional cooling, the nomads of the Saudi Arabian desert roll up the inside walls of their tents; in more humid (and more tropical) India, screens are aligned to catch any cooling breeze.

Now let us consider the houses with which you are more familiar. In our discussion of the human body we used the concept of heat balance, which may also be applied to the typical contemporary house. As with the body, there are four types of flows to and from the roof and walls of a house. Instead of—but analogous to—

metabolism, the house contains an artificial heating (and/or cooling) system. An artificial heating system usually channels heat from the central interior toward the walls and roof of the building. The evaporative cooling part of the house is normally designed with a sloping roof that removes surface water or snow as rapidly as possible. However, most houses have the important function of reducing the amount of convectional cooling by wind; they provide shelter from the wind and facilitate control of the inside climate via the heating system.

Radiant heating and cooling applies to the outside of a house much as it applies to the body, and architects take it into consideration in several ways. If they want to use radiation for heating the house by means of the greenhouse effect, they can ensure that the windows of the building are open to the direct rays of the sun. Sometimes they use a partial shade to cut out direct rays coming from high sun angles in the summer, when heating is not required; but rays from low sun angles bypass the shade and supply winter heating. Thus, by adapting their designs to the operation of the four factors of the heat balance system, architects can provide our houses with a small environment capable of adjustment to the changing seasonal needs of our bodies.

URBAN MICROCLIMATES

For the most part, the marvels of modern architectural design and engineering work well as small, individual open systems. But in today's world, where about 50 percent of humankind resides in metropolitan areas, individual buildings do not function as single entities but as a group.

Mass, Energy, and Heat in the City

The city is an extraordinary processor of mass and energy and has its own metabolism. A daily input of water, food, and energy of various kinds is matched by an output of sewage, solid refuse, air pollutants, energy, and materials that have been transformed in some way. The quantities involved can be enormous. Each day, directly or indirectly, the average U.S. urbanite consumes about 600 liters (156 gal) of water, 2 kg (4.4 lb) of food, and 8 kg (17.6 lb) of fossil fuel. This is converted into roughly 500 liters (130 gal) of sewage, 2 kg of refuse, and more than 1 kg of air pollution. Multiply these figures by the population of your metropolitan area, and you will get some idea of what a large processor it is, even without including its industrial activities. Many aspects of this energy use affect the atmosphere of the city, particularly in the production of heat.

In winter, the values of heat produced by a city and its suburbs can equal or surpass the amount of heat available from the sun. All the heat that warms a building eventually transfers to the surrounding air, a process that is quickest where houses are poorly insulated. But an automobile produces enough heat to warm an average house in winter; and if a house were perfectly insulated, one adult could also produce more than enough heat to warm it. Therefore, even without any industrial production of heat, an urban area tends to be warmer than the adjacent countryside.

The burning of fuel, such as by cars, is not the only source of this increased heat. Two other factors contribute to the higher overall temperatures in cities. The first is the heat capacity of the materials that constitute the cityscape, which is dominated by concrete and asphalt. During the day, heat from the sun can be conducted into these materials and stored—to be released at night. But in the countryside, materials have a significantly lower heat capacity because a vegetative blanket prevents heat from easily flowing into and out of the ground. The second factor is that radiant heat coming into the city from the sun is trapped in two ways: (1) by a continuing series of reflections among the numerous vertical surfaces that buildings present, and (2) by the **dust dome** (dome-shaped layer of polluted air) that most cities spawn. Just as in the greenhouse effect, shortwave radiation from the sun passes through the pollution dome more easily than outgoing longwave radiation does; the latter is absorbed by the gaseous pollutants of the dome and reradiated back to the urban surface (Fig. 21-4).

Urban Heat Islands

These are the reasons why the city will be warmer than its surrounding rural areas, and together they produce the phenomenon known as the **urban heat island**. If we regard isotherms (lines of constant temperature) as analogous to contour lines of elevation on a map, then the distribution of temperatures within a city gives the general impression of an area of higher land—or an island of higher temperatures—set above a more uniform plain. This effect is clearly seen in Fig. 21-5, which maps the distribution of average winter low (nighttime) temperatures in the heart of the metropolis anchored by Washington, D.C.

Note that the mildest temperatures on this surface are recorded above downtown Washington, D.C. and neighboring Arlington and Alexandria, Virginia (a pattern also marked by the micromaritime influence of the Potomac River and its tributary, the Anacostia). From the "high ground" of this island, in every direction, temperatures decline away from downtown across the city and the Maryland/Virginia suburban ring toward the edge of the built-up metropolitan frontier—which lay approximately along the −2.8°C (27°F) isotherm when this map was constructed in 1964. By the mid-1990s, of course, three decades of subsequent rapid growth had expanded the edges of the urban heat island up to 40 km (25 mi) deeper into the countryside.

Heat islands develop best under the light wind conditions associated with anticyclones, but in large cities they can form at almost any time. The precise configuration of a heat island depends on several factors. The island can be elongated away from the prevailing wind;

FIGURE 21-4 A classic cross-sectional view of an urban pollution dome, low enough to the ground that the tallest skyscraper penetrates the clear air above it. This is not smoggy Los Angeles or Denver, but the center of another western U.S. city: Seattle (with yet another dome—the professional football/baseball Kingdome—located on the apron of the downtown skyline). For a look at the Los Angeles dust dome, *see* Fig. 8-7 on p. 89.

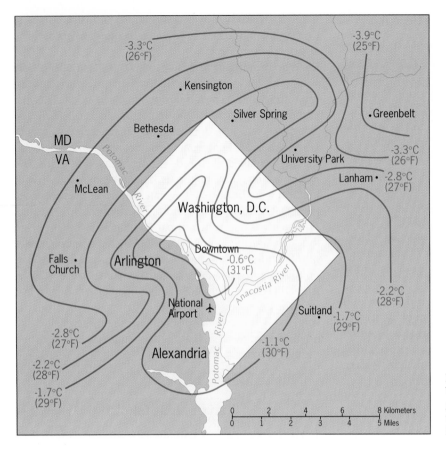

FIGURE 21-5 The urban heat island of Washington, D.C. and its Maryland and Virginia inner suburbs, as represented by isotherms of mean low winter temperatures, in degrees Celsius and Fahrenheit.

pools of cold air can be found over unbuilt parkland within the city; and sometimes tongues of warmer air follow the course of major rivers (as seen south and east of downtown Washington, D.C. in Fig. 21-5). When the heat island is well developed, microscale variations can be extreme. In winter, busy streets in cities can be 1.7°C (3°F) warmer than the side streets; the areas near traffic lights can be similarly warmer than the areas between them because of the effect of idling cars. Even in small towns, the existence of a structure like a large dairy processing plant can also be detected in the form of higher temperatures.

The maximum difference in temperature between neighboring urban and rural environments is called the *heat-island intensity* for that region. In general, the larger the city-suburban complex, the greater its heat-island intensity; the actual level of intensity depends on such factors as the physical layout, population density, and productive activities of a metropolis. The consequences of higher temperatures in urban areas should not be overlooked. During a typical recent heat wave in the U.S. Midwest and Northeast, the cities suffered the most. In St. Louis, the number of deaths attributed to the heat rose dramatically in temperatures above 32.2°C (90°F); when the temperature reached 32.8°C (91°F), 11 people died, but mortalities rose to 73 when the temperature reached 35°C (95°F). The total death toll from

the hot spell reached 500 in St. Louis, and more than 1100 in both New York City and the state of Illinois.

Peculiarities of City Climates

The surface-atmosphere relationships inside metropolitan areas understandably produce a number of climatic peculiarities. For one thing, the presence or absence of moisture is affected by the special qualities of the urban surface. With much of the built-up landscape impenetrable by water, even gentle rain runs off almost immediately from rooftops, streets, and parking lots. Thus city surfaces, as well as the air above them, tend to be drier between precipitation episodes; with little water available for the cooling process of evaporation, relative humidities are usually lower.

Wind movements are also modified by the cityscape because buildings increase the friction on air flowing over and around them. This tends to slow the speed of winds (by as much as 80 percent at a height of 30 m [100 ft]), making them far less efficient at dispersing pollutants when they travel above large cities. At certain locations within the metropolis, on the other hand, air turbulence increases in association with high-velocity, skyscraper-channeled airflows and wind eddies on street corners. Other unique aspects of city climates originate in the artificially modified air over the urban landscape. As we

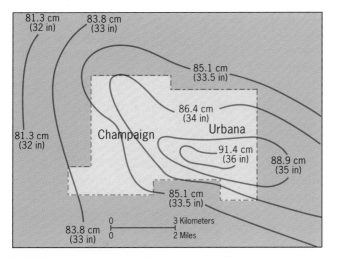

FIGURE 21-6 Wind circulation in an urban dust dome.

know, a dust dome forms, which often gives rise to the wind circulation cells shown in Fig. 21-6. In the box on p. 88 in Unit 8, we discuss the impact of dust domes on the urban radiation balance; here we examine the dome's effects on the atmospheric moisture of cities.

Within the dust dome, there are far more airborne particles that can act as condensation nuclei than in corresponding rural areas, and this significantly modifies the moisture content of the urban atmosphere. Although little water vapor rises from the city surface, horizontal flow in the atmosphere brings just as much moisture to the city as to the nearby countryside. Because of the greater number of particles in the air above a city, there is a greater propensity for condensation and the formation of small droplets of liquid mist, fog, and cloud. As a result, fog is more frequent in cities than in surrounding areas, an effect heightened when there are local moisture sources such as lakes and rivers within the metropolitan area.

Not surprisingly, the presence of a metropolis also increases the amount of rainfall. Most cities seem to have about 10 percent more precipitation than the surrounding area. But some, such as the urban areas of Cham-

paign–Urbana, Illinois (Fig. 21-7) and St. Louis, Missouri, exhibit larger increases. The cause appears in part to be greater turbulence in the urban atmosphere, the result of hot air rising from the built-up surface. However, as we will see in the following section, artificially produced pollution can also play a major role.

AIR POLLUTION

The most important climate modifications of urban areas are summarized in Table 21-1, and many of them can be attributed to the pollutants that cities discharge into the air. We say that air is polluted when its composition departs significantly from its natural composition of such gases as nitrogen and oxygen. However, we might call cigarette smoke pollution but not the aroma of a charbroiled steak. We are therefore concerned with factors that are in some way detrimental to, or uncomfortable for, human life.

TABLE 21-1 THE EFFECT OF CITIES ON CLIMATIC ELEMENTS

ELEMENT	COMPARISON WITH RURAL ENVIRONMENT
Radiation	
Global	15–20% less
Ultraviolet, winter	30% less
Ultraviolet, summer	5% less
Sunshine duration	5–15% less
Temperature	
Annual average	0.5°–1.0°C more
Winter low (average)	1°–2°C more
Contaminants	
Condensation nuclei and particles	10 times more
Gaseous mixtures	5–25 times more
Wind speed	
Annual average	20–30% less
Extreme gusts	10–20% less
Calms	15–20% more
Precipitation	
Totals	5–10% more
Days with less than 5 mm	10% more
Snowfall	5% less
Cloudiness	
Cover	5–10% more
Fog, winter	100% more
Fog, summer	30% more
Relative humidity	
Winter	2% less
Summer	8% less

FIGURE 21-7 Average yearly precipitation (in centimeters and inches) in Champaign–Urbana, Illinois, 1949–1967.

Source: Adapted from Landsberg, 1970. © World Meteorological Organization.

The Nature of Air Pollution

We can group airborne pollutants into two categories: primary and secondary. **Primary pollutants**, which may be gaseous or solid, come from industrial and domestic sources and the internal combustion engines of motor vehicles. The principal gaseous primary pollutants are carbon dioxide, water vapor, hydrocarbons, carbon monoxide, and oxides of sulfur and nitrogen—particularly sulfur dioxide and nitrogen dioxide. The effluents toward the end of this list are sometimes called "status-symbol" pollutants because they are especially associated with the industrially developed countries. The leading solid primary pollutants are iron, manganese, titanium, lead, benzene, nickel, copper, and suspended coal or smoke particles. Except where coal or wood is burned in homes, these pollutants also emanate mainly from industrial sources.

Secondary pollutants are produced in the air by the interaction of two or more primary pollutants or from reactions with normal atmospheric constituents. There also are two kinds of secondary pollution. The first is the *reducing type*. An example occurs when sulfur dioxide changes to sulfur trioxide during combustion or in the atmosphere. The sulfur trioxide then combines with atmospheric water to form droplets of sulfuric acid. This acid is corrosive, irritating, and attracts water, thereby enhancing the development of rain droplets—and acid precipitation (*see* Perspective: Acid Precipitation in Unit 19). The other kind of secondary pollution is called the *oxidation type*. Here the effects of sunlight—known as photochemical effects—play a role. An example is nitrogen dioxide, the source of the brown color of urban dust domes (Fig. 21-4), when it reacts with sunlight to form nitrogen monoxide and one odd oxygen atom. The freed oxygen atom can then combine with normal oxygen (O_2) to form ozone (O_3), which acts as an irritant.

Air pollution is an old, persistent, and costly problem. Indeed, a treatise on London's polluted air was published as long ago as the late seventeenth century. An oft-quoted death toll of 4000 during the infamous London smog episode of December 1952 (Fig. 21-8), as well as similar widely publicized mid-century tragedies in Belgium and Donora, Pennsylvania, are evidence enough of the dangerous effects of air pollution. New data, published in conjunction with every major smog outbreak, indicate that substantial increases in respiratory disease accompany the severely polluted air. The costs of air pollution are difficult to gauge accurately, but they are surely astronomical. Three decades ago it cost $2 million to reface New York's City Hall; since then, the cost of deterioration of that city's public and private property owing to air pollution has been reckoned at well over $500 million annually.

In Unit 8, we discuss the conditions that lead to the formation of smog and pollution domes over cities, particularly the influence of local temperature inversions.

FIGURE 21-8 St. Paul's Cathedral in central London, almost engulfed by smog at the height of that city's 1952 air pollution disaster.

These inversions can prevent urban air from rising more than a few hundred meters, thereby acting as an atmospheric "lid" that traps airborne pollutants and greatly increases their concentration in the city's surface layer. Horizontal flushing by winds can play a significant role in relieving air pollution in an urban area. But this means that using the skies above a city as a dumping ground quickly becomes someone else's problem as the effluents are transported downwind, often for considerable distances when pollutants are discharged via smokestacks taller than 300 m (1000 ft). We now take up this matter by considering the macroscale effects of airborne pollution.

Larger-Scale Air Pollution

The problem of air pollution is no longer a local one, and its effects are increasingly being felt over wide areas. Just as individual houses tend to interact within a city, so cities have significant effects beyond their immediate vicinities, thereby contributing to pollution on a broader scale. We are already familiar with the concept of the urban dust dome (*see* Fig. 21-6). When the prevailing wind is greater than 13 kph (8 mph) the dome begins to detach itself from the city, and the airborne pollutants stream out above the surrounding countryside as a **pol-**

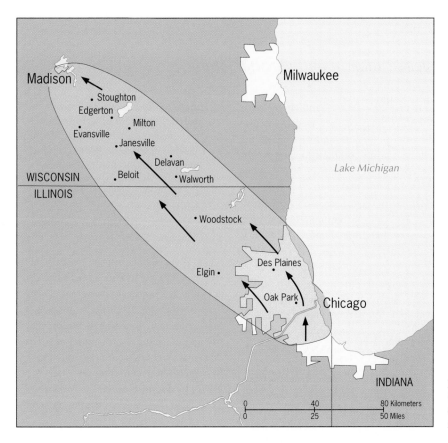

FIGURE 21-9 The pollution plume of the Chicago metropolitan area, extending downwind to Madison, nearly 250 km (150 mi) to the northwest.

lution plume. As Fig. 21-9 shows, the plume emanating from metropolitan Chicago has been observed to stretch as far as Madison, Wisconsin under the influence of a southeasterly wind—a distance of 240 km (150 mi). Similarly, population clusters around the world—from Fargo, North Dakota to Hong Kong—are affected by pollution plumes that originate in industrial cities.

As it advances, the heat- and dust-bearing plume brings with it many of the climatic characteristics associated with cities. For instance, the amount of shortwave solar radiation received on a clear day in the environs of Toronto, Canada depends markedly on which way the wind is blowing out of the central city. Pollution plumes affect more than the particulate content and heat levels of the atmosphere. La Porte, Indiana is 50 km (30 mi) downwind from the Chicago–Gary heavy-industrial complex, but its precipitation variations appear to be determined by the latter's level of manufacturing activity (*see* Perspective: Weather Modification in an Industrial Region).

Another factor reinforcing the problem of macroscale air pollution is that many of our big cities are concentrated in certain regions. For example, the metropolitan areas of the U.S. northeastern seaboard— as well as many in Western Europe and Japan—are geographically so close to one another that their pollution plumes merge and act together rather than individually. These coalesced, multimetropolitan regions (called *conurbations* by urban geographers) are now so large that they function as pollution sources on a continental scale. Thus the release of sulfur dioxide over southeastern England or western Germany's Ruhr Valley leads to an increase in the acidity of precipitation in Norway and Sweden (similar North American pollution flows are mapped in Fig. 19-6).

HUMAN ACTIVITIES AND THE GLOBAL CLIMATE MACHINE

At the global scale, human climatic impacts may be reaching a level capable of interfering with certain natural processes of the atmosphere. We discuss such hypotheses in other units, particularly in the Perspective boxes on ozone depletion (Unit 6), deforestation (Unit 17), desertification (Unit 17), and acid precipitation (Unit 19). Because little is being done worldwide to counteract these human-induced environmental problems, they are likely to intensify in the future. They might even be joined by more serious crises. The most dramatic of these is *nuclear winter*, the hypothesized global cooling caused by massive smoke plumes that would be generated by the extensive fires of a nuclear war.

Weather Modification in an Industrial Region

In the spring of 1968, the magazine *Saturday Review* ran an article entitled "Home-Brewed Thunderstorms of La Porte, Indiana." And shortly thereafter, *Newsweek* picked up the same story and catapulted the weather of this obscure Midwestern town into a leading national news item. Climatologist Stanley Changnon, writing in a prestigious meteorological journal, presented some of the startling facts behind the headlines.

Beginning in about 1925, La Porte experienced a dramatic increase in rain, thunderstorms, and hail. Weather stations in nearby towns did not show evidence of the same changes. Some people charged that air pollution from steel plants 50 km (30 mi) upwind in Chicago and Gary had caused the increase in local storms. Others blamed errors by weather observers who recorded the data.

But the micrometeorological changes soon proved to be real, and atmospheric scientists believed they were related to the heat, pollutants, and vapor from the manufacturing complex located to the west of La Porte. During the 50 preceding years, La Porte had been drenched by a 30 to 40 percent increase in precipitation. Especially heavy rainfall occurred in years of heightened steel production in the Chicago area. From 1949 to 1965, La Porte experienced 38 percent more thunderstorms than the surrounding towns. Between 1920 and 1965, La Porte averaged 59 percent more hail than neighboring communities; and from 1951 to 1965, the figure was 246 percent more hail.

This strikingly altered weather pattern, according to Changnon, was probably the result of

> an interaction between two major problem areas of meteorology: weather modification and atmo-

spheric pollution. . . . La Porte is located downwind of a major industrial complex capable of sizeable increase in heating, moisture content, and condensation nuclei and freezing nuclei, and the temporal fluctuations in smoke days and steel production compare favorably with those of the precipitation conditions (Changnon, 1968, 10).

La Porte stands out as a prime example of weather modification by human activities, a phenomenon associated with many major manufacturing regions. As a final comment, it is worth noting that precipitation and storm patterns in La Porte have become more similar to those of surrounding communities since the 1970s. By no coincidence, the years since 1980 have seen a sharp decline in steel production in the Chicago–Gary area—a connection that further supports Changnon's hypothesis.

Indeed, studies of large forest fires (often ignited by humans) have already demonstrated significant filtering of solar radiation by smoke particles (Fig. 21-10). Moreover, natural versions of nuclear winters have probably occurred in the past. A growing number of scientists claim that just such an event, approximately 65 million years ago, led to the extinction of the dinosaurs and many other species. The cause of this catastrophe was a large asteroid or comet (*see* Perspective: When Worlds Collide in Unit 4) that smashed into the earth, disinte-

FIGURE 21-10 The edge of one of the massive wildfires that swept across Yellowstone National Park in mid-1988, here threatening several tourist facilities. It is not hard to envision that thick plumes of smoke generated by such blazes can screen out significant amounts of solar radiation. In Kuwait, for example, the 1991 oil-well fires set by the fleeing Iraqi army (just before they were expelled by U.S. military forces) burned for months, producing huge clouds of smoke sufficient to lower daytime ground temperatures by more than 10°C (18°F).

grated, and cast up a globe-girdling layer of thick dust that lingered for at least a year—wiping out numerous life forms on the temporarily frigid surface below.

Despite the impressive advances of late-twentieth-century science, we still understand very little about the longer-term consequences of atmospheric pollution and its possible linkage to the forces of climatic change. Recent experience has shown that monitoring and forecasting methods need to be improved, because instead of slow, incremental change, certain pollutants may build up silently for years. Then, only after they surpass a crit-

ical mass, do they produce rapid and potentially far-reaching environmental change.

With scenarios such as these to contend with, soon it may no longer be possible to regard the earth's climate as a finely tuned natural machine in long-term equilibrium, constantly correcting itself to maintain exactly the right balance of warmth and moisture required to sustain the range of life on this planet. Today we may be beginning to disturb the delicate workings of that machine, and what that portends is one of the leading new pursuits of the atmospheric, life, and earth sciences.

KEY TERMS

dust dome (p. 231)

microclimate (p. 228)

pollution plume (p. 235)

primary pollutants (p. 234)

secondary pollutants (p. 234)

urban heat island (p. 231)

REVIEW QUESTIONS

1. Describe the various sources of energy and heat received by and lost from the human body.

2. Describe how an urban heat island develops and the meteorological consequences of this climatic modification.

3. What are primary pollutants? What are secondary pollutants? Give examples of both.

4. What is a pollution plume and how does it develop?

REFERENCES AND FURTHER READINGS

BRYSON, R. A., and ROSS, J. E. "The Climate of the City," in Detwyler, T. R., et al., *Urbanization and the Environment: The Physical Geography of the City* (North Scituate, Mass.: Duxbury Press, 1972), 51–68.

CHALONER, W. G., and MOORE, P. D. *Global Environmental Change* (Cambridge, Mass.: Blackwell, 1995).

CHANGNON, S. A., JR. "The La Porte Weather Anomaly: Fact or Fiction?," *Bulletin of the American Meteorological Society*, Vol. 49, No. 1, 1968, 4–11.

DETWYLER, T. R., et al. *Urbanization and the Environment: The Physical Geography of the City* (North Scituate, Mass.: Duxbury Press, 1972).

DOUGLAS, T. *The Urban Environment* (London: Edward Arnold, 1984).

ELSOM, D. *Atmospheric Pollution: A Global Problem* (Cambridge, Mass.: Blackwell, 2nd ed., 1992).

GOUDIE, A. S. *The Human Impact on the Natural Environment* (Cambridge, Mass.: Blackwell, 4th ed., 1993).

LAMB, H. H. *Weather, Climate and Human Affairs: A Book of Essays and Other Papers* (London/New York: Routledge, 1988).

LANDSBERG, H. E. "Climates and Urban Planning," in *Urban Climates* (Geneva: World Meteorological Organization, No. 254, Technical Paper 141, Technical Note No. 108, 1970), 364–374.

LEIGHTON, P. A. "Geographical Aspects of Air Pollution," *Geographical Review*, Vol. 56, 1966, 151–174.

LOCKWOOD, J. G. *World Climatic Systems* (London: Edward Arnold, 1985).

MANNION, A. M. *Global Environmental Change: A Natural and Cultural Environmental History* (New York: Wiley/Longman, 1991).

PIELKE, R., and COTTON, W. R. *Human Impacts on Weather and Climate* (New York: Cambridge University Press, 1994).

SEINFELD, J. H. "Urban Air Pollution: State of the Science," *Science*, Vol. 243, February 10, 1989, 745–752.

SILVER, C. S., and DE FRIES, R. S. *One Earth, One Future: Our Changing Global Environment* (Washington, D.C.: National Academy Press, 1990).

SIMMONS, I. G. *Changing the Face of the Earth: A History of the Human Impact* (New York: Basil Blackwell, 1989).

TURCO, R. P., et al. "The Climatic Effects of Nuclear War," *Scientific American*, August 1984, 33–43.

TURNER, B. L., II, ed. *The Earth as Transformed by Human Action: Global and Regional Changes in the Biosphere over the Past 300 Years* (New York: Cambridge University Press, 1990).

United Nations Environmental Programme. *Urban Air Pollution in Megacities of the World* (Cambridge, Mass.: Blackwell, 1992).

UNIT 22
CLIMATE, SOIL, PLANTS, AND ANIMALS

UNIT 23
FORMATION OF SOILS

UNIT 24
PHYSICAL PROPERTIES OF SOIL

UNIT 25
CLASSIFICATION AND MAPPING OF SOILS

PART THREE

THE BIOSPHERE

UNIT 22

Climate, Soil, Plants, and Animals

OBJECTIVES

◆ To expand our concept of physical geography to include biotic systems operating at the earth's surface.

◆ To relate biotic systems to our understanding of global climates.

◆ To link physical geography to the more general topic of conservation.

In Part Two, we focused on one major aspect of physical geography—climate. After studying the processes that take place in the atmosphere, our inquiry concentrated on a geographic interpretation of the results: the earth's present climate regions (Fig. 16-3 summarizes the global distribution of macroclimatic regions). Part Two concluded with a discussion of the dynamics and human interactions of climate.

As we delve further into the physical geography (and with it, the natural history) of our planet in Part Three, we should remember that processes and resulting patterns are always changing. Just a few thousand years ago, leafy forests and green pastures stood where desert conditions prevail today; animals now extinct roamed countrysides that are now empty and barren. A mere 10,000

years ago, glaciers were melting back after covering much of the U.S. Midwest and the heart of Europe.

If you have been following the news lately, you have heard about other changes in our terrestrial environments. One such example is the hole in the atmosphere's ozone layer that recently opened above Antarctica. As we note in the Perspective box in Unit 6, it may portend a future of greater cancer risk and unforeseeable effects on plants and animals. In short, maps of climates—as well as other elements of the environment—are still pictures of a changing world.

UNIT OPENING PHOTO: Sowing a ricefield in India, a vivid demonstration of interrelationships among climate, soil, plants, animals—and human societies. (Authors' photo)

240

In Part Three, we discuss several aspects of the environment that are closely related to climate. Geographers are most interested in interrelationships—between climate and soils, soils and plants, natural environments and human societies, and many other ecological connections. More often than not, the prevailing climate, past or present, is a key to our understanding of such relationships. Again and again, when we study the distributions of natural phenomena such as soils, plants, and animals, we see the pattern of climate.

NATURAL GEOGRAPHY

The study of soils, plants (*flora*), and animals (*fauna*) in spatial perspective is a branch of physical geography (*see* diagram, p. 6), although this aspect of the discipline might be better designated *natural* geography—the geography of nature. The geography of soils is a part of the science of **pedology**, a term that derives from the ancient Greek word *pedon*, meaning ground. The geography of plants and animals defines the field of **biogeography**, a combination of biology and geography. Biogeography is further divided into two subfields: **phytogeography**, the geography of plants, and **zoogeography**, the geography of animals.

Geography of Soils

The next three units (23–25) of Part Three examine the development, properties, classification, and regionalization of soils. On the landmasses, soils and vegetation lie at the *interface* between the lithosphere and the atmosphere, at the plane of interaction between rocks and their minerals on the one hand, and the air with its moisture and heat on the other. Soil is the key to plant life, containing mineral nutrients and storing water. But the vegetation itself contributes to its own sustenance by adding decaying organic matter to the soil's surface, which is absorbed and converted into reusable nutrients (Fig. 22-1). The processes that go on at this interface are intricate and complicated.

It is often said that life as we know it would not exist on this planet without the presence of water. One reason is that without water there would be no soil. The moon's lifeless surface of rock fragments and pulverized rubble holds no water—and therefore no soil. Water is the key to the chemical and physical processes that break down rocks, thus triggering the process of soil formation. As the soil develops or matures, water sustains its circulatory system, promotes the necessary chemical reactions, transfers nutrients, helps decompose organic matter, and ensures the continued decay of rocks below the evolving soil layer. Soils capable of supporting permanent vegetation can develop in a short period (between one and two centuries), but a soil may require thousands of years

FIGURE 22-1 Dead vegetative matter on the floor of a forest, a raw material in the soil's ability to sustain living plants with vital nutrients.

to mature fully (Fig. 22-2). Over this period of time the soil absorbs and discharges water, inhales and exhales air through its pore spaces, takes in organic matter and dispenses nutrients, and becomes inhabited by organisms of many kinds. In every sense of the term, *the soil is a living entity.*

As with all living things, the soil's well-being can be threatened, and soils can actually die. When allowed to

FIGURE 22-2 Cutaway view of a mature soil, whose structure has slowly evolved over several thousand years.

develop and mature, a soil will achieve a state of equilibrium with the prevailing climate, with the vegetation it supports (and that supports it), and with other elements of the surrounding environment. Many conditions can threaten a soil, some natural, others artificial. Climatic change may also lead to change in the natural vegetation, which, in turn, can expose a soil to increased erosion. Farming may overtax the soil, giving it insufficient time to recover from plowing, planting, and harvesting year after year (Fig. 22-3). Domesticated animals may trample vegetation and topsoil alike, weakening the soil and subjecting it to removal by water and wind.

The geographic study of soils, therefore, not only involves us in learning about the development and maturation of this critical cloak of life, but also leads us to investigate the state of its health—and ways to protect and conserve it. When it comes to the components of the earth's biosphere, the more we know about nature the better we are prepared to help sustain it. In Part Three, therefore, we explore the development of a typical soil, view a mature soil in cross-section, and discuss the processes that go on within and between the layers of a soil. Next we study ways of classifying soils, and from this classification emerges regionalization, the map of world soils (*see* Fig. 25-14). A comparison between this map and other global maps covered in this text (such as climate [Fig. 16-3], precipitation [Fig. 12-11] , and terrain [Fig. 52-1]) reveals some of the spatial relationships between soil regions and other elements of the natural environment. Moreover, comparing the soil map and a map

of world population distribution (*see* Fig. 2-5) indicates the critical importance of certain soil types for food production.

Biogeography

Another major topic, addressed in Units 26–28, is biogeography, or the geography of flora and fauna. Several of the world's climates are named after the vegetation that characterizes them (e.g., savanna, steppe, and tundra). When climate, soil, vegetation, and animal life reach a stable adjustment, vegetation constitutes the most visible element of the ecosystem. When you fly southward in Africa from the central Sahara (ca. 25°N) to the equatorial heart of the Zaïre (Congo) Basin, you cross over a series of ecosystems ranging from desert to rainforest. You cannot see the climate changing, nor can you clearly see the soil most of the way. But the vegetation tells the story: the barren desert gives way to shrub; the grasses gradually become denser; trees appear, widely spaced at first, then closer together; and finally, the closed canopies of the tropical rainforest come into view. The best indicator of the succession of prevailing ecosystems was vegetation (Fig. 22-4).

As we learn in Unit 27, however, the earth's great vegetation assemblages consist not of just one or two but of literally millions of species. Biologists suggest that there may be as many as 30 million species alive today, of which only about *1.5 million* have been identified and classified (*see* Perspective: Biodiversity Under Siege). Geographers are especially interested in the distribution of the known species, as well as in the relationships between plant and animal communities and their natural environments. Biogeographers seek explanations

FIGURE 22-3 Poor conservation practices led to this dramatic example of severe soil erosion and gullying in Madagascar, the large island country located off the coast of southeastern Africa.

FIGURE 22-4 A vegetation boundary zone at the base of the Chugach Range in southeastern Alaska. At this northern limit of tree growth, the thinning coniferous forest steadily gives way to the ecosystem of the hardier tundra.

Biodiversity Under Siege

Biodiversity, shorthand for *biological diversity*, refers to "the variety of [Earth's] life forms, the ecological roles they perform, and the genetic diversity they contain" (Wilcox, 1984). The cataloguing of life forms according to species has proven to be the most practical approach (spatial variations are discussed in Unit 26 in the Perspective box on p. 287). A **species** may be defined as a population of physically and chemically similar organisms within which free gene flow takes place.

How many living species does the global environment contain? The renowned biologist Edward O. Wilson, a leading expert, indicates that approximately 1.5 million species have been identified and described. About 750,000 of these are insects, 250,000 are plants, 41,000 are vertebrate animals, and the remainder are accounted for by invertebrates, fungi, algae, and microorganisms (Wilson, 1988, pp. 3–5). However, biologists unanimously agree that these subtotals constitute only a small proportion of the total number of terrestrial species. Wilson estimates that total far exceeds the number of known species, and lies somewhere between 5 and 30 million.

Biodiversity may be regarded as one of our planet's most important resources, and is a rising concern today because it is under siege. In Unit 17 we note that tropical rainforests are especially threatened by the mass burning and cutting of trees, which *annually* remove a woodland area no less than the size of Washington State. These forests also are home to about 80 percent of all living species, and the rate of loss today is about 100 species per day (36,500 per year). Other fragile biological communities are besieged as well (particularly coastal zones and wetlands), and the main source of these exterminations is loss of habitat caused by the expansion of such human activities as deforestation, agriculture, urbanization, and air and water pollution.

As these uses—and misuses—of the global environment multiply, there is growing evidence that decline in biodiversity may soon become a universal phenomenon. An ominous example is the decrease in field observation, since the mid-1980s, of fully 70 percent of the bird species that are known to summer in the eastern United States. This decrease is consistent with Wilson's finding that perhaps 20 percent of all bird species have disappeared in modern times and another 10 percent are endangered. Although biodiversity is one of the life sciences' newest research arenas, we already know enough to draw the following conclusion: as human technology continues to advance, it is increasingly accompanied by the largest extinction of natural species the world has undergone since the mass extinctions that occurred at the time the dinosaurs disappeared 65 million years ago.

for the distributions the map reveals. For example, why are certain species found in certain areas but not in others that also seem suitable for them? And why do species exist cooperatively in some places but competitively in others? What is the effect of isolation on species and their interrelationships?

The founder of biogeography as a systematic field of study was Alexander von Humboldt (1769–1859), who traveled much of the world in search of plant specimens. When Humboldt reached South America's Andes, he recognized that altitude, temperature, natural vegetation, and crop cultivation were interrelated. After returning to Europe, he produced a monumental series of books that formed a basis not only for biogeography but also for many other fields of the natural sciences.

The separation of biogeography into phytogeography and zoogeography occurred after the appearance of Humboldt's writings. Probably the most important book on zoogeography to appear during the nineteenth century was written by Alfred Russel Wallace and entitled *The Geographical Distribution of Animals* (1876). Wallace was particularly interested in the complicated distribution of animals in Southeast Asia and Australia. Australia is the last major refuge of the marsupials (animals whose young are born very early in their development and then carried in an abdominal pouch); the kangaroo, koala, and wombat are three Australian marsupials. While a few marsupials survive in other areas of the world, such as the opossum in North America and the lemur in Madagascar, Australia's fauna are unique, and Wallace wanted to establish the zoogeographical boundary line between Southeast Asia's very different animal assemblage and that of Australia. When he did his fieldwork, he discovered that Australian fauna existed not only in Australia itself but also in New Guinea and even on some eastern islands of what is today Indonesia. So Wallace drew a line between Kalimantan (Indonesian Borneo) and Sulawesi (Celebes), and between the first island (Bali) and second island (Lombok) east of Java (Fig. 22-5). *Wallace's Line* soon became one of the most hotly debated zoogeographic boundaries ever drawn, and the debate continues to this day.

As we note in Unit 28, zoogeography is a field of many dimensions and challenges. Vegetation regions can be seen from the air and can be mapped from remotely sensed data. But animals move and migrate, their *range* (area of natural occurrence) changes over time and even seasonally, and detection can present problems as well. Mapping faunal distributions for large animals is difficult enough; for smaller species it is often far more complicated.

CONSERVATION AND THE BIOSPHERE

The more we learn about the earth's biosphere, the more concerned we become for the future. The overuse and erosion of soils are worsening global problems (Fig. 22-6). A large part of the sediment load carried oceanward by the world's great rivers comes from slopes where cultivators have loosened the topsoil, which is carried away by rainstorms. The destruction of tropical rainforests is accelerating in South America, Africa, and Asia. Botanists estimate that as many as one-quarter of all the presently living plant species may become extinct during our lifetime. Animals large and small are also facing extinction, from the great African elephant to tiny insects.

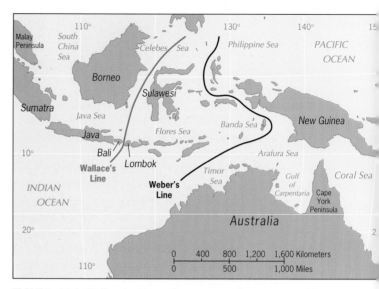

FIGURE 22-5 Wallace's Line, the presumed zoogeographic boundary between the faunal assemblages of Southeast Asia and Australia. One of Wallace's many challengers, the proposer of the alternative Weber's Line (*see* Unit 28), placed that boundary farther to the east.

People pose many threats to animals, only one of which is their invasion of the last refuges of wildlife. The horn of the African rhinoceros, for instance, is prized in

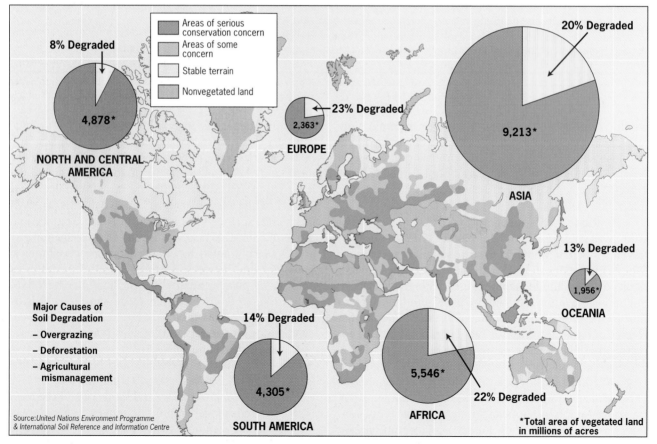

FIGURE 22-6 The global geographic pattern of human-induced soil degradation since 1945. Over the past half-century, this environmental abuse has destroyed one-eighth of the earth's vegetated land.

FIGURE 22-7 Three decades ago, southwestern Germany's Black Forest was still one of the world's most beautiful expanses of woodland. Today, much of its landscape lies decimated by pollutants such as acid rain.

Saudi Arabia as a dagger handle, and in East Asia it is regarded as an aphrodisiac in powdered form. As a result, rhinos are killed by poachers so quickly that only comparatively few survive. In 1990, the price of pow-

dered rhino horn on the Asian market was four to five times higher by weight than gold. So Africa's rhinos are threatened less by competition from poor, land-hungry farmers than by the whims of the distant wealthy.

Whether we focus on soils, plants, or animals, our geographic perspective soon underscores the reality that the earth's biosphere is under severe stress. All too often, the changes that take place are irreversible. Topsoil that has taken hundreds or thousands of years to develop is washed away in a few months. Plant species never even inventoried are lost forever; we will never know what role they might have played in combating disease. As Africa's elephant herds dwindle in the face of ivory poachers, we are witnessing the extinction of a legacy of hundreds of millions of years. The explosive growth of the earth's human population as well as the profligate spending behavior of consumers in the world's wealthiest countries have combined to put the biosphere under stress as never before. From fishing grounds to mountain pastures, from rainforest to desert margin, the evidence is everywhere (Fig. 22-7).

How can this tide of destruction be slowed or reversed? Knowledge and awareness are powerful allies in any such endeavor. For years after the communists took power in China in 1949, Chinese leaders extolled the virtues of population expansion, at home and abroad, as a means of furthering their ideological objectives. Then China became aware that economic development would be stymied by the needs of so many millions of additional citizens, and its policies were changed (Fig. 22-8). Not only population control but also soil conservation and reforestation programs were made high priorities. But the People's Republic of China, unlike India and many

FIGURE 22-8 Over the past 20 years, China has attempted to regulate its population growth. Among other approaches, exhortations to the masses dot the public landscape, as here in the western city of Chengdu. Nonetheless, after considerable success during the 1980s, birth rates are creeping upward again in the 1990s. (Authors' photo)

other countries of the world, has a powerful, highly centralized government that can impose such measures rather effectively. Elsewhere, change must come through education and voluntary cooperation, which is far more difficult to achieve.

Although Part Three concentrates on the biosphere, we should remember that the principles and practices of conservation apply to more than soils, plants, and animals. **Conservation** entails the careful management and use of natural resources, the achievement of significant social benefits from them, and the preservation of the environment. For more than half a century, courses in conservation were a cornerstone of an education in geography.

One of the incentives behind this practice was the terrible experience of the 1930s, a phenomenon now known simply as the *Dust Bowl*. In the U.S. Great Plains, where raising wheat was the mainstay of the region's farmers, below-average rainfall was recorded for several years running. Wheat had been sown on land that was only marginally suitable for grain cultivation, and when the rains failed the loosened soil fell prey to the ever-present wind. Great clouds of dust soon blackened the skies as millions of tons of topsoil were blown away (Fig. 22-9). Dunes formed, some more than 3 m (10 ft) high,

and the landscape was transformed. Thousands of farmers abandoned their land and moved westward, hoping to make a new start in California. The physical destruction of much of the Great Plains and the social dislocation that attended it made an indelible mark on America, and the idea of conservation took hold. Since geographers studied climates, soils, agriculture, and related topics, many students came to geography because of their interest in conservation.

The administration of President Franklin D. Roosevelt (1933–1945), too, launched programs to counter future dust-bowl experiences. Congress passed legislation in support of land-use planning and soil protection, and the federal government created several offices and agencies to implement these initiatives. The Soil Conservation Service was among these agencies, as were the Natural Resources Board and the Civilian Conservation Corps. One of the most important agencies to be founded during this period was the Tennessee Valley Authority (TVA), a massive regional project begun in 1933 to control destructive floods, assist farmers, improve navigation, and create electrical power sources in the Tennessee River basin (Fig. 22-10). The TVA, still operating today, includes parts of seven states: Alabama, Georgia, Kentucky, Mississippi, North Carolina, Tennessee, and

FIGURE 22-9 A classic photograph taken during the height of the Dust Bowl in the mid-1930s. Here a massive dust storm is about to engulf a western Oklahoma village, which, according to the Bettmann Archives notes that accompanied this photo, was reduced to a "blot of dust" a few minutes later. These notes further indicate that "merciless black blizzards [of dust] raged across the Plains five months of each year, sweeping out crops, burying pastures, and driving housewives to distraction by filtering their homes with dust picked up from the dry land."

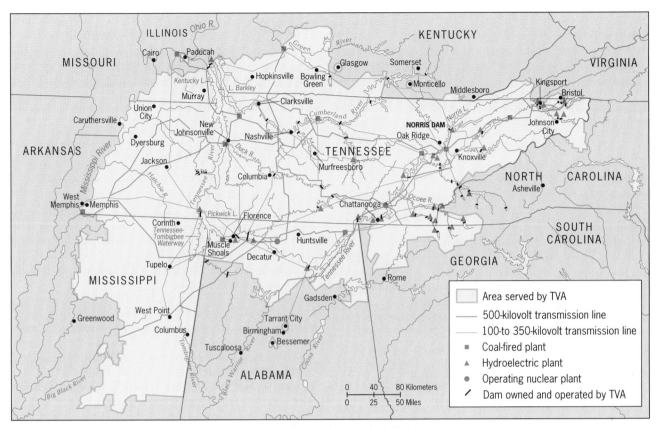

FIGURE 22-10 The region developed by the Tennessee Valley Authority (TVA), beginning in the mid-1930s.

Virginia. It has transformed the physical and human geography of a large region.

The 1930s, which some have called the golden age of conservation, were followed by a period during which attention in the United States was diverted to other causes: war and global strategic competition. In the meantime, urbanization was increasing and many farmers, unable to afford mechanization and other new technologies required to compete in the postwar marketplace, were forced to sell their land. Fortunately, some farsighted people continued to support conservationist ideas and programs with their own money. Among privately funded conservation organizations were the Conservation Foundation (founded in 1948) and Resources for the Future (1952). Such organizations sponsored research and publications on conservation issues and practices, and helped keep the conservation ethic before the American public.

As environmental problems intensified not only at home but also abroad, the public interest in and concern over conservation matters was revived. The U.S. government finally responded in 1970 by creating the Environmental Protection Agency (EPA). The EPA soon became a high-profile agency that addressed a wider range of conservation and environmental concerns than any of its predecessors. After a number of scandals and exposures of lax enforcement during the 1980s, under the Clinton administration in the mid-1990s the EPA—along with a reinvigorated U.S. Department of the Interior—is again playing a major role in the conservation arena.

As you can see, the geographic study of the biosphere is not just a theoretical exercise. It involves science as well as policy, research as well as application. To make decisions we need basic information, and each of the next six units in Part Three contains essential background to help us do so.

KEY TERMS

biodiversity (p. 243)	conservation (p. 246)	phytogeography (p. 241)	zoogeography (p. 241)
biogeography (p. 241)	pedology (p. 241)	species (p. 243)	

REVIEW QUESTIONS

1. Which subjects are encompassed by *pedology* and *biogeography*?

2. What are the major functions of soil?

3. In what way does physical geography aid in understanding problems of conservation?

4. Why was the Tennessee Valley Authority created in the 1930s?

REFERENCES AND FURTHER READINGS

BRADBURY, I. K.   *The Biosphere* (New York: Wiley, 1991).

CONSERVATION FOUNDATION.   *State of the Environment: An Assessment at Mid-Decade* (Washington, D.C.: The Conservation Foundation, 1984).

CUTTER, S. L., et al.   *Exploitation, Conservation, Preservation: A Geographic Perspective on Natural Resource Use* (New York: Wiley, 2nd ed., 1991).

EBLEN, R. A., and EBLEN, W. R., eds.   *The Encyclopedia of the Environment* (Boston: Houghton Mifflin, 1994).

EYRE, S. R.   *Vegetation and Soils: A World Picture* (London: Edward Arnold, 2nd ed., 1975).

FURLEY, P. A., and NEWEY, W. W.   *Geography of the Biosphere: An Introduction to the Nature, Distribution, and Evolution of the World's Life Zones* (Stoneham, Mass.: Butterworth, 1982).

HIGHSMITH, R. M., et al.   *Conservation in the United States* (Chicago: Rand McNally, 1962).

HOLE, F. D., and CAMPBELL, J. B.   *Soil Landscape Analysis* (Totowa, N.J.: Rowman & Allanheld, 1985).

LOVELOCK, J.   *The Ages of Gaia: A Biography of Our Living Earth* (New York: Norton, 1988).

MAURER, B.   *Geographical Analysis of Biodiversity* (Cambridge, Mass.: Blackwell, 1994).

MCMICHAEL, A. J.   *Planetary Overload: Global Environmental Change and the Health of the Human Species* (New York: Cambridge University Press, 1993).

SCIENTIFIC AMERICAN.   *The Biosphere: A Scientific American Book* (San Francisco: Freeman, 1970).

SEARS, P. B.   *Deserts on the March* (Norman, Okla.: University of Oklahoma Press, 4th ed., 1980).

SHELFORD, V. E.   *The Ecology of North America* (Urbana, Ill.: University of Illinois Press, 1963).

TRUDGILL, S. A.   *Soil and Vegetation Systems* (London/New York: Oxford University Press, 2nd ed., 1988).

WALLACE, A. R.   *The Geographical Distribution of Animals; With a Study of the Relations of Living and Extinct Faunas as Elucidating the Past Changes of the Earth's Surface* (New York: Hafner [reprint of 1876 original], 1962).

WILCOX, B. A.   "In Situ Conservation of Genetic Resources: Determinants of Minimum Area Requirements," in J. A. McNeeley & K. R. Miller, eds., *National Parks, Conservation, and Development: The Role of Protected Areas in Sustaining Society* (Washington, D.C.: Smithsonian Institution Press, 1984), pp. 639–647. Quotation taken from p. 640.

WILSON, E. O.   "The Current State of Biological Diversity," in E. O. Wilson, ed., *Biodiversity* (Washington, D.C.: National Academy Press, 1988), 3–18.

WILSON, E. O.   *The Diversity of Life* (Cambridge, Mass.: Belknap/Harvard University Press, 1992).

Formation of Soils

OBJECTIVES

◆ To understand the components of soil.

◆ To outline the factors affecting soil formation.

◆ To describe and explain a typical soil profile and the processes responsible for the formation of soil horizons.

S oil may be regarded as a living system inasmuch as it supports plants, the organisms responsible for plant decay, and a variety of other life forms. Using more formal terms, the U.S. Soil Conservation Service defines **soil** as "a mixture of fragmented and weathered grains of minerals and rocks with variable proportions of air and water; the mixture has a fairly distinct layering; and its development is influenced by climate and living organisms." *Weathering*, the chemical alteration and physical disintegration of earth materials by the action of air, water, and organisms, is more closely examined in Unit 24; here we will concentrate on the major controls governing soil formation.

The term *soil* actually means different things to different people. Agricultural scientists regard soil as the few top layers of weathered material that plants root and

grow in. But geologists use the term to refer to all materials that are produced by weathering at a particular site. Using this definition, we can still consider as soil those soils that were produced thousands of years ago and are now covered by layers of other material, even though it is impossible to grow plants in them. Alternatively, civil engineers look upon soil as something to build on and, in general, as anything that doesn't have to be blasted away.

Soil is obviously located at the earth's surface and in contact with the atmosphere. It is not, except in small quantities, found in the air, although when dry or unpro-

UNIT OPENING PHOTO: Exposed soil in the tropical rainforest of southeastern Nigeria. This erosional scar is largely the result of local human environmental abuse. (Authors' photo)

FIGURE 23-1 Intricate terracing on steep hillsides has marked the agriculture of Indonesia and many other countries for centuries. Success depends on carefully working with the local terrain and constantly maintaining a harmonious human-environmental relationship.

tected, soil is subject to the action of wind. Nor is it naturally encountered in large quantities in rivers, although erosion and transportation by water may put it there. The general location of soil is at the interface of the atmosphere, hydrosphere, biosphere, and lithosphere. The soil layer, in fact, constitutes one of the most active interfaces among these four spheres of the earth system.

We may not normally think of it as such, but the soil is one of our most precious natural resources. Geographers classify the earth's resources into **renewable resources**, those that can regenerate as they are exploited, and **nonrenewable resources**, such as metallic ores and petroleum, which when consumed at a certain rate will ultimately be used up. Schools of fish, for example, form a renewable resource: harvested at a calculated rate, they will restore themselves continuously. But, as all fishing people know, this balance can easily be disturbed by overfishing. And once a population has been overexploited, the fish may not return for many years—if ever.

So it is with soil. Soil is a renewable resource: we use it and deplete it, but it continues to regenerate. Renewable resources, however, are not inexhaustible. Damaged beyond a certain level, a soil may be lost to erosion and perhaps permanently destroyed. Farmers

know the risks involved in cultivating steep slopes by the wrong methods. Terraces must be created, but more importantly they must be *maintained* (Fig. 23-1). Once neglected, a terraced slope is ripe for soil erosion; once the gullies appear, the process may be irreversible (Fig. 23-2).

THE FORMATION OF SOIL

When the earth first formed and molten rock began to solidify, there was as yet no soil. The surface was barren, and not until an atmosphere evolved could any soil develop. Once the earth acquired its layer of moisture-carrying and heat-transferring air, soil formation progressed—as it continues to do today.

Soil Components

Soil contains four components (Fig. 23-3): minerals, organic matter, water, and air. A soil's minerals make up its tiny rock particles; the organic matter is of various types; water usually clings to the surfaces of the rock particles; and air fills the intervening gaps. We now look at each component more closely.

FIGURE 23-2 When agricultural terraces, such as those depicted in Fig. 23-1, are not well maintained, erosion quickly leads to gullying. That has happened on this particular slope in Nepal in the foothills of the Himalayas. Unless remedial action is taken swiftly, these gullies will multiply in the rainy season and the terraces will soon have to be abandoned.

Minerals are naturally occurring chemical elements or compounds that possess a crystalline structure and are the constituents of rocks. The most common elements in the earth's crust—silicon, aluminum, iron, calcium, sodium, potassium, and magnesium—are the building blocks for many minerals. Actually, the most prevalent element of all is oxygen, and it too is found in countless combinations with other elements. Thus rocks contain assemblages of minerals formed from thousands of such element combinations. When rocks break down into soils, these mineral components become available to plants as nutrients.

Another soil component is *organic matter*, the material that forms from living matter. In the upper layers of the soil, there is an accumulation of decaying and decayed remains of the leaves, stems, and roots of plants. There is also waste matter from worms, insects, and other animals. The decay processes are carried out by an astronomical number of microorganisms, such as bacteria and fungi. All of these contribute to the organic content of the soil.

Soil also contains life-sustaining *water*. There is an electrical attraction between the mineral particles and the

water molecules surrounding them. Normally, water fills much of the space between the minerals. But even in very dry soils, the attraction is so persistent that there may be a thin film of water, possibly one or two molecules thick, around the mineral particles. The water is not pure but exists as a weak solution of the various chemicals found in the soil. Without water, the many chemical changes that must occur in the soil could not take place.

Air fills the spaces among the mineral particles, organic matter, and water. It is not exactly the kind of air we know in the atmosphere. Soil air contains more carbon dioxide and less oxygen and nitrogen than atmospheric air does.

It is simple enough to deduce that these four soil components came from the other spheres of the earth system. The more interesting task is to discover *how*. Five important factors will direct us to that understanding.

Factors in the Formation of Soil

Some of the factors in soil formation are obvious, whereas others are not so apparent. The rocks of the earth are the parents of soil, and so we give them, and the deposits formed from them, the name **parent material**. The atmosphere provides the water and air for the soil layer, but the processes of the atmosphere vary across the earth's surface; we must therefore look at *climate* as a distinct soil-forming factor. Because of the presence of organic matter, we might also expect vegetation and other *biological agents* to be a factor. Only close observation tells us that soils seem to vary with the ruggedness of the landscape; thus *topography* is not quite so obvious a factor in soil formation. Neither is the last factor, but the formation of soils must depend on *time*. Let us see how all these factors come into play.

Parent Material

When a soil forms directly from underlying rock, the dominant soil minerals bear a direct relationship to the original rock. This, the simplest kind of soil formation, gives rise to what is known as a **residual soil**. Thus, on the Atlantic coast of the United States, thick red soils contain the insoluble residues of the iron oxides and aluminum silicates from the original rock. Such cases are quite common throughout the world. Yet even on the east coast of the United States, there may also be soil differences resulting from the variation of climate along this seaboard.

In a second category of soil, known as **transported soil**, the soil may be totally independent of the underlying solid rock because the parent material has been transported and deposited by one or more of the gradational agents, often far from its original source. During the most recent ice age, large quantities of soils and other materials were transported by glaciers thousands of kilometers and deposited in new areas. These materials

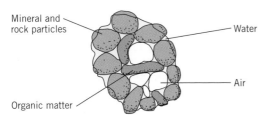

Mineral and rock particles

Water

Air

Organic matter

FIGURE 23-3 The four major components and their relative positions within a tiny clump of soil.

then formed the parent material for new soil formation. Many Eastern and Midwestern states have soils formed of such parent material.

The sediments deposited in stream valleys are another source of parent material. This type of parent material often creates fertile soils, as in the case of the lands bordering the Mississippi River. In still other instances, the wind of the atmosphere carries and eventually deposits thick blankets of fine matter that form the parent materials of new soil. Such deposits of windblown material are called *loess*; much of the central Great Plains region of the United States is underlain by loess deposits that are sometimes 60 m (200 ft) thick.

Climate

In the Piedmont areas of central Georgia and Maryland, the soil has developed from a rock called granite. Yet the resulting soils are not the same because the two states have different climates. The warmer, moister climate of the Southeast has engendered a much more complete pattern of chemical change in the Georgian soils. As far as the development of soils is concerned, the important elements of climate are moisture, temperature, and wind. The amount of soil moisture is determined by the amount of precipitation and evapotranspiration at a particular location. Both moisture and higher temperatures accelerate

FIGURE 23-4 Humus, the blackish layer at the top of the soil, consists of decomposing organic matter.

the chemical reactions of the soil. Thus we could expect to find thick, well-developed soils in the lower, warmer latitudes. Wind is another factor in the formation of some soils. In certain areas, wind action is responsible for the accumulation of sediment that may become the parent material of soil. Elsewhere, especially in areas with sparse vegetation cover, wind may be responsible for removing soil.

Biological Agents

Climate affects the types and amounts of vegetation and other organisms that grow in an area. Outside the tropics, usually the greater the amount of vegetation on the soil the greater the amount of organic matter in the soil. Partially decomposed organic matter called **humus** forms a dark layer at the top of the soil (Fig. 23-4). The most fertile soils are rich in humus. In many respects, humus and vegetation form a closed system. The substances that circulate through this system are plant foods—such as nitrogen compounds, phosphates, and potassium—collectively known as *plant nutrients.*

We would need a microscope to see the bacteria and fungi that are the key to this circulation. Decomposing microorganisms continually change the nutrients into simpler compounds that can enter plants through their roots. At the same time, other bacteria "fix" atmospheric nitrogen so that it too can be absorbed. Plants use these nutrients to grow, and when they die the decomposing bacteria return the nutrients to the soil to continue the cycle. This system turns into an open system when humans or animals remove the vegetation. Then the soil must often be balanced with artificial nutrients or fertilizers.

Several macroorganisms living in the soil also act as biological agents of formation. Earthworms alone can turn up to 6.5 metric tons per hectare (18.1 tons per acre) of soil every year. This mixing activity greatly helps soil formation and fertility. You may have seen advertisements for earthworms, which are a great aid to any healthy garden. Ants, termites, gophers, and scores of other burrowing animals perform the same function, but usually not with the efficiency of worms.

Topography

Another factor that affects the formation of a soil is its location with respect to the earth's terrain. In the case of mountain climates, the aspect of a slope partially determines its receipt of radiation and therefore the amount of moisture evaporating from it. Windward and leeward slopes receive varying amounts of precipitation. Their steepness also affects runoff and thus the amount of moisture that penetrates to the lower layers of the soil. These phenomena, to a large degree, control the amounts of moisture and heat in a soil-forming area. A hillside might have a relatively thin layer of soil in part because of its efficient drainage. If the hillside faces away from the sun, this is even more likely because both heat and moisture are minimized. In contrast, a less-well-drained valley bottom, receiving a large amount of heat and moisture, is an optimal location for the chemical processes of soil formation. We could therefore expect a deep soil layer in a valley.

Time

Whether a soil is deep or shallow, it still may need a long time to form. The processes of soil formation are slow, and thus time becomes an important factor. An example of relatively rapid soil formation comes from the Indonesian volcanic island of Krakatau. In its tropical climate, 35 cm (13.5 in) of soil developed on newly deposited rock materials within 45 years. The same process often takes much longer in colder climates: some of the organic matter in Arctic soils at Point Barrow, Alaska is still not thoroughly decomposed, even though it is nearly 3000 years old. Often the first evidence of a layered soil develops in as little as a century, but a fully mature soil requires thousands of years to evolve. This tells us how serious any damage to a soil can be: the destruction caused in a few years of careless farming (such as on overly steep slopes) can take centuries to repair.

PROCESSES IN THE SOIL

From the foregoing, we may conclude that soil formation results from a set of processes, all occurring simultaneously within the soil's developing layers. It is difficult to generalize about these processes: what goes on in evolving equatorial soils is very different from what happens in higher-latitude or high-altitude soils.

In 1959, the soil scientist Roy W. Simonson published a general theory of soil formation that provides a useful framework for understanding what takes place within a maturing soil. Simonson noted that certain processes occur in all soils during their formation, but that some very active processes present in some soils are nearly dormant in others. He further noted the lack of sharp boundaries between soils: soils tend to change gradually over space, and few *soil bodies* (geographic areas within which soil properties remain relatively constant) have distinct margins. Soils also change vertically, but in a much more abrupt manner so that we normally observe a sequence of layers. Simonson concluded that the development of such layers in soils can be ascribed to four sets of processes:

1. **Addition**: the gains made by the soil through adding organic matter from plant growth, or sometimes when loose surface material moves downslope and comes to rest on the soil. Many soils have a dark-colored upper layer whose appearance results from the addition of organic material. Some soils even develop an entirely organic uppermost layer consisting of decaying vegetative matter.

2. **Transformation**: the weathering of rocks and minerals and the continuing decomposition of organic material in the soil. The breakdown of rocks and minerals proceeds throughout all the layers of a soil, but the processes of weathering tend to be more advanced in the upper layers. Near the base of the soil chunks of yet-unaltered rock still exist, but toward the top of the soil no trace of these can be found.

3. **Depletion**: the loss of dissolved soil components as they are carried downward by water, plus the loss of other material in suspension as the water percolates through the soil from the upper toward the lower layers. While the upper layers are thus depleted, the dissolved and suspended materials are redeposited lower down in the soil.

4. **Translocation**: the introduction of dissolved and suspended particles from the upper layers into the lower ones. In deep soils, for example, in equatorial and certain other tropical areas, the translocation processes redeposit these particles so deep into the soil that plant roots cannot reach them.

These fundamental processes take place within all soils, but not everywhere at the same rate or with the same degree of effectiveness. They depend on the soil-forming factors and on the conditions we have just discussed: parent material, climate, biological agents, topography, and time. As they proceed, the soil is internally differentiated into discrete layers with particular properties.

SOIL PROFILES

We now have some information about the factors involved in soil formation and the major processes that go on within soils. The time has come to dig a hole in the soil in a certain location and see what lies below the surface.

After an hour or so of vigorous digging, we have made a hole or pit about 2 m (6.5 ft) deep. Even before we finished we made a major discovery: the soil consisted of a series of layers revealed by changes in color and by the "feel" or texture of the materials of which they are made. Each of these soil layers is called a **soil horizon**, and the differentiation of soil into distinct layers is called *horizonation*. Some soils have quite sharply defined horizons; in others, horizons are difficult to identify. All the horizons, from top to bottom, are known as the **soil profile**. A soil profile is to a soil geographer what a fingerprint is to a detective. Soil scientists, of course, cannot take the time to dig holes wherever they study soils, so they often use a device called a soil auger, a hollow sharp-edged pipe that is pushed into the soil and retrieves a sample of the upper horizons (or, in a shallow soil, the entire profile).

When we look at a 2-m (6.5-ft) profile of a well-developed, mature soil, it is usually not difficult to identify the horizonation within it. Soil scientists for many years used a logical scheme to designate each horizon in a model profile: they divided the profile into three segments named alphabetically. Accordingly, the **A** horizon lies at the top (often darkened by organic material); the **B** horizon in the middle, often receiving dissolved and suspended particles from above; and the **C** horizon at the bottom, where the weathering of parent material proceeds.

This **A-B-C** designation has been in use for many years, although it has undergone much modification. For instance, some soil profiles include an uppermost horizon—*above* the **A** horizon—consisting entirely of organic material in various stages of decomposition. Where this occurs it is identified as an **O** horizon, an organic horizon separate from the **A** horizon on top of which it lies. Note that a soil with an **A** horizon colored dark from vegetation growing in it is *not* an organic horizon; an **O** horizon consists exclusively of organic material.

As soil scientists learned more about soil profiles, they found that individual horizons were themselves layered. Some upper **A** horizons, for example, are dark-colored but turn light-colored perhaps 30 cm (1 ft) down. At first, the upper **A** horizon was designated A_1 and the lower, lighter horizon A_2; but eventually it became clear that the A_2 horizon had its own distinct qualities. So now many soil profiles have an **E** horizon between the **A** and **B** horizons. At the base of the soil, the horizon where the bedrock is breaking up and weathering into the particles from which soil is being formed is designated the **R** horizon (**R** for *regolith*).

A typical soil profile, representing a soil found in (for example) the moist, midlatitude coastal plain of the U.S. Atlantic seaboard, would look like the one shown in Fig. 23-5. Note that the major horizons (**O**, **A**, **B**, etc.) are sometimes given secondary designations, such as **Oa** and **Bt**. We identify the significance of these combinations as we study this diagram.

The profile shown in Fig. 23-5 contains horizons marked by six capital letters—**O**, **A**, **E**, **B**, **C**, and **R**. These are called the *master horizons*. As the drawing shows, this soil does have an **O** horizon, and it consists of two layers: a bed of leaves and twigs not yet decomposed on top (**Oi**), and beneath this a layer of highly decomposed organic matter (**Oa**). Immediately below this layer is the **A** horizon, the uppermost layer of the soil derived from the parent material below but colored dark by the organic matter from above. Soon, however, the **A** horizon becomes lighter, and about 25 cm (10 in) down it shows evidence of depletion—the removal by percolating water of particles in solution or suspension. It now gives way to the **E** horizon (the **E** standing for the process

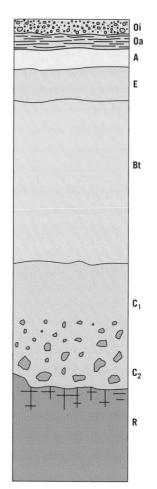

FIGURE 23-5 A soil profile, typical of the humid midlatitudes, showing the various soil horizons.

known as **eluviation**, the general term for such removal; it means "washed out").

Below the **E** horizon comes the **B** horizon. When a soil is very young, it may not yet possess a clearly developed **B** horizon. Figure 23-6 shows the development of soil horizons on a sedimentary parent material: note that a **B** horizon develops only in the second stage of soil evolution. Also note that when the **B** horizon first develops, it is not yet enriched by particles carried downward from the **O** and **A** horizons; eluviation, too, is just beginning. Such weakly developed **B** horizons are marked **Bw**. But after enough time has elapsed, the soil displays an **A** as well as an **E** horizon, translocation is in full force, and the **B** horizon matures. The symbol **Bt** is used to identify such a mature **B** horizon, and the term **illuviation** signifies the deposition of particles carried downward by percolating water. Thus eluviation from the **A** and **E** horizons is matched by illuviation into the **B** horizon.

The **Bt** designation reflects the common presence of calcium carbonate ($CaCO_3$) in parent materials. For instance, the sedimentary materials from which many soils of the U.S. Midwest are formed, consisting of rock

Exchange of Ions

Just how do mineral substances become nutrients for plants? How do mineral particles enter the water that percolates through the soil, and how do they leave these liquids to enter plant roots? We would need a powerful microscope to observe these exchanges in the soil. When minerals and humus break down, they disintegrate into tiny particles no larger than 0.1 micron (0.000001 mm [0.00000001 in]) in diameter. Such tiny fragments are called *colloids*, and they can be observed in suspension in the soil solution, making it look turbid.

The colloid surface is electrically charged, attracting other particles with an opposite charge. These other particles also result from the breakup of humus and minerals, but not into colloidal fragments. Rather, they result when mineral compounds dissolve in the soil water and form atoms or groups of atoms called *ions*. Calcite ($CaCO_3$), for example, breaks up into atoms of calcium and atoms of carbonate.

The ions resulting from the decomposition of minerals also are electrically charged. Positively charged ions, such as those of calcium (Ca^{2+}), are called *cations*. Negatively charged ions, such as the carbonates, nitrates, and phosphates, are known as *anions*.

Now imagine a colloid particle with its negatively charged surface: it will attract (depending on its size) many positively charged ions. These cations come from plant nutrients such as calcium, potassium, magnesium, and sodium. The colloids hold these cations in the soil solution, but let them go when a stronger attraction pulls them into the roots of plants. So the colloids function to make the nutrient-providing ions available to the plants growing in the soil. Without them, these nutrients would be quickly washed away.

pulverized by glaciers, contain lime; so do sediments of the Atlantic coastal plain. When rainwater percolates downward through such lime-containing soils, the $CaCO_3$ is dissolved in the water and carried away in a process called **leaching**. This eventually makes the topsoil acidic, which in turn stimulates mineral weathering. The result of this weathering is the release of many fine particles—ions—of nutrients, which enter the soil solution and greatly contribute to soil fertility (*see* Perspective: Exchange of Ions).

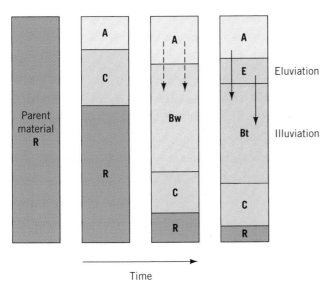

FIGURE 23-6 The stages of soil horizon evolution on a sedimentary parent material.

Another effect is the formation of microscopic clay particles. These clay particles, translocated by eluviation and illuviation from the **A** horizon into the **B** horizon, signal the maturing of the **Bt** horizon. The illuvial accumulation of clay particles transforms a weakly developed **Bw** horizon into a more fully developed **Bt** horizon.

What happens when the parent material contains little or no clay and there is little lime to set the clay-producing process in motion? There will be little clay illuviation into the **B** horizon, and no **Bt** horizon will develop. However, the decaying organic materials in the **O** horizon will acidify rainwater, causing chemical reactions with aluminum (Al) and iron (Fe) in the upper soil. Oxides of Al and Fe are then illuviated into the **B** horizon, creating the red-colored **Bhs** horizon of so many tropical and equatorial soils (**h** for humus; **s** for sesquioxides of aluminum and iron). In the United States, soils with **Bhs** horizons occur in forested areas from Maine to Florida, and they are common in the tropics. Soils with such **B** horizons tend to be of low fertility and low water-retention capacity.

The **C** horizon is the soil layer in which parent material is transformed by weathering into soil particles. The parent materials of soil range widely, of course, from already pulverized glacial sediments and river-deposited alluvium to hard bedrock. Figure 23-5 assumes a hard-bedrock parent material. Accordingly, the **C** horizon is divided into a lower zone (C_2) where pieces of bedrock still lie interspersed with weathered soil material, and an upper zone (C_1) where the weathering process is more advanced.

The **R** horizon denotes the regolith, where solid rock is first affected by soil formation. Here, cracks and other zones of weakness in the bedrock are loosened and opened, and the transformation process is in its beginning stages.

SOIL REGIMES

The soil profile represented in Fig. 23-5 is only one of thousands of such profiles charted by soil scientists around the world. Imagine the range of possibilities: take the complex pattern of global geology, superimpose a map of climates much more intricate than the Köppen regionalization we studied, overlay this with the diversity of biogeography, add local variations of relief and slope, and insert the factor of time. The resulting soil map is infinitely complicated, and understanding it is quite a challenge.

Soil geographers like to use the notion of **soil regimes** to help us comprehend spatial patterns of soil formation. Even if their parent material remained constant throughout the world, soils would differ because they would form under different temperature, moisture, biogeographic, and other conditions or regimes. We can imagine certain of these regimes by recalling some information from Part Two. Among other things, we learned that soils may form under continuously moist, seasonally moist, or dry conditions, and under continuously warm, seasonally warm, continuously cool, seasonally cold, or continuously frigid conditions—or at any point in between. Early on in the study of soils, it became clear that certain soils with particular characteristics prevailed under certain regimes: calcium-rich soils under arid regimes in lower latitudes; silica-rich soils under somewhat moister but much cooler regimes in higher latitudes; and iron- and aluminum-rich soils under moist, warm regimes in tropical areas.

From soils developed under such contrasting regimes, soil scientists learned how soil-forming processes varied in intensity from region to region. Under lower-latitude arid regimes, eluviation is constrained by the limited amount of water available for downward percolation, and calcium and magnesium remain in the upper soil, giving it a characteristic light color (Fig. 23-7). Soils that develop under such regimes are appropriately called *aridisols*.

Soils in higher latitudes, where it is moister than in the subtropical arid zones but also cooler, reveal that a different process prevails there. Under a vegetative cover of pine trees, the soil acquires an **O** horizon made up of a mat of pine needles. As these pine needles decay, they produce a mild organic acid; when rainwater or meltwater seeps through the **O** horizon, it too becomes slightly acidic. This mild acid solution now enters the **A** horizon, dissolving most of its minerals and eluviating

FIGURE 23-7 Calcified soils, characteristically light in color, dominate California's Mojave Desert.

FIGURE 23-8 The gray, ash-like color of the **A** horizon is the signature of the soil type classified as *spodosol*.

them downward—except silica, which will not dissolve. Left behind in the **A** horizon, the silica colors the topsoil a light gray, ash-like color (Fig. 23-8). The Greek word for "wood ash" gave these soils their modern name: *spodosols*.

Quite different soil regimes prevail in moist equatorial and tropical areas, where temperatures are high and moisture is abundant. Bacterial action is rapid, so that organic matter is destroyed quickly, leaving little or no humus. Because there is little humus, not much acid is produced. The water that percolates downward leaches the soil (removes materials in solution), but it leaves behind compounds of iron and aluminum. The result is an often very thick soil poor in nutrient materials but rich in iron and aluminum, which color it with a characteristic reddish tinge. Sometimes the iron and aluminum oxides are so concentrated that they develop into a hard layer called an *oxic* horizon, described by farmers as *hardpan*. This layer can form a severe obstacle to cultivation already made difficult by the infertility of these aptly named *oxisols*.

These are just three combinations of regimes producing characteristic soils—three among thousands. Before we try to make sense of the global and regional distribution of soils (Unit 25), we must examine the more important physical properties that mark them (Unit 24).

KEY TERMS

addition (p. 253)

depletion (p. 253)

eluviation (p. 254)

humus (p. 252)

illuviation (p. 254)

leaching (p. 255)

nonrenewable resources (p. 250)

parent material (p. 251)

renewable resources (p. 250)

residual soil (p. 251)

soil (p. 249)

soil horizon (p. 253)

soil profile (p. 253)

soil regime (p. 256)

transformation (p. 253)

translocation (p. 253)

transported soil (p. 251)

REVIEW QUESTIONS

1. What are the four primary soil components?

2. What are the five major factors in soil formation?

3. What are the four processes of soil formation?

4. Describe the dominant characteristics of the **O, A, B, C,** and **R** soil horizons.

REFERENCES AND FURTHER READINGS

BIRKELAND, P. W. *Soils and Geomorphology* (London/New York: Oxford University Press, 2nd ed., 1984).

BUOL, S. W., et al. *Soil Genesis and Classification* (Ames, Iowa: Iowa State University Press, 3rd ed., 1989).

COURTNEY, F. M., and TRUDGILL, S. A. *The Soil: An Introduction to Soil Study* (London: Edward Arnold, 2nd ed., 1984).

DANIELS, R. B., and HAMMER, R. D. *Soil Geomorphology* (New York: Wiley, 1992).

FANNING, D. S., and FANNING, M.C.B. *Soil: Morphology, Genesis, and Classification* (New York: Wiley, 1989).

FITZPATRICK, E. A. *Soils: Their Formation, Classification, and Distribution* (London/New York: Longman, 1983).

FOTH, H. D. *Fundamentals of Soil Science* (New York: Wiley, 8th ed., 1990).

GERRARD, A. J. *Soils and Landforms: An Integration of Geo-morphology and Pedology* (Winchester, Mass.: Allen & Unwin, 1981).

JENNY, H. *The Soil Resource: Origin and Behavior* (New York/Berlin: Springer-Verlag, 1981).

ROSS, S. *Soil Processes: A Systematic Approach* (London/New York: Routledge, 1989).

ROWELL, D. L. *Soil Science: Methods and Applications* (New York: Wiley/Longman, 1994).

SIMONSON, R. W. "Outline of Generalized Theory of Soil Genesis," *Soil Science Society of America, Proceedings*, Vol. 23, 1959, 152–156.

SOIL SCIENCE SOCIETY OF AMERICA. *Glossary of Soil Science Terms* (Madison, Wis.: Soil Science Society of America, 1984).

STEILA, D., and POND, T. E. *The Geography of Soils: Formation, Distribution, and Management* (Totowa, N.J.: Rowman & Littlefield, 2nd ed., 1989).

Physical Properties of Soil

OBJECTIVES

◆ To introduce terminology used to describe soil characteristics.

◆ To define some important properties that arise out of a soil's physical characteristics.

◆ To illustrate the likely arrangement of soil characteristics in a hypothetical landscape.

The next time you take a daytime highway trip of any substantial length, you can make it much more interesting by taking time to stop at some road cuts to examine the exposed rock and soil. In Parts Four and Five we discuss some of the rock types and structures you may be able to see; if you are lucky, you might even find an unusual mineral or fossil. But even the soil alone makes a stop worthwhile. Especially in relatively fresh road cuts, you may be able to see several soil horizons, regolith, humus, plant roots, even the imprints made by worms and other inhabitants of the soil. However, do not expect to be able to recognize all the soil properties the road cut reveals. It is one thing to understand a soil profile from a textbook, but in the field

those well-defined horizons might not be so evident. But it does help to know some basic terminology, and you can measure some of the soil properties yourself.

If the road cuts through a hill, you may note that the soil's depth at the top of the hill is less (say 1 m or 3.3 ft) than halfway down the slope. Much of the difference may come from the **A** horizon, which is especially exposed to erosion on top of the hill.

UNIT OPENING PHOTO: The fertile wheatfields of the Texas Panhandle near the southwestern edge of the Great Plains' most productive grain farming zone.

SOL AND *PED*

The terms *sol* and *ped* appear frequently in soil study, either alone or in some combined form. We already have encountered both, as in oxisol and, of course, pedology, the science of soils. In examining the classification and regionalization of soils in Unit 25, we identify soil orders by means of a dominant characteristic followed by *sol*. For example, soils forming in dry areas are known as *aridisols*: *arid* + *sol*. Why sol and not soil? The pedologists who devised the classification wanted it to have global application, and Russian scientists were among the world's leading experts in this area. (The Russian word for soil is *sol*.)

Thus, when a comprehensive soil classification was established, the suffix *sol* was adopted. It also appears in the term *solum*. The **solum** of a soil consists of the **A** and **B** horizons, and constitutes that part of the soil in which plant roots are active and play a role in the soil's development. Below the solum, in the **C** horizon, parent material is being weathered. Therefore, when it is reported that "the solum is 1.5 m (5 ft) thick," the reference is to the zone where all the interacting processes of plant life and soil development are taking place.

The *ped* in pedology also appears in the term **pedon**, a column of soil drawn from a specific location, extending from the **O** horizon (if it is present) all the way down to the level where the bedrock shows signs of being transformed into **C** horizon material (Fig. 24-1). In other words, a pedon is a soil column representing the entire soil profile. The term *ped* also is used by itself to identify a naturally occurring aggregate or "clump" of

soil and its properties. We defer discussion of this topic until later in the unit, when we are in a better position to understand why soils exhibit this property of forming natural peds.

SOIL TEXTURE

Descriptions of soils often contain terms such as "sandy clay," "loam," or "silt." These terms are not just general descriptions of the character of soils. In fact, they have very specific meaning and refer to the sizes of the individual particles that make up a soil (or one of its horizons). If you were to rub a tiny clump of soil between your thumb and forefinger, you would be left with the smallest grains that make up that part of the soil from which the clump came. These grains may be quite coarse and look or feel like sand, or they may be very fine, like dust on your fingertip. The size of the particles in soil, or its *texture*, is very important, because it has to do with the closeness with which soil particles can be packed together, the amount of space there is in the soil for air and water, how easily roots can penetrate, and other aspects of its behavior and performance.

Soils, of course, often exhibit several kinds of textures. For instance, that clump on your forefinger may contain some sandy and some much finer particles. In your field notes, therefore, you might call it a *sandy clay*—a mostly fine-grained soil, but with some coarser particles in it. The coarsest grains in a soil are *sand* (not counting even larger gravel, which is sometimes found in soil as well). Sand particles, by the official definition of the United States Department of Agriculture (USDA), range in size from 2 down to 0.05 mm (0.08 to 0.002 in). The next smaller particles are called *silt* (0.5 to 0.002 mm [0.002 to 0.00008 in]). There are still smaller grains than silt, namely, the *clay* particles. The smallest of the clay particles are in the colloidal range and are less than one hundred-thousandth of a millimeter in diameter.

A single soil may contain grains of all three size categories—sand, silt, and clay. Such a soil is called a **loam** if all three are present within specific proportions (Fig. 24-2). Unlike sand, silt, and clay, therefore, the term *loam* refers not to a size category but to a certain combination of variously sized particles. The USDA has established a standard system to ensure that such terms as "sandy clay" and "silt loam" have more than a subjective meaning.

When we rub some soil between thumb and forefinger, we can tell that there are particles of different sizes in the clump—but we can only estimate in what proportions. Accurate determination of these proportions requires additional analysis, and then the numerical results must be checked against the official USDA graph (Fig. 24-2). Note that this triangular chart shows the percentages of all three components. A soil that is about

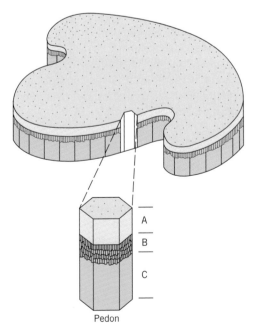

FIGURE 24-1 A complete soil column or pedon.

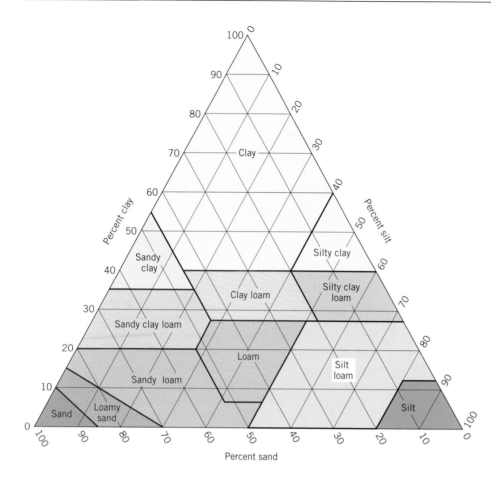
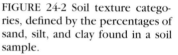

FIGURE 24-2 Soil texture categories, defined by the percentages of sand, silt, and clay found in a soil sample.

one-third sand, one-third silt, and one-third clay falls in the clay loam area. To be called a silty clay, a typical soil would have 45 percent silt, 45 percent clay, and 10 percent sand. You can choose other combinations yourself and see what they are called using the chart.

Soil texture is related to the parent material from which the soil was derived. Some types of bedrock yield sand-rich soils, while others give rise to clayey soils. Most soils contain some combination of components. As we noted earlier, texture is the critical factor determining the pore spaces in soil, and hence its capacity to hold water. This is not difficult to imagine: the very term *clay* seems to imply a waterlogged soil, and sand is usually dry and light. Thus a sandy soil allows water to percolate downward under the influence of gravity, draining (and drying up) rapidly. Plants with roots in sandy soils do not have much opportunity to absorb water, because the soil is quickly drained.

Clay, on the other hand, has a far greater **field capacity** (ability to hold water against the downward pull of gravity). But this characteristic produces a different problem for plants: the pore spaces in clay are so small that drainage is poor, reducing the circulation of nutrient- and oxygen-carrying solutions. Furthermore, the close packing of clay particles may make it difficult for plant roots to penetrate deeply into the soil. Such texture-related properties must be considered when farmers plant crops. The potato plant, for example, handles the

wetness and compactness of clayey soils well, but wheat should not be sown in these soils.

From what we have just learned, it would appear that loams present the best combination of textural properties. Indeed, that is the case. Loams do not stay waterlogged, nor do they yield their water content too rapidly. Pore spaces are large enough to let plant roots find their way downward. Good drainage, which is directly related to soil texture, is a key to successful crop cultivation. Indeed, many farmers say it is more important than nutrient content. Nutrients can be supplemented artificially, but soil texture cannot easily be changed.

SOIL STRUCTURE

Earlier we referred to so-called *peds*, naturally occurring clumps of soil that tend to form and stay together unless they are purposely broken up. Again, you will discern this tendency in soil when you examine it: dislodge a bit of soil and it will not disintegrate, like loose sand, but form small clumps. Only when you rub these peds in your hand will they break up into the individual grains described previously. Peds develop because soil particles are sometimes held together by a thin film of clay, which is deposited during soil formation by circulating solutions. Other peds may develop because of molecular attraction among the particles they contain. Either way,

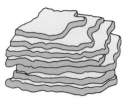

a

b

FIGURE 24-3 Platy soil structure.

peds give soil its structure, and they are quite important because they affect the circulation of water and air throughout the soil. Soil structure, that is, the nature of its peds, affects the soil's vulnerability to erosion, its behavior under cultivation, and its durability during dry periods when cohesive forces weaken.

Soils exhibit four basic structures: *platy, prismatic, blocky,* and *spheroidal* (*see* Figs. 24-3 through 24-6).

1. **Platy structure** (Fig. 24-3), as the term suggests, involves layered peds that look like flakes stacked horizontally. If a soil you were examining had platy structure, you would recognize it immediately because the individual plates often are as much as 1 to 2 cm (0.4 to 0.8 in) across, and occasionally even larger.

2. **Prismatic structure** (Fig. 24-4) reveals peds arranged in columns, giving the soil vertical strength. In Unit 49, we discuss a wind-deposited material called *loess* on which very fertile soils develop. In their deep **B** horizons, loess soils have a well-developed prismatic structure and can form, without collapsing, bluffs many meters high. Individual peds in soils with prismatic structure range from 0.5 cm (0.2 in) to as large as 10 cm (4 in).

3. **Blocky** or **angular structure** (Fig. 24-5) consists of irregularly shaped peds. These peds, however, have straight sides that fit against the flat surfaces of adjacent peds, giving the soil considerable strength.

b

a

b

FIGURE 24-4 Prismatic soil structure.

b

FIGURE 24-5 Blocky (angular) soil structure.

a

b

FIGURE 24-6 Spheroidal (granular) soil structure.

4. **Spheroidal** or **granular structure** (Fig. 24-6) displays peds that are usually very small and often nearly round in shape, so that the soil looks like a layer of bread crumbs. Such soils are very porous, and with the peds so small and cohesion very weak, they are more susceptible to erosion.

As we have noted, not only the shape of the peds (i.e., the structure alone) but also their size is important. Descriptions of soil structure include observations on whether the peds are coarse, medium, or fine. There are no hard rules governing these size categories, but previously we have referred to dimensions that range from a fraction of 1 cm to 10 cm (note the scale of each of the photographs in Figs. 24-3 through 24-6).

Another indicator of soil properties is what pedologists call *soil consistence*. This is a rather subjective measure of a moist or wet soil's stickiness, plasticity, cementation, and hardness. It is a test done in the field by rolling some moist soil in the hand and observing its behavior. After subjecting a bit of soil to this test, some of it will have stuck to the skin (indicating its stickiness). The rolled-up soil may form a small rope and then break up, or it may attain a thin, twine-like shape, as moist clay would. This reveals its plasticity. The greater the clay content, the more tightly the soil particles bind together because of the cohesiveness of clay, and the longer the rope- or worm-like roll will be. Conversely, the greater

the sand content, the more likely the soil is to crumble and fall apart. Soil hardness often varies along its profile as well. The soil may crumble easily in the **A** horizon, but parts of the **B** horizon may resist even a knife. All these attributes relate to soil consistence.

SOIL COLOR

Soils generally exhibit a range of colors. Only rarely are soils encountered that do not display some color variation along their profile. Even a typical equatorial or tropical soil, dominated by the rusty redness of iron and aluminum oxides, exhibits lighter and darker shades of red and orange (*see* Fig. 25-12). Soil color is a most useful indicator of the processes that prevail in its maturation. A dark brown-to-black upper layer reflects the presence of the organic material, humus. A soil's color may range from nearly black at the top of the profile to brown in the lower **A** and **B** horizons to beige in the lower **B** horizon—indicating the decrease in humus content with depth.

Soil color tends to change with the degree of wetness. Wetting causes soil colors to become much more vivid. Obviously, the range of hues in soils is almost infinite, and soil scientists use the Munsell Soil Color Chart to describe soil colors objectively. This chart, which takes into account many possible conditions, contains hundreds of colors, each with a letter and number code. The color of a sample of soil can therefore be codified. If the job is done carefully, we can be certain that a coded soil from the **B** horizon of a U.S. Great Plains pedon is exactly the same color as one from the **B** horizon in Ukraine without having to put the samples side by side.

SOIL ACIDITY AND ALKALINITY

Soil colloids are associated with the presence of ions in the soil (*see* Perspective box in Unit 23). Hydrogen (H^+) cations are very common, and their dominance in the soil solution defines an *acid* condition. Conversely, a relative absence of H^+ cations and the presence of hydroxyl (OH^-) anions plus sodium (Na^+) and other associated cations make the soil *alkaline* (or "basic") in nature. The acidity of soil is measured by its pH value. A pH value of 7.0 is considered neutral, which occurs when H^+ and OH^- ions are balanced and present at relatively low levels. Lower values (normally between 4.0 and 7.0) indicate acidic soils, whereas higher values (7.0 to 11.0) indicate alkaline soils.

The acidity of a soil is closely related to its fertility because acids are necessary to make nutrients available to plants. However, extreme acidity or alkalinity is detrimental to plant growth. In dry climate regimes, where

Fertilizer

The global population explosion raises a painful question—can we feed all of these nearly 6 billion people? Our limited amount of agricultural land must produce more and more food. One way to meet the challenge is to use fertilizer, which enriches the existing soil and produces much larger harvests. Some eastern Native Americans increased soil fertility by planting a fish with each seed. In other cultures, farmers apply organic fertilizer—primarily manure. But organic fertilizer cannot meet our needs. Today, over 50 million tons of nutrients are added to the soil each year. Chemical fertilizers supply the primary nutrients: nitrogen, phosphorus, and potassium.

Although pesticides, new hybrid seeds, and mechanization contribute to higher crop yields, fertilizers bring the most substantial rewards. For each kilogram (2.2 lb) of fertilizer a farmer spreads on the soil, 10 extra kilograms of grain may be harvested. By using the proper combination of nutrients to complement those available in the soil, a farmer may grow 2 or 3 times as many crops. By 2000, we may need 120 million tons of nutrients to support the world's 6.2 billion people. This means that one 45-kg (100-lb) bag of chemical fertilizer will be used for each person on earth.

Understanding the composition of the soil will allow us to fertilize more ef-ficiently, but there are some undesirable side effects. These include nitrate contamination of groundwater and accelerated *eutrophication* (excessive growth of organic matter through overfertilization) of water bodies, especially lakes, resulting from surface runoff containing nitrogen and phosphorus. Animal feedlot runoff, dairies, and sewage effluents also contribute to eutrophication. Over-enrichment of nutrients and excessive algae blooms signal the rapid aging, ecological degradation, and eventual demise of affected water bodies.

---

alkaline soils often occur, one remedy is to treat them with compounds containing sulfur. Conversely, the most common treatment for too much acidity is to apply lime to the soil. (These artificial manipulations of soil quality, of course, remind us that fertilizers have long been an integral part of the human use of the earth [*see* Perspective: Fertilizer].) Different plants and microorganisms are adapted to varying degrees of acidity. The variation of acidity often bears a relationship to both climate and parent material, and is associated with the different soil-forming processes.

SOILS OF HILLS AND VALLEYS

Topography strongly influences soil formation. On gently undulating (rolling) countryside, soil profiles tend to develop fully and would look much like the profile examined in Fig. 23-5. When the landscape flattens out, the soil reflects this by developing a thick **B** horizon. A flat surface promotes leaching, and we would probably encounter a dense clay layer in the **B** horizon. When the landscape becomes hilly, soils tend to become thinner. As a rule, hilltop soils have a thin **A** horizon, and the **B** horizon will be shallow as well. Rapid draining and exposure to surface erosion inhibit the development of mountaintop soils. Where the land is poorly drained, as in meadows and bogs, the soil profile may show a lack of contrast between **A** and **B** horizons. The regional land-

forms, therefore, are an important guide to what we should expect to find in the soils.

On hillsides in southern Sudan in Africa, as in many other parts of the world, soils have a characteristic arrangement from the top to the bottom of the slopes. As Fig. 24-7 shows, lateritic soils and a capping of hard *laterite* (the name sometimes given to a very hard iron oxide horizon) are found on the top of the hill. Some of the soil particles washed downhill during overland flow and sheet erosion come to rest farther down the slope. They form a material known as *colluvium*, and soils that develop from this are called colluvial soils. Most of the ma-

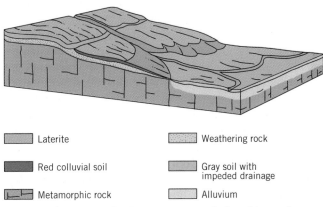

Laterite	Weathering rock
Red colluvial soil	Gray soil with impeded drainage
Metamorphic rock	Alluvium

FIGURE 24-7 A type of soil catena commonly found in southern Sudan.

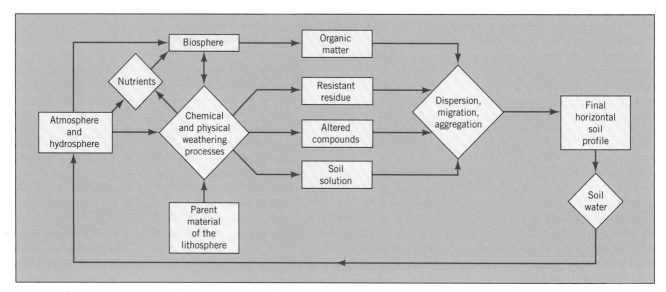

FIGURE 24-8 The structure and flows of the soil-development system.

terial brought down valleys by river water ends up as *alluvium* on the valley floor, and alluvial soils may develop from it (these are often poorly drained). When the same parent material results in an arrangement of different soil types, say along a hillside, this is called a soil **catena**. This term (pronounced "kuh-TEENA") comes from a Latin word meaning chain or series. A catena is usually defined as a sequence of soil profiles appearing in regular succession on landform features of uniform rock type.

THE SOIL-DEVELOPMENT SYSTEM

We are now close to putting together all the factors relating to soil development. They will come together to form a coherent picture if the total process and the soil profile are regarded as a system, as shown in Fig. 24-8. An overall view of this system reinforces the statement that the soil serves as an interface among the atmosphere, biosphere, lithosphere, and hydrosphere.

These four main spheres are the starting point of the soil-development system. The atmosphere and hydrosphere provide heat and moisture, while the lithosphere and biosphere furnish the materials. Then a variety of chemical and physical weathering processes break down and transform the parent material and organic matter. The biosphere provides not only organic material, but also new chemicals, especially acids, resulting from the breakdown of this material. These can contribute to further weathering of the inorganic matter of the lithosphere. At the same time, the chemical and physical weathering processes can release nutrients to the plants of the biosphere. The interaction between the biosphere

and the soil of the lithosphere is therefore both reciprocal and vital.

As a result of the weathering process and the decay of the plants of the biosphere, the soil system ends up with four ingredients. First, there is *organic* matter. Second there is the *resistant residue* that cannot be altered in any way by the weathering processes; this often takes the form of silica, such as the quartz particles we find in most soils, particularly sandy soils. Third, there is a whole host of newly *altered chemical compounds*, such as oxides and carbonates. These include the various clay minerals, which by ion exchange react with the fourth component, the *soil solution*. The soil solution contains many of the minerals extracted from the original parent material.

These four components of the soil system are then subjected to various processes of dispersion, migration, and aggregation. Some of the more important of these processes are the flow of water through the soil under the influence of gravity, the upward movement of water by capillary action, and the evapotranspiration of water from the soil. Water is clearly important. Although the diagram does not include such events as the carbon dioxide and nitrogen cycles (*see* Unit 6), the groundwater part of the hydrologic cycle is illustrated. The final result of these and many more processes is usually the soil profile, with its distinctive horizons.

We need all our powers of observation and deduction to gain some understanding of soils, because all parts of the soil-development system often operate simultaneously. Soils are a particularly dynamic component of the physical world, and they also exhibit important spatial variations. These are considered in Unit 25, which focuses on the geography of soils.

KEY TERMS

blocky (angular) structure (p. 261)

catena (p. 264)

field capacity (p. 260)

loam (p. 259)

pedon (p. 259)

platy structure (p. 261)

prismatic structure (p. 261)

solum (p. 259)

spheroidal (granular) structure (p. 262)

REVIEW QUESTIONS

1. What is meant by the terms *sol* and *ped*?

2. Identify the size ranges for grains of sand, silt, and clay.

3. Define *field capacity*.

4. What is the difference between an acid soil solution and an alkaline soil solution?

5. What is a soil *catena*?

6. What is fertilizer, and why is it so highly prized by farmers the world over?

REFERENCES AND FURTHER READINGS

BIRKELAND, P. W. *Soils and Geomorphology* (London/New York: Oxford University Press, 2nd ed., 1984).

BRADY, N. C. *The Nature and Properties of Soils* (New York: Macmillan, 9th ed., 1984).

COURTNEY, F. M., and TRUDGILL, S. A. *The Soil: An Introduction to Soil Study* (London: Edward Arnold, 2nd ed., 1984).

FANNING, D. S., and FANNING, M.C.B. *Soil: Morphology, Genesis, and Classification* (New York: Wiley, 1989).

FOTH, H. D. *Fundamentals of Soil Science* (New York: Wiley, 8th ed., 1990).

GERRARD, A. J. *Soils and Landforms: An Integration of Geomorphology and Pedology* (Winchester, Mass.: Allen & Unwin, 1981).

PITTY, A. F. *Geography and Soil Properties* (London: Methuen, 1978).

ROSS, S. *Soil Processes: A Systematic Approach* (London/New York: Routledge, 1989).

ROWELL, D. L. *Soil Science: Methods and Applications* (New York: Wiley/Longman, 1994).

THOMPSON, L. M., and TROEH, F. R. *Soils and Soil Fertility* (New York: McGraw-Hill, 4th ed., 1978).

U.S. GOVERNMENT PRINTING OFFICE. *Soils and Men: Yearbook of Agriculture, 1938* (Washington, D.C.: U.S. Department of Agriculture, 1938).

WHITE, R. E. *Introduction to the Principles and Practice of Soil Science* (New York: Wiley, 2nd ed., 1987).

Classification and Mapping of Soils

OBJECTIVES

◆ To present a brief history of pedology and highlight the difficulties of deriving soil classification schemes.

◆ To outline the current system of soil classification, the Soil Taxonomy.

◆ To survey the 11 Soil Orders in the Soil Taxonomy and examine their regional patterns on the world map.

I n this unit, we focus on the geographical perspective in order to discover how soils are distributed across the United States and the rest of the world. Unit 16 points out the problems associated with the classification of climates and their spatial representation. It is necessary to establish criteria to distinguish climate types from one another—criteria not ordained by nature but established by scientists. Certainly there are justifications for those lines in Fig. 16-3 (pp. 180–181): climatologists found transition zones between different climate regimes that also were marked by vegetation changes and often by other modifications as well.

As noted at the beginning of Part Three, the map of world climates keeps reemerging as we discuss soils, vegetation, and animal life on the earth's surface. In many respects, classifying and regionalizing soils is even more difficult than classifying climates. A small area may contain a bewildering variety of soils of different profiles, thicknesses, textures, and structures. For more than a century, pedologists have been working to devise an ac-

ceptable system of classifying soils on which a map of world soil distribution could be based.

CLASSIFYING SOILS

The study of soils has for more than a century been dominated by Russian and American scientists. Up to about 1850, soils were believed to be simply weathered versions of the underlying bedrock. The Russian scholar Vasily V. Dokuchayev (1846–1903) was the first pedologist to demonstrate that soils of the same parent material develop differently under different environmental conditions. Dokuchayev began an elaborate description of Russian soils based on their profiles, and after 1870 he and his colleagues produced several successive soil classifications. In 1914, Dokuchayev's student Konstantin D.

UNIT OPENING PHOTO: A boundary between two soil regions marks the edges of the irrigated Rio Sumbay Valley in Arequipa Province on the dry coastal plain of southernmost Peru.

Glinka (1867–1927) published a book, which was translated into English, that summarized the achievements of the school of thought Dokuchayev had started. This work made a great impression on soil scientists in the United States, who had been grappling with many of the same problems the Russians had faced.

During the first half of the twentieth century, the most productive American soil scientist undoubtedly was Curtis F. Marbut (1863–1935). He laid the groundwork for the first genetic soil classification, in which he used Russian ideas and Russian terminology (hence the large number of Russian terms in American pedology). Marbut was director of the Soil Survey Division of the U.S. Department of Agriculture (USDA), and the system he and his colleagues devised was first published in 1938. The Marbut System, as it became known, was frequently revised and modified during the 1940s. But as time went on, it became clear that the Marbut System had fundamental flaws that no amount of modification would solve.

Soil scientists were troubled by a number of points. Many soils have been altered from their natural state by either agricultural practices or other events. Soil scientists felt that the Marbut System laid too much emphasis on the original soil-forming factors. In the final analysis, it was the soils—not their formation factors—that were to be classified. Thus classification criteria must be stated in terms of the actual soils and not in terms of, say, climate. They also argued that the origin of a soil was sometimes unknown and that such a soil clearly could not be classified genetically. The effort to build a new system was begun during the 1950s by the USDA's Soil Conservation Service. The Soil Survey staff had assembled an enormous mass of data on soils in the United States, and now an attempt was made to create the first all-encompassing classification system based on the soils themselves. This initial effort was followed by a series of revisions until, in 1960, the seventh version of it was approved and adopted. This came to be called the Seventh Approximation, shorthand for the *Comprehensive Soil Classification System (CSCS)*.

THE SOIL TAXONOMY

Even this version, however, was not the last word on soil classification. The CSCS grouped the earth's soils into six categories, identifying 10 Soil Orders, 47 Suborders, 185 Great Groups, nearly 1000 Subgroups, 5000 Families, and more than 10,000 Series. But even as this scheme came into general use, it was modified further. Soil scientists kept identifying additional Series, and it became necessary to expand the number of Families and Subgroups too. Before long, the CSCS was no longer a "seventh approximation," and the term became obsolete as further revision continued.

After 1975, the name *CSCS* also began to fade from use as soil scientists increasingly came to prefer the simpler term **Soil Taxonomy**. By 1990, the number of recognized Series had grown to about 16,000 (10,000 in the United States alone!). While heightened knowledge understandably expanded the number of lower order Great Groups, Subgroups, and Series, few pedologists anticipated that even the 10 primary Soil Orders of the old CSCS would have to be modified. But in 1990 this did happen, and an eleventh Soil Order (the andisols) was added. The current categorical organization of the Soil Taxonomy is shown in Table 25-1.

For our purposes we need consider only the 11 Soil Orders, three of which (aridisols, spodosols, and oxisols) were already encountered in Unit 23. In the Soil Taxonomy, a **Soil Order** is a very general grouping of soils with broadly similar composition, the presence or absence of specific diagnostic horizons, and similar degrees of horizon development, weathering, and leaching. Our objective now is to study the general distribution of these Soil Orders.

Soil Distribution on a Hypothetical Continent

When we discussed the world distribution of climates, we postulated a "model continent" and mapped the ex-

TABLE 25-1 ORGANIZATION OF THE SOIL TAXONOMY

LEVEL	DESCRIPTION
Order (11)	The most general class. Soils of a given Order have a similar degree of horizon development, degree of weathering or leaching, gross composition, and presence or absence of specific diagnostic horizons, such as an oxic horizon in an oxisol.
Suborder (53)	Distinguished by the chemical and physical properties of the soil. At this level formative and environmental factors are taken into account.
Great Group (230)	Distinguished by the kind, array, or absence of diagnostic horizons.
Subgroup (about 1000)	Determined by the extent of development or deviation from the major characteristics of the Great Groups.
Family (about 5000)	Soil texture, mineral composition, temperature, and chemistry are distinguishing features.
Series (about 16,000)	A collection of individual soils that might vary only in such items as slope, stoniness, or depth to bedrock.

Source: U.S. Department of Agriculture.

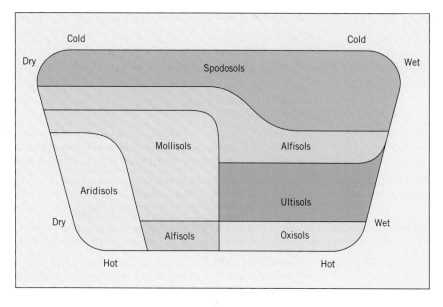

FIGURE 25-1 The distribution of Soil Order regions on a hypothetical Northern Hemisphere continent.

pected climatic distributions on it (*see* Fig. 16-2). One way to approach the Soil Taxonomy is to consider the spatial distribution of Soil Orders on another such hypothetical continent (Fig. 25-1). The mapped contents of this model should strike us as somewhat familiar because we know that climate is a soil-forming factor. Thus the southeastern region would be dominated by **mT** air masses, the southwestern by **cT** air, and so forth (*see* Fig. 13-10). Remembering what we learned about soil-forming processes, we can conclude that the warm, moist southeast would experience the kind of leaching that produces oxisols; the hot, dry southwest should be marked by aridisols; and the cool to cold north should yield a region of spodosols.

But we should be careful in drawing these conclusions. Whereas the regional climates might indicate such a distribution, parent materials could change the picture considerably. For our hypothetical continent, therefore, we need to make some far-reaching assumptions, including a uniform parent material (with calcareous and mixed mineral content and a loam texture), gently rolling relief without groundwater in the solum, and a long period of soil evolution (say, 100,000 years). Given these idealized conditions, the resulting generalized Soil Order map might look as in Fig. 25-1, but it does put our hypothetical continent at some distance from the complexities of the real world.

The Soil Orders

Before looking more closely at Fig. 25-1, let us acquaint ourselves in some detail with the properties of all 11 Soil Orders. These are listed, along with their Suborders, in Table 25-2. Notice how frequently the USDA gives climatic factors under the heading "characteristics." As you read the text that follows, you should be consulting Perspective: Soil Taxonomy—What's in a Name? (p. 271)

which explains how the names of the Soil Orders were derived. Be aware that the first five Soil Orders (entisols, histosols, vertisols, inceptisols, and andisols) involve soils that for some reason have not developed the kinds of distinct and complete horizons we associate with most soil profiles. They also may possess some particular property that makes them quite distinct. After these we come to the aridisols and thus to more familiar territory.

1. Entisols

When the **entisol** Soil Order was established, it was intended, literally, as the order that would contain all the soils that would not fit into any of the other Orders. Such soils tend to be of recent origin, often developed in unconsolidated material (alluvium, for example) but also found on hard bedrock. Entisols are even younger than inceptisols and may have a thin **A** horizon overlying a **C** or **R** horizon; otherwise, they show little development (Fig. 25-2). Because there are numerous reasons for the absence of well-developed horizons, entisols are found in many different environments. Climate, therefore, is not a strong influence in their distribution.

2. Histosols

Histosols are organic soils that are often water-saturated for most of the year. We would recognize them in such common terms as bogs, peats, and moors. Organic material tends to dominate in the **O** horizon and clay in the **C** horizon (Fig. 25-3). Histosols are unique among the Soil Orders because they can be totally destroyed over time or altered by drainage, natural or artificial. It is difficult to generalize about the geographic distribution of these soils because bogs are so widespread. Poor drainage in low or flat-lying topographic locales explains their occurrence better than climate or prevailing vegetation. When they can be drained, intensive cultivation of such crops as cabbage, carrots, potatoes, and other root crops is possible.

3. Vertisols

You may have seen areas where clay soils develop large cracks in the dry season and swell with moisture when rain returns (Fig. 25-4). In the Soil Taxonomy, these are called **vertisols**. More than 35 percent of their content is clay, which causes the swelling and cracking. The clay particles are normally derived from the parent material, so vertisols are found where the clay-producing materials are available—in mesothermal or tropical climates with periodic dry and wet seasons. Their Suborders are closely related to climatic divisions, as Table 25-2 indicates. They are most extensive in Australia, India, and Sudan. Vertisols are hard to use for most human purposes, particularly construction. When they shrink and crack, fences and telephone poles may be thrown out of line. Pavements, building foundations, and pipelines can all be damaged by the movement of these "turning" soils.

4. Inceptisols

Inceptisols take their name from the Latin word for beginning; that is, they have the beginnings of a **B** horizon. They form rather quickly but are generally older than entisols. They have a weakly developed **B** horizon noted

FIGURE 25-3 Cross-section of a histosol in Michigan. In many parts of the world, such organic soil is cut as peat to be used for fuel.

for its reddish color, which lacks strong clay development and the accumulation of other compounds. But inceptisols do contain significant amounts of organic matter and/or evidence that the parent material has been weathered to a certain extent. Inceptisols are generally found in humid climates, but they occur from the Arctic to the tropics and are often found in highland areas as

FIGURE 25-2 These sand dunes lining Oregon's Pacific coast constitute an entisol developed in recently deposited, unconsolidated materials.

FIGURE 25-4 The sequence of wetting and drying in this irrigated field creates the pattern of cracks characteristic of vertisols.

TABLE 25-2   ORDERS AND SUBORDERS OF THE SOIL TAXONOMY

ORDER	SUBORDER	CHARACTERISTICS
1. Entisol	Aquent	Shows evidence of saturation at some season
	Arent	Lacks horizons because of plowing or other human activity
	Fluvent	Formed in recent water-deposited sediments, as in floodplains
	Orthent	Occurs on recent erosional surfaces, such as high mountains
	Psamment	Occurs in sandy areas, such as sand dunes
2. Histosol	Fibrist	Occurs in poorly drained areas, slightly decomposed
	Folist	Occurs in poorly drained areas, mass of leaves in early stage of decomposition
	Hemist	Occurs in poorly drained areas, intermediate stage of decomposition
	Saparist	Occurs in poorly drained areas, highly decomposed
3. Vertisol	Torrert	Occurs in arid climates
	Udert	Occurs in humid climates
	Ustert	Occurs in monsoon climates
	Xerert	Occurs in Mediterranean climates
4. Inceptisol	Aquept	Wet with poor drainage
	Ochrept	Freely drained, light in color
	Plaggept	Has a surface layer more than 50 cm (20 in) thick resulting from human activity, such as manuring
	Tropept	Freely drained, brownish to reddish, found in the tropics
	Umbrept	Dark reddish or brownish, acidic, freely drained, organically rich
5. Andisol	Aquand	Occurs in wet areas
	Cryand	Occurs in cold areas, including high altitudes
	Torrand	Occurs in warm, arid climates
	Udand	Occurs in humid climates
	Ustand	Occurs in monsoon climates
	Vitrand	Formed in association with volcanic glass
	Xerand	Occurs in Mediterranean climates
6. Aridisol	Argid	Has an illuvial horizon where clays have accumulated to a significant extent
	Orthid	Has an altered horizon, a hard layer (called a hardpan or duripan), or an illuvial horizon of water-soluble material
7. Mollisol	Alboll	Has a surface layer that covers a white horizon from which clay and iron oxides have been removed, and a layer of clay accumulation below
	Aquoll	Saturated with water at some time during the year
	Boroll	Cool or cold, relatively freely drained
	Rendoll	Occurs in humid climates, developed from parent materials rich in calcium carbonate
	Udoll	Not dry for as much as 60 consecutive or 90 cumulative days per year
	Ustoll	Occurs in monsoon climates
	Xeroll	Occurs in Mediterranean climates
8. Alfisol	Aqualf	Periodically saturated with water
	Boralf	Freely drained, found in cool places
	Udalf	Brownish to reddish, freely drained
	Ustalf	Partly or completely dry for periods longer than 3 months
	Xeralf	Occurs in dry climates
9. Spodosol	Aquod	Associated with wetness
	Ferrod	Has an iron-enriched sesquioxide horizon
	Humod	Has a humus-enriched sesquioxide horizon
	Orthod	Has significant amounts of humus and iron in the sesquioxide horizon
10. Ultisol	Aquult	Occurs in wet places
	Humult	Freely drained, rich in humus
	Udult	Freely drained, poor in humus
	Ustult	Occurs in warm regions with high rainfall but a pronounced dry season
	Xerult	Freely drained, found in Mediterranean climates
11. Oxisol	Aquox	Formed under the influence of water
	Perox	Always moist, with a high humus content
	Torrox	Occurs in arid climates, may have formed under a different climate from that now existing in that location
	Udox	Occurs in places with a short or no dry season, other than aquoxes
	Ustox	Occurs in humid climates with at least 60 consecutive dry days per year

Sources: U.S. Department of Agriculture; Foth, pp. 281–282.

Soil Taxonomy—What's in a Name?

Science is based in part on classification systems. Biologists organize the living world into kingdoms. Chemists classify types of changes in matter and energy. Anthropologists identify types of kinship systems and family structures. New systems are constantly being put forward, yet not all survive.

What defines the difference between a useful system and one that does not stand the test of time? A useful system is flexible enough to accommodate new data and phenomena, yet rigid enough so that its underlying principles can be ap-plied again and again. Because of the increasing internationalization of science, names should have some significance in a variety of languages. And finally, a system should organize its information in ways that show useful relationships.

How does the Soil Taxonomy, which is surveyed in this unit, measure up according to these standards? When the USDA first proposed the CSCS in 1960, its major goals were: (1) to replace entirely the terms used in older soil name systems; (2) to produce words that could be remembered easily and that would suggest some of the properties of each kind of soil; and (3) to provide names that would have some meaning in all languages derived from either Latin or Greek. (The majority of modern European and American languages are so derived.) In fact, the Department called upon classical scholars from Belgium's University of Ghent as well as the University of Illinois to provide the Latin and Greek roots to match the soil characteristics. The etymology of the names finally chosen is as follows:

ORDER NAME	DERIVATION	ROOT/KEY
1. Entisol	None: coined for this purpose	rec**ent**
2. Histosol	*histos*, Greek for "tissue"	**hist**ology
3. Vertisol	*verto*, Latin for "turn"	in**vert**
4. Inceptisol	*inceptum*, Latin for "beginning"	**incept**ion
5. Andisol	*ando*, volcanic ash	**and**esite (volcanic mineral)
6. Aridisol	*aridus*, Latin for "dry"	**arid**
7. Mollisol	*mollis*, Latin for "soft"	**molli**fy
8. Alfisol	*al* from aluminum; *ferrum* is Latin for "iron"	ped**alf**er (former soil name)
9. Spodosol	*spodos*, Greek for "wood ash"	**pod**sol (Russian term)
10. Ultisol	*ultimus*, Latin for "last"	**ulti**mate
11. Oxisol	*oxide*, French for "containing oxygen"	**oxi**de

well. These soils most frequently develop under forest cover, but can be found under tundra (where they dominate) or grass (Fig. 25-5). Once more, the Suborders in Table 25-2 indicate their environmental conditions.

5. Andisols

The newly added Soil Order of **andisols** was established to include certain parent-material-controlled soils, notably those developed on volcanic ash. For example, the Suborder of andepts was moved out of the inceptisol Order to become part of the new andisol Soil Order. Altogether, as Table 25-2 shows, the andisols can be differentiated into no fewer than seven Suborders. These weakly developed soils lie principally in the "Pacific Ring

FIGURE 25-5 Cross-section of an inceptisol, exhibiting physical features described in the text.

FIGURE 25-6 Cross-section of an andisol atop layered volcanic ash on the island of Hawaii.

of Fire" (*see* Unit 32), Hawaii, and other volcanic zones (Fig. 25-6). They occupy less than 1 percent of the world's land surface, and are so finely distributed that they cannot be seen on the small-scale maps employed later in this unit. Andisols also contain much organic matter and hold water well; they are quite fertile.

None of these first five Soil Orders can be specified geographically with precision, which is why they do not appear in Fig. 25-1. As Fig. 25-7 indicates, they are primarily related to earth spheres other than the atmosphere. But the remaining six Orders are definitely associated with particular climates. This is especially clear in the soils of dry areas of the world, which we consider next.

6. Aridisols

Aridisols cover a larger area of the world's land surface (19.2 percent) than any other soil. These soils are usually

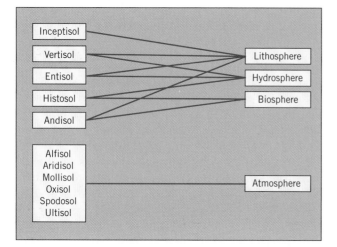

FIGURE 25-7 Primary relationships among the four spheres of the earth system and the Soil Orders of the Soil Taxonomy.

dry unless they are irrigated, whether naturally or artificially. They usually have a thin, light-colored horizon at the surface that is low in organic carbon. Moreover, these soils often contain horizons rich in calcium, clay, gypsum, or salt minerals, as shown in Fig. 25-8. Table 25-2 describes the two Suborders of aridisols. Large expanses of aridisols are found in deserts such as northern Africa's Sahara and eastern Asia's Gobi. Areas with these soils can be used for grazing or intensive production of crops with the aid of irrigation. Desert shrubs and grasses are the main form of vegetation, and overgrazing of the land is often a serious problem. The soil-forming process of calcification is common in aridisols but is even more pronounced in the next Order.

7. Mollisols

You can find **mollisols** in climates that normally have dry seasons, but temperatures can range from microthermal to tropical. Rainfall may be sufficient to leach these soils, but calcification is far more common. Mollisols are the soils of the steppes, the grass-covered plains of south-central Russia and the U.S. Great Plains that lend their name to the **BS** climatic zone. The dominant attribute of a mollisol is a thick, dark surface layer (Fig. 25-9), high in alkaline content and at least 50 percent saturated

FIGURE 25-8 Cross-section of an aridisol adjacent to Mormon Mesa, Nevada.

FIGURE 25-9 Cross-section of a mollisol, the dominant soil that underlies the grasslands of the Great Plains.

between the moist equatorial region and the Sahara and Kalahari Deserts of the subtropical latitudes.

9. Spodosols

Soils that result when organic soil acids associated with pine-needle decay cause the depletion of most **A** horizon minerals, like the one in Fig. 23-8 (p. 256), are **spodosols**. Such soils develop mainly in northerly Northern Hemisphere latitudes. They are characterized by a horizon with an illuvial accumulation of sesquioxides (oxides with 1½ oxygen atoms to every metallic atom). Usually, this horizon shows rounded or subangular black or very dark-brown, iron-rich pellets the size of silt. A characteristic ash-gray **A** horizon—the signature of silica, which, unlike other soil minerals, is resistant to dissolving by organic acids—often marks the topsoil of spodosols (*see* Fig. 23-8) but is not itself a defining feature. Spodosols are found only in humid regions, mostly with coniferous forest covers; the four Suborders are listed in Table 25-2.

Because of the association of forest cover with spodosols, lumbering is one of the most important human activities that occurs on these soils. Agriculture is mini-

with basic cations (calcium, magnesium, potassium, sodium). This layer has a ratio of carbon to nitrogen of less than 17 percent (13 percent if cultivated) and a moderate to strong stable structure. Mollisols occasionally are found under water-loving plants or deciduous hardwood forests, but the vast majority are found under tall or short grasslands. Mollisols are often associated with large-scale commercial grain production (*see* Fig. 17-10) and livestock grazing. Corn is the predominant grain when precipitation is sufficient, but sorghum is increasingly grown on such soils in the western United States. Drought is the most common problem facing the agriculturalist who farms on mollisols.

8. Alfisols

Alfisols are found in moister, less continental climatic zones than the mollisols. They are high in mineral content and are usually moist. As Fig. 25-10 indicates, alfisols lack the dark surface horizon of the mollisols, but they do exhibit noteworthy clay accumulation in the **B** horizon. Alfisols are usually found under higher-latitude forests or middle-latitude deciduous forests, but occasionally can be found in areas of vegetation adapted to dryness. Climatic factors are apparent in the Suborders of alfisols, as shown in Table 25-2. Areas of alfisols are notable for some of the most intensive forms of agriculture. Part of the Corn Belt of the United States lies on these soils; oats, soybeans, and alfalfa are also widely grown there. The largest area of alfisols (Suborder ustalfs) lies in tropical Africa, within two wide zones (one in the Northern and one in the Southern Hemisphere)

FIGURE 25-10 Cross-section of an alfisol, a soil associated with some of the most productive farming areas in the U.S. Midwest.

mal, although corn, wheat, oats, and hay are now being grown in upstate New York and other places where such soils occur.

10. Ultisols

Warmer, wetter climatic zones host a soil distinguished by a **B** horizon of strong clay accumulation (Fig. 25-11). The native vegetation of these **ultisols** may have been forest or savanna grassland. Ultisols have at least a few minerals that may be subject to weathering. They may well be alfisols that have been subjected to greater weathering. In the eastern United States, ultisols lie to the south of the southern border of Pleistocene glaciation, and alfisols lie to the north. Therefore, ultisols are older, some being pre-Pleistocene, whereas alfisols are of Pleistocene age or younger. There are five Suborders of ultisols, as Table 25-2 indicates, and all are highly favorable to cultivation. Ultisols are often associated with the farming of cotton and peanuts.

11. Oxisols

As noted in Unit 23, **oxisols** are restricted to tropical areas with high rainfall. Organic matter is quickly destroyed, and downward-percolating water leaches the soil, leaving behind compounds of iron and aluminum. It is hard to distinguish the horizons in such soils. Oxisols are therefore characterized only by a horizon with a large part of the silica, previously combined with iron and aluminum, removed or altered by weathering. This often produces a hard layer called an *oxic horizon* (Fig. 25-12), which is bright red or orange in color, caused by a high concentration of clay-sized minerals in the **B** horizon, mainly sesquioxides. The five Suborders of oxisols are listed in Table 25-2. The natural vegetation on most oxisols is tropical rainforest or savanna, and the soils tend to be of low fertility (as we explain in Unit 24), except where they develop from alluvial or volcanic deposits.

The low fertility of oxisols has given rise to a pattern of *shifting cultivation*: the land is farmed for only a year or two and is then left for many years to naturally renew its nutrients. Population pressures in some parts of the world, such as Nigeria, have altered this agricultural system, and the rapid deterioration of this soil's already limited fertility is usually the result (*see* unit opening photo in Unit 23).

THE SPATIAL DISTRIBUTION OF SOILS

Let us now return to the hypothetical continent we devised (Fig. 25-1) to model the distribution of the Soil Orders. Note that between the oxisols of the southeast, the aridisols of the southwest, and the spodosols of the

FIGURE 25-11 Cross-section of an ultisol in Tennessee, exhibiting a **B** horizon rich in clay.

FIGURE 25-12 Oxisols, when fully developed, are deep (often dozens of meters thick) and characteristically red-colored because of the preponderance, in the oxic horizon, of iron and aluminum. This upper part of the profile of an oxisol on the South Pacific island of Viti Levu in Fiji shows the virtual absence of humus in these soils. Note that there is little color change from near the very top of the soil downward. (Authors' photo)

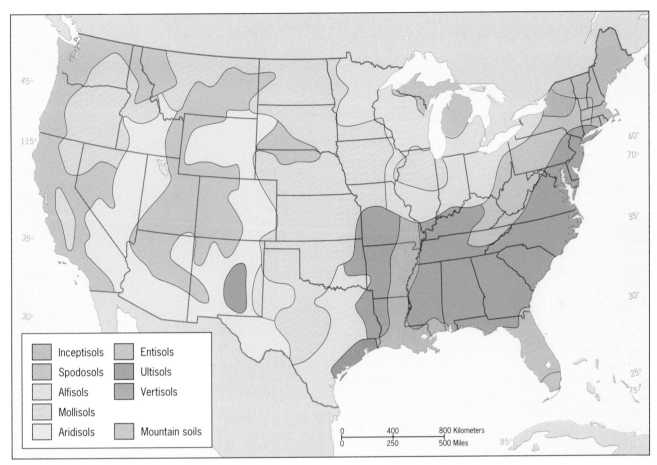

FIGURE 25-13 Generalized spatial distribution of soils within the conterminous United States.

north, there lie zones of mollisols, ultisols, and alfisols (the last developing under different climatic conditions in two separate areas).

The U.S. Pattern

The distribution of soils in the real world is more complicated. Figure 25-13 shows the spatial pattern for the conterminous United States. The climatic relationships are quite clear with soils such as spodosols and aridisols, but complications arise because other soils, such as inceptisols, are caused by different factors. Thus Appalachian soils, because of their highland environment, are poorly developed, whereas the inceptisols of the lower Mississippi Valley are at an incipient stage of development because of their youth. Terrain can complicate the actual pattern even further. The Rocky Mountain region, for instance, is a complex mosaic of aridisols, mollisols, and especially mountain soils. It is therefore difficult to place all the Soil Orders that are not directly related to climate on a map of this scale. Yet, as we should keep in mind, all Soil Orders except the oxisols exist within the continental United States.

The World Soil Map

At the global scale there are many distinct relationships between soils and climate, but there also are some differences in detail as Fig. 25-14 demonstrates. The relationship between soils and climates is quite clear in central Eurasia. Soil types vary from south to north, changing from aridisols to mollisols, then to alfisols, and finally to spodosols. These variations parallel changes in climate from desert through steppe and microthermal to polar climates (*see* Fig. 16-3). A similar progression may be observed in corresponding parts of North America. Perhaps the most obvious relationship at the global scale is that between the desert climates and the aridisols and entisols. Notice, too, that in classifying soils on a world scale, mountain areas (as is also the case at the continental scale of Fig. 25-13) and ice-covered terrain have been grouped into separate categories.

Despite the clear relationships between climate and soil types at this macroscale, similar climates do not always correspond to similar soil types. The humid subtropical climate (**Cfa**) is a case in point. This climate type prevails in the southeastern parts of the five major con-

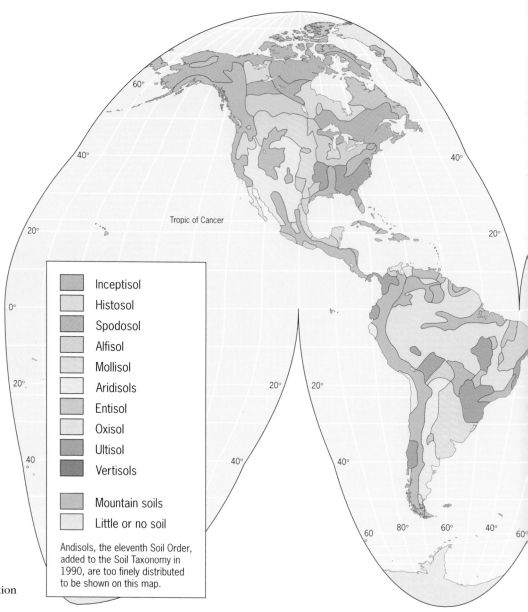

FIGURE 25-14 The global distribution of soils.

Legend:
- Inceptisol
- Histosol
- Spodosol
- Alfisol
- Mollisol
- Aridisols
- Entisol
- Oxisol
- Ultisol
- Vertisols
- Mountain soils
- Little or no soil

Andisols, the eleventh Soil Order, added to the Soil Taxonomy in 1990, are too finely distributed to be shown on this map.

tinents (refer to Fig. 16-3), but another look at Fig. 25-14 shows that the soil types in these areas vary. Only in the southeastern United States and southeastern China do we find ultisols. Southeastern parts of South America around the Rio de la Plata exhibit mollisols, whereas southeastern Africa possesses alfisols. And east-central Australia, an area of predominantly humid subtropical climate, has

alfisols, vertisols, and mountain soils. If you continue to compare Figs. 25-14 and 16-3, you will be able to pick out additional spatial discrepancies between climate and soils. This underscores that other factors in soil formation, such as parent material and the way people use soil, also help determine the prevailing type of soil.

KEY TERMS

alfisol (p. 273)   entisol (p. 268)   mollisol (p. 272)   Soil Taxonomy (p. 267)   ultisol (p. 274)
andisol (p. 271)   histosol (p. 268)   oxisol (p. 274)   spodosol (p. 273)   vertisol (p. 269)
aridisol (p. 272)   inceptisol (p. 269)   Soil Order (p. 267)

REVIEW QUESTIONS

1. What is the Soil Taxonomy?
2. List and briefly describe the distinguishing characteristics of the 11 Soil Orders.
3. Which Soil Orders are associated with particular climatic zones?

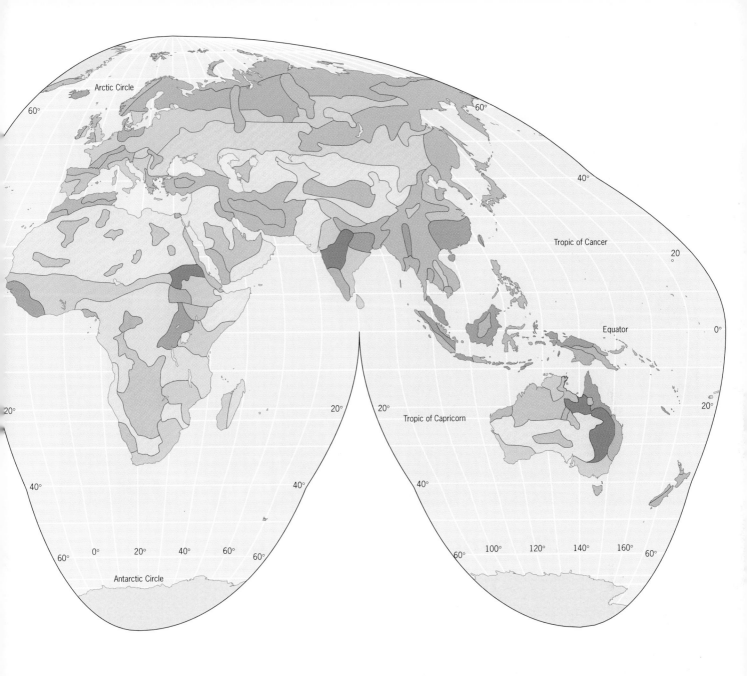

REFERENCES AND FURTHER READINGS

BASILE, R. M.   *A Geography of Soils* (Dubuque, Iowa: Wm. C. Brown, 1971).

BRIDGES, E. M.   *World Soils* (London/New York: Cambridge University Press, 2nd ed., 1978).

BRIDGES, E. M., and DAVIDSON, D. A.   *Principles and Applications of Soil Geography* (London/New York: Longman, 2nd ed., 1986).

BUOL, S. W., et al.   *Soil Genesis and Classification* (Ames, Iowa: Iowa State University Press, 3rd ed., 1989).

CLARKE, G. R.   *The Study of Soil in the Field* (London/New York: Oxford University Press [Clarendon], 5th ed., 1971).

DENT, D., and YOUNG, A.   *Soil Survey and Land Evaluation* (Winchester, Mass.: Allen & Unwin, 1981).

FANNING, D. S., and FANNING, M.C.B.   *Soil: Morphology, Genesis, and Classification* (New York: Wiley, 1989).

FITZPATRICK, E. A.   *Soils: Their Formation, Classification, and Distribution* (London/New York: Longman, 1983).

FOTH, H. D.   *Fundamentals of Soil Science* (New York: Wiley, 8th ed., 1990).

OLSON, G. W.   *Soils and the Environment: A Guide to Soil Surveys and Their Applications* (London: Methuen, 1982).

PITTY, A. F.   *Geography and Soil Properties* (London: Methuen, 1978).

ROWELL, D. L.   *Soil Science: Methods and Applications* (New York: Wiley/Longman, 1994).

STEILA, D., and POND, T. E.   *The Geography of Soils: Formation, Distribution, and Management* (Totowa, N.J.: Rowman & Littlefield, 2nd ed., 1989).

U.S. DEPARTMENT OF AGRICULTURE.   *Soil Taxonomy: A Basic System of Soil Classification for Making and Interpreting Soil Surveys* (Washington, D.C.: USDA, Soil Conservation Service, Handbook No. 436, 1975 [reprinted 1988 by Krieger Publ. Co.]).

Biogeographic Processes

OBJECTIVES

◆ To discuss the process of photosynthesis and relate it to climatic controls.

◆ To introduce the concept of ecosystems and highlight the important energy flows within ecosystems.

◆ To outline the factors influencing the geographic dispersal of plant and animal species within the biosphere.

The soil is located at the base of the biosphere. Not only is the soil itself a living, maturing entity; life exists within the soil layer in many forms. In this unit and the next, we study the most obvious evidence of the earth's "life layer"—the natural vegetation. Biogeography, as we noted in Unit 22, consists of two fields: *phytogeography* (the geography of plants) and *zoogeography* (the geography of animals). We begin with the plants.

DYNAMICS OF THE BIOSPHERE

The story of the development of life on earth parallels that of the formation of the atmosphere. The earliest atmosphere, more than 3 billion years ago, was rich in gases such as methane, ammonia, carbon dioxide, and

water vapor. About 45 years ago, a scientist named Stanley Miller filled a flask with what he believed may have been a sample of this early atmosphere. He then subjected the contents to electrical discharges (to simulate lightning) and to boiling (much of the crust was volcanic and red-hot). This resulted in the formation of amino acids, the building blocks of protein that are, in turn, the constituents of all living things on earth. Another scientist tried freezing the same components, and out of that experiment came organic material that forms one of the ingredients of deoxyribonucleic acid or DNA, a key to life. It is believed that these experiments replicated what

UNIT OPENING PHOTO: These alligators are emblematic of the teeming life of subtropical Okefenokee Swamp in Georgia's southeastern corner.

actually happened on Earth over 3 billion years ago, events that led to the formation of the first complex molecules in the primitive ocean. The earliest forms of life—single-celled algae and bacteria—came from these molecules.

Photosynthesis

When the first algae floated on the surface of the earth's evolving ocean, a process could begin that would be essential to advancing life: *photosynthesis.* This process requires solar energy, carbon dioxide, and water. In those ancient oceanic algae, the first conversion of water (H_2O) and carbon dioxide (CO_2) under solar energy yielded carbohydrate (an organic compound and food substance) and oxygen (O_2). After more than 2500 million years of slow photosynthesis, the atmosphere became sufficiently rich not only in oxygen but also in higher-altitude ozone (O_3), so that life forms could forsake the oceans and emerge onto the landmasses. The first land plants and insects colonized the earth more than 400 million years ago and were followed by higher forms of life (Fig. 26-1).

A fundamental requirement for photosynthesis is a green pigment, *chlorophyll,* at the surface of the part of the plant where the process is taking place. The color of this pigment in part ensures that the wavelength of light absorbed from the sun is correct for photosynthesis. This produces the dominant green color of plants. If photosynthesis required a different wavelength of light, the earth's vegetated landscapes might look blue or orange! As to the need for light or solar energy, plants constantly compete for the maximum exposure to this essential ingredient of life. Some plants, as we will note, have adapted to survival in low-light environments. But mostly we see every leaf turned toward the sun. Where solar energy arrives on earth in the greatest quantities, the tall trees of the equatorial and tropical forests soar skyward, spreading their leafy crowns in dense canopies as they vie for every ray of the sun (Fig. 26-2).

Photosynthesis, therefore, is critical to life on earth. It removes carbon dioxide from the atmosphere and substitutes oxygen. Humans could not live as they do in an oxygen-poor atmosphere, and had photosynthesis not altered the earth's primitive envelope of gases, the evolution of life would have taken a very different course. Thus we depend on plants in a very direct way. If the great tropical rainforests are destroyed, as is continuing to happen (*see* first Perspective box in Unit 17), the atmosphere will sustain damage of a kind we cannot accurately predict.

One result of such destruction could possibly involve not only oxygen depletion but also, through the relative increase of carbon dioxide, a general warming

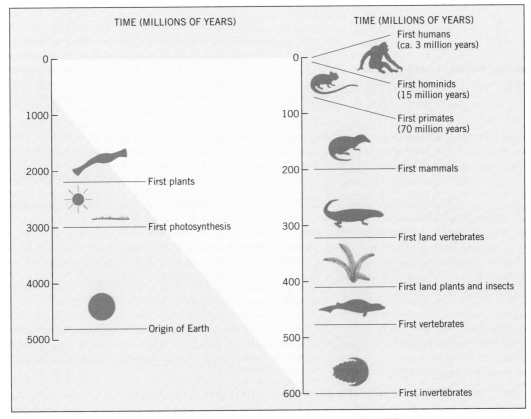

FIGURE 26-1 The development of the biosphere on Earth: the time scale of life.

FIGURE 26-2 A continuous canopy of leafy vegetation marks the tropical rainforest from the air. The tall trees have spread their crowns until they interlock, competing for each ray of sunlight. This is the vast equatorial forest of the Zaïre (Congo) Basin in west-central Africa, the world's second largest tropical rainforest after that of the Amazon Basin.

of the planet—a further enhancement of the greenhouse effect. Photosynthesis also produces carbohydrates, which nourish plants, and consequently animals and people. The natural production of organic food substances would cease entirely without photosynthesis. In addition, our fossil fuels (coal, oil, natural gas) were produced from carbohydrates originally formed during photosynthesis.

Limitations of Photosynthesis

Combining what we have learned about climatic patterns with what we know about photosynthesis, we can deduce where the process goes on most productively. Solar energy is most abundant in the tropical latitudes, and there, in general terms, photosynthesis is most active. However, some significant limitations must be recognized. In equatorial and tropical latitudes, the high solar energy also generates heat, and heat increases the plants' rates of *respiration*. Respiration runs counter to photosynthesis because it breaks down available carbohydrates, combines them with oxygen, and yields carbon dioxide, water, and biochemical energy that sustains life. When leafy plant surfaces become hot, therefore, heightened respiration diminishes the effectiveness of photosynthesis.

Another limiting factor is the availability of water. Carbon dioxide from the atmosphere is not available for photosynthesis until it dissolves in water at the plant surface. Thus continued photosynthesis requires moisture. Small holes in the leaf surface, called *stomata*, are openings for the water that arrives from the roots and stem

of the plant. The more water there is at the stomata, the greater the amount of carbon dioxide that can be dissolved and the greater the production of plant food. Our knowledge of the map of moisture (*see* Fig. 12-11) is an important aid in understanding the distributional pattern of photosynthesis.

Several other processes occur at or near the leaf surface as photosynthesis and respiration proceed. Obviously, when heat and moisture are present, evaporation will occur. Evaporation from the leaf surfaces has a drying effect, which in turn affects photosynthesis. Plants also lose moisture in the same way we humans do, through transpiration. These two processes in combination, as we learned in Unit 12, are referred to as *evapotranspiration*. As we also know from that discussion, an intimate relationship exists between the amount of moisture lost in evapotranspiration (*see* Fig. 12-10) and the production of living organic plant matter.

Phytomass

The total living organic plant matter produced in a given area is referred to as its **biomass**. Technically, biomass refers to all living organic things, including an area's fauna, so the plant matter should be called the **phytomass**. But biomass is expressed in terms of weight—that is, in grams per square meter—and the weight of plant matter is so dominant that the terms *biomass* and *phytomass* tend to be used synonymously. Figure 26-3 maps, in very general terms, the global distribution of annual biomass productivity. Note the overall similarities between certain climatic patterns and this configuration. Biomass productivity is greatest in well-watered equatorial and tropical lowlands; it is at a minimum in desert, highland, and high-latitude zones. But this is true only for the *natural* vegetation, not for cultivated crops. As the map suggests, forests contain the largest biomass; replace the tropical forest with a banana plantation, and photosynthesis and biomass will decline precipitously.

Ecosystems and Energy Flows

An **ecosystem** is a linkage of plants (or animals) to their environment in an open system as far as energy is concerned: solar energy is absorbed, and chemical and heat energy are lost in several ways. You can observe part of an ecosystem in action in a sunlit freshwater pond (Fig. 26-4). Energy from sunlight is taken up in photosynthesis by microscopic green plants called *phytoplankton* (you may be able to see these in the aggregate as a greenish sheen on the water), which in turn produce carbohydrate, the food substance. Such food-producing plants are called *autotrophs*, and they provide sustenance for small larvae and other tiny life forms in the pond collectively called *zooplankton*. Zooplankton is eaten by small fish, and these fish are later consumed by larger fish.

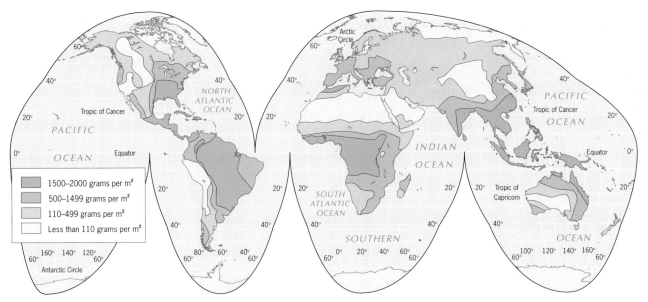

FIGURE 26-3 The global distribution of annual biomass productivity (in grams per square meter of dry biomass matter).

Meanwhile, plants and animals die in the pond and decay, thereby releasing chemicals back into the water to be used once more by autotrophs in the production of food. Thus food energy passes from organism to organism in the ecosystem represented by our pond, gener-

FIGURE 26-4 Crucial part of an ecosystem. The sun's energy reaches the surface of a lake in upstate New York, where phytoplankton (microscopic green plants) convert it through photosynthesis into carbohydrate, a food substance. Tiny life forms, zooplankton, feed on this carbohydrate, and small fish in turn eat the zooplankton. Larger fish feed on the smaller fish, and the food chain continues.

ating a **food chain**. Food chains exist in all ecosystems, on land as well as in the oceans.

Ecological Efficiency

It is very difficult to measure precisely the production and consumption of energy in a food chain. Research has shown, however, that only a fraction of the food produced by autotrophs—as little as 15 percent—is actually consumed by the next participants in the food chain. For example, the zooplankton in our pond consumes only about 15 percent of the food energy yielded by the phytoplankton. A still smaller percentage of the zooplankton is eaten by the smallest fish. Each of these groups—the phytoplankton, zooplankton, small fish, and larger fish—along the food chain is called a **trophic level**. On an African savanna, the autotrophs are the grasses and other plants; at the next trophic level are the animals that eat these plants, the **herbivores**. In turn, the animals that eat herbivores (as well as other animals), the **carnivores**, form a still higher trophic level.

Central to the way the biosphere operates is the loss of chemical food energy at each trophic level. We have already noted that a mere 15 percent of the food energy generated by the autotrophs in the pond is passed on to the herbivores (the zooplankton). Only 11 percent of the zooplankton reaches the carnivores (the smallest fish), and just 5 percent of these go on to the larger fish when they eat the smaller ones. These numbers indicate a very low efficiency, but they reflect efficiency at *each trophic level* in the pond. We also can estimate efficiency for entire ecosystems, averaging the efficiencies at different levels. Desert ecosystems have the lowest efficiency of all, with values of less than 0.1 percent. Swamps in trop-

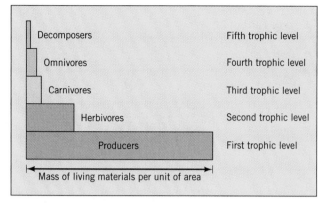

FIGURE 26-5 Mass of living materials per unit of area in different trophic levels of an ecosystem.

ical areas have the highest, but even they average as little as 4 percent.

Several important consequences arise from the various efficiencies in ecosystems. First, there must always be a large number of producers to support smaller quantities of herbivores and even fewer carnivores. The masses of living material at each trophic level therefore stack up like the pyramid in Fig. 26-5. Because only about 10 percent of the energy produced in the form of food is passed from one stage to another, to obtain enough food the animals at the higher trophic levels must have large territorial areas that provide enough of the species at the lower trophic level. This explains why large carnivores such as lions require a wide territorial range.

Another consequence is that, because food and energy move along a chain in only one direction, the whole system collapses if earlier links of the chain are broken. Thus the removal of the autotrophs (grass) by too many rabbits in Australia caused the breakdown of an entire ecosystem. A related consequence of the chain-like structure is the fact that undesirable materials can be passed along and concentrated by the ecosystem. The insecticide DDD was applied in Clear Lake, California to kill gnats. It was sprayed onto the water at a concentration of 0.02 parts per million (ppm). The DDD density was 5 ppm in the plankton, 15 ppm in the herbivores feeding on the plankton, 100 ppm in the fish, and 1600 ppm in the grebes (birds) that ate the fish. The grebes died. It is therefore important to understand the nature of food energy flow through ecological systems. We now turn to look at a vegetation system that changes, and we find once again that energy is crucial.

PLANT SUCCESSIONS

The flow of energy through a food chain illustrates the dynamic nature of the biosphere. Indeed, the vegetation layer is as dynamic as the atmosphere, soil, or crust of the earth. Some of the changes occur within a stable eco-

system, but sometimes one type of vegetation is replaced by another. This is called a **plant succession**, of which three types exist.

A **linear autogenic succession** occurs when the plants themselves initiate changes in the land surface that consequently cause vegetational changes. "Linear" indicates that the order of succession in any one place is not normally repeated. Figure 26-6 shows how the growth of vegetation in an area that has been a lake (Stage I) is part of a linear autogenic succession. As the lake gradually fills with sediments, the water becomes chemically enriched; mosses and sedges as well as floating rafts of vegetation build up (Stage II). Plant productivity increases in the lake, and other plants encroach around its edges (Stage III). After the lake has completely filled with organic debris, plants and trees may finally take over (Stage IV).

Sometimes one kind of vegetation is replaced by another, which is in turn replaced by the first. Or possibly the original vegetation follows a series of two or three others. This is a **cyclic autogenic succession**. An example can be found at the northernmost limit of tree growth in Alaska, where permafrost (permanently frozen ground) lies beneath tundra vegetation of grasses, sedges, and bare ground. The permafrost melts to a sufficient depth in summer to allow colonization by willow scrub and later by spruce trees. Gradually, this forest becomes denser and forms a layer of litter. The permafrost, thus insulated, gradually rebuilds. The forest degenerates and eventually gives way to the original tundra vegetation. The cycle is then complete and ready for another sequence.

A third type of succession occurs where vegetation changes because of some outside environmental force. This kind of succession is termed an **allogenic succession** because the agent of change comes from outside the plant's immediate environment. Devastation through nuclear radiation could be one such force, but disease is more common. An epidemic of chestnut blight in the eastern United States created oak and oak-hickory forests where oak-chestnut forests had once existed.

In all types of plant successions, the vegetation builds up through a series of stages, as indicated in Fig. 26-6. When the final stage is reached, the vegetation and its ecosystem are in complete harmony with the soil, the climate, and other parts of the environment. This balance is called a **climax community** of vegetation. The major vegetation types, or *biomes* (described in Unit 27), all represent climax communities, which also are characterized by ecosystems with stable amounts of accumulated energy. Solar energy is taken in and energy is lost through respiration and other processes, but the stored energy of the biomass is relatively constant. During a plant succession, however, the amount of energy stored in the biomass increases as Fig. 26-7 shows. Input of energy must exceed losses for the plant succession to develop.

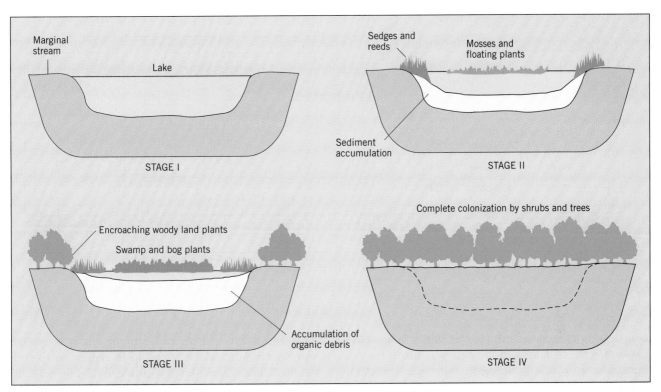

FIGURE 26-6 An idealized sequence of a linear autogenic plant succession by which a lake is eventually colonized by shrubs and trees.

GEOGRAPHIC DISPERSAL

Geographers are interested in the spatial distribution of the species of the biosphere, a topic treated in Units 27 and 28. Here we investigate the factors that determine the spread and geographic limits of any particular group of organisms. The limiting factor may be either physical or biotic. Each species has an *optimum range* where it can survive and maintain a large healthy population, as shown in Fig. 26-8. Beyond this range, a species increasingly encounters *zones of physiological stress*. Although

it can still survive in these zones, the population is small. When conditions become even more extreme, in the *zones of intolerance*, the species is absent altogether except possibly for short, intermittent periods.

Physical Factors

Temperature

A common limiting factor for both plants and animals is temperature. Unit 16 mentions the correspondence, suggested by Köppen, of the northern tree line and certain

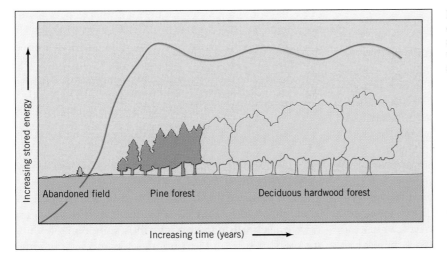

FIGURE 26-7 The increase in stored energy of the biomass in a typical secondary autogenic succession. In this case, a deciduous hardwood forest takes over from an abandoned field over a period lasting about 175 years.

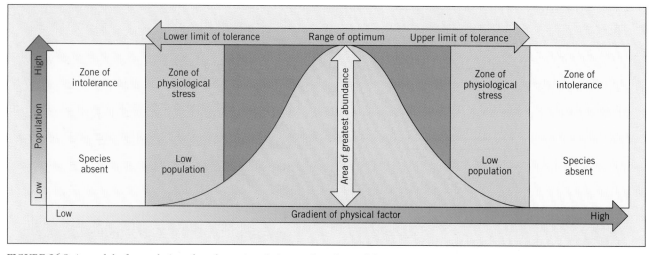

FIGURE 26-8 A model of population abundance in relation to the physical factors in the environment of a species.

temperature conditions. Another example of temperature's effect on spatial distributions was discovered by the British ecologist Sir Edward Salisbury in the 1920s. One species he studied was a creeping woody plant known as the wild madder (*Rubia peregrina*). He found that the northern boundary of the wild madder in Europe coincided closely with the January 4.5°C (40°F) isotherm. This temperature was critical because in January the plant formed new shoots, and lower temperatures would inhibit their development and subsequent growth. Indeed, temperatures play such a large role in determining plant distributions that plants are sometimes classified according to their propensity for withstanding heat. Plants adapted to heat are called *megatherms*; those that can withstand low temperatures are designated *microtherms*; and those with a preference for intermediate temperatures are called *mesotherms*.

Availability of Water

Another vital factor—the availability of water—limits the spread of plants and animals throughout the physical world. Water is essential in photosynthesis and in other functions of plants and animals. Several plant classifications take water availability into account.

Plants that are adapted to dry areas are called *xerophytes*, and desert plants have evolved many fine adjustments. Stomata are deeply sunken into the leaf surface to reduce water loss by evapotranspiration. Roots often reach 5 m (16 ft) or more into the ground in search of water, or more commonly spread horizontally for great distances. Plants that live in wet environments are classified as *hygrophytes*, with rainforests, swamps, marshes, lakes, and bogs the habitats of this vegetation. The aquatic buttercup, a curious example of a hygrophyte, produces two kinds of leaves: it develops finely dissected leaves when in water and simple entire leaves when exposed to the air. Finally, plants that develop in areas of

neither extreme moisture nor extreme aridity are called *mesophytes*. Most plants growing in regions of plentiful rainfall and well-drained topography fall into this category.

In tropical climates with a dry season, flowering trees and plants drop their leaves to reduce water loss during the dry season. This phenomenon spread to plants of higher latitudes, where the formation of ice in winter sometimes causes a water shortage. Trees and other plants that seasonally drop their leaves are called *deciduous*, and those that keep their leaves year-round are called *evergreen*.

Other Climatic Factors

Several other factors related to climate play a role in the dispersal of plants. These include the availability of light, the action of winds, and the duration of snow cover. The position of a species within a habitat shared by other species determines the amount of light available to it. In deciduous forests of the middle latitudes, many low shrubs grow intensely in spring before the leaves of taller trees filter out light. The amount of available light is further determined by latitude and associated length of daylight. Growth in the short warm season of humid microthermal (**D**) climates is enhanced by the long daylight hours of summer, and plants can mature surprisingly fast (*see* Fig. 19-2). Wind influences the spread of plants in several ways. It can limit growth or even destroy plants and trees in extreme situations; but it also spreads pollen and the seeds of some species.

Distribution of Soils

Another factor that affects plant dispersal is the distribution of soils. Factors concerned with the soil are known as **edaphic factors**, the most important of which are soil structure and texture, the presence of nutrients,

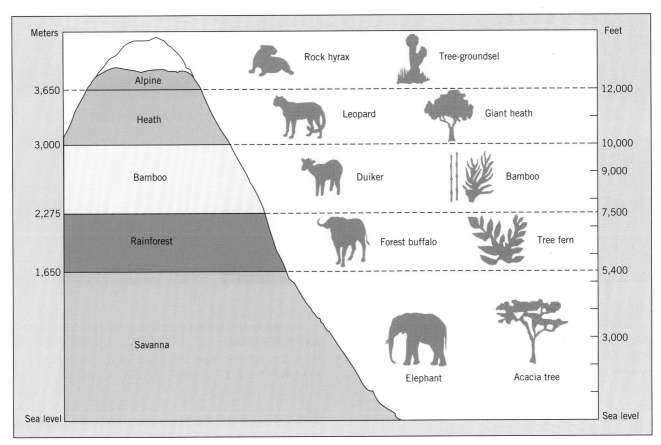

FIGURE 26-9 Zones of vegetation and animal life on the flanks of Mount Kenya in East Africa, which lies directly on the equator.

and internal quantities of air and water. Soil structure and texture affect a plant's ability to root. Nutrients, to some extent, determine the type of vegetation. Grasses, for example, need large quantities of calcium, so they are more likely to be found in dry climates where downward-percolating water is limited, which allows calcium compounds to concentrate in the upper soil. The presence of soil water depends not only on precipitation but also on the porosity of the soil. Thus, given the same amount of rainfall, a grassland might exist over porous soils and a forest over nonporous soils.

Landforms

The final physical factor, landforms, controls vegetation distribution in many ways. On a large scale, as we note in Unit 19, vegetation changes with altitude. On the flanks of Mount Kenya in East Africa (Fig. 26-9), there is a transition from savanna grassland below 1650 m (5400 ft) to alpine vegetation above 3650 m (12,000 ft). Landforms have a small-scale effect as well. Steep slopes foster rapid drainage and may lead to a lack of soil water. Furthermore, the aspect of a mountain slope (the direction it faces) controls the amount of incoming radiation and determines the degree of shelter from the wind, thereby influencing the local distribution of plant and animal species.

Biotic Factors

Competition

As might be expected from the interactions in any one ecosystem, many biotic as well as physical factors affect the spatial distribution of plants and animals. The biosphere is seldom static: individual species may compete with one another, be suppressed by other species, be predators or prey, or live in intimate cooperation with other species. Competition for food and space plays a strong part in plant and animal distributions. Sometimes new species compete for resources so well that they eliminate old species. For example, in the intermountain western United States, such as in Utah, native bluebunch wheatgrass (*Agrophyron spicatum*) has been almost totally replaced by European cheatgrass (*Bromus tectorum*). One reason for the success of the cheatgrass is that it produces 65 to 200 times as many seeds as the native species does.

Amensalism

Another form of biological interaction, the inhibition of one species by another, is called **amensalism**. In the coastal hills of Southern California, sage shrubs grow on slopes and grasses inhabit the valleys. Sometimes, however, the sage shrubs occur in the grassland zone. As Fig.

FIGURE 26-10 Sage shrubs suppressing grass growth along the California coast.

26-10 shows, the shrubs are usually surrounded by grass-free bare ground. Foraging by birds and small animals clears some of the ground, but a major cause is the cineole (a liquid with camphor-like odor) and camphor oil emitted by the sage. Both are toxic to grass seedlings; the sage, therefore, suppresses the growth of grass.

Predation

Anyone who saw the ravaged landscapes of Australia before rabbit control began (Fig. 26-11) cannot doubt the efficiency of predation as a factor in the distribution of vegetation. But examples of one species' eating all members of another species tend to be rare. It is in the best interest of predators in balanced ecosystems to rely on a number of prey species so that their food will never be exhausted.

In this more natural situation, predation affects plant distribution mainly by reducing the pressure of competition among prey species. In general, the presence of predators tends to increase the number of species in a given ecosystem. Charles Darwin suggested that ungrazed pasture in southern England was dominated by fast-growing tall grasses that kept out light. Consequently, the ungrazed areas contained only about 11 species while the grazed lands possessed as many as 20.

Mutualism

Yet another biological interaction is termed **mutualism**, the coexistence of two or more species because one or more is essential to the survival of the other(s). Many examples of mutualism, or *symbiosis*, can be found in

FIGURE 26-11 The rabbit-warren-infested landscape of the Western Australia outback. The destruction of vegetation is clearly visible from 1200 m (4000 ft) above the surface.

A Biogeographic Puzzle: The Species-Richness Gradient

For over a century, it has been known that the number of species per unit area decreases with latitude. A single square kilometer of tropical rainforest contains thousands of plant and animal species; a square kilometer of tundra may contain only a few dozen. This phenomenon is known as the **species-richness gradient**, and many theories have been proposed to explain it.

Associated with the species-richness gradient is the relationship between the most abundant and important species on the one hand and lesser species on the other. This is the principle of *species dominance*. In tropical rainforest environments, where the number and diversity of species are very large (*see* Perspective box in Unit 22), it is often the case that no species is clearly dominant. But in higher-latitude environments, for example, where oak-hickory or spruce-fir forests prevail, a few species (such as an evergreen tree or a large herbivore) often predominate.

We know that species diversity in biotic communities increases with evolutionary time, with environmental stability, and with favorable (warm, humid, biologically productive) habitat. In general, the territorial ranges of individual species are comparatively small in equatorial and tropical environments, and much larger in the higher latitudes. Biogeographers have concluded that the spatial patterns of species richness indicate that niche differentiation is the main active process in the small, favorable low-latitude microclimates, whereas adaptation to the rigors of high-latitude environments is the selective force acting on (and restricting) species there.

But there may be more to it. In their large ranges at high latitudes, species must adapt to wide fluctuations in temperature and other environmental conditions. Animals roam far and wide; plants are hardy and durable. In tropical areas, however, species tend to be closely bound to a narrow set of environmental conditions; the mosaic of microclimates reflects the limited range of many species. Such species cannot adapt to even a small change in their environment. Plants may be fragile and vulnerable. Migrating animals quickly find themselves at a disadvantage.

The species-richness gradient may thus be explained in terms of adaptation, species success and dispersal, and environmental limitations. Some ecologists argue that biological resources can be shared more finely where the environment is relatively constant; specialization is more possible under such circumstances. Considering the relationship another way, we observe that the high input of solar energy near the equator may provide more placement scope for specialization and less at higher latitudes. An early theory, still popular among some biogeographers, is that tropical rainforests are older than temperate and high-latitude communities, and thus have had more time to accumulate complex biotic communities. But newly gained knowledge about the Pleistocene epoch seems to run counter to this idea.

Recent research by an ecologist, George Stevens, introduces still another factor into the debate. In any particular environment there are always very successful, abundant, well-adapted species alongside less-well-adapted, "marginal" species. When large numbers of a successful species disperse into a less suitable environment, they should succumb to competition from established species in that environment. Often, however, they manage to survive because their numbers are constantly replenished by the continuing stream of migrants. This is known as the *rescue effect*.

Migrants in small tropical microclimates quickly move beyond their optimal environmental niches (niches are discussed on p. 299). But in high-latitude environments, where species ranges are larger and where species are adapted to wider environmental ranges, dispersing individuals of successful species are much less likely to stray into unsuitable habitats. As Stevens points out, the rescue effect is much less influential in higher-latitude areas than in the tropics, contributing to the geographic pattern of the species-richness gradient.

Various theories combine different ideas about the species-richness gradient in various ways, but no single theory has won general acceptance. And so one of biogeography's grandest global designs continues to evoke discussion and debate.

equatorial and tropical forests. When certain species in the rainforest are cut down and removed, their leaf litter and other organic remains no longer decay on the forest floor. Since equatorial oxisols are so infertile, this leaf litter keeps other plants supplied with nutrients. Remove them, and the remaining plants may die. Such human intervention, of course, reminds us that another species—*Homo sapiens*—can often significantly shape the geography of plants and animals as well.

Although researchers have not been able to fully explain every major geographic variation (*see* Perspective: A Biogeographic Puzzle: The Species-Richness Gradient), these limiting physical and biotic factors contribute a great deal to our understanding of the distribution of plant and animal species. In Unit 27, we see how the biogeographic processes discussed in this unit are expressed spatially on the world map of vegetative associations.

KEY TERMS

allogenic succession (p. 282)

amensalism (p. 285)

biomass (p. 280)

carnivores (p. 281)

climax community (p. 282)

cyclic autogenic succession (p. 282)

ecosystem (p. 280)

edaphic factors (p. 284)

food chain (p. 281)

herbivores (p. 281)

linear autogenic succession (p. 282)

mutualism (p. 286)

phytomass (p. 280)

plant succession (p. 282)

species-richness gradient (p. 287)

trophic level (p. 281)

REVIEW QUESTIONS

1. Which subjects are encompassed by *phytogeography* and *zoogeography*?

2. Describe the process of photosynthesis, including consideration of requirements, limitations, and related processes.

3. What are food chains, and how efficient are they at transferring energy to higher trophic levels?

4. Describe how a lake might undergo linear autogenic succession.

5. Describe the physical and biotic factors that affect the geographic dispersal of plants and/or animals.

REFERENCES AND FURTHER READINGS

BARBOUR, M. G., et al. *Terrestrial Plant Ecology* (Menlo Park, Calif.: Benjamin/Cummings, 1980)

BEGON, M., et al. *Ecology* (Oxford, U.K.: Blackwell Scientific, 1988).

BIRCH, B. *Biogeography and Soils* (Cambridge, Mass.: Blackwell, 1995).

BRADBURY, I. K. *The Biosphere* (New York: Wiley, 1991).

BROWN, J. H., and GIBSON, A. C. *Biogeography* (St. Louis: Mosby, 1983).

COX, C. B., and MOORE, P. D. *Biogeography: An Ecological and Evolutionary Approach* (Cambridge, Mass.: Blackwell, 5th ed., 1993).

DANSEREAU, P. M. *Biogeography: An Ecological Perspective* (New York: Ronald Press, 1957).

DARLINGTON, P. J., JR. *Biogeography of the Southern End of the World* (New York: McGraw-Hill, 1975).

JARVIS, P. J. *Plant and Animal Introductions* (New York: Blackwell, 1989).

MACARTHUR, R. H. *Geographical Ecology: Patterns in the Distribution of Species* (Princeton, N.J.: Princeton University Press, 1984).

MAURER, B. A. *Geographical Population Analysis: Tools for the Analysis of Biodiversity* (Cambridge, Mass.: Blackwell, 1994).

MIELKE, H. W. *Patterns of Life: Biogeography of a Changing World* (Winchester, Mass.: Unwin Hyman, 1989).

MYERS, A. A., and GILLER, P. S., eds. *Analytical Biogeography: An Integrated Approach to the Study of Animal and Plant Distributions* (New York: Chapman & Hall, 1988).

NELSON, G., and PLATNICK, N. *Systematics and Biogeography: Cladistics and Vicariance* (New York: Columbia University Press, 1981).

RICKLEFS, R. E., and SCHLUTER, D., eds. *Species Diversity in Ecological Communities: Historical and Geographical Perspectives* (Chicago: University of Chicago Press, 1993).

SIMMONS, I. G. *Biogeographical Processes* (Winchester, Mass.: Allen & Unwin, 1983).

TAYLOR, J. A., ed. *Biogeography: Recent Advances and Future Directions* (Totowa, N.J.: Barnes & Noble, 1985).

TIVY, J. *Biogeography: A Study of Plants in the Ecosphere* (London/New York: Longman, 2nd ed., 1982).

WILSON, E. O. *The Diversity of Life* (Cambridge, Mass.: Belknap/Harvard University Press, 1992).

WILSON, E. O. "Threats to Biodiversity," *Scientific American,* September 1989, 108–116.

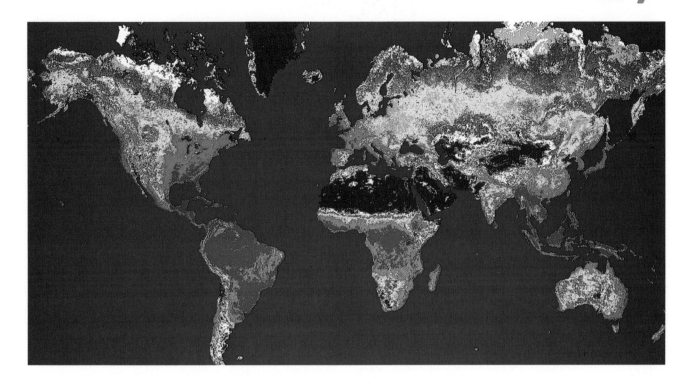

The Global Distribution of Plants

OBJECTIVE

◆ To briefly survey the principal terrestrial biomes.

The classification and mapping of climate poses many geographical challenges. It was necessary to establish a set of measures to form the basis for Fig. 16-3, the map of world climates. That map shows the distribution of *macro*climates—climatic regions on a broad global scale. Embedded within those macroclimates, as we noted, are local microclimates that do not always conform to the established criteria. The global map, therefore, is only a general guide to what we should find in particular places.

Mapping vegetation is in some ways an even more difficult problem. On one of your field trips, stop near any vegetated area and note the large number of plants you can identify, probably ranging from trees and grasses to ferns or mosses. Part of the local area you examine may be tree-covered; another part of it may be open grassland; some of it may be exposed rock, carrying mosses or lichens. How can any global map represent this intricate plant mosaic? The answer is similar to that for climates and soils: plant geographers (phytogeographers) look for the key to the largest units of plant association. In the case of climate, this was done by analyzing temperatures, moisture, and seasonality. The

great Soil Orders (*see* Fig. 25-14) were based on a determination of the most general similarities or differences in soil horizons and formative processes. In classifying the earth's plant cover, biogeographers use the concept of the *biome* to derive their world map.

BIOMES

A **biome** is the broadest justifiable subdivision of the plant and animal world, an assemblage and association of plants and animals that forms a regional ecological unit of subcontinental dimensions. We can readily visualize the most general classes of vegetation—forest, grassland, desert, and tundra—but such a classification would not be useful to geographers. For example, there are tropical forests and high-latitude coniferous forests, and they are so different that grouping them within a

UNIT OPENING PHOTO: Vegetation patterns across the landmasses. Based on the Global Vegetation Index devised by NOAA, colors range from black for barren land to dark green for tropical rainforests.

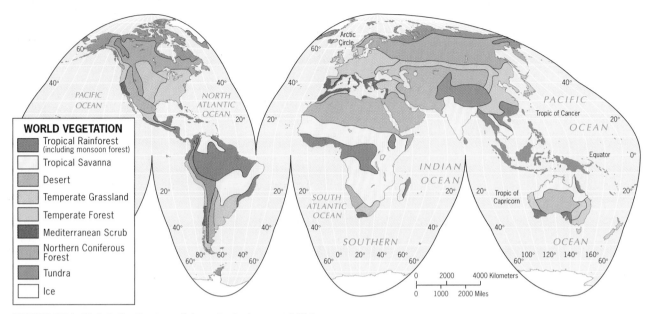

FIGURE 27-1 Global distribution of the principal terrestrial biomes.

single ecological unit would serve little or no purpose. However, if we subdivide forests into tropical, monsoon, temperate (midlatitude), northern (coniferous), and Mediterranean, we have identified true biomes. At this level, the classification is quite useful.

Just as geographers continue to debate the merits of the global climatic map, so biologists are not in total agreement concerning the earth's biomes. In fact, if you were to look at maps of biomes in biology textbooks, you would find quite a range of interpretations. Moreover, biologists include both *terrestrial* and *marine* biomes in their classifications (in biogeography, we tend to concentrate on the terrestrial biomes). Even the identity of terrestrial biomes is not uniformly accepted, so you should view Fig. 27-1 as one justifiable representation of the earth's land biomes, but not the only possible one.

We also must remember that biomes are not so sharply defined regionally that their limits are clearly demarcated on the surface. The lines separating biomes on the world map, therefore, represent broad transition zones. If you were to walk from one biome region to another, you would observe a gradual change as certain species thin out and vanish while others make their appearance and become established. To move from one fully developed biome to its neighbor might require several days of walking.

The distribution of biome regions results from two major factors—climate and terrain. As you will see, both the climatic map (Fig. 16-3) and the physiographic map (Fig. 52-1) are reflected on the map of global vegetation. The key *climatic factors* are: (1) the atmosphere and its circulation systems, which determine where moisture-carrying air masses do (and do not) go, and (2) the energy source for those circulation systems, solar radiation. (The sun's energy not only drives atmospheric movements, but it also sustains photosynthesis and propels the

endless march of the seasons.) The main *terrain factors* are: (1) the distribution of the landmasses and ocean basins, and (2) the topography of the continents.

Some biomes that are widely distributed in the Northern Hemisphere hardly ever occur south of the equator because the Southern Hemisphere does not include large landmasses at comparable latitudes. The varied topography and elevation of the landmasses disrupt much of the regularity the map might have shown had the continents been flatter and lower. Accordingly, the orientations of the earth's great mountain ranges can clearly be seen on the world map of these biomes (Fig. 27-1). Interestingly, many of these same patterns exist at the continental scale as well (*see* Perspective: North America's Vegetation Regions).

We should therefore conclude that the latitudinal transition of biomes from the hot equatorial regions to the cold polar zones can also be observed along the slope of a high mountain. After all, mountain-slope temperatures decrease with altitude, and precipitation increases with altitude. Mount Kenya, in East Africa, stands with its foot on the equator and is capped with snow. So it is, too, with Ecuador's Mount Chimborazo in northwestern South America; this mountain is remembered in this connection because it was there that the great naturalist Alexander von Humboldt first recognized the relationships not only between vegetation and altitude, but also between altitude and latitude. These relationships, applied to the vertical sequencing of biomes, are illustrated in Fig. 27-3. Note that it is not just mountains that have a tree line, a zone above which trees will not grow: the entire earth exhibits a tree line, poleward of which the northern coniferous forests are replaced by the stunted plants of the frigid tundra.

Another way to gain a perspective on the factors influencing the global distribution of biomes is represented

North America's Vegetation Regions

At the continental scale, or the level of spatial generalization below that which is used on the world map (Fig. 27-1), we see in more detail the results of the forces and processes that shape vegetation. The map of North America's vegetation displayed in Fig. 27-2, constructed by biogeographer Thomas Vale, parallels our world map of biome regions, most no-

FIGURE 27-2 Distribution of natural vegetation in North America.

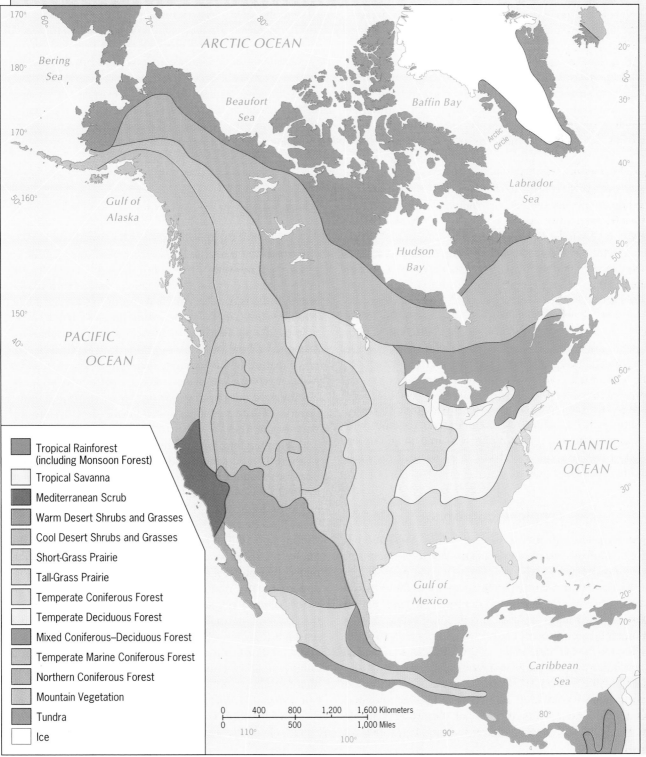

Tropical Rainforest (including Monsoon Forest)

Tropical Savanna

Mediterranean Scrub

Warm Desert Shrubs and Grasses

Cool Desert Shrubs and Grasses

Short-Grass Prairie

Tall-Grass Prairie

Temperate Coniferous Forest

Temperate Deciduous Forest

Mixed Coniferous–Deciduous Forest

Temperate Marine Coniferous Forest

Northern Coniferous Forest

Mountain Vegetation

Tundra

Ice

tably in its broad distinctions between forest, grassland, tundra, and desert zones. There are some differences too, but most are attributable to the downward cartographic shift from the global to the continental scale. In other words, Vale's map is a more detailed portion of Fig. 27-1, the detail made possible by the larger scale of the North American map.

This comparison of spatial frameworks again suggests the possibility of multiple regionalization schemes, with the shift to the North American scale raising a host of familiar problems concerning regional boundaries. An instance is the forest-desert boundary in the U.S. Pacific Northwest, which results from the rain shadow effect east of the Cascade Mountains (*see* Unit 13). On the world map (Fig. 27-1), that fairly rapid west–east change can be represented only by a line dividing two adjacent biomes. But on the North American map (Fig. 27-2) that now-fuzzy ''line'' has become a transition zone, and Vale finds it necessary to introduce a narrow, intermediate *mountain vegetation* region to contain the phytogeographical changes that occur across the Cascade range.

The Cascades example also reminds us that the geography of vegetation at the continental scale results from the same factors that operate at the biome region level—climate and terrain. Solar radiation and atmospheric circulation patterns are again the key climatic factors. Accordingly, much of the North American map can be related to the actions of processes that are discussed in Part Two, especially those shaping north–south temperature gradients and east–west moisture variations.

Terrain differences (which are treated in the survey of North American physiography in Units 52 and 53) markedly influence vegetation as well, and such major topographic features as the Cascades and the ranges of the Rocky Mountains are prominently visible in the western third of the United States in Fig. 27-2. To the east, topography appears to play a lesser role, but the southwestward-pointing prong of the *mixed coniferous-deciduous forest* region near the central northeastern seaboard clearly reflects the presence of the Appalachian Mountains. It also suggests the existence of yet another finer-scale mosaic of vegetation regions at the next lower level of spatial generalization.

in Fig. 27-4. This scheme has latitude increasing along the left side of the triangle and moisture decreasing along the base. Notice that forest biomes are generated within three latitudinal zones: tropical, temperate, and subarctic. Also note that the tropical forests develop more than one distinct biome before giving way to savanna and ultimately desert.

PRINCIPAL TERRESTRIAL BIOMES

Let us now examine the earth's major biomes, keeping in mind the tentative nature of any such regional scheme. Using Fig. 27-1 as our frame of reference, together with the accompanying photographs, we can gain an impression of the location and character of each of the eight biomes discussed below.

Tropical Rainforest

The **tropical rainforest biome**, whose vegetation is dominated by tall, closely spaced evergreen trees, is a teeming arena of life that is home to a great number and diversity of both plant and animal species. Alfred Russel Wallace wrote of the tropical rainforest in 1853: ''What we may fairly allow of tropical vegetation is that there is a much greater number of species, and a greater variety of forms, than in temperate zones.'' This was an understatement. We now know that more species of plants and animals live in tropical rainforests than in all the other world biomes combined. In fact, the rainforest often has as many as 40 species of trees per hectare (2.5 acres), compared to 8 to 10 species per hectare in temperate forests.

True climax tropical rainforest lets in little light. The crowns of the trees are so close together (*see* Fig. 26-2) that sometimes only 1 percent of the light above the forest reaches the ground. As a result, only a few shade-tolerant plants can live on the forest floor (*see* Fig. 17-2). The trees are large, often reaching heights of 40 to 60 m (130 to 200 ft). Because their roots are usually shallow, the bases of the trees are supported by buttresses. Another feature is the frequent presence of epiphytes and lianas. *Epiphytes* are plants that use the trees for support, but they are not parasites; *lianas* are vines rooted in the ground with leaves and flowers in the canopy, the top parts of the trees.

What organic matter there is decomposes rapidly, so there is little accumulation of litter on the rainforest floor. Although it is easy to walk through the true climax tropical rainforest, many areas contain a thick impenetrable undergrowth. This growth springs up where humans have destroyed the original forest. The areas where shifting agricultural practices are common are especially likely to exhibit such second growth. It has been estimated that most of the true tropical rainforest may disappear by the middle of the next century. Moreover, where natural vegetation is cleared from the oxisols of this biome, hardpans frequently develop through the extreme leaching of the upper soil that produces high concentrations of iron and aluminum compounds. Agriculture is difficult and minimally productive in such untillable soil.

Monsoon rainforests are included in this biome, even though they differ slightly from tropical rainforests. Mon-

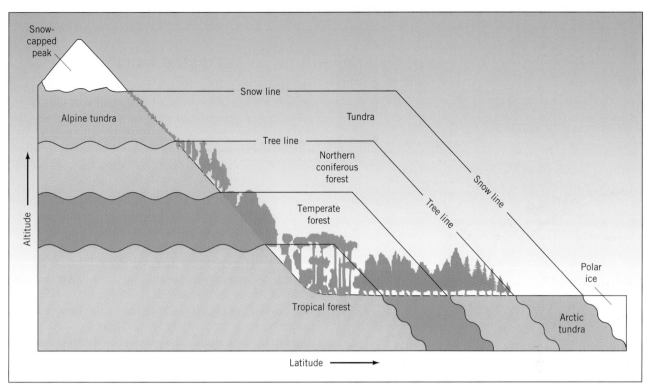

FIGURE 27-3 Vegetation changes with latitude and altitude. Temperature, which affects vegetation, decreases as one travels up a mountain or away from the equator, so that if there is plenty of moisture, vegetation is similar at high altitudes and at high latitudes, as shown here.

soon rainforests are established in areas with a dry season and therefore exhibit less variety in plant species. This vegetation is lower and less dense, and it also grows in layers or tiers composed of species adjusted to various light intensities.

Until recently, tropical rainforests covered almost half the forested area of the earth (Fig. 27-1). The largest expanses are in the Amazon River Basin in northern South America and the Zaïre (Congo) River Basin in west-equatorial Africa. A third area includes parts of

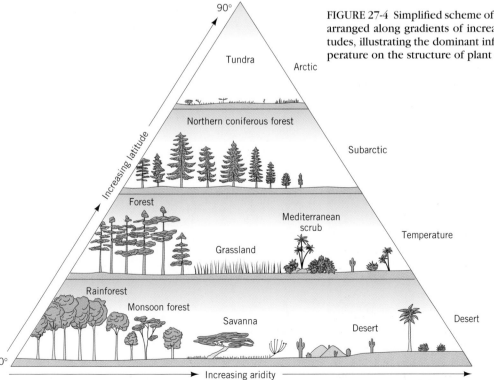

FIGURE 27-4 Simplified scheme of the major terrestrial biomes, arranged along gradients of increasing aridity at different latitudes, illustrating the dominant influence of moisture and temperature on the structure of plant communities.

FIGURE 27-5 The umbrella-like acacia rises above the lower vegetation in this area of East African savanna. Note the abundant grasses. The savanna thus supports grazers of grasses and browsers of trees—and other species, as the large eagle's nest in the central tree proves. (Authors' photo)

southeastern Asia, northeastern Australia, and the Indonesian island territories between them. Human destruction of rainforests (*see* first Perspective box in Unit 17) continues at an alarming rate, with devastating consequences for the tropics, and possibly far beyond.

FIGURE 27-6 The diversified vegetation in a desert biome. Despite the often thin and unproductive aridisols prevailing under desert climatic conditions, the occasional rains sustain a variety of plants, as in this area just north of Tucson, Arizona. (Authors' photo)

Tropical Savanna

The **savanna biome** encompasses a transitional environment between the tropical rainforest and the desert, consisting of tropical grassland with widely spaced trees (Fig. 27-5). Bulbous plants are abundant, but thorn forests, characterized by dense, spiny, low trees, are more obvious. The most common trees of the savanna are deciduous, and they lose their leaves in the dry season. These include the acacia and the curious water-storing, fat-trunked baobab tree. Grasses in the savanna are usually tall, sometimes growing to 5 m (16 ft), and have stiff, coarse blades.

Savanna vegetation has developed primarily because of the seasonally wet and dry climate (**Aw**) in large areas of Africa, South America, northern Australia, and India as well as in parts of Southeast Asia. However, periodic burning plays a significant role in limiting tree growth. In some places, the grasses form a highly inflammable straw mat in the dry season. This can be ignited through natural causes, but humans may play an integral role. Typical of the food chain of this biome are large herds of grazing animals (*see* Fig. 17-6).

Desert

The **desert biome** is characterized by sparse vegetation or even its complete absence (*see* Fig. 17-7). Whereas the grasses of the savanna are *perennials*, persisting from year to year, many of the desert plants are *emphemerals*, completing their entire life cycle in a single growing season. These emphemerals often grow quickly after the short but intense seasonal rains, covering open sandy or rocky areas in a spectacular display (Fig. 27-6). The seeds of these ephemerals often lie in the soils for many years and then germinate rapidly after a rainstorm. The perennial plants in the desert biome, such as cactuses and euphorbias (spurges), are dormant much of the year. Some, with fleshy water-storing leaves or stems, are known as *succulents*. Others have small leathery leaves or are deciduous. Woody plants have very long roots or are restricted to localized areas of water.

The sparse vegetation of desert ecosystems can support only small animals of the higher trophic levels. These animals are well adapted to the arid conditions. Rodents live in cool burrows; insects and reptiles have waterproof skins that help them retain water. The desert biome, of course, coincides with areas of arid (**BW**) climates. Aridisols and entisols are the most common soil orders associated with the desert biome.

Temperate Grassland

The **temperate grassland biome** generally occurs over large areas of continental interiors. Perennial and sod-forming grasses like those shown in Fig. 27-7 are dominant. The temperate grassland biome, like the savanna,

FIGURE 27-7 Bison grazing on prairie grass (temperate grassland biome) in a wildlife refuge in eastern Oklahoma.

is inhabited by herds of grazing animals and their predators. In North America, the short-grass prairie of the central Great Plains gradually gives way toward the east to the moister, richer, tall-grass prairie (Fig. 27-2) of what is now the Corn Belt. This transition in vegetation is accompanied by a transition in the soil layer from mollisols to alfisols (*see* Fig. 25-13).

This biome has been highly susceptible to human influence. As the U.S. example suggests, large areas have been turned over to crop and/or livestock farming. This is true of the interior areas of North America, the Pampas grasslands of Argentina, and the steppes of southern Rus-

sia, where the biome is most widespread (Fig. 27-1). Temperate grasslands maintain a delicate ecological balance, and mismanagement or climatic change quickly turns them into temperate forests or deserts (desertification is discussed in Unit 17).

Temperate Forest

There are several varieties of temperate forest. The **temperate deciduous forest biomes** occur in the eastern United States (Fig. 27-8), Europe, and eastern China. These forests of broadleaf trees are shared by herbaceous plants, which are most abundant in spring before the growth of new leaves on the trees. An outstanding characteristic of temperate deciduous forests is the similarity of plants found in their three locations in the Northern Hemisphere. Oak, beech, birch, walnut, maple, elm, ash, and chestnut trees are all common. As with the temperate grasslands, large areas of this forest type have been cleared and turned over to agricultural activities.

Temperate evergreen forest biomes are found on western coasts in temperate latitudes where abundant precipitation is the norm. In the Northern Hemisphere, they take the form of needleleaf forests. The coastal redwoods and Douglas firs of the northwestern coast of North America are representative (Fig. 27-9). The podocarps of the temperate evergreen rainforest on the western coast of New Zealand exemplify the broadleaf and small-leaf evergreen forests of the Southern Hemisphere.

Mediterranean Scrub

The Mediterranean climate (*see* Unit 18) is characterized by hot dry summers and cool moist winters. Such climates prevail along the shores of the Mediterranean Sea, along the coast of California, in central Chile, in South

FIGURE 27-8 Deciduous broadleaf trees near the peak of their fall foliage ring a quiet lake in northern Connecticut.

FIGURE 27-9 The lush Douglas fir forest of the U.S. Pacific Northwest.

FIGURE 27-10 The landscape of Mediterranean vegetation along the Italian coast.

Africa's Cape Province, and in southern and southwestern Australia (*see* Fig. 16-3). The **Mediterranean scrub biome** corresponds to these **Csa** and **Csb** climates. The vegetation of this biome consists of widely spaced evergreen or deciduous trees (pine and oak) and often dense, hard-leaf evergreen scrub. Thick waxy leaves are well adapted to the long, hot, dry summers.

Mediterranean vegetation creates a very distinctive natural landscape (Fig. 27-10). Even though this biome's regions are widely separated and isolated from each other, their appearance is quite similar. In coastal California, the Mediterranean landscape is called *chaparral*; in the Mediterranean region of southern Europe, it is referred to as *maquis*; in Chile it is known as *mattoral*; and in South Africa it is called *fynbos*. Unless you know species that are unique to each of these locales, you could easily mistake one regional landscape for the other.

Mediterranean regions are among the world's most densely populated and most intensively cultivated. Human activity has profoundly altered the Mediterranean biome through the use of fire, through grazing, and through agriculture. Today vineyards and olive groves have replaced countless hectares of natural chaparral and maquis. Great stands of trees have been removed, and many species of fauna have been driven away or made extinct. Before the rise of ancient Greece, the hills of the Greek peninsula were covered with oaks and pines that had adapted to this climatic regime. Only a few of the legendary cedars of Lebanon now survive. The cork oak, another example of adaptation, still stands in certain corners of the Mediterranean lands. From these remnants, we deduce that the Mediterranean biome has been greatly modified, but it has not lost its regional identity.

Northern Coniferous Forest

The upper-midlatitude **northern coniferous forest biome** goes by many different names. In North America it takes a Latin name to become the *boreal forest*. In Russia it is called the *snowforest* or *taiga*. The most common *coniferous* (cone-bearing) trees in this biome are spruce, hemlock, fir, and pine. These needleleaf trees can withstand the periodic drought resulting from long periods of freezing conditions. The trees are slender and grow to heights of 12 to 18 m (40 to 60 ft); they generally live less than 300 years but grow quite densely (Fig. 27-11). Depressions, bogs, and lakes hide among the trees. In such areas, low-growing bushes with leathery leaves, mosses, and grasses rise out of the waterlogged soil. This type of growth assemblage, combined with stunted and peculiarly shaped trees, is known as *muskeg*.

All the biomes discussed here could be differentiated even more precisely. For instance, the boreal forest of Canada can be divided into three subzones: (1) the main boreal forest, which is characterized by the meeting of the crowns of the trees; (2) the open boreal woodland, marked by patches where the trees are broken up by open spaces of grass or muskeg; and (3) a mixture of woodland in the valleys and tundra vegetation on the ridges, called forest tundra, that is found along the polar margins of this biome.

Tundra

The **tundra biome** is the most continuous of all the biomes, and it occurs almost unbroken along the poleward margins of the northern continents (Fig. 27-1). It is also

FIGURE 27-11 The *taiga* (snowforest) blankets more than half of Russia, the world's largest country in territorial size. This view is across the heart of the Yenisey Basin in western Siberia.

found on the islands near Antarctica and in alpine environments above the tree line on mountains at every latitude. Only cold-tolerant plants can survive under harsh tundra conditions. The most common are mosses, lichens, sedges, and sometimes, near the forest border, dwarf trees (*see* Fig. 19-7). Ephemeral plants are rare; the perennial shrubs are pruned back by the icy winter winds and seldom reach their maximum height. Nor can plant roots be extensive in this biome because the top of the permafrost is seldom more than 1 m (3.3 ft) below the surface. The permafrost also prevents good surface drainage.

During the short summer, shallow pools of water at the surface become the home of large insect populations.

In the Northern Hemisphere, birds migrate from the south to feed on these insects. The fauna is surprisingly varied, considering the small biomass available. It consists of such large animals as reindeer, caribou, and musk ox and such small herbivores as hares, lemmings, and mouse-like voles. Carnivores include foxes, wolves, hawks, falcons, owls, and, of course, people.

This survey of the earth's principal terrestrial biomes has concentrated on the vegetation that dominates their landscapes. A biome, however, consists of more than plants; the animals that form part of its biological community also must be considered. In Unit 28 we look at the geography of fauna: zoogeography.

KEY TERMS

biome (p. 289)

desert biome (p. 294)

Mediterranean scrub biome (p. 296)

northern coniferous forest biome (p. 296)

savanna biome (p. 294)

temperate deciduous forest biome (p. 295)

temperate evergreen forest biome (p. 295)

temperate grassland biome (p. 294)

tropical rainforest biome (p. 292)

tundra biome (p. 296)

REVIEW QUESTIONS

1. What is a biome?
2. What climatic and terrain features influence the distribution of terrestrial biome regions?
3. List and briefly describe each of the earth's eight principal terrestrial biomes.

REFERENCES AND FURTHER READINGS

BARBOUR, M. G., and BILLINGS, W. D. *North American Terrestrial Vegetation* (London/New York: Cambridge University Press, 1988).

COLLINSON, A. S. *Introduction to World Vegetation* (Winchester, Mass.: Unwin Hyman, 2nd ed., 1988).

EYRE, S. R. *Vegetation and Soils: A World Picture* (London: Edward Arnold, 2nd ed., 1975).

GOOD, R. *The Geography of the Flowering Plants* (London: Longman, 4th ed., 1974).

HENGEVELD, R. *Dynamic Biogeography* (London/New York: Cambridge University Press, 1990).

KELLMAN, M. C. *Plant Geography* (New York: St. Martin's Press, 2nd ed., 1980).

KÜCHLER, A. W. *Vegetation Mapping* (New York: Ronald Press, 1967).

KÜCHLER, A. W., and ZONNEVELD, I. S., eds. *Handbook of Vegetation Science 10: Vegetation Mapping* (Hingham, Mass.: Kluwer, 1988).

MAURER, B. *Geographical Analysis of Biodiversity* (Cambridge, Mass.: Blackwell, 1994).

MORIN, N., chief ed. *Flora of North America North of Mexico* (New York: Oxford University Press, 14 projected vols., 1992–2004).

POLUNIN, N. V. *Introduction to Plant Geography and Some Related Sciences* (New York: McGraw-Hill, 1960).

RILEY, D., and YOUNG, A. *World Vegetation* (London/New York: Cambridge University Press, 1966).

SAUER, J. D. *Plant Migration: The Dynamics of Geographic Patterning in Seed Plant Species* (Berkeley: University of California Press, 1988).

VALE, T. R. *Plants and People: Vegetation Change in North America* (Washington, D.C.: Association of American Geographers, Resource Publications in Geography, 1982).

VANKAT, J. L. *The Natural Vegetation of North America: An Introduction* (New York: Wiley, 1979).

VAVILOV, N. I. *Origin and Geography of Cultivated Plants* (New York: Cambridge University Press, 1992).

WALTER, H. *Vegetation of the Earth and Ecological Systems of the Geo-Biosphere* (New York: Springer-Verlag, 2nd ed., 1984).

WILSON, E. O. *The Diversity of Life* (Cambridge, Mass.: Belknap/Harvard University Press, 1992).

WOODWARD, F. I. *Climate and Plant Distribution* (London/New York: Cambridge University Press, 1987).

Zoogeography: Spatial Aspects of Animal Populations

OBJECTIVES

◆ To briefly outline the theory of evolution and related principles influencing the spatial distribution of animals.

◆ To give a brief history of zoogeography.

◆ To relate zoogeography to the larger context of environmental conservation.

The principal terrestrial biomes discussed in Unit 27 are based primarily on the distribution of dominant vegetation. But the distribution of fauna (animals) is closely associated with plants, and, as we noted earlier, with soils that sustain the flora. A biome, therefore, actually is an interacting set of ecosystems that extends over a large area of the earth. Its establishment and maintenance depend not only on climate and soils but also on plants and animals—animals ranging from the tiniest bacteria to the largest mammals.

PROCESSES OF EVOLUTION

To appreciate the work of zoogeographers (their field is pronounced *ZOH-oh-geography*), we should take note of some aspects of the theory of evolution. This theory has led to our understanding that both variety and order

UNIT OPENING PHOTO: A member of the world's most famous near-extinct species roaming through a ravine in a wildlife preserve in southern China.

mark life on this planet. Basic to evolutionary theory is the concept of natural variation within a single species. This natural variation stems mainly from the reproductive process. Information carriers from each parent, called *genes*, join in such a way as to combine a small degree of randomness (chance) with a high degree of specification (stability). Thus a human child may have blue or green eyes, but very likely only two arms and legs.

The mechanism of specification (which confers continuity on the species) sometimes breaks down, and when this happens the exact message of heredity is not passed on. The result is a **mutation** (an inheritable change in the DNA of a gene), and a new species may originate from such an occurrence. Another important part of evolutionary theory holds that a species will produce more offspring than can survive to reproduce. In all species, many immature individuals die by accident, disease, or predation. We encountered this idea in studying food chains, where we saw that autotrophs are eaten by herbivores and herbivores are consumed by carnivores. But while individuals are often killed, the entire population continues to evolve. No matter how many antelope are eaten by lions on the African plains, the herds return every year.

An important related principle has to do with the place where a species can best sustain itself and thrive. This is referred to as its **ecological niche** (or, in the zoogeographic context, simply *niche*), the environmental space within which an organism operates most efficiently. Some niches are very large, perhaps coinciding with entire biomes or continental parts of biomes. The

ecological niche of the South American jaguar is a substantial part of the Amazonian rainforest. Other niches are very small. A specialized (or very small) niche reduces competition from other species, but it also increases the risk of total annihilation, perhaps resulting from a change in the natural environment. A large niche may overlap other niches, causing competition, but it has the advantage of permitting adjustment in the event of environmental change.

In zoogeography, we focus mainly on the larger ecological niches. These are so complex that they are better termed habitats. The **habitat** of a species is the environment it normally occupies within its geographic range. A habitat usually is described in quite general terms, such as "grassland" or "seashore" or "alpine." Obviously, each of these habitats contains many smaller ecological niches.

Now we come to a zoogeographic-evolutionary principle of great importance. It holds that some offspring are better able to adapt to their habitat or niche than others. One familiar example of this idea is the evolution of the giraffe's long neck. We may assume that neck length varied in previous giraffe populations. The animals with the longest necks could reach higher into the trees and thereby obtain more food than giraffes with shorter necks. Shorter-necked individuals had to compete for food near the ground with several other species. These shorter-necked giraffes were less successful, and over time the longer-necked giraffes came to prosper and dominate the species (Fig. 28-1). This implies, then, that better-adapted organisms are more likely to survive and

FIGURE 28-1 In the evolution of giraffes, the longer-necked animals were the most prosperous and produced more offspring, who passed the genetic information concerning the long neck to later generations.

reproduce, whether they are giraffes or camouflaged insects. Through their genes, they pass on favorable aspects of their adaptation to successors.

This evolutionary process necessarily requires time, and adaptive changes come slowly. But, as we know, while adaptation takes place, environments are also changing. When the earth's climates change, as they have many times during its history, habitats are modified or relocated. Some species are unable to adapt further or migrate to new abodes, whereas others achieve new successes. The process is endless, and the evidence exists everywhere in the natural landscape.

A good example of complex adaptations to a shared habitat is the Serengeti Plain of East Africa, an area of savanna grassland grazed by many species. The herbivores are so finely adapted that they use different portions of the Serengeti's vegetation at different times. During part of the year, mixed herds graze on the short grass, satisfying their protein requirements without requiring them to use too much energy in respiration while obtaining the food. Eventually, the short grass becomes overgrazed. Then the largest animals, the zebras and buffaloes, move into areas of mixed vegetation—tall grasses, short grasses, and herbs. Zebras and buffaloes eat the stems and tops of the taller grasses, and trample and soften the lower vegetation. The wildebeests, which resemble the American buffalo somewhat in appearance, can then graze the middle level of vegetation, trampling the level below. Next, the Thompson's gazelles move into the softened area to eat the low leaves, herbs, and fallen fruit at ground level.

The pastoral indigenous people of the Serengeti have long been part of this well-balanced ecosystem. Some scientists believe that their regular use of fire has helped to maintain the short grass best adapted to the native animals. In its totality, the Serengeti ecosystem perfectly illustrates the arrangement of plants and animals that has developed under the rules governing both energy flow and evolution. An understanding of these rules can explain both the nature of species and their geographic spread throughout the biosphere.

EMERGENCE OF ZOOGEOGRAPHY

The field of zoogeography began to develop following the publication of works by the great naturalist-explorer Alexander von Humboldt (1769–1859) and the biologist Charles Darwin (1809–1882). Von Humboldt's maps and drawings of plants in their environmental settings and of animals encountered on his adventurous explorations formed the first hard evidence for early zoogeographic theories. Darwin's momentous 1859 work on evolution, *The Origin of Species*, spurred ideas about environment and adaptation. Through his remarkable field

studies, Darwin focused attention on some unique zoogeographic areas that have remained at the center of research ever since, especially the Galapagos Islands in the Pacific off Ecuador. During the middle part of the nineteenth century, information about the distribution of species and ideas about habitats and food chains began to crystallize quite rapidly.

As more became known about animals and plants in places distant from the centers of learning, urgent zoogeographic questions presented themselves. One of the most interesting involved the transition of fauna from Southeast Asia to Australia. Somewhere in the intervening Indonesian archipelago, the animals of Southeast Asia give way to the very different animals typical of Australia. Australia is the earth's last major refuge of *marsupials*, animals whose young are born very early in their development and are then carried in a pouch on the abdomen. (Kangaroos, wallabies, wombats, and koalas are among Australia's many marsupials.) Australia also is the home of the only two remaining egg-laying mammals, the platypus (Fig. 28-2) and the spiny anteater (*echidna*). Marsupials are found in New Guinea and on islands in the eastern part of Indonesia, but nonmarsupial animals (such as tigers, rhinoceros, elephants, and primates) prevail in western Indonesia. Where, then, lies the zoogeographic boundary between these sharply contrasting faunal assemblages?

One answer to this still debated question was provided by Alfred Russel Wallace in an important work published in 1876, *The Geographical Distribution of Animals*. Wallace mapped what was then known about the animals of Indonesia. He showed how far various species had progressed eastward along the island stepping stones between mainland Southeast Asia and continental Australia; he also mapped the westward extent of marsupials. On the basis of these and other data, Wallace drew a line across the Indonesian archipelago, a zoogeographic boundary between Southeast Asia's and Australia's fauna (Fig. 28-3). *Wallace's Line* lay between Borneo and Sulawesi (Celebes), and between the first (Bali) and second (Lombok) islands east of Java. This line became one of the most hotly debated zoogeographic delimitations ever drawn, and to this day it remains one of the more intensely discussed boundaries in all of physical geography.

As more evidence was gathered concerning the complex fauna of the Indonesian archipelago, other zoogeographers tried to improve on Wallace's Line. In the process, they proved how difficult the problems of regional zoogeography can be. The zoologist Max Weber argued that Wallace's Line lay too far to the west. His alternative line (Fig. 28-3) was placed just west of New Guinea and north-central Australia, making virtually all of Indonesia part of the Southeast Asian faunal region. The substantial distance between the Wallace and Weber

FIGURE 28-2 The platypus, one of the earth's two egg-laying mammal species. It is found only in Australia, which is home to a number of unusual fauna.

boundaries underscores that zoogeographical data can be interpreted differently.

Many biogeographers would agree that the delimitation of vegetative regions presents fewer problems than faunal regions do. It is one thing to draw maps of stands of rainforest, an expanse of desert, or the boundary between taiga and tundra. But to do the same for African elephants, American jaguars, or Indian tigers is quite another matter. Animals move and migrate, their range (region of natural occurrence) varies, and they may simply be difficult to locate and enumerate.

THE EARTH'S
ZOOGEOGRAPHIC REALMS

A global map of zoogeographic realms, therefore, is an exercise in generalization, but it does reflect evolutionary centers for animal life as well as the work of natural barriers over time (Fig. 28-4). Note that zoogeographic realm boundaries in several areas coincide with high mountains (Himalayas), broad deserts (Sahara, Arabian), deep marine channels (Indonesia), and narrow land bridges (Central America).

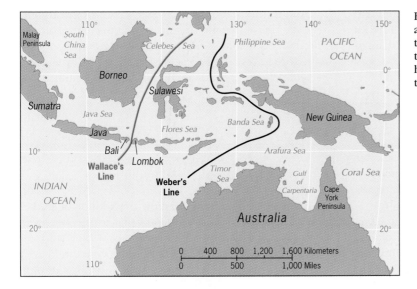

FIGURE 28-3 Wallace's Line across the Indonesian archipelago. This controversial boundary between the faunal assemblages of Southeast Asia and Australia was challenged by Max Weber, who placed his alternative Weber's Line much closer to Australia and New Guinea.

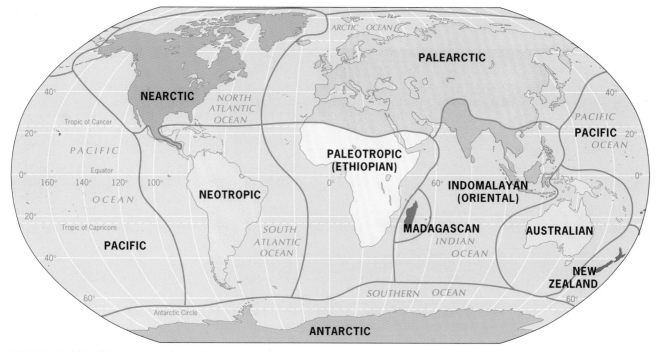

FIGURE 28-4 World zoogeographic realms.

By many measures, the *Paleotropic (Ethiopian) realm* contains the earth's most varied fauna, an enormously rich assemblage of animals, many of which are unique to this realm. Some species, however, have relatives in other realms; for example, the lion and the elephant also occur in the *Indomalayan (Oriental) realm*. Nevertheless, the Indomalayan realm has less diverse fauna. The fauna of the island of *Madagascar*, shown on our map as a discrete realm, differs quite strongly from that of nearby Africa. While Madagascar is home to the marsupial lemur, a small primate, the island has nothing to match East Africa's herd animals, lions, or even poisonous snakes. Just how an island so near the Paleotropic realm became so distinctive zoogeographically remains a problem without a satisfactory solution.

The *Australian realm's* faunal assemblage exhibits the consequences of prolonged isolation and separate evolution. This is the realm of marsupials such as the kangaroo and wombat; of the platypus and the Tasmanian "tiger." While some biogeographers include New Zealand in the Australian realm, others map *New Zealand*, like Madagascar, as a discrete realm. There is ample geographic reason for doing so: New Zealand's fauna include no mammals, very few terrestrial vertebrates of any kind, and nothing to match the assemblage of Australia—except when it comes to birds. New Zealand has a rich of variety of bird life, with an unusual number of flightless species (represented in Africa by the large ostrich and in Australia by the smaller emu).

The *Neotropic realm* also has a rich and varied faunal assemblage that includes such species as the pig-like tapir, the jaguar, and the boa constrictor. In both plants and animals, biogeographers can discern evidence of the theory of **convergent evolution**, which holds that organisms in widely separated biogeographic realms, though descended from diverse ancestors, develop similar adaptations to measurably similar habitats.

The two remaining realms, the *Nearctic* and the *Palearctic*, are much less rich and much less diverse than the other major zoogeographic realms. Some biogeographers prefer to map these together as a single realm, but evidence of long-term isolation (the Bering land bridge notwithstanding) is much stronger in the North American Nearctic than in the Eurasian Palearctic realm. Remarkable adaptations exist, such as the polar bear (both realms), the Siberian tiger and giant panda (Palearctic), and the bison and mountain lion (Nearctic).

Figure 28-4 presents the global zoogeographic map at a high level of generalization, and we should remember that this is a map of *realms*. Embedded within each realm are numerous subdivisions or zoogeographic *regions*. Let us use the Australian realm as an example. The fauna of mainland Australia, despite commonalities with the nearby large Pacific island of New Guinea, nonetheless differs sufficiently to make New Guinea and Australia separate zoogeographic regions within the Australian realm.

FURTHER STUDIES IN ZOOGEOGRAPHY

As interest in zoogeography grew, new kinds of studies were undertaken. It may be said that in the twentieth century zoogeography became more ecological and less cartographic. Zoogeographers became concerned with changing environments, past faunal migration routes, and present migratory habits, such as the flight paths of

302

birds between summering and wintering grounds. You can study this change in the field by comparing Wallace's book, published more than a century ago but reprinted in 1962, with a volume carrying a very similar title published in 1957, *Zoogeography: The Geographical Distribution of Animals,* by Philip J. Darlington, Jr. Darlington posed four questions for zoogeographers: (1) What is the main pattern of animal distribution? (2) How has this pattern been formed? (3) Why has this pattern developed as it has? (4) What does animal distribution—past and present—tell us about lands and climates? Note that the first question is only the beginning, not the entire objective as was the case in Wallace's work.

By the time Darlington did his research, much more was known about ice ages, shifting landmasses, and changing sea levels than was known in Wallace's time. It was not long before *ecological zoogeography,* the study of animals as they relate to their total environment, became a leading theme. A book published in 1965, *The Geography of Evolution,* by G. G. Simpson, deals with such topics as adaptation, competition, and habitats.

Island Zoogeography

Biologists believe that there may be as many as 30 million species of organisms on our planet, of which only 1.5 million have been identified and classified. Obviously, many species live in the same habitat, often sharing ecological niches or parts of niches. What determines how many species can be accommodated in a specific, measured region? This is a central question in modern zoogeography. Where ranges and niches overlap, as on large landmasses, it is not practical to calculate the number of species an area can sustain because there are no controllable boundaries. But an island presents special opportunities.

Small islands can be inventoried completely, so that every species living on one is accounted for. When zoogeographers did this work, their research had some expected and some unexpected results. First it was learned that the number of species living on an island is related to the size of that island. Given similar environmental conditions, the larger the island, the larger the number of species it can accommodate. We might have anticipated this, except that the larger island does not have a wider range of natural environments, so that it is the size, not the internal variability that might come with it, that affects the species total. Moreover, it was discovered that an island of a given size can accommodate only a limited number of species. New species may arrive, brought by birds from far away or entering via debris washing up onshore. If the island already has a stable population, these new species will not succeed unless another species first becomes extinct.

This conclusion, that an island's capacity to accommodate species has a numerical limit, was tested on some small islands off the coast of Florida. Zoogeographers counted the species of insects, spiders, crabs, and other arthropods living on each island. Then they sprayed the islands, wiping out the entire population. After a few years, the islands were reinhabited. The same number of species had returned to them, although the number of individuals was proportionally different. That is, while there were crabs and spiders before the extinction, now there were more crabs and fewer spiders. This indicates not only that a certain number of species is characteristic of an island in a certain environment, but also that the kinds of species making up this number varies and may depend on the order in which they arrived to fill the available ecological niches.

Biogeographers can also learn about niches and habitats when new land is formed—for example, following the formation of a volcanic island. In 1883, the volcanic island of Krakatau in western Indonesia exploded, denuding the entire island and creating a wholly new and barren landscape. Soon new soil began to form on the fertile lava base, and plant seeds arrived by air and on birds. New vegetation could be seen within a few years (Fig. 28-5), and soon the animal repopulation be-

FIGURE 28-5 Plants manage to establish themselves quite soon after new rocks are created by volcanic eruptions, as evidence from Krakatau and Mount St. Helens has confirmed. This small fissure in a recent eruption in Tanzania, East Africa was exploited within a few years by seeds that took hold at its base, where weathered particles and moisture supplied the essentials for growth. (Authors' photo)

The African Stowaways*

Around 1929, a few African mosquitoes arrived in Brazil. They had probably stowed away aboard a fast French destroyer in the West African city of Dakar. Once in Brazil, the immigrants established a colony in a marsh along the South Atlantic coast. Although the residents of a nearby town suffered from an unusual outbreak of malaria, nobody seemed to notice the presence of the foreign mosquitoes for a long time. Meanwhile, the insects settled comfortably in their new environment, and during the next few years they spread out over about 320 km (200 mi) along the coast. Then, in 1938, a malaria epidemic swept across most of Northeast Brazil. A year later the disease continued to ravage the region; hundreds of thousands of people fell ill, and nearly 20,000 died.

Brazil always had malaria-carrying mosquitoes but none quite like the new African variety. Whereas the native mosquitoes tended to stay in the forest, the foreign pests could breed in sunny ponds outside the forest, and they quickly made a habit of flying into houses to find humans to bite. With the mosquito transplantation problem finally understood, the Brazilian government, aided by the Rockefeller Foundation, hired more than 3000 people and spent over $2 million to attack the invaders. After studying the ecological characteristics of the "enemy," they sprayed houses and ponds. Within three years the battle wiped out the African mosquitoes in South America. Brazil also initiated a quarantine and inspection program for incoming aircraft and ships to keep out unwanted stowaways.

High-speed transportation developed by humans has inadvertently carried many plants and animals to other continents. In their new environment, the immigrants may react in one of three ways: they may languish and die; they may fit into the existing ecosystem structure; or they may tick away like a time bomb and eventually explode like the African mosquitoes.

Another accidental African arrival that could not be controlled in Brazil, the so-called African "killer bee," threatens the United States today. These bees escaped from a Brazilian research facility in 1957 and have been slowly spreading out across the Americas ever since. By 1980, the bees had entered the Central American land bridge and were steadily progressing northward. By the early 1990s, they had crossed all of Mexico and penetrated Texas across the Rio Grande. Although joint U.S.-Mexican efforts to slow the advance of the bees were moderately successful in the late 1980s, the U.S. Department of Agriculture was forced to abandon hope that the invaders could be repelled, and USDA is now busy implementing strategies to cope with this spreading pest on American soil.

*The source for most of this box is Elton, 1971.

---

gan. The arrival of new species was monitored, adding to our knowledge about the occupation of available niches and habitats. Such knowledge permitted optimistic predictions about the revegetation and recovery of Mount St. Helens in Washington State following its destructive eruption of 1980. Just three years after this eruption, 90 percent of the plant species that originally inhabited the devastated slope had already reestablished themselves.

What has been learned about island zoogeography also applies to other isolated places, such as the tops of buttes and mesas (*see* Fig. 42-6) where steep slopes act as barriers to invasion. The stable populations atop such landforms resemble island populations in some ways, but in other ways they relate more closely to surrounding ecosystems. By studying these hilltop faunal assemblages, zoogeographers come still closer to an understanding of the complex relationships that characterize habitats and niches generally. We now turn to how the distribution of animal life is influenced by humans, who,

even by their inadvertent actions (*see* Perspective: The African Stowaways), can produce widespread negative consequences.

ZOOGEOGRAPHY AND CONSERVATION

Research in zoogeography often has important implications for the survival of species. Knowledge of the numbers, range, habits, and reproductive success of species is crucial to their effective protection. The explosive growth of the earth's human population and the destructive patterns of consumption by prospering peoples have combined to render many species of animals extinct, endangered, or threatened. Awareness of the many threats to the remaining fauna has increased in recent decades, and some species have been salvaged from the brink of extinction. But for many others hope is fading.

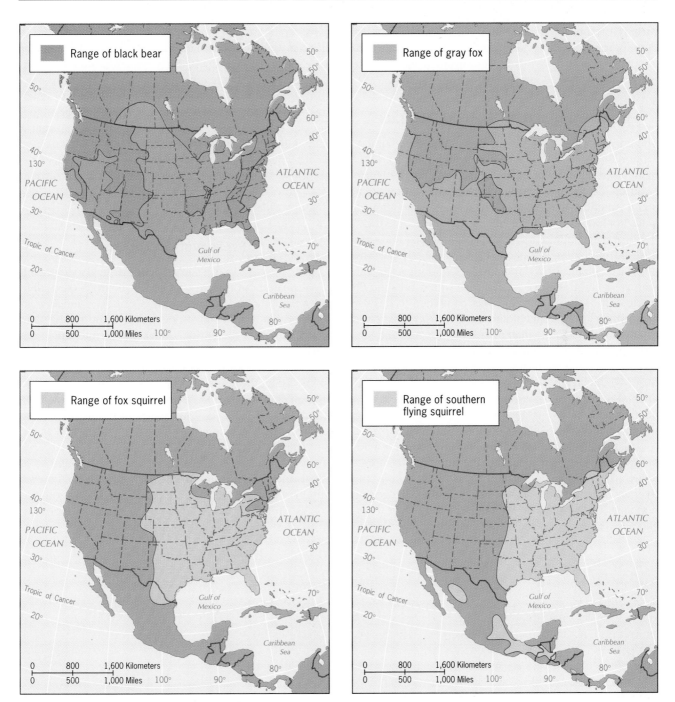

FIGURE 28-6 Ranges of some North American mammals.

Animal Ranges

An important zoogeographic contribution to research in support of species conservation lies in detailed studies of the spatial properties of *animal ranges*. This concept is discussed in Unit 26, and the ranges of four North American mammals are mapped in Fig. 28-6. It is not just the size of a range, however, but also its geographic pattern that will determine whether a species can be expected to survive. The case of the spotted owl in the U.S.

Pacific Northwest is a good example. What this owl needs is old forest, not the trees of new forestation. It needs soft and rotting wood so that it can establish nests in the trunks; those neatly reforested slopes do not have such old decaying trees. The range of the spotted owl is therefore restricted to the dwindling stands of older forest, and the map conservationists need will show where those patches of remaining old forest are located. The pattern of these patches—how large they are, how densely forested, how close together—will reveal more about the prospects of the spotted owl than any estimate of the total size of its range.

Human Impact on Animal Habitats

The effects of human encroachment on animal habitats is another important sphere of zoogeographic study. In many parts of the world—in the foothills of the Himalayas, on the plains of East Africa, in the forests of tropical South America—people and animals are competing for land. Not long ago, substantial parts of India teemed with herds of wild buffalo, many kinds of antelope and deer as well as their predators, lions and tigers, and large numbers of elephants and rhinoceros. The South Asian human population explosion, together with a breakdown in the conservation program have combined to destroy one of the earth's great wildlife legacies. The Indian lion is virtually extinct; the great tiger is endangered and survives in only small numbers in remote forest areas. In East Africa, population pressure in the areas fringing the major wildlife reserves, together with poaching, are decimating the fauna. In the Amazon Basin, rainforest habitat destruction (*see* Perspective box in Unit 17) threatens the survival of many species of animals as well as plants.

All this is reminiscent of what happened to the wildlife of North America following the arrival of the Europeans. The American conservation effort during the twentieth century has concentrated on the salvageable remnants of this realm's wildlife, and there have been successes as well as setbacks. The protection of wildlife

poses complex problems. Special-interest groups such as hunters sometimes demand the right to shoot particular species, endangering what may be a delicate balance in the ecosystem. Others object to closing wilderness areas to the public for any purpose, even sightseeing, when this is necessary to allow an endangered area or animal species to recuperate. And wildlife is mobile, so that it may migrate beyond the boundaries of a wildlife refuge or national park, thus becoming endangered. The protection of birds, whose migratory habits may carry them across the length of the continent, is made especially difficult by their mobility.

Preservation Efforts

Conservationists in the United States have also learned a great deal about the most effective methods of wildlife preservation. This is not simply a matter of fencing off an area where particular species exist, but involves the management of the entire habitat, the maintenance of a balance among animals and plants—and, sometimes, people. A series of congressional actions and commissions during the 1960s and 1970s served to facilitate and implement the newly developed conservation practices. There have been several successes, including the revival of the bison, the wild turkey, and the wolf as well as the survival of the grizzly bear and bald eagle.

In European countries, as well as in the countries of the former Soviet Union to the east, the story of conservation efforts is largely one of remnant preservation. Europe's varied wildlife fell before the human expansion; it was lost even earlier than America's wildlife. Conser-

FIGURE 28-7 A wildlife refuge in southern Kenya. This richness of animal life prevailed across much of East Africa before the arrival of the European colonialists, who upset the ecological balance among native people, their livestock, and the region's wildlife. (Authors' photo)

vation is an expensive proposition, afforded most easily by the wealthier developed countries. European colonial powers carried the concept to their colonies, where they could carve wildlife refuges from tribal lands with impunity. In many areas of Africa and Asia, the Europeans found the indigenous population living in harmonious balance with livestock and wildlife. The invaders disturbed this balance, introduced the concept of hunting for profit and trophies, and then closed off huge land tracts as wildlife reserves with controlled hunting zones (Fig. 28-7). That the African wildlife refuges have for so long survived the period of decolonization (some are now in danger) is testimony to a determination that was not present when orgies of slaughter eliminated much of North America's wildlife heritage. Americans may have advanced knowledge of conservation theory and management practices, but are in no position to proselytize.

Zoogeography, therefore, has many theoretical and practical dimensions. The spatial aspects of ranges, habitats, and niches require research and analysis. The results of such investigations are directly relevant to those who seek to protect wildlife and who make policy to ensure species survival.

KEY TERMS

convergent evolution (p. 302)   ecological niche (p. 299)   habitat (p. 299)   mutation (p. 299)

REVIEW QUESTIONS

1. What is an *ecological niche* and how is it related to a habitat?

2. How were Alexander von Humboldt and Alfred Russel Wallace instrumental in developing the field of zoogeography?

3. In what ways are animal and plant conservation similar?

4. In what ways do animal and plant conservation differ?

5. What have biogeographers learned about ecological niches and habitats from their fieldwork on islands and land newly emerged from the sea?

REFERENCES AND FURTHER READINGS

DARLINGTON, P. J., JR. *Zoogeography: The Geographical Distribution of Animals* (New York: Wiley, 1957 [reprinted 1980 by Krieger]).

DASMANN, R. F. *Environmental Conservation* (New York: Wiley, 5th ed., 1984).

DE LAUBENFELS, D. J. *A Geography of Plants and Animals* (Dubuque, Iowa: Wm. C. Brown, 1970).

ELTON, C. S. "The Invaders," in T. R. Detwyler, ed., *Man's Impact on Environment* (New York: McGraw-Hill, 1971), pp. 447–458.

HENGEVELD, R. *Dynamic Biogeography* (London/New York: Cambridge University Press, 1990).

ILLIES, J. *Introduction to Zoogeography* (New York: Macmillan, 1974).

JARVIS, P. J. *Plant and Animal Introductions* (New York: Blackwell, 1988).

MACARTHUR, R. H., and WILSON, E. O. *The Theory of Island Biogeography* (Princeton, N.J.: Princeton University Press, 1967).

MAURER, B. *Geographical Analysis of Biodiversity* (Cambridge, Mass.: Blackwell, 1994).

NEWBIGIN, M. I. *Plant and Animal Geography* (London: Methuen, 1968).

NOWAK, R. M., and PARADISO, J. L. *Walker's Mammals of the World* (Baltimore: Johns Hopkins University Press, 2 vols., 4th ed., 1983).

SIMPSON, G. G. *The Geography of Evolution: Collected Essays* (Philadelphia: Chilton Books, 1965).

WALLACE, A. R. *The Geographical Distribution of Animals; With a Study of the Relations of Living and Extinct Faunas as Elucidating the Past Changes of the Earth's Surface* (New York: Hafner [reprint of the 1876 original], 1962).

WHITMORE, T. C. *Wallace's Line and Plate Tectonics* (London/New York: Oxford University Press [Clarendon], 1981).

WILSON, E. O. *The Diversity of Life* (Cambridge, Mass.: Belknap/Harvard University Press, 1992).

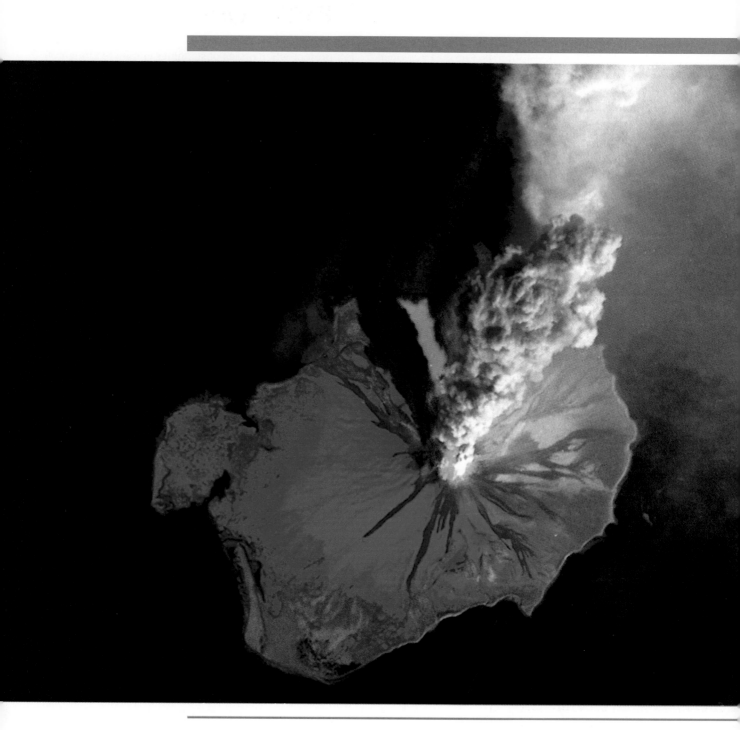

UNIT 29
PLANET EARTH IN PROFILE:
THE LAYERED INTERIOR

UNIT 30
MINERALS AND IGNEOUS ROCKS

UNIT 31
SEDIMENTARY AND METAMORPHIC ROCKS

UNIT 32
PLATES OF THE LITHOSPHERE

PART FOUR

THE RESTLESS CRUST

UNIT 33
PLATE MOVEMENT: CAUSES AND EFFECTS

UNIT 34
VOLCANISM AND ITS LANDFORMS

UNIT 35
EARTHQUAKES AND LANDSCAPES

UNIT 36
SURFACE EXPRESSIONS OF SUBSURFACE
STRUCTURES

UNIT 29

Planet Earth in Profile: The Layered Interior

OBJECTIVES

◆ To outline the relevant properties of the earth's five internal layers and to discuss some of the evidence leading to their discovery.

◆ To introduce the salient properties of the earth's lithosphere, the nature of the crust, and the underlying mantle.

◆ To provide an outline of the gradational processes responsible for continually creating and removing relief elements of the earth's crust.

In other parts of this book we examine various aspects of the earth's environments. Several of the *spheres* of our planet are introduced in these parts, including the atmosphere and hydrosphere (Part Two) and certain properties of the biosphere (Part Three). The atmosphere, hydrosphere, biosphere, and lithosphere are the major visible layers of our planet. Above the effective atmosphere are additional layers of thinner air and different chemical composition; their physical and chemical properties are well known because detailed data about them have been collected by balloons, high-flying aircraft, and space vehicles.

Much less is known however, about the layers that make up the internal structure of the earth. Even the directly observable crust of the planet, which we will examine later in this unit, is not well known. No instruments have been sent down very far into the crust. More than a quarter-century after people first set foot on the moon, the deepest boreholes have penetrated barely 12 km (7.5 mi) into the lithosphere. Since the radius of the

UNIT OPENING PHOTO: A fountain of lava erupting at the top of Kilauea volcano, Hawaii. The temperature of the liquid rock is approximately 1200°C (2200°F).

planet is 6370 km (3959 mi), we have penetrated less than one five-hundredth of the distance to the center of the earth.

Nonetheless, scientists have established the fact that the interior of the earth is layered like the atmosphere, and they have deduced the chemical composition and physical properties of the chief layers below the crust. This research is of importance in physical geography because the crust is affected by processes that take place in the layer below it, and this layer in turn may be influenced by conditions deeper down. So it is important to understand what is known about the earth's internal structure and how this information has been acquired.

EVIDENCE OF THE EARTH'S INTERNAL STRUCTURING

Evidence that supports the concept of an internally layered earth comes from several sources. The crust affords a glimpse of the nature of rocks normally hidden from scientists because rocks formed very deep below the surface occasionally have been elevated to levels in the crust where they can be reached by boreholes. From these samples, as well as from analyses of more common rocks, it is possible to deduce the overall composition of the crust and its average density.

By studying the wavelengths of light emanating from the sun, it is possible to determine the elements the sun contains and their proportions, because certain specific wavelengths correspond to particular elements. Geophysicists have concluded that these proportions will be the same for the earth. However, such abundant elements as iron, nickel, and magnesium are relatively depleted in the crust. This suggests that those heavy elements are concentrated in deeper layers of the planet.

This conclusion is strengthened by two pieces of evidence. First, rock samples taken from great depths do contain higher concentrations of iron and magnesium than "average" crustal rocks. Second, when the mass and size of the earth are measured, the resulting figure is 5.5 grams per cubic centimeter (g/cm^3). This is about double the density of rocks found in the continental crust. Again, the heavier, denser part of the earth should be in the deep interior.

Earthquakes

Further evidence for the internal structuring of the earth comes from the planet's magnetic field and from the high temperatures and pressures known to prevail at deeper levels. Not only does molten rock sometimes flow onto the surface through volcanic vents (*see* unit opening photo), but also it is possible to determine the temperatures at which rocks now solid were once liquefied. But the most convincing body of evidence is derived from

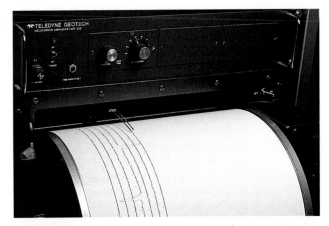

FIGURE 29-1 A seismograph at the earthquake monitoring station at the University of California, Berkeley. The pen plotter continuously records the earth's tremors and shocks as a series of wiggly lines on the slowly rotating, paper-covered drum whose motion is precisely regulated by a clock. This printout is called a seismogram.

the analysis of **earthquakes**—the shaking and trembling of the earth's surface caused by sudden releases of stress within the crust.

Earthquakes occur in many areas of the crust, and their causes are discussed in Unit 35. For the present, we should note that earthquakes generate pulses of energy called **seismic waves** that can pass through the entire earth. A strong earthquake in the Northern Hemisphere will be recorded by *seismographs* in the Southern Hemisphere. Today, thousands of seismographs continuously record the shocks and tremors in the crust (Fig. 29-1), and computers help interpret these earthquake data. This source has given us a picture of the interior structure of our planet.

Seismic waves take time to travel through the earth. In general terms, the speed of an earthquake wave is proportional to the density of the material through which it travels. The denser the material, the faster the speed. Seismic waves, like light waves and sound waves, also change direction under certain circumstances. When a seismic wave traveling through a less dense material reaches a place where the density becomes much greater, it may be bounced back; this is known as *seismic reflection* (Fig. 29-2a). If the contrast in densities between the adjacent layers is less severe, the wave may be bent rather than reflected; here, *seismic refraction* changes the course of the seismic wave (Fig. 29-2b).

Types of Seismic Waves

Seismic waves behave differently as they propagate through the earth. One type of wave travels along the surface of the crust and is termed a *surface wave* (or *L wave*). Two other types of waves travel through the interior of the earth and are referred to as **body waves**. The body waves are of two kinds, known as **P** waves

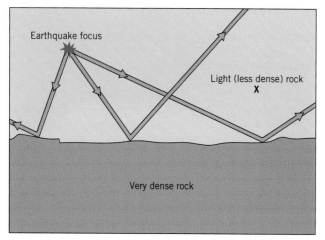

a

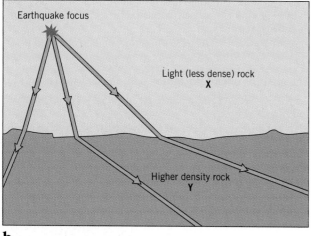

b

FIGURE 29-2 When seismic waves travel through the interior of the earth, several things happen. When they reach a plane where the rock material becomes much denser, they may be *reflected* back (a). If the contrast in rock density is less, they may be *refracted* (b). Their velocities are also affected. Speeds would be less in the layers marked **X** and greater in layer **Y**.

and **S** waves. The **P** *waves* are compressional waves, sometimes called push waves. As they propagate, they move material in their path parallel to the direction of movement. They even travel through material in the liquid state, although their impact is then much reduced. The **S** *waves* are also called shear or shake waves. These waves move objects at right angles to their direction of motion. They do not propagate through liquid material. This is of great importance, because if **S** waves fail to reach a seismograph in an opposite hemisphere, it may be concluded that liquid material inside the earth stopped their progress.

When an earthquake occurs, seismographs nearest its point of origin begin to record the passing of a sequence of waves. The seismogram (*see* Fig. 29-1) will reveal the passage of the **P**, **S**, and **L** waves in a near-continuous sequence that may reflect great destruction of structures in the area. Farther away, the different

speeds of propagation begin to show on the seismogram, and reflected **P** and **S** waves from interior earth layers make their appearance. Eventually, seismographs around the world will record the earthquake. But some stations will not register any **P** or **S** waves, and others will record only **P** waves. From these data significant conclusions about the interior of the earth can be drawn, as we shall discover.

THE EARTH'S INTERNAL LAYERS

The paths of seismic waves, illustrated in Fig. 29-3, reveal the existence of a layer beneath the crust that ends at a boundary where **S** waves (shown by white arrows) are not propagated. If an earthquake occurs at zero degrees, **P** as well as **S** waves are recorded by seismographs everywhere to 103 degrees from its source (a distance of 11,270 km [7000 mi]). Then, from 103 to 142 degrees, the next 4150 km (2600 mi), neither **P** nor **S** waves are recorded (except for **P** waves propagated along the crust). But from 142 to 180 degrees (15,420 to 19,470 km [9,600 to 12,140 mi] distant from the quake), **P** waves—always shown by black arrows—reappear. From this evidence, it is concluded that the earth possesses a liquid layer that begins about 2900 km (1800 mi) below the surface. At the contact between this liquid layer and the layer above it, **S** waves cease to be propagated and **P** waves are refracted.

But some **P** waves that arrive on the far side of the earth, between 142 and 180 degrees, have not been refracted just once or twice, but *four* times! Moreover, their speed has increased. This means that the **P** waves that reach the seismographs located antipodally to the earthquake source (i.e., on the exact opposite point of the spherical earth) must have traveled through a very dense mass *inside* the liquid layer. Confirmation of the existence of such a dense mass at the core of the earth comes from the fact that many **P** waves are reflected back at its outer edge (Fig. 29-4). From the travel times of these reflected **P** waves and the seismograms inside 142 degrees (Fig. 29-3), it is concluded that the earth has a solid inner core—a ball of very heavy, dense material. This may well be where the iron and nickel, depleted from the upper layers, is concentrated.

On the basis of seismic and other evidence, therefore, the interior earth is believed to have four layers: a solid inner core, a liquid outer core, a solid lower mantle, and a partially molten upper mantle (Fig. 29-5). On top of all this lies the crust, still very thin, and in places it is active and unstable.

Solid Inner Core

The solid **inner core** has a radius of just 1220 km (760 mi). Its surface lies 5150 km (3200 mi) below sea level.

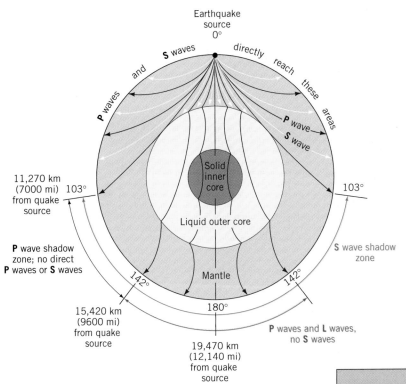

FIGURE 29-3 Imagine that a strong earthquake occurs at the North Pole (zero degrees on the drawing). This diagram shows the paths of the radiating **P**, **S**, and **L** waves as they travel through the planet. Note that no **P** waves are received over a large *shadow zone* in the Southern Hemisphere, between 103 and approximately 142 degrees from the quake's source at zero degrees. This allows us to identify the depth at which the solid mantle yields to the liquid outer core. From the refraction of the **P** waves, we can deduce the contrast in density between mantle and outer-core materials.

Iron and nickel exist here in a solid state, scientists believe, because pressures are enormous—so great that the melting-point temperature is even higher than the heat prevailing in the inner core.

Liquid Outer Core

The liquid **outer core** forms a layer 2250 km (1400 mi) thick. Its outer surface lies at some 2900 km (1800 mi) below sea level, just slightly less than halfway to the center of the planet. The liquid outer core may consist of essentially the same materials as the solid inner core, but because pressures here are less, the melting-point temperature is lower and a molten state prevails. The density of the inner and outer core combined has been calculated as 12.5 g/cm³, which compensates for the lightness of the crust (2.8 g/cm³) and accounts for the density of the planet as a whole (5.5 g/cm³).

Solid Lower Mantle

Above the liquid outer core lies the solid **lower mantle** (Fig. 29-5). Although the scale of our diagram cannot properly convey it, the contact between mantle and core is not smooth and even, but has relief—rather like the uneven upper surface of the crust. The lower mantle has a thickness of about 2230 km (1385 mi). Seismic data show that the lower mantle is in a solid state. Geologists believe that this layer is composed of oxides of iron, magnesium, and silicon.

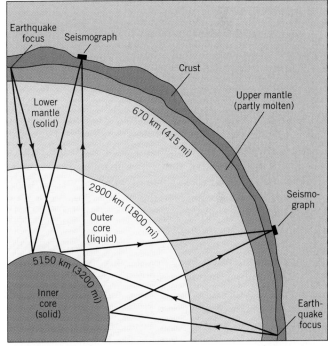

FIGURE 29-4 Certain **P** waves are reflected back toward the crust when they reach the outer edge of the solid inner core. From their travel times, the position of the contact between solid inner core and liquid outer core can be deduced. *Note*: the refraction of these waves, as they traverse the interior of the earth, is not shown.

Upper Mantle

The partially molten **upper mantle** is of great interest to geologists as well as physical geographers because it interacts with the overlying crust in many ways. The upper mantle, however, still is not well understood, al-

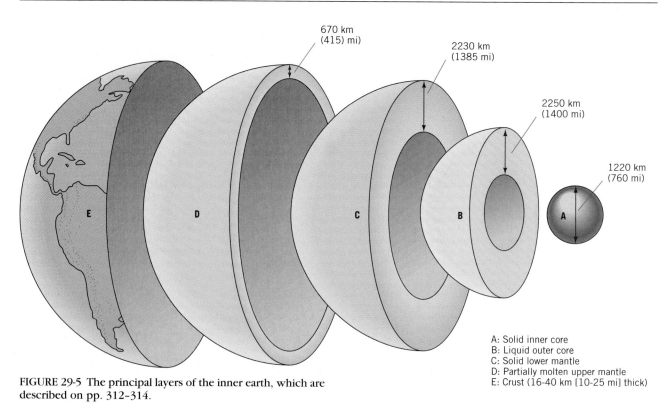

A: Solid inner core
B: Liquid outer core
C: Solid lower mantle
D: Partially molten upper mantle
E: Crust (16-40 km [10-25 mi] thick)

FIGURE 29-5 The principal layers of the inner earth, which are described on pp. 312-314.

though it extends from the base of the crust to a depth of just 670 km (415 mi). The upper mantle is differentiated from the lower mantle on the basis of mineral composition and the state of the rock material: the lower mantle is solid, but the upper mantle's material is viscous (like thick syrup, capable of flowing slowly). However, the zone of the upper mantle just beneath the crust is solid. This portion of the upper mantle also contains pockets of molten rock, some of which feeds chambers from which lava pours onto the surface of the crust. This solid part of the upper mantle together with the crust is referred to as the *lithosphere.*

The interior earth still holds many of its secrets, but our comprehension increases steadily. New knowledge about the upper mantle continues to emerge, and it may well be that this subcrustal layer can itself be subdivided into a number of additional layers. In any case, our interpretation of what we observe at the surface must begin with an understanding of what lies below. As we continue this unit with an examination of the lithosphere, we focus on further aspects of this uppermost layer of the interior earth.

THE EARTH'S OUTER LAYER

Now that we are familiar with the earth's interior, we can turn to the surface layer itself, the lithosphere upon which all else—air, water, soil, life—rests. The crust, as just noted, lies directly above the mantle (Fig. 29-5). The uppermost parts of the crust are the only portions of the

solid earth about which scientists have direct, first-hand knowledge. The rocks that make up the outer shell of our planet have been analyzed from the surface, from mine shafts, and from boreholes. Even the deepest boreholes, however, only begin to shed light on what lies below.

For many years, it was believed that temperatures and pressures below the crust would be so great that the earth material there would be in a completely molten state. Another subsequent theory, based on the study of earthquake waves, suggested that the mantle beneath the crust was solid. Quite recently, more detailed, computer-assisted analyses of the paths and speeds of earthquake waves through the earth's interior have revealed the existence of a broad viscous layer within the upper mantle. Undoubtedly, there will be more revelations in the future.

Structural Properties of the Crust

One of the most significant discoveries relating to the earth's crust occurred in 1909. In that year, the Croatian scientist Andrija Mohorovičić concluded from his study of earthquake waves that the density of earth materials changes markedly at the contact between the crust and mantle. This contact plane has been named the **Mohorovičić discontinuity**, or **Moho**, an abbreviation of his name. Despite nearly a century of far more sophisticated analyses and interpretations, Mohorovičić's conclusion has proven correct: a density discontinuity does indeed mark the base of the earth's crust.

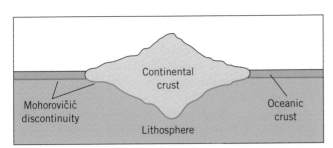

FIGURE 29-6 The Mohorovičić discontinuity (Moho) marks the base of continental as well as oceanic crust. As the sketch shows, it lies much closer to the crust's surface under the oceans than beneath the land.

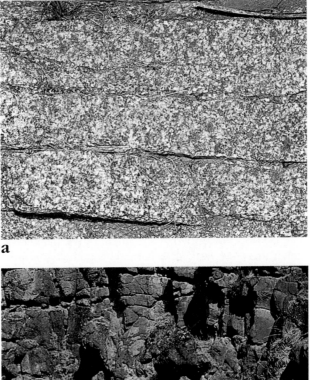

a

b

FIGURE 29-7 A closeup view of sialic continental rock, dominated by light-colored granites (a). Alternatively, the simatic rocks of the ocean floor are dominated by dark-colored basalts (b).

This information made it possible to calculate the thickness of the crust. Earthquake waves speed up at the Moho discontinuity, indicating that the crust is less dense than the mantle below. In some places, this happens a mere 5 km (3 mi) down from the surface; elsewhere, the change in earthquake-wave velocity does not come until a depth of 40 km (25 mi) or even more has been reached. This proved that the crust is not of even thickness. It also showed that the crust is thinner than the shell of an egg relative to the planet's diameter.

When the Moho was mapped, it was found to lie much closer to the surface under the ocean floors than under the continental landmasses (Fig. 29-6). This confirms a conclusion also drawn from gravity measurements: the continents have crustal "roots" that create, in a rough way, a reverse image of the topography at the surface (a matter explored in Unit 33). Under the oceans, the crust averages only 8 km (5 mi) in thickness; under the exposed continental surfaces, the average depth is about 40 km (25 mi).

For many years, it was not realized that a fundamental difference between continental and oceanic crust might account for these differences. This was so, in part, because rocks brought to the surface from the offshore continental shelves resembled those found on the continents themselves; boreholes in shallow water off the coast produced no hint of what was to come. But then technology made possible the drilling of the continental slope farther out to sea. Those rocks, it turned out, were darker and somewhat heavier than the rocks of the continental landmasses. Advances in drilling technology also raised the possibility of directly punching a borehole through the Moho into the mantle itself—a project earth scientists can still only dream about.

In any case, there are fundamental differences between continental crust and oceanic crust. The rocks that make up the continental landmasses have the lowest density of all, so that the continents are sometimes described as "rafts" that float on denser material below. These low-density rocks have come to be known as *sialic rocks*, or just **sial** (from the chemical symbols of their dominant mineral components: *si*lica and *al*uminum [Fig. 29-7a]). Granite is a common sialic rock, and

its density is about the same as the density of the continental landmasses as a whole (2.8 g/cm³). As a result, you will see continental crust referred to as "granitic" crust—although the landmasses are made up of many other rocks as well, some of which have names we commonly use, such as shale and marble. These are the rocks that, because of their different capacities to withstand weathering and erosion, create the diversity of landscapes we will study later.

Oceanic crust, on the other hand, consists of higher-density rocks collectively called the *simatic rocks*, or **sima** (for *si*lica, of which they contain much less than continental rocks, and *ma*gnesium, a heavy dark-colored component [Fig. 29-7b]). Here the dominant rock is the heavy, dark-colored basalt. Oceanic crust, therefore, is often referred to as basaltic crust, although many other rocks also form part of oceanic crust. In combination, rocks of the oceanic crust have a density of about 3.0 g/cm³.

Despite these overall differences between low-density, granitic, continental crust and higher-density, basaltic, oceanic crust, there are places *on* the continents where oceanic-type basalt can be found. How did this supposedly oceanic rock get there? The answer is that continental crust sometimes cracks open, allowing molten rock from deep below to penetrate to the surface. The basalt that has come to the surface through these fissures proves that heavier, denser rocks exist below the continents—so the notion of the landmasses as rafts on a simatic sea is not so far-fetched!

The Lithosphere

The crust terminates at the Moho, but rocks in the solid state do not. The uppermost segment of the mantle, on which the crust rests, also is rigid. Together, the crust and this solid uppermost mantle are called the **lithosphere**, the sphere of rocks. Below the lithosphere, the upper mantle becomes so hot that it resembles hot plastic: it can be made to change shape, it can be molded.

This soft plastic layer in the upper mantle is called the **asthenosphere** (Fig. 29-8). Like the Mohorovičić discontinuity, the asthenosphere begins at a much deeper level below the continental landmasses than under the ocean floor. Beneath the landmasses, it begins at a depth averaging 80 km (50 mi) below the surface. Beneath the ocean floors, it lies only about 40 km (25 mi) below the surface (i.e., of the seafloor, not the water).

Discovery of the existence of the asthenosphere was important to our understanding of what happens in and on the crust. Because the asthenosphere is in a hot plastic state, the lithosphere can move over it. This movement of the crust, which is related to the formation of mountains and even the movement of whole continents, takes place because heat sources deep inside the mantle keep the athenosphere in motion. Unlike the Moho, the contact zone between the rigid lithosphere and the soft asthenosphere is not abrupt. Rather, it is gradual, so that material can pass from one state to the other as it moves vertically as well as horizontally. For many years, geologists wondered what forced molten rock material into the crust, and even through it as lava. As more became known about the asthenosphere, the causes of such processes came to be better understood.

Much more remains to be learned about the lithosphere and its uppermost layer, the crust. Geophysicists today can create small earth tremors where earthquakes normally do not originate, and they can study the behavior of the resulting waves. In the 1980s, scientists began to realize that the so-called bright spots revealed by these artificial waves were more common than they had believed since the first one was noticed in 1975. These bright spots are zones in the crust, usually 15 to 20 km (10 to 13 mi) below the surface, where the seismic waves are being reflected more than elsewhere in the lithosphere. One theory suggests that there is a transition zone of rock from a brittle state to a soft state within the crust, a kind of mini-Moho. The proof may not come for decades, and the discovery of the bright spots is a reminder of the limited state of our knowledge—even of the crust on which we live.

Lithospheric Plates

The crust varies in thickness and is also a discontinuous layer. To us, living on the landmasses, the idea that the crust is not a continuous, unbroken shell is difficult to grasp because we see no evidence of cracks or fractures in it. In fact, the crust and the rest of the lithosphere is fragmented into a number of segments called **lithospheric plates** (or sometimes *tectonic plates*). These plates move in response to the plastic flow in the hot asthenosphere. Many of the earth's mountains are zones where the moving plates have come together in gigantic collisions. This aspect of the lithosphere is so important to our later study of landscapes and landforms that it is treated in a pair of units (32 and 33).

THE CRUSTAL SURFACE

The earth's crust is subject to tectonic forces from below. The rocks that form the crust are pushed together, stretched, fractured, and bent by the movement of the lithospheric plates. These forces tend to create great contrasts at the crustal surface—jagged peaks and sharp crests, steep slopes and escarpments, huge domes and vast depressions. Before we begin an in-depth exami-

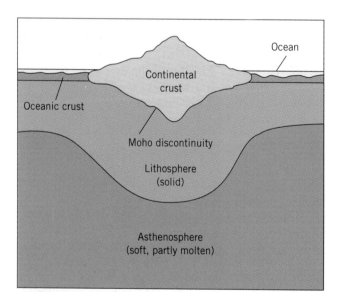

FIGURE 29-8 The position of the asthenosphere in the earth's mantle. The boundary between the asthenosphere and the lithosphere is a transition zone rather than a sharp divide. Its depth beneath the surface is about twice as great under the continents as under the oceans.

Hills, Mountains, Plains, Plateaus

To distinguish regional patterns of crustal relief more sharply, these land surface features have been classified by the geographers Glenn Trewartha, Arthur Robinson, and Edwin Hammond.

Hills and mountains are defined as terrain of less than 50 percent gentle slope. In terms of local relief, *hills* exhibit variations of 0 to 300 m (0 to 1000 ft), *low mountains* variations of 300 to 900 m (1000 to 3000 ft), and *high mountains* variations in excess of 900 m (3000 ft).

Plains and plateaus are land surfaces of more than 50 percent gentle slope. *Plains* are low-lying areas that exhibit less than 90 m (300 ft) of local relief; *flat plains* exhibit more than 80 percent gentle slope and less than 30 m (100 ft) of local relief. *Plateaus*, or tablelands, display more than 90 m (300 ft) of local relief, and more than 50 percent of their gentle slope occurs in the lower half of their elevational ranges. In addition, plateaus almost always are bounded on at least one side by a sharp rise or drop in elevation.

Despite these attempts at definitional precision, the global map is littered with *toponymy* (place names) that disregards the geographer's sense of physiographic order. Plains and plateaus usually fare better than upland areas. A glaring exception would be South America's large central Andean plateau of Peru and Bo-livia that lies more than 3600 m (12,000 ft) above sea level, which is called the "High Plain" (*Altiplano* in Spanish).

When studying hills and mountains on the landscape, therefore, map readers should be forewarned that place namers have taken some blatant liberties. Two examples from the northeastern United States will suffice. In western Massachusetts, the Berkshires are called hills even though they contain several peaks above 1000 m (3300 ft). Alternatively, a sand-dune formation adjacent to the central New Jersey coast is called the Forked River Mountains—which "tower" all of 25 m (80 ft) above the surrounding Atlantic coastal plain!

nation of those forces in the remaining units of Part Four, it is useful to take another look at the surface of the continents and their varied relief.

Topographic Relief

The term **relief** refers to the vertical difference between the highest and lowest elevations in a given area. Thus a range of tall mountains and deep valleys, such as the Rocky Mountains, is an area of *high relief* (Fig. 29-9). A coastal plain is an area of *low relief* (Fig. 29-10). An area of low relief can lie at a high elevation: a nearly flat plateau with an elevation of 3000 m (10,000 ft) has lower relief than a mountainous area with peaks no higher than 2000 m (6600 ft) and valleys at 500 m (1650 ft). Attempts by physical geographers to classify land surfaces notwithstanding, great liberties are often taken with such features on the map (*see* Perspective: Hills, Mountains, Plains, Plateaus).

When we view the continental landmasses even at a small scale (*see* Fig. 2-3), it is evident that they have areas of high relief and other areas of low relief. North Amer-

FIGURE 29-9 The aptly named Sawtooth wilderness, a zone of very high relief in the Rocky Mountains of central Idaho.

FIGURE 29-10 The rural landscape of the flat coastal plain of southern New Jersey, an area of decidedly low relief.

ica, for example, has large areas of low relief, especially in central and eastern Canada, the interior United States, and the coastal plain bordering the Atlantic Ocean and Gulf of Mexico. High relief prevails in the western third of the continent, from the Rocky Mountains westward. In the east lies an area of moderate relief in the Appalachian Mountains.

The two types of relief just identified represent two kinds of continental geology. The earth's landmasses consist of two basic geologic components: *continental shields* and *orogenic belts*. In North America, the region centered on Hudson Bay is a continental shield, expressed topographically as a plain of low relief. The Rocky Mountains represent the topographic results of a period of mountain building and constitute an orogenic belt.

Continental Shields

All the continental landmasses contain shields as well as orogenic belts. The **continental shields** are large, stable, relatively flat expanses of very old rocks, and they may constitute the earliest "slabs" of solidification of the molten crust into hard rocks. This happened more than 3 billion (in some areas more than 4 billion) years ago, and ever since these shields have formed the nuclei of the landmasses.

The shield in northern North America is called the *Laurentian* (or *Canadian*) *Shield* (Fig. 29-11). It is larger than the area of ancient rocks presently exposed because it is covered by water in the north and by sediments along its southern flank. In South America, there are two major shield zones: the *Guyana* (*Venezuelan*) *Shield*

and the *Brazilian Shield*. These shield areas, unlike the Laurentian Shield, are uplands today and present the aspect of low-relief plateaus rather than plains.

Eurasia has three major shields: the *Scandinavian Shield* in the northwest, the *Siberian Shield* in the north, and the *Indian Shield* in the south. The world's largest shield presently exposed is the *African Shield*, a vast region of ancient rocks that extends into the Arabian Peninsula at its northeastern extremity. Some of the oldest known rocks have been found in the *Australian Shield*, which occupies the western two-thirds of that continent. And under the ice in eastern Antarctica lies the *Antarctic Shield*. Wherever these shield zones form the exposed landscape, they exhibit expanses of low relief (Fig. 29-12).

Orogenic Belts

In contrast, the **orogenic belts**—series of linear mountain chains—are zones of high relief. The term *orogenic* derives from the ancient Greek word, *oros* (mountain). As we note later, the earth during its 4.6-billion-year lifetime has experienced several periods of mountain building. These episodes are marked on the topographic map by linear mountain chains, such as the old Appalachians and the younger Rockies in North America. The Andes Mountains in South America, the Alps and Himalayas in Eurasia, and the Great Dividing Range in Australia all represent orogenic activity, when rocks were bent and crushed into folds like a giant accordion (Fig. 29-13). Ever since, processes in the atmosphere have been wearing down those structures; but they persist to the present day, bearing witness to past orogenies.

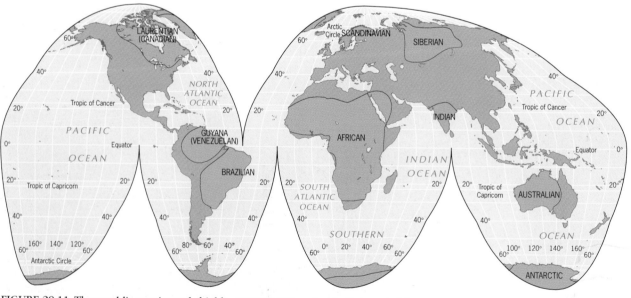

FIGURE 29-11 The world's continental shields, representing materials that cooled from the earliest molten surface of the earth.

FIGURE 29-12 A canyon incised into the otherwise monotonous, low-relief surface of the Canadian Shield. This long view is from above Ouimet Canyon Provincial Park, just north of Lake Superior, near the far western Ontario town of Thunder Bay.

Gradational Processes

If the earth had no atmosphere and no moisture, those shields and orogenic belts would stand unchallenged, destroyed only by new tectonic forces. But the earth does have an atmosphere, and as a result the geologic build-up is attacked by a set of processes that work to wear it down. These are called **gradational processes**, which is the focus of several units in Part Five.

The rocks of which the crust is composed are subject to various forms of *weathering*—the physical, chemical, and even biological processes that operate to distintegrate rocks, break them apart, and make them ready for removal. The force of gravity plays an important role in this removal as the motivating force for *mass movement* of soil and rock, which also takes several forms. (A land-

FIGURE 29-13 The extreme high relief of the Andes in western South America, one of the world's mightiest orogenic belts. These impenetrable mountains of central Peru form a natural

fortress, which led the Incas to locate their fabled city of Machu Picchu here (the light green patch below the center of the photo where the main road ends).

slide, for instance, is a form of mass movement.) But this mass movement does not carry loosened rock material very far. That requires *erosion*—which also includes the longer-distance removal of weathered materials. Great rivers transport rock grains thousands of kilometers from mountain slopes in the deep interiors of continents to their deltas on the coasts. There, and in their valleys upstream, the process of deposition fills the lowlands with the very material removed from the highlands. Even glaciers, wind, and ocean waves participate in this downwearing of the land thrust up by tectonic forces.

It is a continuous contest that affects some areas of the crust more strongly than others at different times in earth history. This conclusion is based on the geologic record: where great mountains once rose, only their roots now remain. Areas once tectonically active are today quiescent. But other zones now appear to be on the verge of great tectonic activity and will subsequently be attacked by gradational forces. It is all part of the continued recycling of earth materials within the dynamic lithosphere.

KEY TERMS

asthenosphere (p. 316)
body waves (p. 311)
continental shield (p. 318)
earthquake (p. 311)
gradational processes (p. 319)
inner core (p. 312)

lithosphere (p. 316)
lithospheric plate (p. 316)
lower mantle (p. 313)
Mohorovičić discontinuity (Moho) (p. 314)
orogenic belt (p. 318)
outer core (p. 313)

relief (p. 317)
seismic wave (p. 311)
sial (p. 315)
sima (p. 315)
upper mantle (p. 313)

REVIEW QUESTIONS

1. How do earthquakes and their seismic waves suggest a layering of the earth's interior?

2. Give the approximate thickness of the inner core, outer core, mantle, and crust.

3. What is the Mohorovičić discontinuity, and what is its significance?

4. What are the differences between oceanic and continental crust?

5. What is the significance of the asthenosphere? How does it relate to the concept of lithospheric plates?

6. Which gradational processes contribute to the recycling of the earth's materials?

REFERENCES AND FURTHER READINGS

BLOOM, A. L. *The Surface of the Earth* (Englewood Cliffs, N.J.: Prentice-Hall, 1969).

BOLT, B. A. *Inside the Earth: Evidence from Earthquakes* (New York: Freeman, 1982).

BOTT, M. H. P. *The Interior of the Earth: Its Structure, Constitution, and Evolution* (London: Edward Arnold, 1982).

DAVIS, G. H. *Structural Geology of Rocks and Regions* (New York: Wiley, 1976).

"The Dynamic Earth," *Scientific American,* September 1983 (special issue).

ERNST, W. G. *Earth Materials* (Englewood Cliffs, N.J.: Prentice-Hall, 1969).

GARLAND, G. D. *Introduction to Geophysics: Mantle, Core, and Crust* (Toronto: Holt, Rinehart, & Winston, 1979).

GASS, I., et al., eds. *Understanding the Earth* (Cambridge, Mass.: MIT Press, 1971).

KING, P. B. *The Evolution of North America* (Princeton, N.J.: Princeton University Press, 1977).

MIYASHIRO, A., et al. *Orogeny* (New York: Wiley, 1982).

POWELL, C. S. "Peering Inward," *Scientific American,* June 1991, 100–111.

PRESS, F., and SIEVER, R. *Earth* (San Francisco: Freeman, 2nd ed. 1978).

RAYMO, C. *The Crust of Our Earth: An Armchair Traveler's Guide to the New Geology* (Englewood Cliffs, N.J.: Prentice-Hall, 1983).

TREWARTHA, G. T., et al. *Elements of Geography* (New York: McGraw-Hill, 5th ed., 1967). Classification discussed on pp. 262–266.

YORK, D. *Planet Earth* (New York: McGraw-Hill, 1975).

Minerals and Igneous Rocks

OBJECTIVES

♦ To understand the relationship between rocks and their constituent minerals.

♦ To briefly investigate the important properties of minerals and to provide an elementary scheme for their classification.

♦ To discuss some important aspects of igneous rocks and their influence on landscape form.

The earth's outermost solid sphere is the crust, and the crust is the upper layer of the lithosphere. The crust consists of many different types of rocks, which range from concrete-like hardness to soap-like softness. When subjected to pressure, some fracture; others bend and warp. When heated to high temperatures, some melt and flow while others remain solid. When exposed to the forces of weathering and erosion, some withstand these conditions better than others. To understand what we see in the physical landscape, it is essential to comprehend the properties of the underlying building materials—the rocks.

MINERALS AND ROCKS

In chemistry we learn that all known matter is made up of about 100 elements. Elements are the most basic substances; they consist of atoms arranged in characteristic structures. An element cannot be broken down further, either by heating or by chemical reaction. Minerals can

UNIT OPENING PHOTO: The 1980 eruption of Mount St. Helens in Washington State's Cascade Range. The column of ash spewed by the volcano is shaped into a thick plume by the upper-air westerly winds.

FIGURE 30-1 A closeup view of large crystals of malachite embedded in calcite rock.

consist of a single element, such as diamond (which is pure carbon) and gold (a metallic element), or they can be combinations of elements, such as quartz, a compound of silicon and oxygen. Minerals are *crystalline*; that is, their atoms are arranged in regular, repeating patterns. It is often impossible to see these patterns in hand samples, but microscopes and X-ray machines reveal them. Sometimes, however, nature displays these crystal structures in spectacular fashion (Fig. 30-1). Thus minerals have distinct properties, given to them by the strength and stability of the atomic bonds in their crystals. To summarize all its characteristics, a **mineral** is a naturally occurring inorganic element or compound having a definite chemical composition, physical properties, and, usually, a crystal structure.

Many minerals can be quickly recognized by the shape of their crystals as well as their color and hardness. For crystals of a given mineral to form, there must be time for their atoms to arrange themselves into the proper pattern. As the formation time increases, so does the size of the mineral structures. Imagine a reservoir of molten rock contained somewhere deep inside the crust. This mass of molten rock cools slowly, and its various component crystals have time to develop their structures quite fully; the atoms, therefore, have the opportunity to arrange themselves in the regular patterns of various minerals. When the mass finally hardens, its various component minerals will be large, well formed, and easily recognizable.

But what happens if that mass of molten rock does not cool slowly deep inside the crust, but is instead poured out through a fissure or vent onto the surface of the crust? Now cooling takes place very rapidly, and there is little time for the atoms to arrange themselves into orderly patterns. What results is a solid rock in which

the atoms are randomly arranged and mineral structure is hardly discernible. A glass-like form of lava called obsidian is an example of such rapid hardening (Fig. 30-2). **Rocks**, therefore, are composed of mineral assemblages. A few rocks consist of only one mineral, such as quartzite, which is completely quartz. But most rocks contain several minerals, and these minerals have much to do with the way rocks break or bend, weather, and erode.

Mineral Properties

As stated above, all minerals exhibit specific properties that enable them to be identified and differentiated. These properties include chemical composition, hardness, cleavage or fracture, color and streak, and luster.

Chemical Composition

Every chemical element is identified by a one- or two-letter symbol. Aluminum, for example, is *Al*; iron is *Fe*. This does not mean, however, that all minerals made from each element are the same. Take, for example, the element identified as *C*, carbon. The mineral *diamond*, one of the hardest substances known, is pure carbon. But so is *graphite*, the soft "lead" in a pencil. Chemically, they are the same, but their crystalline structures differ. In a diamond, all atoms are bonded strongly to each other. In graphite, certain bonds are weaker, creating

FIGURE 30-2 Obsidian is a shiny volcanic glass formed when lava cools very rapidly, such as when it flows into a body of water.

a

b

FIGURE 30-3 Four diamond specimens, highlighting this valuable mineral's hardness, clarity, and transparency (a). By contrast, the graphite form of carbon exhibits the opposite characteristics (b) and is so inexpensive that it is commonly manufactured into pencil lead.

TABLE 30-1	MOHS' HARDNESS SCALE

MINERAL	HARDNESS
Diamond	10
Corundum	9
Topaz	8
Quartz	7
Potassium feldspar	6
Apatite	5
Fluorite	4
Calcite	3
Gypsum	2
Talc	1

By way of comparison, here are some everyday items ranked according to their approximate hardness: pocketknife blade, 5-6; glass, 5; copper penny, 3.5; fingernail, 2.5.

sheets that are easily pulled apart. Note, too, that while diamond is clear, transparent, and very hard, graphite is opaque, black, and soft—the very opposite qualities (Fig. 30-3). Other minerals are even softer than graphite, and rocks containing such soft minerals are more quickly broken down by weathering and erosion than rocks containing only harder minerals.

Hardness

Hardness, therefore, is an important property of minerals. This quality can be useful in identifying minerals in the field, and it can tell us much about the overall hardness of the rock in which they occur. As long ago as 1822, a German mineralogist named Friedrich Mohs noticed that certain minerals could put a scratch into other minerals, but not vice versa. Diamond, the hardest mineral of all, will scratch all other natural mineral surfaces, but cannot be scratched by any of the others. So Mohs established a hardness scale ranging from 1 to 10, with diamond (10) the hardest. Mohs determined that talc (the

base mineral of talcum powder) was the softest naturally occurring mineral, and this he numbered 1. Mohs' Hardness Scale (Table 30-1) continues to be used to this day. Notice that quartz, a commonly appearing mineral, ranks 7 and is quite hard.

Cleavage/Fracture

The crystal form of minerals quickly identifies them in some cases, but not many crystals can grow unimpeded to the full form shown in Fig. 30-1. More useful is the property of *cleavage*, the tendency of minerals to break in certain directions along bright plane surfaces, revealing the zones of weakness in the crystalline structure. When you break a rock sample across a large crystal of a certain mineral, the way that crystal breaks may help to identify it. Sometimes minerals do not break as cleanly as this, however. Instead they *fracture* in a characteristic way. That glass-like obsidian mentioned earlier (Fig. 30-2) has a way of fracturing in a shell-like fashion when broken.

Color/Streak

A mineral's color is its most easily observable property. Some minerals have very distinct colors, such as the yellow of sulfur and the deep blue of azurite. Other minerals have identical colors or occur in numerous colors, and therefore cannot be differentiated according to this property. However, the color of a mineral's *streak* (the mineral in powdered form when rubbed against a porcelain plate) can sometimes help identify it. For example, although both galena and graphite are metallic gray in color, their streaks are gray and black, respectively. Like color, streak is often unhelpful in that most minerals have white or colorless streaks.

TABLE 30-2 COMPOSITION OF THE EARTH'S CRUST

ELEMENT	PERCENTAGE (BY WEIGHT)
Oxygen (O)	46.6
Silicon (Si)	27.7
Aluminum (Al)	8.1
Iron (Fe)	5.0
Calcium (Ca)	3.6
Sodium (Na)	2.8
Potassium (K)	2.6
Magnesium (Mg)	2.1
Total	98.5

Luster

A mineral also displays a surface sheen or *luster*, which, along with color, can be a useful identifying quality. For instance, the difference between real gold and a similar-looking but much less valuable mineral, pyrite (FeS_2), can be detected by their comparative lusters. Not surprisingly, pyrite is called fool's gold for good reason!

Mineral Types

Although as many as 103 chemical elements are known in nature, only 8 make up more than 98 percent of the earth's crust by weight (*see* Table 30-2). Moreover, the two most common elements in the crust, silicon and oxygen, constitute nearly 75 percent of it. Geologists divide the minerals into two major groups, the *silicates* and the *nonsilicates*. Each group is in turn subdivided. This classification is a central concern of the field of mineralogy.

The *silicates*, as their name suggests, are the compounds containing silicon (*Si*) and oxygen (*O*) and, mostly, other elements as well. The *nonsilicates* include the carbonates, sulfates, sulfides, and halides. Among these, the *carbonates* are of greatest interest in physical geography. All carbonates contain carbon and oxygen (CO_3). With calcium they form calcite, the mineral of which limestone is made. Limestone is fairly widely distributed, and it creates unusual landforms under both humid and arid conditions (*see* Unit 44). Add magnesium (*Mg*) to the formula, and the mineral dolomite is formed. Dolomite, too, creates distinctive landforms.

The *sulfates* (SO_4) all contain sulfur and oxygen. The calcium sulfate, gypsum, in some places lies exposed over sufficiently large areas to be of geomorphic interest. The *sulfides* (SO_3), on the other hand, occur in veins and ores, and do not build or sustain landforms themselves. Pyrite (FeS_2) is such a mineral. The *halides* consist of metals combined with such elements as chlorine, fluo-rine, and iodine. The most common is halide, a compound of sodium (*Na*) and chloride (*Cl*), the substance that makes ocean water salty; but halite rock salt also can create landforms.

Finally, there are the oxides and natural elements. The *oxides* are formed by a combination of metal and oxygen—nothing more. This kind of crystallization takes place in veins or ore chambers, and the result may be an economically important deposit of, for example, hematite (Fe_2O_3) or magnetite (Fe_3O_4). Oxidation also can take place as a result of the intrusion of liquid water or water vapor into concentrations of iron or aluminum. The *natural elements* are those rare and prized commodities that are among the most valuable on earth: gold (*Au*), silver (*Ag*), platinum (*Pt*), and sometimes copper (*Cu*), tin (*Sn*), and antimony (*Sb*).

CLASSIFICATION OF ROCK TYPES

From what has been said about the minerals that make up the crustal rocks, it is evident that the diversity of rock types is almost unlimited. Still, when rocks are classified according to their mode of origin, they all fall into one of three categories. One class of rocks forms as a result of the cooling and solidification of **magma** (molten rock), and this process produces **igneous rocks**. Because they solidified first from the earth's primeval molten crust (*see* Perspective: The World's Oldest Rocks), igneous rocks are known as *primary* rocks. The deposition and compression of rock and mineral fragments produce **sedimentary rocks**, and when existing rocks are modified by heat or pressure or both, they are transformed into **metamorphic rocks**. Because they are derivatives of preexisting rocks, sedimentary and metamorphic rocks are called *secondary* rocks. Igneous rocks are treated in the remainder of this unit; secondary rocks are the subject of Unit 31.

IGNEOUS ROCKS

The term *igneous* means "origin by fire," from the Latin word *ignis* (fire). The ancient Romans, upon seeing the flaming lava erupt from Mounts Etna and Vesuvius, undoubtedly concluded that fire stoked the rock-forming ovens inside the earth. But igneous rocks actually form from cooling—the lowering of the temperature of molten magma or **lava** (magma that reaches the earth's surface). This can happen deep inside the crust or on the surface. Magma is not only a complex melt of many minerals; it also contains gases, including water vapor. It is a surging, swelling mass that pushes outward and upward, sometimes forcing itself into and through existing layers of rocks in the crust. If its upward thrust ceases before it

The World's Oldest Rocks*

Geologists have known for decades that the earth was formed about 4.6 billion years ago, condensing from a rotating, gaseous mass along with the other terrestrial planets of the inner solar system. At first our planet consisted entirely of molten rock, but eventually its heavier constituent elements (iron and nickel) settled to form a solid core. Above that inner core, as Fig. 29-5 indicates, three concentric layers emerged as the earth continued to cool: the outer core, the lower mantle, and the upper mantle. Until a few years ago, scientists had little information as to when the upper mantle began to develop a crust, a process they liken to the formation of the crust atop

boiling pea soup. Recent discoveries, however, have begun to shed light on the events associated with the birth of the crust—and have produced field specimens that may rank among the first rocks ever formed.

Over the past decade, the age of the oldest known rocks has been pushed back by at least 300 million years. Chunks of granite obtained near the Arctic Circle in far northern Canada have now been dated at 3.96 billion years, and neighboring rocks may be as old as 4.1 billion years. That is also the age of individual mineral grains found in sedimentary rocks discovered in Australia during the 1980s. Clearly, the ancient granite spec-

imens prove the existence of continental crust almost 4 billion years ago, whereas the sedimentary grains strongly suggest that the cycle of transformation that affects all rocks (*see* discussion of the rock cycle on p. 335) was operating even earlier than that. These monumental findings have not only filled major gaps in our knowledge of the earliest stage of earth history: they are also alerting scientists to field research opportunities that, before the 1990s, few could even imagine existed.

*The source for much of this box is Hilts, 1989.

reaches the surface, the resulting rocks formed from the cooled magma are called **intrusive igneous rocks**. If it penetrates all the way to the surface and spills out as lava or ash, the rocks formed from these materials are called **extrusive igneous rocks**.

As we noted previously, intrusive igneous rocks tend to have larger mineral crystals than faster cooling extrusive ones. Intrusives such as granite and gabbro are coarse-grained, with mineral crystals as much as 1 cm (0.4 in) or more in diameter. For intrusive rocks with exceptionally large crystals, some ranging from 2 to 3 cm (0.8 to 1.2 in) long, the cooling process obviously was unusually slow. It is concluded that this occurred at unusual depth and that the magmatic mass must have been very large. Such coarse-grained intrusive rocks are called *plutonic* igneous rocks, another term of Roman origin (Pluto was the Roman god of the underworld).

The color of igneous rocks can tell us much about their origins. When the original magma is rich in silica (*acidic* or *silicic*), it yields rocks rich in feldspar and quartz. Such rocks are light-colored, with beige-colored feldspar and glassy quartz dominating. Magma that was poorer in silica (*basic* or *mafic*) yields darker rocks, both intrusives and extrusives. For example, a light-colored, coarse-grained granite formed as an intrusive rock; had the same magma spilled out onto the surface, it would have yielded a light-colored but much finer grained rhyolite. But a dark-colored, coarse-grained gabbro came from a basic magma; had it penetrated to the surface as

an extrusive rock, it would have become a fine-grained basalt or perhaps even a black obsidian (*see* Fig. 30-2).

Intrusive Forms

In the analysis of landscape, the form an intrusion (a mass of intrusive rock) takes becomes an important factor. Magmas vary not only in composition, but also in viscosity (fluidness). Thick, viscous masses will remain compact; if the magma is very fluid, it can penetrate narrow cracks in existing rock strata and inject itself between layers. Sometimes the pent-up gases in a magma will help force it through its chamber walls. Other magmas, containing less gas, are calmer.

In general terms, intrusions may be **concordant** if they do not disrupt or destroy existing structures or **discordant** if they cut across previously formed strata. For instance, a **batholith** is a massive *pluton* (a body of plutonic rock) that has destroyed and melted most of the existing structures it has invaded; a **stock** also is discordant but smaller (Fig. 30-4). Sometimes magma inserts itself as a thin layer between strata of existing rocks without disturbing these older layers to any great extent; such an intrusion is called a **sill** (Fig. 30-4). But magma can also cut vertically across existing layers, forming a kind of barrier wall called a **dike** (Fig. 30-4). The sill is a concordant intrusion; the dike obviously is discordant. An especially interesting concordant intrusive form is the **laccolith**. In this case, a magma pipe led to a chamber

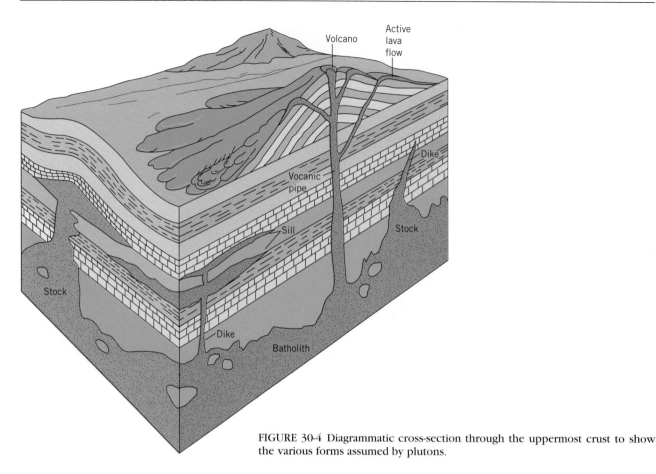

FIGURE 30-4 Diagrammatic cross-section through the uppermost crust to show the various forms assumed by plutons.

that grew, dome-like, pushing the overlying strata into a gentle bulge without destroying them (Fig. 30-4).

Jointing and Exfoliation

Igneous rocks such as granite and basalt display a property that is of great importance in their breakdown under weathering and erosion. **Jointing** is the tendency of rocks to develop parallel sets of fractures without any obvious movement (such as faulting). Granite often exhibits a rectangular joint pattern so that it breaks naturally into blocks (Fig. 30-5). Basalt, on the other hand, usually possesses a columnar joint system that produces hexagonal forms. Jointing in igneous rocks appears to be related to the cooling process of the magma: the contraction of the material produces planes of weakness and

FIGURE 30-5 Jointing in igneous rock is revealed and exploited by wave erosion on this wave-cut platform on the coast of the Tasman Peninsula, Tasmania, Australia. At one time, the waves rolled over this surface, planing it down. Later, it was uplifted to rise above mean sea level. But at high tide, water still washed over it. (Authors' photo)

FIGURE 30-6 Batholiths and stocks, formed deep below the surface, eventually rise as erosion removes the overburden. This release of weight leads to exfoliation, and the outer shells of the rock mass peel off, leaving the rounded landforms shown here towering over the urban landscape of Rio de Janeiro, Brazil. (Authors' photo)

separation—the *joint planes*—but jointing is not confined to igneous rocks. Sedimentary and metamorphic rocks also display forms of jointing.

A special kind of jointing, found in certain kinds of granite, produces a joint pattern resembling a series of concentric shells—not unlike the layers of an onion. The outer layers, or shells, peel away progressively, leaving the lower layers exposed (Fig. 30-6). This phenomenon, called **exfoliation**, is caused by the release of confining pressure. These granite domes were at one time buried deep inside the crust and under enormous pressure. As erosion removed the overlying rocks, the pressure was reduced and these rock masses expanded. The outer shells, unable to resist this expansion force, cracked and peeled along hidden joint planes.

Igneous Rocks in the Landscape

Igneous rocks tend to be strongly resistant to weathering and erosion. Intrusive igneous structures often form characteristic landforms when their overburdens are removed through weathering and erosion. For example, when the strata overlying a laccolith (Fig. 30-4) are eroded away, the granitic core stands as a mound above the landscape, encircled by low ridges representing remnants of the softer sedimentary cover. The intrusive sill, which long ago squeezed between sedimentary layers (Fig. 30-4), resists erosion longer than the softer sedimentary rocks do. Eventually, such a sill is likely to cap a table-like landform called a *mesa* (*see* diagram on p. 433), a remnant of the intrusion. A dike (Fig. 30-4), which is also more resistant than its surroundings, will stand out above the countryside as a serpentine ridge (*see* photo of New Mexico's Ship Rock on p. 432). Exfoliation also can be seen in progress in many places (some spectacular examples are the domes in California's Yosemite National Park).

The most spectacular landforms associated with igneous rocks undoubtedly are shaped by extrusive structures, especially the world's great volcanoes. A famous one is Devil's Tower in Wyoming (Fig. 30-7), a columnar

FIGURE 30-7 Devil's Tower in northeastern Wyoming, a columnar basalt structure 265 m (865 ft) high.

structure of basaltic rock formed from an ancient eruption. The violent eruption of Mount St. Helens in the U.S. Pacific Northwest in 1980 (*see* unit opening photo) provided physical geographers and other scientists with an opportunity to witness volcanic processes and their consequences. Vesuvius, the great volcano that looms over the Italian city of Naples (*see* photo on p. v), is the most legendary of all such mountains. The processes and landforms of volcanism are investigated in Unit 34. But we now continue our survey of the two remaining major rock types in Unit 31.

KEY TERMS

batholith (p. 325)

concordant (p. 325)

dike (p. 325)

discordant (p. 325)

exfoliation (p. 327)

extrusive igneous rock (p. 325)

igneous rock (p. 324)

intrusive igneous rock (p. 325)

jointing (p. 326)

laccolith (p. 325)

lava (p. 324)

magma (p. 324)

metamorphic rock (p. 324)

mineral (p. 322)

rock (p. 322)

sedimentary rock (p. 324)

sill (p. 325)

stock (p. 325)

REVIEW QUESTIONS

1. What are minerals?

2. How are minerals related to elements and rocks?

3. How are intrusive and extrusive igneous rocks different, and how can they generally be distinguished?

4. How is a *sill* different from a *dike*?

5. How is a *batholith* different from a *laccolith*?

REFERENCES AND FURTHER READINGS

BARKER, D. S.   *Igneous Rocks* (Englewood Cliffs, N.J.: Prentice-Hall, 1983).

COX, K. G., et al.   *The Interpretation of Igneous Rocks* (Boston: Allen & Unwin, 1979).

DEER, W. A., HOWIE, R. A., and ZUSSMAN, J.   *An Introduction to Rock Forming Minerals* (New York: Wiley, 2nd ed., 1992).

DIETRICH, R. V., and SKINNER, B. J.   *Gems, Granites, and Gravels: Knowing and Using Rocks and Minerals* (New York: Cambridge University Press, 1990).

DIETRICH, R. V., and SKINNER, B. J.   *Rocks and Rock Minerals* (New York: Wiley, 1979).

EHLERS, E. G., and BLATT, H.   *Petrology: Igneous, Sedimentary, and Metamorphic* (New York: Freeman, 1982).

HESS, P. C.   *Origins of Igneous Rocks* (Cambridge, Mass.: Harvard University Press, 1989).

HILTS, P. J.   "Canadian Rock, At 4 Billion Years, Is Called Oldest," *New York Times,* October 5, 1989, 8.

HURLBUT, C. S., JR.   *Minerals and Man* (New York: Random House, 1969).

KLEIN, C., and HURLBUT, C. S., JR.   *Manual of Mineralogy* (New York: Wiley, 21st ed., 1993).

MACKENZIE, W. S., et al.   *Atlas of Igneous Rocks and Their Textures* (New York: Wiley/Halsted, 1982).

POUGH, F.   *A Field Guide to Rocks and Minerals* (Cambridge, Mass.: Riverside Press, 3rd ed., 1960).

PRINZ, M., et al.   *Simon and Schuster's Guide to Rocks and Minerals* (New York: Simon & Schuster, 1978).

Sedimentary and Metamorphic Rocks

OBJECTIVES

♦ To discuss the circumstances under which sedimentary and metamorphic rocks form.

♦ To identify common sedimentary and metamorphic rock types.

♦ To discuss some observable structures within sedimentary and metamorphic rock masses.

The igneous rocks have been called the earth's primary rocks—the first solidified material derived from the molten mass that once was the primeval crust. The other two great classes of rocks could therefore be called secondary, because they are derived from preexisting rocks. These are the sedimentary and metamorphic rocks.

SEDIMENTARY ROCKS

Sedimentary rocks result from the deposition and compaction (*lithification*) of rock and mineral grains derived from other rocks. These grains are broken away from existing rocks by the action of water, wind, and ice, processes explored in Part Five. Again, the ancient Roman scholars understood what they saw: *sedimentum* is the Latin word for "settling." Many sedimentary rocks begin their existence as loose deposits of sand or gravel at the bottom of a sea or lake, on a beach, or in a desert (Fig.

31-1). Later, the sediment is lithified—compressed into a rock.

As successive layers of sediment accumulate, the weight of the sediments expels most of the water between the grains. Pressure caused by the weight of the overlying materials will compact and consolidate lower strata. The grains are squeezed tightly together, especially in fine-grained sediments such as clays and silts. This is the process of **compaction** (Fig. 31-2a). Compaction rarely takes place alone. Most sedimentary material has some water in the pore spaces between the grains, and this fluid contains dissolved minerals. This mineral matter, such as silica or calcite, is deposited in thin films on the grain surfaces, which has the effect of gluing them together. This is the process of **cementation** (Fig. 31-2b). Together, compaction and cementa-

UNIT OPENING PHOTO: Intricate layering marks a sedimentary rock formation in southern Utah.

FIGURE 31-1 This accumulation of poorly sorted sediment, resulting from stream erosion and short-distance transportation in a valley near the Gila River in eastern Arizona, would become a conglomerate if compaction and cementation followed. More likely, future rainstorms and floods will wash much of this loose deposit downslope. (Authors' photo)

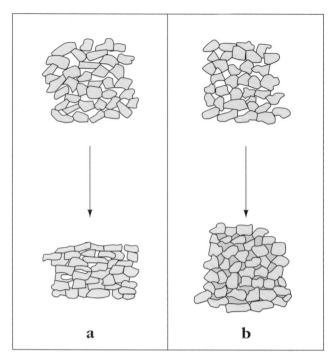

FIGURE 31-2 Compaction and cementation in sedimentary rocks. In *compaction* (a), the grains are packed tightly together by weight from above. In *cementation* (b), the spaces between the grains are filled through the deposition of a cement, such as silica or calcium carbonate.

tion can transform a bed of loose sand into a layer of cohesive sedimentary rock.

Clastic and Nonclastic Sedimentary Rocks

The range of agents and materials that combine to produce sedimentary rocks is wide, and as a result the structure and texture of these rocks also vary greatly. Even the finest wind-blown dust can become lithified. The same is true for a mixture of boulders, cobbles, pebbles, and sand swept down by a river and subsequently compacted and cemented. Sedimentary rocks made from particles of other rocks are referred to as **clastic**, from the ancient Greek *klastos*, meaning broken. The vast majority of sedimentary rocks are clastic. **Nonclastic** sedimentary rocks form from chemical solution by deposition and evaporation, or from organic deposition.

Clastic sediments are most conveniently classified according to the size of their fragments, which can range from boulder to fine clay particles. The coarsest grained sedimentary rock is the **conglomerate**, a composite rock made of gravels, pebbles, and sometimes even boulders. An important property of conglomerates is that the pebbles or boulders tend to be quite well-rounded. This characteristic is evidence that they were transported by water for some distance, perhaps rolled down a river valley or washed back and forth across a beach. A large pebble may reveal the source area from where it was

removed, perhaps telling us something about ancient drainage courses. Sometimes pebbles are elliptical in shape, and in the conglomerate a significant number of them lie cemented with their long axes in the same direction. Such information helps reveal the orientation of the coastline where the sediment accumulated.

When pebble-sized fragments in a conglomerate are not rounded but angular and jagged, it is called a **breccia** (Fig. 31-3). The rough shape of the pebbles indicates that

FIGURE 31-3 This mound of breccia (pronounced BRETCH-uh), made up of pebbles and small boulders, many of them broken, was cemented by a matrix hard enough to retard erosion in this area south of Globe, Arizona. (Authors' photo)

Oilfield Formation

Petroleum, one of the world's leading fuels, became a full-fledged energy resource in the second half of the nineteenth century when oil-drilling technology was pioneered. Drilling began with a few barrels a day drawn from a single well in northwestern Pennsylvania in 1859. From that humble beginning, global production has risen to more than *60 million* barrels a day in the 1990s, a staggering output of approximately 22 billion barrels per year. Because oil-producing countries can count on a substantial income, the search for additional petroleum deposits continues all over the world.

Petroleum is found in the sedimentary rocks of nonshield zones (Fig. 29-11) where conditions have favored the development of geologic structures capable of containing oil reservoirs. The formation of petroleum itself involved large, shallow bodies of water where, scientists believe, microscopic plant forms (such as diatoms) contained minute amounts of it. At death, these tiny plants released this substance, so that it became part of the sediments accumulating on the seabed.

Millions of years later, a thick accumulation of sediments—now transformed into sedimentary rock layers—might contain a large quantity of oil. Then, when these rock layers were subsequently compressed and bent into arching structures called folds (*see* Unit 36), the accumulated oil would be squeezed into a reservoir as shown in Fig. 31-4. Such a reservoir might be an upfold in the rock layers, or a dome capped by an impermeable stratum. (Note that natural gas often forms above such an oil pool—the two energy resources frequently occur together—and that the oil also floats above any groundwater that may lie below the upfold in the porous, reservoir rock layer.) There the petroleum deposit remains under pressure until its existence is discovered by exploration. Then a well is drilled, the black liquid is pumped to the surface, and the world's oil production capacity is recorded as having increased.

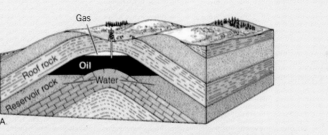

Structural traps

FIGURE 31-4 An *oil pool* (a body of rock in which oil occupies all the pore spaces) trapped in an upward-arching layer of reservoir rock. These curving rock structures are known as folds, and constitute the most important of all oil traps.

little transport took place prior to lithification. When compaction and cementation occur after a landslide, for instance, the result is a breccia. Again, the properties of the fragments can constitute a key to the past.

Another common and important sedimentary rock is **sandstone**. In a sandstone the grains, as the name implies, are sand-sized, and they usually are quartz grains. Some sandstones are very hard and resist erosion even in humid climates. This is because the cementing material in such sandstones is silica. But other sandstones are less compacted and are cemented by calcite or even iron oxide. Such sandstones are softer and more susceptible to weathering and erosion. Thus sandstones are also a key to the past. They may have rounded or angular grains, depending on the distance they have traveled.

As with conglomerates, the size of sandstone grains may vary. Rounded, even-sized grains indicate long-term deposition. Variations in particle size and irregular shapes mean poor sorting and rapid deposition. Sandstones also have economic importance: because they are porous, they can contain substantial amounts of water and even oil. Under certain structural circumstances, such water or oil can form a reservoir suitable for exploitation (*see* Perspective: Oilfield Formation).

A sedimentary rock even softer than most sandstones is **shale**, the finest grained clastic sedimentary rock. Shale is compacted mud. Whereas sandstone contains quartz grains that are often visible to the naked eye, shale is made from clay minerals, and the individual mineral grains cannot be seen. Shale has a tendency to split into thin layers, making this already soft rock even more susceptible to weathering and erosion (Fig. 31-5). In many places (such as the Appalachian Mountains in the eastern United States), the low valleys are often underlain by soft shale and the higher ridges by other rocks, including hard sandstone.

One of the most interesting sedimentary rocks, both because of the way it forms and its response to weathering and erosion, is **limestone**. Limestone can form from the accumulation of marine shell fragments on the ocean floor, which qualifies it as a special kind of clastic sedimentary rock. Most limestone, however, results from

FIGURE 31-5 An outcropping of shale in South Africa. This soft, thin-layered sedimentary rock is easily weathered and eroded.

the respiration and photosynthesis of marine organisms in which calcium carbonate is distilled from seawater. This calcium carbonate ($CaCO_3$) then settles on the ocean floor, and accumulations may reach hundreds of meters in thickness. Limestones can vary in composition and texture, but much of it is finely textured and, when exposed on the continental landscape, hard and resistant to weathering. Limestone, however, is susceptible to solution, and under certain environmental conditions it creates a unique landscape both above and below the ground (*see* Unit 44).

None of the nonclastic sedimentary rocks plays a significant role in the formation of landscape, other than limestone in its chemical form. *Evaporites* form from the deposits left behind as water evaporates. Such evaporites as halite (salt), gypsum, and anhydrite have some economic importance, but their areal extent is small. Biological sediments include the carbonate rocks formed by coral reefs, cherts formed from silica skeletons of diatoms and radiolarians (marine microorganisms), and, technically, the various forms of coal.

Sedimentary Rocks in the Landscape

A sequence of sedimentary rocks in the landscape is unmistakable because it displays variations in texture, color, and thickness of the various layers (*see* unit opening photo). This layering, or **stratification**, reminds us that conditions changed as the succession of rock beds or **strata** were being deposited. Often, distinct surfaces between strata, or *bedding planes*, are evident. Sometimes it is apparent that the sequence was interrupted and that a period of deposition was broken by a period of erosion before the deposition resumed. Where such an interruption is evident in the **stratigraphy** (order and arrangement of strata) of sedimentary rocks, the contact between the eroded strata and the strata of resumed deposition is called an **unconformity** (Fig. 31-6).

The texture and color of the sedimentary layers allow us to deduce the kinds of environments under which they were deposited. Sedimentary rocks, therefore, are crucial in the reconstruction of past environments. Even more importantly, sedimentary rocks contain fossils (Fig. 31-7). Much of what is known about earth history is based on the fossil record. Interpretations from the fossil record, as well as conclusions drawn from the stratigraphy of sedimentary rock sequences far removed from one another, make possible correlations that provide further evidence for reconstructions of the past.

Sedimentary rocks can be observed as they accumulate today, which provides further insight into similar conditions in the distant geologic past. You may have seen ripples in the sands of a beach or in a desert area. These ripples can be created by the wash of waves or by the persistent blowing of wind. Most of the time they are erased again, only to reform later. But sometimes they are cemented and preserved in lithifying rock as *ripple marks*. Millions of years ago, ripple marks were formed that have become exposed by erosion today—providing evidence of wind or wave direction in the distant past.

Features of Sedimentary Strata

When originally formed, most sedimentary strata are layered horizontally. Another form of layering, **cross-bedding**, consists of successive strata deposited at vary-

FIGURE 31-6 This exposed cliff on the Italian island of Ischia reveals an eventful sedimentary history. Unconformities mark the lower strata (below the church on the flank of the cliff). Note the contrasting angle of dip of the light-colored sandstone strata (upper right). Clearly, these layers were deposited during times of much-interrupted deposition *and* tectonic activity. (Authors' photo)

FIGURE 31-7 Fossils contained in sedimentary rocks provide valuable clues about the geologic past. This easily recognizable shoal of fossilized fish represents important evidence in the archeological interpretation of marine life forms.

ing inclines. Like ripple marks, this forms on beaches and in dunes. The sand layers do not lie flat, but at angles caused by wind and water-current action over an irregular bed. We can see this happening today, and we can compare angles of repose and other aspects of the process to crossbedded layers in old sedimentary rocks.

As noted previously, all rocks have jointing properties. Not only are sedimentary rocks layered—with their bedding planes often a factor in weathering and ero-

sion—but they also are jointed. Joints are produced by a variety of processes, ranging from desiccation (drying) in sedimentary rocks to unloading by erosion in igneous rocks. Furthermore, over time sedimentary rocks may be folded, faulted, and otherwise deformed (*see* Unit 36). All these circumstances contribute to the rate of erosion in areas where sedimentary rocks dominate the landscape, and they create the sometimes spectacular, multicolored, and varied scenery of such places (Fig. 31-8).

FIGURE 31-8 Where the Colorado Plateau plunges into the Grand Canyon, the Colorado River's erosion has excavated tens of millions of years of sedimentary accumulation. Note the color variations in the depositional sequence; also (for later reference), note the sharp, cliff-like drop from the level plateau's upper surface into the canyon.

METAMORPHIC ROCKS

Metamorphic rocks are rocks that have been changed by heat and pressure. The term *metamorphic* comes from a Greek word meaning "change," but the complex processes involved in rock metamorphism have only begun to be understood in modern times. All rocks may be subject to metamorphism. Granites can be remelted and recrystallized. Sandstones can be fused by heat and pressure into much harder rocks. And metamorphic rocks themselves can be changed again.

All this happens through tectonic action in the crust (*see* Unit 33) or through volcanic action (*see* Unit 34). Zones of the earth's crust are pushed down to deeper levels; other segments of the crust rise. The rocks making up these crustal zones are subjected to changing temperature and pressure conditions, and are modified as a result. When intrusive action by magma occurs, rocks in the zone near the batholith or dike will be affected (Fig.

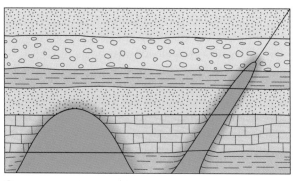

a

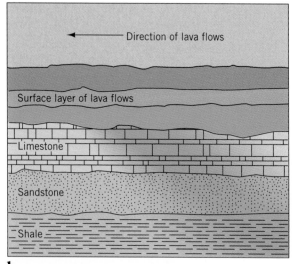

b

FIGURE 31-9 Two examples of contact metamorphism. The effects of two kinds of magmatic intrusions on the existing sedimentary rock strata: metamorphism radiates deeply into these layers from both the batholith and the dike (a). The effects of extrusion, in this case repeated lava flows, on underlying sedimentary layers (b).

31-9a). (This process is known as *contact metamorphism.*) When a sheet of lava flows out over the surface, its heat changes the rocks it covers. Imagine a sedimentary sequence of limestone, sandstone, and shale being interrupted by repeated lava flows from a fissure. The heat and weight of the lava will create metamorphic rocks out of the sedimentary rocks immediately below it (Fig. 31-9b). This also happens when a sill forms, with contact metamorphism occurring in the overlying and underlying rocks.

Metamorphic Rock Types

Some metamorphic rocks have quite familiar names. Sandstone, made of quartz grains and a silica cement, becomes **quartzite**, a very hard rock that resists weathering. Limestone is converted into much denser and harder **marble**, used by sculptors for statues that can withstand exposure to the elements for thousands of years. Shale may be metamorphosed into **slate**, a popular building material; slate retains shale's quality of breaking along parallel planes.

Sometimes metamorphism alters the preexisting rocks so totally that it is not possible to determine what the previous form of the rock may have been. A common metamorphic rock is **schist**. This rock is fine-grained, and it breaks along roughly parallel planes (but very unevenly, unlike slate). If schist was, at least in part, shale in its premetamorphic form, there is little resemblance left (*see* Fig. 31-5).

When schist is seen outcropping on the landscape, it displays wavy bands such as those shown in Fig. 31-10. These bands show that the minerals in the preexisting rock were realigned during the metamorphic process. This realignment indicates that metamorphism did not completely melt the older rocks, but made them viscous enough for the minerals to orient themselves in parallel strips. This process gives certain metamorphic rocks their unmistakable banded appearance, or **foliation**. One of the best and most frequently seen examples is in **gneiss**, the metamorphic rock derived from granite (Fig. 31-11).

Metamorphic Rocks in the Landscape

Because metamorphic rocks have been subjected to heat and pressure, we might conclude that they would be the most resistant of all rocks to weathering and erosion. But even metamorphic rocks have their weak points and planes (Fig. 31-11). Slate, for example, is weak at the surfaces along which it breaks because water can penetrate along these planes and loosen the rock slabs. Schist often occurs in huge masses, and therefore seems to resist weathering and erosion quite effectively. But schist is weak along its foliation bands and breaks down quite rapidly. Even gneiss is weakest along those foliation planes, especially where dark minerals such as micas

FIGURE 31-10 The characteristically shiny surface of this hand sample of schist is caused by the small, elongated flakes of mica that have been oriented in the same direction by the foliation process. We found it among metamorphic rocks in the Superstition Mountains of southern Arizona. (Authors' photo)

have collected. In some areas where gneiss is extensive, the dark minerals have been weathered so effectively that vegetation has taken hold in these bands. From the air one can follow the foliation bands by noting the vegetation growing in the weakest ones.

The earth's first rocks were igneous rocks—rocks solidified from the still-molten outer sphere some 4 billion years ago. Ever since, existing rocks have been modified and remodified, and there are few remains of these

original, ancient-shield rocks. What we see in the landscape today are only the most recent forms in which rocks are cast by the processes acting upon them. This explains why it is so difficult to piece together the planet's history from the geologic record: subsequent metamorphism has erased much of it.

THE ROCK CYCLE

The earth, therefore, is continuously changing. Plutons form deep in the crust; uplift pushes them to the surface; erosion wears them down; the sediments they produce become new mountains. This cycle of transformation, which affects all rocks and involves all parts of the crust, is conceptualized as the **rock cycle** (Fig. 31-12).

The rock cycle has neither beginning nor end, so you can start following it anywhere on the diagram. High temperature and pressure deep inside the crust melt the crustal (rock) material. Magma rises, either intruding into existing rocks or extruding as lava, creating both igneous and metamorphic rocks. Weathering and erosion attack the exposed rocks; deposition creates sedimentary rocks from the fragments. Crustal forces push sedimentary, igneous, and metamorphic rocks downward, and if they reach the lower levels of the crust they will be melted and the cycle will start anew—or rather, continue. In our lifetimes, we are witness to just a brief instant in a cycle that affects the entire planet in space and its whole history in time.

Now that we have become familiar with the characteristics and cycling of earth materials, we are ready to consider the forces of the restless crust. Our examination of these processes, which contribute importantly to the shaping of surface landscapes, begins in Unit 32 with an overview of the lithospheric plates that fragment the crust.

FIGURE 31-11 Foliation in metamorphic rocks lines up the minerals in parallel bands so that the rocks appear streaked with alternating light- and dark-colored belts. Because such banding can have the effect of concentrating softer minerals in narrow zones, erosion can progress rapidly in these weaker belts. But metamorphic rocks generally are strong and resistant. Here an outcrop of heavily banded gneiss creates rapids in the Tana River in Kenya, East Africa. (Authors' photo)

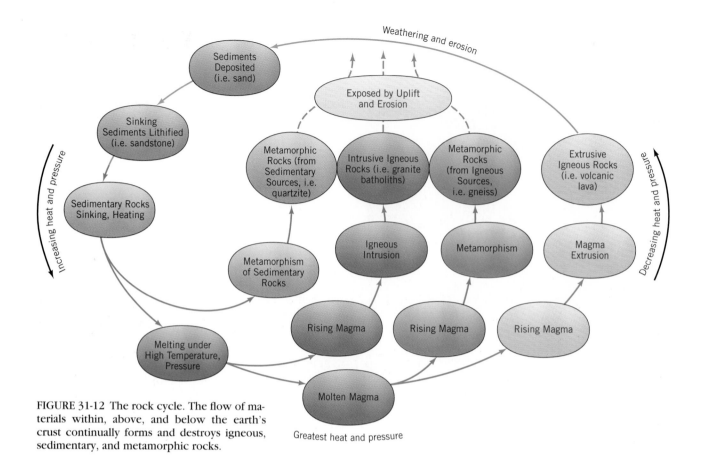

FIGURE 31-12 The rock cycle. The flow of materials within, above, and below the earth's crust continually forms and destroys igneous, sedimentary, and metamorphic rocks.

KEY TERMS

breccia (p. 330)
cementation (p. 329)
clastic and nonclastic sedimentary rocks (p. 330)
compaction (p. 329)
conglomerate (p. 330)
crossbedding (p. 332)
foliation (p. 334)

gneiss (p. 334)
limestone (p. 331)
marble (p. 334)
nonclastic sedimentary rocks (p. 330)
quartzite (p. 334)
rock cycle (p. 335)
sandstone (p. 331)

schist (p. 334)
shale (p. 331)
slate (p. 334)
strata (p. 332)
stratification (p. 332)
stratigraphy (p. 332)
unconformity (p. 332)

REVIEW QUESTIONS

1. How are sedimentary rocks formed?
2. How do clastic sedimentary rocks differ?
3. How are metamorphic rocks formed?

4. What are the metamorphic equivalents of sandstone, limestone, and shale?
5. Briefly outline the major components of the rock cycle.

REFERENCES AND FURTHER READINGS

BLATT, H., et al. *Origin of Sedimentary Rocks* (Englewood Cliffs, N.J.: Prentice-Hall, 2nd ed., 1980).

BOGGS, S. *Principles of Sedimentology and Stratigraphy* (Columbus, Ohio: Merrill, 1987).

COLLINSON, J. D. *Sedimentary Structures* (Boston: Allen & Unwin, 1982).

DEER, W. A., HOWIE, R. A., and ZUSSMAN, J. *An Introduction to Rock Forming Minerals* (New York: Wiley, 2nd ed., 1992).

DIETRICH, R.V., and SKINNER, B. J. *Gems, Granites, and Gravels: Knowing and Using Rocks and Minerals* (London/New York: Cambridge University Press, 1990).

DIETRICH, R.V., and SKINNER, B. J. *Rocks and Rock Minerals* (New York: Wiley, 1979).

EHLERS, E. G., and BLATT, H. *Petrology: Igneous, Sedimentary, and Metamorphic* (New York: Freeman, 1982).

HYNDMAN, D. W. *Petrology of Igneous and Metamorphic Rocks* (New York: McGraw-Hill, 2nd ed., 1985).

MASON, R. *Petrology of the Metamorphic Rocks* (Boston: Allen & Unwin, 1978).

PETTIJOHN, F. J. *Sedimentary Rocks* (New York: Harper & Row, 3rd ed., 1975).

REINECK, H. E., and SINGH, I. B. *Depositional Sedimentary Environments* (New York: Springer-Verlag, 2nd ed., 1980).

Plates of the Lithosphere

OBJECTIVES

◆ To introduce the concepts of continental drift and plate tectonics.

◆ To identify the major plates of the lithosphere.

◆ To discuss the important boundary zones between lithospheric plates in which rifting, subduction, and transform faulting occur.

When Christopher Columbus reached America in 1492, his discovery was recorded in his ship's log—and on the first map to be based on the Atlantic Ocean's western shores. When Columbus returned on his next three voyages, and as others followed him, the Atlantic coastline of the Americas became better known. During the sixteenth century, Portuguese navigators and cartographers mapped Africa's Atlantic coasts all the way to the Cape of Good Hope at the continent's southern tip.

By the early 1600s, the general configuration of the Atlantic Ocean was fairly well known, even though the maps of the time were often inaccurate in detail. Nonetheless, in 1619, Francis Bacon, the great English naturalist, made an observation that contained the kernel of a momentous concept. Looking at the evolving map of the South Atlantic Ocean, Bacon said that the opposite coasts of South America and Africa seemed to fit so well that it looked as though the two continents might at one

time have been joined. It was a notion soon forgotten, however, and not revived until nearly three centuries later. And even then the idea that whole continents could move relative to each other was greeted with skepticism and, in some circles, derision.

CONTINENTAL DRIFT

In 1915 a German earth scientist, Alfred Wegener (1880–1930), published a book that contained a bold new hypothesis. Not just Africa and South America, Wegener suggested, but all the landmasses on earth once were united in a giant supercontinent. This primeval landmass,

UNIT OPENING PHOTO: The island of Surtsey off the south coast of Iceland, located along the Mid-Atlantic Ridge. This is the top of a still-building undersea volcano that emerged from below the waves in 1963. (Authors' photo)

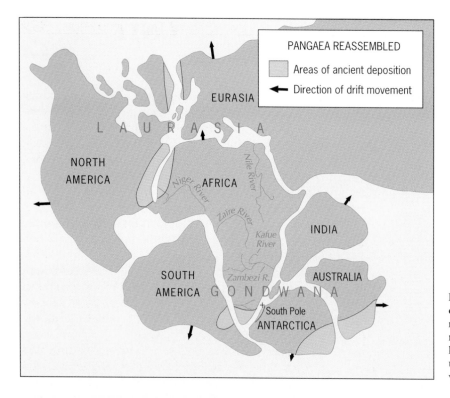

FIGURE 32-1 The breakup of the super-continent Pangaea began more than 100 million years ago. Note the radial movement of its remnants away from Africa and how areas of ancient deposition help us to understand where today's landmasses were once joined together.

FIGURE 32-2 The world ocean-floor map underscores the prominence of the global network of midoceanic ridges. A larger-scale version of this map can be found on pages 20–21.

which he named **Pangaea** (meaning "all-earth"), broke apart, forming the continents and oceans as we know them today. Wegener theorized that Pangaea consisted of two major parts, **Laurasia** in the north and **Gondwana** in the south. Today, Eurasia and North America are the remnants of Laurasia; South America, Africa, India, Australia, and Antarctica form the principal fragments of Gondwana (Fig. 32-1).

Wegener's book, *The Origin of Continents and Oceans* (1915), was not translated into English until the end of the 1920s. By then, Wegener's notion of **continental drift**—the fragmentation of Pangaea and the slow movement of the continents away from this supercontinent—was already a topic of hot debate in geological and physical-geographical circles in many parts of the world. An American geologist, F. B. Taylor, had written a long article in support of continental drift. A South African geomorphologist, Alexander du Toit, also supported Wegener's hypothesis and busied himself in gathering evidence from opposite sides of the South Atlantic Ocean. But most other geologists could not conceive of the possibility that whole continents might be mobile, functioning like giant rafts.

Wegener had marshalled a good deal of circumstantial evidence: fossil plants and animals from widely separated locales; climatic environments (as indicated by sedimentary rocks) unlike those now prevailing; and, of course, the remarkable jigsaw-like "fit" of the continents. Plausible as continental drift was to those who believed this evidence, there was one major problem: the process that could move continents was unknown. There simply was no evidence for a propelling mechanism.

As sometimes happens when a new scientific concept emerges, the hypothesis of continental drift lost credibility among many geologists, in part because of the mechanisms proposed by Wegener himself as well as others. Wegener suggested that the earth's gravitational force, which is slightly lower at the equator, was over time strong enough to pull the continents apart. Taylor proposed that the moon was torn from the earth in what is today the Pacific Basin, and that the continents have been steadily moving into the gap thereby created. Such notions damaged the credibility of the entire continental drift hypothesis, and despite the accumulation of geologic and paleontological evidence in favor of it, few geologists (especially in the United States) were willing to accept the possibility.

Some scientists, however, kept working on the problem. One geologist, Arthur Holmes, as early as 1939 proposed that there might be heat-sustained convection cells in the interior of the earth and that these gigantic cells could be responsible for dragging the landmasses along (*see* Unit 33). Others argued that the evidence for continental drift had become so overwhelming that notions of a rigid crust would have to be abandoned: there must be a mechanism, and further research would uncover it.

CONTINENTS AND SEAFLOORS

Geologists and physical geographers had been searching for evidence to support continental drift on the landmasses. But a large part of the answer lay not on the exposed continents, but on the submerged ocean floor. The existence of a Mid-Atlantic Ridge—a submarine mountain range extending from Iceland south to the Antarctic latitudes, approximately in the center of the Atlantic Ocean—had been known for many years. For decades it was believed that this feature was unique to the Atlantic Ocean. But during the 1950s and 1960s, evidence from deep-sea soundings made by a growing number of transoceanic ships carrying new sonar equipment began to reveal the global map of the ocean floors in unprecedented detail. The emerging map decidedly revealed that midoceanic ridges are present in all the ocean basins (Fig. 32-2).

This was a momentous development, but more was to come. When oceanographers and geologists analyzed rocks brought up from the ocean floor and determined their ages, the rocks were found to be much younger than most of those on the landmasses. These basaltic rocks, moreover, were youngest near the midoceanic ridges and progressively older toward the continental margins of the ocean basins.

Further investigation revealed another startling pattern: the midoceanic ridges were not just submarine mountain ranges like those on the landmasses, but constituted a global pipeline for hot upwelling magma. New rock was being formed, soon to be pushed away horizontally by still newer rock forcing its way up from below all along the midoceanic ridge. The process came to be called **seafloor spreading**, involving the creation of new crust and its continuous movement away from its source.

Obviously, if the midoceanic ridges are zones where new crust forms and diverges, the earth's crust is divided into segments—large fragments separated along the ridges. These segments of the crust were called *plates* (also **lithospheric plates** to denote their rigidity and *tectonic plates* to describe their active mobile character). Now, at last, the scientists were one giant step closer to understanding the mechanism needed to explain the drifting movement of continents.

If ocean floors can move and "spread," then continents can also be displaced. Moreover, if the ocean floor spreads outward from the midoceanic ridges, then the crust must be crushed together elsewhere, and parts of it must be pushed downward to make space for the newly forming crust. Thus the plates of which the crust is made are formed in one zone and destroyed in another. This process of destruction occurs where plates moving in opposite directions collide. Earthquakes, volcanism, and mountain building mark such zones of crustal collision.

DISTRIBUTION OF PLATES

When seafloor spreading was first recognized and the map of midoceanic ridges took shape, it appeared that the earth's crust was divided into seven major plates, all but one carrying a major landmass. But as more became known about both ocean floors and landmasses, additional plates were identified. By the 1980s, more than a dozen lithospheric plates had been mapped (Fig. 32-3), but there is still uncertainty as to the exact boundaries and dimensions of several, and others may yet be discovered. The largest plates are as follows:

1. The *Pacific Plate* extends over most of the Pacific Ocean floor from south of Alaska to the Antarctic Plate.

2. The *North American Plate* meets the Pacific Plate along California's San Andreas Fault and related structures. It carries the North American landmass.

3. The *Eurasian Plate* forms the boundary with the North American Plate at the Mid-Atlantic Ridge north of 35°N. It carries the entire Eurasian landmass north of the Himalayas.

4. The *African Plate* extends eastward from the Mid-Atlantic Ridge between 35°N and 55°S. It carries Africa and the island of Madagascar, and it meets the Antarctic Plate and Australian-Indian Plate under the Southern and Indian Oceans, respectively.

5. The *South American Plate* extends westward from the Mid-Atlantic Ridge south of 20°N. It carries the South American landmass.

6. The *Australian-Indian Plate* carries both Australia and the Indian subcontinent. It meets the Eurasian Plate at the Himalayas and the Pacific Plate in New Zealand.

7. The *Antarctic Plate* boundaries encircle the Antarctic landmass (which it carries) under the Southern Ocean.

These seven plates were identified early on. Later, a number of smaller plates were detected (Fig. 32-3). The largest among these is the *Nazca Plate*, wedged between the South American Plate to the east and the Pacific Plate

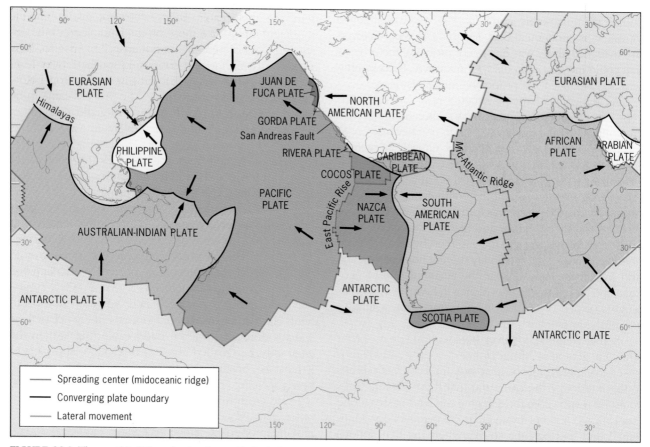

FIGURE 32-3 The earth's lithospheric plates. Each drifts continuously in the direction shown by the arrows. As the legend indicates, plate-boundary movement falls into one of three categories: divergence (spreading), convergence, or lateral motion.

to the west. Immediately to the north of the Nazca Plate, partly separating the South American Plate from the North American Plate, are two smaller plates. The larger of these, the *Caribbean Plate*, supports most of Central America and the southern Caribbean region. West of it (and thus north of the Nazca Plate) lies the *Cocos Plate*, and its northwestern extension (the *Rivera Plate*) off west-central Mexico. The Cocos and Rivera Plates are all oceanic crust and, unlike the Caribbean Plate, do not support a landmass.

At the southern end of South America lies a plate whose boundary arches eastward around an island chain much like the Caribbean Plate's eastern boundary; this is the *Scotia Plate*. Off the western coast of North America, near the point where the U.S.-Canada boundary reaches the Pacific Ocean, lies the *Juan de Fuca Plate*; and to its south, off southern Oregon and northernmost California, lies the smaller *Gorda Plate*.

Other smaller but significant plates that have been recognized include the *Philippine Plate*, located between the Pacific and the Eurasian Plates, and the *Arabian Plate*, which supports the landmass known as the Arabian Peninsula. The Philippine Plate skirts the island arcs off Asia's eastern coast, and these *archipelagoes* (island chains) mark its western boundary zone, a most active part of the earth's crust. The Arabian Plate is more stable at present, although its boundaries are also marked by crustal activity, especially in the south.

It is quite likely that the mid-1990s map of lithospheric plates will be revised again, because some smaller plates may not have been identified yet. It is noteworthy, for example, that all the smaller plates that have been recognized lie in ocean-floor areas of the crust. We know that the sialic landmasses lie on simatic crust that continues beneath them (*see* Unit 29). It is possible that the crust beneath the landmasses is more fragmented than is now known; later in this unit we examine the map of Africa to explore this possibility.

Location of Plate Boundaries

Once the notion of seafloor spreading gained acceptance and the segmented character of the crust became known, the search was on for the location of the boundaries of tectonic plates. Some earlier evidence now acquired new significance: the map showing the global distribution of earthquakes (Fig. 32-4) provided an important clue. Note that this map shows that earthquakes often originate in the midoceanic ridges, in island arcs, and in mountain belts such as the Andes and Himalayas. It was concluded that these linear earthquake zones represented plate boundaries, and thus the early idea that there were seven major lithospheric plates was reexamined.

The map of the world distribution of active volcanoes further supported this theory (Fig. 32-4). Of course, continental-surface patterns of volcanic activity were much better known than submarine volcanism. Volcanic activity is so common in western South and North America, in Asia's island archipelagoes, and in New Zealand that this circum-Pacific belt had long been known as the **Pacific Ring of Fire**. Thus it was at first concluded that the entire Pacific Ocean floor constituted a single, giant tectonic plate. Only later, when knowledge of the ocean-floor topography and geology improved, was the exis-

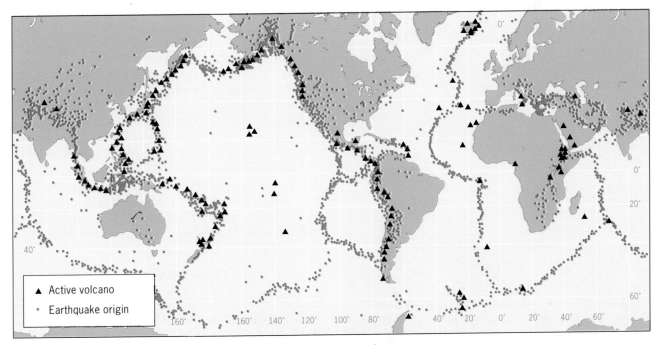

FIGURE 32-4 The global distribution of recent earthquakes and active volcanoes.

tence of smaller plates such as the Juan de Fuca and Cocos recognized.

Figure 32-5 shows the seafloor of the Atlantic Ocean. Note that the midoceanic ridge is not linear and unbroken, but divided into segments that are set off, giving them an overall zig-zag appearance. Right-angle fractures in the crust (*faults*) separate the segments. This relates to the conditions that prevail where new crust is created: rock is hot, molten, and viscous. It rises and spreads, beginning to cool as it moves away from the midoceanic ridge. But the ridge itself is not rigid and stable enough to sustain continuity. Lateral forces of movement in one direction sometimes are stronger than in the other. Thus one part of the Mid-Atlantic Ridge (between 20° and 35°N) is dragged to the west; elsewhere, such as just south of the equator, another part may lag behind or even move slightly to the east. As the rocks harden, a fault develops between the segments, and they are offset. Under the oceans, away from the landmasses, this pattern prevails.

But the map of world lithospheric plates (Fig. 32-3) also shows that the plate boundaries along continental margins take on another form. Along the edge of western South and North America, and off eastern Asia, the plate boundaries mostly appear on the map as solid black lines. The same is true for the longest plate boundary

FIGURE 32-5 The central portion of the Atlantic seafloor, between roughly latitudes 50°N and 20°S, highlighting the topography associated with the Mid-Atlantic Ridge.

known to exist across a landmass: the contact zone between the Eurasian Plate and plates to the south of it. No new crust is being created there; rather, crust is being crushed and pushed downward. From the distribution of these plate-boundary types, we can infer the movement of the lithospheric plates.

MOVEMENT OF PLATES

It is now known that the lithosphere consists of more than half a dozen major plates and at least a dozen smaller ones. These plates move relative to one another, apparently maintaining their prevailing direction of movement for millions of years. The term *tectonic plate* refers to this motion, which has carried the continental landmasses ever since Pangaea was fragmented by it. Indeed, plates and landmasses may have been in motion ever since the earth's crust was formed more than 4 billion years ago; Pangaea existed a mere 200 million years ago. It is quite possible that Pangaea itself resulted from an earlier phase of plate movement during which landmasses coalesced to form a single supercontinent.

The movement of lithospheric plates is directly responsible for many of the earth's major landscapes and landforms. All our studies of geomorphic processes and features must take into account the effect of crustal mobility. As plates migrate and carry landmasses along, they push, drag, tilt, bend, warp, and fracture. Lava pours out of fissures and vents. Rocks laid down as horizontal strata are deformed in every conceivable way. The weight of accumulating sediments pushes part of a plate downward. Where erosion has removed the material turned into sediment, the weight of the upper crust is reduced and the plate will rebound upward—all while movement continues.

Major directions of plate movement can be inferred from the map of the distribution of plates (Fig. 32-3). (Recent research has also provided information on the relative velocities of moving plates, the topic of the Perspective box in Unit 33.) Clearly, the African Plate has moved eastward, and its dominant direction remains eastward today. The South American Plate moves westward. New crust for both these plates is being created along the Mid-Atlantic Ridge. For some other plates, the direction of movement is less clear. What is certain, however, is that plate boundaries take on three kinds of character: (1) divergence or spreading; (2) convergence or collision; and (3) transform or lateral displacement.

Plate Divergence

We have noted that plates *diverge* or separate along the midoceanic ridges in the process called seafloor spreading. Magma wells up from the asthenosphere, new lithosphere is created, and the lithosphere on opposing sides

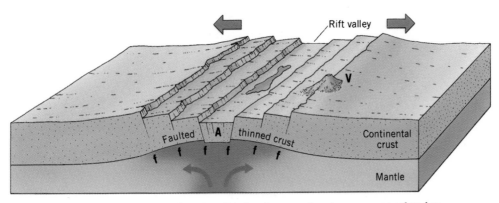

FIGURE 32-6 The development of a rift valley involves tensional movement related to motion in the mantle, faulting (**f**), the collapse of elongated strips of crust (**A**), and crustal thinning. Sometimes lava erupts along the tensional fault planes (**V**). Lakes fill large portions of rift valleys in East Africa (*see* map on p. 380).

of the midoceanic ridges is pushed apart. Here the tectonic forces are tensional, and the crust is so thin that it **rifts** open. Some geographers have pointed out that this process can also affect continental crust. If tensional forces exist beneath a part of a plate where a landmass occurs, both the simatic crust below and the sialic crust above are pulled apart. At the surface, this results in a sometimes spectacular landform called a **rift valley** (Fig. 32-6).

At present, a major system of rift valleys occurs in eastern Africa (*see* Fig. 36-7 on p. 380), and this system may signal the future fracturing of that continent (and the African Plate) along this zone of apparent crustal thinning. The Red Sea represents a more advanced stage of this process (Fig. 36-7): the Arabian Plate has separated from the African Plate, and between them now lies a basalt-floored sea. As time goes on, the Red Sea is likely to widen and become a new ocean.

Thus the geologic term *seafloor spreading* is perhaps better replaced by a geographic one, **crustal spreading**. Even today, not all spreading is confined to the ocean floor. And when Pangaea was a supercontinent, the first fractures in it occurred as rift valleys. Only when the rifts widened and magma filled the now-collapsed, water-filled trenches did the spreading become "seafloor."

The map of Africa, from a physical-geographic standpoint, also cautions us against taking the current map of known plates too seriously. The African Plate may well consist of three or four plates, all moving in the same general direction at this moment in geologic time, but capable of separating (*see* Fig. 36-9 on p. 381)—as the great East African rifts seem to suggest. An East African Plate may exist east of the easternmost (of Africa's) rift valleys; a small Victoria Plate may exist between the eastern and western rifts (*see* Fig. 36-7). The surface evidence further suggests that a plate boundary may exist, or be forming, along a line extending from the Gulf of Guinea toward Lake Chad in west-central Africa. Let us remember that Wegener made effective use of physiographic information to develop his hypothesis of continental drift, thereby paving the way for other scientists. The landscape is still a valuable guide, even in this age of satellite imagery, sophisticated data collection, and computer analysis.

Plate Convergence

If plates form and spread outward in certain areas of the crust, then they must *converge* and *collide* in other zones. The results of such a gigantic collision depend on the type of lithosphere involved on each side. Continental crust, as we know, has a relatively lower density and, in these terms, it is "light" compared to oceanic crust, which is denser, heavier, and more prone to sink or be forced downward where plates collide. When an oceanic plate meets a plate carrying a continental landmass at its leading edge, the lighter continental plate overrides the denser oceanic plate and pushes it downward. This process is termed **subduction**, and the area where it occurs is defined as a *subduction zone* (Fig. 32-7).

A subduction zone is a place of intense tectonic activity. The plate being forced down (subducted) is heated by the asthenosphere, and its rocks melt. Some of the molten rock forces its way upward through vents and fissures to the surface, so that volcanism is common along both oceanic and continental subduction zones. The movement of the plates is comparatively slow, averaging 2 to 3 cm (about 1 in) per year, but this is enough to generate enormous energy; some of that energy is released through earthquakes. Subduction zones are earthquake-prone and are among the world's most dangerous places to live.

Three types of convergent zones exist, only two of which result in significant subduction. Subduction occurs where oceanic crust subducts beneath continental crust and where one oceanic plate subducts beneath another. However, where two plates of continental crust converge, a somewhat different sequence of events follows. Let us examine each of the three cases.

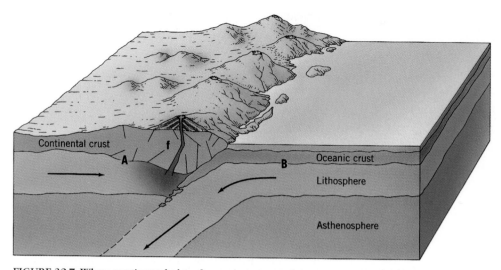

FIGURE 32-7 When continental plate **A**, moving eastward, meets oceanic plate **B**, moving westward, the process of subduction carries the heavier oceanic plate downward beneath the thicker but lighter continental plate. In this process, high relief develops along the coastline, the continental crust is heavily deformed, and magma can penetrate through vents and fissures (**f**) to erupt as lava at the surface.

Oceanic-Continental Plate Convergence

The best example of the kind of oceanic-continental subduction zone just described (Fig. 32-7) lies along South America's western margin, where the oceanic Nazca Plate is subducting beneath the continental South American Plate. The crust in the collision zone is dragged downward into deep oceanic trenches close to shore. A few kilometers to the east, continental rocks are crumpled up into the gigantic mountain ranges of the Andes. Volcanoes tower over the landscape (*see* photo on p. 348); earthquakes and tremors are recorded almost continuously. Sediments are caught in the subduction zone and become part of the hot molten magma. Here we can see the rock cycle (Fig. 31-12) in progress: basaltic crust from the ocean floor and granitic and sedimentary rocks from the landmass are melted and forced downward. They will eventually be carried back to the midoceanic ridges where they emerge and solidify, becoming part of the lithosphere once again.

Oceanic-Oceanic Plate Convergence

Other convergent plate boundaries involve two oceanic plates. The contrast between lithospheric plate densities is not present, and the crust is thrown into huge contortions. One of the plates will override the other, resulting in subduction; deep trenches form and volcanoes protrude, often above sea level (Fig. 32-8). The collision zone between the Pacific, North American, Eurasian, Philippine, and Australian-Indian Plates in the northern and western Pacific Ocean creates island arcs, such as the Aleutian and Japanese archipelagoes, as well as other segments of the Pacific's Ring of Fire.

Continental-Continental Plate Convergence

Where collision involves two continental plates, the situation is different. The best example is a segment of the convergence between the Eurasian Plate and the western flank of the Australian-Indian Plate. The Eurasian Plate is

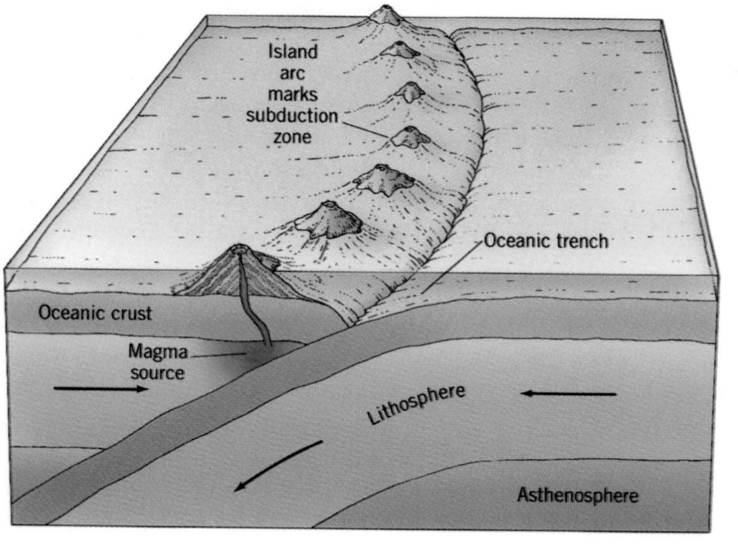

FIGURE 32-8 A convergent plate boundary involving two oceanic plates. Where one oceanic plate is subducted beneath the other, a deep oceanic trench forms. Above the trench, on the margin of the upward-riding plate, an island arc is created by volcanic activity.

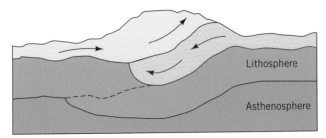

FIGURE 32-9 Two continental landmasses collide at a convergent plate boundary. There is much deformation of the crust, and high relief develops (South Asia's Himalayas mark such a convergent continental plate boundary). But while there is considerable thickening of the crust, less actual subduction occurs than when contrasting continental and oceanic plates converge.

moving southward, and the Indian subcontinent has moved to the north (thus a part of Gondwana was carried into collision with Laurasia). Such continental convergence creates massive deformation and a huge buildup of sialic mass. One continental mass may override the other, but the lower mass is not forced down into the asthenosphere or mantle (Fig. 32-9). Rather, the landmass thickens along the contact zone; earthquakes will attend the process, but not the widespread volcanism that accompanies oceanic-continental collisions.

Lateral Plate Contact

For many years, California's San Andreas Fault was known to be a place of crustal instability, a source of earthquakes, a line of danger on the map of the Golden State. Not until plate tectonics became understood, however, could the real significance of the San Andreas Fault be recognized. The fault marks a plate boundary—not a

FIGURE 32-10 A Southern California orange grove planted astride the San Andreas Fault. The fault's movement was sufficient to displace the originally straight rows of trees. Lateral ground motion was such that the trees in the background (beyond the fault line) moved from left to right with respect to the orange trees in the foreground.

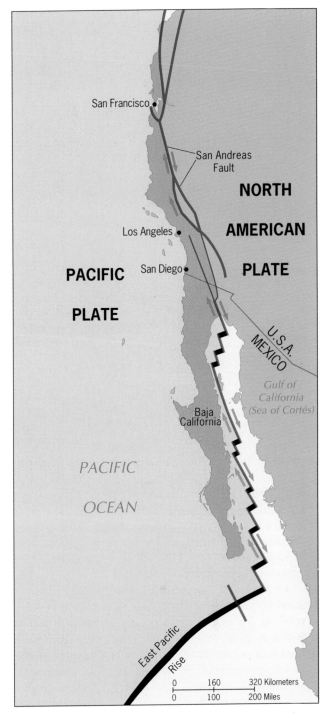

FIGURE 32-11 California's San Andreas Fault in its regional context. This fault separates the Pacific Plate from the North American Plate, which here are sliding past each other.

boundary of divergence or convergence, but a boundary along which two plates are sliding past each other (Fig. 32-10). These boundaries are referred to as **transform faults**. Transform (or *lateral*) movement along such plate boundaries may not have the dramatic topographic consequences displayed by convergent movement, but it, too, is accompanied by earthquakes and crustal deformation.

In effect, the area to the west of the San Andreas Fault is part of the Pacific Plate, whereas the area to the east is part of the North American Plate (Fig. 32-11). The

The Los Angeles Earthquake of 1994

FIGURE 32-12 Even a moderate earthquake can produce major disruptions in a densely developed metropolitan area. This is what remained of the interchange between Interstate 5 and the Antelope Valley Freeway (one of northern L.A.'s busiest) a few hours after the San Fernando Valley quake.

In the predawn darkness of January 17, 1994, the Los Angeles area was rocked by a destructive earthquake centered in Northridge in the city's San Fernando Valley, approximately 35 km (22 mi) northwest of downtown. The quake's magnitude was not particularly severe, but it struck a heavily populated suburban sector, killing 61, injuring 8500, and damaging more than 26,000 buildings as well as three vital freeways (*see* Fig. 32-12 above). When everything was added up weeks later, the destruction totaled a staggering $30 billion, making it the most expensive disaster in the history of the United States.

As geologists descended on the Valley to explain the event, they immediately discovered that the earthquake had oc-

curred along a then-unknown fault—but one that was well within the corridor containing the San Andreas system of transform faults (Fig. 32-11). Ominously, the quake also fit a pair of patterns that suggest this region faces more trouble. For one thing, this earthquake confirmed the existence of yet another fault lying directly beneath the nation's second largest metropolis. When added to the five other such crustal fractures recently identified and mapped, it is now clear that the Los Angeles Basin is honeycombed with faults that are associated with the plate-boundary stress zone lying along its inland margin.

More significantly, the 1994 quake was the latest in a series of noteworthy tremors that have plagued Southern Cal-

ifornia since 1986. This represents a sharp increase in earthquake activity compared to the four previous decades, and worries many scientists because the great San Francisco quake of 1906 was preceded by a similar cycle of more moderate tremors. The theory is that after a big earthquake on the San Andreas system, the moving plates steadily accumulate new stresses that finally build up again to the breaking point. History has shown that California experiences major earthquakes every 75 to 100 years. The last great quake in the Los Angeles area took place in 1857, and several geologists now say there is a 50–50 chance that "The Big One" will strike within the next 20 years.

fault extends southward into the Gulf of California, thus also separating Mexico's Baja California from the Mexican mainland, which is part of the North American Plate as well. At its northern end, the San Andreas Fault enters the Pacific Ocean north of San Francisco. Thus Baja California and Southern California (including metropolitan Los Angeles) are sliding north-northwestward past the North American Plate (Fig. 32-11) because of the northward motion of the Pacific Plate as a whole. This process is going on at a fairly high rate of speed, estimated to average more than 7.5 cm (3 in) per year.

In the case of the San Andreas Fault, earthquakes attend this movement; moreover, the San Andreas is the major fault in a much larger and complex system of transform faults in this corridor (*see* Perspective: The Los Angeles Earthquake of 1994). But lateral movement, for obvious reasons, is comparatively quiescent: no major subduction occurs, and no great volcanoes rise to mark the zone of contact. This means that lateral plate-contact boundaries beneath continental landmasses may yet be undiscovered. Thus the map of tectonic plates is still subject to modification. And Wegener's original vision, stimulated by the observation of geographic patterns, has been proven essentially correct. Accordingly, we continue the discussion of plate dynamics in Unit 33 by asking more questions about the breakup of Pangaea.

KEY TERMS

continental drift (p. 339)	Laurasia (p. 339)	Pangaea (p. 339)	seafloor spreading (p. 339)
crustal spreading (p. 343)	lithospheric plates (p. 339)	rift (p. 343)	subduction (p. 343)
Gondwana (p. 339)	Pacific Ring of Fire (p. 341)	rift valley (p. 343)	transform fault (p. 345)

REVIEW QUESTIONS

1. Briefly describe some of the evidence supporting the notion of continental drift.

2. Briefly describe the global map of lithospheric plates.

3. Where are most of the present-day rifting zones located?

4. How are subduction boundaries different from transform-fault boundaries? Give an example of each.

REFERENCES AND FURTHER READINGS

ANDERSON, D. L. "The San Andreas Fault," *Scientific American,* November 1971, 52–68.

BRIDGES, E. M. *World Geomorphology* (London/New York: Cambridge University Press, 1990).

CONDIE, K. C. *Plate Tectonics and Crustal Evolution* (Elmsford, N.Y.: Pergamon, 2nd ed., 1982).

DU TOIT, A. L. *Our Wandering Continents* (Edinburgh: Oliver & Boyd, 1937).

MARVIN, U. B. *Continental Drift: The Evolution of a Concept* (New York: Random House [Smithsonian Institution Press], 1973).

RAYMO, C. *The Crust of Our Earth: An Armchair Traveler's Guide to the New Geology* (Englewood Cliffs, N.J.: Prentice-Hall, 1983).

TARLING, D. H., and TARLING, M. P. *Continental Drift: A Study of the Earth's Moving Surface* (Garden City, N.Y.: Anchor/Doubleday, 2nd ed., 1975).

WEGENER, A. *The Origin of Continents and Oceans* (New York: Dover Publications, translated from the 1929 4th ed. by J. Biram, [reprinted 1966]).

WINDLEY, B. F. *The Evolving Continents* (New York: Wiley, 2nd ed., 1984).

WYLLIE, P. J. *The Way the Earth Works* (New York: Wiley, 1976).

Plate Movement: Causes and Effects

OBJECTIVES

♦ To briefly outline the mechanism that is believed to drive lithospheric plate movement.

♦ To discuss the evolution of the earth's continental landmasses.

♦ To discuss the concept of isostasy and relate it to the topography of the continental landmasses.

The most recent phase of crustal plate movement has broken up Pangaea, the supercontinent first envisaged fully by Alfred Wegener nearly a century ago (*see* Fig. 32-1). Continental drift proved to be one manifestation of plate tectonics, partial evidence for a process of planetary proportions involving immense amounts of energy. Confirmation of the breakup of Laurasia and Gondwana, however, raises new questions.

The fragmentation of Pangaea, the collision between India and Asia, the separation of Africa and South America—all this has taken place within the last 200 million years of earth history. But that period of time is just the last *4 to 5 percent* of the earth's total existence as a planet. What happened during previous phases of plate movement? Did earlier movements cause Pangaea to form

through a coalescence of landmasses, only to be pulled apart again? Can remnants of former Pangaeas still be found?

Certainly there are rocks older than 200 million years. As we have already noted, the earth formed more than 4.5 billion years ago. The oldest known rocks are from the geologic era known as the Precambrian, and they date from as long as 4 billion years ago (*see* Perspective: The Geologic Time Scale in Unit 37). These are all igneous and metamorphic rocks that today form part

UNIT OPENING PHOTO: A linear sequence of active volcanic peaks atop the backbone of the Ecuadorian Andes, the landscape signature of the subduction of the Nazca Plate beneath the South American Plate.

How Fast Do Drifting Plates Move?

Today it has been ascertained that the plates of the lithosphere move at velocities ranging from 1 to 12 cm (0.4 to 4.7 in) per year. We know this because technological breakthroughs in the past few years have enabled earth scientists to measure plate motion with remarkable precision. Studies of past velocities were first calculated by measuring the magnetic properties of rocks on each plate, and the inference was drawn that these motions have remained constant up to the present time. In the 1980s, that hypothesis was proven to be correct: by bouncing laser beams off new, specially equipped satellites, surface distances could for the first time be measured within an accuracy of 1 cm. Subsequent laser measurements quickly led to the construction of a new world map of current plate velocities.

That map is shown in Fig. 33-1, and its numbers refer to plate motion in centimeters per year. Numbers along the midoceanic ridges are average velocities indicated by (now-confirmed) measurements of rock magnetism. A figure of 11.7, as indicated for the northernmost segment of the East Pacific Rise (south of Mexico), means that the distance between a point on the Cocos Plate and a point on the Pacific Plate to the west annually increases by an average 11.7 cm (4.6 in) in the direction of the arrows that diverge from this midoceanic ridge. The red lines connect stations that measure plate movement by using satellite laser (**L**) technology several times each year. Note that the recorded plate velocities between these stations closely agree with the mean velocities estimated from magnetic measurements (**M**).

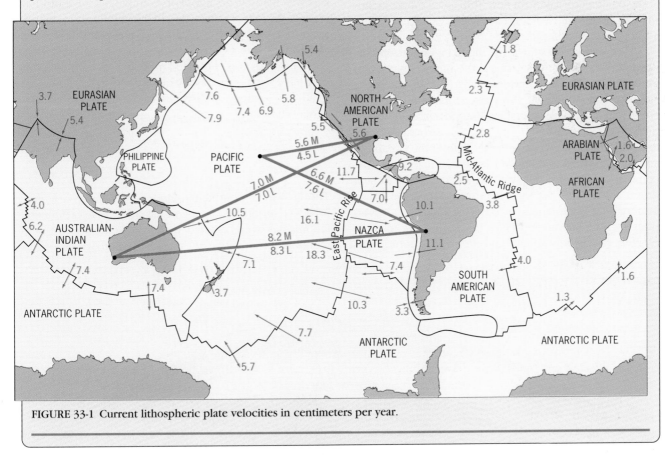

FIGURE 33-1 Current lithospheric plate velocities in centimeters per year.

of the core areas of the landmasses—their shields (*see* Fig. 29-11).

The ancient original shields must have been involved in every phase of crustal movement after their formation. Today, the Canadian (Laurentian) Shield forms the geologic core of North America; but during the Late Precambrian period, it probably lay at the heart of a landmass of very different shape and dimensions. This question is of interest in physical geography because the comparatively inactive continental shields carry landscapes that also are very old and have, in some areas, changed very little over many millions of years. These ancient landscapes, found in shield areas of Africa, South America, and Australia, may provide insights into the physiography of parts of Pangaea before its breakup.

MECHANISM OF CRUSTAL SPREADING

The movement of plates forming the earth's crust has been established beyond a reasonable doubt. Although we can accurately measure plate motion today (*see* Perspective: How Fast Do Drifting Plates Move?), the mechanism that propels the plates, past as well as present, is still not completely understood. The first model, proposed by Arthur Holmes, involves a set of internal convection cells in the earth's mantle and has been refined to account for the new knowledge of seafloor spreading. A model proposed by Harry Hess suggests that hot mantle material rises at the spreading midoceanic ridges. Some of it emerges to form new, thin crust; most remains in a hot plastic state, sliding slowly away from the ridges and cooling in the process.

This would explain why the temperature of the ocean-floor crust is highest near the midoceanic ridges and drops toward the continental margins. By spreading sideways and dragging the crust along, the sublithospheric magma keeps the spreading ridges open. By the time the new crust and the magma carrying it have spread as far as the continental margin of the ocean basin, they have cooled and thickened sufficiently to become so dense and heavy that they are ready to sink down again. This occurs when, at a convergent plate boundary, a continental (or other oceanic) plate overrides it. Now subduction takes place, and the material reenters the asthenosphere and the mantle, where it is heated up on the return journey to the spreading midoceanic ridge. All of these relationships are shown in Fig. 33-2.

If a set of convection cells such as those shown in Fig. 33-2 exists beneath all parts of the earth's crust, several questions arise, some not yet fully explained. First, how many of these cells exist? Are they all the same size, or do they range in size as the plates themselves do? How deeply do they penetrate the mantle? Evidence suggests that oceanic crust can be subducted to depths of 700 km (440 mi), which is far below the lower boundary of the asthenosphere. Some geologists suggest that the entire mantle may be in motion, and not just its upper layer. Does the earth's internal heat sustain the process, or is a heat-generating process, such as radioactive decay, responsible for the energy to keep convection cells in motion?

Most of the answers to such questions remain as speculative today as Wegener's continental drift hypothesis was many years ago. For example, as the earth has cooled, the rate of convection in the mantle and asthenosphere may have slowed down. Thus plate movement before the formation of Pangaea may have been even more rapid than it is today, and convergent boundaries may have been even more violently active. It follows that continental-margin landscapes had still more relief and variety than those of today's Pacific Ring of Fire.

EVOLUTION OF CONTINENTS

The growth of continental landmasses is also a largely unsolved riddle. It has been assumed that continental landmasses were created by the solidification of segments of the primitive crustal sphere, perhaps as long as

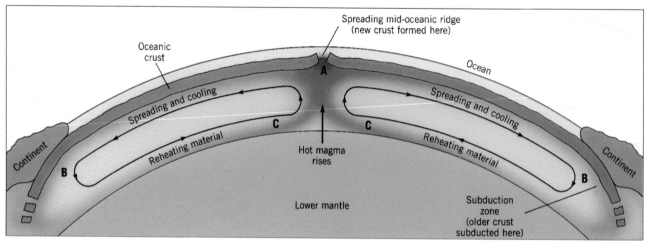

FIGURE 33-2 Convection cells in the mantle may look like this in cross-section. Hot magma rises at **A**, spreads toward **B**, and in the process drags the existing oceanic crust with it. At the spreading midoceanic ridge, new oceanic crust is being created from some of this upwelling magma. As the material below the crust spreads toward **B**, it cools slowly. When it reaches a convergent boundary with a continental landmass, the oceanic crust is subducted. The material in the convection cell now moves toward **C**, reheating at this depth. By the time it has passed **C**, it has enough energy to rise again into the spreading ridge. Speed of movement may be only about 2.5 cm (1 in) per year.

4 billion years ago. This process may have given rise to the igneous shields, which thus have existed ever since as the cores of the continents. When methods of dating rocks became more reliable, studies indicated that the oldest shield regions were indeed flanked by successively younger rock regions. Erosion of the original shield rocks created sedimentary strata around their margins. This would imply not only that the rocks of a continent become progressively younger away from the core shield, but also that the continents have grown, continuously or in stages, ever since their cores were first formed.

Crustal Formation

Other geologic evidence, however, suggests that shield areas have not grown by successive consolidation of magma around the original cores. Rather, it appears that the landmasses were formed from the solidification of the outermost cooling mantle, during a period approximately 2.5 to 3.5 billion years ago. After 2.5 billion years ago, the continental landmasses appear to have retained about the same total volume (if not the familiar shapes) as today. The crust has been recycled ever since, material being lost to subduction at convergent plate boundaries and regained by reformation at the spreading ridges.

Throughout their existence, the continental shields have lost little, because subduction has affected mostly the sedimentary strata accumulated at their margins. Even when a coalesced landmass became subject to crustal spreading, as happened when Africa and South America separated and the Mid-Atlantic Ridge appeared, the shield thereby fragmented lost no part of its mass. This model of the evolution of continents is still a subject of debate, and it may be modified when more becomes known about the subcrustal convection currents.

But even the continents themselves continue to yield their secrets. Very recently, geologists realized that certain parts of landmasses do not, geologically speaking, seem to belong where they are located. Their rocks and geologic histories are so different from their surroundings that the conclusion is inescapable: these chunks of continent must have been moved from faraway locales to their current, foreign positions.

Suspect Terranes

Geologists refer to a region of "consistent" rocks (in terms of age, type, and structure) collectively as a *terrane*. When there is a mismatch—a subregion of rocks possessing properties that sharply distinguish it from surrounding regional rocks—the "exotic" rock mass is called a **suspect terrane**. A prime example exists in North America. In the western part of the continent, three major rock masses have been identified as suspect terranes: an area of south-central Alaska near the Canada-Alaska border; another area stretching north from the

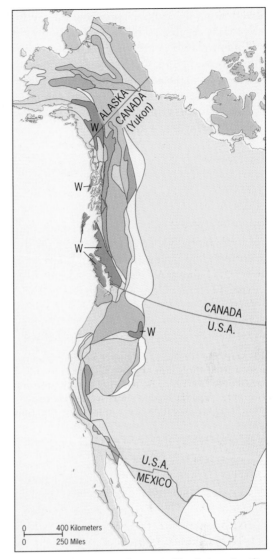

FIGURE 33-3 The terranes of western North America (lines are terrane boundaries). The three suspect terranes that constitute Wrangellia (marked **W**) are highlighted.

conterminous U.S.-Canada border along the Pacific Coast as far as the Queen Charlotte Islands; and a third in the interior of the U.S. Pacific Northwest (Fig. 33-3). These rock masses are collectively called *Wrangellia* (named after southeastern Alaska's Wrangell Mountains). They appear to have been wedged into North America about 100 million years ago, having been moved nearly 6000 km (3700 mi) from some distant southerly source.

How these suspect terranes reached their present locations is still a mystery. The existence of similar suspect terranes in the mountain belts of other landmasses also is known but not explained. It is also known that pieces of continental crust are lodged on the ocean floors, often visible on seafloor topographic maps as submerged plateaus. One example of such a continental rock mass rising to the ocean surface lies in the Seychelles Islands off East Africa in the western Indian Ocean. Such comparatively small rock masses with continental origins may be moved along as seafloor spreading proceeds. They may eventually reach a convergent plate boundary, and parts of them may become wedged into the mountain belt being formed there. Other suspect terranes may be former island arcs, pushed into continental margins and enveloped by mountain building. Clearly, this is one way a continental landmass may grow in present times, even if the Precambrian phase of shield formation has long passed.

ISOSTASY

The upper surfaces of the continents display a high degree of topographic variety. Mountain ranges rise high above surrounding plains; plateaus and hills alternately dominate the landscape elsewhere. Mountain ranges have mass. Because of the law of gravity, they exert a

FIGURE 33-4 The Himalayas form an awesome mountain wall when seen from the south. A view such as this undoubtedly greeted George Everest as he took his gravity measurements. This is central Nepal (a small country wedged between India's Gangetic Plain and the Himalayan massif) in the vicinity of Annapurna, the world's eleventh tallest peak.

certain attraction on other objects. If we were to hang a plumb line somewhere on the flank of a mountain range, we would expect the mountains to attract the plumb line from the vertical toward the range.

More than a century ago, British scientist George Everest, after whom the world's highest mountain is named, took measurements along the southern flanks of the Himalayas in India (Fig. 33-4). He suspended his plumb line and did indeed find that the great Himalayas caused some attraction—but far less than his calculations, based on the assumed mass of the mountain range, led him to expect. Everest and his colleagues soon realized the importance of what they had discovered. If the deviation of the plumb line toward the mountains was less than calculated, there must be rocks of lesser density extending far below the Himalayas, displacing the heavier simatic material that would have caused greater attraction. In other words, the lighter sialic rocks appear to extend far down into the simatic rocks, and mountain ranges seem to have "roots" penetrating downward farthest where their surface elevations are greatest.

This possibility was realized as early as 1855 by George Airy, whose hypothesis of mountain roots is depicted in Fig. 33-5. In Fig. 33-6, the sialic part of the crust is likened to blocks of copper that, because they are less dense, float in the mercury representing the sima. The higher the block stands above the dashed line representing sea level, the deeper the root below pushes into the simulated sima. Thus the blocks, or parts of the earth's crust, reach a kind of balance. Under the Himalayas and other major mountain ranges, the sialic part of the crust is comparatively thick. Under plateaus it is thinner, and under low-lying plains it is thinner still. Thus

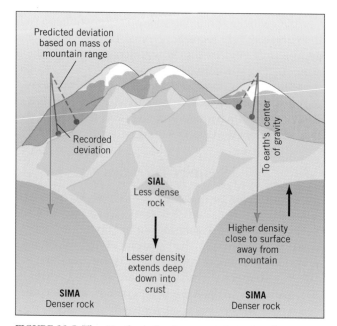

FIGURE 33-5 The Airy hypothesis: mountain ranges have roots of sialic rock that penetrate the denser simatic rock below.

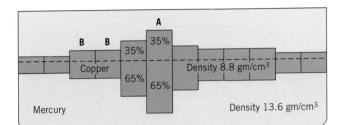

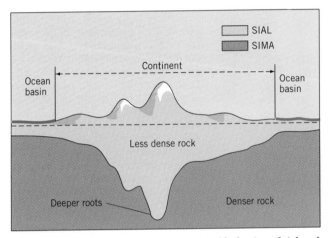

FIGURE 33-6 Isostasy. The distribution and behavior of sial and sima is analogous to blocks of copper floating in mercury. Note that, no matter how thick the block, the same percentage (35 percent/65 percent) floats above and below the surface. Each block is therefore in balance.

the relief of the continental landmasses has a mirror image below. Since the development of tectonic-plate theory, described in Unit 32, we have come to believe that the balance is not a question of sialic "rafts" floating on a simatic "sea." Rather, the balancing movements occur at the base of the lithosphere, far below the Mohorovičić discontinuity (*see* Unit 29).

The vertical changes in the crust are thought to take place for two reasons: (1) the lithosphere floats on the asthenosphere as the copper blocks float on the mercury; and (2) the lithosphere is subjected to changes of density from time to time. This situation of sustained adjustment, as visualized by Airy and modified by others after him, has come to be known as the principle of isostasy. The source of this term is not difficult to determine: *iso* means "the same" or "equal," and *stasy* comes from the ancient Greek word meaning "to stand." Thus **isostasy** is a condition of equilibrium between floating landmasses and the asthenosphere beneath them, maintained despite the forces that tend to change the landmasses all the time.

Isostasy and Erosion

We can use the model shown in Fig. 33-6 to envision what would happen if a high mountain range were subjected to a lengthy period of erosion. If we were to remove the upper 10 percent of the column marked **A**, we

would expect that column to rise slightly—not quite to the height it was before but nearly so. If we were to place the removed portion of **A** on the two columns marked **B**, they would sink slightly, and their upper surface would adjust to a slightly higher elevation than before. Thus column **A** would have a lower height and a shorter root, whereas the columns labeled **B** would have a greater height and a deeper root.

This tendency explains why erosional forces in the real world have not completely flattened all mountain ranges. Scientific experiments have indicated that, at present rates of erosion, the earth's mountain ranges would be leveled in a single geologic period, certainly within 50 million years. But mountains hundreds of millions of years old, such as the Appalachians of the eastern United States, still stand above their surroundings. What seems to happen is that as erosion removes the load from the ridges, isostatic adjustment raises the rocks to compensate. Rocks formed deep below the surface, tens of thousands of meters down, are thereby exposed to our view and to weathering and erosion.

We can also answer some puzzling questions about the deposition of enormous thicknesses of sedimentary rocks. One sequence of sedimentary rocks found in Africa involved the accumulation of nearly 6.5 km (over 20,000 ft) of various sediments followed by an outpouring of great quantities of lava. Other parts of the world have even thicker deposits. It is possible to deduce the environment under which deposition took place from the character of the deposits themselves. In some areas such deposition took place in shallow water. Although thousands of meters of sediments collected over millions of years, the depth of the water somehow remained about the same. In the accumulating sediments near the Bahamas, for instance, rocks formed in shallow or intertidal flats are now 5500 m (18,000 ft) thick.

We may therefore conclude that some cause, or combination of causes, continuously depresses the region of deposition, keeping the surface at about the same level. Such slowly accumulating sediments might be another place to effectively dispose of waste material produced by human activity, which might slowly sink from sight within the sediments. Slow accumulation is now taking place in the Mississippi Delta. For millions of years, the great river has been pouring sediments into its delta, but these deposits have not formed a great pile, nor have they filled in the Gulf of Mexico (Fig. 33-7). Isostatic adjustment constantly lowers the material to make room for more.

Isostasy and Drifting Plates

If isostasy involves a condition of equilibrium, then the contact and collision of drifting plates must greatly affect that situation. When a continental plate meets an oceanic plate, the oceanic plate plunges below the continental plate, causing the deformation and dislocation shown in

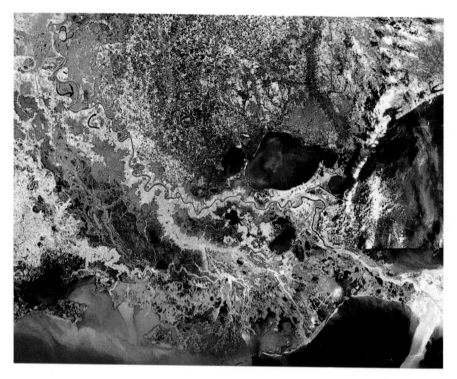

FIGURE 33-7 A satellite mosaic of the Mississippi Delta region, with the river threading its way from the northwest to the southeast corner. This view of the coastal zone of the Gulf of Mexico extends from approximately Lafayette, Louisiana in the west to Biloxi, Mississippi in the east. The almost round water body just to the right of the center of the figure is Lake Pontchartrain; the city of New Orleans lies between its southern shore and the Mississippi. The color changes in the lower-right portion of the figure were caused by the fitting together of slightly different—but contemporaneous—imagery.

Fig. 32-7. Along the leading edge of the continental plate, rocks are crushed and folded, sediments are baked into metamorphic rocks, and magma penetrates and erupts along fissures and from volcanoes. In effect, the sialic mass increases in volume and, according to isostatic principles, rises upward. Right next to the high mountains so formed, the downward thrust of the oceanic plate often creates a deep trench, as deep as or deeper than the highest mountains are high. This is the situation along much of the Pacific's Ring of Fire, where crustal instability and isostatic maladjustment are greatest. Earthquakes occur continuously along this zone as the plates converge and collide (*see* Fig. 32-4).

But the process does not go on forever at the same rate. We still do not know just why, but the geologic record shows that the earth has gone through various rather distinct periods of mountain building and other quieter periods. Eventually, even the Ring of Fire will quiet down, and plate collision and mountain building may start elsewhere. When this happens, erosion becomes dominant and begins to lower the mountains that have been created—but very slowly, because isostatic uplift will occur as mass is removed. The first phase of erosion, however, may be rather rapid.

The sialic mountains along the plate margin may have been pushed beyond the elevations justified by the depth of their roots, and isostatic readjustment will not commence until the overload has been removed. After that, the mountain masses undergo isostatic uplift as mass is eroded, a process that ensures their persistence for a long time. The Appalachians and Southern Africa's Cape Ranges have roots deep enough to ensure their topographic prominence over the past 200-plus million years. Some mountain ranges have indeed been flattened by erosion all the way down to their roots, but these are much older still.

Isostasy and Regional Landscapes

In studying the effects of the isostatic principle, we tend to be preoccupied with mountain ranges, mountain building, plate compression, and associated phenomena. But we should not lose sight of the consequences of isostasy in areas of less prominent, less dramatic relief. Erosion is active on the continents' plains too, and millions of tons of material are carried away by streams, rivers, and other erosional agents. Even moving ice and wind denude and reduce land surfaces. Unlike the mountainous zones, however, the plains are vast in area and slopes are gentler. Rivers erode less spectacularly on the plains than in the mountains as a consequence of several factors (*see* Unit 41). All these circumstances mean that eroded material is removed from the plains at a slower rate.

Plains and Uplands

The sialic crust has a certain rigidity. It does not behave, as in Airy's model, as a series of discrete columns. Therefore, isostasy affects plains and plateaus in phases. For a certain period, the amount of material removed does not trigger isostatic readjustment because the hardness of the crust prevents continuous uplift. But when the plain has been lowered sufficiently for the push of isostatic uplift to overcome the resistance of the crust, a change takes place. Thus, at any given moment, an area may not be in isostatic equilibrium. Instead, it awaits the time when readjustment is forced by the removal of a sufficient mass of landscape.

Scientists suggest that this periodic adjustment may also occur in mountain ranges, especially older ones. In the beginning, when the sialic root is deep, almost continuous isostatic uplift occurs. But as time goes on the root becomes shorter, erosion continues, and comparatively more eroded material must be removed for readjustment to occur. In fact, the Appalachians were probably flattened almost completely and then rejuvenated by a recurrence of isostatic uplift. Now the old ridges are being worn down again, and the whole area may be transformed into a plain before another readjustment occurs (Fig. 33-8).

Icesheets and Isostatic Rebound

Some other manifestations of isostatic change also are of interest. When icesheets spread over continental areas during glaciations, the weight of the ice (which can reach a thickness of several thousand meters) causes isostatic subsidence or sinking of the crust below, just as sediment accumulation has produced in other regions. This is what happened in northern North America and Eurasia during the most recent glaciation, where the last of the great icesheets melted less than 12,000 years ago. During the maximum extent of these continental glaciers, the crust below was depressed by their weight. When the icesheets retreated, isostatic readjustment caused the crust to rebound. The upward readjustment, however, could not keep pace with the relatively rapid melting of the ice. In geologic terms, the melting removed the enormous load of the icesheets almost instantaneously.

Studies show that the ensuing isostatic rebound is still going on; thus the crust is still not in equilibrium. In the heart of the Scandinavian Peninsula in northern Europe, the site of a huge icesheet, the crust is rising at more than 1 m (3.3 ft) per century, a very fast rate. In some coastal parts of Norway, metal rings placed in rocks centuries ago to tie up boats are now much too high above sea level to be of use. Similarly, much of coastal California is flanked by ancient beaches tens of meters above the present-day sea level.

Dams and Crustal Equilibrium

Even human works on the surface of the earth can produce isostatic reaction. When a river dam is constructed, the weight of the impounded water behind it may be enough to produce isostatic accommodation in the crust. Measurable readjustment of this kind has taken place in the area of Kariba Lake, formed upstream of the great dam on the Zambezi River in Southern Africa, and

FIGURE 33-8 A portion of the central Appalachians that suggests these highlands are being eroded into a plain-like surface.

around Lake Mead behind Hoover Dam on the Colorado River in Nevada. We cannot see these changes with the naked eye, but scientific instruments detect them. To us, in our everyday existence, the crust may seem permanent, unchanging, and solid, but even our own comparatively minuscule works can disturb its equilibrium.

The dynamics of plate movement produce much of the restlessness that characterizes the earth's crust. One of the most spectacular surface manifestations of this geologic activity is volcanism, a topic that is the focus of Unit 34.

KEY TERMS

isostasy (p. 353)   suspect terrane (p. 351)

REVIEW QUESTIONS

1. Describe the mechanism that is believed to drive lithospheric plate movement.
2. What are mountain "roots"?
3. How might long periods of erosion trigger uplifting of the landscape?
4. What is meant by the term *suspect terrane*?
5. Describe the process of isostatic uplift.

REFERENCES AND FURTHER READINGS

ALLEGRE, C. *The Behavior of the Earth: Continental and Sea-floor Mobility* (Cambridge, Mass.: Harvard University Press, 1988).

BRIDGES, E. M. *World Geomorphology* (London/New York: Cambridge University Press, 1990).

GLEN, W. *Continental Drift and Plate Tectonics* (Columbus, Ohio: Merrill, 1975).

KING, L. C. *Wandering Continents and Spreading Sea Floors on an Expanding Earth* (Chichester, U.K.: Wiley, 1983).

MIYASHIRO, A., et al. *Orogeny* (New York: Wiley, 1982).

OLLIER, C. D. *Tectonics and Landforms* (London/New York: Longman, 1981).

SULLIVAN, W. *Continents in Motion: The New Earth Debate* (New York: McGraw-Hill, 1974).

SUMMERFIELD, M. A. *Global Geomorphology: An Introduction to the Study of Landforms* (New York: Wiley/Longman, 1991).

TARLING, D. H., and RUNCORN, S. K. *Implications of Continental Drift to the Earth Sciences* (New York: Academic Press, 1973).

UYEDA, S. *The New View of the Earth: Moving Continents and Moving Oceans* (San Francisco: Freeman, 1978).

VITA-FINZI, C. *Recent Earth Movements* (Orlando, Fla.: Academic Press, 1986).

WILSON, J. T., ed. *Continents Adrift and Continents Aground* (San Francisco: Freeman, 1976).

Volcanism and Its Landforms

OBJECTIVES

◆ To relate volcanic activity to plate boundary types.

◆ To discuss typical landforms produced by volcanic eruptions.

◆ To cite some dramatic examples of human interaction with volcanic environments.

olcanism is the eruption of molten rock at the earth's surface, which is often accompanied by rock fragments and explosive gases. The process takes various forms, one of which is the creation of new lithosphere at the midoceanic spreading ridges (about 75 percent of the world's volcanoes are on the seafloor). Along some 50,000 km (31,000 mi) of ocean-floor fissures, molten rock penetrates to the surface and begins its divergent movement (*see* Fig. 33-2).

It is a dramatic process involving huge quantities of magma, the formation of bizarre submarine topography, the heating and boiling of seawater, and the clustering of unique forms of deep-sea oceanic life along the spreading ridges. But it is all hidden from us by the ocean water above, and what we know of it comes from the reports of scientist-explorers who have approached the turbulent scene in specially constructed submarines capable of withstanding the pressure at great depths and the high temperatures near the emerging magma. Volcanism also occurs on the continents in the vicinity of plate boundaries (*see* Figs. 32-3 and 32-4), and leaves a characteristic signature in the form of volcanic landscapes (*see* photo above).

UNIT OPENING PHOTO: Mount Fuji, southwest of Tokyo, is a towering volcanic mountain of almost perfect symmetry. This highest of Japanese peaks rises 3776 m (12,388 ft).

FIGURE 34-1 One of the spectacular volcanic explosions that rocked the Icelandic island of Heimaey in 1973. Incredibly, life soon returned to normal here following the termination of this volcanic episode. But the threat of renewed activity is, of course, ever present.

Islands situated on midoceanic ridges afford a glimpse of a process that is mostly concealed from view. Iceland and smaller neighboring islands lie on the Mid-Atlantic Ridge between Greenland and Norway in an area where the ridge rises to the ocean surface (*see* Fig. 2-6). Iceland and its smaller neighbors all are totally of volcanic origin, and there is continuing volcanic activity there. In 1973 a small but populated and economically important island off Iceland's southwest coast, Heimaey, experienced a devastating episode of midoceanic-ridge volcanic activity. First, Heimaey was cut by fissures, and all of its inhabitants were quickly evacuated. In the months that followed, lava poured from these new gashes in the island, and volcanic explosions rained fiery pieces of ejected magma onto homes and commercial buildings (Fig. 34-1).

Heimaey actually increased in size, but the lava flows threatened to fill and destroy its important fishing port. This threat led to an amazing confrontation between people and nature: the islanders quickly built a network of plastic pipes at the leading edge of the advancing lava. They pumped seawater over and into the lava, aware that by cooling it more quickly than nature could, the lava would form a solid dam that might stop the advance and restrain the lava coming behind it. This daring scheme worked: part of the harbor was lost to the lava, but a

critical part of it was saved. When this volcanic episode ended, life returned to Heimaey. But Iceland and its neighbors lie on an active midoceanic ridge, and volcanism will surely attack them again. What happened above the surface at Heimaey is happening, continuously, all along those 50,000 km (31,000 mi) of submerged spreading ridges.

DISTRIBUTION OF VOLCANIC ACTIVITY

Most volcanism not associated with seafloor spreading is related to subduction zones (*see* Fig. 32-7). As the global map (Fig. 32-4) shows, volcanic activity is concentrated at convergent plate boundaries. Not surprisingly, a majority of the world's active volcanoes lie along the Pacific Ring of Fire. But note that there is some volcanic activity that is associated neither with midoceanic ridges nor with subduction zones. The island of Hawaii, for example, lies in an archipelago near the middle of the Pacific Plate. Lava poured from one of its volcanic mountains, Kilauea, repeatedly during the 1970s and almost continuously during the 1980s and early 1990s. On the African Plate, where West Africa and Equatorial Africa meet (at Africa's so-called armpit), lies Mount Cameroon, another active volcano far from spreading ridges and subduction zones. The map reveals a number of similar examples, both on ocean-floor crust and on continental crust. As we will learn later, this distribution is difficult to explain.

Active, Dormant, and Extinct Volcanoes

Physical geographers differentiate among active, dormant, and extinct volcanoes based, in some measure, on their appearance in the landscape. An *active* volcano is one that has erupted in recorded history (which, geologically speaking, is but an instant in time). A *dormant* volcano has not been seen to erupt, but it shows evidence of recent activity. This evidence lies on its surface: lava tends to erode quickly into gulleys and, on lower, flatter slopes, weather into soils. If a volcano seems inactive but shows little sign of having been worn down, it may be concluded that its latest eruptions were quite recent, and activity may resume. When a volcano shows no sign of life and exhibits evidence of long-term weathering and erosion, it is tentatively identified as *extinct*.

Such a designation is always risky because some volcanoes have come to life after long periods of dormancy. As noted earlier, the great majority of continental volcanoes lie in or near subduction zones. Volcanic activity there is concentrated, and parts of the landscape are dominated by the unmistakable topography of eruptive volcanism. Many of the world's most famous mountains are volcanic peaks standing astride or near plate boundaries: Mount Fuji (Japan), Mount Vesuvius (Italy), Mount

Rainier (U.S. Pacific Northwest), Mount Egmont (New Zealand), and many others. Frequently, such mountains stand tall enough to be capped by snow, their craters emitting a plume of smoke. It is one of physiography's most dramatic spectacles.

Lava and Landforms

The viscosity of magma and lava varies with its composition. Basaltic lavas, such as those flowing from the mid-oceanic spreading ridges, are relatively low in silica and high in iron and magnesium content, and are therefore quite fluid when they erupt. They flow freely and often (on land) quite rapidly. Other lavas are poorer in magnesium and iron but richer in silica—and thus more acidic. These lavas tend to be more viscous, and as a result they flow more slowly. Basaltic lava can flow like motor oil; the less mafic lava moves more like a thick porridge would.

Magma also contains steam and other gases under pressure, with variations again a function of its mineral content. The acidic, silica-rich magmas tend to contain more gases, and when they erupt as lavas these gases often escape explosively. Gobs of lava are thrown high into the air, solidifying as they fall back to the mountain's flanks. Such projectiles, not unreasonably, are called volcanic *bombs*. Smaller fragments may fall as volcanic *cinders* or volcanic *ash*. After the explosive 1980 eruption of Mount St. Helens in Washington State (*see* opening photo in Unit 30), lighter volcanic *dust* fell over a wide area downwind from the mountain. Geologists use the term **pyroclastics** for all such fragments erupted explosively from a volcano.

When lava creates landforms, still other factors come into play. The rate of cooling relates not only to the composition and viscosity of the molten material, but also to the thickness of the flow and the nature of the surface over which the lava spreads. The nature of the *vent*, the opening through the existing crust where the lava has penetrated, is an additional factor. Some eruptions do not come from pipe-shaped vents, but from lengthy cracks in the lithosphere. *Fissure* eruptions do not create

mountains; rather, they release magma to form lava that extends in sheets across the countryside, creating sometimes extensive plateaus (Fig. 34-2). In the United States, the best example is the Columbia Plateau, an area underlain by thick sheets of basalt that erupted in many successive layers and eventually covered 50,000 km² (20,000 mi²) of southern Washington, eastern Oregon, and much of Idaho (Fig. 34-3). Just before Gondwana broke up, great fissure eruptions produced a vast lava plateau of which parts still exist in South Africa, India, South America, and Antarctica.

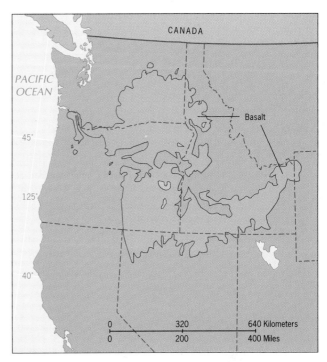

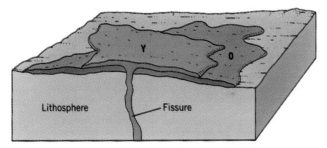

FIGURE 34-2 During a fissure eruption, lava flows onto the surface and spreads out in a sheet rather than forming a dome. When the fissure opens again, a later, younger flow (**Y**) will cover all or part of the older lava sheet (**O**).

FIGURE 34-3 The basalt flows of the Columbia Plateau cover much of the interior Pacific Northwest of the United States (a). The horizontal layering of these successive lava flows is strikingly visible in the landscape (b).

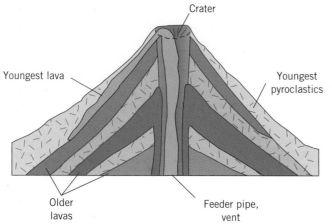

FIGURE 34-4 Simplified cross-section of a composite volcano, showing a sequence of lavas interspersed with compacted pyroclastics.

VOLCANIC MOUNTAINS

The most characteristic product of volcanic eruption is the towering mountain form, represented by such peaks as Fuji (Japan), Rainier (U.S.), Popocatépetl (Mexico), Vesuvius (Italy), and Kilimanjaro (Tanzania). But not all volcanic action, even from pipe-shaped vents, produces such impressive landforms. Some volcanoes are even larger in volume but less prominent in shape. Others are smaller and less durable. Four types of volcanic landforms exist: composite volcanoes, volcanic domes, cinder cones, and shield volcanoes.

Composite Volcanoes

Most of the great volcanoes formed over subduction zones are **composite volcanoes**: they disgorge a suc-cession of lavas and pyroclastics. In cross-section such volcanoes look layered, with lavas of various thicknesses and textures interspersed with strata formed by compacted pyroclastics (Fig. 34-4). Neither the heavier pyroclastics nor the rather viscous lava travel very far from the crater; thus the evolving volcano soon takes on its fairly steep-sided, often quite symmetrical appearance. Many composite volcanoes are long-lived and rise to elevations of thousands of meters. Appropriately, they also are called *stratovolcanoes*.

Composite volcanoes, with their acidic, gas-filled lavas, are also notoriously dangerous. They often erupt with little or no warning, and molten lava is not the only threat to life in their surroundings. Pyroclastics can be hurled far from the crater; volcanic ash and dust create health hazards that choke human and animal life even farther away.

Lahars

On snowcapped volcanoes, the hot ash sometimes melts the snow and ice, which forms a flood of ash, mud, and water rushing downslope. Such a mudflow can be extremely destructive. In 1985, a volcano named Nevado del Ruiz in the Andes of central Colombia erupted, and much of its snowcap melted. In the mudflow that swiftly followed, more than 20,000 people perished. After it was over, the scene at a town in its path at the base of the mountain range was one of utter devastation, a mass of mud containing bodies of people and animals, houses and vehicles, trees and boulders (Fig. 34-5). Such a deposit of volcanic origin is called a **lahar** (pronounced "luh-HARR") once it solidifies.

Nuées Ardentes

Perhaps even more dangerous is the outburst of hot gas and fine dust that may accompany or precede an erup-

FIGURE 34-5 The aftermath of the 1985 mudflow, caused by a nearby erupting volcano, that buried the Colombian town of Armero without warning late at night. This lahar killed at least 20,000 people in the immediate area, amounting to one of the worst natural disasters of the 1980s.

tion. Gas, as we know, is pent up in magma itself, but in some composite volcanoes a large reservoir of gas may accumulate in a chamber below the crater. This happens when the crater becomes clogged by solidified lava, which forms a *plug* in the top of the vent. The gas cannot escape, and building temperatures inside the mountain may exceed 1000°C (1800°F). Pressures finally become so great that the side of the volcano may be blown open, allowing the gas to escape, or the plug may be blown out of the top of the volcano's pipe, enabling the gas to rush from the crater along with heavier pyroclastics and lava. Such an event produces a **nuée ardente** (French for glowing cloud; pronounced "noo-AY ahr-DAHNT"), which races downslope at speeds exceeding 100 kph (65 mph). Everything in its path is incinerated because the force of the cloud of gas plus its searing temperatures ensure total destruction.

In 1902, a *nuée ardente* burst from Mount Pelée on the eastern Caribbean island of Martinique. It descended on the nearby port town of St. Pierre at its base in a matter of minutes, killing an estimated 30,000 persons (the only survivor was a prisoner in a cell deep underground in the town's prison). The "Paris of the Caribbean," as St. Pierre had been described, never recovered and carries the scars of its fate to this day.

Predicting Risk

Among the other dangers posed by volcanoes of the gas-filled, viscous-magma variety are quieter emissions of lethal gases that can reach people and animals at air temperature and kill without warning. Estimating the risk to people living near such volcanoes and predicting dangerous activity have become part of an increasingly exact science, but many volcanoes (especially in developing countries) are not yet subject to such research. The 1980 eruption of Mount St. Helens had been predicted by scientists who had measured the telltale signs of resumed activity, and their warnings probably saved thousands of lives.

Forecasting techniques have been improving since 1980, and a number of volcanoes are now being monitored by scientists in the United States, Japan, Mexico, and elsewhere. Among their latest successes were predictions of the eruptions of Alaska's Mount Redoubt in 1989 and Mount Pinatubo in the Philippines in 1991 (Fig. 34-6). But the magnitude of these eruptions went well

FIGURE 34-6 The massive explosions that tore apart Mount Pinatubo in mid-1991 amounted to nothing less than the largest volcanic event of this century (at least through press time). This particular blast, accompanied by spectacular lightning bolts, occurred late at night near the end of the three-month-long eruption sequence, whereby Pinatubo roared back to life after six centuries of dormancy. Locally, on the major Philippine island of Luzon, this disaster killed nearly 1000 people, rendered 1.2 million homeless, and destroyed more than 80,000 hectares (200,000 acres) of fertile farmland. Beyond the Philippines, the Pinatubo eruptions injected such massive quantities of volcanic ash and dust into the atmosphere that air temperatures around the world were slightly diminished through 1994.

Risking the Wrath of Goddess Pele

Native Hawaiians do not regard Hawaii's volcanism as simply a geologic pheno- menon. They have lived with Hawaii's fountains of fire far longer than the white invaders have been on their islands. The goddess Pele (pronounced ''PAY-lay'') rules here, and she displays her pleasure or wrath through her power over the main island's volcanoes. In accordance with tradition, Hawaiians walk bare- foot on the *aa* lava to the very edges of Kilauea (Fig. 34-8) and Hawaii's other steaming craters, pray to Pele, and leave fern garlands for her.

During the 1980s, plans were an- nounced to tap geothermal energy from the interior of Kilauea. The project was expected to generate enough electricity to some day provide all the Hawaiian Is- lands with power. But Pele's followers argued that such a penetration of the heart of a holy mountain would destroy the goddess herself. A confrontation de- veloped, and it reached the courts of law. Christian missionaries long ago had sup- pressed Pele worship, but the geother- mal project proved that Pele still has many followers.

FIGURE 34-8 The unearthly landscape of Hawaii's Kilauea, a classic shield volcano. An eruption is in progress, one of a still-continuing series that began in the 1970s. Lava emanating here can flow for many kilometers and reach the sea.

Meanwhile, the goddess seemed to prove a point by sending lava into the Royal Gardens subdivision and consum- ing several houses. In 1988, the Pele Defense Fund even bought a full-page advertisement in the *New York Times* to state its case—and the day it appeared, Kilauea put on a volcanic display de- scribed as the most spectacular in more than a year. Native Hawaiians were not surprised.

FIGURE 34-7 Cinder cones are common in areas of East Africa. This photograph shows the characteristic convex configuration of this type of volcanic landform.

beyond what had been anticipated, reminding us that volcanic activity still defies exact prediction.

Volcanic Domes

When acidic lava penetrates to the surface, it may ooze out without pyroclastic activity. This process usually pro- duces a small volcanic mound called a **volcanic dome**. A volcanic dome often forms inside a crater following an explosive eruption, as happened at Mount St. Helens. But volcanic domes can also develop as discrete land- forms in a volcanic landscape. Although some grow quite large, volcanic domes, on average, are much smaller than composite volcanoes.

Cinder Cones

Some volcanic landforms consist not of lava, but almost entirely of pyroclastics. Normally, such **cinder cones**

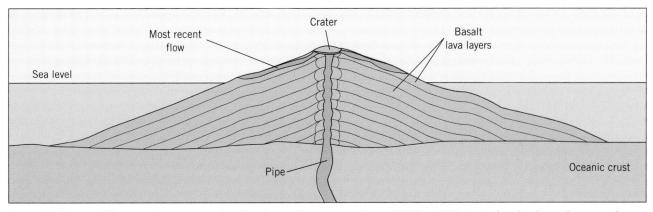

FIGURE 34-9 Simplified cross-section of a shield volcano. Vertical scale is greatly exaggerated. The base of this volcano extends over 320 km (200 mi); its height above the ocean floor is around 13 km (8 mi).

(which may also include fragments larger as well as smaller than cinders) remain quite small, frequently forming during a brief period of explosive activity. Probably the most extensive area of cinder-cone development lies in East Africa, associated with the rift-valley system of that region (Fig. 34-7). In North America, the Craters of the Moon (Idaho) and Sunset Crater (Arizona) often are cited as examples of older cinder cones. Geologists were able to closely observe the growth of a cinder cone in Mexico from 1943, when it was born in a cornfield in Michoacán State (about 320 km [200 mi] west of Mexico City), until 1952. This cinder cone, named Paricutín, grew to a height of 400 m (1300 ft) in its first eight months of activity and remained intermittently active for nearly a decade.

Shield Volcanoes

Shield volcanoes are formed from fluid basaltic lavas. These lavas contain sufficient gases to create a sometimes dramatic "fountain" of molten rock and some cinders, but these are tiny compared to the explosive eruptions at the craters of composite volcanoes. The basaltic lava is very hot, however, and flows in sheets over a countryside being gradually built up by successive eruptions (Fig. 34-9). Compared to their horizontal dimensions, which are very large, the tops of such volcanoes are rather unspectacular and seem rounded rather than peaked. This low appearance has given them the name shield volcanoes.

The most intensively studied shield volcanoes undoubtedly are those on the main island of Hawaii. Mauna Loa is the largest active shield volcano there. It stands on the ocean floor, and from there rises 10,000 m (33,000 ft), with the uppermost 3500 m (13,680 ft) protruding above sea level. To the north, Mauna Kea's crest is slightly higher. And to the east lies Kilauea (pronounced "kill-uh-WAY-uh"), the volcano that has experienced eruptions since the 1970s, and also holds special meaning for native Hawaiians (*see* Perspective: Risking the

Wrath of Goddess Pele). Kilauea's lavas even flowed across a housing subdivision, Royal Gardens, obliterating homes and streets, and reached the ocean, thereby adding a small amount of land to the island.

The volcanic landscape in this area displays some interesting shapes and forms. Lava that is especially fluid when it emerges from the crater develops a smooth "skin" upon hardening. This slightly hardened surface is then wrinkled, as the lava continues to move, into a ropy pattern called **pahoehoe** (Fig. 34-10). Less fluid lava hardens into angular, blocky forms called **aa**, so named (according to Hawaiian tradition) because of the shouts of people trying to walk barefoot on this jagged terrain! In the Hawaiian Islands, it is the big island—Hawaii—that displays the currently active volcanism. But all the islands in this archipelago are formed by shield volcanoes. Why should Hawaii contain the only volcanic activity? And why are there volcanoes at all, here in the middle of a lithospheric plate?

FIGURE 34-10 The "big island" of Hawaii, the product of giant shield volcanism, is a laboratory for the study of volcanic activity as well as landforms. Shown here is a mass of recently erupted pahoehoe lava, its smooth "skin" wrinkled into ropy patterns by continued movement of molten rock below. (Authors' photo)

These questions are answered in part by the topographic map of the northwestern Pacific Basin (Fig. 34-11a). Note that the Hawaiian Islands lie in an arched line from Kauai and its small neighbors in the northwest to Hawaii in the southeast. That line continues northwestward through Midway Island and then bends north-northwestward on the ocean floor as the Emperor Seamounts. It is remarkable that the only pronounced volcanic activity along this entire corridor is on Hawaii—at the very end of the chain.

Another part of the answer lies on the surface. In general, the rocks of Midway and Kauai (Fig. 34-11a) show evidence of much longer erosion than those of Oahu and, of course, Hawaii itself. Midway and Kauai seem to be much older than Oahu and Hawaii, and geologic evidence confirms this. When the rocks of these Pacific islands were dated, those of Oahu were found to be around 3 million years old, those of Midway 25 million years old, and those at the northern end of the Emperor Seamounts 75 million years old.

Hot Spots

Geologists theorize that the Pacific Plate has been moving over a **hot spot** in the mantle, a "plume" of extraordinarily high heat that remains in a fixed location, perhaps stoked by a high concentration of radioactivity. As the Pacific Plate moved over this hot spot, shield volcanoes formed over it (Fig. 34-11b). Thus one location after another would experience volcanic activity, acquiring volcanic landforms as that plate moved.

Today Hawaii is active, but eventually that island will move to the northwest and a new island will be formed southeast of where Hawaii now lies. Already, a large undersea volcano, named Loihi, is being built upward about 35 km (22 mi) southeast of Hawaii; that seamount, which lies 1000 m (3300 ft) beneath the waves, is expected to emerge above the ocean surface in approximately 50,000 years. In the meantime, Hawaii's shield volcanoes will become extinct.

Hot Spots and Plate Dynamics

If the hot-spot theory is correct, and subcrustal hot spots are indeed stationary, then it is possible to calculate the speed and direction of plate movement. As the map (Fig. 34-11a) shows, the Emperor Seamounts extend in a more northerly direction than the Midway-Hawaii chain. Thus the moving Pacific Plate—which today travels toward the

a

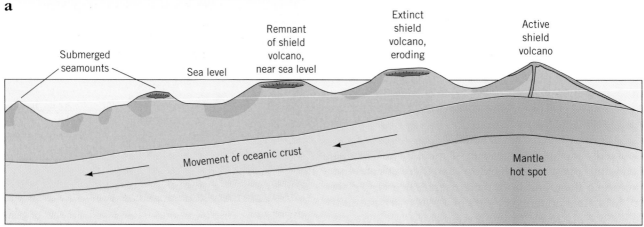

b

FIGURE 34-11 The ocean floor of the northwestern Pacific, dominated by the Emperor Seamounts and Hawaiian Chain (a). Volcanic chain formed by the seafloor moving over a geologic hot spot (b); these volcanic landforms become progressively older toward the left.

northwest (*see* Fig. 32-3)—changed direction about 40 million years ago if the same hot spot is responsible for both the Emperor Seamounts and the Midway-Hawaii chain. Given the age of Midway's lavas (25 million years), we may conclude that the plate traveled some 2700 km (1700 mi) over this period. This works out to about 11 cm (4.5 in) per year, a rate of movement consistent with average rates for other plates (*see* Fig. 33-1).

The volcanic effects of hot spots under comparatively thin oceanic lithosphere can be discerned rather easily and are revealed by the ocean-floor topography (as in Fig. 34-11). But it is likely that hot spots also are active under thicker continental crust, which may account for some of the intraplate volcanism shown on the world distribution map (Fig. 32-4). The effect of hot spots under continental plates is rather more difficult to identify. Hot spots may be responsible for some of those giant fissure eruptions, such as the ones that formed the Columbia Palateau. An isolated zone of volcanic activity (e.g., Mount Cameroon in west-central Africa and the range to the northeast of it) also may be associated with hotspot activity.

CALDERAS

A volcano's lavas and pyroclastics come from a subterranean magma chamber, a reservoir of active molten rock material that forces its way upward through the volcanic vent. When that magma reservoir ceases to support the volcano, the chamber may empty out and the interior of the mountain may literally become hollow. Left unsupported by the magma, the walls of the volcano may collapse, creating a **caldera** (Fig. 34-12). Such an event can occur quite suddenly, perhaps when the weakened structure of the volcano is shaken by an earthquake. A caldera also can result from a particularly violent eruption, which destroys the peak and crater of the volcano. In such cases, however, the magma chamber below is at the peak of its energy and will soon begin to rebuild the mountain.

Calderas are often large and sometimes filled with water. They also are often misnamed, as, for example, Oregon's Crater Lake, which is a circular caldera 10 km (6 mi) across, with walls more than 1200 m (4000 ft) high; a lake 600 m (2000 ft) deep fills this caldera. Another misnamed caldera is Ngorongoro Crater in Tanzania, some 18 km (11 mi) across and 600 m (2000 ft) deep. Ngorongoro contains a small lake, but it is best known for the enormous concentration of wildlife that has occupied it and its fertile natural pastures for many thousands of years (*see* Fig. 17-6).

Phreatic Eruptions

Water, poured on advancing lava, can help cool and consolidate the hot crust and slow down or divert the movement of the flow. But when water penetrates into the magma chamber below a volcano, it has quite a different effect. Just as pouring water on a grease fire only intensifies the blaze, so water entering a superheated magma chamber results in an explosive reaction—so explosive, in fact, that it can blow the entire top off the volcano above. This may be the reason for the gigantic explosions known to have occurred in recorded history, explosions that involved large composite volcanoes standing in water. Such explosions are called **phreatic eruptions**, and their effects reach far beyond the volcano's immediate area. Some phreatic eruptions, described in the following subsections, are believed to have changed the course of human history.

Krakatau

The most recent major phreatic eruption happened in 1883, when Indonesia's Krakatau volcano blew up with a roar heard in Australia 3000 km (1900 mi) away. The blast had a force estimated at 100 million tons of dynamite. It rained pyroclastics over an area of 750,000 km² (300,000 mi²) and propelled volcanic dust through the troposphere and stratosphere to an altitude of 80 km (50 mi). For years following Krakatau's eruptive explosion, this dust orbited the earth, affecting solar radiation and coloring sunsets brilliant red. Krakatau was an uninhabited, forested island located between Java and Sumatra. Although no one was killed by its explosion, great water waves were generated that dealt death and destruction. First, the explosion itself set in motion a giant *tsunami* (a sea wave set off by crustal disturbance), which radiated to the nearby coasts of Java, Sumatra, and other islands. Next, water rushed into the newly opened caldera, setting off further explosions and successive tsunamis. When these waves reached the more heavily populated coasts of Indonesia, an estimated 40,000 people were killed. For an in-depth account of the Krakatau eruption and its reconstruction, see the *Earth* Magazine article on pp. *Earth* 25–32.

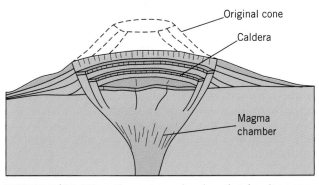

FIGURE 34-12 When the magma chamber that has long supplied an active volcano is somehow deprived of its conduit to the magma source, it empties out, leaving a hollow chamber beneath the cone. Eventually, the structure of the volcano yields and the cone collapses, creating a caldera.

Tambora

Explosive eruptions of Krakatauan dimensions occur suddenly but rarely. Nonetheless, Krakatau's actually was the second eruption of the nineteenth century in its corner of the world. In 1815, the Indonesian volcano Tambora's phreatic eruption was even larger. The volcanic dust it generated interfered so strongly with incoming solar radiation that the weather turned bitterly cold over much of the world, and as a result 1816 was widely described as the "year without a summer." Crops failed to ripen that following year, food and fuel shortages developed, and problems lasted into the ensuing winter and beyond.

Santorini

Certainly phreatic eruptions can affect climate for a time. But possibly no such event had a greater impact on human history than the explosion of the Mediterranean volcano Santorini (Thera), which took place in the middle of the seventeenth century B.C., more than 3600 years ago. The exact date of this cataclysmic event is still being debated, but its dimensions are clear.

The volcanic island of Santorini stood in the Mediterranean about 110 km (70 mi) north of Crete, where the Minoan civilization thrived. On a fateful day sometime around 1645 B.C., Santorini exploded, possibly as a result of seawater entering its magma chamber. So much volcanic ash and dust were blasted into the air that skies were darkened for days. (It has been suggested that this was the event described in the Bible's Old Testament as the act of God in retribution against the pharaoh: "thick darkness in all the land of Egypt for three days.") The Mediterranean Sea turned into a cauldron of tsunamis that lashed the coasts of other islands.

As for Santorini itself, when daylight reappeared its core was gone, replaced by the protruding margins of a vast caldera now filled with seawater (Fig. 34-13). The loss of life cannot be estimated. Fertile, productive, well-located Santorini had towns and villages, ports and farms; fleets of boats carried trade between it and prosperous Crete. Minoan civilization was the Mediterranean's most advanced, and great palaces graced Crete as well as Santorini. Many archeologists and historical geographers have speculated that the eruption of Santorini spelled the end of Minoan culture. Whatever the cause, Minoan civilization went into decline at about the time of Santorini's explosion. It has also been suggested that the legend of a drowned city of Atlantis was born from Santorini's destruction.

Santorini's caldera did not form from collapse; rather, it resulted from explosive forces. Geologically, very little time has elapsed since Santorini's explosion, but considerable volcanic activity and associated seismic (earthquake) activity have taken place. Near the middle of the caldera, probably above the original vent, a new dome is emerging. The volcano remains active, and al-

FIGURE 34-13 A great volcano once towered over this site on the Greek island of Santorini (Thera). In one gigantic explosion, it was pulverized. Enormous blocks of rock fell into the Mediterranean; ash rained on countrysides hundreds of kilometers away; dust encircled the earth for years. Towns, villages, and farms on prosperous Santorini disappeared. Left behind was a huge caldera, open to the sea—the lower rim of the former volcano. Houses perch precariously on the inner wall of the caldera, but Santorini's volcanic activity has not ended, and many houses have been destroyed by associated earthquakes. A new central cone is slowly rising in the middle of the caldera (out of the picture to the right); the fresh lava pouring from the vent will eventually build a new mountain.

though its caldera rim is quite densely populated today, many inhabitants have fled after each volcanic and seismic event. In the meantime, archeologists continue to excavate the ruins of a Bronze Age settlement near Akrotiri on the southern edge of the caldera, hoping to learn more about life on Santorini before its catastrophic interruption. We can only speculate on the course history might have taken if Minoan civilization had spread throughout the archipelago in which Santorini lies, diffused onto the mainland, and given birth to what would have been a very different kind of ancient Greece.

LANDSCAPES OF VOLCANISM

Volcanic activity, especially the mountainous type, creates unique and distinct landscapes. As Fig. 32-4 reminds

FIGURE 34-14 The Cascade volcanoes of the U.S. Pacific Northwest and adjoining Canada.

us, volcanic landscapes are limited in their geographic extent, but they do dominate certain areas. Even a single composite cone, by its sheer size or threat, can dominate physical and mental landscapes over a much wider area.

Mount Vesuvius, which in A.D. 79 buried a ring of Roman towns, including Pompeii (*see* Fig. 1-2) and Herculaneum, under ash and lahars, still towers over the southern Italian city of Naples (*see* photo p. v). Active and monitored anxiously, Vesuvius has erupted disastrously some 18 times since the first century A.D. In the twentieth century alone, it has erupted in 1906, 1929, and 1944, causing destruction and death in each instance. Farther south, on Sicily, the island just off the toe of the Italian peninsula, stands Mount Etna, perhaps Europe's most dangerous active volcano. Again "the mountain," as such dominating volcanic landforms seem to be called wherever they stand, pervades the physiography of the entire area. East Africa's Kilimanjaro, 5861 m (19,340 ft) tall, carries its snowcap within sight of the equator (*see* Fig. 19-11). It has not erupted in recorded history, but its presence is predominant, its form a reminder of what could happen, its soils fertile and intensively farmed.

Volcanic landscapes are most prevalent, as noted earlier, along the Pacific rim of subduction zones, from southern Chile counterclockwise to New Zealand, and in Indonesia from Sumatra through Java and the Lesser Sunda Islands (Fig. 32-4). On the Asian side of the Pacific, these landscapes are almost exclusively associated with island arcs and convergent contact between oceanic plates (Fig. 32-8). Virtually all the material here is vol-

canic, but not all of the topography appears to represent active volcanism. Japan's Mount Fuji, for example, towers as impressively over the area southwest of Tokyo (*see* unit opening photo) as Vesuvius does over Naples. Yet Fuji stands in a zone that is entirely of volcanic origin. Its morphology makes it unusual, because Japan does not consist of a row of snowcapped composite cones. In fact, Japan's eroded interior mountains do not evoke the image of a volcanic landscape.

The combination of orogenic (mountain-forming) and volcanic activity is more dramatically reflected in the topography of the Americas, where oceanic and continental plates are colliding. In Chile, Peru, Ecuador, and Colombia, South America's Andes Mountains are studded with great volcanic cones (*see* opening photo in Unit 33). Central America and southern Mexico are similarly dominated, in their mountain backbones, by volcanic peaks, craters, lakes, and recent lava flows. In North America, the Aleutian Islands and areas of southern Alaska have volcanic landscapes. In the conterminous United States, the Cascade Range from Mount Baker in northern Washington to Lassen Peak in northern California incorporates a line of volcanoes of which Mount St. Helens is only one (Fig. 34-14). Mount Hood was active as recently as 1865, Mount Baker in 1870, Mount Rainier in 1882, and Lassen Peak in 1921. The eruption of Mount St. Helens in 1980 may signal a new round of activity in the Pacific Northwest that could change the volcanic landscape of the most vulnerable region of the United States.

KEY TERMS

aa (p. 363)	composite volcano (p. 360)	*nuée ardente* (p. 361)	pyroclastics (p. 359)
caldera (p. 365)	hot spot (p. 364)	pahoehoe (p. 363)	shield volcano (p. 363)
cinder cone (p. 362)	lahar (p. 360)	phreatic eruption (p. 365)	volcanic dome (p. 362)

REVIEW QUESTIONS

1. In what three geologic settings do volcanoes occur?

2. How do shield volcanoes differ from composite volcanoes?

3. What are *hot spots*, and how do they help explain current volcanic activity in the Hawaiian Islands?

4. What causes a phreatic eruption?

5. Describe Santorini's monumental volcanic explosion and its impact on regional human activity.

REFERENCES AND FURTHER READINGS

BLONG, R. J.   *Volcanic Hazards: A Sourcebook on the Effects of Eruptions* (New York: Academic Press, 1984).

BULLARD, F. M.   *Volcanoes of the Earth* (Austin: University of Texas Press, 2nd ed., 1984).

CHESTER, D.   *Volcanoes and Society* (Sevenoaks, U.K.: Edward Arnold, 1993).

DE BLIJ, H. J., ed.   *Nature on the Rampage* (Washington, D.C.: Smithsonian Institution Press, 1994).

DECKER, R. W., and DECKER, B. B.   *Mountains of Fire: The Nature of Volcanoes* (London/New York: Cambridge University Press, 1991).

DECKER, R. W., and DECKER, B. B.   *Volcanoes* (San Francisco: Freeman, 1981).

Earthquakes and Volcanoes: A Scientific American Book (San Francisco: Freeman, 1980).

FRANCIS, P.   *Volcanoes* (Harmondsworth, U.K.: Penguin, 1976).

GREEN, J., and SHORT, N. M., eds.   *Volcanic Landforms and Surface Features: A Photographic Atlas and Glossary* (New York/Berlin: Springer-Verlag, 1971).

HARRIS, S. L.   *Fire Mountains of the West: The Cascade and Mono Lake Volcanoes* (Missoula, Mont.: Mountain Press, 1991).

LIPMAN, P. W., and MULLINEAUX, D. R., eds.   *The 1980 Eruptions of Mt. St. Helens, Washington* (Washington, D.C.: U.S. Geological Survey, Professional Paper 1250, 1981).

MACDONALD, G. A., et al.   *Volcanoes in the Sea: The Geology of Hawaii* (Honolulu: University of Hawaii Press, 2nd ed., 1983).

MCPHEE, J. A.   *The Control of Nature* (New York: Farrar, Straus & Giroux, 1989).

OLLIER, C. D.   *Volcanoes* (New York: Blackwell, 1988).

RITCHIE, D.   *The Encyclopedia of Earthquakes and Volcanoes* (New York: Facts on File, 1994).

ROBINSON, A.   *Earth Shock: Hurricanes, Volcanoes, Earthquakes, Tornadoes and Other Forces of Nature* (New York: Thames & Hudson, 1993).

SCARTH, A.   *Volcanoes: An Introduction* (College Station, Tex.: Texas A&M University Press, 1994).

WOOD, C. A., and KIENLE, J.   *Volcanoes of North America: United States and Canada* (London/New York: Cambridge University Press, 1990).

WRIGHT, T. L., et al.   *Hawaii Volcano Watch: A Pictorial History, 1779–1991* (Honolulu: University of Hawaii Press, 1992).

Earthquakes and Landscapes

OBJECTIVES

◆ To describe and quantify earthquakes.

◆ To relate the spatial pattern of earthquakes to plate tectonics.

◆ To discuss landscapes and landforms that bear the signature of earthquake activity.

I n Unit 29, we describe how earthquakes in the crust and upper mantle generate seismic waves that travel through the lithosphere as well as the interior of the earth (*see* Fig. 29-4). These waves yield key evidence toward our understanding of the internal structure of the planet. As the database expands and methods of analysis improve, properties of the unseen interior become known.

Earthquakes also have an impact at the surface of the crust of the earth, and they affect physical as well as cultural landscapes. Following a major earthquake, physical evidence of its occurrence can be seen on the ground in the form of dislocated strata, open fractures, new scarps, and lines of crushed rock. Earthquakes also trigger earth movements such as landslides and mudslides.

Moreover, the shocks and aftershocks of an earthquake can do major damage to buildings and other elements of the cultural landscape (*see* photo above, which shows one aftermath of the 1989 Loma Prieta earthquake in the San Francisco Bay Area). Add to this the fact that certain areas of the world are much more susceptible to earthquake damage than others, and no doubt remains that we should study this environmental hazard in a geographic context.

UNIT OPENING PHOTO: The collapsed Nimitz Freeway in Oakland, California after the 1989 Loma Prieta earthquake along the San Andreas Fault. The greatest destruction occurred on unstable landfill areas such as this one alongside San Francisco Bay.

EARTHQUAKE TERMINOLOGY

We have seen how lithospheric plates collide at convergent boundaries, producing fractures in rocks called faults. A **fault** is a fracture in crustal rock involving the displacement of rock on one side of the fracture with respect to rock on the other side. Some faults, such as the San Andreas Fault, are giant breaks that continue for hundreds of kilometers (*see* Fig. 32-11); others are shorter. Along the contact zone between the North American and Pacific Plates, the upper crust is riddled with faults, all resulting from the stresses imposed on hard rocks by the movement of plates. Like volcanoes, certain faults are active whereas others are no longer subject to stress. In the great continental shields that form the cores

of continents lie many faults, fractures that bear witness to an earlier age of crustal instability. Today those faults appear on geologic maps, but they do not pose an earthquake hazard.

When rock strata are subjected to stress, they begin to deform or bend (Fig. 35-1). All rocks have a certain rupture strength, which means that they will continue to bend, rather than break, as long as the stress imposed on them does not exceed this rupture strength. When the stress finally becomes too great, the rocks suddenly move along a plane (the fault) that may or may not have existed before the deformation began. That sudden movement snaps the rocks on each side of the fault back into their original shape and produces an earthquake.

An **earthquake**, therefore, is the release of energy that has been slowly built up during the stress of increasing deformation of rocks. This energy release takes the form of seismic waves that radiate in all directions from the place of movement (Fig. 35-2). Earthquakes can originate at or near the surface of the earth, deep inside the crust, or even in the upper mantle. The place of origin is the **focus**, and the point directly above the focus on the earth's surface is the **epicenter** (Fig. 35-2). Earthquakes range from tremors so small that they are barely detectable to great shocks that can destroy entire cities. This reflects their **magnitude**, the amount of shaking of the ground as the quake passes, as measured by a seismograph (*see* Fig. 29-1).

Magnitude is assessed on the *Richter Scale*, which assigns a number to an earthquake based on the severity of that ground motion. This open-ended scale, developed in 1935 by geophysicist Charles Richter, ranges from 0 to 8+ (Table 35-1). It is logarithmic, so that an earthquake of magnitude 4 causes 10 times as much ground motion as one of magnitude 3 and 100 times as much as a quake of magnitude 2. It should be noted, however, that the original Richter Scale has undergone some recent changes, because more sophisticated equipment and more precise measurements of magnitude have been developed. Thus when the U.S. Geological Survey now announces a "Richter magnitude" following an earthquake, it is referring only to the magnitude of surface seismic waves. Actually, the most widely used measure today (especially by the U.S. media) is the *Moment Magnitude Scale*, based on the size of the fault along which a quake occurs and the distance the rocks around it slip.

Another measure of an earthquake's size is its **intensity**. This measure reflects the impact of an earthquake on the cultural landscape—on people, their activities, and structures. Intensity is reported on the *Mercalli Scale*, which was first developed by the Italian geologist Giuseppe Mercalli in 1905 and modernized in 1931. Now called the *Modified Mercalli Scale* (Table 35-2), it assigns a number ranging from I to XII to an earthquake (Roman numerals are always used). For instance, an earthquake

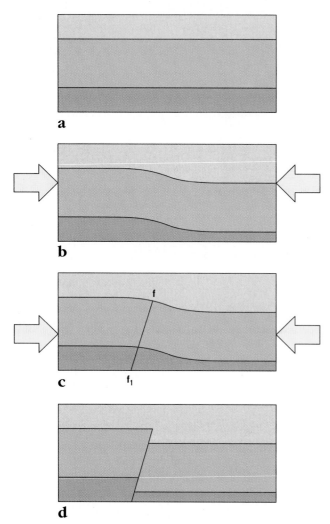

FIGURE 35-1 When stress (in this case compression) affects horizontal rock layers (a), the strata begin to bend, and they continue to do so as long as their rupture strength is not exceeded (b). When this level is exceeded, a fault plane (**f-f₁**) develops (c). Often, more than one such fault plane will be formed. Sudden movement along this fault plane, accompanied by one or more earthquakes, relieves the now-exceeded rupture strength, and the rock strata resume their original (horizontal) position (d).

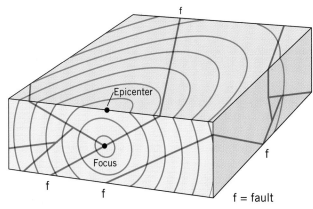

FIGURE 35-2 In an area honeycombed by faults (**f**), movement on a particular fault plane (perhaps at intersecting faults) becomes the focus for energy release. The epicenter is the point at the earth's surface vertically above the point of focus. The energy released at the earthquake's focus radiates outward to other parts of the earth in the form of seismic waves, represented by the rings spreading from the focus and the epicenter.

of intensity IV is felt indoors, and hanging objects swing. Intensity V produces broken windows and dishes, awakens many sleepers, and cracks plaster. Intensity IX damages building foundations and breaks in-ground pipes. At intensity XII, damage is total, and even heavy objects are thrown into the air.

As Table 35-1 indicates, the severest of earthquakes occur on average perhaps once every few years. Tens of thousands of smaller quakes occur annually, many felt

TABLE 35-1 RICHTER SCALE OF EARTHQUAKE MAGNITUDE, COMPARED WITH MERCALLI INTENSITIES

MAGNITUDE	APPROXIMATE MAXIMUM INTENSITY	NUMBER PER YEAR	APPROXIMATE ENERGY RELEASED IN EXPLOSIVE EQUIVALENTS
0			1 lb (0.45 kg) TNT
1		700,000	
2			
2–2.9	II	300,000	
3			
3–3.9		49,000	
4	III		
4–4.9	Minor	6,200	
5	VI		Small atom bomb, 20,000 tons TNT (20 kilotons)
5–5.9	Damaging	800	
6	VII		Hydrogen bomb, 1 megaton
6–6.9	Destructive	120	
7	X		
7–7.9	Major	18	
8	XII		
8–8.6	Great	1 every few years	60,000 1-megaton bombs

Source: After Wyllie, P. J. *The Way the Earth Works* (New York: Wiley, 1976), p. 47.

TABLE 35-2 MODIFIED MERCALLI SCALE OF EARTHQUAKE INTENSITIES

INTENSITY	QUALITATIVE TITLE	DESCRIPTION OF EFFECTS
I	Negligible	Detected by instruments only.
II	Feeble	Felt by sensitive people. Suspended objects swing.
III	Slight	Vibration like passing truck. Standing cars may rock.
IV	Moderate	Felt indoors. Some sleepers awakened. Hanging objects swing. Sensation like a heavy truck striking building. Windows and dishes rattle. Standing cars rock.
V	Rather strong	Felt by most people; many awakened. Some plaster falls. Dishes and windows broken. Pendulum clocks may stop.
VI	Strong	Felt by all; many are frightened. Chimneys topple. Furniture moves.
VII	Very strong	Alarm; most people run outdoors. Weak structures damaged moderately. Felt in moving cars.
VIII	Destructive	General alarm; everyone runs outdoors. Weak structures severely damaged; slight damage to strong structures. Monuments toppled. Heavy furniture overturned.
IX	Ruinous	Panic. Total destruction of weak structures; considerable damage to specially designed structures. Foundations damaged. Underground pipes broken. Ground fissured.
X	Disastrous	Panic. Only the best buildings survive. Foundations ruined. Rails bent. Ground badly cracked. Large landslides.
XI	Very disastrous	Panic. Few masonry structures remain standing. Broad fissures in ground.
XII	Catastrophic	Superpanic. Total destruction. Waves are seen on the ground. Objects are thrown into the air.

Source: After Wyllie, P. J. *The Way the Earth Works* (New York: Wiley, 1976), p. 45.

only by sensitive seismographs. Great earthquakes that cause death and destruction are long remembered; several recent ones are listed in Table 35-3. Probably the most infamous earthquake in North America during the twentieth century was the 1906 San Francisco quake, which is estimated to have had a magnitude of 8.3 and an intensity of XII. A dangerous associated effect of such a severe shock is fire, which rages out of control from many points of origin (especially when gas lines rupture)

and cannot be fought because water supply systems are disrupted (Fig. 35-3).

Much more recently, in early 1995, an earthquake measuring 6.8 in magnitude severely damaged Kobe, Japan's sixth largest city and second leading port. Because the quake's epicenter lay on an island only 25 km (15 mi) from downtown Kobe, the destruction was massive (Fig. 35-4). More than 5500 people were killed and 26,744 were injured; 46,400 buildings were destroyed, thereby rendering nearly 300,000 residents homeless; and the total damage was estimated to exceed 40 billion U.S. dollars. (The costliest natural disaster in the history of the United States was also the result of an earthquake—the devastating shock that struck Los Angeles exactly one year earlier in 1994, which is profiled in the Perspective box in Unit 32.)

EARTHQUAKE DISTRIBUTION

The global distribution of earthquake epicenters, whatever the period of record, indicates that many earthquakes fortunately originate in locations that are less vulnerable than San Francisco, Los Angeles, or Kobe. The heaviest concentration is along the *Circum-Pacific belt* of subduction zones associated with the Pacific and

TABLE 35-3 SOME MAJOR TWENTIETH CENTURY EARTHQUAKES

YEAR	PLACE	RICHTER MAGNITUDE	ESTIMATED DEATH TOLL
1906	San Francisco, California	8.3	700
1908	Messina, Italy	7.5	120,000
1920	Kansu, China	8.5	180,000
1923	Tokyo-Yokohama, Japan	8.2	143,000
1935	Quetta, India	7.5	60,000
1939	Chillan, Chile	7.8	30,000
1962	Northwestern Iran	7.3	14,000
1964	Southern Alaska	8.6	131
1970	Chimbote, Peru	7.8	66,794
1976	Tangshan, China	7.6	242,000*
1985	West-Central Mexico	7.9, 7.5	9,500
1988	Armenia, (then) U.S.S.R.	7.0	55,000+
1989	Loma Prieta, California	7.0	63
1990	Northwestern Iran	7.7	40,000+
1992	Landers, California	7.5	1
1994	Los Angeles, California	6.8	61
1995	Kobe, Japan	6.8	5,527+

*Reports still persist that as many as 750,000 died.

FIGURE 35-3 The height of the raging fire that engulfed much of the city's center following the great San Francisco earthquake of 1906.

FIGURE 35-4 A segment of the Hanshin Expressway near central Kobe following the 1995 earthquake. The shock lasted only 20 seconds, yet it was powerful enough to corrugate the elevated roadway and then collapse the entire structure when 15 supporting pillars snapped at their bases.

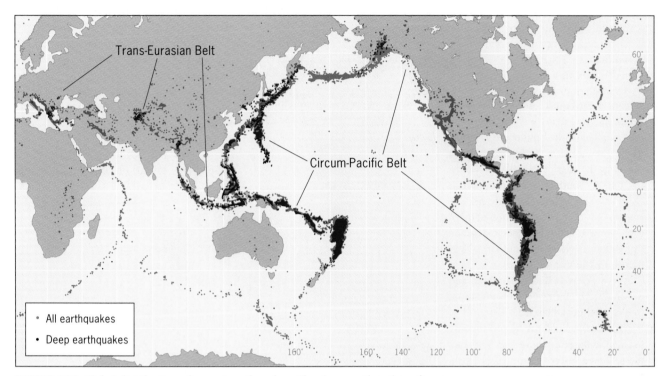

FIGURE 35-5 Global distribution of recent earthquakes. The deep earthquakes, shown by black dots, originated more than 100 km (63 mi) below the surface.

Nazca Plates and their neighboring plates (Fig. 35-5). About 80 percent of all shallow-focus earthquakes (depths of less than 100 km [63 mi]) originate in this belt. As the map indicates, earthquakes with deeper foci in the lower lithosphere are even more heavily concentrated in this zone.

When we compare Fig. 35-5 with a map of the world distribution of population (p. 19), we quickly see where large numbers of people are at risk from earthquakes. All of Japan's 126 million people live in an area of high earthquake incidence and, therefore, high risk. The Philippines and Indonesia also are earthquake-prone and are home to nearly 275 million people. But the earthquake-affected zone then extends into a less populated area of the Pacific Ocean and turns south, lessening in intensity just to the north of New Zealand.

In the north, the Circum-Pacific earthquake zone affects the sparsely populated Aleutian Islands, but also penetrates populated areas of southern Alaska. A severe earthquake struck Anchorage on Good Friday in 1964, causing death and destruction. That earthquake's magnitude was 8.6, and its epicenter was only about 120 km (75 mi) from the coastal city; the damage was caused not only by the shaking ground, but also by the massive ocean waves generated by the shock. As the map shows (Fig. 35-5), between the main body of Alaska and Vancouver Island in southwesternmost Canada there is comparatively little earthquake risk—but the population is also rather sparse. Southward from the U.S.-Canada border in the conterminous U.S. West, earthquake frequency increases again, endangering the large population cen-

ters of California (where the 1994 Los Angeles earthquake is still fresh in people's minds) and posing a constant threat to populous areas in Mexico and Central America. And virtually all of western South America is an active earthquake zone, including major population centers in Colombia, Ecuador, Peru, and Chile.

Another zone of high earthquake incidence is the *Trans-Eurasian belt*, which extends generally eastward from the Mediterranean Sea through Southwest Asia and the Himalayas into Southeast Asia, where it meets the Circum-Pacific belt. The incidence of major earthquakes in this corridor is not as high as it is in parts of the Circum-Pacific belt, but some important population centers are at risk. These include the countries of former Yugoslavia, Greece, Iraq, Iran, and the highlands of Afghanistan, northern Pakistan, northernmost India, and Nepal.

A third zone of earthquakes is associated with the global system of *midoceanic ridges* (compare Figs. 35-5 and 32-2). The earthquakes there are generally less frequent and less severe than they are in the Circum-Pacific belt, and except for population clusters on islands formed by these ridges, this earthquake zone does not endanger large numbers of people.

From Fig. 32-3 it is evident that known contact zones between tectonic plates form the earth's most active earthquake belts. But earthquakes, some of them severe, can and do occur in other areas of the world. Note (in Fig. 35-5) the scattered pattern of epicenters in interior Asia, in eastern Africa from Ethiopia to South Africa, and in North America east of the west-coast subduction zone. The causes of these **intraplate earthquakes** are still not

Earthquake Risk in the Eastern United States

A map of earthquake risk in the United States suggests that the West is not the only portion of the country to face this hazard. Major risk exists in three areas of the East: the middle Mississippi Valley, the coastal Southeast, and the interior Northeast (Fig. 35-6). Moderate risk exists not only in the surroundings of these three areas, but also in a zone extending from eastern Nebraska south to Oklahoma. In the conterminous U.S., only southern Texas as well as southern and northwestern Florida are believed to be free of earthquake hazard.

This assessment (and the accompanying map) is based on more than 300 years of records of Eastern earthquakes. In the East, nearly 30 significant earthquakes have been recorded since 1663, all with an estimated magnitude of 5.0 or more. A dozen have taken place since 1925. Geologists report that an Eastern earthquake could devastate an area 100 times as large as would be affected by an equivalent quake in the West because the shock waves are cushioned much more in the fault-infested crust of the West.

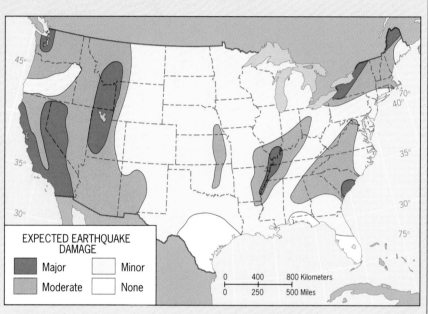

FIGURE 35-6 Earthquake risk in the conterminous United States.

But the map is based on a past that gives little clue to the future. There is no evidence of major earthquake activity in the middle Mississippi Valley or in coastal South Carolina before great earthquakes struck there in 1811–1812 and 1886, respectively. The next large quake could strike an area deemed to have only moderate or minor earthquake risk. The science of earthquake prediction is still in its infancy; the available record is short-term and unreliable. In facing earthquake threats, however, the American West is not alone.

---

well understood, but they can produce severe damage because people in the affected areas are less well prepared for earthquakes. For example, municipal ordinances governing high-rise construction in California cities take seismic hazards into consideration; in the eastern United States, local building codes do not. Yet the most severe earthquake ever experienced in the United States did not occur in California but in the central Mississippi Valley.

In 1811 and 1812 three great earthquakes, all with epicenters near New Madrid in the "bootheel" of extreme southeastern Missouri, changed the course of the Mississippi River and created a 7300-hectare (18,000-acre) lake in the valley. The severity of these quakes has been estimated to have been as high as magnitude 8.5. Had nearby St. Louis and Memphis been major cities at that time, the death toll would have been enormous. In 1886, an earthquake of magnitude 7.0 devastated Charleston, South Carolina, and the ground shook as far away as New York City and Chicago. Thus the eastern United States, too, must be considered a potentially hazardous

seismic zone (*see* Perspective: Earthquake Risk in the Eastern United States).

In East Asia, the earthquake that cost the largest number of lives in this century did not originate in the offshore subduction zone but in eastern China. In 1976, the city of Tangshan was destroyed by an earthquake of magnitude 7.6, whose focus lay directly beneath the urban area. The exact loss of life will never be known, but estimates still range as high as 750,000 (the Chinese government insists the toll was 242,000). Damage occurred over more than an 80-km-long (50-mi) corridor that reached as far as the major port city of Tianjin and the capital, Beijing.

The global map of earthquake distribution should not lead us to the conclusion that areas outside the plate-contact zones are free from seismic hazard. Even apparently stable shields, such as Australia's and Africa's, can be shaken by significant earthquakes. Beyond the subduction zones, other areas of plate contact, and mid-oceanic ridges, the pattern of epicenters is not yet understood.

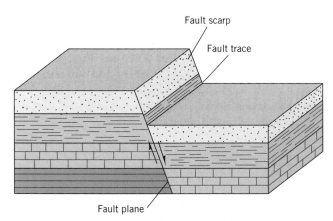

FIGURE 35-7 A fault scarp resulting from the vertical movement of one block with respect to another. The blocks are in contact along the fault plane, whose exposed face is the fault scarp. If the upper block were eroded down to the level of the lower block, the only surface evidence of the fault plane would be the fault trace.

EARTHQUAKES AND LANDSCAPES

Earthquakes, as we noted earlier, are not comparable to volcanic action as builders of a distinct landscape. But earthquakes do *modify* physical and cultural landscapes, and some landforms actually are created by earthquake movements. When movement along a fault generates an earthquake, or when a fault is newly created following long-term deformation of rock strata, the result may be visible at the surface in the form of a **fault scarp**. As Fig. 35-7 shows, a fault scarp is the exposed cliff-like face of the **fault plane**, the surface of contact along which blocks on either side of a fault move.

Not every earthquake produces a scarp, and in some cases the movement is lateral (sideways) rather than vertical. But if one block is raised with respect to another, a fault scarp is produced. Such a scarp may have a vertical extent (or *face*) of less than 1 m (3.3 ft); others may exceed 100 m (330 ft). Repeated vertical movement along the fault scarp can raise the height of the exposed face in stages. As Fig. 35-7 shows, the lower edge of the fault scarp is called the **fault trace**. Imagine that erosion wears down the raised block to the left so that the scarp is no longer visible. The existence of the fault is then only revealed at the surface by the trace. Along the trace may lie a band of crushed, jagged rock fragments called *fault breccia*, evidence of the powerful forces that created the fault.

Earthquakes affect different types of rocks in different ways. When a major earthquake struck Mexico City in 1985, parts of the city were devastated while other areas of it showed little damage. The reason lay in the underlying geology: where the rocks were solid, there was little impact. But where the buildings stood on the soft ground of an old lake bed on which part of Mexico City is built, the damage was great because the earthquake waves were significantly amplified. Many highrise structures collapsed or simply fell over as the ground wobbled like a bowl of gelatin. In Anchorage, the 1964

FIGURE 35-8 The astonishing ferocity of the 1964 earthquake (magnitude 8.6) that struck Anchorage, Alaska is seen in this downtown street, whose very topography was reshaped in the process.

earthquake had similar effect: houses and other buildings that stood on solid rock sustained minor damage, but areas of the city that were underlain by clays or soft and uncemented sedimentary strata slid downslope. Hundreds of houses were carried away and destroyed, and in places the whole topography of Alaska's leading urban area was changed (Fig. 35-8).

The effect of a major earthquake, therefore, is to produce movements of different kinds. In the same affected area, blocks of solid rock will shake but remain stable; loosely compacted sediments, especially on slopes, will slide downhill. When there has been prolonged rain in an area of thick clay, the saturated clay may stay in place—until an earthquake or even a tremor provides the impetus to dislodge it. In snowy mountain terrain, an avalanche may be started by a slight earth tremor when otherwise the snowpack would have stayed in place. Landslides, mudslides, and other forms of mass movement (*see* Unit 39) often result from a combination of circumstances among which a quake can be crucial.

Undoubtedly, as the photos in this unit demonstrate, the impact of a major earthquake on the cultural landscape is its most dramatic manifestation. Such effects range from the offsetting of linear features such as fences and hedges, roads and pipelines, to the devastation of major structures such as skyscrapers, highways, and bridges. An earthquake's capacity to reduce to rubble a structure built of steel and concrete inspires awe and terror. Not surprisingly, earth scientists are intensely interested in developing ways to forecast earthquake activity. An overview of their recent progress is reported in the *Earth* Magazine article on pp. *Earth 32–39.*

TSUNAMIS

The Japanese word for a great sea wave, **tsunami**, has come into general use to identify a seismic sea wave. When an earthquake's epicenter is located on the ocean floor or near a coastline, its shock will generate one or more waves on the water. These waves radiate outward from the point of origin. In the open ocean, they look like broad swells on the water surface, and ships and boats can ride them out quite easily. But a tsunami's vertical magnitude, when it reaches shallow water near shore, may be many times that of an ordinary wind-caused swell. Near the shore it can create huge breakers that smash into coastal towns and villages.

When seismographs in various locales record a severe underwater quake, a tsunami alert is immediately issued. This alert is intended for places along all vulnerable coastlines. But tsunamis travel very fast, reaching 1000 kph (630 mph), and warnings do not always arrive in time. Tsunami waves have reached more than 65 m (200 ft) in height, and they have been known to travel all the way across the Pacific Ocean. A major earthquake in Chile in 1960 generated a tsunami that reached Hawaii about 15 hours later, smashing into coastal lowlands with breakers 7 m (23 ft) high. Ten hours later, the still-advancing wave was strong enough to cause damage in Japan. Situated in the mid-Pacific, the Hawaiian Islands are especially vulnerable to tsunamis. In 1946, a tsunami caused by an earthquake near Alaska struck the town of Hilo on the island of Hawaii, killing 156 persons. The tsunami generated by the 1964 Alaska earthquake that devastated Anchorage flooded the port of Kodiak, Alaska and drowned a dozen people in a California coastal town far to the south.

Seismic sea waves are not tidal waves, as they are sometimes misnamed. Waves with tsunami-like properties are sometimes created by explosive volcanic eruptions, such as that of Santorini (*see* pp. 365–366). When these giant waves break onto an exposed shore, they can modify the coastal landscape significantly. Occasionally, they can do more to change a coastline in a few moments than normal processes do in centuries of erosion and deposition.

This unit has focused on the sudden, dramatic movements of rocks that produce earthquakes and accompanying impacts on overlying landscapes. But the surfaces of the earth's landmasses are also affected by the consequences of less spectacular stresses on their underlying rocks. Unit 36 examines those types of stresses and the rock structures they shape.

KEY TERMS

earthquake (p. 370)	fault plane (p. 375)	focus (p. 370)	magnitude (earthquake) (p. 370)
epicenter (p. 370)	fault scarp (p. 375)	intensity (earthquake) (p. 370)	tsunami (p. 376)
fault (p. 370)	fault trace (p. 375)	intraplate earthquake (p. 373)	

REVIEW QUESTIONS

1. Compare the Richter and Mercalli scales for measuring earthquakes.
2. Where on the earth are earthquakes most frequent? Why?
3. How are fault scarps produced?
4. What is the Circum-Pacific belt of seismic activity?
5. Why are parts of the eastern United States considered to constitute a hazardous earthquake region?

REFERENCES AND FURTHER READINGS

BERKE, P. R., and BEATLEY, T. *Planning for Earthquakes* (Baltimore, Md.: Johns Hopkins University Press, 1992).

BOLT, B. A. *Earthquakes* (New York: W. H. Freeman, 3rd ed., 1993).

CLARK, S. P. *Structure of the Earth* (Englewood Cliffs, N.J.: Prentice-Hall, 1970).

COFFMAN, J. L., and VON HAKE, C. A., eds. *Earthquake History of the United States* (Washington, D.C.: U.S. Department of Commerce, NOAA, 1973).

DE BLIJ, H. J., ed. *Nature on the Rampage* (Washington, D.C.: Smithsonian Institution Press, 1994).

Earthquakes and Volcanoes: A Scientific American Book (San Francisco: Freeman, 1980).

EIBY, G. A. *Earthquakes* (New York: Van Nostrand-Reinhold, 1980).

GERE, J. M., and SHAH, H. C. *Terra Non Firma: Understanding and Preparing for Earthquakes* (New York: Freeman, 1984).

HODGSON, J. H. *Earthquakes and Earth Structure* (Englewood Cliffs, N.J.: Prentice-Hall, 1964).

LOMNITZ, C. *Fundamentals of Earthquake Prediction* (New York: Wiley, 1994).

LOMNITZ, C. *Global Tectonics and Earthquake Risk* (Amsterdam: Elsevier, 1974).

PENICK, J.L., JR. *The New Madrid Earthquakes* (Columbia, Mo.: University of Missouri Press, 2nd ed., 1981).

RICHTER, C. F. *Elementary Seismology* (New York: Freeman, 1958).

RITCHIE, D. *The Encyclopedia of Earthquakes and Volcanoes* (New York: Facts on File, 1994).

ROBINSON, A. *Earth Shock: Hurricanes, Volcanoes, Earthquakes, Tornadoes and Other Forces of Nature* (New York: Thames & Hudson, 1993).

Surface Expressions of Subsurface Structures

OBJECTIVES

◆ To introduce basic terminology used in describing rock structure.

◆ To distinguish between types of fault movements and the landforms they produce.

◆ To discuss the folding of rocks and relate it to the landforms produced.

The physical landscapes of the continental landmasses are sculpted from rocks with diverse properties. We study some of these properties in other units: the hard, resistant, crystalline batholiths formed deep in the crust and exposed by erosion (Unit 30); the flows of lava (Units 30 and 34); the foliated schists (Unit 31); the layers of sedimentary strata (Unit 31). Rocks may also be changed after their formation, metamorphosed from one state to another (Unit 31). The structure of the rocks reveals the nature of the stress that changed them and reflects the way the rocks reacted to this stress. Brittle rocks fracture under stress. Rocks exhibiting plastic behavior can bend or even fold, and when the stress is removed these structures remain permanent.

Stresses of many kinds are imposed on rocks. In zones of convergent plate contact, rock strata in the col-

lision zone are crushed into tight folds. In shield areas, divergent movement in the mantle below can pull segments of rock apart, creating parallel faults. Where sediments accumulate, their growing weight pushes the underlying rocks downward. Where erosion removes rock, the crust has a tendency to "rebound." Rocks are deformed in so many ways that the resulting forms seem endlessly complicated. But, in fact, certain geologic structures occur many times over and can be recognized even on topographic maps that do not contain any stratigraphic information. We have said elsewhere that our understanding of the surface must begin with what lies

UNIT OPENING PHOTO: The eastern front of Wyoming's Grand Teton Range, whose base marks a boundary between two different areas of rock stress in the formation of landscape.

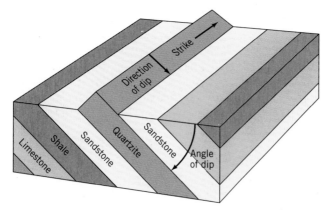

FIGURE 36-1 Strike, direction of dip, and angle of dip.

below, and in this unit we study the relationships between geologic structure and visible landscape.

TERMINOLOGY OF STRUCTURE

If we are to accurately describe the nature and orientation of structures below the surface, we must use consistent terminology. Imagine a ridge of quartzite rising above the surface and extending from northeast to southwest according to your compass (Fig. 36-1). This is the **strike** of that ridge—the compass direction of the line of intersection between a rock layer and a horizontal plane. Thus in your field notes you would report the strike of the ridge as N45°E. The strike of any linear feature always is recorded to range from 0° to 90° east or west.

The ridge of quartzite may consist of a layer of this metamorphic rock that angles downward between softer strata (Fig. 36-1). This is referred to as the **dip** of that layer, which is the angle at which it tilts from the horizontal. If the quartzite layer tilts 30° from the horizontal, its dip is recorded as 30°. In addition, the direction of the dip must be established. If you stand on the ridge and see the quartzite layer tilting downward toward the southeast, its dip is 30°SE. As the diagram indicates, the direction of dip is always at a right angle to the strike.

A prominent feature such as our quartzite ridge is called an *outcrop*, a locality where exposed rock occurs. Here, the hardness of the rock and its resistance to erosion have combined to create a prominent topographic feature whose length and straightness indicate tilting, but little or no other deformation. Where rock strata are bent or folded, the determination of strike and dip can become difficult, and the field map may have to record many variations in strike and dip along segments of the outcrop.

FAULT STRUCTURES

The rocks of the earth's crust are honeycombed by fractures. Some areas, such as the zones near the subduction

of tectonic plates, are more strongly affected by faulting than other locales. But detailed geologic maps of even the "stable" shields of the continental cores reveal numerous fractures, legacies of earlier periods of stress. We have encountered faults in our study of convergent plate boundaries and in connection with earthquakes. Here we identify several types of faults and see how they are expressed at the surface of the crust in the landscape.

A **fault** is a fracture in crustal rock involving the displacement of rock on one side of the fracture with respect to rock on the other side. A fracture without displacement is called a *joint*, so slippage along the fault plane is key evidence for faulting. Faulting results when brittle rocks come under stress, cannot bend or fold, and therefore break. But even rocks that exhibit plastic behavior can be subjected to such severe stress that they, too, can fracture (*see* Fig. 35-1).

As explained in Unit 35, sudden slippage along a fault plane generates an earthquake or a tremor, depending on the amount of energy released. Slippage may involve a few millimeters (less than a quarter of an inch) or 100 m (330 ft) or more. At the surface, a fault may be barely visible—or it may be marked by a tall scarp. Sometimes, when a sequence of sedimentary layers is faulted, it is possible to locate the same bed on opposite sides of the fault plane, and the amount of vertical displacement can be determined. But if the blocks on either side of the fault consist of the same rock (say, a mass of granite), it may not be possible to measure displacement.

Our increased understanding of plate tectonics has contributed to better comprehension of faulting, its causes, and its effects. Where plates converge and collide, *compressional* stresses are strong and rocks are crushed tightly together. The lithosphere is forced to occupy less horizontal space, and rocks respond by breaking, bending, folding, sliding, and squeezing upward and downward. Where plates diverge, and where the crust is subjected to spreading processes elsewhere, the stress is *tensional* and rocks are being pulled apart. Some plastic rocks respond by thinning, but others break and faults result. Where a series of parallel faults develop, blocks of crust between the faults may actually sink down, having been left with less support from below than existed previously. And where plates slide past each other, or where forces below the crust are lateral, the stress is *transverse*. Rock masses that slide past each other also are subject to faulting and associated earthquakes. Let us now examine these phenomena in detail.

Compressional Faults

Where crustal rock is compressed into a smaller horizontal space, shortening of the crust is achieved by one block riding over the other along a steep fault plane between them (Fig. 36-2). Such a structure is called a **reverse fault**. When vertical movement occurs during faulting, blocks that move upward (with respect to ad-

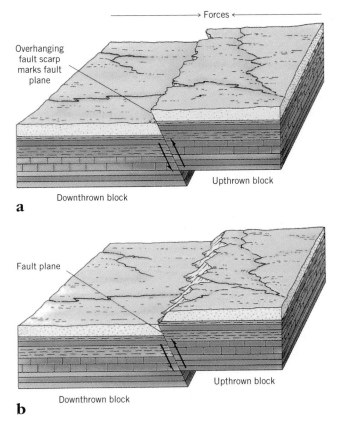

FIGURE 36-2 In a reverse fault, the overhanging scarp of the upthrown block (a) soon collapses, and a new slope, which lies at an angle to the original, is produced by erosional forces (b).

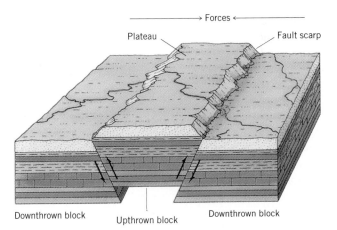

FIGURE 36-3 A plateau formed as an upthrown block between two reverse faults that run parallel to each other where they intersect the surface.

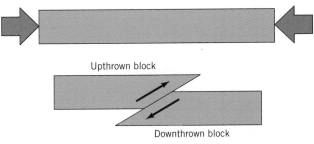

FIGURE 36-4 Side view of a thrust fault resulting from compression. The low angle of the fault plane produces substantial overriding by the upthrown block.

jacent blocks) are referred to as *upthrown*, while blocks that move downward are termed *downthrown*. In the case of reverse faults, note that the upthrown block creates an initial scarp that overhangs the downthrown block (Fig. 36-2a). Such an overhanging scarp soon collapses under the effects of weathering, erosion, and gravity. Landslides and other mass movements (*see* Unit 39) are associated with such scarps, and these processes soon produce a slope at an angle to the original fault plane (Fig. 36-2b). Therefore, when we see a fault scarp in the field we cannot conclude without further investigation that it represents the dip of the fault plane from which it has resulted.

Compressional forces sometimes produce series of nearly parallel (or *echelon*) faults. A strip of crustal rock positioned between reverse faults may assume the dimensions of a block-like plateau (Fig. 36-3). When the angle of a fault plane in a compressional fault is very low, the structure is referred to as a **thrust fault** (sometimes *overthrust fault*) (Fig. 36-4). Note that the overriding block slides almost horizontally over the downthrown block, covering much more of it than is the case in a reverse fault. Following erosion, the resulting scarp would be lower and less prominent.

Tensional Faults

Tensional stresses pull crustal rock apart, so that now the situation is the opposite: there is more horizontal space for crustal material, not less as in compressional faults. The result is one or more **normal faults** (Fig. 36-5). The typical normal fault has a moderately inclined fault plane separating a block that has remained stationary, or nearly so, from one that has been significantly downthrown. As Fig. 36-5 shows, the resulting fault scarp initially reflects the dip angle of the fault plane, although erosion may modify it over time. Sometimes the fault scarp has been eroded for so long that it has retreated from its original position. In such instances, the fault trace (*see* Fig. 35-7) reveals the original location of the scarp. The eastern front of the Grand Teton Mountains near Jackson, Wy-

FIGURE 36-5 Unlike compressional stresses, tensional stresses pull the crust apart. Normal faults result, and the fault scarp often reflects the dip angle of the fault plane.

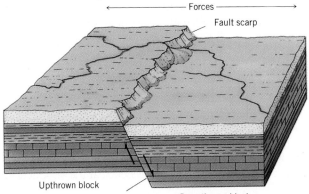

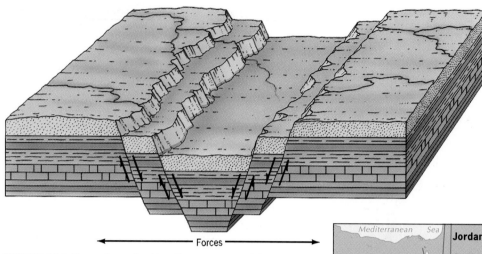

FIGURE 36-6 Formation of a rift valley as tensional forces generate parallel normal faults between which crustal blocks slide downward.

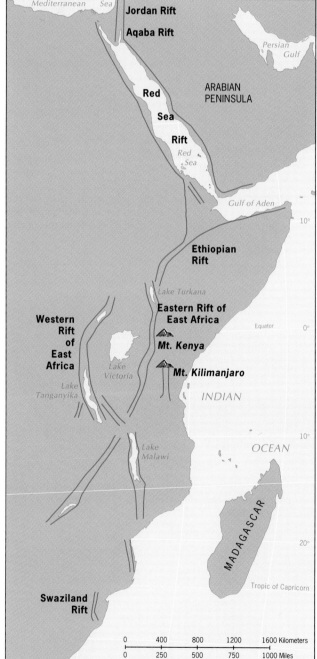

FIGURE 36-7 The African rift-valley system extends from beyond Africa in the north (the Jordan-Aqaba segment) to Swaziland in the south.

oming (*see* unit opening photo), and the western front of the Wasatch Mountains behind Salt Lake City, Utah, are excellent examples of large fault blocks produced by tensional stresses on the crust.

As tensional forces continue to stress the affected area, slippage may continue along a major normal fault. The lower (downthrown) block will continue to sink, and the upper (upthrown) block may rise. But the tensional forces are likely to generate additional normal faults, generally parallel to the original one. This creates conditions favorable to the development of rift-valley topography (Fig. 36-6). The midoceanic ridge system, as noted elsewhere, is essentially a vast rift system, where tensional forces are pulling the crust apart.

On the continental landmasses, the best example of rift-valley topography lies in eastern Africa (Fig. 36-7). You do not need a geologic map to see this magnificent example of crustal rifting, because East Africa's Great Lakes fill the rift valley over much of their length. The system actually extends northward, through the Red Sea, into the Gulf of Aqaba and the Jordan River Valley beyond. In eastern Africa, it crosses Ethiopia and reaches Lake Turkana, the northernmost large lake in the rifts. South of the latitude of Lake Turkana the system splits into two gigantic arcs, one lying to the east of Lake Victoria and the other to the west. These arcs come together just to the north of Lake Malawi. The system then continues southward through the valley in which that lake lies, and it does not end until it has crossed Swaziland.

As long ago as 1921, the British geomorphologist J. W. Gregory offered an interpretation of the rift valleys of East Africa. He concluded that the floors of the valleys, in places thousands of meters below the adjacent uplands, were **grabens** (sunken blocks) between usually parallel normal faults (Fig. 36-8). This theory of a tensional origin of East Africa's rift valleys was later upheld for the arc that lies east of Lake Victoria. But the western arc, in which Lake Tanganyika lies, proved to have compressional origins. The fault scarps looked similar to

FIGURE 36-8 Aerial view of the East African rift valley in southern Ethiopia. The tensional stresses pulling the crust apart here are evident in the fault scarps and the downthrown crustal strips at their base. This scene provides a striking example of the landscape illustrated in Fig. 36-6.

those of the eastern arc, but they had begun as overhanging scarps (Fig. 36-2a). One of Africa's greatest mountains, Ruwenzori, proved to be a **horst**, a block raised between reverse faults (Fig. 36-3).

If the East African rift valleys originated, at least in part, from tensional forces, we should expect volcanism

to be associated with them, just as volcanism affects the spreading midoceanic ridges. These expectations are confirmed in that composite cones, lava domes, cinder cones (Fig. 34-7), and fissure eruptions all mark the region. Volcanic activity still continues along segments of the East African rifts. Seismic activity, too, is stronger there than in other areas of the great African shield (Fig. 35-5). Some geologists now believe that the African Plate may not be a single tectonic unit, and that an East African Plate (or Somali Plate, as it is also known) is separating from Africa just as the Arabian Plate did earlier. In that case, millions of years from now a Red Sea-like body of water may invade Africa from near the mouth of the Zambezi River and penetrate northward into Lake Malawi—and a Madagascar-like chunk of the continent will move off the new East African coastline (Fig. 36-9).

Transverse Faults

Where blocks of crustal rock move laterally, motion along the fault plane is horizontal, not vertical. Thus there are no upthrown or downthrown blocks. The fault is **transcurrent**—that is, movement is in the direction of the fault (Fig. 36-10). When we discussed lateral plate contact, the San Andreas Fault was identified as the contact plane between the Pacific and North American Plates

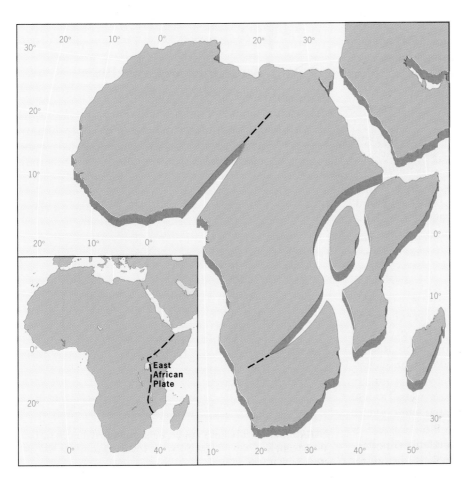

FIGURE 36-9 The possible configuration of Africa, ca. 10 million years from now, if the East African Plate (inset map) completes its separation from the rest of the continent.

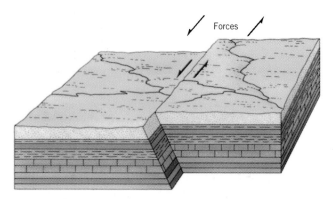

Forces

FIGURE 36-10 A horizontal or transcurrent fault, also known as a strike-slip fault.

(*see* Fig. 32-11). The San Andreas Fault is a special case of transcurrent faulting, a **transform fault** that marks the contact boundary between plates. Because movement at a transcurrent fault takes place along the strike of the fault, transcurrent faults are also known as **strike-slip faults**. To describe the fault completely, the direction of movement also is given. This is determined by looking *across* the fault and stating the direction in which the opposite block is moving. For example, consider the San Andreas Fault again. Stand on the North American Plate side: as you look across to the west, the Pacific Plate is sliding past to the right (*see* Fig. 32-10). Accordingly, the San Andreas Fault is a right-lateral strike-slip fault.

Field Evidence of Faulting

These three types of faults—reverse, normal, and transcurrent—exhibit many variations, and regional fault patterns and structures can become very complicated. Even in the field it is not always easy to distinguish a scarp formed by a fault from one formed by erosion. If the upthrown block contains water-bearing rock, water may pour from the scarp face in springs, probable evidence of faulting. The movement of the rocks along the fault plane may also produce smooth, mirror-like surfaces on the scarp face, and these *slickensides* are further evidence that faulting created the scarp. If the fault trace is marked by a breccia (a shattered rock layer), faulting is indicated as well. Most often, a combination of evidence provides the surest basis for interpretation.

FOLD STRUCTURES

When rocks are compressed, they respond to the stress by **folding** as well as by faulting. All rocks—even a sill of granite—have some capacity to bend before fracturing. But the fold structures we discuss here are most characteristic of layered sedimentary rocks. Folds, again like faults, come in all dimensions. Some are too small to see; others are road-cut size; still others are the size of entire mountain ridges (Fig. 36-11).

Unlike faults, however, folds are primarily compressional features. In areas where crustal rocks are under

FIGURE 36-11 The folded sedimentary strata of this hillside in central Arizona were laid down horizontally and later deformed. (Authors' photo)

stress, the crust may bend into basins hundreds of kilometers across in association with normal faulting. But the accordion-like folds in the Andes Mountains, the Appalachians, and parts of Eurasia's Alpine Mountain system (such as the Zagros Mountains of Iran) are the result of intense compression. Not surprisingly, folding and faulting generally occur together: just as even brittle rocks can bend slightly, so the most plastic rocks have a limited capacity to fold.

Anticlines and Synclines

Folds are rarely simple, symmetrical structures. Often they form a jumble of upfolds and downfolds that make it difficult to discern even a general order. But when we map the distribution of the rock layers and analyze the topography, we discover that recognizable and recurrent

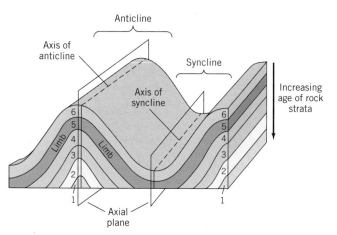

FIGURE 36-12 Anticlines are arching upfolds, while synclines are trough-like downfolds. Note the relative age of rock layers within each type of fold.

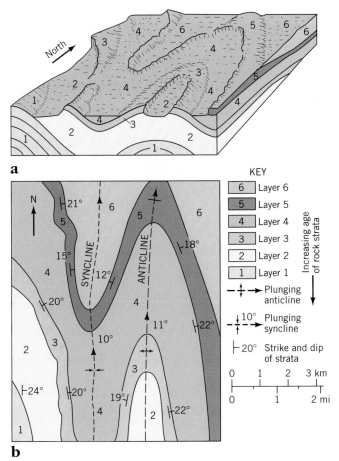

a

b

FIGURE 36-13 An eroded anticline-syncline landscape of surface outcrops with parallel strikes (a). The age sequence of these rock formations can be interpreted by referring to the geologic map (b) below the block diagram.

KEY

6	Layer 6
5	Layer 5
4	Layer 4
3	Layer 3
2	Layer 2
1	Layer 1

Increasing age of rock strata

Plunging anticline

Plunging syncline

Strike and dip of strata

structures do exist. The most obvious of these are up-folds, or **anticlines**, and downfolds, or **synclines**. An anticline is an arch-like fold, with the limbs dipping away from the axis (Fig. 36-12, left). A syncline, on the other hand, is trough-like, and its limbs dip toward its axial plane (Fig. 36-12, right).

When flat-lying sedimentary strata are folded into anticlines and synclines, the cores of the synclines are constituted by younger rocks. Imagine that erosion removes the upper parts of both anticlines and synclines. What is left are a series of surface outcrops with parallel strikes and a sequence of rocks showing a succession of ages. From these data, it is possible to interpret the geologic structures here—based on our understanding of anticlines and synclines (Fig. 36-13).

Plunging Folds and Landscapes

Folds are rarely as symmetrical as shown in Fig. 36-12, however; nor are their axes usually horizontal. Anticlines and synclines often *plunge*, which means that their axes dip. You can easily demonstrate the effect of this by taking a cardboard tube and cutting it lengthwise. Hold it horizontally and you have a symmetrical anticline (and a syncline, if you place the other half adjacent). Now tilt

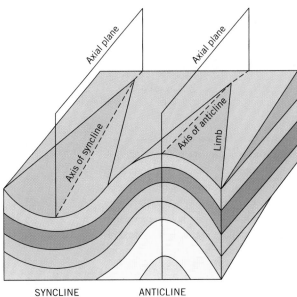

FIGURE 36-14 An anticline-syncline structure, plunging in opposite directions.

the tube, say, about 30 degrees from the horizontal, by the edge of a table or other flat surface. Draw the horizontal line, and cut the tube along it. The second cut represents the outcrop of the youngest rocks of the anticline. Note how they form a "nose" at one end and open progressively toward the other end. In the field, such an outcrop signals an eroded, plunging anticline; the adjacent syncline would plunge in the other direction (Fig. 36-14).

In the intensely folded central Appalachian Mountains, plunging anticlines and synclines adjoin each other in wide belts, eroded into attractive scenery by streams over many millions of years. But as the Appalachians' topography suggests, there is considerable regularity and even symmetry to the pattern (Fig. 36-15). That is not always the case in areas of intense compression and folding. Europe's Alps, for instance, are much more severely distorted. There, folds are not only anticlinal and synclinal: many of the anticlines have become **recumbent** and even **overturned** (Fig. 36-17). Add to this the presence of intense faulting and erosion by streams and glaciers, and complex spectacular alpine landscapes are produced.

At this point, an important distinction must be made between what some geomorphologists call *primary* landforms (the structures created by tectonic activity) and *secondary* landforms (the products of weathering and erosion). Sometimes geologic and topographic maps reveal amazing contradictions. An anticline of soft sedimentary rocks may be eroded into a low valley, while a nearby synclinal structure, composed of more resistant rocks, stands out as a ridge or upland. Thus the geologic upfold forms a geographic lowland and vice versa! The relationship between structure, rock resistance, and erosional processes forms a major theme that runs through many units in Part Five.

The Decline and Fall of Southeast Texas

If you believe that such matters as crustal tilting and warping are merely textbook topics, consider what is happening in Baytown and nearby areas of coastal Southeast Texas (Fig. 36-16). Baytown, Texas lies near Trinity Bay, an extension of Galveston Bay on the Gulf of Mexico. With its offshore sandbars and lagoons (Galveston itself is located on such a sandbar), this coastline would seem to be one of emergence and accumulation (topics discussed in Unit 51). Instead, it is an area of subsidence, and in places very rapid subsidence.

The ground here is tilting downward, and the sea is invading the land. The rate of decline is rapid indeed—15 cm (6 in) per year in some areas. Seawalls that once stood high above the water now barely hold back the waves. Areas that were once high and dry are now swampy and wet. Some people have already abandoned their homes and simply left, having given up the battle against the course of nature. And it is not just Baytown that is threatened. The whole Houston region of coastal Southeast Texas is subsiding, tilting southward and downward, and what the people in Baytown are facing is only the beginning.

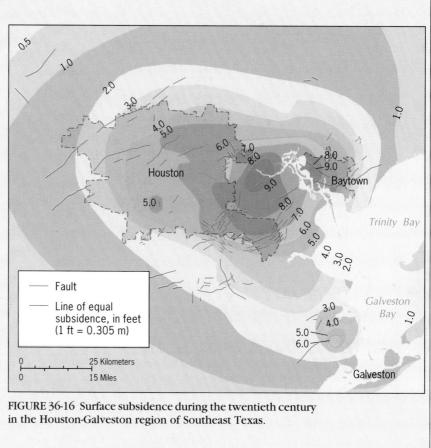

FIGURE 36-16 Surface subsidence during the twentieth century in the Houston-Galveston region of Southeast Texas.

REGIONAL DEFORMATION

Not all types of crustal deformation are encompassed by the various types of faulting and folding discussed in this unit. Over large areas of stable lithosphere, especially in the major continental shields, the crust undergoes slight deformation *without* being faulted or folded. Geologists in the past gave various terms to these and other crustal movements. *Diastrophism* was one, and *crustal warping*

FIGURE 36-15 Satellite image of the intensely folded Appalachian topography of south-central Pennsylvania. This area is dominated by often parallel ridges (reddish brown) and valleys (lighter colors) produced by the differential erosion of sedimentary rock layers of varying resistance. The large river at the upper right is the Susquehanna; the city of Harrisburg, Pennsylvania's capital, is located on the east bank of the river where it emerges from the ridge-and-valley terrain.

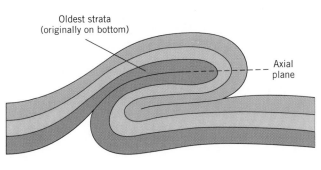

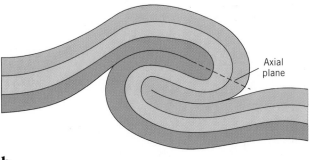

a

b

FIGURE 36-17 Extreme compression can produce folds that double back upon themselves with axial planes that approach the horizontal. The anticline in (a) is termed recumbent: the anticline in (b) is classified as overturned.

was another. Yet another term, still in occasional use, is **epeirogeny**; it refers to the vertical movement of the crust over very large areas, involving little or no bending or breaking of the rocks.

Such rising or sinking of the upper crust undoubtedly is related to the movement of lithospheric plates over the mantle's convection cells, and in that sense it is subsumed under plate motion. But in physical geography these very slight regional movements are extremely important. Even the slightest change in the slope of a large region can have an enormous impact on an entire drainage basin, rates of erosion and deposition, and other aspects of the regional physiography (*see* Perspective: The Decline and Fall of Southeast Texas). We should remember, therefore, that while this unit focuses on the tectonically active zones of the world (present and past), more subtle but still highly significant deformation also affects the more stable sectors of the landmasses.

KEY TERMS

anticline (p. 383)

dip (p. 378)

epeirogeny (p. 385)

fault (p. 378)

folding (p. 382)

graben (p. 380)

horst (p. 381)

normal fault (p. 379)

overturned fold (p. 383)

recumbent fold (p. 383)

reverse fault (p. 378)

strike (p. 378)

strike-slip fault (p. 382)

syncline (p. 383)

thrust fault (p. 379)

transcurrent fault (p. 381)

transform fault (p. 382)

REVIEW QUESTIONS

1. What is the basic difference between normal and reverse faulting?

2. What evidence of faulting might one encounter when looking at an isolated rock outcrop?

3. Differentiate between *anticlines* and *synclines*, and describe how plunging affects their orientation.

4. What is the difference between *primary* and *secondary* landforms?

5. Differentiate between the concepts of *strike* and *dip* in describing a ridge on the earth's surface.

6. What is a rift valley? Why are these formations so widespread in eastern and northeastern Africa?

REFERENCES AND FURTHER READINGS

BENNISON, G. M. *An Introduction to Geological Structures and Maps* (London: Edward Arnold, 5th ed., 1990).

BILLINGS, M. P. *Structural Geology* (Englewood Cliffs, N.J.: Prentice-Hall, 3rd ed., 1972).

DAVIS, G. H. *Structural Geology of Rocks and Regions* (New York: Wiley, 1984).

HOBBS, B. E., et al. *An Outline of Structural Geology* (New York: Wiley, 1976).

HILLS, E. S. *Elements of Structural Geology* (New York: Wiley, 2nd ed., 1972).

MIYASHIRO, A., et al. *Orogeny* (New York: Wiley, 1982).

OLLIER, C. D. *Tectonics and Landforms* (London/New York: Longman, 1981).

POWELL, D. *Interpretation of Geological Structure Through Maps: An Introductory Practice Manual* (New York: Wiley/Longman, 1992).

RAGAN, D. M. *Structural Geology: An Introduction to Geometrical Techniques* (New York: Wiley, 3rd ed., 1984).

WEYMAN, D. *Tectonic Processes* (Boston: Allen & Unwin, 1981).

WINDLEY, B. *The Evolving Continents* (New York: Wiley, 2nd ed., 1984).

SCULPTING
THE SURFACE

UNIT 37

The Formation of Landscapes and Landforms

OBJECTIVES

◆ To focus attention on the gradational processes that produce secondary landforms.

◆ To introduce three primary gradational processes: weathering, mass movements, and erosion.

◆ To recognize the role of degradation and aggradation in the formation of landscapes.

What we see in the earth's landscapes is the temporary result of many processes. As the units in Part Four detail, the continents are affected by the horizontal and vertical movement of tectonic plates. These movements result in the folding, faulting, tilting, and warping of the crust, and are often accompanied by earthquakes and volcanic eruptions. These movements and processes create **primary landforms**. A volcano, for example, can be worn down over a long period of time by various agents of erosion such as water, wind, and even ice in certain places. But the original mountain is a primary landform created by volcanic activity.

When erosion sculpted it into the form we see today, a **secondary landform** was produced. The same is true,

over a larger area, in a subduction zone. We know that the crustal plates move several centimeters per year. The enormous energy involved is expressed in part by the crushed, folded, and faulted rocks that mark the scenery of these subduction zones (*see* Fig. 29-13). Here it may be more difficult to differentiate between primary and secondary processes because both happen at the same time. Even as the crust buckles and breaks, erosion attacks; in combination, these processes produce characteristic terrain.

UNIT OPENING PHOTO: Limestone tower topography in southern China around the city of Guilin. Chemical action played a major role in sculpting this unusual landscape.

388

All parts of the landmasses, even relatively stable areas unaffected by subduction or other severe deformation, are subject to vertical and horizontal movement that affect the evolution of their landscapes. As the plates carrying the continents move over the mantle, the landmasses are pushed, dragged, pressed, and stretched. Geologists still are not certain about the exact nature of tectonic activity beneath the landmasses. Hot spots such as those we note for the Pacific Plate (Unit 34) also exist beneath continent-bearing plates; the Yellowstone National Park region is a good example (*see* Fig. 40-11).

As noted in Part Four, at least some of the extensive continental plates may, like the oceanic plates, consist of segments. These segments may exhibit convergent, divergent, and lateral contact. The effect of these movements on the landmasses is not yet completely understood. However, there can be no doubt that they continuously deform the continental crust—and thus its surface, the landscape. In Part Five, we concentrate on the secondary processes that mold the scenery of the continents, but we should keep in mind that the processes sculpting the surface we observe are the result of forces from below as well as from above.

LANDSCAPES AND LANDFORMS

As the title of this unit indicates, our concern is with the surface configuration of the exposed land, not the submerged seafloor. Thus we refer frequently to particular types of *land*scapes and *land*forms. A **landform** is a single and typical unit that forms part of the overall shape of the earth's surface. A solitary mountain (such as a composite cone), a hill, a single valley, a dune, and a sinkhole all are landforms. A **landscape** is an aggregation of landforms, often the same types of landforms. A volcanic landscape, for example, may consist of a region of composite cones, lava domes, lava flows, and other features resulting from volcanic activity, all modified, to a greater or lesser extent, by erosion. A dune may be part of a desert landscape or a coastal landscape. So the term *landform* often refers to the discrete product of a set of processes; a *landscape* is the areal (or regional) expression of those processes.

GRADATION

When a dirt road is "graded," it is leveled off to a smooth horizontal or sloping surface, usually by a machine designed to perform this task. The forces of erosion, too, work to lower and level the surface of landmasses, to wear them down to a smooth and nearly level plane. Landscapes and landforms represent the progress made by the agents of weathering and erosion as they transform the surface of the landmasses rising above sea level.

Landscapes reveal the nature of these erosional agents, the hardness and resistance of the rocks, the geologic structures below, and the tectonic activity affecting the crust.

The key force is that of the earth's *gravity*. When soil and loose rock on a hillside become waterlogged, the force of gravity pulls it downslope where stream water (also responding to gravity) removes it (Fig. 37-1). When a mineral grain on a rock face is loosened, gravity causes it to fall to the ground below and the wind may carry it away. Streams flow and glaciers move down their valleys under the laws of gravity: steeper slopes mean faster movement. Even winds and waves are subject to the earth's gravitational pull. Landslides, avalanches, waterfalls—all reveal the ever-present force of gravity in the wearing down of the earth's landmasses.

The result is a reduction in the relief of the landmasses—a lowering of the mountains, a filling of the valleys, a planing of the continents down to as close to sea level as the laws of physics permit. Millions of tons of sediment are annually carried downstream by the great rivers of the world (Fig. 37-2). This material comes from the interior uplands, and it may be deposited in deltas or on the ocean floor. Some of it is laid down in the valleys of the rivers themselves, so that many streams in their lower courses no longer flow in rock-floored channels but over their own sediments.

FIGURE 37-1 A talus cone or scree slope (foreground) in the central Rocky Mountains of western Colorado. These sizeable rock fragments, many as big as large boulders, have been loosened by mechanical weathering and have been moved downslope by the force of gravity. They have not, however, been moved far from their bedrock sources, so erosion is not yet a major agent in their removal.

FIGURE 37-2 The chocolate-colored waters of the sediment-rich Chang Jiang, China's mightiest river. This is the spectacular Three Gorges section of the Chang, just before the river emerges from the highlands onto its lower plain, the earth's most heavily populated region. Grandiose dam-building schemes have been initiated in these gorges; when implemented, they will forever alter the hydrography—and human-environmental relationship—in this part of the world.

The wearing down of the landmasses goes on through a combination of processes collectively referred to as **degradation**. This term implies that a landmass is being lowered, reduced, smoothed. When a river cuts an ever-deeper channel, or a glacier scours its valley, or waves erode a beach, degradation prevails. But the opposite also occurs: a combination of processes removes material from one area and deposits it elsewhere. When a river builds a delta at its mouth, or when a valley fills with sediment, or when a glacier melts and deposits the ground-up rocks it is carrying along, **aggradation** takes place. Aggradation also contributes to the lowering of relief by reducing the height differences between the high points and low places in an area.

Degradational Processes and Landscapes

The processes and agents that work to lower the uplands of the continents range from the quiet disintegration of regolith (loose rock material) to the violent cascade of boulder-filled mountain streams. In Part Five, we study the multiple processes that combine to destroy and remove the rocks of the continents and the landscapes they create. In a very general way, the gradational processes can be divided into three categories: (1) *weathering*, (2) *mass movements*, and (3) *erosion*. We devote entire units each to weathering and mass movements, and several to the agents and processes of erosion.

A good way to understand how these degradational processes differ is to consider the *distances* traveled by loosened rock particles. In weathering a rock disintegrates, but the loose grains do no more than fall to the

ground below under the force of gravity (Fig. 37-1). In mass movements (such as landslides), rock material travels farther downslope over distances that can reach 20 km (12 mi) or more. And erosional processes can carry material hundreds, even thousands of kilometers.

As the term implies, **weathering** is the breakdown of rocks in situ, that is, their disintegration or decomposition without distant removal of the products. As discussed in Unit 38, rocks can be weakened in many ways: by exposure to temperature extremes, by chemical action, and even by the growing roots of plants. The processes of weathering prepare the rock for later removal.

Mass movement is the spontaneous downslope movement of earth materials under the force of gravity. Anyone who has witnessed or seen a picture of a landslide or mudslide knows what such slope failure involves. Houses, roads, even entire hillsides tumble downslope, often after the weathered surface material has become waterlogged (Fig. 37-3). In Unit 34 we report on the great mudflow (or *lahar*) that occurred on Colombia's volcano, Nevado del Ruiz, when it erupted in 1985; saturated by abruptly melted snow from the volcano's peak, this mass movement traveled over 50 km (30 mi) down valleys from the base of the mountain and buried at least 20,000 people living in its path (*see* Fig. 34-5). Thus mass movements (treated in Unit 39) can involve large volumes of material, sometimes with disastrous consequences. But other forms of mass movement are slow and almost imperceptible.

Most of our attention, however, is devoted to the fascinating processes of **erosion**. Here, the particles resulting from weathering and mass movement are carried away over long distances. During this process of transportation, additional breakdown occurs, both of the transported particles *and* of the rocks of the valleys through which they travel. For example, a boulder first loosened in a mass movement is removed by a stream and bounces along the valley as the water transports it.

FIGURE 37-3 The aftermath of a mudslide in the foothills of the Wasatch Mountains in northeastern Utah.

It breaks into smaller pieces that become rounded into pebbles, and the parts that are broken off during this process form smaller particles. As these and other boulders, pebbles, gravel, and sand move along, they chip pieces of rock off the valley floor and sides. In this way, stream erosion involves both transportation and breakdown.

Running Water

Running water is by far the most effective and significant agent of erosion, as noted in Units 40 to 43. Streams are complex gradational systems whose effectiveness relates to many factors, including the gradient (slope) of the valley, the volume of water, and the form of the valley. Few places on earth are unaffected by the erosional power of water. Even in deserts, water plays a major role in shaping the landscape. When we study maps of streams, it is clear that many river systems consist of a major artery that is joined by *tributaries*, smaller streams that feed into the main river (Fig. 37-4). The spatial pattern of these systems can reveal much about the rock structures below. Certain patterns are associated with particular structures and landscapes. Often the configuration of a regional stream system can yield insights about the underlying geology and geomorphic history of an area, as explained in Unit 42.

Glaciers

Streams are not the only agents of erosion; ice, in the form of glaciers, is also an important agent of degradation. During earlier stages of the Pleistocene ice age, great glaciers covered much of northern North America and Eurasia. These icesheets modified the landscape by scouring the surface in some areas and depositing their erosional "load" elsewhere. At the same time, valleys in high mountain areas (such as the Alps and Rockies) were filled by glaciers that snaked slowly downslope. These mountain glaciers widened and deepened their containing valleys, carrying away huge loads of rock in the process. When the global climate warmed up and the glaciers melted back, the landscape below had taken on an unmistakable and distinctive character. We examine the evidence in Units 45 to 48.

FIGURE 37-5 Wind is a frequent landscape sculptor in desert environments. These are some of the massive dunes in Namibia's Namib Desert, which parallels the southwestern coast of Africa.

Wind

Wind, too, is a gradational agent. Globally, compared to running water and moving ice, wind is an insignificant factor in landscape genesis. In certain areas of the world, however, wind *is* an effective modifier of the surface. Wind can propel sand and dust at high velocities, wearing down exposed rock surfaces in its path. Wind is also capable of moving large volumes of sand from one place to another, creating characteristic dune landscapes (Fig. 37-5). The processes and landscapes of wind action are treated in Unit 49.

Coastal Waves

Coastal landscapes are of special interest in physical geography. Where land and sea make contact, waves attack and rocks resist, and the resulting scenery often is spectacular as well as scientifically intriguing. As in the case of wind action, it is not the waves alone that erode so effectively. When waves contain pieces of loosened rock, they are hurled against the shore, sometimes with enormous impact. The erosional results are represented by cliffs and other high-relief landforms (Fig. 37-6). Coastal landscapes and landforms are the focus of Units 50 and 51.

Chemical Solution

A special case of degradational action involves the removal of rock not by physical breakdown but by chemical solution. As we observe in Unit 44, this process produces a landscape of highly distinctive surface and near-surface features. When soluble rocks, principally limestone, are layered in a suitable way and subjected to humid climatic conditions, the limestone dissolves. Par-

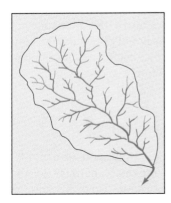

FIGURE 37-4 A stream system in its drainage basin. The main river is built up by the smaller tributary streams that empty into it.

FIGURE 37-6 Coastal landforms are shaped by several forces, including waves, winds, streams at maximum volume, and tectonic action. As a result, coastlines present some of the world's most dramatic—and occasionally enigmatic—scenery. This particular combination of landforms, dominated by a spectacular natural bridge, occurs on the Pacific coast of Oregon. (Authors' photo)

tially carried away in solution, the limestone strata are honeycombed by caves and underground channels. As the ground above collapses into these subsurface cavities, the landscape is pocked by sinkholes, evidence of the efficacy of chemical solution in sculpting the earth's surface.

Aggradational Processes and Landforms

Streams, glaciers, wind, and waves are erosional, degradational agents—but they also have the capacity to build, to deposit, and thus to create aggradational landscapes and landforms. As we study the degradational power of rivers, we note that what is removed as boulders and pebbles from interior highlands may be laid down as mud or silt in a coastal delta. Those Pleistocene glaciers that scoured the surface of much of Canada deposited their ground-up load of sediments to the south, much of it in the U.S. Midwest. Mountain glaciers excavated the upper valleys, but they filled their lower valleys with deposits. Wind action wears rocks down in some places but builds dunes in others. Waves that cut cliffs create beaches in other locales and under different circumstances. So the degradational agents, the agents of erosion, also are aggradational agents, or agents of deposition. The key is transportation: the removal of rocks, their long-distance conveyance, and their pulverization (grinding) in the process. All of this is part of the rock cycle (see Fig. 31-12), which is driven by gravity and sustained by the hydrologic cycle (see Fig. 12-2), and serves to reduce the relief of the landmasses.

EROSION AND TECTONICS

If the continental landmasses were static (that is, if they were not subject to tectonic forces as parts of lithospheric plates), they would indeed eventually be worn down to near sea level. Moreover, in the context of the earth's lifetime, it would not take very long. To prove this, geomorphologists first calculated the total mass of the continents presently existing above sea level. Next, they estimated the quantity of sediment that all the world's rivers carried to the oceans in an average year. By dividing the total mass by the amount annually removed, and by factoring in the reduced efficiency of streams as the continents are (theoretically) lowered, it was possible to estimate the number of years it would take for the continental landmasses to be planed to near sea level. The surprising result: 270 million years, just a fraction (about 6 percent) of the earth's total history!

Obviously, this calculation is not consistent with observed reality. Continental landmasses probably have existed for well over 3 billion years, and in approximately their present form (if not shape) for at least 650 million years. The oldest sedimentary rocks are from the Precambrian era (see Perspective: The Geologic Time Scale); thus we know that degradation and aggradation were taking place at least 570 million years ago. We must therefore conclude that the continents are being rejuvenated tectonically as they are worn down by erosion.

The principle of isostasy (Unit 33) relates to this rejuvenation. According to isostatic principles, the removal of a large volume of rock (or other burden, such as the melting of ice) from an area of the crust leads to an upward "rebound" of that part of the crust. Thus downward erosion is compensated for by upward rebounding of lithosphere. If this is true, then it follows that erosion will go on indefinitely, because the crust will always replace what has been lost through degradation.

This also has implications for the study of landscapes. It may mean that landscapes retain their relief properties much longer than had previously been believed. For instance, a topography such as that of the Appalachian highlands may retain its ridge-and-valley scenery (see Fig. 53-4) for many millions of years after erosion *should* have worn the steep-sided ranges into low, undulating hills. Compensating isostatic uplift may do more than preserve continental landmasses: it also may preserve landscapes. We return to this theme several times later in Part Five, but, as noted elsewhere, it is important to keep in mind the tectonic forces from below when assessing the erosional forces above.

REGIONAL LANDSCAPES

Finally, Part Five includes the application of a fundamental geographic theme—regional analysis—to the distribution of physical landscapes. Units 52 and 53

The Geologic Time Scale

When geologists began to realize that there was a degree of order in the massive accumulations of rocks that made up the earth's crust, it became necessary to establish a timetable for permanently labeling newly identified rocks. Accordingly, the **geologic time scale**, a chronicle of earth history, was developed more than a century and a half ago by British geologists. They called the oldest sequence of rocks the *Primary*, and the next two the *Secondary* and the *Tertiary*. Later, the youngest rocks, such as sediments now accumulating in river valleys and deltas, were separately identi-

fied as *Quaternary*. Soon it became necessary to subdivide these original major units. Bit by bit, the geologic time scale became more exact—and more complicated.

As scientists learned more about fossils (traces of plants and animals preserved in the rocks), they were also correlating their discoveries with those made in distant countries. These discoveries about the evolution of life largely determined the names in the geologic time scale now in use (Fig. 37-7); the dates are still being refined as new evidence comes to light. Note that Primary

has become **Paleozoic** (era of ancient life), and Secondary is now **Mesozoic** (era of medieval life). We can still find **Tertiary** and **Quaternary** in the modern classification, but they are now subdivisions of a third great era, the **Cenozoic** (era of recent life).

The time scale in Fig. 37-7 shows these three major eras divided into *periods*, which, in the case of the Tertiary and Quaternary, are further subdivided into *epochs*. These names refer to the evolution of all life on earth, not human life alone, which did not appear fully developed until the latest period of the Ce-

FIGURE 37-7 The geologic time scale, showing the emergence of various life forms, calibrated according to dates established during the mid-1990s.

nozoic era—the Quaternary. There are also rocks predating the Paleozoic era (older than about 570 million years). This era is called the **Precambrian** because it is older than the oldest period of the Paleozoic era, the Cambrian. The shield zones of the earth (*see* Unit 29) are in this oldest of age categories.

The simplified geologic time scale in Fig. 37-7 also shows that more is known about relatively recent geological times than about older periods. Subdivisions in the right-front quarter of the chart are more numerous, and they cover shorter time intervals. This is not surprising because the youngest rocks tend to be closest to the surface. Layers left in the northern United States and Canada during a recent glaciation are fresh and have hardly changed, geologically speaking. Other rocks are being deposited at this very moment. We can extrapolate from the processes we see to the rocks buried deep down, which must have been deposited under similar circumstances.

We will have occasion to refer to this time scale much as we refer to the years and months of our own lifetimes. To better understand the magnitudes of the time periods involved, let us relate the earth's life span to your age. If you are 20 years old and we take as the age of the earth the time of formation of the oldest rocks we can find (more than 4 billion years), consider the following:

1. One *year* of your life equals 230 million years of the earth's. That puts you in the late Paleozoic just one year ago.

2. One *month* of your life equals just over 19 million years of the earth's. The Rocky Mountains formed just 3½ months ago.

3. One *week* of your life equals nearly 5 million years of the earth's. The Pleistocene ice age began three days ago.

4. One *day* of your life equals about 630,000 years of the earth's. Human evolution was still in its early stages just yesterday at this time.

5. One *hour* of your life equals over 26,000 years of the earth's. In that single hour, the human population grew from less than 100,000 to almost 6 billion, and the major civilizations developed.

Where will we be one hour from now?

---

concentrate on North America and the United States, respectively. Landscapes often display a certain sameness or homogeneity over large areas. We acknowledge this in our everyday language by referring to such areas as the "Great Plains" or the "Rocky Mountains." But if we are to map such regional landscapes with any precision, we must establish criteria on which to base the boundaries drawn. This leads us into the field of *regional physiography*, where everything we have learned in this book comes together: climate, soil, vegetation, topography, terrain, and relief. In combination, these elements of physical geography allow us to view our continent and country regionally, and to better understand its many faces, from sea to shining sea.

KEY TERMS

aggradation (p. 390)	geologic time scale (p. 393)	Mesozoic (p. 393)	Quaternary (p. 393)
Cenozoic (p. 393)	landform (p. 389)	Paleozoic (p. 393)	secondary landform (p. 388)
degradation (p. 390)	landscape (p. 389)	Precambrian (p. 394)	Tertiary (p. 393)
erosion (p. 390)	mass movement (p. 390)	primary landform (p. 388)	weathering (p. 390)

REVIEW QUESTIONS

1. How does a secondary landform differ from a primary landform?
2. Briefly describe the processes of gradation.
3. What is meant by the term *weathering*?
4. Describe the major time divisions of the geologic time scale.

REFERENCES AND FURTHER READINGS

BRADSHAW, M., et al. *The Earth's Changing Surface* (New York: Wiley, 1978).

BUTZER, K. W. *Geomorphology from the Earth* (New York: Harper & Row, 1976).

CURRAN, H. A., et al. *Atlas of Landforms* (New York: Wiley, 1984).

EICHER, D. L. *Geologic Time* (Englewood Cliffs, N.J.: Prentice-Hall, 2nd ed., 1976).

HARLAND, W. B., et al. *A Geologic Time Scale 1989* (London/New York: Cambridge University Press, 1990); *A Geologic Time Scale 1989—Wallchart* (New York: Cambridge University Press, 1990).

KING, L. C. *The Morphology of the Earth: The Study and Synthesis of World Scenery* (New York: Hafner, 1967).

OLLIER, C. D. *Tectonics and Landforms* (London/New York: Longman, 1981).

PITTY, A. F. *Landforms and Time* (New York: Blackwell, 1989).

RITTER, D. F. *Process Geomorphology* (Dubuque, Iowa: Wm. C. Brown, 2nd ed., 1986).

SNEAD, R. E. *World Atlas of Geomorphic Features* (Huntington, N.Y.: Krieger, 1980).

THORNBURY, W. D. *Principles of Geomorphology* (New York: Wiley, 2nd ed., 1969).

Weathering Processes

OBJECTIVES

◆ To differentiate the major categories of weathering—mechanical, chemical, and biological.

◆ To introduce and briefly discuss common weathering processes.

◆ To note the general environmental controls over weathering processes.

On the northeast coast of Brazil, about 1300 km (800 mi) north of Rio de Janeiro, lies the city of Salvador, capital of the state of Bahía. Two centuries ago, Salvador was a wealthy, thriving place. Agricultural products from nearby plantations and whales from offshore waters yielded huge profits. The Salvadorans, under the tropical sun, built themselves one of the world's most beautiful cities. Ornate churches, magnificent mansions, cobble-stoned streets, and manicured plazas graced the townscape. Sidewalks were laid in small tiles in intricate patterns, so that they became works of art. And speaking of art, the townspeople commissioned the creation of many statues to honor their heroes.

But times eventually changed. Competition on world markets lowered incomes from farm products, and over-exploitation eliminated the whale population. There was no money to maintain the old city; people moved away, and homes stood abandoned. Salvador had to await a new era of prosperity. But when things improved, the better-off residents preferred to live in high-rise luxury buildings on the coast or in modern houses away from the old town. Old Salvador lay unattended, exposed to the elements (Fig. 38-1). The equatorial sun beat down

UNIT OPENING PHOTO: Sandstone formation, Dinosaur National Monument, Utah. The expanding roots of this juniper tree, through mechanical action, are a force in the weathering of these rocks.

FIGURE 38-1 The historic city of Salvador, located on Brazil's northeastern coast, has been exposed to the elements and has been in decay ever since its time of prosperity centuries ago. Ornate buildings show the effects of weathering—the same weathering that affects natural structures. (Authors' photo)

FIGURE 38-2 This high-mountain granite boulder has been split in two by the repeated freezing and thawing of water that penetrated its central joint plane. So far, the boulder has not been dislodged from the place where it was wedged apart.

on roofs and walls; moisture seeped behind plaster; mold and mildew formed as bright pastel colors changed to ugly gray and black. Bricks were loosened and fell to the ground, and statues began to lose their features as noses, ears, and fingers wore away. People who knew the old city in its heyday were amazed at the rapid rate of decay that was evident everywhere.

The fate of old Salvador reminds us of nature's capacity to attack and destroy—not just with swirling rivers, howling winds, or thundering waves, but quietly, persistently, and intensively. Nature's own structures are affected just as much as buildings and statues are. As we detail in Unit 23, the first steps in soil formation involve the breaking down of rocks into smaller particles. The same processes that destroy abandoned buildings also contribute to the degradation of landforms. This quiet destruction proceeds in many ways. When the sun heats an exposed wall, materials in that wall expand. The bricks may expand more than the mortar that binds them. Day after day, this repeated, differential expansion will destroy the bond between brick and mortar. On vulnerable corners, bricks will fall.

The same process affects rocks, and when various minerals expand at different rates, they will be loosened. Not only temperature but also moisture plays a critical role. Just as water seeps behind plaster, it can invade rocks, for example, along cracks and joints. Once there, water exploits the rocks' weaknesses, opens cracks wider, and allows moist air to penetrate. Soon the rock is broken into pieces, and the next rainstorm carries the smaller fragments away.

Other forces also come into play in the silent, unspectacular breakdown of rocks. In combination, these processes are called *weathering*—a good term, because it signifies the impact of the elements of weather examined in Part Two of this book. Weathering goes on continuously, not only at exposed surfaces but also beneath the ground and within rock strata. It is the first stage in that series of processes we have called degradation. Several kinds of weathering processes can be recognized. The three principal types are: (1) *mechanical* weathering; (2) *chemical* weathering; and (3) *biological* weathering. The terms are self-explanatory, and we examine each in turn. But it must not be assumed that these processes are mutually exclusive or that if one occurs the others do not. In fact, weathering processes usually operate in some combination, and it is often difficult to separate the effects of each process.

MECHANICAL WEATHERING

Mechanical weathering, also called *physical* weathering, involves the destruction of rocks through the imposition of certain stresses. A prominent example is **frost action**. We are all aware of the power of ice to damage roads and sidewalks and split open water pipes (water increases in volume by about 9 percent as it freezes). Similarly, the water contained by rocks—in cracks, joints, even pores—can freeze into crystals that shatter even the strongest igneous masses. This ice can produce about 1890 metric tons (2100 tons) of pressure for every 0.1 m² (1 ft²).

Of course, frost action operates only where winter brings subfreezing temperatures. In high-altitude zones, where extreme cooling and warming alternate, the water that penetrates into the rocks' joint planes freezes and thaws repeatedly during a single season, wedging apart large blocks and boulders and loosening them completely (Fig. 38-2). Then, depending on the local relief, these pieces of rock may either remain more or less where they are, awaiting dislodgement by wind or precipitation, or they may roll or fall downslope and collect at the base of the mountains.

When the pieces of rock accumulate near their original location, they form a **rock sea** (Fig. 38-3). Some-

FIGURE 38-3 A *felsenmeer* (rock sea) has developed from the mechanical weathering of quartzite on this slight slope. The blocks have been pried apart by weathering but have moved very little since being loosened. Now they form an expanse of angular fragments of many sizes, covering an area of Montana countryside. (Authors' photo)

FIGURE 38-5 Extreme chemical weathering has obliterated most of the detailed features on these statues high atop one of the facades of the Bayeux Cathedral in the northern French *département* of Normandy.

times other terms are used for this, such as *blockfield* or the German word *felsenmeer*. When the rock fragments roll downslope, they create a *scree slope* or **talus cone**. If you have been in the Rocky Mountains or other mountainous areas in the western United States, you have probably seen such piles of loose boulders. They often lie at steep angles and seem ready to collapse (Fig. 38-4).

Rocky soils can also reflect the action of frost. For a long time physical geographers wondered what produced the remarkable geometrical structures of stone that mark the soils in Arctic regions (*see* Fig. 48-6). These *stone nets* are created when ice forms on the underside of rocks in the soil, a process that tends to wedge the rock upward and sideways. Eventually, the rocks meet and form lines and patterns that look as though they were laid out by ancient civilizations for some ceremonial purpose.

In arid regions, too, mechanical weathering occurs. There the development and growth of salt crystals has an effect similar to that of ice. When water in the pores of such rocks as sandstone evaporates, small residual salt crystals form. The growth of those tiny crystals pries the rocks apart (in a process called *salt wedging*) and weakens their internal structure. As a result, caves and hollows form on the face of scarps, and the wind may remove the loosened grains and reinforce the process (*see* Perspective: The Dust Bowl).

The mineral grains that rocks are made of have different rates of expansion and contraction in response to temperature changes; therefore, the bonds between them may be loosened by continual temperature fluctuations. Although impossible to replicate in the laboratory, this kind of weathering may play a role in the mechanical disintegration of rocks. Over many thousands of years, daily temperature fluctuations could well have a weakening effect on exposed rocks. But other weathering processes also enter the picture: as the contacts between the grains are loosened, moisture enters and promotes decay of the minerals.

CHEMICAL WEATHERING

The minerals that rocks are made of are subject to alteration by **chemical weathering**, as the statues in Fig. 38-5 illustrate. Some minerals, such as quartz, resist this alteration quite successfully, but others, such as the calcium carbonate of limestone, dissolve easily. In any rock made up of a combination of minerals, the chemical breakdown of one set of mineral grains leads to the disintegration of the whole mass. In granite, for instance, the quartz resists chemical decay much more effectively than the feldspar, which is chemically more reactive and weathers to become clay. Often you can see a heavily

FIGURE 38-4 Coalesced talus cones form a thick, continuous apron along the base of the ridge lining the far shore of this lake in the Canadian Rockies' Banff National Park.

pitted granite surface. In such cases, the feldspar grains are likely to have been weathered to clay and blown or washed away. The quartz grains still stand up, but they may soon be loosened too. So even a rock as hard as granite cannot withstand the weathering process forever.

Hydrolysis

Three kinds of mineral alteration dominate in chemical weathering. When minerals are moistened, **hydrolysis** occurs, producing not only a chemical alteration but expansion in volume as well. This expansion can contribute to the breakdown of rocks. Hydrolysis, it should be noted, is not simply a matter of moistening: it is a true chemical alteration, and minerals are transformed into other mineral compounds in the process. For example, feldspar hydrolysis yields a clay mineral (silica) in solution, and a carbonate or bicarbonate of potassium, sodium, or calcium in solution. The new minerals tend to be softer and weaker than their predecessors. In granite boulders, hydrolysis combines with other processes to cause the outer shells to flake off in what looks like a miniature version of exfoliation (*see* Fig. 30-6). This is **spheroidal weathering** (Fig. 38-6), and it affects other igneous rocks besides granite.

Oxidation

When minerals in rocks react with oxygen in the air, the chemical process is known as **oxidation**. We have plenty of evidence of this process in the reddish color of soils in many parts of the world and in the reddish-brown hue of layers exposed in such places as the Grand Canyon (*see* Fig. 31-8). The products of oxidation are compounds of iron and aluminum, which account for the reddish colors seen in so many rocks and soils (*see* Fig. 25-12). In tropical areas, oxidation is the dominant chemical weathering process.

FIGURE 38-6 These two small boulders of dark-colored granite, from Swaziland in Southern Africa, are undergoing spheroidal weathering. (Authors' photo)

Carbonation

Various circumstances may convert water into a mild acid solution, thereby increasing its effectiveness as a weathering agent. With a small amount of carbon dioxide, for instance, water forms carbonic acid, which in turn reacts with carbonate minerals such as limestone and dolomite (a harder relative of limestone, a carbonate of calcium and magnesium). This form of chemical weathering, **carbonation**, is especially vigorous in humid areas, where limestone and dolomite formations are often deeply pitted and grooved, and where the evidence of solution and decay are prominent. This process even attacks limestone underground, contributing to the formation of caves and subterranean corridors (Unit 44). In arid areas, however, limestone and dolomite stand up much better, and although they may show some evidence of carbonation at the surface, they appear in general to be much more resistant strata.

Chemical weathering is the more effective agent of rock destruction in humid areas because moisture promotes chemical processes. Physical or mechanical weathering prevails to a greater extent in dry and cold zones. But water plays an important role in the dry as well as the moist environments. The growth of destructive salt crystals in porous rocks of arid areas, for instance, takes place only after some moisture has entered the pores and then evaporated, thereby triggering the crystals' growth. As we will discover, the role of water in other rock-destroying processes is also paramount.

BIOLOGICAL WEATHERING

We should remind ourselves at this point of the role of weathering in the formation of soils (*see* Unit 23). It is through the breakdown of rocks and the accumulation of a layer of minerals that plants can grow—plants whose roots and other parts, in turn, contribute to the weathering processes. But it is likely that the role of plant roots in forcing open bedding planes and joints is somewhat overestimated. The roots follow paths of least resistance and adapt to every small irregularity in the rock (*see* unit opening photo). Roots certainly keep cracks open once they have been formed. More importantly, however, areas of roots tend to collect decaying organic material that is involved in chemical weathering processes.

One of the most important aspects of **biological weathering** is the mixing of soil by burrowing animals and worms. Another interesting aspect is the action of lichens, a combination of algae and fungi, that live on bare rock. Lichens draw minerals from the rock by ion exchange (a mechanism discussed on p. 255). The swelling and contraction of lichens as they alternately get wet and dry may also cause small particles of rock to fall off.

Technically, humans are also agents of biological weathering. Human activity contributes to various forms

The Dust Bowl

A dust storm blows up early in the novel, *The Grapes of Wrath*. Author John Steinbeck describes how the wind in the Dust Bowl swept away the soil and the livelihood of an Oklahoma farming community in the 1930s:

The air and the sky darkened and through them the sun shone redly, and there was a raw sting in the land, dug cunningly among the roots of the crops.

The dawn came, but no day, and the wind cried and whimpered over the fallen corn.

Men and women huddled in their houses, and they tied handkerchiefs over their noses when they went out, and wore goggles to protect their eyes.

*When the night came again it was black night, for the stars could not pierce the dust to get down. Now the dust was evenly mixed with the air, an emulsion of dust and air. Houses were shut tight, but the dust came in so thinly that it could not be seen in the air, and it settled like pollen on the chairs and tables and dishes. The people brushed it from their shoulders. Little lines of dust lay at the door sills.**

In the story, as in real life, the farmers could not grow enough crops to survive.

Thus the "Okies" packed up their families and headed for California. How did this tragedy occur? After Congress passed the Homestead Act in 1862, families poured onto the Great Plains following the Civil War, anxious to establish farms of their own. They plowed up the thick protective layer of vegetation to plant crops or put out cattle to graze on the dense grass cover.

The region usually receives sparse rainfall, marginal for agriculture, but for a few years heavier rain fell. The farmers moved quickly to cash in on the unexpected moisture, and they plowed more fields to plant wheat or put more cattle out to graze. They gambled that the extra rain would continue and nourish their farms, but they quickly lost their bet.

In the mid-1930s, the weather turned exceptionally dry. The wheat did not survive; the cattle devoured the remaining grass and then went hungry; farmers soon abandoned their plots and moved west. Absentee landlords ignored their land and hoped for better times. The land now lay exposed without any plant cover to hold down the soil, and when the wind began to blow, layers of topsoil were stripped off the earth and whisked away in duststorms (*see* Fig. 22-9).

The natural drought conditions, therefore, were aggravated by the extensive expansion of wheat farming and grazing. By 1938, the Dust Bowl conditions finally began to subside and heavier rainfall returned. The government encouraged the remaining farmers to use conservation techniques to protect the soil. Some land was retired from cultivation and planted with cover crops. Shelter belts of trees were planted to inhibit the wind. Fewer cattle pressured the grassland.

The Dust Bowl is part of the past—or is it? Some farmers have cut down their shelter belts to plant a few more rows of wheat. Ranchers have added a few more head of cattle to their herd to increase their meager profits. Speculative farming on the Great Plains, overgrazing, and a few dry years could bring us to the brink of a dust bowl again (as almost happened in the mid-1950s). Only great care by farmers can prevent the return of these conditions

*Excerpts from John Steinbeck, *The Grapes of Wrath* (New York: Viking Press, 1939), pp. 2–3.

---

of weathering in a number of ways: (1) by polluting the air with various substances, we greatly accelerate some chemical weathering, especially in and around large urban centers (*see* Unit 21); (2) by quarrying and mining, we accelerate mechanical, chemical, and biological weathering through exposure of deep strata to these processes (Fig. 38-7); and (3) by farming and fertilizing, we influence soil-formation processes, sometimes destructively.

GEOGRAPHY OF WEATHERING

As we have noted, particular weathering processes are more prevalent and effective in certain areas than in others. In very general terms, soils are much thicker in equa-

FIGURE 38-7 The Bingham Canyon open-pit copper mine just southwest of Salt Lake City, Utah is reputed to be the largest hole on the earth's surface.

torial and tropical areas than in the polar and subpolar latitudes, a contrast that reflects the comparative intensities of weathering in those locales. The heat, high humidity, and often copious rainfall of low-latitude zones are conducive to particularly active weathering. But certain processes occur only under specific conditions. For example, the wedging effect of frost action occurs only in higher latitudes (the **D** climates with their strong seasonal and diurnal temperature contrasts are a good indicator) and at high, frost-affected altitudes. Again, stone nets of various kinds are formed only where a special set of conditions prevails: this happens in Arctic latitudes, yet not all Arctic areas have stone nets.

It would be impractical to devise a small-scale global map of weathering incidence because the various processes are not confined to specific regions, and different processes are at work in the same areas. But the maps of world temperature (Fig. 8-11), world precipitation (Fig. 12-11), world climate (Fig. 16-3), and world soils (Fig. 25-14) provide some indication of what we should expect to be happening in particular places. Where moisture (as indicated by the rainfall map) and temperature are high, all forms of weathering are intense. Where tem-

perature ranges are high and moisture is low (as in **BW** and **BS** climate regions), mechanical weathering takes the more prominent role, and chemical weathering and biological weathering are less effective. In the higher latitudes, the dominant form of mechanical weathering is frost action.

The regional geologic map also must be consulted. In regions where precipitation is low, limestone and dolomite are quite resistant to weathering. But where temperatures are moderate to high and where moisture is ample, chemical weathering of limestone and dolomite can be so effective that the whole landscape is transformed by it; this special case of chemical weathering forms the basis of Unit 44. In our study of the mass removal of loose earth material, we focus on the great rivers that sweep vast amounts of rock fragments and particles downstream (Unit 41). Let us not forget that much of that material was first loosened by the quiet, relentless processes of weathering. The intermediate force that acts on weathered materials prior to their removal is gravity, and its important influence on slope stability is explored in Unit 39.

KEY TERMS

biological weathering (p. 398)
carbonation (p. 398)
chemical weathering (p. 397)
frost action (p. 396)

hydrolysis (p. 398)
mechanical weathering (p. 396)
oxidation (p. 398)

rock sea (p. 396)
spheroidal weathering (p. 398)
talus cone (p. 397)

REVIEW QUESTIONS

1. Why is frost action such an aggressive mechanical weathering agent?

2. How does spheroidal weathering occur?

3. Under what environmental conditions could chemical weathering be most aggressive?

4. What is meant by the term *Dust Bowl*?

5. What is biological weathering?

REFERENCES AND FURTHER READINGS

BIRKELAND, P. W. *Pedology, Weathering, and Geomorphological Research* (London/New York: Oxford University Press, 2nd ed., 1984).

CARROLL, D. *Rock Weathering* (New York: Plenum, 1970).

KELLER, W. D. *The Principles of Chemical Weathering* (Columbia, Mo.: Lucas Brothers, 2nd ed., 1962).

OLLIER, C. D. *Weathering* (London/New York: Longman, 2nd ed., 1984).

REICHE, P. *A Survey of Weathering Processes and Products* (Albuquerque: University of New Mexico Publications in Geology No. 3, 1962).

RITTER, D. F. *Process Geomorphology* (Dubuque, Iowa: Wm. C. Brown, 2nd ed., 1986).

SELBY, M. J. *Hillslope Materials and Processes* (London/New York: Oxford University Press, 1982).

STATHAM, I. *Earth Surface Sediment Transport* (London/New York: Oxford University Press, 1977).

WHALLEY, W. B., et al. *Weathering* (Cambridge, Mass.: Blackwell, 1995).

YATSU, E. *The Nature of Weathering* (Tokyo: Sozosha, 1988).

YOUNG, A. *Slopes* (London/New York: Longman, 2nd ed., 1975).

Mass Movements

OBJECTIVES

♦ To demonstrate the role of gravity in promoting mass movements in weathered materials.

♦ To discuss the various types of mass movements and the circumstances under which they usually occur.

In 1985, when more than 20,000 people were buried by mudflows from Colombia's active volcano Nevado del Ruiz (*see* Fig. 34-5), the world was again reminded of the hazards posed by unstable slopes. Instantly-melted snow from this erupting volcano's peak saturated the loose material on its upper slopes, and suddenly a huge mass of lubricated material thundered downward, sweeping away all in its path. Whole towns and villages and herds of livestock disappeared beneath the deep, muddy mixture. The disaster triggered by Nevado del Ruiz was, admittedly, a rather spectacular example of the hazard posed by mass-movement processes. In all mountain zones, rock material moves downslope in quantities ranging from individual boulders to entire hillsides. It happens at different rates of speed; some rocks actually fall downslope, bounding along, and occasionally even landing on highways. Other material moves slowly, almost imperceptibly.

There are times when the signs of a coming collapse are unmistakable, and people at risk can be warned. At other times it occurs in an instant. In 1970, a minor earthquake loosened a small mass of rock material high atop Mount Huascarán in the Peruvian Andes, and this material fell about 650 m (2000 ft) down the uppermost slopes. It normally would have done little damage, but at an elevation of about 6000 m (20,000 ft) it happened to land on a steep slope of loose material on the mountainside. Both the falling mass and this loose material then combined and roared downslope. According to later calculations, the combined debris took only two minutes to move 14.5 km (9 mi), in the process reaching a speed of 435 kph (270 mph). By the time it stopped, it had traveled 65 km (40 mi). The loss of life is estimated to have been between 25,000 and 45,000 persons. No trace was found of entire communities, particularly

UNIT OPENING PHOTO: Rockslides are a constant hazard in high-relief terrain. This blocked highway is located in the rugged Hindu Kush Mountains of northern Pakistan.

FIGURE 39-1 The 1970 debris avalanche that destroyed the Peruvian town of Yungay (left of center) and its neighbors.

Yungay (Fig. 39-1), which alone accounted for the loss of 17,000 lives.

The force of gravity plays a major role in the modification of landscape. Gravity pulls downward on all materials. Solid bedrock can withstand this force because it is strong and tightly bonded. But if weathering loosens a fragment of bedrock, that particle's first motion is likely to be caused by gravity. Slopes consist of many different kinds of material. Some are made of solid bedrock; others consist of bedrock as well as loose rock fragments; still others are sustained by soil and regolith. A key factor is how well these materials are held together. Another factor is the steepness of the existing slope.

You can experiment with this yourself. Take a flat surface, the size of a cafeteria tray, and cover it with a layer of sand about 2.5 cm (1 in) thick. Now tilt the tray slightly, say to an angle of 5 percent. The material is unlikely to move. Tilt it slowly to a steeper angle, and at a given moment movement will begin. At first you may be able to observe some small particles moving individually, but soon the whole mass will move downslope because

you have exceeded the *angle of repose*, the maximum angle at which granular material remains at rest. All slopes have an angle of repose. For solid bedrock, that angle may approach 100 percent (90 degrees); for loose sand, it is less than 10 percent. For a slope consisting of regolith (loose rock material) and soil it may, depending on its composition, be 15 to 30 percent.

In this experiment you "oversteepened" the slope by tilting the tray to a higher angle. Slopes become oversteepened in nature too. When a stream cuts its valley, the valley sides will become steeper and their angles of repose are exceeded. Then loose material from the valley sides falls into the stream and is carried away. In general, it may be said that the steeper the slope, the stronger the downslope pull or *shearing stress*; counteracting this shearing stress is the *friction* of the loose material on the slope (Fig. 39-2).

This is easy to imagine: when there is little friction between particles, they will not lie on a high-angle slope. For example, had you used ball bearings or marbles in your experiment, they would have rolled off the tray at the slightest angle of slope. Another factor in the downslope movement of material is water or, more technically, fluid. Material that is stable below its critical angle of repose may become unstable when saturated with water. The water has the effect of adding significant weight and reducing friction. Heavy rain on a slope known to be near the angle of repose, therefore, can spell trouble—if not disaster—in a populated area (Fig. 39-3).

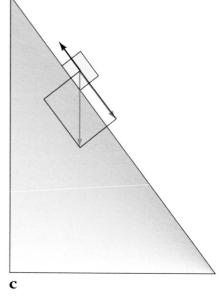

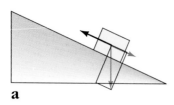

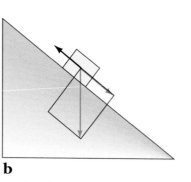

a   b   c

FIGURE 39-2 Slope angle and slide potential. In this drawing the length of the green arrow represents the amount of shearing stress on a block of rock placed against a slope. On the low-angle slope (a), the shearing stress is low, and is not enough to dislodge the rock. On the medium-angle slope (b), the shearing stress is greater, and the downslope pull is nearly enough to move the rock. Note how long the green arrow has become on slope (c): it exceeds the friction that held the rock against the other slopes, and the rock will tumble down. Thus the greater the slope angle, the stronger the shearing stress or downward pull on particles large and small. (In each diagram the black arrow indicates friction and the red arrow gravity.)

FIGURE 39-3 A small stream near Soledad, California has over-steepened a slope and caused a sudden landslide. Not even the dense Mediterranean-type vegetation (*see* Unit 27) could prevent the collapse of this hillside. (Authors' photo)

FIGURE 39-4 An avalanche of snow and ice thundering down a slope in the Rocky Mountains near Silverton, Colorado.

MASS MOVEMENT

The movement of earth materials by the force of gravity is called **mass movement**. From this term it is clear that the materials involved are moved *en masse*—in bulk. We refer to such movements in our everyday language: a landslide, for instance, is a form of mass movement. So is an avalanche, involving not rock debris but snow and ice (Fig. 39-4). Another form of mass movement is the imperceptible downslope creep of soil.

You will sometimes see the term *mass wasting* employed to describe these various, gravity-induced movements. But our interest focuses in large part on the nature of the *movement* that displaces the materials—whether this movement is fast or slow, whether it occurs in a valley or is unconfined, and how it relates to the surface over which movement takes place. Hence, the term *mass movement* is preferable.

Mass movements, for obvious reasons, are important in the breakdown of rocks. If weathering processes loosen a joint block from the face of a granite escarpment, it is the force of gravity that pulls that block down the slope. Thus a new target for weathering is exposed. If mass movements did not occur or were less effective, the entire degradational system would be slowed down.

It is difficult to find a satisfactory typology of mass movements. One way to distinguish them is by rate of speed. Accordingly, we recognize: (1) *creep* movements, the slowest and least perceptible; (2) *flow* movements, in which water often plays a crucial role; (3) *slide* movements, in which fluids may be less important but speed is great; and (4) *fall* movements.

Creep Movements

Mass movements produce characteristic marks and scars on the landscape. Sometimes this evidence is so slight that we have to assume our detective role to prove that mass movement is indeed taking place. The slowest mass movement—**creep**—involves the slow, imperceptible motion of the soil layer downslope. We cannot observe this movement as it actually happens, but it is quite common. Soil creep reveals itself in trees, fence posts, and even gravestones that tilt slightly downslope, as shown in Fig. 39-5. The upper layer of the soil moves faster than the layers below, and vertical objects are rotated in the process.

Soil creep (and *rock creep*, where upturned weathered rock strata are similarly affected) appears to result from the alternate freezing and thawing of soil particles or from alternate periods of wetting and drying. The particles are slightly raised or expanded during freezing and wetting, and then settle slightly lower along the slope when they are warmed or dried. The total change during a single freeze-thaw or wetting-drying sequence is minuscule, but over periods of years the results are sub-

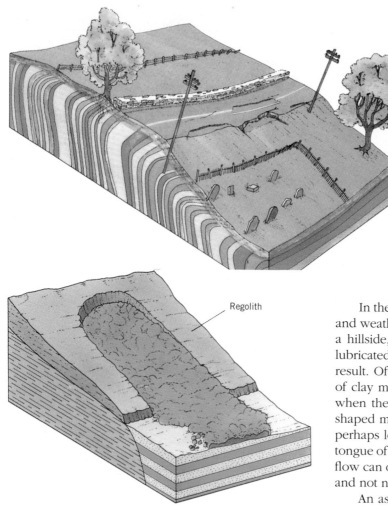

FIGURE 39-5 Effects of soil creep in the cultural and physical landscape: objects tilting downslope.

Regolith

FIGURE 39-6 Earth flow. This lobe-shaped mass leaves a small scarp at its upper end and pushes a tongue of debris onto the valley floor in the foreground.

stantial. Creep is faster on less vegetated surfaces and steep slopes.

A special kind of soil creep is called **solifluction**, a term from Latin and French that literally means "soil flow." In solifluction, the soil and rock debris are saturated with water and flow as one mass. The process is especially common in subpolar regions, where the ground below the thaw zone is impermeable because of permafrost (*see* photo p. 491). The results of solifluction on the landscape look as if a giant painter had painted it with a brush that was far too wet.

Flow Movements

The impact of soil creep on the total landscape is obviously slight over a short period of time. The form of a hill changes hardly at all, even though soil creep may be occurring on it. Perhaps the valley between one hill and the next is slightly modified by some infilling, with soil creeping downslope from both sides. Although far less widespread, flow movements have a much stronger impact on the landscape. Once again, the critical ingredient—besides gravity, of course—is water.

In the case of **earth flows**, a section of soil or weak and weathered bedrock, lying at a rather steep angle on a hillside, becomes saturated by heavy rains until it is lubricated enough to flow. Figure 39-6 diagrams the result. Often this lubrication is helped by the presence of clay minerals, which are slippery and promote flow when they become wet. The earth flow forms a lobe-shaped mass that moves a limited distance downslope, perhaps leaving a small scarp at the top and pushing a tongue of debris into the valley below. Although an earth flow can occur quite suddenly, it is normally rather slow and not nearly so dramatic as, say, a landslide.

An associated feature of hillside flow is **slumping**. In 1950, a spectacular example of this motion blocked a main highway just outside Oakland, California (Fig. 39-7). Again, saturation and lubrication caused these

FIGURE 39-7 This dramatic photograph of slumping, which halted traffic on a major road in the San Francisco Bay Area, appeared in newspapers all over the United States in 1950.

404

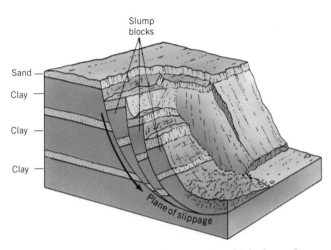

FIGURE 39-8 An example of slumping in which three slump blocks have moved downslope. Note the backward rotation of each slump block, the scarp at the head of the slump cavity, and the forward flow at the toe of the mass.

rock masses to weaken and slump. However, unlike earth flows, slumping involves major sections of regolith, soil, or weakened bedrock that stand in near-vertical scarps and move as block-like units. As Fig. 39-8 shows, slumping includes a downslope motion and a backward rotation of the slump block. Slumping is normally a rather slow process, faster than creep but much less rapid than slides.

Another form of flow is the **mudflow**—a stream of fluid, lubricated mud. The mudflow originates when heavy rainfall comes to an area that has long been dry and where weathering has loosened a large quantity of relatively fine material. Very little of the rainwater can percolate downward, and therefore virtually all of it runs down the valleys, carrying the loose particles downslope. But so much material is brought with it that, instead of a stream, the channel soon contains a porridge-like mass of mud.

When the water supply is ample, the mudflow is rather thin and moves rapidly. At other times there is an overload of debris (the mudflow soon accumulates rocks and boulders as well), and then it becomes stiff and moves rather slowly, eventually coming to a stop. But before this happens the mudflows from several valleys may join in a larger valley, forming an enormous volume of material that is pushed along like advancing molten lava. As we have noted, this sort of mudflow causes great destruction when it reaches populated areas (see Fig. 34-5). Mudflows can move for several kilometers, emerging from mountainous areas and extending far into adjacent flatlands.

Slide Movements

Whenever rock and soil lie at a high angle on a steep slope there is a possibility that a segment of such material will break away in a **landslide** or **rockslide**. The differ-

ence between landslides and the previously discussed mass movements is the speed with which the slide occurs. Landslides move much faster than flows and sometimes roar downslope with a thunderous sound and great force. Here water plays a lesser role, although it does help in the weathering and loosening that has gone before. A landslide is in effect a collapse, and no lubricants are necessary, although the weight of water saturation certainly does contribute. Many of Southern California's feared landslides are triggered by heavy rains that overload slopes that are already near collapse (see Fig. 39-3).

Landslides are produced in other ways as well. Earthquakes and earth tremors sometimes provide the vibrations necessary to set the downward slide in motion. A river cutting into the side of its valley can oversteepen the slope and initiate a collapse. People, too, play a role in starting landslides by cutting into mountainsides to construct roads and other works (see unit opening photo)—thereby weakening and eventually destroying the support needed to keep the slope intact (see Perspective: The Human Factor). In a study conducted in 1976, the Japanese Ministry of Construction located 35,000 sites that were potentially dangerous because of land development. Torrential summer rains that year on Japan's Izu Peninsula, near Tokyo, had caused almost 500 landslides, killing nine people in a popular holiday resort and leaving nearly 10,000 homeless.

The destructiveness of landslides makes them perhaps the best known of all mass movements. Tens of millions of tons of rock debris may break loose at once and in minutes bury entire towns. The 1970 Huascarán landslide in Peru (see Fig. 39-1) involved as much as 50 million m³ (1.7 billion ft³) of debris, snow, and ice.

Landslides and rockslides differ from creep and flow movements not only in terms of the speed of movement but also, frequently, in the nature of the material carried down. In a landslide, everything goes down—bedrock as well as overburden. Weathering loosens pieces of bedrock by attacking along joint planes and fracture zones. Any steep slope carries within it the parallel planes of weakness that may eventually fail and yield (Fig. 39-9). As soon as the landslide has occurred, exposing a fresh scarp or slope, the cycle begins again. New joints and fractures are attacked, and gradually the slide threat returns.

Fall Movements

The last of the mass movements to be discussed is the fastest of all: free **fall**, or the downslope rolling of pieces of rock that have become loosened by weathering. Such rock fragments and boulders usually do not go far, ending at the base of the slope or cliff from which they broke away (Fig. 39-11). They form a *talus cone*, an accumulation of rock fragments, large and small, at the foot of the slope, as we saw in Figs. 37-1 and 38-4. Such an

The Human Factor

Mass movements most frequently are caused by natural forces and conditions, but human activities also contribute. Sometimes people fail to heed nature's warnings, as in the case of the disastrous Langarone landslide in northern Italy in 1963. The materials on the slopes of the Langarone Valley were known to have low shear strengths. Nevertheless, a dam was constructed across the valley, and a large artificial lake (Lake Vaiont) was impounded behind it. The water rose and lubricated the inundated valley sides, adding to the risk of landslides. When heavy rains pounded the upper slopes in the summer and early autumn of 1963, major landslides seemed to be inevitable.

In early October, the severity of the hazard was finally realized, and it was decided to drain the dam. But it was too late, and on October 9 a mass of rock debris with a surface area of 3 km² (1.2 mi²) slid into Lake Vaiont, causing a huge splash wave in the reservoir. The wave spilled over the top of the draining dam and swept without warning down the lower Langarone Valley, killing more than 2600 persons.

Yet, despite this tragic lesson, the building of houses and even larger structures on slide-prone slopes continues in many countries. This includes the United States, which contains numerous areas that are susceptible to landslide risk (Fig. 39-10).

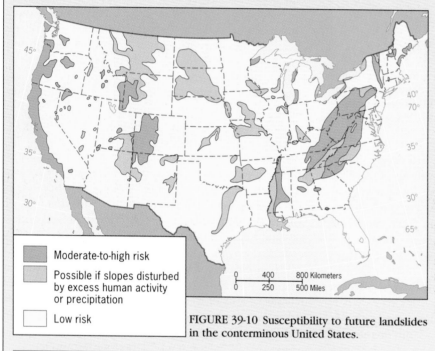

Moderate-to-high risk

Possible if slopes disturbed by excess human activity or precipitation

Low risk

FIGURE 39-10 Susceptibility to future landslides in the conterminous United States.

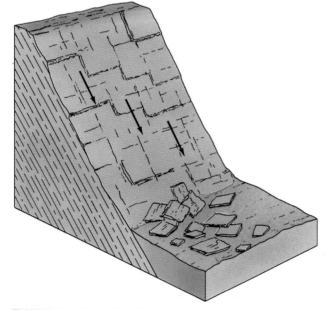

FIGURE 39-9 Rockslide. Note the steep slope and the parallel planes of weakness in the rockface.

accumulation, also known as a *scree slope*, lies at a surprisingly high angle—as high as 35 degrees for large fragments. When an especially large boulder falls onto the slope, the talus may slide some before attaining a new adjustment.

THE IMPORTANCE OF MASS MOVEMENTS

It is easy to underestimate the importance of mass movements among the processes of erosion. Mass movement often occurs almost imperceptibly. But it is taking place along the banks of every stream, and a large portion of the materials swept out to sea by the world's rivers is brought to these water channels through mass movements. Figure 39-12 compares the volumes of rock eroded directly by a stream and the volume first moved by mass movement. Even slow, sluggish, meandering

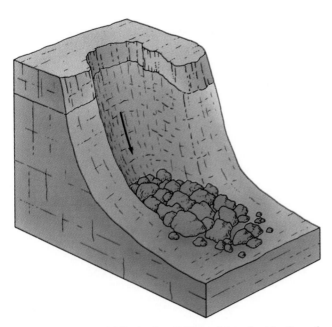

FIGURE 39-11 Rockfall: the free falling of detached bodies of bedrock from a cliff or steep slope. The loosened boulders usually come to rest at the base of the slope from which they fell.

streams undercut their banks, causing the collapse of materials into their waters. Rapidly eroding streams in highland areas and streams in deep canyon-like valleys cause

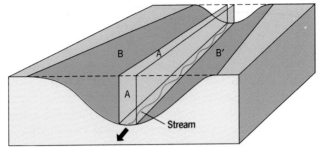

FIGURE 39-12 A comparison of the amount of rock eroded directly by a stream (A) and that first moved to the stream by various kinds of mass movement (B, B').

much oversteepening and constant collapse along their valley walls.

We have dealt with four major categories of mass movement: creep, flow, slide, and fall. Within these categories there is sometimes more than one type of movement. These movements can be summarized in terms of the rate of downslope movement and the amount of water involved. Figure 39-13 shows how the different types of mass movements relate to one another with respect to these two parameters.

Why should we spend so much time examining the sliding and flowing earth? A knowledge of mass move-

FIGURE 39-13 Common types of mass movement classified according to rate of movement and water content.

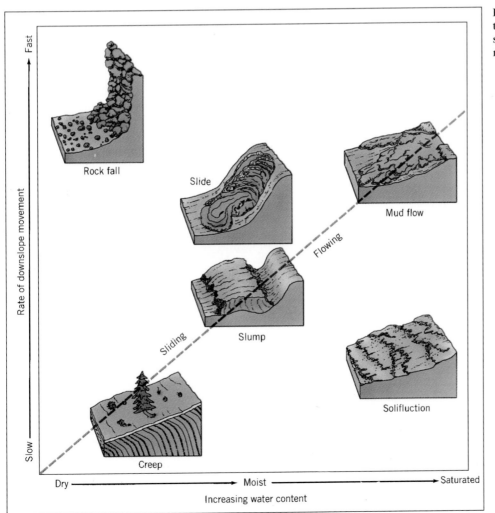

ment can help in planning and policymaking. It is quite costly to use public money to build highways in unstable areas (*see* Fig. 39-7). Moreover, many people build houses in locations chosen for their spectacular scenic views, with little or no regard for the stability of the land. Some of these residents have had rude awakenings, as Fig. 39-14 shows.

Physical geographers are interested in mass movement for other reasons. In sculpting the land surface, mass movements also perform the important job of exposing new bedrock to the forces of weathering. By the time the material sags and collapses in a flow or slide, weathering has become much less effective in attacking deeper rock layers. But the slide removes the weathered material. Thus mass movements are vitally important in the total complex of erosional processes.

FIGURE 39-14 Scenic locations for homebuilding can prove to be temporary if the underlying land is unstable. This house was built too close to the eroding shoreline of Lake Michigan in a suburb north of Chicago. Shortly after the photo was taken, this precariously balanced structure was successfully moved a few dozen meters inland—where lakeside mass movements could jeopardize it again in the future.

KEY TERMS

creep (p. 403)

earth flow (p. 404)

fall (p. 405)

landslide (p. 405)

mass movement (p. 403)

mudflow (p. 405)

rockslide (p. 405)

slumping (p. 404)

solifluction (p. 404)

REVIEW QUESTIONS

1. What is the angle of repose, and why is it significant?
2. What fundamental force acts to restrain mass-movement processes?
3. How does creep differ from a flow type of movement?
4. Where (in a tectonic setting) would one expect the highest frequency of large landslides?
5. What is solifluction? In what environment is it most common?

REFERENCES AND FURTHER READINGS

ALLISON, R. *Rock Slopes* (Cambridge, Mass.: Blackwell, 1995).

BRABB, E. E., and HARROD, B. L., eds. *Landslides: Extent and Economic Significance* (Rotterdam: Balkema, 1989).

CARSON, M., and KIRBY, M. *Hillslope Form and Process* (London/New York: Cambridge University Press, 1972).

CROZIER, M. J. *Landslides: Causes, Consequences and Environment* (London/New York: Methuen, 1986).

HAYS, W. W., ed. *Facing Geologic and Hydrologic Hazards* (Washington, D.C.: U.S. Geological Survey, Professional Paper 1240-B, 1981).

SCHUMM, S. A., and MOSLEY, M. P., eds. *Slope Morphology* (Stroudsburg, Pa.: Dowden, Hutchinson and Ross, 1973).

SCHUSTER, R. L., and KRIZEK, R. J., eds. *Landslides: Analysis and Control* (Washington, D.C.: National Academy of Sciences, 1978).

SELBY, M. J. *Hillslope Materials and Processes* (London/New York: Oxford University Press, 1982).

SHARPE, C.F.S. *Landslides and Related Phenomena: A Study of Mass Movements of Soil and Rock* (Paterson, N.J.: Pageant Books, 2nd ed., 1960).

VOIGHT, B., ed. *Rockslides and Avalanches* (Amsterdam: Elsevier, 2 vols., 1978).

YOUNG, A. *Slopes* (London/New York: Longman, 2nd ed., 1975).

ZARUBA, Q., and MENCL, V. *Landslides and Their Control* (Amsterdam: Elsevier, 1969).

Water in the Lithosphere

OBJECTIVES

◆ To discuss the various paths water may take on and within the surface of the lithosphere.

◆ To introduce fundamental aspects of river flow.

◆ To outline basic concepts related to groundwater hydrology.

Water is the essence of life on earth. Humanity's earliest civilizations arose in the valleys of great rivers. Our ancestors learned to control the seasonal floods of these streams, and irrigation made planned farming possible. Today we refer to those cultures as *hydraulic civilizations* in recognition of their ability to control and exploit water. Our current dependence on water is no less fundamental. Society's technological progress notwithstanding, water remains the earth's most critical resource. We could conceivably do without oil, coal, or iron, but we cannot survive without

water. Thus the historical geography of human settlement on this planet is in no small part the history of the search for, and use of, water.

The operations of the hydrologic cycle are introduced in Fig. 12-2 (p. 128) in the unit that treats the part of the hydrosphere found in the atmosphere (water vapor). In this unit, the focus is on the part of the hydro-

UNIT OPENING PHOTO: A high waterfall is one of the most spectacular features of water flow on the lithosphere. Angel Falls, Venezuela, the world's highest.

sphere contained in the lithosphere. In Units 41 to 43 we study rivers and streams, the primary agents of landscape formation. But before water collects in streams it must travel over or through the surficial layer. Arriving as precipitation (mostly as rainfall), some of it evaporates. Some of it falls on plants; some of it moistens the upper soil layers; some of it seeps deeper down and collects in underground reservoirs. And part of it runs downslope and collects, first in small channels, then in larger creeks, to become part of the volume of permanent streams.

WATER AT THE SURFACE

Our inquiry starts with a raindrop that arrives at the earth's surface. What happens to it depends on the nature and state of the surface. In some cases, raindrops never actually reach the ground: they fall on vegetation and evaporate before they can penetrate the soil. Figure 40-1 shows where this **interception** occurs.

The amount of water intercepted by vegetation depends on the structure of the plants involved. In Australia, for example, eucalyptus trees intercept only 2 to 3 percent of the rain. Hemlock and Douglas fir forests in California possess a different structure and intercept as much as 40 percent of the rain. If the rainwater reaches the ground, it can be absorbed by the surface. Such surfaces as concrete roads or granite outcrops that do not permit water to pass through them are said to be *impermeable*. Most natural surfaces absorb a portion of the water that falls on them and are considered *permeable*, although the degree of permeability depends on many factors.

The flow of water into the earth's surface through the pores (spaces between particles) and openings in the soil mass is called **infiltration**. The infiltration rate depends on several factors: (1) the physical characteristics of the soil, (2) how much moisture is already in the soil, (3) the type and extent of the vegetation cover, (4) the slope of the surface, and (5) the nature of the rainfall.

FIGURE 40-1 The interception of raindrops by springtime grass.

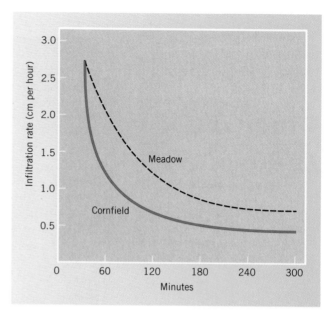

FIGURE 40-2 Infiltration rates for two different vegetated areas. Both show higher rates at the beginning than at the end of the rainfall, but the meadow has higher rates throughout.

The most evident characteristics of soil that affect infiltration are its structure and the closeness of the soil particles. For the most part, water can infiltrate more rapidly into coarse, sandy soils than into clay soils. In a much-studied valley in Switzerland, forest soil absorbs 100 mm (4 in) of water in two minutes, but a pasture where cattle graze requires three hours to take in the same amount of water.

A soil that is already wet will allow less infiltration than a dry soil because the soil surface becomes compacted by the rain. The soil also may swell, closing small openings. Small particles wash into the surface openings, decreasing the porosity, and the existing pores become filled with water and cannot accept any more. Infiltration rates are usually highest in vegetated areas because the vegetation prevents raindrops from compacting the soil, and the roots act to increase the porosity of the surface layer. Moreover, organic litter provides a home for burrowing animals, whose activities further loosen the soil.

The type of vegetation also affects infiltration. The infiltration rate of a bluegrass meadow is decidedly greater than that of a tilled cornfield. Figure 40-2 shows the typical shape of an infiltration curve for two kinds of vegetation. Infiltration rates for both kinds are high at the start of a rainfall event and then decrease over time.

Steep slopes may encourage runoff before water can be absorbed, so the infiltration rate is likely to be somewhat higher on gentler slopes and flat surfaces. The characteristics of local rainfall are also critical to the infiltration rate. Rainfall can vary in intensity, duration, and amount. The *intensity* is the amount of water that falls in a given time. A rainfall intensity of 5 mm (0.2 in) per hour is quite heavy; if it kept up for a *duration* of 10 hours, it would produce a total *amount* of 50 mm (2 in) of rain.

If the rainstorm has a great enough intensity and duration, it will exceed the infiltration rate. Water begins to accumulate into small puddles and pools. It collects in any hollow on a rough ground surface, detained behind millions of little natural dams. This situation is called *surface detention*. When there is more rainwater than the small detention hollows can hold, the water flows over the land as surface **runoff** (or *overland flow*). After a short period of detention, any rainfall that does not infiltrate the soil runs off it. Thus, as precipitation continues and infiltration rates decrease, runoff rates increase.

Once surface runoff has started, the water continues to flow until it reaches a stream or river or an area of permeable soil or rock. An area of land on a slope receives all the water that runs off the higher elevations above it. The longer the total flow path, assuming that the runoff flows across uniform material, the greater the amount of runoff across the area. Therefore, the largest runoff rates occur at the base of a slope, just before the runoff enters a river. This emphasizes the need for sound land management practices. If water is needed for raising crops, then the less that runs off the better. Because infiltration and runoff rates are inversely related, runoff depends on all the factors that affect infiltration. By employing appropriate farming techniques, high infiltration rates can be maintained and runoff rates reduced (*see* Fig. 23-1).

FIGURE 40-3 The downward rush of water in a high-gradient mountain stream. Roaring Fork, Great Smoky Mountains National Park, Tennessee.

WATER FLOW IN RIVERS

For some, looking down at a flowing river holds the same kind of fascination as looking into the flames of a fire. The physical geographer's interest in stream flow is less romantic but just as fascinating. A study of river flow helps us to make predictions about such matters as pollution and floods. From the smallest flow in a tiny rivulet to the largest flow in a major river, certain general rules apply.

River Channels

One fundamental property of a river channel is its **gradient**, or **slope**, the difference in elevation between two points along the stream course. Gradients can be measured for the entire course of a river or over a short stretch of it. When a gradient is high, the flow of water is very turbulent, as in a mountain stream that has rapids and falls in its valley (Fig. 40-3). By contrast, the lower portions of such rivers as the Mississippi and the Amazon have very low gradients. The Amazon River, for instance, falls only about 6 m (20 ft) over its final 800 km (500 mi).

Not all the water in a river channel moves at the same rate of speed, or **velocity**. A stream tends to move fastest in the center of its channel, just below the surface of the water, and slowest along the bottom and sides of the

valley where the resistance to flow is greatest (Fig. 40-4). The **discharge** of a river is the volume of water passing a given cross-section of its channel within a given amount of time, and is measured as average water ve-

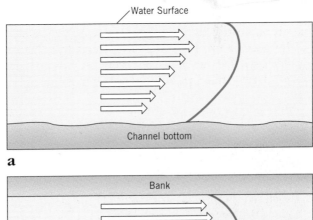

FIGURE 40-4 Velocity variation in a river, viewed from the side (a) and from above (b).

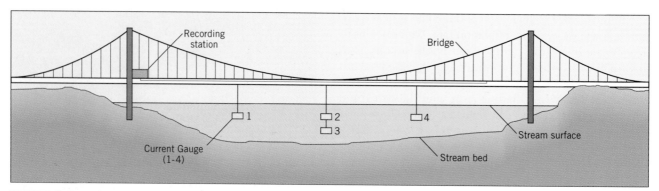

FIGURE 40-5 A stream-gauging installation.

locity multiplied by the cross-sectional area. Long-term records of channel discharge are very important in planning dam construction and flood control strategies. Discharge records also tell us how much material the river can carry downstream in suspension (that is, kept in the flow by its turbulence and movement and not deposited). This reflects the river's effectiveness as an erosional agent.

River Flow

Measurements of river flows and sediment loads produce some surprises. For example, some sparkling mountain "torrents" have lower velocities than the placid lower segments of the Mississippi River! This happens because mountain streams often flow in circular *eddies*, whose backward movement is almost as great as their forward movement. Only careful measurement can prove the eye wrong. Hydrologists have devised various instruments to obtain some measure of the properties of rivers. Some of these instruments are designed to be hung from bridges or other structures, and once installed they begin to record the behavior of the stream (Fig. 40-5).

The most important aspect to be measured, of course, is the river's discharge, from which other properties can be deduced. But discharge varies—by season, by year, and over longer periods. A graph of a river's discharge over time is called a **hydrograph** (Fig. 40-6), and the one shown here records a 10-day period during which the discharge was affected by a storm in Nova Scotia's St. Mary River drainage basin. This is only one event in that river's life; the longer the hydrographic record, the more accurate our knowledge of a river basin's hydrology.

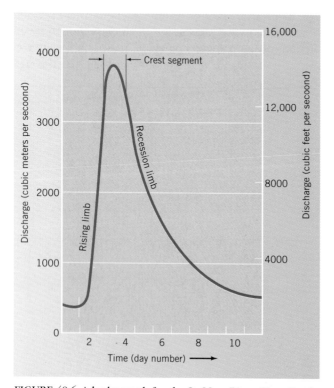

FIGURE 40-6 A hydrograph for the St. Mary River, Nova Scotia, Canada, showing the discharge associated with an individual storm.

WATER BENEATH THE SURFACE

Despite the beauty and importance of rivers, they contain only 0.03 percent of all the fresh water in the world. Twice as much is stored as soil moisture, and 10 times as much is held in lakes. By far the greatest proportion of the world's fresh water (75 percent) is locked in glaciers and icesheets. But about a quarter of the total freshwater supply is available only under certain conditions: this water is hidden beneath the ground, within the lithosphere, and is called **groundwater**. A raindrop that falls to the earth may remain above ground as we have seen, but it may also infiltrate into the soil. Two zones within the ground may hold this water. The upper zone, usually unsaturated except at times of heavy rain, is called the **zone of aeration** or the *vadose zone*; below this is the **zone of saturation**, sometimes called the *phreatic zone* (Fig. 40-7). The raindrop infiltrates into the vadose zone first.

Soil Moisture in the Zone of Aeration

Once rainwater has infiltrated into the soil it is called *soil moisture*, and any further movement is by processes

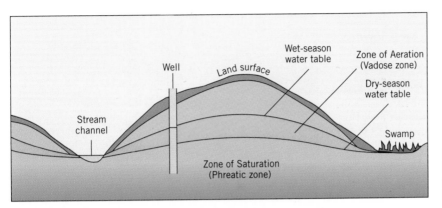

FIGURE 40-7 The two subsurface water-holding zones. The term *groundwater* applies to water lying below the water table—in the zone of saturation.

other than infiltration. The downward movement of water through the pores and spaces in the soil under the influence of gravity, called *percolation*, is the most common method of water movement in a soil. In a contrary motion, described in Unit 23, water may also move upwards, like liquid in a straw, through *capillary action*. Moisture can also move around within the soil through evaporation, movement of water vapor, and recondensation onto new surfaces.

The texture and structure of the soil determine the amount of water it can hold. **Field capacity** is the maximum amount of water that a soil can possibly contain without becoming waterlogged. Theoretically, soil moisture can fall to zero. But it takes a large amount of drying to reach this limit because a thin film of water clings tenaciously to most soil particles. This *hygroscopic* water is unavailable to plant roots. A more practical lower limit to soil moisture is termed the *wilting point*, and it varies for different soils and crops. Below this point a crop dries out, suffering permanent injury. The total *soil storage capacity* for agricultural purposes is the product of the av-

erage depth in centimeters to which roots grow and the water storage per centimeter for that soil type.

Groundwater in the Zone of Saturation

If you dig a hole far enough into the earth, chances are that sooner or later you will come to a layer that is permanently saturated with water. This is the top of the phreatic zone, a surface called the **water table** (Fig. 40-7). Instead of lying horizontally, the water table tends to follow the outline of the land surface, as Fig. 40-8 indicates. Where it intersects the surface, a spring, stream, river, or lake is exposed (Fig 40-8 shows the location of a stream in relation to the water table).

The materials of the lithosphere below the water table can be classified according to their water-holding properties, and these are also exhibited in the diagram. Porous and permeable layers that can be at least partially saturated are called **aquifers**. Sandstone and limestone often are good aquifers. Other rock layers, such as mudstone and shale, consist of tightly packed or interlocking

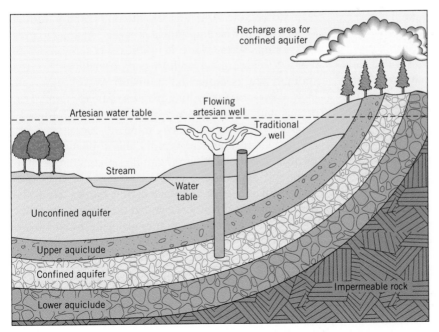

FIGURE 40-8 Aquifers, aquicludes, and their relationship to the water table and wells.

particles and, therefore, are usually quite impermeable and resist groundwater infiltration. These are known as **aquicludes** (sometimes called *aquitards*).

You might imagine aquifers and aquicludes in the earth as layers of sponge-like material set between layers of plastic. An *unconfined aquifer* obtains its water from local infiltration, as if the sponge were at the surface. But a *confined aquifer* exists between aquicludes, and it often obtains its water from a distant area where the rock layer of the aquifer eventually reaches the surface (as shown in the upper right-hand portion of Fig. 40-8).

Wells and Springs

In many parts of the world, settlements are located near wells and springs. In such cases, geology directly influences human locational decision making. Wells can be dug wherever an aquifer lies below the surface, and in most cases springs are formed when an aquifer intersects the surface.

There are two kinds of wells—the traditional and the artesian. The *traditional well* is simply a hole in the ground that penetrates the water table. Water is then drawn or pumped to the surface. Traditional wells usually are sunk below the average level of the water table. The actual level of the water table may vary because of droughts and storms. So the deeper the well is sunk below the water table, the less chance there is of the well's becoming dry when the water table falls.

The French region of Artois is blessed with a confined aquifer whose water supply is recharged from a remote location. Wells sunk into this aquifer produce water that flows under its own natural pressure to the surface (Fig. 40-8). Artois has lent its name to this type of well, and now **artesian wells** are common in many parts of the world. The pressure in artesian wells can sometimes be quite strong. One dug in Belle Plaine, Iowa, spouted water over 30 m (100 ft) into the air and had to be plugged with 15 wagonloads of rock before it could be brought under control.

Wells sometimes suffer from side effects that limit their use. In some cases, the water is withdrawn faster than it can be replaced by water flowing through the aquifer. When this happens, the local water table directly surrounding the well drops, forming a *cone of depression* like that in Fig. 40-9a. The amount of the drop in the local water table is called the *drawdown*. When it drops, energy must be used to bring the water to the surface. In addition, saltwater can intrude into a well near a coastline. Excessive pumping gradually moves denser, saline seawater into the well, as diagrammed in Fig. 40-9b. Wells, therefore, must be used with care.

A surface stream of flowing water that emerges from the ground is called a **spring**. Springs can be formed in a number of ways. Most commonly, an aquiclude stops the downward percolation of water, which is then forced

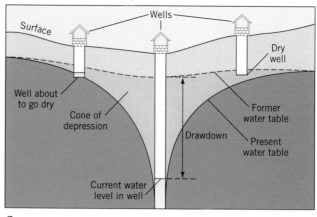

a

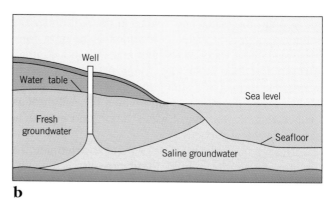

b

FIGURE 40-9 The potential disadvantages of wells. When water is pumped out faster than it can be replaced, the local water table drops in the form of a cone of depression (a). Near a coastline, excessive pumping can lead to the intrusion of seawater into the well (b).

to flow from a hillside, as indicated in Fig. 40-10a. Occasionally, as this diagram also shows, the aquiclude leads to the formation of a separate water table, called a **perched water table**, at a higher elevation than the main water table. Sometimes, water finds its way through joints in otherwise impermeable rocks, such as granite, and springs form where it reaches the surface. Often, faulting rearranges aquifers and aquicludes so that springs form, as shown in Fig. 40-10b. And exceptionally high water tables after long periods of heavy rainfall can occasionally raise the water level high enough to cause temporary springs, as shown in Fig. 40-10c.

An interesting variation is the formation of *hot springs*. The water flowing from these has a temperature averaging above 10°C (50°F), and often comes from springs that overlie portions of the earth's crust that contain magma chambers close to the surface (as in Fig. 40-10d). Hot springs are common in the mountains of Idaho and, of course, in nearby Yellowstone National Park (Fig. 40-11). Near Rotorua, New Zealand, hot springs are used for cooking and laundry, and steam from them is used for generating electricity; similarly, *geothermal* steam powers turbines in the hills north of

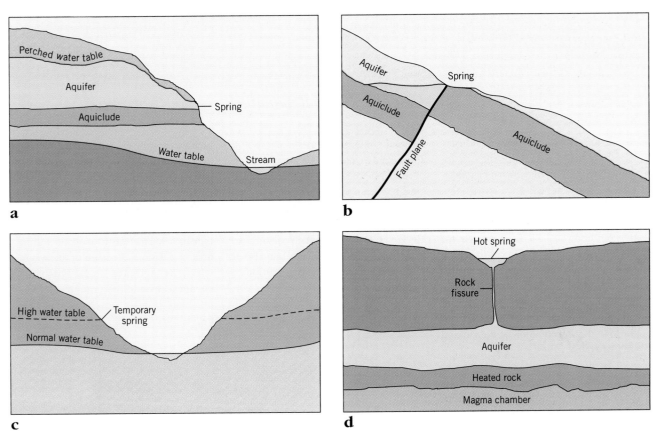

FIGURE 40-10 Conditions leading to the formation of springs.

San Francisco that produce electricity for more than 1.5 million homes. Much of Reykjavik, the capital of Iceland, is heated by water from hot springs—as are parts of the city of Boise, Idaho.

Hot springs containing large quantities of minerals dissolved from the surrounding rocks are sometimes called *mineral springs*, and the mineral water may be used for medicinal purposes. Although it is very pleasant to sit in a warm-water bath of a tapped mineral spring, this is one of the less important uses of water. A far more significant human impact is the growing pollution of subsurface water supplies (*see* Perspective: Groundwater Contamination).

The movement of water on and through the surface of the earth has important consequences for the erosion, transportation, and deposition of surficial materials. In the following units we highlight the various pathways this water takes, the environmental controls determining their relative magnitude, and the resultant geomorphic work that is being done. We begin our survey in Unit 41, which explores the processes of stream erosion.

FIGURE 40-11 Old Faithful geyser in Yellowstone National Park, which erupts approximately once an hour. A *geyser* is a hot spring that periodically expels jets of heated water and steam.

Perspectives on THE HUMAN ENVIRONMENT

Groundwater Contamination

Residents of highly developed, industrialized countries have long taken it for granted that when they turn on a faucet, safe, potable water will flow from the tap. Throughout most of the rest of the world, however, the available water—particularly groundwater—is usually unfit for human consumption. Although natural dissolved substances make some of this water undrinkable, the more common situation is that these groundwater supplies have become contaminated through the introduction of human and industrial wastes at the surface.

The most common source of water pollution in wells and springs is sewage. Drainage from septic tanks, malfunctioning sewers, privies, and barnyards widely contaminates groundwater. If water contaminated with sewage bacteria passes through soil and/or rock with sizeable openings, such as coarse gravel or cavity-pocked limestone, it can travel considerable distances while remaining polluted.

Vast quantities of human garbage and industrial waste products are also deposited in shallow basins at the land surface. When such a landfill site reaches its capacity, it often is covered with earth and revegetated. Many of the waste products, now buried below ground, are activated by rainwater that percolates downward through the site, carrying away soluble substances. In this manner, harmful chemicals slowly seep into aqui-

fers and contaminate them. The pollutants migrate from landfill sites as plumes of contaminated water, steered by the regional groundwater flow regime, and they are dispersed at the same rates as the moving subsurface water (Fig. 40-12). Moreover, these effluents frequently are toxic not only to humans but also to plants and animals in the larger biotic environment.

Yet another hazard is posed by toxic chemicals. Each year, pesticides and her-

bicides are sprayed in massive quantities over countless farm fields to improve crop quality and productivity. As is well known, some of these chemicals have been linked to cancers and birth defects in humans; other toxic substances of this type have led to disastrous declines in animal populations. Because of the way in which they are spread, toxic agricultural chemicals invade the groundwater system beneath huge areas as precipitation flushes them into the soil.

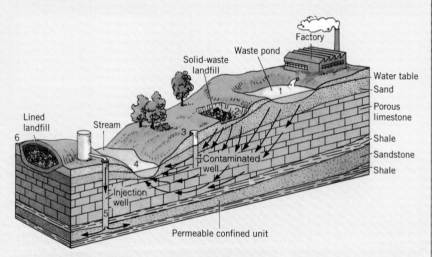

FIGURE 40-12 A groundwater system contaminated by toxic wastes. Toxic chemicals in an open waste pond (1) and an unlined landfill (2) percolate downward and contaminate an underlying aquifer. Also contaminated are a well downslope (3) and a stream (4) at the base of the hill. Safer, alternative approaches to waste management include injection into a deep, confined rock layer (5) that lies well below aquifers used for water supplies, and a carefully constructed surface landfill (6) that is fully lined to prevent downward seepage of wastes. Since neither of the latter approaches is completely reliable, constant monitoring at both sites would be necessary.

KEY TERMS

aquiclude (p. 414)   gradient (slope) (p. 411)   interception (p. 410)   velocity (p. 411)

aquifer (p. 413)   groundwater (p. 412)   perched water table (p. 414)   water table (p. 413)

artesian well (p. 414)   hydrograph (p. 412)   runoff (p. 411)   zone of aeration (p. 412)

discharge (p. 411)   infiltration (p. 410)   spring (p. 414)   zone of saturation (p. 412)

field capacity (p. 413)

REVIEW QUESTIONS

1. What general factors influence infiltration characteristics within a soil mass?

2. What is a hydrograph?

3. How does the water table relate to the zones of aeration and saturation?

4. Describe an aquifer and an aquiclude.

5. What is an artesian well?

REFERENCES AND FURTHER READINGS

BALDWIN, H. L., and MCGUINNESS, C. I. *A Primer on Ground-water* (Washington, D.C.: U.S. Geological Survey, 1963).

FREEZE, R. A., and CHERRY, J. A. *Groundwater* (Englewood Cliffs, N.J.: Prentice-Hall, 1979).

KROMM, D. E., and WHITE, S. E., eds. *Groundwater Exploitation in the High Plains* (Lawrence, Kan.: University Press of Kansas, 1992).

LEOPOLD, L. B. *A View of the River* (Cambridge, Mass.: Harvard University Press, 1994).

LEOPOLD, L. B. *Water: A Primer* (San Francisco: Freeman, 1974).

MATHER, J. R. *Water Resources: Distribution, Use, and Management* (New York: Wiley/Winston, 1984).

MATHEWS, O. P. *Water Resources, Geography and the Law* (Washington, D.C.: Association of American Geographers, Resource Publications in Geography, 1984).

MAURITS LA RIVIÈRE, J. W. "Threats to the World's Water," *Scientific American,* September 1989, 80–94.

MILLER, D. H. *Water at the Surface of the Earth: An Introduction to Ecosystem Hydrodynamics* (New York: Academic Press, 1977).

TODD, D. K. *Ground Water Hydrology* (New York: Wiley, 2nd ed., 1980).

TODD, D. K., ed. *The Water Encyclopedia: A Compendium of Useful Information on Water Resources* (Port Washington, N.Y.: Water Information Center, 1970).

U.S. Department of Agriculture. *Water: The 1955 Yearbook of Agriculture* (Washington, D.C.: U.S. Government Printing Office, 1955).

Slopes and Streams

OBJECTIVES

◆ To discuss the processes involved in the erosion of hillslopes.

◆ To outline the factors influencing the erosional activity of rivers and to discuss the mechanisms of stream erosion and sediment transport.

◆ To characterize the river as a system and to describe the tendencies of this system.

S treams are the most important sculptors of terrestrial landscapes. Elsewhere in Part Five we study the landscapes carved by glaciers, molded by the wind, and shaped by waves. However, none of these erosional agents comes close to flowing water as the principal creator of landforms and landscapes on our planet. Even where glaciers once prevailed and where deserts exist today, water plays a major role in modifying the surface.

The ancient Latin word for river was *fluvius*, from which is derived the term *fluvial* to denote running water. Thus fluvial processes are the geomorphic processes associated with running water, and fluvial landforms and landscapes are produced by streams. We note in Unit 37 that rivers degrade (erode) and aggrade (deposit). Hence, the landscape contains *degradational* or *erosional* landforms, created when rock is removed, and *aggradational* or *depositional* landforms, resulting from the accumulation of sediment. The Grand Canyon of Arizona is essentially an erosional landscape; the delta of the Mississippi River is an assemblage of depositional landforms. After we have studied fluvial processes, we will examine the landscapes they create.

UNIT OPENING PHOTO: One of the earth's most active zones of streamflow across steeply sloped terrain: the southern flank of the Himalayas in central Nepal. These southward-flowing rivers all empty into northern India's mighty Ganges.

EROSION AND THE HYDROLOGIC CYCLE

In Unit 12 we examine the hydrologic cycle, the global system that carries water from sea to land and back again (*see* Fig. 12-2 on p. 128). This unceasing circulation of water ensures the continuation of fluvial erosion because the water that falls on the elevated landmasses will always flow back toward sea level. As it does so, it carries the products of weathering with it.

Rainfall comes in many different forms, from the steady, gentle rain of a cloudy autumn day to the violent heavy downpour associated with a midsummer afternoon thunderstorm. During a misty drizzle, the soil generally is able to absorb all or most of the water because its **infiltration capacity** (the rate at which it is able to absorb water from the surface) is not exceeded. This means that no water collects at the surface. But when rain falls at higher intensities, it may quickly saturate the soil and exceed the infiltration capacity, resulting in *run-off*. Many small, temporary streamlets form, and with such runoff comes erosion.

Large, heavy raindrops dislodge soil particles in a process called *splash erosion* (Fig. 41-1). Once loosened, these grains are quickly carried away in the streamlets that form during an intense rainstorm. If the surface is flat, much of the loosened soil may be deposited nearby, and the area suffers little net loss of soil. But if the exposed soil lies on a slope, splash erosion results in a downslope transfer of soil. The steeper the slope, the faster this degradation proceeds. If the rate of erosion, over the long term, exceeds the rate of soil formation, the slope will lose its soil cover and suffer **denudation**. The term *denudation* is often used to describe the combined processes of weathering, mass movement, and erosion, and a reduction in the relief of the landscape.

Vegetation plays an important role in restraining erosional forces. Leaves and branches break the fall of raindrops and lessen their erosive impact, and the roots of plants bind the soil and help it resist removal. Leaf litter and/or grass cover forms a cushion between raindrop impact and the soil. Wherever vegetation cover is dense and unbroken, erosion is generally slight. But where natural vegetation is removed to make way for agriculture, or perhaps decimated by overgrazing, accelerated erosion often results (*see* Fig. 22-3).

STREAMS AND BASINS

Rain that is not absorbed by the soil runs off as *sheet flow* (sometimes called *sheet wash*). Sheet flow is a thin layer of water that moves downslope without being confined to channels. This thin film of water can cause considerable erosion as it removes fine-grained surface materials; such **sheet erosion** is an important degradational process in certain areas, especially in deserts where the ground is bare and unable to absorb water rapidly. Continued runoff causes the initiation and growth of small channels called *rills* that may merge into larger *brooks* or *creeks*, and these in turn coalesce into more permanent streams. This water carries sediment with it, and both water and sediment become part of the river system that is in the process of shaping the regional landscape.

A river system consists of a *trunk* river joined by a number of *tributary* streams that are themselves fed by smaller tributaries. These branch streams diminish in size the farther they are from the trunk river. The complete system of the trunk and its tributaries forms a network that occupies a region known as a **drainage basin**. One of the best defined drainage basins in the world is the Mississippi Basin in North America: the Mississippi River

FIGURE 41-1 Splash erosion. The impact of a raindrop throws soil particles into the air. If this occurs on a slope, the particles

often fall downslope from their original position. The "crater" left after impact is about four times larger than the raindrop.

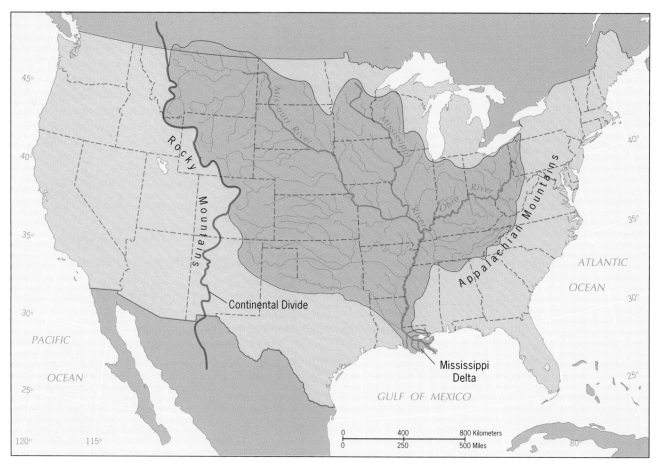

FIGURE 41-2 Drainage basin of the Mississippi River. It includes all the areas of the central United States drained by the Mississippi (trunk stream) and its chief tributaries, the Missouri and Ohio Rivers.

is the trunk stream in that basin, and the Missouri and Ohio Rivers are its chief tributaries (Fig. 41-2).

A drainage basin, therefore, is defined by the orientation of water flow. Within it, all streams flow into other streams that ultimately join the trunk stream. In the United States, a drainage basin is also sometimes referred to as a *watershed*. Drainage basins are separated by topographic barriers called *divides*. One of the most prominent divides in the world is the Continental Divide in the western United States, where it follows the spine of the Rocky Mountains (*see* Fig. 45-1b). Water from the eastern slope of the central and northern Rocky Mountains flows into the Mississippi drainage basin and reaches the Atlantic Ocean via the Gulf of Mexico (Fig. 41-2). Water falling on the western slope becomes part of the Colorado or Columbia River system and eventually flows into the Pacific Ocean.

To assess the erosional activity of a river system in a defined basin, physical geographers take a number of measurements. An obvious one is the quantity of sediment that passes a cross-section of the trunk stream at or near its mouth. The measurement of the total volume of sediment leaving a stream basin is the *sediment yield*. But this may not reveal all the erosional work that goes on inside the drainage basin, because some sediment is de-

posited as **alluvium** in the floodplains of streams and thus does not reach the lowest course of the trunk river.

Another set of measurements identifies every stream segment of the network and the sedimentary load carried by each. This research has yielded some interesting results. We might expect, for example, that the larger a drainage basin is, the larger will be the sediment load carried out of it at the mouth of its trunk river. But this is not always the case. The vast Amazon Basin of equatorial South America, for instance, is more than six times as large as the basin of India's Ganges River. But its annual sediment load, measured at the Amazon's mouth, is less than one-fifth that of the Ganges. Other comparisons also indicate that basin size alone is not a reliable indicator of the amount of erosion that occurs within a drainage basin.

What, then, are the conditions that influence the rate of erosion? Obviously, the amount of *precipitation* is a factor: the more water available, the more erosion can take place. But the cover of *vegetation* also plays an important role because it inhibits erosion. The dense forests in the Amazon Basin undoubtedly help slow down erosion; but as these tropical rainforests are removed—at a prodigious pace, as we explain in Unit 17—the rate of erosion also increases. The *relief* in the drainage basin is

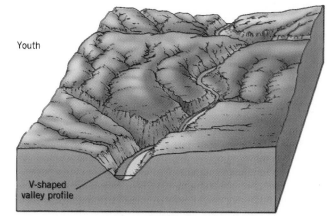

FIGURE 41-3 The Aswan High Dam, built during the late 1960s in the Nile Valley of upper (southern) Egypt, is a classic example of human interference in the river erosion process. In this 1988 photo, taken by an astronaut aboard the space shuttle, the area is dominated by a huge reservoir (Lake Nasser) that has filled behind the dam (top center, where the lake ends). Besides enhancing erosion all around the new lake, the dam has caused increased deposition within the reservoir, as well as an entirely new flow regime in the river below the dam all the way from Aswan north to the densely populated Nile Delta (*see* Fig. 41-5).

important as well. In a basin where the relief is generally low, slopes are less steep, water moves more slowly, and erosion is less active than in a basin where the relief is high. Another factor is the underlying *lithology*, or rock type; soft sedimentary and weak metamorphic rocks are degraded much more rapidly than are hard crystalline rocks. We must also consider *human impact* to be a factor influencing stream basin erosion rates. Although human activity can affect the fluvial system in many ways, the usual result is an increased rate of regional degradation (Fig. 41-3).

STREAM FUNCTIONS AND VALLEY PROPERTIES

As streams modify the landscape, they perform numerous functions. These functions can be grouped under three headings: *erosion* (degradation), *transportation* (transfer), and *deposition* (aggradation). Before we examine these processes, let us consider what happens to the river valley itself. A river is not static—it changes continuously. In fact, one conceptual model used by geographers describes the life cycle of a river in organic terms, suggesting that it evolves through stages of youth, maturity, and old age (Fig. 41-4).

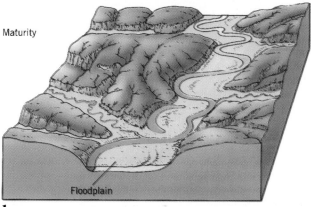

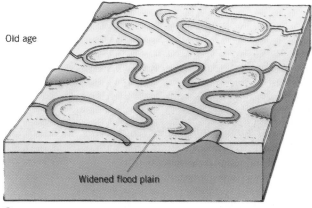

FIGURE 41-4 Three model stages in the life cycle of a river. In youth (a), the river's energy concentrates on deepening its valley, which in cross-section tends to be **V**-shaped. In maturity (b), the river's energy is directed laterally, and a floodplain forms. In old age (c), the floodplain is much wider, and the river bends or meanders form huge loops. Diagrams a and b drawn by Erwin Raisz (copyright © Arthur N. Strahler); reproduced by permission of Arthur N. Strahler.

As the river works, the properties of its valley change. In the area where the river has its origin, streamlets merge to form the main valley. Over time, the "head"

FIGURE 41-5 An astronaut's eye view of the Nile Delta, with the Mediterranean Sea to the north (left) and the Egyptian capital, Cairo, on the east (upper) bank of the river at the southern apex of the delta (right). This near-perfect deltaic triangle supports highly intensive agriculture and a huge population—in stark contrast to the empty desert that surrounds the delta and narrow valley watered by this lowest course of the Nile. In this eastward-facing view, the dark narrow zone beyond the Nile near the upper right-hand corner is the Red Sea.

or source of the valley is extended upslope in a process called **headward erosion**. This has the effect of lengthening the stream valley, but it is not the only way a river's course grows longer. At the other end of the valley, the stream reaches the coast of a sea or lake and may deposit its sedimentary load in a **delta** (Fig. 41-5). Now the river must flow across its own deltaic deposits, which lengthens the distance to its mouth. And between source and mouth, a river will develop bends and turns that make its course longer still. Valley *lengthening*, therefore, affects river valleys as the streams within them perform their erosional and depositional functions.

Streams tend to flow rapidly, even wildly, in mountainous areas. Many streams arise in the mountains and rush downslope in deep valleys whose cross-sections have a shape resembling a **V** (Fig. 41-4a). The high velocity of the water, its growing volume, and the sediments being swept along all contribute to valley *deepening*.

Erosion by Rivers

As we have noted, streams are the most important and efficient of all agents of erosion, continuously modifying their valleys and changing the landscape. Stream erosion takes place in three ways: *hydraulic action, abrasion,* and *corrosion.*

Hydraulic Action

The work of the water itself, as it dislodges and drags away rock material from the valley floor and sides, is referred to as its **hydraulic action**. You can feel the force of the water by wading across a shallow mountain stream; even in water little more than a foot deep, you may have trouble keeping your balance. Large volumes of fast-moving water can break loose sizeable boulders and move them downstream. In its rough-sided valley, the river develops numerous eddies and swirls that can gouge out potholes and other depressions.

Abrasion

Abrasion refers to the erosive action of boulders, pebbles, and smaller grains of sediment as they are carried along the river valley. These fragments dislodge other particles along the stream bed and banks, thereby contributing to the deepening and widening process. Gravel- and sand-sized particles tend to scour the valley bottom, wearing it down while evening out the rough edges. Abrasion and hydraulic action most often function in combination; without abrasive action, hydraulic action would take much longer to erode a valley.

Corrosion

In terms of the volume of rock removed, **corrosion** is the least important form of erosion by streams. It is the process by which certain rocks and minerals are dissolved by water. Limestone, for instance, is eroded not only by hydraulic action and abrasion, but also through corrosion. A sandstone held together by a calcite matrix will be weakened and removed because the water dissolves its cement. In Unit 44, we discuss the special landscapes formed when solution is the dominant form of erosion, but they are unusual and develop only under certain environmental conditions.

Transportation by Rivers

Erosion and transportation go hand in hand. The materials loosened by hydraulic action and abrasion, as well as dissolved minerals, are carried downstream. Along the way, different transportation processes are dominant. In the mountains, streams carry boulders and pebbles along in a high-velocity rush of water; near the coast, slow-moving water is brown or gray with sediment. Let us now identify the four processes at work when a stream transports its load: *traction, saltation, suspension,* and *solution.*

Traction

Traction refers to the sliding or rolling of particles along the river bed. This is accomplished by hydraulic action, as large pieces of rock are literally dragged along the stream bottom (Fig. 41-6). Traction breaks down boul-

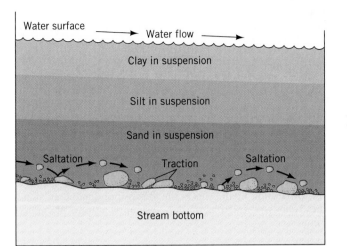

FIGURE 41-6 Heavier sand and gravel fragments are dragged along the stream bed by traction, while lighter fragments in the bed load advance downstream by saltation. Finer silt and clay particles are carried above in the suspended load, maintained there by the stream's turbulence until the water flow nears a standstill.

ders and pebbles into gravel- and sand-sized fragments, which begin to bounce along the stream bed.

Saltation

The speed of the water lifts these fragments off the river bottom, and they bounce along in a process called **saltation** (from a Latin word meaning jump). Saltation, therefore, is a combination of traction and suspension: the particles make contact with the stream bed, but they also are briefly suspended as they move along (Fig. 41-6).

Suspension

Very fine sediment, now ground down to silt- and clay-sized particles, is carried within the stream by a process known as **suspension** (Fig. 41-6). When suspension dominates, stream water becomes muddy, and even when it moves very slowly it can carry huge amounts of fine sediment toward its mouth. Material in suspension does not make contact with the river bottom except when the water is slowed to a near standstill.

Solution

As we noted earlier, some rock material dissolves in stream water and is carried downriver in **solution**. Corrosion is not confined to calcium-rich rocks. Numerous other minerals can be partially dissolved, and even a clear mountain stream contains ions of sodium, potassium, and other materials.

Deposition by Rivers

In combination, these four transportation processes move hundreds of millions of tons of earth materials an-

nually from the interior of the landmasses toward the coasts, where deposition takes place (as, for example, in deltas). In the upper reaches of drainage basins, sediment particles in streams are generally large. With increasing distance downstream, particle size decreases, and the stream does not need as much erosional power. As a result, aggradational processes begin to dominate. This depositional work of rivers is discussed in Unit 43; but first we need to explore the factors that govern the effectiveness of streams as they erode and transport their loads.

FACTORS IN STREAM EROSION

Stream Power

Physical geographers use the term **stream capacity** to denote the maximum load of sediment that a stream can carry with a given *discharge*, or volume of water. It is rather obvious that a stream with a large discharge has a larger capacity than a smaller stream. Another way to measure the erosional effectiveness of a stream is by determining its *competence*, which depends on the speed, or velocity, of the water's movement. The faster a stream flows, the greater is its ability to move large boulders in its channel. A combination of large volume *and* high velocity give a stream substantial capacity and competence.

The *velocity* of a stream, therefore, is a critical factor in its ability to erode and transport. As a rule, water velocity is greatest in the middle of the channel, but near the banks and on the river bottom the velocity is lowest because friction between water and rock slows the water's motion (as shown in Fig. 40-4). Under ideal conditions, the highest velocity prevails in the middle of the channel where the stream is deepest (this deepest part of the stream channel is called the *thalweg*). But even in a straight portion of a river's course, the thalweg tends to shift sideways from the middle, which initiates the swings that eventually lead to bends or meanders. In Fig. 41-7, the red line represents the thalweg; note that it lies on the outside of the bends in the channel. Stream erosion is most active on the outside of those bends, enlarging them over time.

The *gradient*, or slope, is the key factor in a stream's velocity. Again, this is obvious: a stream plunging down a steep mountain slope has a much higher velocity than a river crossing a coastal plain. What is less obvious, however, is the effect of even a slight change in gradient on the velocity (and thus the erosional power) of a river. Sometimes, the region across which a river flows is tilted slightly upward by tectonic forces. We might not notice this in the landscape, but a stream whose gradient is increased immediately begins to erode its valley with new vigor.

The *channel* of the river also affects velocity and

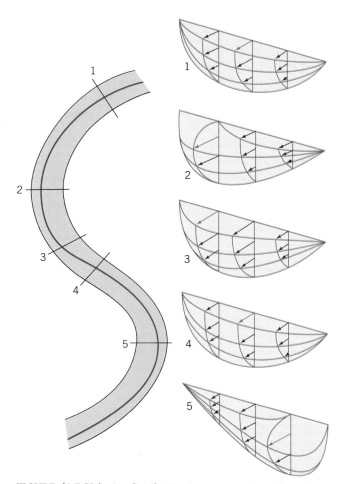

FIGURE 41-7 Velocity distribution in cross-sections through a curving channel (lengths of arrows indicate relative flow velocities). The zone of highest velocity (red arrow), associated with the thalweg, lies near the surface and toward the middle of the stream where the channel is relatively straight (cross-sections 1 and 4). At bends, the maximum velocity swings toward the outer bank and lies below the surface (cross-sections 2 and 5).

erosional capacity. An open, bowl-shaped, smooth valley cross-section promotes rapid water flow and efficient action. A wide, shallow channel generates too much friction and reduces both velocity and erosional capacity. And a stream valley that has irregular sides and a rock-strewn floor also slows water movement.

Stream Floods

Streamflow volume varies through time as a result of changing water inputs from adjacent upland areas and tributaries. Abnormally high rainfall or abrupt seasonal snowmelt can produce a **flood** in the trunk river. (A flood occurs when a stream overflows its channel.) These floods often result in aggressive scouring of stream beds and banks, the reshaping of the channel, the undercutting of valley sides, and many other consequences. Although the erosional power of large floods is visually striking and temporarily devastating, their infrequent occurrence limits their effect over long time periods. Far more erosional work is done under moderate flow conditions, which have a greater cumulative impact.

FIGURE 41-9 Floodplains are frequently endowed with highly fertile soils, which encourages many people to undertake the risk of farming them. This is the Mississippi floodplain, along the Tennessee-Arkansas border, just upriver from Memphis, Tennessee.

Not all floods are exclusively erosional, of course. Rivers, particularly in their lower courses, often rise above their average levels and sometimes overflow their normal channels. During such episodes the rivers inundate their **floodplains**, the low-lying ground adjacent to the stream channel (Figs. 41-4b and 41-4c). Floodplains are built by successive floods as sediment is deposited as *alluvium*; each flood adds more sediment to the plain. Fine-grained and mineral-rich, the alluvium gives rise to highly productive soils, often attracting dense human settlement. Although floodplain dwellers know the risks involved in living on the low ground near flood-prone streams (*see* Perspective: The Hazards of Floodplain Settlement), the fertility of the soils has made that gamble worth taking (Fig. 41-9).

Base Levels

We can learn much about the behavior of streams by studying their longitudinal *profiles*, their downward curve from the interior highlands to the coast. As a stream erodes and modifies its valley by deepening, widening, and lengthening its channel, it also tends to create a smooth, downward profile. A critical factor in this process is the river's **base level**—the level below which a stream cannot erode its bed. When a river reaches the ocean, its capacity to erode ends (although muddy currents offshore are known to be capable of some degradation). For our purposes, a river's **absolute base level** lies no more than a few meters below sea level. There, the stream slows down, deposits its sedimentary load, and its erosional work is terminated.

Some streams, however, do not reach the ocean. A stream that flows into a lake does not relate to global sea level. The effective base level is the lake level, at whatever altitude it may lie; the lake, therefore, becomes the river's **local base level**. Thus, for a stream that originates in the mountains and flows onto a plateau containing a lake, that surface is its local base level (Fig. 41-10). On occasion, a stream that erodes downward reaches an especially hard, resistant rock barrier, perhaps in the form

The Hazards of Floodplain Settlement

Episodes of abnormally high stream discharge—known more commonly as *floods*—can have a major impact on the cultural as well as the physical landscape in a river basin (*see* opening photo in Unit 43). About one-eighth of the population of the United States now resides in areas of potential flooding. To these people, the advantages of living in a flood-prone area outweigh the risks (just as people continue to live on the slopes of active volcanoes and near the San Andreas Fault).

The advantages include the fertility of the soils in these floodplain zones and the flatness of the land. Crops grow bountifully, and the land is easy to farm and develop. The risks, however, are great, and they are not only financial. Recent flood losses in the United States (apart from those incurred during the great Midwestern flood of 1993) have exceeded $1 billion per year, and dozens of human lives are lost annually.

The decision to live on or avoid a floodplain is not strictly rational. The geographers Gilbert White, Robert Kates, and Ian Burton have shown that human adjustments to the known dangers of flooding do not increase consistently as the risk becomes greater, because people tend to make an optimistic rationalization for continuing to live in a potential flood zone.

There are other decisions and trade-offs associated with floodplain settlement and development. Floods are natural events, but we sometimes try unnatural methods to prevent their worst effects. Dams and high artificial banks or *levees* like those in Fig. 41-8 can be built, but such structures sometimes have some unwanted side effects. Dams, although they help control flooding, can prevent the natural replacement of fertile alluvial soil because the sediment collects behind the dam instead. The Aswan High Dam in Egypt (*see* Fig. 41-3), stretching across the middle Nile Valley, holds back sediment that would otherwise be deposited on the highly productive agricultural land in the lower Nile floodplain and delta. Furthermore, artificial levees in floodplains can create a false sense of security.

These levees are seldom designed to cope with the greatest possible flood because they would be too expensive to construct. But often the higher the levee, the more people who will live in the area and the greater the disaster will be when a major flood does occur. Therefore it is essential for us to have an understanding of floodplains as well as other natural phenomena of the earth. Otherwise we cannot make rational decisions about where to live or the adjustments we should make in living there. One day you may have to vote on issues such as these.

FIGURE 41-8 An oblique aerial view of the artificial levee system that lines the banks of the Mississippi in southern Louisiana. This is part of the regional water-control system built by the U.S. Army Corps of Engineers.

of a dike (*see* Unit 30). That barrier may keep the river from developing a smooth profile until the stream has penetrated it. This creates a **temporary base level**, which temporarily limits further upstream channel incision.

THE RIVER AS A SYSTEM

The river is one of the simplest and most easily understood examples of a system that we have on the earth's surface (systems are discussed in Unit 1). It is an open system, with both matter and energy flowing through it, as diagrammed in Fig. 41-11. The most obvious material flowing through the system is water, entering as precipitation on any part of the drainage basin and leaving by evaporation and flow into the ocean.

Energy and Work in a River System

The water generates kinetic energy as it flows downslope. Through the action of the hydrologic cycle (*see*

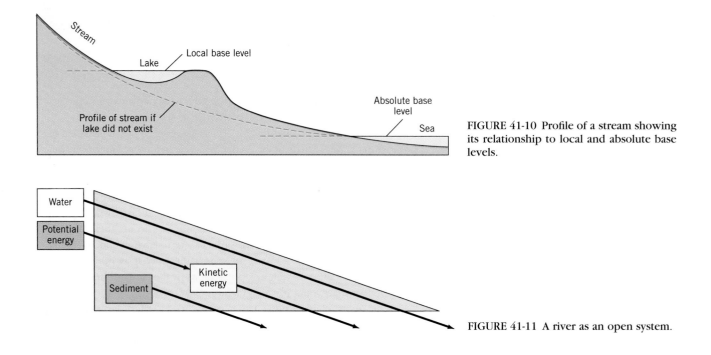

FIGURE 41-10 Profile of a stream showing its relationship to local and absolute base levels.

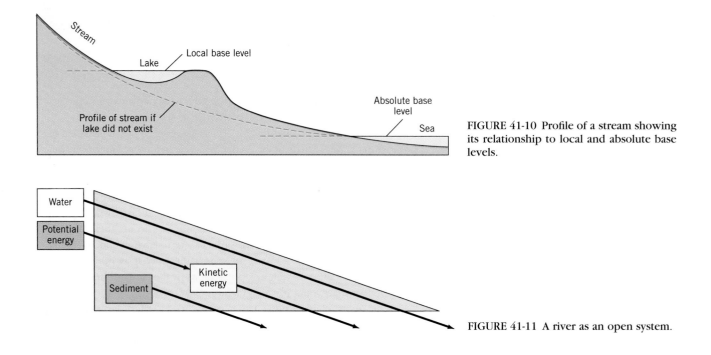

FIGURE 41-11 A river as an open system.

Fig. 12-2)—powered by energy from the sun—water vapor and, after condensation, liquid water are given potential energy. Any object possesses potential energy by virtue of being raised above the earth's surface and of work having been done against gravity. The motion of water in a river represents the transformation of potential energy into kinetic energy, the energy of movement. The water uses kinetic energy to carry its load and move itself. By the time the water has reached its base level, which is usually the sea, there is no more potential energy available. Thus no kinetic energy can be generated, and the river is unable to do further work.

Another kind of material, sediment, enters and leaves the fluvial system. This sediment is produced by weathering processes, soil erosion, mass movement, and scouring of the stream bed and banks. Its movement through the fluvial system is facilitated by the kinetic energy of the flowing water. The two "flows," water and sediment, are used to define the behavior of the river system.

A river is often characterized as a *steady-state* system—one in which inputs and outputs are constant and equal—at least with regard to water and energy. Physical laws suggest that energy and matter in a steady-state system must move in a particular way. First, there is a tendency for the least work to be done. If we imagine a river where all the water starts at the top of one tributary, the least-work profile would be a waterfall straight down to sea level. More practically, the least-work profile would be steep near the head and close to horizontal near the mouth of the river, as demonstrated in Fig. 41-12.

Another tendency in a steady-state system is for work to be uniformly distributed. A river in which this occurred would get wider downstream but have a nearly constant slope, as in the uppermost curve in Fig. 41-12. A *graded* stream profile is a compromise between the principles of least work and uniform distribution of work, but the graded profile is possible only when the channels are in a material that can be degraded and aggraded. The fact that this profile is indeed typical of many streams suggests that the graded river represents a steady-state system, and it also requires us to look more closely at the concept of grade.

A Graded River System

Rivers ultimately establish a longitudinal profile that allows the stream's load to be transported with neither degradation nor aggradation at any part of the profile. If this happens, the river is said to be **graded**. A graded stream represents a balance among valley profile, water volume, water velocity, and transported load. Many fac-

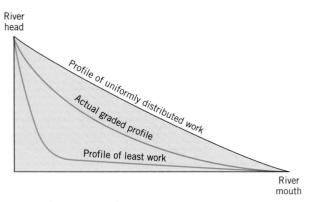

FIGURE 41-12 Formation of a graded profile as a compromise between the principles of uniformly distributed work and least work. This profile is reached if the material of the river channel is adjustable (i.e., capable of being degraded and aggraded).

TABLE 41-1 FACTORS INVOLVED IN THE TENDENCY OF A RIVER TO ACHIEVE GRADE.

INDEPENDENT	SEMIDEPENDENT	DEPENDENT
Discharge	Channel width	Slope
Sediment load	Channel depth	
Ultimate base level	Bed roughness	
	Grain size of sediment load	
	Velocity	
	Meander/braid tendency	

Source: After A. L. Bloom, *The Surface of the Earth* (Englewood Cliffs, N.J.: Prentice-Hall, 1969).

tors are involved in the tendency for a river to attain a graded state, and these factors may be classed as either independent, semidependent, or dependent (Table 41-1).

Independent factors are those over which the river has little or no control and to which it must simply adjust. The ultimate base level is a good example; the river must adjust to its ultimate base level because at this point it no longer has any potential or kinetic energy. The *semidependent* factors are partly determined by the three independent factors in Table 41-1, but also partly interact among themselves. The roughness of the river bed and the grain size of the sediment load, for example, interact. The bed roughness is partly determined by the size of the sediment grains and, in turn, partially determines the degree of mixing, or *turbulence*, in the river. Sediment size generally decreases downstream because of abrasion. But only finer particles remain in suspension in the lower part of the river, where turbulence decreases. Thus sediment grain size and bed roughness interact via the turbulence factor.

There is only one variable in the table that appears to be *dependent* on all others, and this is the gradient of the river. The river has virtually complete control of this by either depositing or scouring away material as necessary. It was the propensity of a river to change its slope that first set investigators to thinking about the concept of grade. Not until later was it completely realized that many other factors were involved in grade.

The modern definition of grade was established by J. Hoover Mackin in 1948 and has been subsequently modified by Luna B. Leopold, Thomas Maddock, and Marie Morisawa. The consensus is that a graded stream is one in which, over a period of time, stream slope, channel characteristics, and flow volumes are delicately adjusted to provide just the velocity required for the transportation of the load supplied from the drainage basin. A graded stream is a steady-state system. Its diagnostic characteristic is that any change in any of its controlling factors causes a displacement of the equilibrium in a direction that tends to absorb the effect of the

change. So if you tried to dam a graded stream with rocks, the stream would immediately begin to wear the rocks away.

Not all rivers are graded, and not all graded rivers assume a smooth longitudinal profile. A newly uplifted tectonic landscape would present a chaotic drainage system, with profiles far from graded. The semidependent variables are usually the first to adjust themselves to the process of moving material downstream. Furthermore, a river usually establishes grade in its lower reaches first, and then the graded condition is slowly extended upstream toward the river head. Because of differences in rock resistance and other factors, it is quite possible for grade to exist in isolated segments of the complete river profile. Brandywine Creek, located in southeastern Pennsylvania and northern Delaware, manifests an irregular profile; yet, it has been shown to have achieved grade, as Fig. 41-13 demonstrates.

The casual observer can assess whether or not a river is graded by judging the stability of a channel. If the channel has clearly remained in the same place for years and years, it is likely that the river is graded. But if there is evidence that the river occasionally changes its channel in some way, it is unlikely that grade has yet been achieved. The establishment of grade has important consequences. Any human alteration of a river system, such as lining a channel with concrete or building a dam, must take into account the state of grade. Otherwise, the dam or new channel may quickly become silted, and the expensive engineering work will have been wasted.

Rivers degrade, transport, and aggrade. Here we have studied the processes of river erosion, and in Units 42 and 43 we examine the products of rivers' work in the landscape. These products range from spectacular canyons and deep gorges to extensive deltaic plains. As we study these landforms, we will learn still more about the functions of rivers, those great sculptors of scenery.

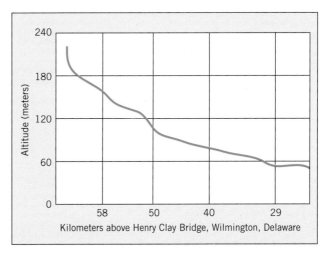

FIGURE 41-13 The longitudinal profile of Brandywine Creek in Pennsylvania and Delaware, a graded river without a smooth profile.

KEY TERMS

abrasion (p. 422)

absolute base level (p. 424)

alluvium (p. 420)

base level (p. 424)

corrosion (p. 422)

delta (p. 422)

denudation (p. 419)

drainage basin (p. 419)

flood (p. 424)

floodplain (p. 424)

graded river (p. 426)

headward erosion (p. 422)

hydraulic action (p. 422)

infiltration capacity (p. 419)

local base level (p. 424)

saltation (p. 423)

sheet erosion (p. 419)

solution (p. 423)

stream capacity (p. 423)

suspension (p. 423)

temporary base level (p. 425)

traction (p. 422)

REVIEW QUESTIONS

1. How does overflowing water accomplish erosion? What processes are involved?

2. Describe the arrangement of streams within a drainage basin.

3. What factors determine the erosional activity of streams within a drainage basin?

4. Describe the three stream functions.

5. What is base level, and how is it related to the development of the longitudinal profile?

6. What are the primary independent controls on the development of a graded river system?

REFERENCES AND FURTHER READINGS

BAKER, V. R., et al., eds. *Flood Geomorphology* (New York: Wiley, 1988).

BLOOM, A. L. *Geomorphology: A Systematic Analysis of Late Cenozoic Landforms* (Englewood Cliffs, N.J.: Prentice-Hall, 2nd ed., 1990).

BURTON, I., KATES, R. W., and WHITE, G. F. *The Environment as Hazard* (New York: Oxford University Press, 1978).

DINGMAN, S. L. *Fluvial Hydrology* (New York: Freeman, 1983).

GORDON, N. D., MCMAHON, T. A., and FINLAYSON, B. L. *Stream Hydrology: An Introduction for Ecologists* (New York: Wiley, 1992).

GREGORY, K. J., and WALLING, D. E. *Drainage Basins: Form, Process and Management* (Cambridge, Mass.: Blackwell, 1995).

HERSCHY, R. W. *Streamflow Measurement* (New York: Elsevier, 1985).

HOYT, W. G., and LANGBEIN, W. B. *Floods* (Princeton, N.J.: Princeton University Press, 1955).

KIRKBY, M. J., ed. *Hillslope Hydrology* (New York: Wiley, 1978).

KNIGHTON, D. *Fluvial Forms and Processes* (London: Edward Arnold, 1984).

LEOPOLD, L. B. *A View of the River* (Cambridge, Mass.: Harvard University Press, 1994).

LEOPOLD, L. B., et al. *Fluvial Processes in Geomorphology* (San Francisco: Freeman, 1964).

MORISAWA, M. *Streams: Their Dynamics and Morphology* (New York: McGraw-Hill, 1968).

MOSLEY, M. P., and MCKERCHAR, A. I. "Streamflow," in D. R. Maidment, ed., *Handbook of Hydrology* (New York: McGraw-Hill, 1993), pp. 8.1–8.39.

PETTS, G., and FOSTER, I. *Rivers and Landscapes* (London: Edward Arnold, 1985).

SCHUMM, S. A. *River Morphology* (Stroudsburg, Pa.: Dowden, Hutchinson & Ross, 1972).

SMITH, D. I., and STOPP, P. *The River Basin* (New York: Cambridge University Press, 1978).

STATHAM, I. *Earth Surface Sediment Transport* (London/New York: Oxford University Press, 1977).

WARD, R. *Floods: A Geographical Perspective* (New York: Macmillan, 1978).

Degradational Landforms of Stream Erosion

OBJECTIVES

◆ To outline the roles of geologic structure, lithology, tectonics, and climate in influencing fluvial erosion.

◆ To introduce descriptive terms for characterizing drainage networks and controls on fluvial erosion.

◆ To briefly consider how landscapes might change or evolve through time in response to river erosion.

When rivers erode the landscape, many processes occur simultaneously. Rock material is being removed, transported, pulverized, and deposited. Stream valleys are widened and deepened. Mass movements are activated. Slopes are being flattened in some places, steepened in others. Long-buried rocks are exhumed and exposed. Endlessly, the work of streams modifies the topography.

The regional landscape—and individual landforms comprising parts of it—reveals the erosional process or processes that dominate in particular areas. Certain fluvial landforms of desert areas, for example, are not found in humid environments. Wide, open floodplains, occupied by curving stream channels, are not found in mountainous areas. Concave slopes and sharp, jagged ridges are more likely to be found in arid areas; round, convex slopes and rounded hilltops reflect moister conditions. We can draw many conclusions about the processes shaping the landscape from simple observational evidence.

UNIT OPENING PHOTO: One of North America's most dramatically eroded landscapes: the Badlands of southwestern South Dakota.

FACTORS AFFECTING STREAM DEGRADATION

Among the many factors influencing fluvial erosion, four have special importance: *geologic structure, bedrock type* (or *lithology*), *tectonic activity,* and *climate.*

Geologic Structure

Geologic structure refers to features such as synclines, anticlines, domes, and faults that were formed originally by geologic processes (*see* Unit 36). These geologic structures are sculpted by streams into characteristic landforms. One of the simplest examples is a landform that results from a high-angle intrusion, a dike.

Hogbacks and Cuestas

The crystalline rock layer of the dike dips at a high angle and penetrates softer surrounding rocks (Fig. 42-1a). Originally, this dike might not have reached the uppermost of the sedimentary rock layers through which it penetrated, but stream erosion has degraded the countryside and exposed the much harder rock of the dike. *Geologically*, it remains a dike; *geographically*, it stands

out as a prominent, steep-sided ridge called a **hogback**. Stream erosion exposes the differences in resistance between the crystalline rock of the dike and the sedimentary rocks surrounding it.

Other less prominent ridges may form when a sequence of sedimentary strata of varying hardness lies at a low-angle dip. This is seen in Fig. 42-1b, where erosion by streams removes the softest rock strata first, leaving a low ridge with one fairly steep and another very gentle slope. This landform, known as a **cuesta**, is not usually as pronounced in the landscape as a hogback, but cuestas can be hundreds of kilometers long.

Ridges and Valleys

When structures become more complicated, so do the resulting landforms. The erosion of synclines and adjacent anticlines produces a series of parallel ridges and valleys that reveal the structures below. If the axes of the folds plunge, as is frequently the case, the result is a terrain of zigzag ridges (*see* Fig. 36-13a on p. 383); note that the steep face of the resistant layer (numbered 3 on this diagram) faces *inward* on the anticline and *outward* on the syncline, providing us with preliminary evidence of the properties of the folds below the surface. The Appalachian Mountains provide many examples of this topography (*see* Figs. 36-15 and 53-4).

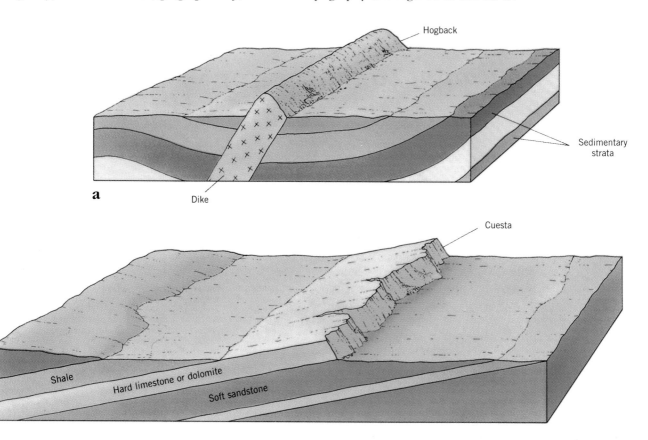

FIGURE 42-1 Hogback (a) and cuesta (b) landforms. Note that hogbacks dip at a high angle whereas cuestas dip at a low angle (the cuesta in this drawing is associated with a dry climate).

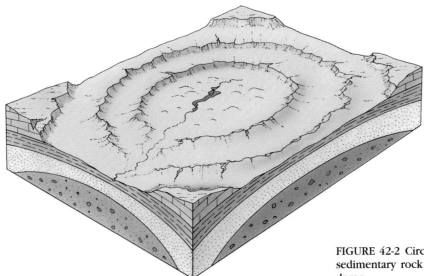

FIGURE 42-2 Circular cuestas produced by stream erosion of sedimentary rock layers pushed upward to form a structural dome.

Domes

In areas where sedimentary strata have been pushed upward to form a dome, stream erosion also produces a characteristic landscape in which the affected layers form a circular pattern, as shown in Fig. 42-2. One of the world's best examples of this type of landscape lies in the Black Hills in the area of the South Dakota–Wyoming border. There, the crystalline core of the dome has already been exposed by erosion, but the overlying sedimentary layers still cover its flanks in all directions. As a result, the regional topography consists of a series of concentric cuestas of considerable prominence, separated by persistent valleys. Not only physical geographic configurations but also the human geographic features reveal the dominance of this concentric pattern: ridges, rivers, roads, and towns all exhibit circular patterns.

Faults

Faults, too, are exposed and sometimes given relief by stream erosion. Of course, normal and reverse faults create topography by themselves because one block moves upward or downward with respect to another. But after the faulting episode, stream erosion begins and the fault becomes a geographic as well as a geologic feature. In fact, physical geographers distinguish between a *fault scarp* (Fig. 42-3a), a scarp (cliff) created by geologic action without significant erosional change, and a *fault-line scarp*, a scarp that originated as a fault scarp but that has been modified, even displaced, by erosion (Fig. 42-3b). And where the geologic structures are formed by numerous parallel (echelon) faults, as happens in regions where lithospheric plates are affected by convergent movements, the fault-generated terrain is also modified by stream erosion. Even prominent upthrust blocks may be worn down to a reduced relief, and valleys fill with sediment derived from these uplands. The overall effect is to lower the regional relief, as shown in Fig. 42-4.

Extrusive Igneous Structures

In Unit 34, we discuss volcanoes as geologic phenomena and as landforms. Created by volcanic action, volcanic landforms are quickly modified by stream erosion. Extinct and long-dormant volcanic cones reveal their inactivity through numerous, often deep stream channels carved into their slopes. Eventually, the entire mountain

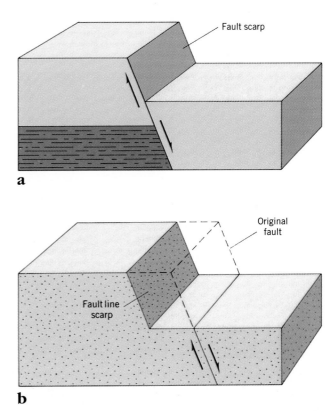

FIGURE 42-3 A fault scarp (a) originates as a result of geologic activity without much erosional modification. A fault-line scarp (b) originates as a fault scarp but then undergoes significant erosional change.

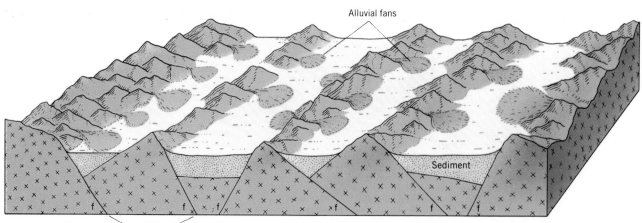

Alluvial fans

Sediment

Faults

FIGURE 42-4 The erosion of upthrown blocks leads to the filling of their intervening valleys with sediment. In the southwestern United States, parallel faults create a so-called basin-and-range topography in which the fault-formed valleys are filling up with sediment from the ranges. Short streams flowing off the uplands produce alluvial fans—fan-shaped sedimentary landforms where stream velocity is sharply reduced when the streams enter the flat valleys (*see* Fig. 43-1).

may be worn down to its plug and radiating dikes (Fig. 42-5). Fissure eruptions create hard sills that may become the caprocks of plateaus or smaller landforms known as **mesas** and **buttes** (Fig. 42-6), although these two landforms may develop in sedimentary rocks as well.

Intrusive Igneous Structures

The sedimentary, fault-dominated, and volcanic structures discussed so far have quite characteristic erosional forms. From topographic and drainage patterns, we can often deduce what lies beneath the surface. But vast areas of the landmasses are underlain by granitic and metamorphosed rocks that do not display such regularity. We

FIGURE 42-5 Rising hundreds of meters above the desert floor, Ship Rock in northwestern New Mexico is all that remains of a towering volcano. Radiating from the vent-filled plug (or volcanic neck) are several dikes that still defy the forces of weathering and erosion. Eventually, however, the agents of erosion will erase this silent witness to another geologic time.

note in Part Four that large batholiths formed within the crust have been uplifted and exposed by erosion and that large regions of metamorphosed crystalline rocks form the landscapes of the ancient shields. Some batholiths now stand above the surface as dome-shaped mountains, smoothed by weathering and erosion. The great domes that rise above the urban landscape of Brazil's Rio de Janeiro are such products of deep-seated intrusion and subsequent erosion (*see* Fig. 30-6), as is Stone Mountain near Atlanta, Georgia, a granite dome now rising nearly 200 m (650 ft) above the surrounding plain (Fig. 42-7).

Metamorphic Structures

Metamorphic rocks often display regional foliation (*see* Unit 31), and sometimes this tendency to form belts of aligned minerals is reflected in the drainage lines that develop upon them. Some metamorphic rocks, furthermore, are much more resistant than others, so that zones of quartzite, for instance, are likely to stand above softer rocks such as weak slates and schists. Gneiss, on the other hand, is quite hard and often supports uplands reminiscent of those formed on granite batholiths.

Geologic structure, therefore, is an important factor in stream erosion (a relationship that cannot be overlooked when human technology attempts to influence river courses [*see* Perspective: Controlling Rome's River]). It also affects the development of drainage systems; later, we will discover that the actual pattern of drainage lines can reveal the nature of the underlying structural geology.

Bedrock Type (Lithology)

Rock type, for obvious reasons, strongly influences landscape evolution. This is true not only because different rock types have different properties of hardness and resistance against erosion, but also because they exhibit

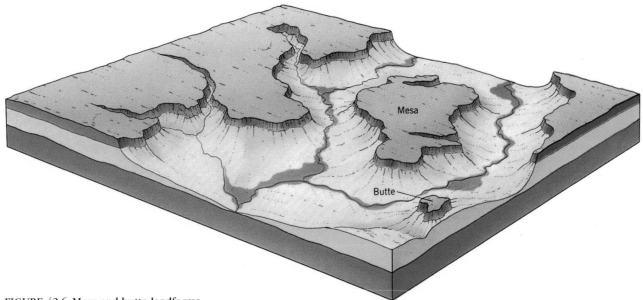

FIGURE 42-6 Mesa and butte landforms.

varying capacities to form slopes. A resistant crystalline rock mass, for example, can form and maintain vertical scarps (hence a slope angle of 90 degrees), but weak shales may, under certain circumstances, support a slope angle of not more than 35 degrees. In Fig. 42-9, weak shales lie between more resistant sandstone layers, creating a segmented slope in which the sandstone forms small cliffs while the shale forms low-angle slopes. If the bedrock shown in this diagram had been uniform, no such alternation would have developed, and the valley would exhibit an unbroken **V**-shape, the angle of the sides depending on the nature of the rock.

Bedrock type also influences the sculpting of landforms, such as the zigzag pattern on plunging anticlines and synclines described earlier. If the strata folded into those structures were uniform, then the resulting landscape would not display the variation shown in Fig. 36-13 (p. 383). Again, when a fault has the effect of thrusting soft sedimentary rocks upward adjacent to hard crystallines, the weak sedimentaries will soon yield to erosion and the resulting landscape may *reverse* the geologic imprint.

Tectonic Activity

Stream erosion is significantly affected by tectonic activity, especially in areas affected by convergent plate movement. We know that the landmasses are continuously influenced by the movement of lithospheric plates. Not only are peripheral areas of convergent plate contact deformed, but also larger regions of the continents are subject to slight but important warping and tilting. All this has a significant effect on drainage systems and erosional effectiveness. Even a very slight increase in regional "tilt" can greatly increase a stream's effectiveness. Sometimes, we can see from the profile of a stream valley that the river has been *rejuvenated*, that is, its energy increased. This may be the result of several factors, among which tectonic activity is the most obvious. Uplift results in increased velocity of streamflow and enhanced erosional capacity.

Climate

Climate plays a role in the evolution of landscape, but this role remains open to debate. From the landscapes we see, the answer would seem to be obvious: landscapes of humid areas tend to be rounded and domi-

FIGURE 42-7 Stone Mountain, Georgia, now all but engulfed by the burgeoning eastern suburbs of Atlanta, is one of the best-known granitic domes in the United States.

Controlling Rome's River*

Rivers flow through most of the world's large cities, and trying to make these water courses behave has been a priority since people first clustered alongside them. Europe's great cities, in particular, have had long experience in attempting to control local river channels, and the famous stone walls and embankments that line the Seine in Paris and the Thames in London are testimony to the fair degree of success that has been achieved. No European city, however, has been engaged in this battle longer than Rome, and the latest Roman skirmish with its Tiber River is a reminder of the limitations that constrain the efforts of hydrologists and engineers.

Taming the Tiber is a struggle that dates back more than 2000 years to the time of Julius Caesar, when stone bridges (still in use today) were built across the river and "improvements" made to its banks. Nevertheless, because the Tiber is a relatively short stream that emanates from the nearby, rugged Appennine Mountains, serious floods continued to regularly bedevil its valley. By the late nineteenth century, following modern Italy's unification and the restoration of Rome as the country's capital, the new government decided to act and ordered that the Tiber be corseted by stone walls to protect the city's treasured riverside monuments (Fig. 42-8). Unfortunately, to save money, the construction program avoided erecting heavy structures that could have significantly enhanced the walls' stability. Not surprisingly, parts of the stonework collapsed a few years later during an especially bad flood. It was

FIGURE 42-8 The Castel Sant'Angelo (Hadrian's Tomb), located just to the east of Vatican City, overlooks the walled embankments of the Tiber as it wends its way through the heart of Rome.

quickly (and cheaply) rebuilt—but this patchwork mentality prevailed until just a few years ago.

Finally realizing that they tamper with the river at their own peril, Rome's engineers are now marshalling the resources to try to achieve a longer-term solution. But first they must overcome an additional problem. Upstream from Rome, the Tiber has recently been dammed to the extent that not enough silt and sand are available to replenish the riverbed. With the removal of most of these deposits, the Tiber's erosive power has concentrated on cutting into bedrock, which threatens to undermine bridges, stone walls, and buildings adjacent to them. To meet the challenge, the most vulnerable bridges are being re-

anchored to the river bottom, huge concrete slabs are being inserted below the waters to trap silt and gravel to reinforce bridge foundations, and the heavy structures to stabilize the walls, dismissed by the builders of a century ago, have at last reached the planning stage.

Whatever happens, human ingenuity and technology will not succeed in controlling the Tiber or any other major river. That realization has been a long time in coming to the so-called Eternal City, which only after two millennia has begun to work more harmoniously with nature to do what is possible to minimize the Tiber's hazards.

*The source for most of this box is Tagliabue, 1993.

nated by convex slopes, whereas landscapes of desert environments are stark, angular, and dominated by concave slopes. But detailed geomorphological research has not confirmed this contrast—or at least it has not confirmed that climate is the key factor.

Part of our impression, geomorphologists say, has to do with the cloak of vegetation in humid areas, as op-

posed to the barrenness of arid zones. Strip away this vegetation, they suggest, and the contrast between humid and arid landscapes may be less pronounced than we expect. Moreover, we should take account of climate change. Areas that are humid today were dry just a few thousand years ago, and vice versa. Elsewhere, glaciers dominated the topography of mountains now being

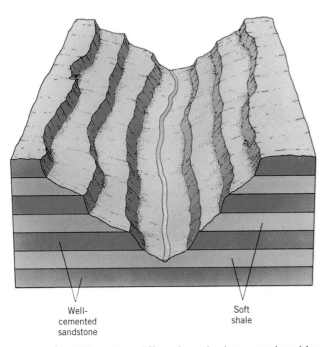

Well-
cemented
sandstone

Soft
shale

FIGURE 42-9 Alternating cliffs and gentle slopes produced by the differential erosion of resistant sandstone and weak shale.

eroded by rivers. There are solution caves in arid areas, indicating the presence of much more water than may currently be the case. And there are valleys in high mountains that were gouged out by glacial ice, not streams. Over the long term, therefore, climate certainly plays a role in forging the landscape. But just how climate and hillslopes are related remains a contentious issue.

DRAINAGE PATTERNS

The structures and rock types that are sculpted by streams into characteristic landforms strongly influence the drainage patterns that develop on them. The drainage pattern often reveals much about the geology that lies below. By examining the way a river system has evolved in a certain area, we can begin to unravel the origins of the regional landscape. Before we study actual drainage patterns, we should acquaint ourselves with the concept of drainage density.

The effectiveness of a drainage system as an erosional force is directly related to the **drainage density**— the total length of the stream channels that exist in a unit area of a drainage basin. The higher the drainage density, other things being equal, the greater is the erosional efficiency of the system. More stream channels in an area mean that more erosion can take place and that more slopes are directly affected by degradation. You might expect drainage density to be higher in humid areas than in arid locales, so that upland surfaces, virtually unaffected by stream erosion, are more likely to exist in dry than in humid regions. In reality, drainage density is influenced by other factors besides climate, so this simple generalization is not always accurate.

Drainage density can be stated quantitatively, such as in meters of stream channel per square kilometer of basin area, but usually it is simply given as low, medium, or high. On a Pacific volcano, in the path of moist winds, drainage density is likely to be high; on a mountain range in the arid interior of Asia, it will be low. When the drainage density is medium to high, drainage patterns become clear. To see a drainage pattern most clearly, it is best to study a map that shows *only* streams, nothing else. The pattern alone will be useful in later interpretations of structures and rocks.

A **radial drainage** pattern, for instance, shows drainage of a conical mountain flowing in all directions (Fig. 42-10a). We can tell that the drainage in this example flows outward in all directions from the way tributaries join. Except under the most unusual circumstances, tributaries join larger streams at angles of less than 90 degrees, and often at much smaller angles. Radial patterns of the kind shown in Fig. 42-10a develop most often on volcanic cones.

Another highly distinctive drainage type is the **annular drainage** pattern, the kind that develops on domes like South Dakota's Black Hills structure. Here the concentric pattern of valleys is reflected by the positioning of the stream segments, which drain the interior of the excavated dome (Fig. 42-10b).

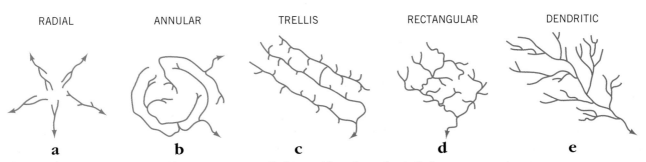

RADIAL   ANNULAR   TRELLIS   RECTANGULAR   DENDRITIC

a   b   c   d   e

FIGURE 42-10 Five distinctive drainage patterns. Each provides clues about their underlying geologic structures.

One of the most characteristic patterns is the **trellis drainage** pattern, in which streams seem to flow in only two orientations (Fig. 42-10c). This pattern appears very regular and orderly, and often develops on parallel-folded or dipping sedimentary rocks of alternating degrees of hardness. The main courses are persistent, but tributaries are short and join the larger streams at right angles.

The **rectangular drainage** pattern also reveals right-angle contacts between main rivers and tributaries, but the pattern is less well-developed than in the case of trellis drainage (Fig. 42-10d). What the diagram cannot show is that rectangular patterns tend to be confined to smaller areas, where a joint and/or fault system dominates the structural geology. Trellis patterns, on the other hand, usually extend over wider areas.

The tree-limb-like pattern shown in Fig. 42-10e is appropriately termed the **dendritic drainage** pattern because it resembles the branches of a tree. It is the most commonly developed drainage pattern, and it is typical on extensive batholiths of generally uniform hardness or on flat-lying sedimentary rocks. The entire drainage basin is likely to slope gently in the direction of the flow of the trunk river.

The five drainage patterns shown in Fig. 42-10 are not the only patterns that may be recognized. Other, less common patterns also develop, and some representative names are *convergent* (or *centripetal* streams flowing into a central basin), *contorted* (disorderly drainage in an area of varied metamorphic rocks), *parallel* (streams flowing down a steep slope or between elongated landforms), and *deranged* (where the drainage has yet to become organized, such as on glacial debris). Physical geographers employ this nomenclature to convey the prevalent character of regional drainage systems, often as a first guide to the interpretation of landforms.

OVERCOMING GEOLOGIC STRUCTURE

The drainage patterns we have just discussed would suggest that structure exercises powerful control over river systems. Certainly the underlying geology influences the development of the patterns we saw in Fig. 42-10, but there are places where rivers seem to ignore structural trends. In some areas, rivers actually cut across mountain ranges when they could easily have flowed around them. What lies behind these discordant relationships?

Superimposed and Antecedent Streams

In many such instances, the river channel first developed on a surface that lay *above* the still-buried mountain range. In Fig. 42-11a, note that the stream has developed on sedimentary rocks that covered the under-

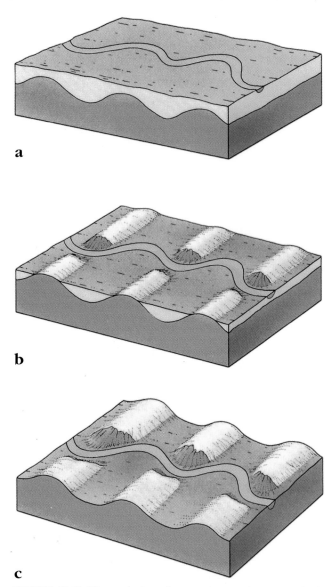

FIGURE 42-11 The evolution of a superimposed stream, which carves water gaps where it slices through three emerging mountain ridges.

lying ranges. If the drawing showed the entire drainage pattern, it probably would be dendritic, but we show only the trunk stream for simplicity. As the stream erodes downward, it reaches the crests of the mountain ranges. In Fig. 42-11b, the emerging range probably causes some waterfalls to develop where the stream cuts across its harder rocks, but a **water gap** (a pass in a ridge or mountain range through which a stream flows) is developing and the stream stays in place. As more of the overlying sedimentary material is removed by the river and its tributaries, the ranges rise ever higher above the stream valley (Fig 42-11c). This process is referred to as stream *superimposition*, and the river itself is called a **superimposed stream** because it maintains its course regardless of the changing lithologies and structures encountered.

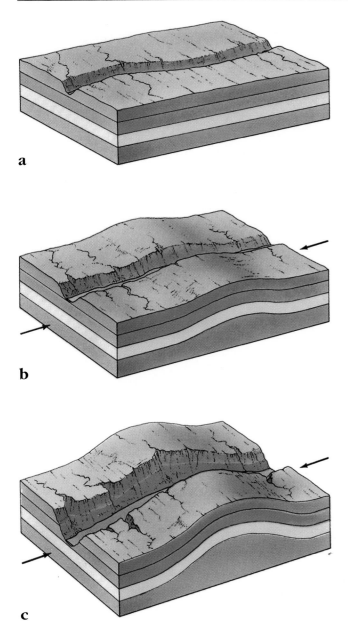

a

b

c

When we see a river slicing across a mountain range, however, we should not conclude without further research that the cause is necessarily superimposition. There are other possible explanations. The ridge through which the river now flows may not have been buried but may have been pushed up tectonically. A river flowing across an area so affected (Fig. 42-12a) may have been able to keep pace, eroding downward as rapidly as the ridge was being formed (Figs. 42-12b and c). This river, therefore, predates the ridge, and is referred to as an **antecedent stream**.

Stream Capture

One of the most intriguing stream processes involves the "capture" of a segment of one stream by another river. Also called **stream piracy**, this process diverts streams from one channel into another, weakening or even eliminating some river courses while strengthening others. This is not just a theoretical notion, nor does it affect only small streams.

Looking at a map of south-central Africa, you will see the great Zambezi River. Flowing toward the Zambezi from the north is the Upper Kafue River. But at Lake Iteshi the Kafue River makes an elbow turn eastward, joining the Zambezi as shown in Fig. 42-13. What happened here is a classic case of stream capture. The Upper Kafue River once flowed south-southwestward, reaching the Zambezi as shown by the dashed line (Fig. 42-13). But another river, the Lower Kafue, was lengthening its valley by headward erosion toward the west. When the upper channel of the Lower Kafue intersected the Upper Kafue and its continuation in the area of Lake Iteshi, the

FIGURE 42-12 The evolution of an antecedent stream, which kept flowing (and eroding downward) as the mountain ridge was being tectonically uplifted across its path. Tectonic forces are shown by the black arrows.

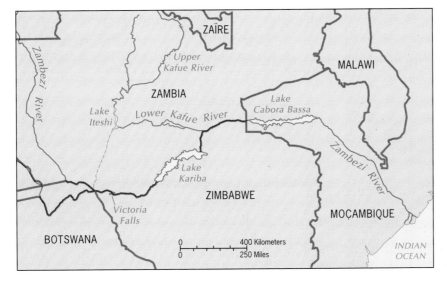

FIGURE 42-13 Stream piracy enabled the headward-eroding Lower Kafue River to capture the Upper Kafue in the vicinity of Lake Iteshi. Previously, the Upper Kafue River drained into the Zambezi via the (now abandoned) connecting channel, shown by the dashed line.

flow of water from the Upper Kafue began to divert into the Lower Kafue, and its original connecting channel to the Zambezi was abandoned. Today, a small stream (appropriately termed *underfit*) occupies the large valley once cut by the Kafue below Lake Iteshi. Its waters have been pirated.

This is a large-dimension case of capture, and many smaller instances can be found on maps of drainage systems. It is important, however, to realize the effect of capture. The capturing stream, by diverting into its channel the waters of another river, increases its capacity to erode and transport, and is, in that sense, rejuvenated. From the valley profile of the Lower Kafue River we can see how this river's competence was increased when its piracy was successful.

REGIONAL GEOMORPHOLOGY

Physical geographers perform research on landforms and drainage systems, erosional processes, and stream histories. Like other scientists, however, they also want to understand the "grand design"—the overall shaping of the landscape, the sculpting of regional geomorphology, and the processes that achieve this. At first, it would seem that this is merely a matter of the sum of the parts. If we

FIGURE 42-14 William Morris Davis, founder of the Association of American Geographers in 1904, was one of this century's most distinguished physical geographers.

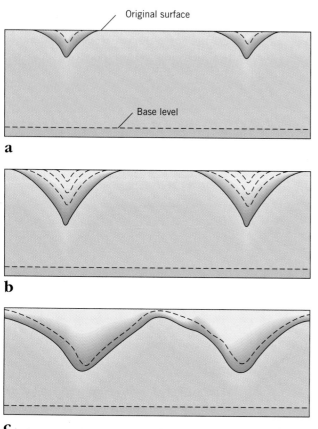

FIGURE 42-15 The Davisian peneplanation model. Streams erode a flat upland area (a), which is gradually transformed into mountainous terrain (b and c). These highlands are subse-quently lowered until only the hilly interfluves remain (d and e). Finally, these hills are eroded away, leaving the peneplain (f).

know about the factors affecting erosional efficiency (such as those discussed in Units 40 and 41), then shouldn't the evolution of landscape surely be understood? The answer is not yet. Physical geographers today still debate some very basic issues concerning regional geomorphology.

Slopes and Plains

One particular debate has been going on for more than a century. It was started by William Morris Davis (1850–1934), pioneer physical geographer and professor at Harvard University from 1878 to 1912 (Fig. 42-14). Davis proposed that a **cycle of erosion** would affect all landscapes. This cycle had three elements: geologic *structure*, geographic *process*, and time or *stage*. Every landscape, Davis argued, has an underlying geologic structure; it is being acted upon by streams or other erosional processes, and it is at a certain stage of degradation. A high mountain range was thought to be in an early stage. It would eventually be worn down to a very nearly flat surface, a "near-plain" Davis called a **peneplain**. This sequence of events is illustrated in Fig. 42-15. In Figs. 42-15a and b, an upland is being attacked by a network of streams. The upland slopes gradually attain lower angles until the upland is transformed into mountainous terrain (Fig. 42-15c). Now the mountains are lowered until they are little more than convex hilly **interfluves** (Figs. 42-15d and e). Finally, even these hills are eroded away, leaving a nearly flat plain (peneplain) with a few remnants on it (Fig. 42-15f). Davis called the most prominent not-yet-eroded remnants **monadnocks**, after Mount Monadnock in southern New Hampshire, an example of the phenomenon (Fig. 42-16).

In Davis's view, slopes are worn *down*: that is, they become increasingly convex in appearance, then are flattened. For many years this idea was generally accepted, although many slopes were seen *not* to have a convex form. Nevin Fenneman and Grove Karl Gilbert, in the first decades of this century, were among the first to suggest that the Davisian model did not fit all landscapes. But not until the German geographer Walther Penck published his doubts (his book, published in the 1920s, was not translated into English until the 1950s) was an alternative theory concerning slopes proposed.

This theory holds that all slopes exhibit the same elements and that they do not wear *downward*, but *backward*. According to a group of scholars, including Alan Wood, Lester C. King, and John T. Hack, slope retreat is the process whereby highlands are reduced to plains—not peneplains but **pediplanes**, plains at the foot of mountains. (Because the comparison is to a flat geometric surface, the term "pedi*plane*" was preferred.) This sequence of events is illustrated in Fig. 42-17. Note that the interfluves in this model retain their near-vertical

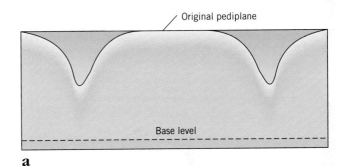

a

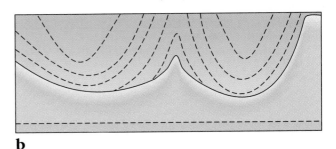

b

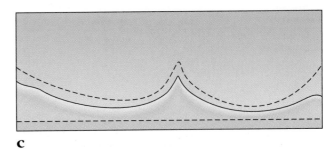

c

FIGURE 42-17 The pediplanation model, whereby slope retreat reduces highlands to plains. Note that the steep slope angles of the interfluvular uplands are retained throughout the entire erosional sequence.

FIGURE 42-16 New Hampshire's Mount Monadnock, which inspired William Morris Davis to name the upland remnants on a peneplain after it.

slopes and that the uplands are essentially unaffected until the retreating slopes intersect.

Landscapes in different areas seem to support both interpretations, and it was not long before some physical geographers began to link theories of slope change to climate. In humid climates, it was suggested, slopes tend to become convex; it is therefore not surprising that Davis reached such conclusions, because most of his field-work was done in humid locales. In arid climates, they held, slopes would develop concave properties because drainage density was much lower, and weathering and mass movement played more important roles. But when these ideas about *climogenetic* or *morphogenetic* landscapes (purportedly sculpted by geomorphic processes resulting from certain climatic conditions) were put to the test, they failed to account for so many exceptions that they were abandoned. And so the debate about the evolution of landscapes continues.

KEY TERMS

annular drainage (p. 435)

antecedent stream (p. 437)

butte (p. 432)

cuesta (p. 430)

cycle of erosion (p. 439)

dendritic drainage (p. 436)

drainage density (p. 435)

geologic structure (p. 430)

hogback (p. 430)

interfluve (p. 439)

mesa (p. 432)

monadnock (p. 439)

pediplane (p. 439)

peneplain (p. 439)

radial drainage (p. 435)

rectangular drainage (p. 436)

stream piracy (p. 437)

superimposed stream (p. 436)

trellis drainage (p. 436)

water gap (p. 436)

REVIEW QUESTIONS

1. How can lithology influence stream erosion?
2. What kinds of landforms are most likely to exhibit a radial drainage pattern?
3. Contrast a dendritic drainage pattern with a trellis pattern.

4. How might stream piracy result in the rejuvenation of a stream?
5. Briefly describe the cycle of erosion as envisioned by William Morris Davis.

REFERENCES AND FURTHER READINGS

CHORLEY, R. J., ed.   *Introduction to Fluvial Processes* (London: Methuen, 1971).

CHORLEY, R. J., et al.   *Geomorphology* (London/New York: Methuen, 1984).

DAVIS, W. M.   *Geographical Essays* (New York: Dover, reprint of 1909 original, 1954).

DERBYSHIRE, E., ed.   *Climatic Geomorphology* (New York: Harper & Row, 1973).

GORDON, N. D., MCMAHON, T. A., and FINLAYSON, B. L.   *Stream Hydrology: An Introduction for Ecologists* (New York: Wiley, 1992).

KING, P. B., and SCHUMM, S. A.   *The Physical Geography of William Morris Davis* (Norwich, England: GeoBooks, 1980).

KNIGHTON, D.   *Fluvial Forms and Processes* (London: Edward Arnold, 1984).

LEOPOLD, L. B.   *A View of the River* (Cambridge, Mass.: Harvard University Press, 1994).

LEOPOLD, L. B., et al.   *Fluvial Processes in Geomorphology* (San Francisco: Freeman, 1964).

MORISAWA, M.   *Streams: Their Dynamics and Morphology* (New York: McGraw-Hill, 1968).

PETTS, G., and FOSTER, I.   *Rivers and Landscapes* (London: Edward Arnold, 1985).

RICHARDS, K. S.   *Rivers: Form and Process in Alluvial Channels* (London/New York: Methuen, 1982).

RICHARDS, K. S., ed.   *River Channels: Environment and Process* (New York: Blackwell, 1987).

SCHUMM, S. A.   *The Fluvial System* (New York: Wiley, 1977).

SMITH, D. I., and STOPP, P.   *The River Basin* (New York: Cambridge University Press, 1978).

TAGLIABUE, J.   "Still Trying to Make the Tiber Behave," *New York Times,* September 16, 1993, A6.

Aggradational Landforms of Stream Erosion

OBJECTIVES

◆ To highlight a number of landforms built by rivers.

◆ To discuss the formation and development of the stream floodplain.

◆ To discuss the evolution of river deltas.

In Unit 42, we examine streams as sculptors, carvers, and cutters, and we explain how streams degrade and how running water can denude countrysides. In this unit, we focus on rivers as builders. Although the world's rivers disgorge hundreds of millions of tons of sediment annually into the oceans, part of their transported load does not reach the sea but is laid down on floodplains, in deltas, and elsewhere on land. Whereas the landforms of stream aggradation are not as spectacular as those in high mountains and incised plateaus, they are nevertheless interesting and important. Often, the deposits along a river's lower course reveal the history of the stream's upper course, and can help us unravel the complexities in understanding the evolution of a regional drainage basin.

ALLUVIAL FANS

Let us begin with a special case. In certain areas, especially in arid zones of the world, streamflow is discontinuous. Rainstorms in the mountains produce a subsequent rush of water in the valleys, and turbulent, sediment-laden streams flow toward adjacent plains. Emerging from the highlands, the water slows down and deposits its sedimentary load. Much of this water infiltrates the ground, and evaporation in the desert heat diminishes the rest. Meanwhile, it has stopped raining in the moun-

UNIT OPENING PHOTO: Deposition is the dominant landform sculptor in this scene of the inundated floodplain of Myanmar's (Burma's) Irrawaddy, one of Southeast Asia's great rivers.

FIGURE 43-1 Aerial view of a classic alluvial fan, formed where a stream has emerged from a canyon in the Black Mountains onto the famous eastern California lowland known as Death Valley.

tains, and soon the stream runs dry. A stream that flows intermittently like this is called an **ephemeral stream** (as opposed to a permanent stream).

Ephemeral streams often construct alluvial fans where they emerge from highland areas. As the term suggests, an **alluvial fan** is a fan-shaped deposit consisting of alluvial material, located where a mountain stream emerges onto a plain (Fig. 43-1). In this situation, the stream is not part of the regional drainage basin that ultimately leads to the ocean. Streams that form alluvial fans are unlikely to flow far beyond the edge of the fan. And when the mountain rains are below average, the stream may not even reach the outer margin of its own deposits.

Fan-shaped deposits are not unique to desert areas, although they are best developed there. They can also be found in areas where glaciation has taken place and where streams now carry heavy loads of debris to the edges of glacier-steepened mountains. Others are located on the flanks of steep-sided volcanoes, where streams loaded with volcanic material reach a flat plain. But the typical alluvial fan is primarily a desert landform, a product of stream aggradation in an arid environment.

The alluvial fan attains its conical or semicircular shape because the stream that emanates from the often canyon-shaped mountain valley tends to have an impermanent course on the fan surface. The sediment-clogged water, when it surges from the mouth of the canyon, quickly slows down, so that it must drop part of its load. Figure 43-2a shows what happens.

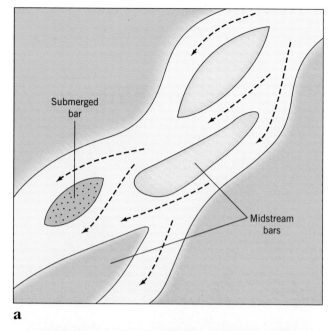

a

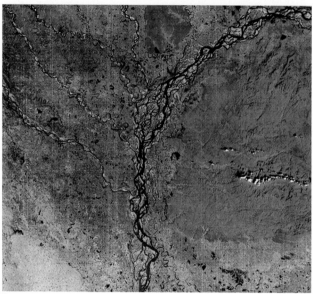

b

FIGURE 43-2 Braided streams begin with the deposition of midstream bars (a). These obstacles split the river into many channels, greatly widening the stream. Some braided rivers can reach widths of 8 km (5 mi), as is the case with the Brahmaputra River in northeastern India as it flows out of the Himalayas on its way to its delta in Bangladesh (b).

First, a single **midstream bar** forms, and water flows around this obstruction; soon more deposition takes place, and many bars, some submerged, develop. Now the stream divides into many smaller channels that intertwine with one another to form a **braided stream** (a process that in certain large rivers can reach several kilometers in width [Fig. 43-2b]). Under such conditions, the stream obviously cannot erode a deep valley; in fact, it is flowing on deposits that may build to an elevation above the rest of the fan surface. The next time rains generate a stream surge, the water may seek a different direction.

FIGURE 43-3 Desert pavement marks the floor of an arid valley in southeastern California. This coarse layer of gravel—varnish-like in appearance—protects finer sediments, lying just below the surface, from wind erosion.

An alluvial fan, therefore, consists of a series of poorly stratified layers, thickest near the mountain front and progressively thinning outward. The coarsest sediments are normally located nearest the apex of the fan and finer-grained material toward the outer edges. Since the fan lies on bedrock and has layers of greater and lesser permeability, infiltrating water can be contained within it. Many alluvial fans in the southwestern United States and throughout the world are sources of ground-water for permanent settlements.

When conditions are present for alluvial fan development, a mountain front may have not just one or two but dozens of larger and smaller alluvial fans, coalescing across the *pediment* (the smooth, gently sloping bedrock surface that underlies the alluvial cover and extends outward from the foot of the highlands). When this happens, the cone shapes of individual fans may be difficult to distinguish, and the landform is called an *alluvial apron* or a **bajada** (from the Spanish word for slope). Such an assemblage of alluvial fans can exhibit many features. Where streamflow has increased, gulleys have been cut into the fans, and thus the older, upper parts of the fan surface are no longer subject to the shifting stream process described above. These older areas become stable, may support vegetation, and often develop a varnished appearance on the weathered gravel surface. This is referred to as **desert pavement** (Fig. 43-3).

RIVERS TO THE SEA

Most streams, and virtually all of the world's major rivers, are parts of drainage systems that ultimately flow into the oceans rather than into a closed desert basin. As stream capacity increases in its lower course, a river ceases degrading and begins aggrading and building alluvial landforms. This is manifested in the development and widening of its floodplain, and ultimately the formation of the coastal equivalent of the alluvial fan—the delta.

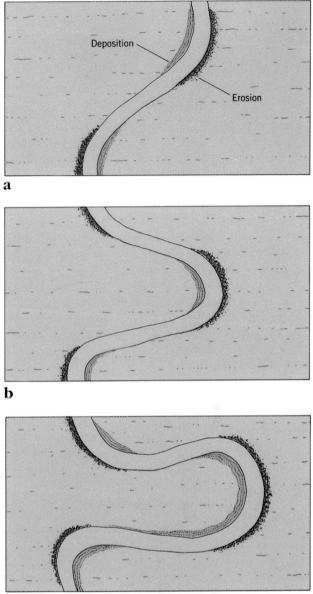

a

b

c

FIGURE 43-4 The development of meanders in a stream involves both deposition and erosion. In (a) a stream bend displays evidence of deposition on the inside of the bend and erosion on the outside curve. In (b) the river's bends are close to developing meander properties. Erosion on the outside of the now-sharper bend is increasing, while deposition is filling the inside. In (c) the process has advanced to the meander stage. The river is building its floodplain; the inside of the meander is a growing alluvial deposit. On the outside curve of the meander, erosion is strongest toward the lower part (the southeast in this sketch), resulting in the down-valley migration of the meander itself.

The course of the river exhibits several changes as its depositional function gains strength. Whereas the river fully occupied its **V**-shaped valley upstream (*see* Fig. 41-4a), the channel now begins to erode laterally and for the first time the valley becomes slightly wider than the stream channel. Bends in the river channel, called **meanders**, are increasingly evident (Fig. 43-4). Erosion on the outside of these meanders is greater than on the inside; as the meanders grow larger, erosion on the inside stops altogether and deposition begins there, as Fig. 43-4 illustrates.

The Floodplain

When meanders develop in a stream channel, they move in two directions. First, as Fig. 43-4 shows, they erode laterally and increase in size; second, they migrate downstream. These two motions, the lateral swing and the downstream shift, have the effect of widening the river's valley and creating an extensive **floodplain**—the flat, low-lying ground on either side of the stream channel that is inundated during periods of unusually high water (Fig. 43-5). Note that the channel of the stream still fills almost the entire valley at first (Fig. 43-5a), but becomes an ever-smaller part of the valley as time goes on (Figs. 43-5b and c). Also, observe that the bottom of the stream channel, swinging back and forth across the valley, creates the base of the floodplain in the underlying bedrock. But by now, the channel is flanked by deposits laid down by the stream itself.

Figure 43-5c shows the development of two small, crescent-shaped lakes. These are **oxbow lakes**, which form when a meander is cut off, as shown in Fig. 43-6. Such cutoffs can occur, as seen in the diagram, when the downstream movement of one meander "catches up"

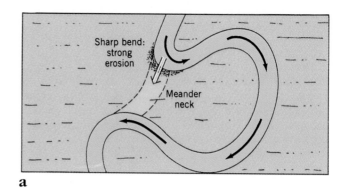

a

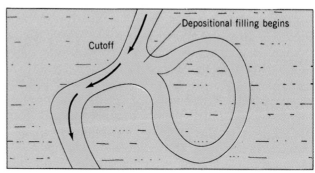

b

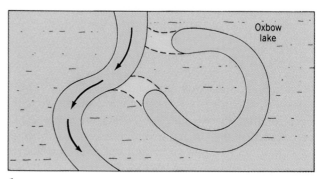

c

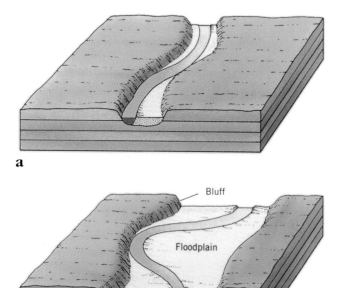

a

b

c

FIGURE 43-5 Floodplain formation and widening as meanders grow, erode laterally, and migrate downstream.

d

FIGURE 43-6 Oxbow lake formation as a result of meander neck cutoff.

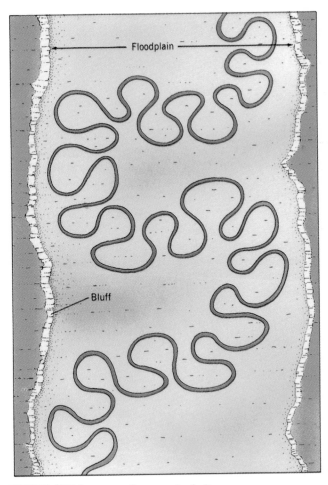

FIGURE 43-7 Segment of a meander belt.

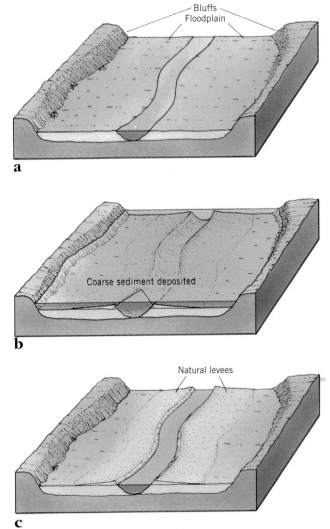

FIGURE 43-8 The relationship between floods and natural levee development. The river's coarsest deposits are laid down closest to the overflow (b). After repeated flooding, these deposits are built up as levees that contain the stream channel (c).

with the lagging movement of the bend immediately below it (Figs. 43-6b and c). Another way this can happen is during a flood, when a meander neck, such as that shown in Fig. 43-6a, can be swept away, and the channel deepened along the course of the dashed lines. When the flood subsides, the new cutoff may be deep enough to have the same effect as in Fig. 43-6d, and an oxbow lake will form.

The alluvium-filled floodplain is bounded by *bluffs*, created by the meandering stream (Fig. 43-5). The floodplain over time may become many kilometers wide, and *meander belts* themselves may form giant meanders, as shown in Fig. 43-7. If you draw a line through the approximate center of the meanders shown, you will see that the pattern is repeated. And, as in the case of individual meanders, the entire meander belt tends to move down the gradient. As a result, a floodplain is full of evidence of previous meander, meander-belt, oxbow-lake, and other positions. These are referred to as scars, so that a dried-up oxbow lake becomes a *meander scar*, a place where a meander once existed.

As we noted earlier, a floodplain is so named because this plain, between the bluffs, is subject to frequent flooding. Annual floods, during which the river overflows its banks, are a normal part of the floodplain's development. These floods deposit sediments that build the river's **natural levees**, broad ridges that run along both sides of the channel (Fig. 43-8). As the river spills out of its channel, the coarsest material it is carrying is deposited closest to the overflow, hence along the levees. When the river contracts after the flood, it stays within its self-generated levees (Fig. 43-8c).

But not all floods are so regular and productive. Infrequently—perhaps once per century—a river may experience a flood of such magnitude that its floodplain is greatly modified. Water up to several meters deep may inundate the entire floodplain, destroying submerged levees, eroding bluffs, and disrupting the entire system. Such floods have cost millions of lives in the densely populated floodplains of Asia's major rivers; they also occur in the Mississippi Basin of the central United States, where the damage, too, can be awesome (*see* Perspec-

The Great Midwestern Flood of '93

Within the space of 17 months, between August 1992 and January 1994, the United States suffered the three costliest natural disasters in its history: Hurricane Andrew, the Midwestern floods, and the Los Angeles earthquake. The hurricane and earthquake wreaked havoc in relatively small areas, notably Miami's southern suburbs and the San Fernando Valley; but the massive flooding that for weeks on end ravaged the lower Missouri and central Mississippi basins at one point inundated an area half the size of Maine. Loss of life was fortunately modest (about 50), but the rest of the toll was truly astounding: property damage that surpassed $7 billion; crop losses in the range of $10 billion; damaged or destroyed

homes numbering over 55,000; failure of more than half of the region's 1400 artificial levees; and, in hundreds of towns and cities, flood crests higher than any previously recorded.

Even for this flood-prone region, the floods of late spring and early summer 1993 were extraordinary. Hydrologists quickly concluded this event was a *500-year flood*, meaning that, statistically, a flood of this magnitude should happen only once every five centuries. The cause of the great flood was a record-shattering deluge of rain that soaked the Upper Midwest for most of May, June, and July. With the ground in both river basins saturated, the only place this water could go was into the Missouri and Mississippi.

The worst flooding occurred where these two mighty rivers join just north of St. Louis (Fig. 43-9). At the Gateway Arch on the riverfront in downtown St. Louis, the Mississippi crested at an all-time high level of 15.1 m (49.58 ft), more than 6 m (20 ft) above flood stage and 11 m (36 ft) higher than its normal midsummer flow. Although it is human folly to deny a river its floodplain, people and activities (with widespread government support) were busily reestablishing themselves in the inundated zones a year after the great flood—blithely confident that such a disaster would not repeat itself in the foreseeable future.

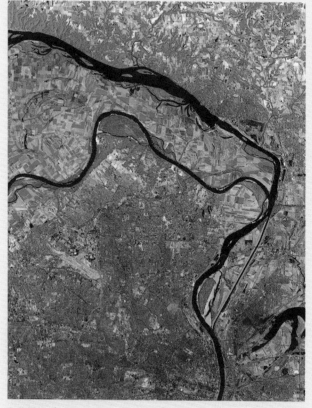 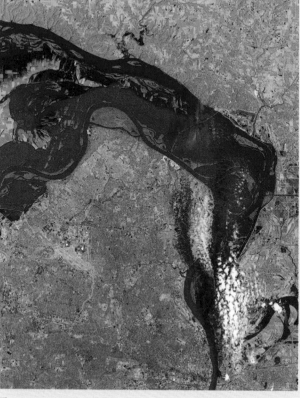

FIGURE 43-9 Satellite images of the confluence zone of the Mississippi and Missouri Rivers just to the north of the city of St. Louis. The image on the left, taken on July 4, 1988, shows the river channels in their normal early-summer positions. The image on the right, taken on July 18, 1993, shows the awesome extent of the mid-1993 floodwaters, which peaked at a slightly higher level on August 1.

tive: The Great Midwestern Flood of '93). No reinforcement of natural levees or construction of **artificial levees** (*see* Fig. 41-8) can withstand the impact of such an onslaught.

Terraces

After a flood, a river returns to pre-flood volume and functions. But what happens in a floodplain if a stream is rejuvenated? Increased volume (through stream capture or longer-term climatic change in the drainage basin) or tectonic events may increase a meandering stream's capacity to erode.

Let us first consider an instance in which a stream cuts into its own alluvial deposits (Fig. 43-10). For simplicity, levees and other features have been left off the diagram. The rejuvenated stream cuts downward from its original level on the floodplain (a) to a new level (b), where it stabilizes and meanders develop. Soon, a new floodplain within the older floodplain develops (c), complete with a new set of bounding bluffs.

Remnants of the older floodplain stand above these newer bluffs as **terraces**. These terraces reveal the two-stage evolution of the valley, and may be correlated with other information about the climate, base level, or tectonic uplift in the region involved. The terraces shown in Fig. 43-10 are *paired terraces*; that is, they lie at the same elevation on each side of the rejuvenated stream. Sometimes terraces are not paired as a result of a combination of valley deepening and lateral (sideways) erosion. This can destroy one side of a set of paired terraces, sometimes making studies of valley history quite difficult.

Conceivably, the rejuvenated stream will remove all of its alluvial base in the floodplain, leaving a **rock terrace** rather than creating an alluvial terrace. Technically, such a rock terrace is a degradational landform, but its genesis relates to an earlier phase of floodplain aggradation. Under certain circumstances, such as the uplifting of the land surface above base level, whole meander belts can be incised into hard bedrock from overlying floodplain topography. These incised or **entrenched meanders** can produce some spectacular scenery, as in the San Juan and Colorado River valleys of southern Utah (Fig. 43-11).

DELTAS

About 2500 years ago, the ancient Greek scholar Herodotus, studying the mouth of the Nile River, found that the great North African river forms a giant fan-shaped deposit where it reaches the Mediterranean Sea. Noting the triangular shape of this area of sedimentation, he called it a **delta**, after the fourth letter of the Greek alphabet, which is written as Δ. Ever since, river-mouth deposits

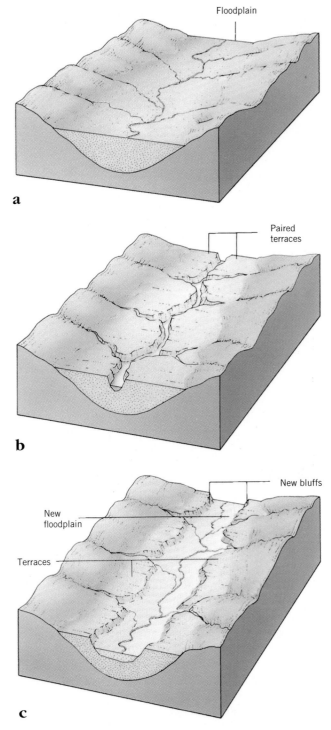

FIGURE 43-10 Paired terraces emerge as the rejuvenated stream cuts downward into its own floodplain (b). Eventually a new floodplain embedded within the older one develops, bordered by newer bluffs formed by the inner slopes of the paired terraces (c).

have been called deltas, even when they have a different shape.

Near its mouth, a river comes close to base level and slows down markedly. Even the finest sediment being carried in suspension is deposited, so that the river

FIGURE 43-11 Entrenched meanders mark the sharply incised Colorado River, here at the confluence with the Green River, as it winds its way through southeastern Utah.

mouth becomes clogged. As a result, the stream channel breaks down into smaller channels that flow over the accumulated sediment. These channels, which begin at the apex of the delta and carry the river's water in several directions over the surface, are called **distributaries**. Thus a trunk river receives *tributaries* in its drainage basin and develops *distributaries* where it forms a delta.

As the map of the Nile delta (Fig. 43-12a) shows, the Nile forms a few prominent distributaries and many smaller ones. When Herodotus did his fieldwork, the Nile Delta was an uninhabited swampy area. Today, it is an area of dense rural settlement and intensive cultivation (Fig. 43-13). Control of the Nile's distributaries and land reclamation have made this transformation possible.

The exact form of a delta is determined by (1) the volume of the stream and the amount of sediment it carries, (2) the configuration of the offshore continental shelf near the river mouth, and (3) the strength of currents and waves. Notice, on a world map, that the Nile and Mississippi Rivers have large deltas, but that the Zaïre

(Congo) River of west-equatorial Africa does not. The Nile and Mississippi flow into relatively quiet waters, and offshore depths increase gradually. The Zaïre River, however, flows into deeper water immediately offshore, coastal currents are strong, and its sediment supply is less than either the Nile's or the Mississippi's. As the map of the Nile Delta (Fig. 43-12a) shows, the Mediterranean Sea is not without coastal currents, but these have the effect of creating sandbars and lagoons rather than destroying the advancing delta.

As the delta grows seaward, the *deltaic plain* (the flat landward portion of the delta) stabilizes. The distributaries of many deltas are today dredged and controlled, affecting the process of formation. In the case of the Mississippi Delta, for example, many distributaries that would have become blocked by sediment are kept open, creating a deltaic form known as a *birdfoot delta* (Fig. 43-12b). This shape differs considerably from that shown in Fig. 43-12c, where the seaward edge of the delta displays none of the finger-like extensions of the birdfoot delta. Here, currents and waves cannot prevent the formation of the Niger Delta on the coast of West Africa, but they do sweep sediment along the shoreline rather than allowing the evolution of birdfoot characteristics.

The Delta Profile

No two deltas form in exactly the same way, and the process is extremely complicated. A simplified version of delta formation is illustrated in Fig. 43-14. The finest deposits to be laid down are the *bottomset beds*. The river in our diagram is depositing its finest-grained material ahead of the delta, where (we may assume in this instance) the water is quiet and such deposition can occur. In the meantime, the river is adding to the *topset beds* of the delta, the horizontal layers that underlie the deltaic plain. As the delta grows outward, the *foreset beds* are built from the leading edge of the topset beds. Later, the newly accumulated foreset beds will be covered by extended topset beds.

The thickness and resulting weight of deltaic sediments can depress the coastal crust isostatically, complicating the process of delta development still further. Areas of surrounding coastland, but not part of the delta structure itself, may be affected by such subsidence, with serious consequences. The Mississippi Delta (*see* Fig. 33-7 on p. 354), where it enters the Gulf of Mexico in southeastern Louisiana, is a classic example of this phenomenon, and much land has been lost in recent years.

Deltas are among the largest aggradational features of fluvial erosion. Those boulders and pebbles that were dragged down mountain valleys now lie as fine grains on the coast—pulverized, transported, and deposited by rivers, those great sculptors and builders of the landscape.

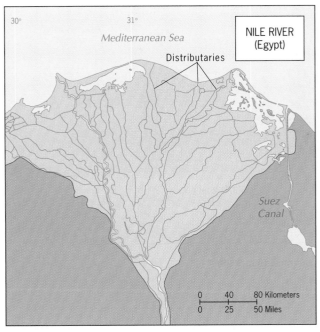

a

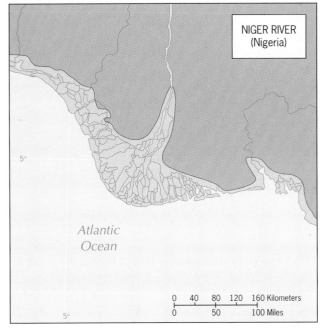

c

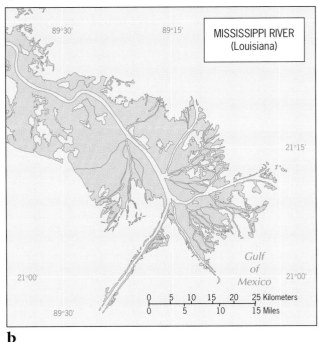

b

FIGURE 43-12 The spatial form of a delta depends on the quantity of sediment carried by the river, the configuration of the continental shelf beyond the river mouth, and the power of waves and currents in the sea. The Nile Delta (a) exhibits the classic triangular shape. The Mississippi Delta (b), exhibiting the birdfoot shape, results from large quantities of sediment carried into quiet water. The Niger Delta (c) is shaped by strong waves and currents that sweep sediment along the coast.

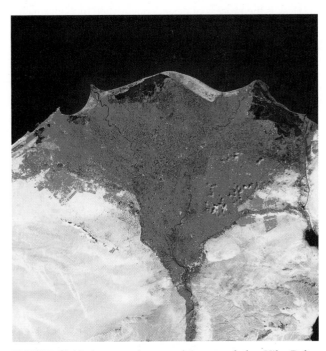

FIGURE 43-13 A remotely sensed image of the Nile Delta, whose intensively cultivated croplands show up in brilliant red. This image should be compared to Fig. 41-5, which shows the same region (rotated 90 degrees to the east) in its true colors.

We have now completed our survey of the geomorphic processes associated with running water and the fluvial landscapes and landforms they shape. But we are not yet ready to turn our attention away from water as a geomorphic agent. A special case of degradational action involves the removal of rock not by physical breakdown but by chemical solution. This process produces landscapes of highly distinctive surface and near-surface features, which we examine in Unit 44.

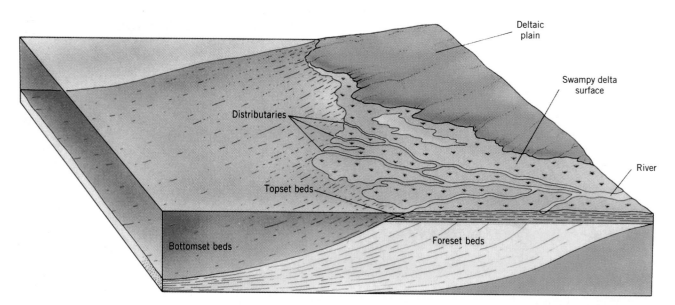

FIGURE 43-14 Internal structure of a small delta. Large deltas, such as those mapped in Fig. 43-12, are far more complex in their composition.

KEY TERMS

alluvial fan (p. 442)

artificial levee (p. 447)

bajada (p. 443)

braided stream (p. 442)

delta (p. 447)

desert pavement (p. 443)

distributaries (p. 448)

entrenched meander (p. 447)

ephemeral stream (p. 442)

floodplain (p. 444)

meander (p. 443)

midstream bar (p. 442)

natural levee (p. 445)

oxbow lake (p. 444)

rock terrace (p. 447)

terrace (p. 447)

REVIEW QUESTIONS

1. Describe the circumstances under which alluvial fans form.

2. How is the process of meandering related to the formation of floodplains?

3. Describe the stages of formation of an oxbow lake.

4. What are stream terraces, and how are they formed?

5. Describe the similarities between alluvial fans and river deltas.

REFERENCES AND FURTHER READINGS

CARLING, P. A., and PETTS, G. E., eds. *Lowland Floodplain Rivers: Geomorphological Perspectives* (New York: Wiley, 1992).

CHANGNON, S. A., ed. *The Great Flood of 1993* (Boulder, Colo.: Westview Press, 1995).

CHORLEY, R. J., ed. *Introduction to Fluvial Processes* (London: Methuen, 1971).

CHORLEY, R. J., et al. *Geomorphology* (London/New York: Methuen, 1984).

DE BLIJ, H. J., ed. *Nature on the Rampage* (Washington, D.C.: Smithsonian Institution Press, 1994).

DURY, G. H., ed. *Rivers and River Terraces* (London/New York: Macmillan, 1970).

GORDON, N. D., MCMAHON, T. A., and FINLAYSON, B. L. *Stream Hydrology: An Introduction for Ecologists* (New York: Wiley, 1992).

GRAF, W. L. *Fluvial Processes in Dryland Rivers* (New York/Berlin: Springer-Verlag, 1988).

HOYT, W. G., and LANGBEIN, W. B. *Floods* (Princeton, N.J.: Princeton University Press, 1955).

KNIGHTON, D. *Fluvial Forms and Processes* (London: Edward Arnold, 1984).

LEOPOLD, L. B. *A View of the River* (Cambridge, Mass.: Harvard University Press, 1994).

LEOPOLD, L. B., et al. *Fluvial Processes in Geomorphology* (San Francisco: Freeman, 1964).

MORISAWA, M. *Streams: Their Dynamics and Morphology* (New York: McGraw-Hill, 1968).

PETTS, G., and FOSTER, I. *Rivers and Landscapes* (London: Edward Arnold, 1985).

RACHOCKI, A. H., and CHURCH, M., eds. *Alluvial Fans: A Field Approach* (New York: Wiley, 1990).

RICHARDS, K. S. *Rivers: Form and Process in Alluvial Channels* (London/New York: Methuen, 1982).

RICHARDS, K. S., ed. *River Channels: Environment and Process* (New York: Blackwell, 1987).

SCHUMM, S. A. *River Morphology* (Stroudsburg, Pa.: Dowden, Hutchinson & Ross, 1972).

SCHUMM, S. A. *The Fluvial System* (New York: Wiley, 1977).

SMITH, D. I., and STOPP, P. *The River Basin* (New York: Cambridge University Press, 1978).

WARD, R. *Floods: A Geographical Perspective* (New York: Macmillan, 1978).

Karst Processes and Landforms

OBJECTIVES

♦ To discuss the general environmental conditions that favor the formation of karst landscapes.

♦ To analyze the landforms that are the signature of karst landscapes.

♦ To relate karst processes to the development of extensive underground cave systems.

As we note elsewhere in Part Five, water erodes rocks of all kinds, sculpting the surface into many distinctive landscapes. Under certain special conditions, however, water dissolves soluble rocks and minerals, carrying them off in solution. Water performs this function both at and *below* the surface. When it dissolves rocks beneath the surface, it may remove soluble layers while leaving overlying as well as underlying strata in place. This leads to the formation of caves and associated subterranean features.

Caves occur in many areas of the world, and some are so large and spectacular that they have become quite famous. Mammoth Cave, located in west-central Kentucky, is a network of underground chambers and passages totaling over 500 km (310 mi) in length. Carlsbad

Caverns in New Mexico (photo above) has more than 37 km (23 mi) of explored chambers and tunnels. Other major cave systems lie in the Appalachians, in Indiana, Missouri, Texas, Utah, and elsewhere in the United States; they also can be found in many parts of Europe, in China, Australia, Africa, and South America. In short, caves have developed wherever the conditions for their formation were favorable, and such conditions exist, or have existed, in thousands of places beneath the earth's surface.

Although this unit focuses on caves and other features formed by solution, we should note that not all

UNIT OPENING PHOTO: The unearthly formations in Carlsbad Caverns, New Mexico, created by dripping water rich in calcium carbonate.

caves are sculpted this way. Caves are also carved by waves along shorelines; they can be created by tectonic movements; and they can even result from large-scale eluviation processes. Here, however, as this unit's title indicates, we focus on *karst* topography, which is associated with *limestone*. Other caves and caverns, including those formed in other materials, are not karst features.

Archeologists have discovered that caves were purposely occupied hundreds of thousands of years ago, and some caves contain valuable evidence about their occupants. In Lascaux Cave in southwestern France, the cave walls were decorated by artists, from whose drawings we can deduce what kinds of animals were hunted and how the inhabitants may have lived. Other caves, such as the Sterkstroom Caves in South Africa, have yielded skeletons whose dimensions helped anthropologists unlock the secrets of the chain of human evolution. Caves, therefore, are more than mere physiographic and archeological curiosities. They are expressions of a particular set of geomorphic processes that also produce many additional related landforms.

FIGURE 44-1 The depression-pocked surface of the Pennyroyal Plain, lying to the east of Bowling Green, Kentucky, is a classic midlatitude karst landscape.

KARST

Except for its entrance, a cave cannot be seen from the surface. But as we have noted, the processes that form caves also produce visible landforms and, indeed, entire landscapes. If rock removal by solution can go on beneath the ground, it obviously can also take place at the surface. When this happens, the landscape takes on a distinctive, sometimes unique appearance (Fig. 44-1). Such scenery is called **karst** landscape, a term that has its origin in the area of east-central Europe where Slovenia and Croatia (newly independent countries that were components of former Yugoslavia) and Italy meet. There, in a zone bordering the Adriatic Sea, an arm of the Mediterranean, lie some of the most spectacular and characteristic of all karst landscapes. Surface streams disappear into subsurface channels, steep-sided and closed depressions dot the countryside, and stark limestone hills rise above a seemingly chaotic topography. This terrain extends to the Adriatic coast itself, and where sea and limestone meet the shore becomes a monument of natural sculpture.

Karst landscapes are not always as spectacular as this, but some karst areas are world famous for their angular beauty. Perhaps the most remarkable of all lies in southeastern China, centered on the city of Guilin (*see* Unit 37 opening photo). Here, the Li River winds its way through a landscape that has for millennia inspired artists and writers. And to the west, in the province of Yunnan, lies another unique manifestation of karst processes, the fantasy-like Stone Forest (Fig. 44-2).

Karst terrain is widely distributed across the earth, occurring on all the continents in hundreds of localities.

More than a century ago, in 1893, a Croatian scholar named Jovan Cvijic (pronounced "yoh-VAHN SVEE-itch") produced the first comprehensive study of karst processes and landscapes, under the title that translates as *The Karst Phenomenon*. Ever since, the term *karst* has been in use. Cvijic also described and gave names to many landforms resulting from karst processes. How-

FIGURE 44-2 Erosion in limestone areas can produce some unusual landforms. The so-called Stone Forest in Yunnan Province, China is an aptly named aggregation of limestone columns, remnants of a once-continuous, thick layer of jointed limestone attacked by solution from above (rainwater) as well as below (rising and falling groundwater). The Stone Forest is studded with small lakes whose rise and fall is marked on the column walls. (Authors' photo)

ever, he was not aware of the numerous places where karst topography also existed, and later additional karst phenomena were identified and named. As these studies progressed, karst geomorphology became an important part of physical geography.

KARST PROCESSES

Karst landforms and landscapes are the products of a complex set of geomorphic processes, conditions, and lithology. These include the stratigraphy, local relief, surface drainage, and groundwater. Karst landscape develops only where certain particularly soluble limestones, rich in calcite ($CaCO_3$), form all or part of the stratigraphy. Another soluble rock, *dolomite* ($CaMg[CO_3]_2$), can also form karst topography.

But, as we have noted earlier, not all limestones dissolve easily. Calcite-poor limestones are much less soluble. Also, where metamorphism has created marble from limestone, solution proceeds more slowly. In arid areas, limestone may resist weathering, solution, and erosion more than other rocks do, and thus may form ridges and plateaus rather than depressions and caves.

The texture and structure of the affected rocks also influence the solution process. The greater the permeability and porosity (water-holding capacity) of the rock, the more susceptible it is to solution and removal. When there are many joints, faults, and cracks in the limestone or dolomite strata, subsurface water can flow more easily. This, in turn, increases the rate of erosion.

The Role of Water

Even when the lithology and stratigraphy are suitable, karst topography may not fully develop unless other conditions also prevail. The most important of these other conditions are water and drainage. In karst areas, three kinds of water movement contribute to erosion: (1) surface streams, (2) underground drainage flows, and (3) groundwater. Surface streams may be poorly developed in karst terrain, but they supply the underground system. The underground network of interconnected channels is the most important solution agent. The fluctuations of groundwater and the water table also influence karst processes.

Together, these waters, moving across and through the limestone-layered rocks, create karst landforms. But the solution process is not uniform. Rainwater absorbs atmospheric gases as it passes through the atmosphere. Thus water (H_2O) plus a small quantity of carbon dioxide (CO_2) combine to become a weak acid (carbonic acid [H_2CO_3]). In this form, and depending on its acidity, rainwater becomes an effective solvent for limestone.

Another factor has to do with the soil and vegetation present in a karst area. They contribute to the presence of carbon dioxide, which is critical to the karstification process. More carbon dioxide dissolves into water in the soil because it is released during the decomposition of dead plants. Therefore, if soil water seeps into underground channels it will increase the acidity of the water in those channels and thus the water's capacity to dissolve limestone. This helps to explain why more fully developed karst topography is found in warmer as well as moister climatic regions. Higher temperatures promote biogenic action, and this in turn enhances the effectiveness of the available water (in the form of carbonic acid) as an agent of erosion.

Relief

The formation of karst landscape is further promoted when the area of limestone and/or dolomite strata affected lies under at least moderate relief. Where the surface is flat or nearly so, and where surface streams have not succeeded in creating some local relief, underground drainage and erosion are slowed, and karst formation is inhibited. Research has shown that in this respect, at least, surface rivers and subsurface streams have something in common.

We know that increasing water velocity at the surface increases a river's capacity to carry loads and perform erosion. Below the surface, water in tunnels also retains its erosional capacity longer if it moves rapidly, and loses it if it is slowed down. Sluggishly moving water soon becomes saturated with dissolved limestone and thereby loses its capacity to dissolve more of it. Dipping strata plus moderate relief combine to favor speedy subsurface water movement. Another condition favoring solution is the substantial uplift of the affected area. This allows underground streams to descend from one level to the next. Many cave networks lie on several levels, indicating that uplift and/or dropping water tables played a role in the evolution of the system (Fig. 44-3).

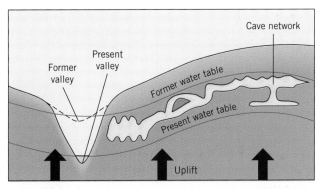

FIGURE 44-3 Cave network formed by water that entered limestone fractures and enlarged them below the (former) water table. When that water table dropped—partly due to uplift and partly due to the nearby stream's deepening of its valley—the cave system filled with air.

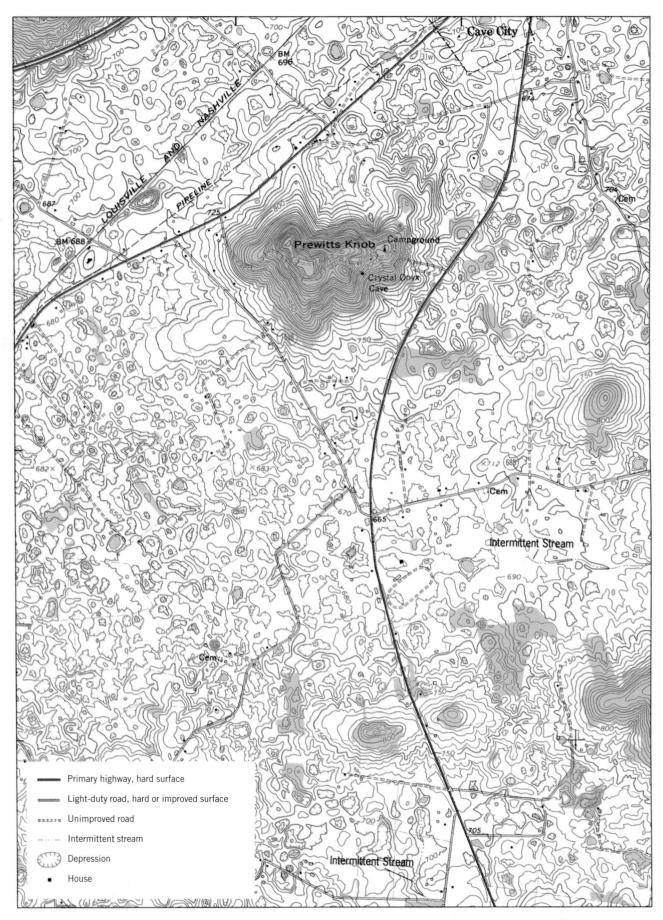

FIGURE 44-4 Karst landscapes exhibit unusual relief and drainage characteristics. This topographic map shows the surface near Mammoth Cave in west-central Kentucky, one of the most prominent areas of temperate karst in the United States. Note: (1) the myriad small basins (symbolized by ⊛) lacking outlets; (2) the pair of intermittent, disappearing streams; and (3) the steep slopes of Prewitts Knob rising from the nearly flat surrounding plain.

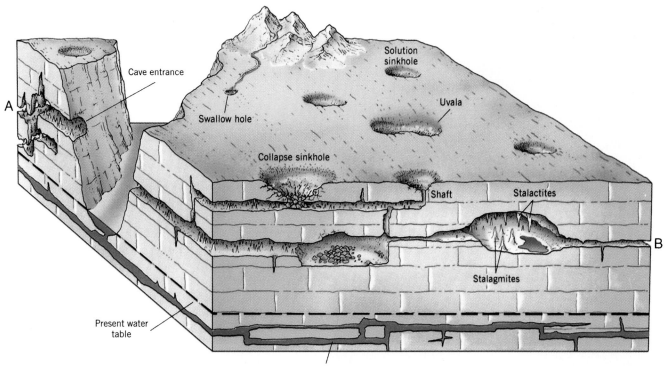

Cave entrance

A

Solution sinkhole

Swallow hole

Uvala

Collapse sinkhole

Shaft

Stalactites

B

Stalagmites

Present water table

Underground water channel

FIGURE 44-5 Surface and underground features of temperate karst. When the groundwater table was higher, the solution features at levels A and B formed. Now the water table is lower and solution proceeds. A new cave network will form when the next drop in the water table occurs.

Groundwater

Below the water table, porous rock is saturated, its pores and other open spaces occupied by water. The water table fluctuates seasonally, and rises and falls locally after rainstorms and during prolonged dry periods. But karst conditions are different because the rocks in limestone areas are soluble. Thus the groundwater does more than simply occupy openings: in the form of weak carbonic acid, it contributes to underground erosion by changing the calcium carbonate in limestone into calcium bicarbonate, which is removed in solution. Studies have shown that karst areas contain *perched aquifers*, pockets of groundwater situated above the level of the local water table. Water in these perched aquifers is confined, just like water under artesian conditions, and sometimes it emerges in natural springs (*see* Fig. 40-10a).

By tracing the movement of water in underground drainage networks, scientists have made significant discoveries about cave systems. One way to accomplish this is by putting colored dyes in surface water where it disappears below the ground, and then to check for the appearance of the colored water in certain accessible cave locations and at springs. This research has indicated that a cave-riddled karst area may have several underground drainage systems that are separate, not interconnected.

In a single karst area, therefore, there may be as many as half a dozen subsurface drainage systems, functioning at different levels but not linked to one another. This means that when we enter a honeycombed mountain and follow the tunnels leading inward from a cave entrance, there are other tunnels that are not directly accessible above as well as below. Thus water entering the subsurface from one location on the surface becomes part of one underground system, while water from another ground-level location joins a different and separate network below.

KARST LANDFORMS AND LANDSCAPES

A topographic map of a karst area quickly reveals the unusual character of relief and drainage (Fig. 44-4). It seems that all the rules we have learned so far are broken: surface streams are interrupted and stop (or shrink) in mid-valley, contour lines reveal basins without outlets and without streams entering or leaving them. Steep-sided hills rise from flat plains without displaying slope characteristics familiar to us.

When we examine the maps more closely, it becomes clear that the signature of karst topography is not the same everywhere. Geomorphologists identify three types of karst landscape: temperate, tropical, and Caribbean.

1. **Temperate karst**, of which the "type" area in Slovenia and Croatia is an outstanding example, forms more slowly than tropical karst. Disappearing streams, jagged rock masses, solution depressions, and extensive cave networks mark temperate karst (Fig. 44-5).

2. **Tropical karst** develops rapidly as a result of the

FIGURE 44-6 A satellite image of east-central Florida, a classic Caribbean karst region. The large number of round lakes, formed in sinkholes, is the signature of this landscape. The triangular promontory on the Atlantic coast at the right is Cape Canaveral, which is covered by NASA's rocket-launch complex. The greenish-tinged large lake at the same latitude, located just southwest of the center of the image, is Lake Apopka; the Orlando metropolitan area lies to its southeast.

higher amounts of rainfall and humidity, biogenic action, and organic acids in the soil and thus the subsurface water. Steep-sided hills tend to be vegetation-covered, and solution features are larger than in temperate karst landscapes.

3. **Caribbean karst** is a special case found only in a few locations. In the type area of central Florida (Fig. 44-6), nearly flat-lying limestones are eroded underground, although they lie barely above sea level. Water comes from hill country to the north, seeps through the limestones, and leaves the system through offshore submarine springs. A similar situation exists in Mexico's Yucatán Peninsula; there, as in Florida, the roofs of the subsurface conduits have in many places collapsed, creating those characteristic depressions in the ground (*see* Fig. 44-1).

Disappearing Streams and Sinkholes

Let us now examine some landforms we might encounter in any karst landscape. Where a surface stream "disappears," flowing into an underground channel, the place of descent is called a *swallow hole* (Fig. 44-5, top left). This may occur at a fault or an enlarged joint that has been widened by solution and leads to a subsurface system. Interrupted drainage of this sort is a general indicator of karst conditions. Another common karst landform is the surface depression, ranging in size from small hollows to large basins. The dominant process of formation, as we have noted, is solution. Depressions also can form from the collapse of part of the roof of an underground stream conduit. Logically, the former are referred to as solution sinkholes and the latter as collapse sinkholes (*see* Fig. 44-5).

Solution sinkholes (Fig. 44-5, top right) range in size from a bathtub to a football stadium. A single area of karst topography may contain tens of thousands of them, some old and established, others just starting to form. Their relative location probably has to do with the configuration of the terrain when karstification began; low-lying places became solution hollows first. These subsequently expanded, and others formed as the surface was lowered overall. In cross-section (*see* sinkhole at front edge center of Fig. 44-5), solution hollows resemble funnels, water seeping down the sides to the approximate center of the basin. There a *shaft* leads downward, and the water joins the regional groundwater (or subsurface conduit) through this outlet.

Collapse sinkholes (Fig. 44-5, left front corner), as we noted, are created by the collapse or failure of the roof or overlying material of a cave, cavern, or underground channel. We should remember that the solution process also created these subsurface tunnels, so even a collapse sinkhole ultimately owes its origin to the solution process. Physical geographers distinguish between a *collapse sink*, in which the rock ceiling of the sinkhole collapses into the underground solution hole, and a *suffosion sink*, created when an overlying layer of unconsolidated material is left unsupported. Where such loose material lies on top of limestone strata affected by the karst process, lower parts of this material are drawn downward into the enlarging karst joints. This creates a void in the lower stratum of this unconsolidated layer, and when the void becomes large it will collapse. That can happen quite suddenly (*see* Perspective: North America's Most Famous Sinkhole), unlike the slow development of solution sinkholes. Obviously, building in a zone of collapse or suffosion sinkholes can be a hazardous proposition!

Occasionally, two or more neighboring sinkholes join to become an even larger depression, called an **uvala** (Fig. 44-5, top center). Uvalas can reach a diameter of more than 1.5 km (1 mi). Some have rough, uneven floors that are dry and vegetated; others are filled with water and form scenic lakes. Studies of sinkhole terrain indicate that this kind of karst landscape develops best where the environment is humid, so that there is plenty of water to sustain the underground drainage system. It also appears that the rise and fall of the water table may have something to do with the distribution of collapse sinkholes. Where groundwater levels rise and fall rapidly and substantially, this process seems to trigger more frequent collapses.

North America's Most Famous Sinkhole

Cultural landscapes often experience sudden change in areas of collapse and/or suffosion sinkholes. In recent times, the most famous disruption of this kind occurred in 1981 when, without warning, a large sinkhole materialized near the center of Winter Park, an elegant suburb of Orlando in central Florida's karst region. Within moments of its appearance, the yawning abyss swallowed a three-bedroom house, half a municipal swimming pool, several motor vehicles (including five new Porsches standing at the rear of a dealership), and part of a nearby street. In all, the damage amounted to more than $2 million in property losses—as well as the disappearance of some prime real estate.

This scene is documented in the photo above (Fig. 44-7), taken on the day of the collapse. The dimensions of the new sinkhole were a diameter longer than a football field (107 m [350 ft]) and a depth of at least 40 m (130 ft). The sinkhole soon stabilized and remains essentially unchanged today. The only difference is that it has mostly filled with water to become one of the thousands of small round lakes that pockmark this part of peninsular Florida (*see* Fig. 44-6).

FIGURE 44-7 North America's most famous sinkhole lies in the middle of Winter Park, Florida, a suburb of Orlando. Appearing quite suddenly in the spring of 1981, it swallowed parts of an automobile dealership and a city pool in the process (this photo was taken only hours after the collapse).

Thanks to high-profile television coverage by the news media at the time, the Winter Park sinkhole remained an object of curiosity for years after its formation. Within a week, local authorities had surrounded the collapse with a chain-link fence, and the public just kept coming to see it. It became a landmark complete with souvenir shops, a popular stop for many of the tourists who yearly stream into the Orlando area to visit its theme parks and related attractions. Perhaps it was even on your itinerary the last time you went to Walt Disney World?

Karst Towers

If sinkholes are the signature landforms of temperate karst areas, then the dominant feature in tropical karst regions is the **tower**. Sinkholes and towers have various other names, but there is no need to complicate our terminology. A tower is a cone-shaped, steep-sided hill that rises above a surface that may or may not be pocked with solution depressions. Even when many such sinkholes are also present, the towers, sometimes hundreds of meters tall, dominate the landscape (*see* photo, p. 388). In tropical karst zones, the contrast between towers and depressions is so sharp that the whole scene is referred to as **cockpit karst**, the term *cockpit* referring to the irregular, often steep-sided depressions between the towers.

Tropical karst, studded with such towers, is found in such locales as Puerto Rico, Jamaica, Cuba, and Vietnam (Fig. 44-8). Exactly what determines the location and distribution of the towers is still uncertain. The towers are remnants of a thick bedrock sequence consisting of limestone and/or dolomite layers. Before the karst topography developed, and following regional uplift (or the lowering of sea level), the original surface presumably developed a soil cover and plants took hold. This initial pattern of soil and vegetation (thick and well developed in some places, thin and sparse in others) probably was determined by the original terrain.

Where the surface was low, moisture collected and soil soon formed, but higher places stayed barren. Eventually, these higher places became the tops of karst towers as the intervening hollows grew ever deeper. China's Guilin area (p. 388) and Yunnan Stone Forest (p. 452) may have originated in this way, although fluctuating groundwater also was a factor in the process. In fact,

FIGURE 44-8 A low aerial view of central Jamaica's "Cockpit Country," a classic, tower-studded tropical karst landscape.

limestone towers in the Stone Forest still rise from lakes whose levels vary seasonally.

KARST AND CAVES

We began this unit with a look at caves as part of the karst phenomenon. Now we are in a better position to examine these remarkable features in more detail. Technically, any substantial opening in bedrock that leads to an interior open space is a cave. The word "substantial" here has a human connotation: it is generally agreed that a cave, in order to be called a cave, must be large enough for an average-sized adult person to enter. Thus even a vacated swallow hole, where a river once flowed into an underground channel, is a cave. This is one way a *cave shaft* forms, and many unsuspecting animals and people have fallen into such vertical cave entrances.

Cave Features

A fully developed cave consists of an entrance (portal) and one or more chambers, passages, and terminations. A *termination*, in accordance with the definition above, marks the place beyond which a person cannot crawl any farther along an underground passage or conduit. Passages in a fully developed cave system form a network of interconnected conduits. The pattern of this network depends on the stratigraphy, faulting, and jointing of the bedrock sequence. It may consist of one major subsurface artery (the *linear* form); it may look like the branches of a tree (*sinuous*); or, if block jointing is well developed, it may have an *angulate* (right-angle, stepped) form. Given the complexity of karst features, various other, more detailed models have been developed, accounting also for the overall structure of the maze of caves, caverns, and conduits.

Where passages grow exceptionally large, chambers or "rooms" develop. These chambers, some with the dimensions of a large hall, contain many fascinating forms. Lakes stand in some of them, and drops of water falling from the ceiling create eerie musical echoes in the dark void. Streams may even flow through them, with the magnified sound of a waterfall.

Dripping water that is saturated with calcium carbonate ($CaCO_3$) precipitates its calcite in the form of the mineral *travertine*. This water, entering the cave, contains calcium bicarbonate. If the air in the cave contains less carbon dioxide than it could, then there will be excess carbon dioxide in the water solution dripping from the cave ceiling. This results in the transfer of some carbon dioxide from the water solution to the air, which means, of course, that some of the calcium bicarbonate that contains the carbon dioxide will have to switch back to calcium carbonate. When this happens, the form of $CaCO_3$ that is deposited is *not* calcite but travertine, a less soluble form.

This complicated process leaves icicle-like **stalactites** hanging from the ceiling and **stalagmites** standing, sentinel-like, on the floor (Fig. 44-9). These white opposing pinnacles can become several meters tall and often coalesce to form **columns** (Fig. 44-9). They have been compared to the pipes of a huge organ and have been seen as the teeth of a lurking giant. Small wonder that caves attract so many visitors.

Cave Networks

To physical geographers, however, caves present other mysteries. In 1988, divers for the first time penetrated the

FIGURE 44-9 More wonders of New Mexico's Carlsbad Caverns—column, stalagmite, and stalactites highlighted. These particular formations, called the *Totem Pole* and the *Chandelier*, are major features of a huge cavern known as The Big Room.

water-filled tunnels of a cave system beneath northern Florida's Woodville Karst Plain near Tallahassee. They entered a sinkhole lake and followed a flooded passage, using battery-powered motors and floodlights. The passage went 75 m (240 ft) below the surface, and they followed it for more than 2500 m (1.5 mi) until they saw daylight above and returned to the surface through another sinkhole. The cave passages, they reported, were as much as 30 m (100 ft) wide, but also narrowed considerably. They saw side passages joining the main conduit, and realized that they were seeing only a fraction of a very large and unmapped network. These Florida cave systems, now below sea level and filled with water, were formed more than 35 million years ago during a period of lower sea level.

Much remains to be learned about caves. Their formation is generally understood, but many details remain unclear. How important is abrasion by underground streams? What role does groundwater play? How do the underground processes combine to produce such extensive cave systems? What happens when those systems are submerged? As yet there is no general agreement on such issues, which proves that physical geography still holds some dark secrets.

Karst terrain and associated caves, as we noted at the beginning of this unit, are widely distributed across the earth. Some of the world's most impressive karst regions are only now becoming known and understood (such as the karst structures of Australia's Kimberley region in the northern part of Western Australia). Karst topography has been submerged by coastal subsidence, and it also has been uplifted into high mountains (there is karst terrain under Canadian ice and on frigid Andean slopes in South America). Sometimes we encounter karst landforms where solution is not an important contemporary process, for example, in semiarid New Mexico. Thus the map of karst and cave distribution contains valuable evidence for geologic as well as environmental change. The signature landscape of karst is one of the most distinctive elements in the entire mosaic of physiography.

KEY TERMS

Caribbean karst (p. 456)

cockpit karst (p. 457)

collapse sinkhole (p. 456)

column (p. 458)

karst (p. 452)

solution sinkhole (p. 456)

stalactite (p. 458)

stalagmite (p. 458)

temperate karst (p. 455)

tower (p. 457)

tropical karst (p. 455)

uvala (p. 456)

REVIEW QUESTIONS

1. Which rock types are prone to karst development?

2. Describe the general climatic conditions that favor karst development.

3. What chemical weathering process is instrumental in the development of karst landscapes?

4. What is the difference between temperate and tropical karst?

5. How do towers form in karst regions?

REFERENCES AND FURTHER READINGS

BECK, B. F., ed. *Engineering and Environmental Impacts of Sinkholes* (Rotterdam, The Netherlands: Balkema, 1989).

BECK, B. F., and WILSON, W. L., eds. *Karst Hydrogeology: Engineering and Environmental Applications* (Rotterdam, The Netherlands: Balkema, 1987).

FORD, D. C., and WILLIAMS, P. W. *Karst Geomorphology and Hydrology* (Winchester, Mass.: Unwin Hyman, 1989).

GILLIESON, D. *Caves* (Cambridge, Mass.: Blackwell, 1995).

HERAK, M., and SPRINGFIELD, V. T. *Karst Regions of the Northern Hemisphere* (Amsterdam: Elsevier, 1977).

JAKUCS, L. *Morphogenetics of Karst Regions* (New York: Wiley/Halsted, 1977).

JENNINGS, J. N. *Karst Geomorphology* (New York: Blackwell, 1985).

LAFLEUR, R. G. *Groundwater as a Geomorphic Agent* (Winchester, Mass.: Allen & Unwin, 1984).

MOORE, G. W., and NICHOLAS, G. *Speleology: The Study of Caves* (Boston: Heath, 1964).

SWEETING, M. M. *Karst Geomorphology* (Stroudsburg, Pa.: Dowden, Hutchinson & Ross, 1981).

SWEETING, M. M. *Karst Landforms* (New York: Columbia University Press, 1972).

TRUDGILL, S. A. *Limestone Geomorphology* (London/New York: Longman, 1986).

WALTHAM, T. *Caves* (New York: Crown, 1975).

WHITE, W. B. *Geomorphology and Hydrology of Karst Terrains* (London/New York: Oxford University Press, 1988).

Glacial Degradation and Aggradation

OBJECTIVES

◆ To discuss the different categories of glaciers.

◆ To give a brief history of how glaciation has influenced the earth's surface.

◆ To outline how glaciers form, move, and erode the landscape.

A **glacier** is a body of ice, formed on land, that is in motion. This motion is not readily apparent over short time periods, however, and to an observer glaciers appear to be mere accumulations of ice, snow, and rock debris. Yet glaciers do move, and steadily erode their valleys. Scientists realized this centuries ago, and in the Swiss Alps they calculated glacial movement by putting stakes in the ice and in the rock on the sides of the valley and measuring the annual downslope advance of the ice. But just how mountain glaciers move and how they modify the rocks below (Fig. 45-1) continues to be a subject of debate and ongoing research. In many ways, glaciers and glacial activity are more difficult to understand than rivers because of the difficulty in observing processes within and beneath the flowing ice.

The glaciers of Switzerland and Alaska are **mountain (alpine) glaciers**. These glaciers are confined in valleys that usually have steep slopes (they are sometimes called *valley glaciers* as well). However, not all glaciers occur in valleys: some glaciers consist of huge masses of ice that are not confined to valleys but that bury whole countrysides beneath them. These glaciers are called **continental (sheet) glaciers**. Today Antarctica, a continent almost twice as large as Australia, is almost completely covered by such a vast icesheet, and so too is Greenland, the world's largest island. Continental

UNIT OPENING PHOTO: Switzerland's Matterhorn, the best-known peak of Europe's Alps, is a vivid example of the sculpting power of glaciers.

a

b

FIGURE 45-1 When mountain glaciers invade an area of high relief originally sculpted by rivers, they bury most of the terrain and fill the valleys with ice, as shown for the Alaska Range (a). (Authors' photo) After the glaciers have melted away, the topography reveals their work in a variety of landforms, including the sharp-edged ridges, deep steep-sided valleys, and jagged mountain peaks that can be seen along the Continental Divide in Colorado's Rocky Mountains (b).

glaciers move, but generally even more slowly than mountain glaciers. Accordingly, they conform to our definition: they are bodies of ice and they exhibit motion. In this unit and in Units 46 to 48 we study the movement and erosive power of glaciers and the landscapes they create.

GLACIERS OF THE PAST

As noted in Unit 20, the earth has periodically experienced ice ages. An **ice age** is a stretch of geologic time during which the earth's average atmospheric temperature is lowered, resulting in the expansion of glacial ice in high latitudes and the growth of glaciers at high altitudes in lower latitudes. During an ice age, which may endure for millions of years, stages of global cooling alternate with stages of warming. As a result, glacial ice respectively expands (advances) and contracts (recedes) over periods measured in tens or hundreds of thousands of years.

A cooling period, during which the ice expands, is known as a **glaciation**. During such a time, icesheets become continental in size and gain many hundreds or even thousands of meters in thickness. At the same time, mountain valleys fill with glacial ice, often replacing rivers that formerly flowed there. After the cooling period has reached its peak and the glaciers have expanded as far as they can, the climate begins to warm up. Now the glaciers start melting and receding in a phase known as **deglaciation**. After deglaciation, the global climate may stabilize for some tens of thousands of years as the earth awaits a new cooling episode. This period between the most recent deglaciation and the onset of the next glaciation is referred to as an **interglaciation**.

The earth now is comparatively warm, glaciers have withdrawn to the coldest of the polar (and mountainous) regions, and areas once covered by continental icesheets are dominated by other geomorphic processes. In other words, we are presently experiencing an interglaciation. Just 12,000 to 15,000 years ago, however, Canada was almost entirely covered by continental glaciers, and these icesheets reached as far south as the Great Lakes. Over the past 10 millennia, the earth has warmed up and the glaciers have receded—but the present interglaciation is unlike any other this planet has witnessed. During the interglaciation now in progress, the world's human population has grown explosively. Geologically, these last 10,000 years constitute the Holocene Epoch (*see* Fig. 37-7). Geographically, the Holocene has witnessed the transformation of the planet—not only by climatic change but also by human activity.

The present interglaciation is unprecedented because, for the first time in the earth's history, humans have become an agent of environmental change. On the basis of what is known about the patterns of previous glaciations, we may today assume that another cooling episode lies ahead and that the glaciers will once again expand and advance. But human interference in the composition of the atmosphere may affect the course of events. Some scientists now warn of the human contribution to the intensification of the atmosphere's greenhouse effect. This might lead to further warming of the earth, thereby causing additional melting of ice in polar regions, a rise in the global sea level, and widespread flooding of low-lying areas. It also may contribute to a sudden "trigger effect" when the next glaciation occurs, again with an unpredictable impact.

The ice age of the present is often called the *Pleistocene Ice Age* because it has seemed to coincide almost exactly with the Pleistocene Epoch of the Cenozoic Era (*see* the geologic time scale, p. 393). But geologists now know that the current ice age began during the Pliocene Epoch, the epoch preceding the Pleistocene, probably between 2.5 and 3 million years ago. In fact, there is evidence of even earlier cooling, so their preferred name for this ice age is the **Late Cenozoic Ice Age**. (However, the Pleistocene Epoch remains closely identified with this whole episode.)

The Late Cenozoic Ice Age is only the latest in a series of such events in the earth's environmental history. For example, there is no longer any doubt that the great supercontinent of Gondwana (*see* Unit 32) experienced an ice age before it broke apart. During the Permian Period of the Paleozoic Era, the Dwyka Ice Age spread great icesheets over the polar regions of Gondwana. More than 240 million years ago, these continental glaciers left ample evidence of their activity. When Gondwana split apart, its several fragments (Africa, South America, India, Australia, Antarctica) all carried this evidence in their landscapes and underlying rock strata. When geologists discovered it, they had a major clue to the former existence of Gondwana—as well as its polar orientation during Permian times. Moreover, much older rocks from West Africa indicate an even earlier ice age dating probably to the Silurian Period about 425 million years ago. Ice ages, therefore, have affected our planet repeatedly. Even though its causes are uncertain (*see* Perspective: What Causes Ice Ages?), the present ice age is unusual—but hardly a unique event in earth history.

When we study glaciers, present or past, it is important to remember their significant connections to global environments. Even today, when the ice is of comparatively limited areal extent, the glaciers of Greenland and Antarctica influence the radiation and heat balances of the planet. Continental glaciers contain huge volumes of fresh water and thereby affect the global water balance as well. When a glaciation begins, precipitation in the form of snow is compacted into glacial ice. Therefore it is not returned to the oceans (remember the hydrologic cycle, diagrammed in Fig. 12-2), so that sea level drops as glaciation proceeds.

Later, when deglaciation begins, the melting glaciers yield their large volumes of water and sea level rises again. Thus glaciation and deglaciation are accompanied, in turn, by falling and rising sea levels. During an interglaciation such as the present one, continental shelves (*see* Fig. 2-6) are inundated. When the glaciers expand again, the flooded continental shelves will be exposed once more. Taking the long view of the future occupation of the earth by humankind, it is therefore true that land lost in the high latitudes to glacial advance will be partly compensated for by land exposed by lowered sea level.

THE FORMATION OF GLACIERS

Glaciers consist of ice, and this ice is formed from compacted, recrystallized snow. But not all snow, not even in mountainous areas, becomes part of a glacier. When you travel through (or over) a high mountain area such as the Rocky Mountains, you will observe that there is a *snow line*, a line above which snow remains on the ground throughout the year. Below this snow line the winter's accumulation of snow melts during the next summer, and none of it is converted into ice. But above the snow line—also known as the *firn line*—the snowpack thickens over time. There, some permanent snow survives the summer and contributes to the growing thickness of the snowpack. Where summer snow loss is less than winter gain, conditions favorable to the formation of glacial ice exist.

Snow is converted into ice in stages. Newly fallen snowflakes are light and delicately structured crystals. A layer of freshly fallen snow generally has a low density. Some melting of the outer "points" of the crystals may take place, changing them into irregular but more spherical grains (Fig. 45-2a). Or a later snowfall might compress the layer below it, packing the crystals more tightly together and destroying their original structure. All this has the effect of increasing the density of the lower layer and reducing its open spaces, or porosity. In areas where periodic melting occurs, fluffy snow can be converted into dense granular snow in a matter of days.

But this first stage does not yet yield glacial ice. The granular, compacted snow—called **firn**—undergoes further compression and recrystallization (Fig. 45-2b). That takes time—more time in cold polar areas than in moister temperate zones. This is so because in the temperate areas, where melting occurs, percolating meltwater fills the remaining pore spaces, refreezes there, and adds to the weight of the snowpack. Glaciologists calculate that the transformation from firn to ice in temperate areas may require less than 50 years. In polar areas, it may take 10 times as long. This means that a snowpack in temperate areas needs to be less thick to be converted into glacial ice.

A glacier in coastal Alaska may need a firn less than 15 m (50 ft) deep for ice to form. On the other hand, in

What Causes Ice Ages?

It is known that the earth has experienced repeated glaciations. The Late Cenozoic Ice Age is only the most recent. When Gondwana was still a supercontinent, it experienced a prolonged glacial age (the Dwyka Glaciation), and there is evidence of still earlier ice ages. Much is known, too, about icesheets and mountain glaciers, and their erosional and depositional work. But scientists remain unsure about the causes behind nature's grand design. Why are the glacial ages periodic? Do they come at regular intervals? Are they caused by terrestrial conditions, or are they the result of conditions in the solar system and planetary orbits? Several theories have been formulated to account for what we know about ice ages, but none has yet gained general acceptance.

One theory attributes ice ages to *continental drift*. This theory holds that when plate tectonics moves landmasses into the polar latitudes, their elevation, combined with polar coldness, generates icesheets—like Antarctica's today. But what is known about past movements of landmasses does not completely support this idea. Nor does it explain why ice ages are marked by alternating periods of cooling (glaciations) and warming (interglaciations).

A second theory links ice ages to *crustal bulging* resulting from plate collisions. The Late Cenozoic Ice Age, for example, is thought to have its origins in the vertical uplift of Asia's Himalayas and adjacent Tibetan Plateau over the past 20 million years, and the contemporaneous uplift of North America's Sierra Nevada

and southern Rocky Mountain ranges. These raised crustal segments, it is argued, would interfere with jet streams and other atmospheric windflows, combining the coldness from the elevation with a latitudinal shift of air circulation, thereby creating hemispheric cooling. A problem here is the absence of such a landmass-generated cooling in the Southern Hemisphere, except in the case of Antarctica—which can be explained by other means.

Yet another theory relates glaciations to episodes of *volcanic activity*. Certain periods in earth history have been marked by intense volcanism. The dust spewed into the atmosphere might, according to this notion, interfere with solar radiation to such an extent that the volcanically derived cloud cover cools the surface enough to spawn a glaciation.

Still another set of theories attributes glacial cooling to changes in the earth's *atmosphere* and *hydrosphere*. Fluctuations in carbon dioxide in the atmosphere could cause alternating warming (intensified greenhouse conditions) and cooling. When global vegetation is abundant, more carbon dioxide is consumed and its presence in the atmosphere is reduced. This would lead to cooling and glacial conditions; but when the vegetation dies, more carbon dioxide is released into the atmosphere, and greenhouse warming resumes. One problem with this idea, however, is that evidence for the short-term vegetation changes required for the model is lacking.

Changes in oceanic circulation are related to the tectonic movement of con-

tinents, and it is believed by some scientists that the inflow of warm Atlantic water into the basin of the Arctic Ocean would melt part of that ocean's ice cover, thereby releasing moisture for snow-bearing air masses. Huge amounts of snow would then accumulate in high-latitude North America and Eurasia—just where the great continental glaciers of the Pleistocene formed. At present, with Greenland and North America located as they are, warm Atlantic water cannot enter the Arctic Basin in large quantity; thus the Arctic Ocean remains frozen most of the time, and the supply of snow is much reduced. The obvious problem with this theory is that it fails to explain the rapid alternations between glaciations and interglaciations.

Additional theories look beyond the earth and suggest that *earth-sun relationships* and *planetary orbits* are ultimately responsible for ice ages (as noted in Unit 20). Over many millions of years, the distance from the earth to the sun during orbits changes slightly. Moreover, the angle of the earth's axis to the plane of the ecliptic also undergoes some variation. In combination, these changes affect the amount of solar radiation received by all areas on the earth's surface. Data from various sources now suggest that this may be the fundamental cause of ice ages, including the short-term advances and withdrawals of the ice of the Pleistocene. Therefore, the intensity and duration of each global ice age probably are determined by orbital variation, plus some of the conditions on which other theories are based.

---

the colder and drier Antarctic, 100 m (330 ft) of firn would be required to produce ice of the same density. Such data are useful in determining the age of the great icesheets. Snow accumulation on the Antarctic icesheet is very slow, but the firn is of enormous depth. Obviously, this continental glacier required a long time (probably more than 4 million years) to achieve its present dimensions.

THE GLACIER AS A SYSTEM

We can visualize a glacier as an open system, as shown in Fig. 45-3. If the glacier is in equilibrium, it will gain as much matter in the form of precipitation in its **zone of accumulation** as it loses through various processes in the **zone of ablation**. The term *ablation* denotes all

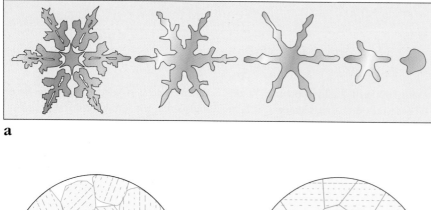

a

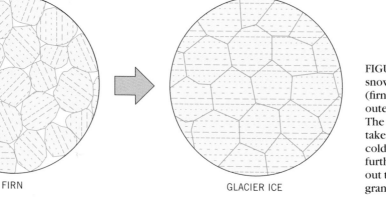

FIRN   GLACIER ICE

b

FIGURE 45-2 The transformation of a snowflake into a granule of old snow (firn) can take several weeks as the outer "points" of the crystal melt (a). The conversion of firn into glacier ice takes decades, even centuries, in the coldest climates, as recrystallization and further compression slowly squeeze out the open spaces between individual granules (b).

forms of loss at a glacier's lower end, including melting and evaporation. Material is moved continuously down-slope from the zone of accumulation to the zone of ablation.

In this system, therefore, matter enters in the solid state as snow, undergoes two changes (to granular and then to crystalline form), and leaves the system in a liquid or vaporized state. Under conditions of equilibrium, a glacier neither grows nor shrinks. But equilibrium conditions rarely exist and never over a long period of time. Glaciers therefore tend to exhibit evidence of fluctuation, especially in the zone of ablation. When the *mass balance*—the gains and losses of matter in the system—is positive, the glacier advances and its leading edge is steep and icy. When the balance is negative, the front end of the glacier is less pronounced, and the gray melting ice is smudged by rock debris being released from it.

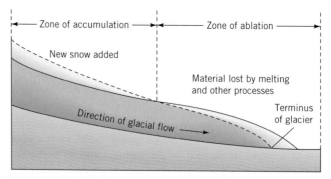

FIGURE 45-3 A glacier is an open system, with new snow added in its upper zone of accumulation and material lost in its lower zone of ablation.

GLACIAL MOVEMENT AND EROSION

It is certainly true that glacier movement, even in high-relief mountain zones, is not as rapid as streamflow. But the erosional power of glaciers is enormous. When mountain glaciers have melted away and vacated their valleys, they leave exposed some of the world's most spectacular scenery. Valley sides are sheer and scarp-like; waterfalls plunge hundreds of meters onto flat, wide valley floors (Fig. 45-4). Whole mountainside spurs, once rounded by a winding river before the glacier occupied

FIGURE 45-4 The glories of Yosemite Valley in California's Sierra Nevada, whose magnificent scenery was carved by a mountain glacier of awesome erosional power.

its valley, are sheared off as if by a giant knife, as the ice straightened and smoothed the valley's course. Lakes are formed behind natural dams made from glacial debris. Angular peaks and ridges rise above the landscape. The glaciers may be gone, but the landscape bears the dramatic imprint of their work.

Temperature and Glacial Erosion

As we have learned, much is known about the erosional functions of rivers. Measuring streamflow, sediment load and character, valley morphology, and other properties of streams is easy when compared to glaciers. But observations of the contact plane between ice and bedrock—where erosion takes place—are difficult to make, and movement within various parts of the glacier cannot be measured simply by hanging a gauge from a bridge. In our discussions of stream erosion, no reference was made to the temperature of the water as a factor in stream velocity or erosional capacity. Temperature, however, is a critical factor in glacial erosion. Indeed, the temperature of the ice at the base of a mountain glacier, and its melting point, may be the most important factor of all in that glacier's capacity to erode its valley.

The temperature of glacial ice does not decrease steadily with depth. Various factors, including the pressure exerted by the weight of the ice and the temperature in the bedrock below, affect the temperature of the lowest ice layer, the **basal ice**. When the basal ice is at the melting temperature, the glacier moves faster, erodes more effectively, and transports a larger sedimentary load than when the basal ice is colder. Mountain glaciers in temperate zones, therefore, erode more strongly than similar glaciers in very cold polar areas, where the temperature of the basal ice is much lower. Continental glaciers such as those covering Greenland and Antarctica are of massive dimensions, but their movement is very slow. Their erosional work, compared to their size, is much less effective than that of lower-latitude mountain glaciers.

The Movement of Ice

Glaciers move slowly. The great continental glaciers move as little as 2 to 3 cm (1 in) per day, and even some cold-area mountain glaciers move just a few centimeters daily. In a rapidly moving alpine glacier, the daily advance may amount to as much as 4 or 5 m (15 ft) or even more. Occasionally, a mountain glacier develops a **surge**, a rapid movement of as much as one meter per hour or more, sustained over a period of months, producing an advance of several kilometers in one season.

A profile through the center of a mountain glacier reveals that its upper layer consists of rigid, brittle ice that is often cut by large cracks called **crevasses**; below this rigid layer, the ice takes on the properties of a plastic material (Fig. 45-5). When the glacier moves downslope,

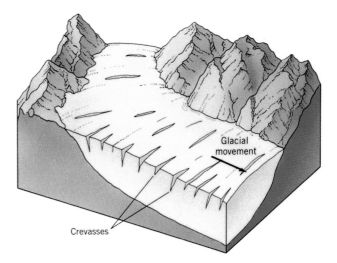

FIGURE 45-5 A glacier's brittle upper layers are studded with large crevasses; down below, plastic flow predominates.

its center advances most rapidly and the sides the slowest (Fig. 45-6). In vertical cross-section, the upper surface moves fastest while the basal ice moves more slowly.

Geomorphologists report that glaciers move in two different ways. The first, called **glacial creep**, involves

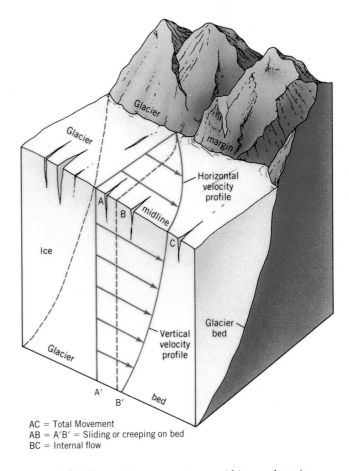

AC = Total Movement
AB = A'B' = Sliding or creeping on bed
BC = Internal flow

FIGURE 45-6 Differential movement rates within an advancing glacier. Vertically, the upper surface moves fastest and the basal ice slowest. Horizontally, the center moves fastest and the sides slowest.

the internal deformation of the ice, with crystals slipping over one another as a result of the downslope movement just described. The second flow mechanism, called **glacial sliding**, is the movement of the entire glacier over the rocks below it. We are most interested in this second movement because it directly involves the glacier's erosional work. It is generally conceded that glacial sliding is enhanced by the existence of a thin film of water between the basal ice and the bedrock floor. This film of pressurized water is just millimeters thick and is probably discontinuous. But it is enough to lubricate the contact plane between glacier and bedrock, and speeds the glacier's movement downslope.

Glacial Erosion

Erosion by glaciers can take place through **plucking** (also called *quarrying*), a complicated process in which the properties of the bedrock (rather than the transported rock debris) are most significant. In plucking, blocks or fragments of bedrock beneath the glacier are pulled from the surface as the ice moves forward. Glacial erosion also can occur by **abrasion**, the scraping process produced by the impact of rock debris carried in the ice upon the bedrock below. Despite its appearance, ice is not a hard substance; on the Mohs hardness scale (*see* Table 30-1) it would rate only about 1.5. Thus ice by itself cannot be an effective erosional agent. Abrasion, therefore, must be performed by the rock fragments being dragged along the bedrock floor (and the submerged valley sides) by the moving ice.

Abrasion

Some of these rock fragments are, of course, quite soft themselves and do not have much effect in glacial erosion. Such soft material is soon pulverized and becomes part of the dark streaks visible on the glacier's surface. Harder fragments, however, do have a powerful impact on the bedrock floor beneath the glacier. The enormous weight of the glacier pushes a boulder downward while dragging it along, and this combination can create rapid degradation. Other factors also come into play: the rate of movement of the glacier, the temperature of the basal ice, and the character of the underlying bedrock.

How fast do glaciers degrade? Various measurements have been taken, but it is not possible to generalize from these. In one area of temperate-zone glaciers, average erosional rates ranged up to 5 mm (0.2 in) per year, but in another area a rate nearly seven times as high was recorded. The effectiveness of the abrasion process is quite variable. Abrasion can produce several telltale features in the landscape. When the abrading debris consists of fine but hard particles (quartz grains, for instance) and the underlying bedrock also is quite hard, abrasion produces a polished surface that looks as though the

bedrock surface has been sandpapered. But when the rock fragments are larger, the underlying surface may be scratched quite deeply. These scratches, made as the boulder or pebble was dragged along the floor, are called glacial **striations** (Fig. 45-7). They often are meters long and centimeters (but more often millimeters) deep. They can be useful indicators of the direction of ice movement where the topography provides few clues, because striations tend to lie parallel to the direction of ice advance.

Plucking

Plucking, a process diagrammed in Fig. 45-8a, also leaves evidence in the landscape. The most common landform associated with glacial plucking is the **roche moutonnée** (Fig. 45-8b). This characteristic, asymmetrical mound appears to result from abrasion to one side (the side from which the ice advanced) and plucking on the leeward side. A complicated process allows the glacier to quarry this leeward side, lifting out and carrying away loosened parts of the *roche*. Studies suggest that jointing in the bedrock, and probably frost-caused fracturing as well, contribute to the glacier's ability to "pluck" the

FIGURE 45-7 This foreground surface of hard bedrock was planed by a glacier, and the surface thus created was grooved by striations. These deep scratches, made when rocks at the base of the ice were dragged over this surface, now reveal the direction of ice movement in this part of the Swiss Alps (toward the Matterhorn at the rear).

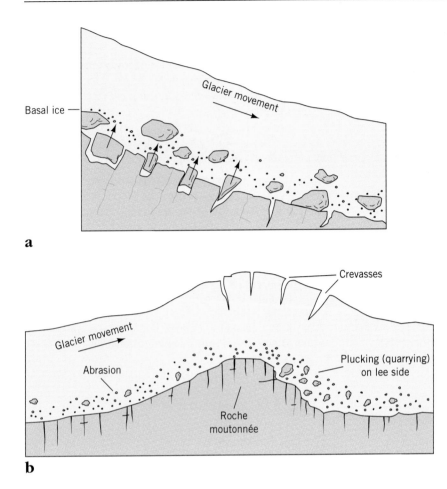

Basal ice —

Glacier movement

a

Crevasses

Glacier movement

Abrasion

Plucking (quarrying) on lee side

Roche moutonnée

b

FIGURE 45-8 Plucking occurs when glacial-bed rock fragments are torn loose (by ice freezing into cracks that are subsequently enlarged by wedging), rotated upward, and carried away downslope embedded in the basal ice flow (a). Larger mound-like landforms called *roches moutonnées* are also created by this process, with the plucking found on the leeward side (b).

mound over which it passes. Whatever the nature of the process, *roches moutonnées*, like striations, help us reconstruct the path of the glacier.

As in the case of rivers, glaciers deposit (aggrade) as they erode (degrade). The degradation of their mountainous source areas is matched by aggradation in their lower valleys and at their terminal edges. Like rivers, glaciers grind up the sedimentary loads they carry; but unlike rivers, fragments of hard rock that have taken the long trip encased in glacial ice appear at the glacier's end as angular boulders. The debris carried downslope by alpine glaciers tends to concentrate in certain zones of

the glacier, and appears on the surface as a series of parallel bands. Once deposited, this material leaves no doubt as to its origin: rounded fragments are a sign of fluvial action, whereas angular fragments signify the work of ice. Continental glaciers, too, degrade their source areas and aggrade where their advance is slowed or stopped. Much of the topography of the area of the Great Lakes and surrounding states is underlain by glacial debris scoured by continental glaciers from the Canadian Shield and deposited far to the south. We turn next to the landforms and landscapes created by the great icesheets of the past.

KEY TERMS

REVIEW QUESTIONS

1. Where do continental glaciers presently exist?

2. What is meant by the term *Late Cenozoic Ice Age?*

3. How does snow become transformed into glacial ice?

4. Describe how the mass balance of a glacier controls the glacier's movement.

5. How is a *roche moutonnée* formed?

REFERENCES AND FURTHER READINGS

ANDREWS, J. T. *Glacial Systems: An Approach to Glaciers and Their Environments* (North Scituate, Mass.: Duxbury Press, 1975).

BROECKER, W. S., and DENTON, G. H. "What Drives Glacial Cycles?," *Scientific American,* January 1990, 48–57.

DYSON, J. L. *The World of Ice* (New York: Knopf, 1962).

EYLES, N., ed. *Glacial Geology* (Elmsford, N.Y.: Pergamon, 1983).

FLINT, R. F. *Glacial and Quaternary Geology* (New York: Wiley, 1971).

FLINT, R. F. *The Earth and Its History: An Introduction to Physical and Historical Geology* (New York: Norton, 1973).

HAMBREY, M., and ALEAN, J. *Glaciers* (New York: Cambridge University Press, 1992).

IMBRIE, J., and IMBRIE, K. P. *Ice Ages: Solving the Mystery* (Short Hills, N.J.: Enslow, 1979).

JOHN, B. S. *The Ice Age: Past and Present* (London: Collins, 1977).

MATSCH, C. L. *North America and the Great Ice Age* (New York: McGraw-Hill, 1976).

PATERSON, W.S.B. *The Physics of Glaciers* (Tarrytown, N.Y.: Pergamon, 3rd ed., 1994).

POST, A., and LACHAPELLE, E. R. *Glacier Ice* (Seattle: The Mountaineers, 1971).

SHARP, R. P. *Living Ice: Understanding Glaciers and Glaciation* (London/New York: Cambridge University Press, 1988).

Landforms and Landscapes of Continental Glaciers

OBJECTIVES

◆ To discuss contemporary continental glaciers and their former extent during the Late Cenozoic Ice Age.

◆ To discuss typical landforms produced by continental glaciers.

uring an ice age, the earth's surface is transformed. Great icesheets form over landmasses situated at high latitudes. Whole regions are submerged under ice—mountains, plateaus, plains, and all. The weight of the ice, which may reach a thickness of more than 3000 m (10,000 ft), pushes the underlying crustal bedrock downward isostatically (*see* Unit 33). So much water is converted into snow (and subsequently into glacial ice) that sea level drops many meters. Large areas of continental shelf are exposed; coastlines are relocated accordingly, and continental outlines change shape. As the ice expands, thereby expanding the region of polar-type temperatures, global climatic zones are compressed toward the lower latitudes. Midlatitude lands that were previously temperate become cold, barren, and subpolar in character; vegetation belts shift equatorward.

In this unit, we focus on the great continental glaciers, present and past, and on the landforms they create. Like mountain glaciers, the continental icesheets migrate, erode by abrasion, and create characteristic landforms

UNIT OPENING PHOTO: One of the most forbidding surfaces on our planet: the massive icesheet that covers virtually all of Antarctica. The mountaintops protruding above this continental glacier are known as nunataks.

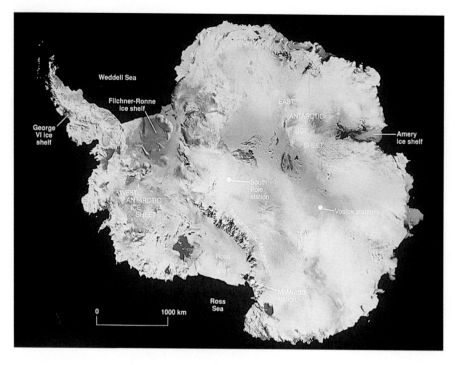

FIGURE 46-1 Satellite view of Antarctica. Except for parts of the Antarctic Peninsula, this continent is covered by an icesheet the size of Canada plus the conterminous United States.

through degradation as well as aggradation. But whereas mountain glaciers tend to increase the relief, continental glaciers have the opposite effect. Mountain glaciers gouge and excavate; continental glaciers scour and fill. In terms of total area affected, continental glaciers have the larger impact by far. It is estimated that glacial deposits laid down by icesheets cover nearly 9 percent of the North American continent. Add to this the vast areas scoured bare by the icesheets, and the significant role of continental glaciers is even more evident.

THE ANTARCTIC ICESHEET

The present climate is relatively warm compared to the atmospheric conditions of the past few million years, and glaciers have receded from many areas. But two representative icesheets persist to this day—in Antarctica and Greenland.

The *Antarctic Icesheet* has existed throughout the entire Late Cenozoic Ice Age and is unlikely to melt in Holocene times. This massive icesheet allows us to measure and observe the properties of continental glaciers and to better understand how they affected the now-deglaciated areas of the Northern Hemisphere. The Antarctic Icesheet (Fig. 46-1) is of a size comparable to that of the icesheets that covered Canada and the northern United States repeatedly during the Late Cenozoic Ice Age. It covers an area of about 13 million km² (some 5 million mi²), constituting almost 9 percent of the "land" area of the globe.

Beneath this great glacier lies an entire continental landmass, including an Andes-sized mountain range and a vast plateau. In places on top of this plateau—particularly the region to the right of the South Pole in Fig.

46-1—the ice is more than 4000 m (13,200 ft) thick. A few of the highest mountain peaks protrude through the ice and snow; such exposed tips are called **nunataks** (*see* unit opening photo). Except for the outer parts of the Antarctic Peninsula, the Antarctic Icesheet prevails from coast to coast. In schematic profile it looks like a giant dome, resting on the landmass below (Fig. 46-2). Because of this great ice accumulation, Antarctica has the highest average altitude of all the continents.

Volume and Weight of the Icesheet

The volume and weight of the Antarctic Icesheet are perhaps best illustrated by the following data. About 65 percent of all the fresh water on earth is presently locked up in the Antarctic ice. If this icesheet were to melt, the global sea level would rise by some 60 m (200 ft) and possibly more, thereby drowning many of the world's great cities (the effects of such a rise on central North America are shown in Fig. 46-3). The weight of the Antarctic Icesheet is so great that the landmass below it has sunk isostatically by an estimated 600 m (2000 ft). Thus, if the ice were to melt, the Antarctic landmass would rebound upward by about 600 m as a result of the removal of this load, further contributing to global sea-level rise.

At present, the Antarctic glacier system experiences very little mass input in the interior zone of accumulation. Average annual snowfall in the interior amounts to less than 10 cm (4 in) of water equivalent, which qualifies this region as a desert according to our climate classification scheme (*see* Unit 17). But over this vast area, even that meager amount of snow is sufficient to keep the great icesheet flowing outward at rates varying from 1 to 30 m (3 to 100 ft) per year.

470

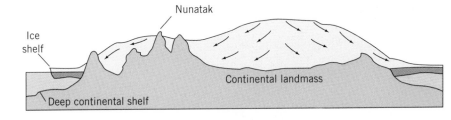

FIGURE 46-2 The Antarctic Icesheet forms a gigantic dome that depresses the landmass below. In a few places mountain peaks (nunataks) rise above the icecap. Vertical scale is markedly exaggerated.

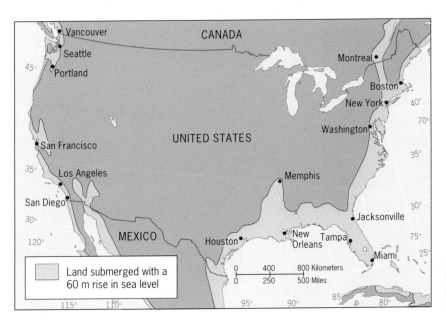

FIGURE 46-3 The effects of a 60-m (200-ft) rise in sea level on central North America, showing coastal zones and some of the major cities that would be submerged.

Features of the Antarctic Icesheet

The Antarctic Icesheet exhibits several features that are useful in our study of other, extinct continental glaciers. One of these is the division of the ice dome into **flow regimes** (Fig. 46-4). The ice does not move outward in a simple radial manner. Rather, it flows seaward in several discrete regions, each of which has its own rates of snow accumulation, ice formation, and velocity.

Another interesting feature of the icesheet is its behavior in Antarctica's marginal areas. Here the ice thins out, and the underlying topography plays a much more important role than it does under the thicker ice of the continental interior. In places, the icesheet fans out into valley glaciers and *ice tongues* (outlet glaciers that extend into the sea). These marginal glaciers are fed in part by the icesheet and in larger part by heavy snows falling on their local areas. As a result, they move faster than the main body of the icesheet. Moreover, recent research suggests volcanism is a factor here as well, which produces ground warmth that melts sufficient basal ice to keep the marginal glaciers flowing seaward atop a slippery layer of mud.

Still another feature of the Antarctic Icesheet is the formation of **ice shelves**. These smaller icesheets are floating extensions of the main glacier, which remain attached to the continental icesheet as they protrude from land into the frigid seawater (*see* far left edge of Fig. 46-2). Antarctica presently has two prominent ice shelves—the Ross Ice Shelf in the Ross Sea and the Filch-ner-Ronne Ice Shelf in the Weddell Sea (shown in Fig. 46-1)—plus many smaller ice shelves. The Ross Ice Shelf has an area of 535,000 km² (205,000 mi²) and is larger than France; the Filchner-Ronne Ice Shelf is nearly 400,000 km² (150,000 mi²), about the size of California.

The thickness of the ice in these shelves declines with distance from the mainland. At the coast, the Ross Ice Shelf is 300 m (1000 ft) thick; at its seaward edge, about 600 km (375 mi) from shore, thickness is reduced to about 180 m (600 ft). At this outer edge, the ice shelves break up into huge tabular icebergs (Fig. 46-5) in a process called **calving** (which is therefore another form of ablation). These flat-topped icebergs generally range in size from a few hundred meters to about 35 km (20 mi) across, although some are several times larger.

The tabular iceberg is characteristic of Antarctic waters. Icebergs, whatever their appearance or origin, have a slightly lower density than the cold water in which they float. Only about one-sixth of the mass of an iceberg appears above the water. A peaked iceberg in northern waters may have an underwater base extending far beyond its exposed form, and many ships have collided with icebergs that still appeared to be a safe distance away.

Surrounding the great Antarctic Icesheet and its zone of tabular icebergs lies a zone of floating sea ice that mostly covers the water's surface. This **pack ice** (Fig. 46-4) does not derive from the icesheet: it forms from the

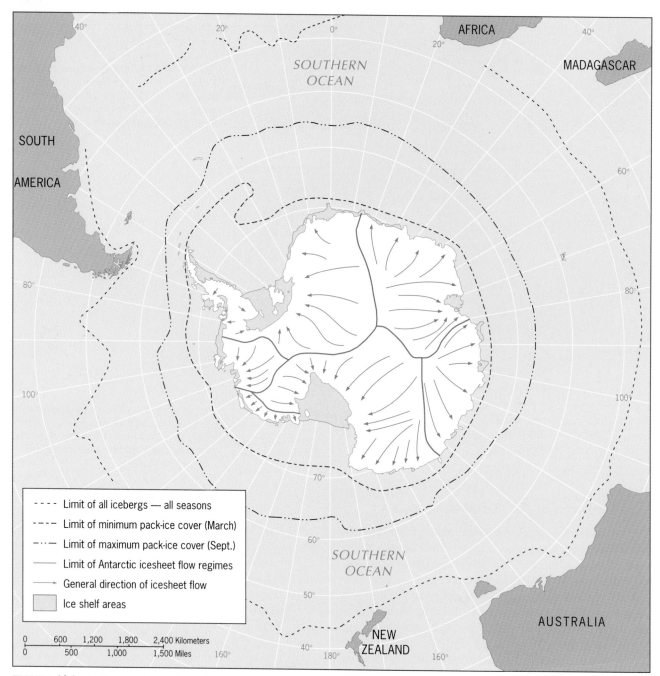

FIGURE 46-4 The flow regimes of the Antarctic Icesheet and ice limits in the surrounding Southern Ocean.

freezing of water in the adjacent Southern Ocean. During the winter, the belt of pack ice thickens and spreads; thus from March to November, Antarctica lies encircled by a nearly solid zone of floating ice so wide that it nearly doubles the "land mass" of the continent. After November the pack ice rapidly breaks up, and *leads*, or channels of open water, form through it which allow ships carrying supplies and equipment to reach the continent's coastal research stations.

FIGURE 46-5 With a thunderous roar, a 30-story-high slab of ice calves from the seaward face of the Hubbard Glacier in southern Alaska's Yakutat Bay.

THE GREENLAND ICESHEET

In total area, the *Greenland Icesheet* is about one-eighth as large as Antarctica's continental glacier, covering about 1.7 million km² (670,000 mi²) of surface (Fig. 46-6). Indeed, this icesheet creates most of the surface of Greenland, the largest of all the world's islands. But in terms of volume, the Greenland Icesheet is far less important, and it contains only about 11 percent of the world's freshwater supply. Like the Antarctic Icesheet, Greenland's glacier exhibits the shape of a dome (*see* Fig. 46-2), reaching its highest elevation (over 3000 m/ 10,000 ft) in the east-central part of the island. From there the surface drops quite rapidly toward the coasts, where the thickness of the ice is reduced to 150 m (500 ft) and even less.

The icesheet also leaves about 12 percent of coastal Greenland uncovered, which is proportionally more than in Antarctica. It appears to be quite stable, ablation in the coastal zone approximating accumulation in the interior. But there are no ice shelves here comparable to those of the Antarctic Icesheet. Indeed, the only ice-shelf feature in the Arctic lies not in Greenland but along the northern coast of adjacent Ellesmere Island to the northwest. Where the Greenland Icesheet reaches the edge of the ocean, it extends seaward in valley outlet glaciers. From these, large masses of ice calve and float as icebergs into the Arctic and North Atlantic Oceans.

Unlike the Southern Hemisphere, the Northern Hemisphere does not have a polar continent. Thus, while the South Pole lies almost in the geometric center of Antarctica (Fig. 46-1), the North Pole lies encircled by, but not on, land (*see* Fig. 46-8). We can therefore conclude that true sheet glaciers do not form over water: they form only on high-latitude landmasses. In the Northern Hemisphere, there is no major landmass at a latitude higher than that of Greenland, and the polar region is an ocean—the Arctic Ocean. This ocean surface is filled with pack ice, thick enough so that it can be traversed, but subject to crushing, buckling, and breaking (Fig. 46-7). In the summer, when the open leads of water form, ships venture into the Arctic, as many wooden boats did in the centuries when the (futile) search was on for a "northwest passage" (around North America) from the Atlantic to the Pacific. Many a ship has been doomed by the rapid closing of the pack ice at the onset of the long Arctic winter.

AGE OF THE PRESENT ICESHEETS

The Antarctic and Greenland Icesheets have survived the warm interglaciation of the Holocene. But were they permanent throughout the present ice age? How old are they? These questions can now be answered because new research methods have been developed to solve the problems of ice-age chronology. As in the case of con-

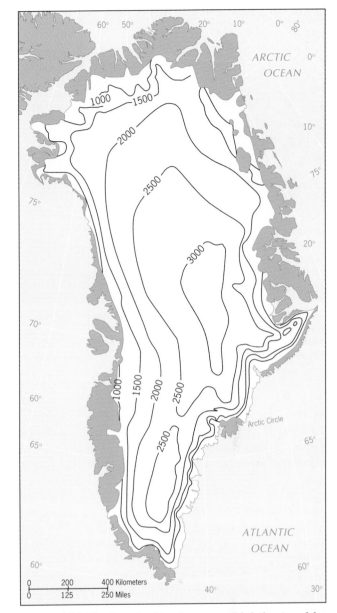

FIGURE 46-6 The Greenland Icesheet, one-eighth the size of the Antarctic Icesheet (Fig. 46-1). Ice thickness contours are given in meters.

tinental drift and crustal spreading, the oceans provided crucial evidence.

Episodes of cooling and warming occur during an ice age, and these stages are recorded by the microorganisms that become part of the deep-sea sediments. These tiny organisms lived at the surface of the sea, like plankton, then died and sank to the ocean floor. Cores of such sediments, obtained by deep-sea drilling, now allow geologists and marine scientists to study the sequence and length of ice-age glaciations, analyze paleomagnetic data, and search for the presence of certain atoms (isotopes) of oxygen in proportions that reveal the times of arrival of meltwater and the removal of seawater during the successive deglaciations and glaciations.

In combination, these data sources have produced evidence that the present ice age began well before the onset of the Pleistocene 2 million years ago. According

473

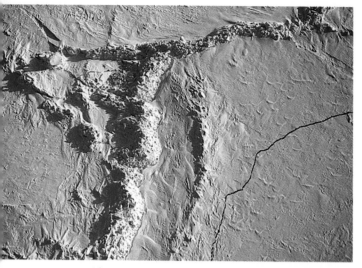

FIGURE 46-7 A close-up aerial view of the surface of the floating pack ice that fills most of the Arctic Ocean. The large crack and network of pressure ridges are testimony to the crushing and breaking that takes place on this constantly churning surface.

to the scientists engaged in this exciting research, there have been more than 30 glaciations since the ice age began. Glaciations appear to occur an average of 90,000 to 100,000 years apart. Within individual glaciations, there are fluctuations when the advancing ice stalls or temporarily recedes before resuming its forward progress.

As for the Antarctic Icesheet, the evidence indicates that cooling and ice formation in Antarctica began long before the Late Cenozoic Ice Age commenced. The great continental glacier may have been growing as long as 15 million years ago, and probably reached its full extent (somewhat larger than its present size) about 3.5 million years ago. It has survived all of the interglaciations of the Late Cenozoic Ice Age, as it is now surviving the warm Holocene.

OTHER LATE CENOZOIC ICESHEETS

The Antarctic and Greenland continental glaciers are the two Holocene survivors of a larger group of Pleistocene icesheets. During Late Cenozoic glaciations, larger and smaller continental glaciers formed in the middle latitudes of both hemispheres, but to a greater extent in the Northern Hemisphere where there is far more land.

Former Northern Hemisphere Icesheets

Biggest of all was the *Laurentide Icesheet*, the vast continental glacier that occupied Canada east of the Rocky Mountains and that expanded repeatedly to cover the North American continent as far south as the Ohio and Missouri River Valleys (Fig. 46-8). To the west of the Laurentide Icesheet, the Rocky Mountains were covered by

a smaller regional ice mass called an **icecap**. (Icecaps differ from icesheets in that they overlay subcontinental areas totaling less than 50,000 sq km [20,000 sq mi].)

In Eurasia, the largest continental glacier was the *Scandinavian (Fennoscandian) Icesheet*. This glacier, centered on the eastern segment of the Scandinavian Peninsula, extended southward into central Europe; southwestward, it coalesced with an icecap situated atop the British Isles (Fig. 46-8). The Scandinavian Icesheet was only slightly less extensive than the Laurentide Icesheet. Elsewhere in Europe, a separate icecap also developed over the Alps, and a smaller one over the Pyrenees; the island of Iceland, too, was buried under ice.

A third major continental glacier in the Northern Hemisphere is believed to have developed with its center approximately over Novaya Zemlya, the elongated island north of central Russia (Fig. 46-8). This icesheet is variously known as the *Siberian* or *Barents Icesheet* (some glaciologists believe it was largely an ice shelf); it may have coalesced with the Scandinavian glacier to the west. So far it is less well known than the Laurentide or Fennoscandian Icesheets, but its mass has been calculated to match that of the Scandinavian glacier. As for the rest of Eurasia, major icecaps also developed over the Himalaya Mountains and the adjacent highlands of central Asia (Fig. 46-8).

Former Southern Hemisphere Icecaps

In the Southern Hemisphere, only the Antarctic Icesheet reached continental proportions. The smaller extent of high-latitude landmasses in the Southern Hemisphere led to the formation of smaller icecaps. The largest of these lay in southern South America, covering the southern Andes Mountains and much of Patagonia to the east. Another icecap developed on the South Island of New Zealand, where substantial mountain glaciers still exist (*see* Unit 47).

NORTH AMERICA'S GLACIATION: THE FINAL FOUR

Before the days of deep-sea drilling and oxygen-isotope analysis, researchers had to rely on stratigraphic evidence to unravel the complicated glacial past. By mapping the surface geology and constructing cross-sections, they tried to determine the succession of glaciations and interglaciations. Dating these in absolute terms was not yet possible, so chronologies were based on what was known about the rates of accumulation of glacial deposits, on the depth of soils that developed between glacial episodes, and on related data.

Under these circumstances, this field research achieved some remarkably good results. In North America, it was concluded that there were four major advances

FIGURE 46-8 The Late Cenozoic continental glaciers of the Northern Hemisphere. Arrows indicate the general direction of ice flow. Today's coastlines are shown as dashed lines; the Pleistocene coastlines on the map developed when global sea level was about 100 m (330 ft) lower than at present. The pack ice covering the Arctic Ocean at that time extended much farther south, reaching the North Atlantic.

of the Laurentide Icesheet, of which the *Wisconsin glaciation* was the most recent (Table 46-1). In the meantime, European glaciologists had also identified four major, corresponding glaciations. Thus it was theorized that the Pleistocene Glaciation was a four-stage sequence, involving a total period variously estimated to have lasted from 0.5 to 1.5 million years.

TABLE 46-1	FOUR NORTH AMERICAN LATE CENOZOIC GLACIATIONS AND THEIR EUROPEAN EQUIVALENTS

NORTH AMERICA	INTERGLACIATION	EUROPE
	Holocene	
Late		
WISCONSIN		WURM
Early		
	Sangamonian	
ILLINOISAN		RISS
	Yarmouthian	
KANSAN		MINDEL
COMPLEX		COMPLEX
	Aftonian	
NEBRASKAN (Pre-Nebraskan)		GUNZ (Pre-Gunz)

The current state of knowledge is that these four glaciations actually represent the last *dozen* of the more than 30 glacial episodes of the Late Cenozoic Ice Age. The Wisconsin, it is now clear, consists of two major advances, not just one: the Early and Late Wisconsin. In Europe, the Mindel Glaciation is now called the Mindel Complex because there is evidence of repeated glacial advances. The pre-Gunz (pre-Nebraskan) glaciation also is now recognized. Nonetheless, the four-stage sequence (Table 46-1) is a remarkable approximation, given that it was based on evidence that exists only in the landscapes of Europe and North America. They form the record of perhaps the last 400,000 to 450,000 years of the Late Cenozoic's 3 million years of rhythmic global cooling.

When the technique of radiocarbon dating of carbon-bearing substances in the most recent (Wisconsin) glacial deposits became possible, the story of this last pre-Holocene deglaciation emerged. The final advance of the Wisconsin ice was so rapid in some areas that the leading edge of the icesheet toppled trees and encased them. When deglaciation began, these tree trunks were deposited along with the glacier's rock debris. Radiocarbon dating revealed the age of the trees when they were engulfed by the ice. This was about 12,000 years ago, and the final deglaciation that led directly to the present Holocene interglaciation is barely 10 millennia behind us.

LANDSCAPES OF CONTINENTAL GLACIERS

In Unit 45, we study the ways glaciers erode and describe two kinds of degradational features: polished and striated surfaces, and *roches moutonnées*. These landforms are neither prominent nor very common. But sheet glaciers do create extensive landscapes by degradation. They acquire their enormous sedimentary load by scouring huge parts of their source areas clear of soil, regolith, and loosened rock. Where the underlying topography has valleys roughly parallel to the direction of ice movement, a sheet glacier can even behave like a mountain glacier, deepening and widening such valleys. This is what happened in the Finger Lakes region of upstate New York (Fig. 46-9).

More often, when a continental glacier melts away, what is left is a vast ice-scoured plain marked by depressions, which are filled with water where the icesheet did its gouging and scouring. Much of the surface of Canada's Laurentian Shield and northern Europe's Scandinavian Shield (*see* Fig. 29-11) display such landforms. Erosion has created extensive continental-glacier landscapes.

FIGURE 46-9 A satellite view of western New York State's Finger Lakes region, long a famous center of tourism and wine-making. Note the parallel alignment of the elongated lakes and their relationship to the area's drainage patterns. Lake Ontario is at the top, and its southernmost indentation lies just to the north of metropolitan Rochester; the Syracuse urban area is located in the extreme upper right.

Glacial Lakes

Glacial lakes, including those just described, are of much importance and interest in the study of glaciated landscapes. In addition to the lakes formed in depressions sculpted at the bottom of sheet glaciers, lakes also formed when glacial deposits blocked the path of outflowing meltwater at the leading edge of icesheets. Such dammed-up lakes formed during the recession of valley glaciers as well as sheet glaciers. These lakes, born on the margins of melting glaciers, contain sediments that are layered in a characteristic pattern.

A cross-section of such lake-bottom sediments reveals pairs of layers: each pair consists of a light-colored band of silt and a dark-colored, finer textured band of clay. Together, these two bands represent one year's deposition. The coarser silt was washed into the lake during the summer, when the ice was melting and sediment entered the lake in quantity; this material settled quickly on the lake floor. During the ensuing winter, the lake surface froze and no meltwater or new sediment arrived. But finer material, still in suspension, now settled slowly on top of the silt. This finer material consisted of clay particles and organic matter, and they created the dark band.

The paired layers, one light and one dark, each constitute a *varve*. By counting varves, geomorphologists can calculate the life span of a glacial lake much as tree rings can be used to date the age of a tree. Elaborate systems of correlations were developed to extend the varve counts from lake to lake. Such research made an important contribution to early estimations of glacial timing.

Glacial Lake Bonneville

As the map of the maximum extent of the Laurentide Icesheet and its adjoining Rocky Mountains Icecap indicates, the continental glacier never reached as far as Utah, Nevada, Oregon, or California (Fig. 46-8). Still, the glaciations had far-reaching effects there. Today, the basins in this area of the Far West are arid. During the glaciations, however, precipitation in this region was substantially higher than at present, and more than 100 lakes, known as **pluvial lakes**, developed as a result (Fig. 46-10).

The largest of them, Lake Bonneville, was the forerunner of Utah's Great Salt Lake. Glacial Lake Bonneville at one stage was about as large as Lake Michigan is now. It reached a maximum depth of 300 m (1000 ft) and overflowed northward through Idaho into the Snake River and hence into the Columbia River. Although Glacial Lake Bonneville has now shrunk into the Great Salt Lake, its former shorelines still can be seen on the slopes of the mountains that encircled it. Today, only a few of the pluvial lakes still contain water, and this water is saline; the Great Salt Lake is the largest of them. As for the other pluvial lakes, they have evaporated away, leaving only physiographic evidence of their former existence.

The Great Lakes and Their Evolution

The Great Lakes, too, owe their origin to the Late Cenozoic glaciers. The area occupied by the present five

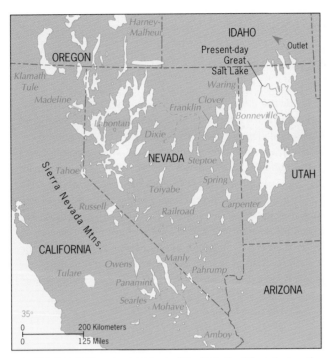

FIGURE 46-10 The pluvial lakes of the U.S. Far West during the glaciation stage. Blue dashed lines represent overflow channels.

Great Lakes (the largest cluster of freshwater lakes in the world) lay at a low elevation and had low relief when the glaciers advanced over it. The ice excavated a set of shallow but very extensive basins. When the ice receded, deposits left along its leading edge blocked the outflow of meltwater southward except through a few channels, and the first stage in the evolution of the Great Lakes began.

The major outlet to the south led from near the southern end of present-day Lake Michigan across Illinois into the Mississippi River (Fig. 46-11, Route 5). What began as a set of separate, marginal lakes grew into an interconnected group of major water bodies. Not only was their overflow channeled southward into the Mississippi Basin, but they also drained eastward into the Hudson (Route 14) and St. Lawrence Valleys (Routes 12 and 13).

The Great Lakes reached their maximum extent in the early Holocene interglaciation, when Lake Huron far exceeded its present size and flowed through the Ottawa River into the St. Lawrence Valley (Fig. 46-11, Route 12). Then, with the meltwater sources gone, the lakes began to shrink. Lowered water level closed the Lake Michigan outflow; crustal rebound closed the Ottawa River exit. But the lakes continued to flow into each other, and the St. Lawrence River and its estuary eventually became the only outlet for Great Lakes water (Fig. 46-11, Route 13).

The future of the Great Lakes is uncertain. The lowering of the lake levels exposed large areas of fertile soils, and these lakeshore zones now constitute one of the continent's most densely populated areas. Major cities have evolved on the shores of the Great Lakes, including Chicago, Detroit, and Toronto. Urban and agricultural pollution have had a severe impact on the lakes, and Lake Erie in particular has been gravely threatened in recent decades. In the meantime, lake levels are kept up by the considerable precipitation received in the lakes region. Over the longer term, however, continued crustal rebound following the recession of the Wisconsin ice, plus a changing water budget, will continue to modify the map of the Great Lakes and their tributary region.

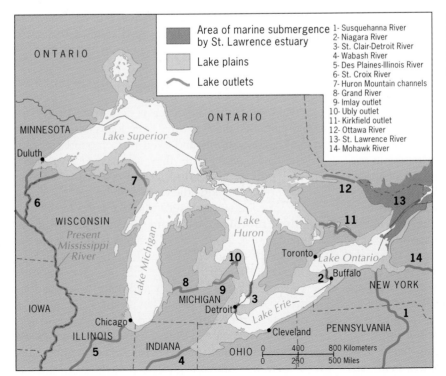

FIGURE 46-11 The lake plains (former lake bottoms) of the late Pleistocene Great Lakes and the major outlets used by these lakes during their evolution into the present-day Great Lakes.

The Driftless Area

Time and again, the continental glaciers of the Late Cenozoic moved southward in North America, covering older landscapes with glacial drift from Nebraska to New England. But one small area never was buried (*see* Fig. 46-13). In southwestern Wisconsin lies a 400-km (250-mi) stretch of scenic landscape of moderate relief, supported by exposed bedrock and exhibiting landforms that could not have survived the overriding of icesheets.

Surrounded on all sides by the characteristic scenery of glacial drift, Wisconsin's *Driftless Area* provides a unique glimpse of the landscape now concealed from view throughout the Midwest. Economically, this region's poorer soils have deflected the Corn Belt to the south (even though the climate is hospitable), and local agriculture is dominated by less-crop-intensive dairying.

There are, in fact, two interpretations of the history of this unique region. One

theory holds that the Driftless Area was never glaciated, escaping the icesheets' advance because of its location midway between the valleys that became Lakes Superior and Michigan. Thus the main mass of the advancing ice was repeatedly diverted around the area. But other glaciologists believe that the Driftless Area escaped only the most recent glaciation. Earlier advances may have covered the area, but the evidence (glacial drift) was removed by subsequent stream erosion.

AGGRADATIONAL LANDFORMS OF ICESHEETS

A continental glacier transports huge amounts of rock debris as it thickens and expands, scouring and sculpting the surface beneath it. When the Laurentide Icesheet moved from the hard crystalline rocks of the Canadian Shield to the softer rocks of adjacent areas, its sedimentary load increased even more. If we were able to take a view in profile of such a sediment-charged glacier, we would note that virtually all of the sediment load is carried near the bed of the glacier. Some of the rock material would be fine-grained, but much of the load would be pebble and boulder-sized or even larger. Laden with all this debris, the icesheet inches forward, depositing some of its load in places along the way and eroding material elsewhere.

Glacial Drift

When deglaciation begins, deposition increases in several ways. Solid material carried at the base of the glacier is dropped as an unsorted mass called **till**, which consists of fragments ranging in size from fine clay particles to boulders. Other rock material carried by the ice is moved some distance by the meltwater coming from the ablating glacier. This material is to some degree sorted during the transportation by water, and shows some layering by size (as clay, silt, gravel, or pebbles). This is known as **stratified drift**. Together, unsorted till and stratified drift are referred to as **glacial drift**.

Over a long period of repeated ice advances and recessions, glacial drift can become very thick, completely burying the underlying bedrock topography.

Much of the U.S. Midwest is covered by glacial drift of varying thickness. Some 30 m (100 ft) below the present landscape of Illinois, for instance, lies a very different buried landscape, covered now by glacial drift accumulated during repeated glaciations and deglaciations. Neighboring Iowa's glacial drift deposits are even thicker, reaching a depth of 60 m (200 ft); but southwestern Wisconsin was bypassed by the icesheets and exhibits a rather different landscape (*see* Perspective: The Driftless Area). The dominant aggradational landscape of continental glaciation, therefore, is a flat to undulating plain underlain by heterogeneous material, often studded with boulders, called *erratics*, that were transported far from their source area by the glacier.

Moraines

Icesheet topography, however, is not always of low relief. Continental glaciers carry large loads of rock debris in their leading fronts and even push mounds of such debris ahead of them as they advance (Fig. 46-12a). When progress stops, this material is left as a curving irregular ridge, marking the outline of the farthest extent of the ice lobe (Fig. 46-12b). Such a ridge is called a **terminal moraine**, and many such moraines can be mapped in the north-central United States. From the properties of the glacial drift and the positions and relationships of the terminal moraines, geomorphologists have been able to reconstruct and map the glaciations shown by the drift margins in Fig. 46-13.

The term **moraine** is applied to many kinds of glacial features. As noted in Unit 47, even material still being carried on a mountain glacier is referred to as a moraine. In the case of continental glaciers, a *terminal moraine* is distinguished from a **recessional moraine**. A terminal moraine marks the farthest advance of the glacier, but a recessional moraine develops when an already receding glacier becomes temporarily stationary. Thus a receding glacier can form several recessional moraines, but it will leave only one terminal moraine. The term *moraine* also is applied to that extensive blanket of unsorted till that is laid down at the base of a melting sheet glacier (Fig. 46-12b): in this case it is called a **ground moraine**.

Drumlins

A special landform associated with icesheet movement is the **drumlin**. This is a smooth hill, modest in size, elliptical in shape, rising from the till plain (Fig. 46-14). The long axis of a drumlin lies parallel to the direction of ice movement. It is believed that drumlins are created when an icesheet overrides and reshapes preexisting till, creating a coated, plastered-looking mound. These occur in groups, rising like the backs of a school of whales above the ocean surface. Indeed, drumlins are sometimes called *whalebacks*.

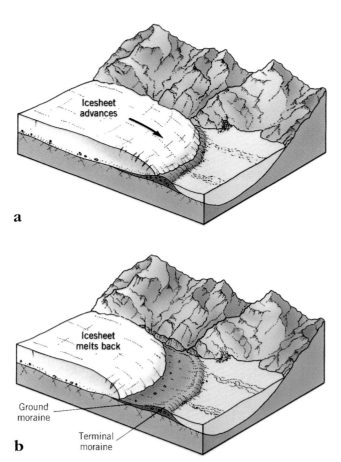

FIGURE 46-12 A terminal moraine is formed from bulldozed debris at an advancing icesheet's front edge (a). When forward glacial movement ends, this debris remains as a ridge after the ice melts back (b).

FIGURE 46-14 The unmistakable topography of drumlins, here in northeastern Saskatchewan Province, Canada. No wonder these cigar-shaped landforms are colloquially referred to as whalebacks.

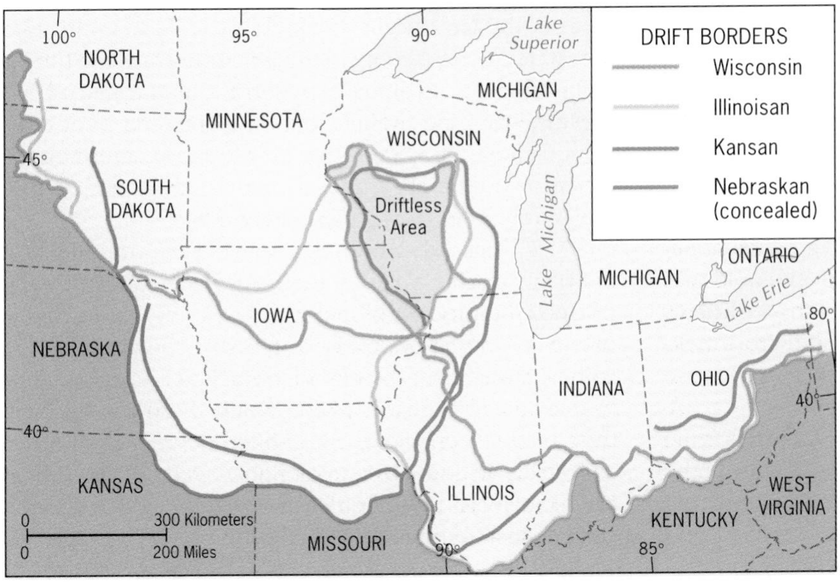

FIGURE 46-13 The drift margins of the north-central United States. The positions of these terminal moraines indicate that in each of the final four glaciations the continental icesheet reached a different line of maximum advance.

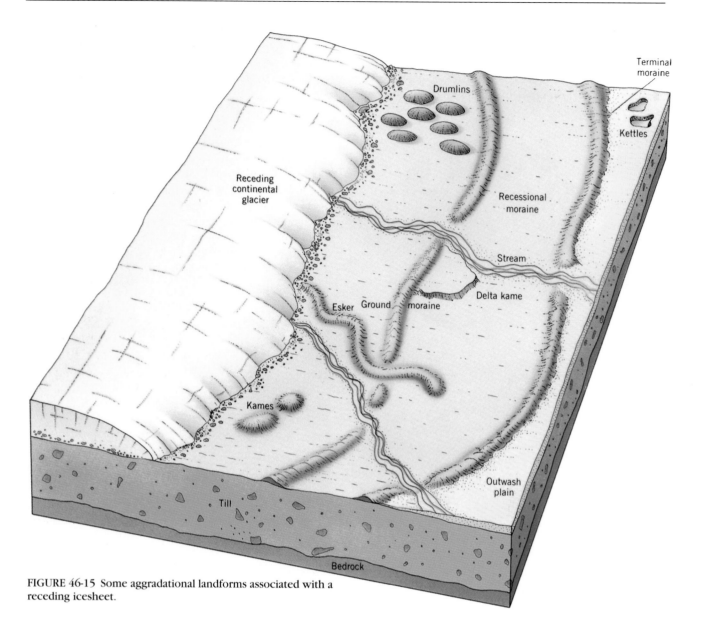

FIGURE 46-15 Some aggradational landforms associated with a receding icesheet.

Glacial Meltwater Deposits

Terminal, recessional, and ground moraines, as well as drumlins, are ice-deposited. But when meltwater becomes part of the depositional process, the landform picture becomes more complicated. Meltwater flows in channels from beneath, on top of, and in front of the ice. In the process, it carries rock debris away from the glacier and sorts it to varying degrees (depending on such factors as the distance of transportation, the volume and speed of the water, and the sedimentary load carried). In the field, unsorted till is easily distinguished from meltwater-deposited layers of sand and gravel, which are called **glacial outwash**.

One of the most interesting outwash landforms is the **esker**. Under certain circumstances, water flowing at the base of a melting sheet glacier forms a long tunnel that begins well (even kilometers) inside the icesheet and

leads to the ice margin. Rock debris collects in this tunnel, sorted considerably by the water even as the tube is being clogged. Eventually, when the entire glacier has melted away, the tunnel's outline is marked by a long, sinuous ridge, the esker, that may look like a terminal or recessional moraine—until its sorted and stratified profile is examined (Fig. 46-15, lower center).

Most outwash landforms, however, form ahead of, not beneath, the stalled or receding glacier. An **outwash plain** is the product of meltwater carrying (and to some extent sorting) rock debris from the wasting ice front. In some places, large blocks of unmelted ice are initially buried in the outwash plain. When these blocks melt, they create steep-sided, water-filled depressions known as **kettles** (Fig. 46-15, upper right). Occasionally, there are so many of these kettles that the affected portion of the outwash plain is described as *pitted*.

We noted earlier that lakes form when meltwater is dammed up by terminal or recessional moraines. Often these marginal lakes do not survive for very long, but they do leave evidence of their brief existence in the landscape. The deposits that constitute this evidence are called *glacio-lacustrine* (glacial-lake) deposits. An example is the **kame** and various landforms associated with it. As in any location where sediment-laden water flows into standing water, the glacial streams form deltas when they pour into marginal glacial lakes. These glacial deltas are made of characteristically coarse-textured material. When the lake drains away or dries up, the deltaic material is left standing in the morainal landscape (Fig. 46-15, right of center). It is typically flat-topped, as the delta surface was. Such a flat-topped hill, sometimes flanked by terraces marking the lowering of the lake level, is a kame.

These landforms and landscapes associated with continental glaciers represent the main features created by icesheet aggradation. As you can imagine, many other features could be identified and their origins analyzed. By understanding the processes that formed these features, we are able to interpret the sequence of events that prevailed during the several glaciations of the Late Cenozoic Ice Age. The landforms of continental glaciers may not be as scenic as those associated with mountain glaciers (which we examine in Unit 47), but they are valuable indicators of past processes and environments.

KEY TERMS

calving (p. 471)	glacial outwash (p. 480)	kettle (p. 480)	pluvial lake (p. 476)
drumlin (p. 479)	ground moraine (p. 479)	moraine (p. 478)	recessional moraine (p. 479)
esker (p. 480)	ice shelf (p. 471)	nunatak (p. 470)	stratified drift (p. 478)
flow regime (p. 471)	icecap (p. 474)	outwash plain (p. 480)	terminal moraine (p. 478)
glacial drift (p. 478)	kame (p. 481)	pack ice (p. 471)	till (p. 478)

REVIEW QUESTIONS

1. What is the approximate size (area and thickness) of the Antarctic Icesheet?

2. Name the three Northern Hemisphere Late Cenozoic icesheets and their major geographic dimensions.

3. How might continental glaciation produce lakes on the landscape?

4. How do pluvial lakes relate to glaciation?

5. What is the difference between till, moraines, and erratics?

REFERENCES AND FURTHER READINGS

DAVIES, J. L.   *Landforms of Cold Climates* (Cambridge, Mass.: MIT Press, 1969).

DAWSON, A.   *Ice Age Earth* (London/New York: Routledge, 1991).

DREWRY, D.   *Glacial Geological Processes* (London: Edward Arnold, 1986).

EMBLETON, C., and KING, C. A. M.   *Glacial Geomorphology* (New York: Wiley/Halsted, 2nd ed., 1975).

FLINT, R. F.   *Glacial and Quaternary Geology* (New York: Wiley, 1971).

GOLDTHWAIT, R. P., ed.   *Glacial Deposits* (Stroudsburg, Pa.: Dowden, Hutchinson & Ross, 1975).

HAMBREY, M., and ALEAN, J.   *Glaciers* (New York: Cambridge University Press, 1992).

HOUGH, J. L.   *Geology of the Great Lakes* (Urbana, Ill.: University of Illinois Press, 1958).

MATSCH, C. L.   *North America and the Great Ice Age* (New York: McGraw-Hill, 1976).

PATERSON, W. S. B.   *The Physics of Glaciers* (Tarrytown, N.Y.: Pergamon, 3rd ed., 1994).

PIELOU, E. C.   *After the Ice Age: The Return of Life to Glaciated North America* (Chicago: University of Chicago Press, 1991).

SHARP, R. P.   *Living Ice: Understanding Glaciers and Glaciation* (London/New York: Cambridge University Press, 1988).

SUGDEN, D. E.   *Arctic and Antarctic: A Modern Geographical Synthesis* (Totowa, N.J.: Rowman & Littlefield, 1982).

SUGDEN, D. E., and JOHN, B. S.   *Glaciers and Landscape: A Geomorphological Approach* (New York: Wiley, 1976).

THEAKSTONE, W. H., et al.   *Glaciers and Environmental Change* (Sevenoaks, U.K.: Edward Arnold, 1994).

Landforms and Landscapes of Mountain Glaciers

OBJECTIVES

◆ To examine the current distribution of mountain glaciers and to comment on the Late Cenozoic extent of these glaciers.

◆ To discuss the landforms produced by mountain glacier erosion and deposition.

The global climate during the Cenozoic was generally mild until the onset of the Late Cenozoic Ice Age. Before the ice age began, even Antarctica was mostly ice-free; on high mountains, streams—not glaciers—sculpted the landscape. The landforms associated with river erosion and deposition also characterized such major mountain ranges as the Rockies, Alps, Andes, and even the Himalayas. Hilltops displayed rounded forms, valleys were **V**-shaped, stream courses meandered, and most tributary junctions were concordant. The mountain regions of the world looked much like today's Great Smoky Mountains (southeastern United States), Atlas Mountains (northwestern Africa), or

Great Dividing Range (eastern Australia), except that higher relief generally prevailed.

When the first cooling episode occurred and the altitude of the snow line dropped, the formation of glaciers began. The Antarctic Icesheet probably was the first continental glacier to develop because of Antarctica's polar location and its high overall elevation. Gradually, on the other continents, permanent ice formed on higher mountain slopes. Snow accumulated above the firn line, and *alpine glaciers* descended down the high valleys.

UNIT OPENING PHOTO: Two converging alpine glaciers join forces to flow as a single unit. Kaskawulsh Glacier, Alaska.

Mountain Glaciers: The View From Space

The relationship between alpine glaciers and their surrounding landscapes is strikingly visible in satellite imagery. The photo above (Fig. 47-1) is a particularly fine example and should be contrasted against the space view of the continental-scale Antarctic Icesheet shown in Fig. 46-1.

The subject of Fig. 47-1, Bylot Island, is located off the northern coast of Canada's Baffin Island. More than two dozen active mountain glaciers can readily be seen, and there are several excellent examples of individual valley glaciers coalescing to form a single main (trunk) glacier. All of them emanate from the island's interior uplands (whose elevation exceeds 600 m [2000 ft]), and move outward and downslope toward the sea. Because of Bylot's high latitude (73°N), the snow line lies at a relatively low elevation and a number of glaciers come close to reaching the coast (two in the northwest actually do). Overall, the island's area is 10,880 sq km (4200 sq mi); its longest east-west dimension measures approximately 145 km (90 mi), and its widest north-south extent is about 115 km (70

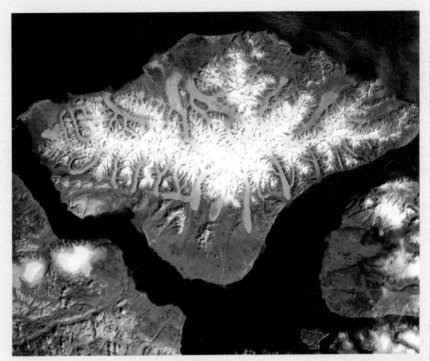

FIGURE 47-1 A vertical satellite view of the alpine-glacier complex that blankets most of Bylot Island in the Canadian Arctic.

mi). Surprisingly, given Bylot's remote location and inhospitable setting, humans have made inroads here: the wide, low-lying, ice-free southern peninsula is occupied by several thousand Inuit people each spring and summer.

These glaciers occupied valleys first carved by rivers, and glacial erosion now replaced stream erosion. Permanent ice appeared on high mountains even in equatorial locales, and the earth was indeed transformed. In this unit, we examine glaciers that form on mountains and erode and deposit in these alpine settings. They differ from the icesheets of continental glaciation in that they are generally confined to valleys, and their behavior is influenced by the topography they inhabit (*see* Perspective: Mountain Glaciers: The View from Space). In other respects, they are quite similar to sheet glaciers.

MOUNTAIN GLACIERS TODAY

In Unit 46, we note that two large icesheets survive to the present (the Antarctic and Greenland Icesheets) and that several other major continental glaciers and icecaps

wasted away with the onset of the Holocene interglaciation. Mountain glaciers, too, were larger and much more prevalent before the current interglaciation began. Many mountain glaciers melted away and vacated their valleys as the firn line rose and ice formation in their source areas diminished or ceased. But despite the warmth of the present global climate, numerous mountain glaciers endure. Some have not disappeared, but have receded up their formerly occupied valleys, leaving abundant evidence of their former advances.

Global Distribution

Every major landmass on earth except Australia contains alpine glaciers. It has been estimated that there are as many as 100,000 individual mountain glaciers in the world today, ranging in size from huge bodies to small narrow ribbons of ice. Ironically, the largest mountain glacier lies in Antarctica's Queen Maud Mountains, where it feeds the continental icesheet. This is the Beardmore Glacier, and it remains a mountain glacier for more than 200 km (125 mi) before it merges into the Antarctic Icesheet. The Beardmore Glacier is as much as 40 km (25 mi) wide, vastly larger than anything seen in Alaska or the Alps.

The remote Beardmore Glacier has been observed by comparatively few people, but other glaciers lie in more accessible locales. Cruise ships visit the Alaskan glaciers where they enter the Gulf of Alaska, and tourists can see active mountain glaciers from their cable-car gondolas in the French and Swiss Alps. Landscapes produced by alpine glaciation are among our most dramatic, and underscore the significance of ice in shaping the surface of the earth.

North America

Many alpine glaciers lie on the islands that encircle the Arctic Ocean, such as Spitsbergen, Ellesmere Island, and Novaya Zemlya (*see* Fig. 47-1). In North America, major clusters of mountain glaciers lie in southeastern Alaska, Canada's Yukon Territory and Coast Mountains of British

FIGURE 47-2 Active mountain glaciers can be seen in several areas of North America, particularly southeastern Alaska. The rugged, steep-sided, angular topography of alpine glaciation is shown in this aerial view of the Alaska Range near Gulkana. (Authors' photo)

Columbia, and in the Canadian Rocky Mountains along the Alberta-British Columbia border (Fig. 47-2).

Equatorward of 50°N latitude, glacier formation becomes more exceptional. South Cascade Glacier near Mount Rainier in western Washington State is an example of a temperate glacier in a moist maritime environment, where abundant orographic snowfall sustains a glacier despite relatively high summer temperatures. In Colorado, where there is much less snowfall in the Rockies, strong winds create high-elevation snowdrifts deep enough to produce ice formation and maintain small remnant glaciers.

South America

In South America, there are large mountain glaciers in the southern Andes of Chile. These begin just south of latitude 45°S (about the latitude of Minneapolis in the Northern Hemisphere) and become progressively larger at higher latitudes. At the Strait of Magellan, where the South American mainland ends, glacial ice reaches the sea. As the world map shows, the Antarctic Peninsula reaches toward South America, but much of this peninsula is not covered by the Antarctic Icesheet. However, large mountain glaciers descend from the highland backbone of the peninsula (an extension of the Andes Mountains) toward its coasts.

Africa

Africa has high mountains in the far northwest (the Atlas massif) and south (the Drakensberg), but neither of these ranges carries glaciers, although they do receive winter snows. All of Africa's glaciers lie on two soaring mountains near the equator, Mount Kilimanjaro (5861m/19,340 ft) and Mount Kenya (5199 m/17,058 ft). Glaciers descend from near the summits of both mountains, and on Mount Kilimanjaro they reach as far down as 4400 m (14,500 ft) on the south side (*see* Fig. 19-11). There are a dozen small glaciers on Mount Kenya, of which two still are quite substantial, although, like the others, receding.

Australia-New Zealand

Australia's mountains are too low to support glaciers, but higher latitude, higher elevation, and greater precipitation combine to sustain large valley glaciers on the South Island of New Zealand. The Southern Alps, as the mountain backbone of this island is called, reach their highest point in Mount Cook (3764 m/12,349 ft). In the vicinity of Mount Cook lie another 15 mountains over 3000 m (10,000 ft) high. The whole mountain range was covered by an icecap in glacial times, and several major glaciers survive in the area of Mount Cook. Among the best known are the Franz Josef Glacier and the Fox Glacier. The Fox Glacier is receding quite rapidly, and its withdrawal is marked by signposts in its lower valley along the road leading to the present glacial margin (Fig. 47-3).

Europe

Most of the world's mountain glaciers lie in two major clusters on the Eurasian landmass. Of these two, the European Alps is undoubtedly the most famous, and the south-central Asian zone is by far the largest (*see* Fig. 46-8). Much of what is known about the degradational and aggradational work of glaciers was learned through research performed in the European Alps. Europe's Alps extend in a broad arc from southeastern France through the area of the Swiss-Italian border into central Austria (Fig. 47-4). Mont Blanc, 4807 m (15,771 ft) high, is the tallest peak, but several other mountains exceed 4000 m (13,200 ft). Active glaciers abound in the Alps, and virtually every erosional and depositional landform associated with glaciation is found there. From the deposits

FIGURE 47-3 The Fox Glacier flows westward off New Zealand's Southern Alps. Near its source area this still is an impressive mountain glacier, but reduced snowfall in its catchment area (and possibly other factors as well) has led to retreat in its lower valley. This withdrawal has been monitored for decades, and the glacial valley is marked by signposts indicating previous positions. Today, the terminus of the glacier shows morainal material in a wide **U**-shaped valley, and rock rubble in and on top of the melting ice. A small stream emanates from the base of the glacier, feeding glacial lakes downstream. (Authors' photo)

FIGURE 47-4 Central Europe's Alps form a gigantic crescent of spectacular mountain ranges. The large lowland just below the center is northern Italy's Po Plain; the Italian peninsula, with its Appennine Mountains backbone, extends seaward from the southern margin of the Po Valley.

FIGURE 47-5 A space view of the Himalayas and the Qinghai-Xizang (Tibetan) Plateau, rising above the plains of northern India (lower left). The snow-and-ice-covered Himalayas and the vast plateau to their north (right), located in the center of the broad southern Asian highland belt, constitute the earth's most prominent upland zone.

left by the repeatedly advancing glaciers, European scientists deduced the glacial sequence presented in Table 46-1.

Asia

The glacial topography of the European Alps is dwarfed, however, by the vast expanse of glacial landscape that extends across the soaring highlands of south-central Asia from Afghanistan to southwestern China (Fig. 47-5). This region was the site of one of the Late Cenozoic's largest icecaps, and tens of thousands of residual glaciers now provide testimony of that phenomenon. The glacial landscape extends from northeastern Afghanistan along the length and breadth of the highlands, marking the boundary between the Eurasian and Australian-Indian Plates (see Fig. 32-3).

Many of the world's highest mountains—including the tallest of all, Nepal's Mount Everest (8848 m/29,028 ft)—lie in this zone. Much of the region still remains buried under ice and snow, and many of the mountain glaciers here are hundreds of meters thick and many kilometers wide. Maximum development occurs in the Himalayan-Tibetan area (Fig. 47-5). Vast as the ice and snow cover is, however, there is abundant evidence that here, too, the glaciers have receded during the Holocene: glacial topography and glacial deposits extend far beyond the margins of the present ice.

Isolated Remnant Glaciers

In addition to the earth's notable clusters of alpine glaciers, there are isolated glaciers in isolated places. We noted the high-elevation glaciers of equatorial East Af-

rica; similar glacier development also occurs on the highest slopes of the Andes Mountains in tropical-latitude Ecuador and Peru. Remnant glaciers also exist in northern Norway, where it is latitude, not altitude, that supports them. Mount Elbrus, the highest peak in the Caucasus Mountains (between the Black and Caspian Seas along Russia's southern flank), reaches 5642 m (18,510 ft) and carries 22 small glaciers. All of these glaciers are surrounded by evidence that they, too, are remnants of larger ones that existed in the past.

DEGRADATIONAL LANDFORMS OF MOUNTAIN GLACIERS

There is no mistaking a landscape sculpted by mountain glaciers, even long after the glaciers have melted away. Mountains, ridges, valleys, and deposits all bear the stamp of the glaciers' degradational or aggradational work. Before examining the major landforms created by glacial action, we should review the nature of alpine glaciers' mass balance as well as their appearance and general morphology. As we note in Unit 45, the zones of mass accumulation of these glaciers lie on high mountain slopes. There snow is compacted into ice, and the ice moves downhill under the force of gravity assisted by basal lubrication to occupy valleys formed earlier by stream erosion. Unlike the surface of an icesheet, which tends to be snow-white or ice-blue, the surface of a mountain glacier normally is streaked by bands of rock debris (see unit opening photo).

Glacial Valleys

A mountain glacier, when it occupies a river valley, immediately begins to change the cross-section and profile of that valley. We have observed that river valleys in mountain areas tend to have **V**-shaped profiles (see Fig. 41-4a), signifying the streams' active downward erosion. Glaciers work to widen as well as deepen their valleys, and the typical cross-section of a glacial valley is **U**-shaped (Fig. 47-6). The **U**-shaped glacial valley, or **glacial trough**, is one of the most characteristic of glacial landforms.

Maps and aerial photographs of glacial troughs also reveal another property of these valleys: they do not turn or bend nearly as tightly as rivers do. Ice is not capable of winding turns as water is, and glaciers are powerful erosional agents that can destroy obstacles in their paths. As a result, many mountainside spurs *around* which rivers once flowed are sheared off by glaciers, thereby straightening the valley course. Such **truncated spurs** are further evidence of glacial action in the landscape after the glaciers have wasted away (Fig. 47-6b).

When a river is joined by a tributary, the water surface of both streams is at the same level and the floors of the two valleys tend to be concordant as well. In other words, if the rivers were to dry up, there would be no

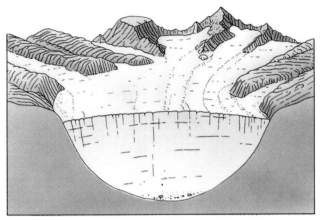

a

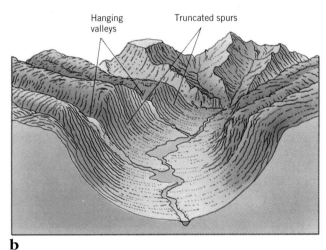

Hanging valleys   Truncated spurs

b

c

FIGURE 47-6 The evolution of a glacial trough. Diagram (a) shows the stage of peak glaciation, with the **U**-shaped trough gouged out by the trunk glacier that is advancing toward the reader. When the valley glacier has melted away, truncated spurs and hanging valleys are readily apparent (b). If the glacial trough is near the coast and has been deepened below (rising) sea level, it will become inundated and a fjord will form (c). Drawn by Erwin Raisz. Copyright © by Arthur N. Strahler. Reproduced by permission of Arthur N. Strahler.

sharp break between the floor of the main stream valley and that of the tributary valley. But when a smaller glacier joins a larger one, the base of the tributary glacier is not nearly as low, nor the valley as deep, as that of the larger trunk glacier. Their ice surfaces will be at about the same level (as shown in Fig. 47-9), but their bedrock floors are discordant, sometimes by hundreds of meters. When both glaciers melt away, the valley of the tributary glacier, as viewed from the floor of the main glacier, seems to "hang" high above. Such a discordant junction is appropriately called a **hanging valley** or hanging trough (Fig. 47-6b), still another sure sign of the landscape's glacial history. A hanging valley is often graced by a scenic waterfall where the stream now occupying the tributary glacier's valley joins the main trough, as illustrated by Yosemite National Park's Bridal Veil Falls (*see* Fig. 45-4, right center).

High-Mountain Landforms

In the mountains above the glacial valleys, the source areas of the glaciers, the landscape also is transformed. A series of three block diagrams in Fig. 47-7 suggests the sequence of events. Initially, the landscape consists of rounded ridges and peaks (Fig. 47-7a). With the onset of glaciation, deep snow accumulations form on the higher slopes, and the icecap subsequently thickens. The ice moves downslope under gravity, and glacial erosion begins (Fig. 47-7b).

In the upper area of continuous snow accumulation, the ice hollows out shallow basins that become the glacier's source area. Not only does the ice excavate such basins: frost wedging on the walls above them, plus undercutting by headward erosion, create distinctive, amphitheater-like landforms that are referred to as **cirques** (Fig. 47-7c). A cirque is a bowl-shaped, steep-sided depression in the bedrock with a gently sloping floor. Two, three, or even more cirques may develop near the top of a mountain; in time these cirques, growing by headward erosion, intersect. Now nothing remains of the original rounded mountaintop in the center of the diagram except a steep-sided, sharp-edged peak known as a **horn** (Fig. 47-8). The Matterhorn in the Swiss Alps (Unit 45 opening photo) is the quintessential example. Thus when the ice melts or the icecap thins out, these horns tower impressively above the landscape (*see* Fig. 45-7).

Other dramatic elements of mountain glacier topography are shown in Fig. 47-7c. One is the large number of razor-sharp, often jagged ridges that rise above the ice-containing troughs. These ridges often separate adjacent glaciers or glacial valleys and are known as **arêtes**. They can develop when two large cirques intersect, but they more frequently form from erosion and frost wedging in two parallel glacial valleys.

Another major feature occurs as the glaciers move downslope: they not only straighten their courses, but also create step-like profiles (valleys emanating forward

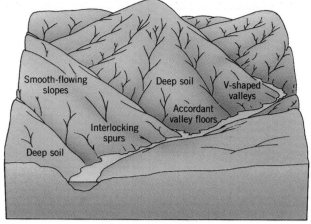

a   **BEFORE GLACIATION**

Labels: Smooth-flowing slopes · Deep soil · V-shaped valleys · Accordant valley floors · Interlocking spurs · Deep soil

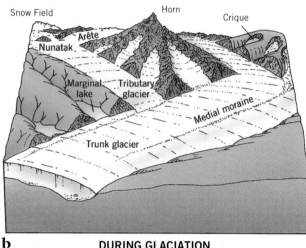

b   **DURING GLACIATION**

Labels: Snow Field · Horn · Crique · Arête · Nunatak · Marginal lake · Tributary glacier · Medial moraine · Trunk glacier

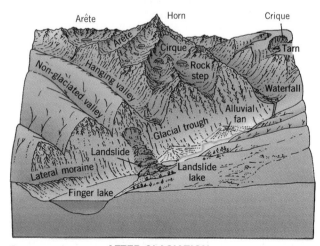

c   **AFTER GLACIATION**

Labels: Arête · Horn · Crique · Arête · Cirque · Tarn · Hanging valley · Rock step · Non-glaciated valley · Waterfall · Alluvial fan · Glacial trough · Landslide · Landslide lake · Lateral moraine · Finger lake

FIGURE 47-7 The transformation of a mountain landscape by alpine glaciation (drawn by Armin K. Lobeck after William Morris Davis). Note how the initial rounded ridges and peaks are sculpted into a much sharper-edged topography by frost wedging and ice excavation.

below the horn in the center of Fig. 47-7c). Such landforms result from a combination of factors. One relates to the differential resistance of various rocks over which

FIGURE 47-8 The terrain of alpine glaciers reveals the capacity of glaciers to sculpt dramatic angular topography. This view of a cirque in the Southern Alps of New Zealand's South Island also reveals a developing horn (right background) and an arête (left foreground). (Authors' photo)

the glacier passes. The jointing of the bedrock (and therefore its susceptibility to frost wedging) also affects this process. In the postglacial landscape, **rock steps** reveal the local effects of such glacial erosion.

Lakes

As in the case of continental glaciers, depressions gouged by alpine glaciers are filled by water during interglaciations such as the present period. Where the climate is warm enough so that even high-altitude cirques are no longer filled with snow, small circular lakes are found on the floors of the cirque basins. These lakes, dammed up behind the "lip" of the cirque, are known as **tarns** (one is labeled in the upper far-right portion of Fig. 47-7c). Lakes also may form on the rock steps previously described, as again shown in those three valleys below the horn in the central portion of Fig. 47-7c. The largest lakes fill substantial parts of glacial troughs, as the lower left portion of this diagram shows. Such lakes may be several kilometers wide and 50 km (30 mi) or more long, and are called **finger lakes** (those in upstate New York are shown in Fig. 46-9). Some of the world's most scenic lakes, from the Alps of Switzerland to the Southern Alps of New Zealand, owe their origins to glacial erosion.

Fjords

Among the most spectacular landforms associated with glacial erosion are fjords. A **fjord** is a narrow, steep-sided, elongated estuary (drowned river mouth) formed from a glacial trough inundated by seawater (Fig. 47-6c). During glaciations, many glaciers reach the ocean. Ice, as we noted elsewhere, has a density about five-sixths that of seawater, so a glacier reaching the ocean can continue to erode a valley many meters below sea level. Thus vigorously eroding glaciers created seaward troughs. When the ice melted, ocean water inundated the glacial valley, creating a unique coastal landscape. Fjords developed mainly in places where glaciated mountains lie near a coastline, such as in southern Alaska and western Norway. Other famous and scenic fjords lie along the southwestern coast of Chile and along the southwestern coast of New Zealand's South Island.

AGGRADATIONAL LANDFORMS OF MOUNTAIN GLACIERS

As components of scenery, the depositional landforms of alpine glaciers are no match for the erosional features just discussed and illustrated. Some of the debris carried downslope is ground into particles so fine that it is called **rock flour**, and when the glacier melts and deposits this rock flour, much of it is blown away by the wind. Larger fragments, as in the case of sheet glaciers, are deposited

at the (stalled) edge of the advancing glacier as *terminal moraines*. Again, as with continental glaciers, stationary periods during a glacier's retreat are marked by *recessional moraines* (*see* Fig. 46-15). Terminal and recessional moraines lie in low ridges across the valley floor. These mounds can form dams that impound meltwater, creating temporary glacial lakes and associated glaciofluvial features (*see* Unit 46).

Moraines

Debris carried by an alpine glacier comes not only from the valley floor it erodes but also from the valley sides above the glacial ice. Frost wedging, the repeated freezing and thawing of water in rock cracks and joints, loosens pieces of bedrock (*see* Fig. 38-2). These fall onto the glacier's surface along the margins of the ice, where they become part of the visible bands of debris called **moraines** (the same term used in relation to sheet glaciers on pp. 478–479, but with a somewhat different meaning here). The vigorously eroding glacier also tends to undercut its valley sides so that mass movement contributes additional material to the glacial surface.

Material that falls from the valley wall first becomes part of the glacier's **lateral moraines**, the moraines situated along the edges of the ice (Fig. 47-9). When a trunk

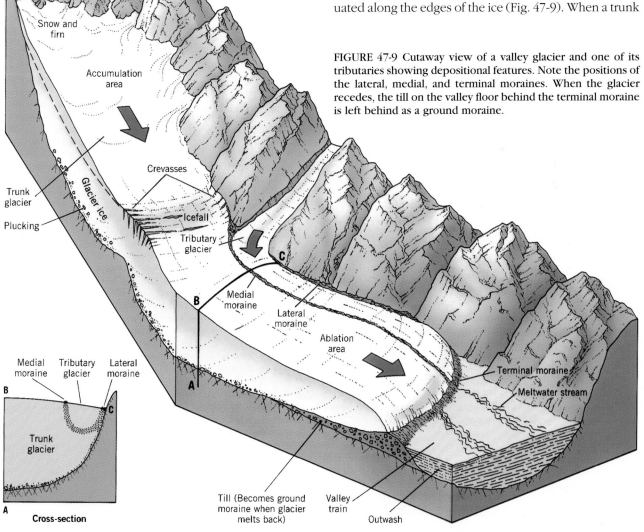

FIGURE 47-9 Cutaway view of a valley glacier and one of its tributaries showing depositional features. Note the positions of the lateral, medial, and terminal moraines. When the glacier recedes, the till on the valley floor behind the terminal moraine is left behind as a ground moraine.

glacier is joined by a substantial tributary glacier, their lateral moraines join to become a **medial moraine** that is situated away from the glacier's edges (Fig. 47-9). This pattern may be repeated several times as additional tributary glaciers enter the main glacier's channel.

Furthermore, as shown in Fig. 47-9, erosion at the base of the valley glacier creates a **ground moraine** (in this case the term has the same connotation as for sheet glaciers). The ground moraine is exposed on the floor of the glacial trough as the ice recedes, and is thickened by the deposition of the contents of the melting, debris-laden glacier. Recession of the valley glacier will also result in the deposition of the lateral and medial moraines, which form irregular ridges and mounds of unsorted material.

Postglacial Landscape Change

River action begins to modify the aggradational landforms in glacial troughs as soon as the glaciers vacate them. When meltwater starts to fill the valley floor with alluvium derived from the morainal material left behind, the new deposit is known as a **valley train** (Fig. 47-9). Modification of the glacial deposits (not the bedrock topography) usually is quite rapid. Glacial lakes are drained, material is sorted and redistributed, and vegetation recovers.

Glacial topography is scenic, and it attracts countless tourists to ski lodges and other highland resorts. But, as the foregoing has made clear, glaciated areas are not stable. The glaciers retreated from valleys with oversteepened walls. Frost wedging loosened huge quantities of rock, much of it perched precariously on steep slopes. Snow accumulations can lead to avalanches that would be harmless in remote terrain, but are often fatal when humans populate these landscapes. Mass movements of various kinds form a significant natural hazard in alpine-glaciated regions, as the large landslide and its dammed lake suggest in the lower central portion of Fig. 47-7c.

KEY TERMS

arête (p. 487)	glacial trough (p. 486)	lateral moraine (p. 489)	rock step (p. 488)
cirque (p. 487)	ground moraine (p. 490)	medial moraine (p. 490)	tarn (p. 488)
finger lake (p. 488)	hanging valley (p. 487)	moraine (p. 489)	truncated spur (p. 486)
fjord (p. 488)	horn (p. 487)	rock flour (p. 489)	valley train (p. 490)

REVIEW QUESTIONS

1. What are the two regions that contain the greatest concentration of mountain glaciers?

2. In what way does a glacial trough differ in shape from a stream valley?

3. What is a hanging valley and how does it form?

4. How would mountain glaciation produce abundant lakes?

5. What is a fjord and how is it formed?

REFERENCES AND FURTHER READINGS

DREWRY, D. *Glacial Geologic Processes* (London: Edward Arnold, 1986).

EMBLETON, C., and KING, C. A. M. *Glacial Geomorphology* (New York: Wiley/Halsted, 2nd ed., 1975).

GOLDTHWAIT, R. P., ed. *Glacial Deposits* (Stroudsburg, Pa.: Dowden, Hutchinson & Ross, 1975).

HAMBREY, M., and ALEAN, J. *Glaciers* (New York: Cambridge University Press, 1992).

IVES, J. D., ed. *Mountains* (Emmaus, Pa.: Rodale Press, 1994).

OWEN, L. A., et al. *Arctic and Alpine Geomorphology* (Cambridge, Mass.: Blackwell, 1995).

PATERSON, W. S. B. *The Physics of Glaciers* (Tarrytown, N.Y.: Pergamon, 3rd ed., 1994).

POST, A., and LACHAPELLE, E. R. *Glacier Ice* (Seattle: The Mountaineers, 1971).

PRICE, L. W. *Mountains and Man: A Study of Process and Environment* (Berkeley: University of California Press, 1981).

PRICE, L. W. *The Periglacial Environment, Permafrost, and Man* (Washington, D.C.: Association of American Geographers, Commission on College Geography, Resource Paper No. 14, 1972).

SHARP, R. P. *Living Ice: Understanding Glaciers and Glaciation* (London/New York: Cambridge University Press, 1988).

SUGDEN, D. E., and JOHN, B. S. *Glaciers and Landscape: A Geomorphological Approach* (New York: Wiley, 1976).

SYVITSKI, J. P. M., et al. *Fjords: Processes and Products* (New York/Berlin: Springer-Verlag, 1987).

WILLIAMS, R. S., JR., and FERRIGNO, J. G. "Cold Beauty: Rivers of Ice," *Earth*, January 1991, 42–49.

Periglacial Environments and Landscapes

OBJECTIVES

◆ To discuss the unique landscapes that develop under near-glacial conditions.

◆ To highlight the important weathering and mass-movement processes that shape periglacial landscapes.

The earth is undergoing an interglaciation at present; yet large regions of the world are anything but warm, even during the summer. Figure 16-3 shows the large expanses of existing **Dfc**, **Dfd**, and **E** climates. Conditions in these high-latitude regions are nearly, but not quite, glacial. The technical term for such environments is **periglacial**: on the perimeter of glaciation. In this unit, we study the processes and landforms that characterize periglacial areas.

Periglacial zones today occupy high polar and subpolar latitudes, almost exclusively in the Northern Hemisphere. No periglacial environments exist in southern Africa or in Australia, although the highlands of Tasmania (off the southeastern coast of Australia) show evidence of recent periglacial conditions. Only small areas of southernmost South America (most notably the island of

Tierra del Fuego) and the Antarctic Peninsula exhibit periglacial conditions. Accordingly, this unit deals almost exclusively with the Northern Hemisphere. Nonetheless, it is estimated that as much as one-quarter of the entire land surface of the earth is dominated by periglacial conditions, and this alone should persuade us to learn more about these cold environments.

We can also assume that periglacial conditions migrate into the middle latitudes when icesheets expand. When the Wisconsin icesheets covered much of northern

UNIT OPENING PHOTO: Soil creep is common on slopes in the periglacial landscape. These solifluction lobes mark the raised beaches at the foot of Mount Pelly on Victoria Island in Canada's Northwest Territories.

491

Humans and the Periglacial Environment

Cold and inhospitable as periglacial environments are, people have lived in and migrated through these regions for many thousands of years. Those who stayed there adapted to the difficult conditions. The Inuit (formerly Eskimo), best known of the Arctic peoples, skillfully exploited the environment's opportunities on both land and sea. Their numbers remained small, their social organization was comparatively simple, and their impact on the fragile periglacial domain was very slight.

But the recent invasion of technologically advanced societies, driven by the search for resources, generated new and major problems and threatened local environments as never before. The ongoing exploitation of oil on Alaska's north coast, the construction of the Trans-Alaska Pipeline, and the catastrophic 1989 oil spill in Prince William Sound underscore this intervention (Fig. 48-1). The periglacial environment poses engineering, construction, and maintenance problems unknown in milder regions. Despite the arrival of modern technology in the subarctic, however, population numbers remain low. Nevertheless, the impact of the new era is felt throughout the region in the form of frontier towns and highways, oil facilities, and military installations.

FIGURE 48-1 North America's most serious high-latitude environmental disaster to date: the fully loaded *Exxon Valdez* disgorging some 42,000,000 l (11,000,000 gal) of crude oil into Prince William Sound shortly after the supertanker ran aground on March 24, 1989.

The periglacial world is a landscape of recent glaciation, of scoured bedrock, of basins and lakes, of thin and rocky soil, and of scattered glacio-fluvial deposits such as kames, eskers, and drumlins. Winter is protracted and bitter; nights are frigid and long. Summer is short and cool, depending on latitude

and exposure. The surface is frozen half the year or more, but when the accumulated snow melts, the ground is saturated. Plants, animals, and indigenous peoples have adapted to a combination of environmental conditions that are delicately balanced and so easily disturbed.

North America, periglacial conditions extended far to the south where the landscape still bears the imprints. This reminds us that the earth's comfortable living space during the next glaciation will be much smaller than the land area not actually covered by ice. Periglacial conditions, extending in a wide belt from the margin of the ice, will restrict the ecumene even more (*see* Perspective: Humans and the Periglacial Environment).

PERMAFROST

It is difficult to exactly define the environmental limits of periglacial regions. Perhaps the most practical way to delimit periglacial conditions is based on a phenomenon

unique to these regions: **permafrost**, permanently frozen ground. In periglacial zones, the ground (soil as well as rock) below the surface layer is permanently frozen. What this means, of course, is that all the water in this subsurface layer is frozen. The permafrost layer (Fig. 48-2) normally begins between 15 cm (6 in) and 5 m (16.5 ft) below the surface.

The upper surface of the permafrost is called the *permafrost table*. The soil above the permafrost table is subject to annual freezing and thawing. This is the *active layer*; it is thickest in the subarctic region and becomes thinner both poleward and southward. Below the permafrost table the frozen ground can be very deep. In North America it averages around 300 m (1000 ft), but in the heart of high-latitude Eurasia permafrost depths of

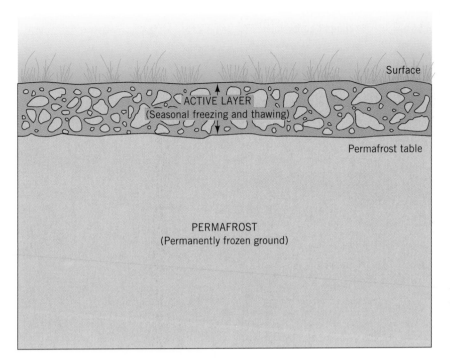

FIGURE 48-2 The subsurface active and permafrost layers characteristic of periglacial zones. Note how the upper active layer contains ice-fractured rock fragments of varying size, which are deposited in the soil as frost wedging in the harsh environment above constantly breaks down boulders and pebbles.

more than 1200 m (4000 ft) have been measured. Like a growing glacier, the permafrost would keep thickening season after season, except that heat from the earth's interior eventually limits this process.

The distribution of permafrost in the Northern Hemisphere is mapped in Fig. 48-3. Note that the map first differentiates between *continuous permafrost*, located in northernmost North America and in a broader zone in northeastern Eurasia, and *discontinuous permafrost*, found as far south as the latitude of Canada's James Bay and northern China. Continuous permafrost, as the term implies, is thick and unbroken, thinning somewhat only under lakes or wide rivers. Discontinuous permafrost is generally thinner and contains unfrozen gaps. Also shown on the map are lower-latitude patches of permafrost. These patches, known as *alpine permafrost*, mostly occur at high elevations in mountains and are remnants of the permafrost of glacial times (*see* dashed-line boundary in Fig. 48-3).

The map of world vegetation (Fig. 27-1) indicates that a certain correspondence exists between the boundary separating continuous and discontinuous permafrost and that delimiting tundra from forest vegetation. While it is not clear whether the vegetation influences the properties of the permafrost or vice versa, it is clear that many factors, including not only vegetation but also precipitation and temperature regimes, affect permafrost development and maintenance.

GEOMORPHIC PROCESSES IN PERIGLACIAL ENVIRONMENTS

Modification of the landscape in periglacial zones takes place in ways that differ from other regions, because freezing and thawing and mass movements play such dominant roles. Water, in the frozen or liquid form, is an important force. When a permanently frozen layer exists below the surface, water cannot drain downward; therefore, it often saturates the active layer. In this upper stratum, boulders are shattered by frost, and fragments are constantly moved by freezing and thawing and the force of gravity. This results in landforms that are unique to periglacial areas.

Frost Action

The presence of water in soil and rock, and its freezing and thawing, is the key disintegrative combination in periglacial environments. **Frost wedging** or *shattering* occurs when the stress created by the freezing of water into ice becomes greater than the cohesive strength of the rock containing it. Research has proven that the more water a rock contains, the greater the power of frost wedging. For instance, porous sedimentary rocks containing water will shatter more rapidly than less porous rocks. Joints and cracks in nonporous crystalline rocks are zones of weakness that are exploited by frost shattering (*see* Fig. 38-2). Frost wedging is capable of dislodging boulders from cliffs, of splintering boulders into angular pebbles, of cracking pebbles into gravel-sized fragments, and of reducing gravel to sand and even finer particles. Thus the active layer consists of a mixture of ice-fractured fragments of all sizes (Fig. 48-2).

The surficial layer of terrain is often characterized by the sorting of fragments by size. This sorting is done by repeated freezing and thawing of the active layer, and produces a phenomenon called *patterned ground* (*see* p. 496). Once rock fragments have been loosened by frost wedging, they are moved by frost heaving. **Frost heaving** causes vertical (upward) displacement when the formation of ice in the ground expands the total mass.

493

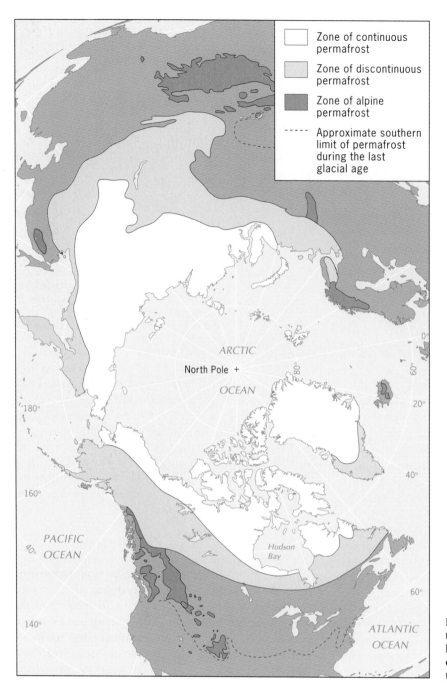

FIGURE 48-3 Distribution of permafrost in the Northern Hemisphere. The southern limit of permafrost during glacial times in eastern Asia has not been established and thus cannot be mapped.

Large fragments are moved upward a greater distance than smaller ones, so that the surface sometimes seems studded with boulders rising above the ground (Fig. 48-4). The same frost heaving that moves boulders upward also pushes concrete blocks, road segments, posts, poles, and other artificial fixtures out of the ground (*see* Fig. 19-3 on p. 208).

In addition to frost heaving, there are processes that move material horizontally. One of these processes is **frost thrusting**. The mechanics of this process are not well understood, but there can be no question that it moves rock fragments horizontally within the active layer. Another process is **frost creep**, the movement of particles in the active layer under the influence of gravity.

A piece of rock brought to the surface by frost heaving will move downslope during the thawing phase. As we will observe shortly, frost wedging, frost heaving, and frost creep combine to produce some remarkable landforms.

Solifluction

Another process closely associated with periglacial conditions is solifluction. **Solifluction** is a form of soil creep, the slow flowage of saturated soil (*see* unit opening photo). Soil in permafrost areas is often saturated because water cannot drain below the permafrost table. In the warm season, such saturated soil begins to move as

FIGURE 48-4 The effects of frost heaving—in the form of boulders thrust upward out of the ground—dominate this high-altitude tundra landscape lying near the summit level of the central Colorado Rockies.

a mass, even when the slope angle is low (Fig. 48-5). The texture of the soil is important, because highly permeable materials such as gravel and sand are not likely to move by solifluction whereas silt-laden soils move quite freely.

Periglacial areas are cold, but they are not without vegetation, which plays a significant role in stabilizing the active layer and in impeding solifluction. Again, human intervention can have devastating effects. When the protective vegetative cover (whether tundra or forest) is removed, binding roots are destroyed, summer thawing reaches a greater depth, and more of the active layer is destabilized. Recovery, in fact, may not occur at all, even after the damaged area is vacated. Periglacial ecologies are particularly fragile.

LANDFORMS OF PERIGLACIAL REGIONS

Landforms in periglacial regions are not as dramatic or spectacular as those of mountain-glaciated areas. They are nonetheless quite distinctive, resulting from a combination of frost action and mass movement. The geomorphic features thus produced often take the form of special patterns that look as though they were designed artificially.

FIGURE 48-6 Aerial view of ice-wedge polygons in the Canadian Arctic.

Ice Wedges

One of these remarkable shapes is created by *ice wedges*. During the frigid Arctic winter, the ground in the active layer (and even the upper permafrost) becomes so cold that it cracks, much as mud cracks form (*see* Unit 38). During the next summer, snow, meltwater, and sediment will fill this crack, creating a wedge of foreign material. The next winter the mix freezes, and the crack, now filled with the ice wedge, opens and widens a little more, and additional water, snow, and sediment enter it. This process is repeated over many seasons.

Some ice wedges reach a width of 3 m (10 ft) and a depth of 30 m (100 ft). They align in patterns that from the air look like an interlocking network, referred to as **ice-wedge polygons** (Fig. 48-6). In some instances, so

FIGURE 48-5 The active layer above the permafrost is frozen in winter but thawed and very wet in summer. Even where slope angles are low, the whole active layer may move slowly downslope. While younger vegetation may not yet reflect this, older trees may lean as a result of this solifluction process.

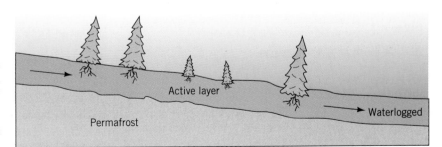

FIGURE 48-7 A ground-level view of a plain filled with rock debris sorted into myriad stone circles 3 to 5 m (10 to 17 ft) in diameter. This striking example of patterned ground was photographed in western Spitsbergen, part of the Arctic Ocean's Svalbard island chain that belongs to Norway.

much fine-grained sediment is available that the wedges become filled with soil. Ice- and soil-wedge patterns are typical of permafrost zones, and when climatic conditions change, they remain imprinted on the landscape.

Patterned Ground

Another characteristic feature of periglacial regions, as we noted earlier, is patterned ground. **Patterned ground** consists of rock and soil debris shaped or sorted in such a manner that it forms designs on the surface resembling rings, polygons, lines, and other repeatedly regular arrangements (Fig. 48-7). Such forms are characteristic of periglacial regions, but unlike ice-wedge features, permafrost is not essential for their formation. Patterned ground results from the frost shattering of bedrock, the lifting and sorting of fragments by frost heaving, and the force of gravity. Stones of various sizes can be moved into circular (or other) arrangements, the smaller fragments accumulating toward the center of the circle or polygon. When it occurs, patterned ground is not just a local feature of limited spatial extent: the phenomenon can persist for kilometers, dominating the landscape and giving it an unmistakable appearance.

Pingos

Another characteristic periglacial landform is a mound called a pingo. **Pingo** is the Inuit (formerly Eskimo) word for hill. Permafrost conditions are prerequisite for the formation of these mounds, which are round or elliptical at the base and can grow quite large. While many are comparatively small and occur in clusters of hundreds, some isolated pingos are as large as 600 m (2000 ft) in diameter and 60 m (200 ft) high (Fig. 48-8). The core of a pingo is made of ice, not rock or soil.

Pingos are believed to form from drained lakes where the permafrost table rises to the surface and bulges upward. As it does so, the saturated overlying lake sediments also are frozen and remain atop the bulging ice. Eventually, only this uppermost active zone remains free of permanent ice and may sustain vegetation, even trees. Pingos are unique to periglacial environments, and thus they are indisputable evidence of former periglacial conditions if they are found as "fossil pingos" in now-temperate zones.

When such fossil pingos are found, they of course represent collapsed features because their icy cores have long since melted. What remains in the landscape is a low circular mound, the remains of the soils of the active layer. In the United States, fossil pingos have been found in northern Illinois, proof that periglacial conditions prevailed there when the Late Cenozoic icesheets lay farther north.

Boulder Fields

So far we have studied periglacial landforms that result mainly from frost action and from the rearrangement of rock debris by freezing and thawing. You will note that we have not discussed river transportation; although rivers do flow through periglacial areas (the northward-flowing streams in Siberia, for example), they do not create distinctive landforms.

The main mover of loosened material in periglacial zones is gravity. Gravity does more than move material in the active layer. In areas of substantial relief, where there are bare bedrock surfaces, large boulders pried off the rock faces by frost wedging are moved into boulder fields. **Boulder fields** are what the name implies: slopes covered by blocky pieces of rock covering a large area (*see* Fig. 38-3). Boulder fields are given several different

FIGURE 48-8 This prominent pingo, located in the Mackenzie River delta of northwesternmost Canada, certainly ranks in the largest size category exhibited by these periglacial landforms.

names; another commonly used term for them is *felsen-meer,* the German word for rock sea.

The boulders or blocks (angular stones) that make up boulder fields are large, as much as 1 to 3 m (3.3 to 10 ft) in diameter. They are loosened from bedrock slopes, cliffs, or other exposed surfaces by frost action, but then gravity takes over. When they are moved down-slope, they accumulate in fields as large as 100 m (330 ft) wide and over 1 km (0.6 mi) long. Exactly how movement takes place is still being investigated. Apparently, it does not occur in a single rock avalanche; there is evidence that slow downslope progress continues, the speed being related to the angle of the slope against which the boulder field lies.

Boulder fields can also be quite deep, and some have been measured at 20 m (65 ft) in thickness. Another interesting aspect is that the long axes of the boulders tend to be lined up approximately parallel to each other, indicating that some sorting does take place. Certain geomorphologists believe that movement may have been aided by a matrix of finer material between the boulders, so that the whole mass was capable of being saturated, thereby facilitating movement. Later, the finer material was eroded away. Others suggest that ice may have filled the openings between the boulders, so that the blocks originally moved as a *rock glacier.*

Whatever the answer, boulder fields are known to occur in periglacial areas and in cold mountainous zones above the timber line. They clearly result from the combination of frost action and mass movement, and although the mechanisms may not be clearly understood, they do provide evidence that near-glacial conditions once prevailed where they exist. We might therefore expect that the Driftless Area of Wisconsin (*see* Perspective in Unit 46), the place the continental glaciers missed but surrounded, ought to be a likely locale in which to find a boulder field. Not surprisingly, this type of landform does indeed occur there.

RESOURCE DEVELOPMENT IN PERIGLACIAL ENVIRONMENTS

Periglacial environments and landscapes are experienced by a very small minority of the earth's population. Our general understanding of these landforms and landscapes is based mainly on what we have learned in the temperate and tropical areas of the world. But now, in our search for resources to satisfy the requirements of the developed countries (especially in fossil-fuel energy), we are invading a realm well known to its indigenous peoples but little understood by outsiders. It is the realm of migrating caribou and reindeer, of musk oxen and wolves, of huge flocks of birds, and of dense swarms of insects. The landscape is one of unfamiliar forms and plants, of mosses and lichens, of needleleaf evergreen trees, and of large patches of barren ground.

So far the modern invasion is limited. The realm is vast, and the invaders are few. But radioactive fallout from the 1986 Chornobyl disaster in the former Soviet Union poisoned the reindeer of Scandinavia's Laplanders, and caribou migration has been adversely affected by Alaska's development since 1970. New economic realities have changed Inuit ways of life. And for all our engineering prowess, developers can never be certain that their impact on subarctic environments will be as they predict. Like Antarctica, the periglacial realm lies open, vulnerable, and fragile in a world of burgeoning demand for what it may contain.

The remote location of most periglacial environments has also severely limited their scientific study. We have repeatedly pointed out knowledge gaps concerning the exact nature of the geomorphic processes operating in these regions. As interest in them continues to grow, there will likely be an expansion of research efforts to better understand this vast, but still superficially explained, landscape.

KEY TERMS

boulder field (p. 496)	frost thrusting (p. 494)	patterned ground (p. 496)	pingo (p. 496)
frost creep (p. 494)	frost wedging (p. 493)	periglacial (p. 491)	solifluction (p. 494)
frost heaving (p. 493)	ice-wedge polygon (p. 495)	permafrost (p. 492)	

REVIEW QUESTIONS

1. Define the term *periglacial* and describe the general spatial distribution of these environments.

2. Which Köppen climatic zones favor the development of periglacial landscapes?

3. Describe permafrost. How is it instrumental in the process of solifluction?

4. How do patterned ground and ice-wedge polygons form?

5. How does the existence of "fossil pingos" help in the reconstruction of climatic conditions?

REFERENCES AND FURTHER READINGS

CLARK, M. J., ed. *Recent Advances in Periglacial Geomorphology* (New York: Wiley, 1988).

DIXON, J. C., and ABRAHAMS, A. D., eds. *Periglacial Geomorphology* (New York: Wiley, 1992).

DREWRY, D. *Glacial Geologic Processes* (London: Edward Arnold, 1986).

EMBLETON, C., and KING, C. A. M. *Periglacial Geomorphology* (New York: Wiley/Halsted, 2nd ed., 1975)

EVANS, D. J. A., ed. *Cold Climate Landforms* (New York: Wiley, 1994).

FRENCH, H. M. *The Periglacial Environment* (London/New York: Longman, 1976).

HARRIS, S. A. *The Permafrost Environment* (Totowa, N.J.: Rowman & Littlefield, 1986).

KING, C. A. M., ed. *Periglacial Processes* (Stroudsburg, Pa.: Dowden, Hutchinson & Ross, 1976).

KRANTZ, W. B., et al. "Patterned Ground," *Scientific American*, December 1988, 68–76.

MATHEWS, J. A. *The Ecology of Recently-Deglaciated Terrain: A Geoecological Approach to Glacier Forelands* (London/New York: Cambridge University Press, 1992).

OWEN, L. A., et al. *Arctic and Alpine Geomorphology* (Cambridge, Mass.: Blackwell, 1995).

PIELOU, E. C. *A Naturalist's Guide to the Arctic* (Chicago: University of Chicago Press, 1994).

PRICE, L. W. *The Periglacial Environment, Permafrost, and Man* (Washington, D.C.: Association of American Geographers, Commission on College Geography, Resource Paper No. 14, 1972).

PRICE, R. J. *Glacial and Fluvioglacial Landforms* (Edinburgh, U.K.: Oliver & Boyd, 1972).

SCHNEIDER, K. "In Aftermath of Oil Spill, Alaskan Sound is Altered," *New York Times*, July 7, 1994, A1, A9.

SUGDEN, D. E., and JOHN, B. S. *Glaciers and Landscape: A Geomorphological Approach* (New York: Wiley, 1976).

WASHBURN, A. L. *Periglacial Processes and Environments* (London: Edward Arnold, 1973).

WILLIAMS, P. J., and SMITH, M. W. *The Frozen Earth: Fundamentals of Geocryology* (London/New York: Cambridge University Press, 1989).

Wind as a Geomorphic Agent

OBJECTIVES

◆ To examine the mechanisms of wind erosion and the landforms produced by this process.

◆ To relate various types of sand dunes to environmental controls.

◆ To note the importance and environmental significance of loess.

The role of **eolian** (wind-related) processes in shaping the earth's surface has been the subject of ongoing debate among geomorphologists and physical geographers. Whereas the effects of running water, flowing ice, and coastal wave action are generally obvious, the role of wind as a geomorphic agent is usually more subtle and difficult to measure. In arid landscapes with little vegetation, wind redistribution of material weathered at the surface takes the form of sand dunes, the morphology of which is controlled by aspects of the local windflow pattern and its strength.

In general, as conditions become more humid, the stabilization of the surface by vegetation diminishes the role of the wind, and other processes become more important in shaping the physical landscape. Complicating this simple assessment is the realization that much of the earth's surface bears the signature of processes that are no longer operating. There is no question that eolian processes have had an important influence on the landscapes of various regions during previous climatic regimes. Furthermore, human activities often destabilize the surface vegetation, and in some areas eolian processes are even becoming more significant (*see* Perspective: The Winds of Interstate-10). Thus we need to examine wind as a geomorphic agent and how it influences the form and dynamics of the landscape.

UNIT OPENING PHOTO: A view across an erg (sand sea) in southwestern Africa's Namib Desert, illustrating the myriad surface sculptures carved by the incessant wind.

499

The Winds of Interstate-10

The arid southwestern corner of the United States is unlike any other desert environment on earth because, over the past half-century, it has been invaded by high-technology civilization and more than three million new urban settlers. Most of this development has occurred within the Interstate-10 corridor, the region's main east-west artery, which crosses the Sonora and Mojave Deserts (in southern Arizona and southern California, respectively) as it threads its lonely way westward for 800 km (500 mi) from Tucson in the east to Los Angeles on the Pacific coast.

To accommodate all this growth, humans have transformed desert landscapes all along I-10. They have forged huge sprawling cities such as Phoenix, which today is the nation's eighth largest; they have built hundreds of resort and retirement communities, of which Palm Springs is the most famous; they have converted millions of hectares of wasteland into productive, irrigated farm fields; and they have created massive water- and power-supply networks to make the whole system work. Harmoniously adapting this still-expanding settlement complex to its harsh habitat is a constant challenge, and the ever-present winds that have sculpted these drylands are a particular concern.

When human-environmental interaction is well conceived, the desert winds can be a boon to regional development. Such is the case around Palm Springs, California, where the local power supply is generated mainly by harnessing the hot, dry winds that blow off the Mojave. A few kilometers northwest of Palm Springs, the narrowing Coachella Valley reaches its apex at the San Gorgonio Pass, which I-10 crosses into the neighboring Los Angeles Basin. Here the local topography funnels the wind through the low-lying (450 m [1500 ft]) pass, and provides the opportunity for

FIGURE 49-1 Converting wind to electrical energy: one of the many windmill farms that line southern California's San Gorgonio Pass, east of Los Angeles.

windmills to convert the moving air into a weak electrical current. By concentrating thousands of windmills in the pass (*see* Fig. 49-1 above), all designed to rotate with the shifting airstream, energy producers are able to generate a nonpolluting power supply sufficient to meet the demands of thousands of local customers.

Unfortunately, throughout the I-10 corridor one is far more likely to encounter examples of misuse of the environment, and where land users have been careless and ignored the effects of desert winds, problems have quickly arisen. Summer duststorms are endemic to this part of the world, and residents must protect their homes, engines, crops, and even their breathing against periodic bombardment by blowing dust. Driving can be especially hazardous because fast-moving duststorms can significantly reduce highway visibility in a matter of moments. On the interstates multiple-vehicle pileups are not uncommon, and the heavily traveled stretch of

I-10 between Phoenix and Tucson contains the scars of dozens of serious chain-reaction collisions.

Geographers Melvin Marcus and Anthony Brazel have studied these accidents and have determined that duststorms here are exacerbated by human modification of the desert surface. The leading problem is the abandonment of farmland adjacent to the I-10 right-of-way, caused initially by groundwater depletion and the disruption of local irrigation networks when the expressway was built in the 1960s. Today, these interstate-paralleling land parcels have largely lost their soil-binding vegetation and have attracted all sorts of surface-disturbing activities, from off-road vehicles to livestock, that make them—and I-10—particularly vulnerable to the windborne movement of large quantities of dust. Belatedly, public agencies are implementing protective land-use practices, but it is still important to be especially alert at the wheel if next summer's travels take you to southern Arizona.

WIND EROSION

In order for wind to be an aggradational agent, it must also be able to erode—to degrade the surface. As in the case of water erosion, the speed of the wind is of primary importance. The higher the velocity of the moving air, the greater is the wind's degradational power. Wind direction also influences the cumulative effect of wind erosion: when the wind blows fairly constantly in the same direction, it erodes more rapidly than it would otherwise.

Moving air alone, however, is not an effective agent of erosion. Only when the wind picks up sand particles does its power to erode become important. These moving particles strike and wear away exposed rock surfaces in a process called **wind abrasion**. If you have ever walked along a dry beach on a windy day, you have probably felt sand particles striking your lower legs; if the wind is strong enough, the experience can even be unpleasant. What this illustrates is that wind abrasion is strongest near the surface and diminishes with height. The largest grains carried by the wind move along in the lower 20 cm (8 in) or so. Almost all wind abrasion takes place within 2 m (6.6 ft) of the ground.

DEGRADATIONAL LANDFORMS

When the wind sweeps along a surface and carries away the finest particles, the process is called **deflation**. We can readily see the results of the deflation process in desert areas today, because the arid landscape often includes shallow basins without outlets (Fig. 49-2). These basins lie in rows parallel to the prevailing wind direction. They begin as small local hollows that are continuously enlarged as the wind removes freshly weathered particles. Eventually, these basins grow quite large (to hundreds of square kilometers in area), and other processes may reinforce their growth.

FIGURE 49-2 A sizeable desert basin shaped by deflation on the floor of California's Death Valley.

FIGURE 49-3 The striking topography of a zone of yardangs dominates the foreground and center of this photo. This locale is near Minab, a coastal town on the Persian Gulf's Strait of Hormuz in southeastern Iran.

Not many landforms can be attributed exclusively to wind erosion. The **deflation hollows** just described as desert basins created by wind erosion undoubtedly result from degradation by moving air. Where wind has removed finely textured material, a surface concentration of closely packed pebbles is left behind as **desert pavement** (*see* Fig. 43-3 on p. 443). Although desert pavement is a residual landform, it may be interpreted to be a feature of wind erosion. The most common product of wind abrasion is the **yardang** (Fig. 49-3). Yardangs are low ridges that form parallel to the prevailing wind direction. They tend to develop in dry sandy areas affected by strong winds, especially where the bedrock is fairly soft and unprotected by vegetation. The yardangs are separated by troughs that are scooped out and smoothed to a polished-looking surface by wind abrasion.

Wind Transportation

The movement of rock particles by wind takes place in three ways. The finest material, such as silt and clay, is carried high in the air in *suspension*. Sometimes this material is carried so high that it enters the upper atmospheric circulation; for example, following the explosive eruption of the Indonesian volcano Krakatau in 1883, an enormous dust cloud encircled the earth for years afterward. Coarser particles are too large to be picked up and carried very far above the ground. The wind still moves them, but during transportation these grains bounce along the ground, sometimes at a high rate of speed. This process is known as *saltation* (Fig. 49-4).

Accordingly, sand-sized particles are swept along in the lowest layer of moving air, mostly below about 20 cm (8 in) but virtually always below 2 m (6.6 ft) above the ground. Sandstorms are the most extreme form of saltation (Fig. 49-5). In contrast to high-altitude-reaching duststorms, in which suspension is the major process (*see*

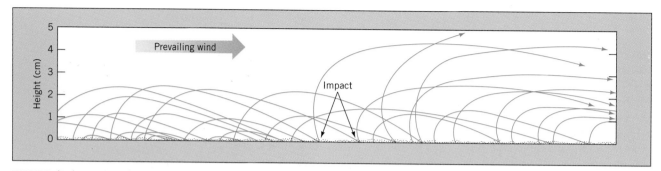

FIGURE 49-4 Wind causes movement of coarse sand grains by saltation. Impacted grains bounce into the air and are carried by the wind as gravity pulls them back to the loose sand surface where they impact other particles, repeating the process.

Fig. 22-9), sandstorms are confined to the lower layers of moving air, and saltation is the dominant mode of movement. Wind also moves larger rock fragments by actually pushing them along the surface in a process called **surface creep**. Pebbles and even small boulders of considerable weight can be moved by strong windstorms.

AGGRADATIONAL LANDFORMS

It is important to remember that wind action is not confined to deserts or semiarid steppelands. Wind also has erosional functions in glacial and periglacial zones, in savannas and humid midlatitude grasslands, and in other areas as well. Anyone who has felt the force of the wind howling through a mountain pass will realize its potential for erosive action. During a severe windstorm in England in 1987, for example, an estimated 1 million trees were toppled.

Nevertheless, wind does its most effective work in dry environments. Deserts and semi-deserts are dominated by eolian processes, and the landforms and landscapes of wind action are best developed in North Africa's Sahara, Southwest Asia's Arabian Desert, the Great Sandy Desert of Australia, and the earth's other extensive drylands. But even in these desert areas, the landforms typically associated with eolian deposition are confined to relatively small sections of the desert. Thus where sand accumulations are large and extensive, we may still not find any prominent eolian landforms. Over large expanses of sandy desert landscape, the dominant feature is the **erg** or *sand sea*. Wind directions may vary seasonally to such an extent that there is no dominantly prevailing airflow, so the sand is continuously moved about. The landscape in such places takes on an undulating (gently rolling) appearance. The surface of the sand may be formed into *ripples* by saltation and surface creep, but otherwise the topography is unremarkable (*see* unit opening photo).

SAND DUNES

The landform most commonly associated with wind deposition is the dune. A **dune** is an accumulation of sand that is shaped by wind action. This definition is clear and concise, but when you look at an aerial image of a dune landscape, it is clear why the many dune formations are difficult to interpret. Dunes come in many shapes and sizes: as straight or curving ridges, as quarter-moon-shaped crescents, as irregular mounds, and more. Physical geographers are interested in three aspects of dunes: (1) whether or not they are stable; (2) what their shape or form is; and (3) how they are arranged in the landscape.

FIGURE 49-5 A ground-hugging sandstorm bearing down on the overgrazed steppeland that marks the Sahel zone of the West African country of Burkina Faso.

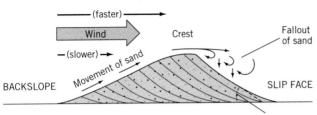

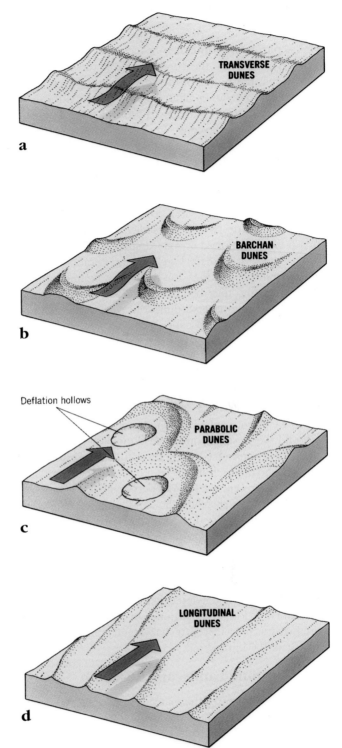

FIGURE 49-6 Cross-section of an active sand dune that is migrating from left to right. The lower-angle, windward backslope is the degradational side of the dune, with surface sand grains pushed upward toward the crest. The steeper, leeward slip face is the aggradational component of the dune, where wind-driven sand is deposited. The accumulation of sand grains on the advancing slip face produces strata inside the dune much like the foreset beds in a delta (see Fig. 43-14).

Dunes in sandy desert areas normally support no vegetation. The wind continually modifies them, removing sand from the windward side and depositing it on the leeward side (Fig. 49-6). This has the effect of moving the dune across the landscape, so that the dune is unstable or *active*. Over time, however, dunes may migrate into moister areas on the desert margin, or a climate change may affect a dune area. Then plants will take hold, and the vegetation will slow or even stop the dune's movement; in such instances a dune is described as stable, or *fixed*.

Dune Features

Every dune has a profile, a cross-section that reveals much about its history. This profile consists of three elements: the windward slope, or **backslope**; the top, or *crest*, of the dune; and the leeward slope, or **slip face** (Fig. 49-6). As this diagram shows, the windward slope (to the left of the crest) has a lower angle than the leeward slope (to the right). The wind drives the sand grains up the length of the backslope, pushes them over the crest, and lets them drop on the slip face. Thus a dune has a degradational and an aggradational side, and from its profile we can determine the prevailing wind direction.

Dune Forms

Given these conditions, we might assume that loose sand influenced by prevailing wind will be arranged into one dominant landform. Unfortunately, that is not the case. Dunes develop various forms, and the exact origin of some of them is not clearly understood.

Barchans

The most common dune form is probably the **barchan**, a crescent-shaped dune. The best way to understand its form is to look at Fig. 49-7b. The convex side of this dune is the windward side, so that its points lie downwind.

FIGURE 49-7 The four most common types of sand dunes. In each diagram the prevailing wind direction is indicated by the arrow. Note that longitudinal dunes lie parallel to the prevailing wind, whereas the others form at right angles to it.

Thus the low-angle backslope faces the wind on the outside, while the steeper slip face lies inside. A cluster of barchans, therefore, immediately indicates the prevailing wind direction in its locality.

Parabolic Dunes

Not all dunes with a crescent shape are barchans, how-ever. A **parabolic dune** also has a crescent shape, but in this type of dune the concave side is the windward side, so that the points of the crescent lie upwind (Fig. 49-7c). Parabolic dunes often develop longer sides than barchans do, and they begin to look like giant horse-shoes rather than crescents. In deserts, they sometimes develop in association with deflation hollows; they also occur frequently along coastlines.

Transverse Dunes

A similar-looking dune is a low sand ridge, called a **transverse dune**, which is usually straight or slightly curved and positioned at right angles to the prevailing wind (Fig. 49-7a). In fact, barchans and parabolic dunes also are transverse dunes, but their sides begin to adjust to the wind and they become rounded. This is not the case with true transverse dunes, however, which look like ripples on the landscape. Indeed, transverse dunes often mark ergs, giving the topography the look of a sandy sea complete with wave crests (*see* unit opening photo).

Longitudinal Dunes

Just as we might confuse barchans and parabolic dunes, so could we mistake transverse dunes for longitudinal dunes. Like transverse dunes, **longitudinal dunes** form lengthy sand ridges, but they lie *parallel* to the prevailing wind direction (Fig. 49-7d), not perpendicular to it. It has been suggested that when the sand supply is plentiful, as in an erg, transverse dunes will develop, but that when the sand availability is limited, longitudinal dunes will develop. Certainly, longitudinal dunes are prominent. Flying from Adelaide on the coast of South Australia to Alice Springs in the heart of the continent, you can see an entire landscape that looks like corrugated cardboard. Thousands of longitudinal dunes over 3 m (10 ft) high, with many as tall as 20 m (65 ft), extend continuously for as long as 100 km (63 mi).

Dune Landscape Research

These are just some of the many dune shapes and forms that have been classified and whose origins are under-stood. Many others remain to be studied. Research on present-day dune landscapes, both in deserts and along coasts, also contributes to the interpretation of the past because dunes offer clues about climate change. Certain areas where dunes now lie are no longer arid, and the dune landscape has become fixed (Fig. 49-8). From the morphology of the dunes, however, conclusions may be drawn about earlier climates not only in general terms but, more specifically, about wind directions and veloci-ties as well.

Another reason to know as much as we can about wind erosion and dune formation is immediate and prac-

FIGURE 49-8 Coastal dunes can attain considerable size and permanence. Dunes may also become anchored by vegetation, as is happening here along the west coast of Australia about 160 km (100 mi) north of Perth. (Authors' photo)

tical. As explained in Unit 17, *desertification* has become a global problem. There even are places along desert margins where advancing dunes are overtaking inhab-ited land. By understanding how dunes migrate and how wind action drives them, we are in a better position to develop ways to stabilize them and to halt their progress (*see* p. 193).

LOESS

Perhaps the most impressive evidence of the capacity of wind to modify the landscape comes from ice-age times. As we describe in Unit 47, glacial erosion produces tre-mendous quantities of finely textured sediment (rock flour), and such fine material became part of the exten-sive outwash deposits formed during glacial recession. Strong winds, which were common in periglacial envi-ronments during the Late Cenozoic, carried away huge quantities of these fine particles. At times, the outwash plains must have looked like the duststorm depicted in Fig. 22-9, as vast clouds of dust darkened the skies and obscured the sun. Because prevailing winds were fairly steady, much of the dust moved in certain specific direc-tions. When the air motion eventually subsided, the fine-grained dust was deposited on the ground, sometimes hundreds of kilometers from its source. In this way, sedi-mentary deposits called **loess** (pronounced "lerss") accumulated.

Distribution of Loess Deposits

Loess was laid down in many areas south of the icesheets in the Northern Hemisphere, and it also occurs in the Southern Hemisphere (Fig. 49-9). In the United States,

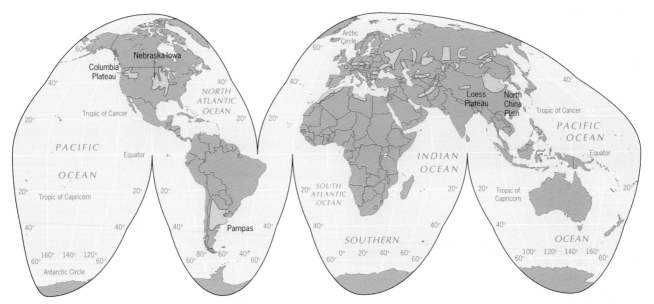

FIGURE 49-9 Major loess deposits of the world.

the most prominent loess deposits extend from the Great Plains to the lowlands of the Mississippi, Ohio, and Missouri River Basins (Fig. 49-10). As this map shows, some of the thickest deposits lie in Nebraska and Iowa, where as much as 60 m (200 ft) of loess has buried the underlying topography. Most of these loess deposits, however, are between 1 m and 30 m (3.3 ft and 100 ft) thick; streams eroded the area after the loess was deposited, and the loess layers can be seen in the walls of many river valleys. Another major deposition of loess occurs on the Columbia Plateau in the Pacific Northwest, near where the states of Washington, Oregon, and Idaho meet (Fig. 49-10).

As the world map (Fig. 49-9) shows, loess deposits are even more extensive in Eurasia than in North America. Loess was first identified in the Rhine Valley as long

ago as 1821, which is how it got its German name (loess means loose in German). It also exists in France's Paris Basin, in the Danube Valley of Eastern Europe, and in large areas of southern and central Russia. But the thickest loess accumulations lie in Asia, especially in east-central China.

Almost the entire surface of the North China Plain consists of loess, and to the west, in the hilly middle basin of the Huang He (Yellow River), lies an even thicker loess deposit. In fact, the Chinese call this region the Loess Plateau, and here the loess averages 75 m (250 ft) thick and reaches as much as 180 m (600 ft) in places (Fig. 49-11).

Loess also occurs over a sizeable area of southern South America, including Argentina's productive Pampas. Other smaller deposits of loess (as well as loess-like

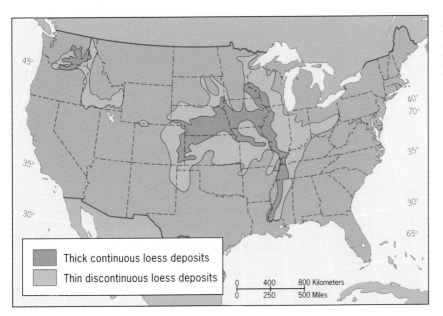

FIGURE 49-10 Loess regions of the United States, highlighting the major deposits of the east-central Great Plains, the Mississippi Valley, and the Pacific Northwest.

FIGURE 49-11 This photo shows the thickness of the loess deposit in China's Loess Plateau, and its tendency to stand in unsupported vertical cliffs even in areas of considerable rainfall. (Authors' photo)

sediments) have been found elsewhere in the Southern Hemisphere. But, as with periglacial phenomena generally, the bulk of the world's loess deposits lie well north of the equator.

Properties of Loess

Ever since loess was first identified, the origins of its various deposits have been debated. Is loess really a windborne sediment, or did water also play an aggradational role? The answer seems to be that loess is indeed a product of wind transportation and deposition: the deposits blanket the terrain below, filling valleys and covering hilltops in ways fluvial deposits do not.

Certainly, there is no argument about the interest in and importance of loess. If you compare the maps of loess deposits, agricultural productivity, and population distribution, you will see that some of the world's most fertile areas—and some great human agglomerations—lie in loess regions. In the United States, the loess of Iowa and Illinois yields massive harvests of corn, soybeans, and other crops. In the drier Great Plains farther west, the wheat of Kansas and Nebraska also comes from loess-derived soils. In terms of population size and density, nothing on earth matches the great human cluster centered on the North China Plain, supported by vast wheatfields on fertile loess-based soils. The most productive farmlands of Russia and Ukraine, too, depend on loess.

Just what makes loess such an unusual sediment? It consists of silt-sized quartz particles along with feldspars, carbonates, clays, and other minerals. Loess contains the whole range of minerals derived from ground-up bedrock, which is about as ready for plant absorption as alluvium is. Moreover, its fertility is not confined to an upper layer; loess is fertile all the way down its profile.

Scrape off the top horizon, and lower layers of it will support plants just as well. But there is more to it.

Technically, we would describe loess as a fine-grained, unstratified, homogeneous, highly porous deposit. After deposition, compaction causes slight shrinkage in the mass, so that vertical passages develop. These vertical passages may take the form of capillaries or may resemble the joints found in harder sedimentary rocks (*see* Unit 31). Water, seeping downward through the loess, dissolves some of its mineral matter and redeposits it, thereby strengthening the walls of capillaries and cleavages. These processes combine to give loess a capacity to stand in upright walls and columns, and to resist collapse when it is excavated.

These qualities (fertility, vertical strength) are on display in what must be the world's most interesting loess region, China's Loess Plateau (Fig. 49-9). This region is neither as large nor as populous as the great North China Plain, having more relief and less water. But here loess is more than fertile soil—loess also serves as living quarters.

In stream valleys, against hillsides, and in the walls of deep road cuts, the local inhabitants have excavated the loess to create dwellings that are sometimes large and elaborate (Fig. 49-12). The entrances to these underground houses are sometimes ornately decorated, but in fact they are caves. As long as the region remains geologically stable, the millions of people living underground are safe. But whereas loess withstands erosion and has vertical strength, it collapses when shaken. In 1920 a severe earthquake struck the Loess Plateau, and

FIGURE 49-12 China's Loess Plateau is more than a thick deposit of wind-blown dust of glacial origin: it is also a region of fertile soils and major agricultural production, dense human population, and considerable danger. The danger involves people's habit of tunneling into the easily excavated loess to build underground dwellings. The entranceways seen here lead to a network of underground tunnels and homes. Even a minor earthquake can collapse these warrens, and countless thousands have lost their lives as a result. Despite the hazard, people continue to burrow underground. (Authors' photo)

an estimated 250,000 people lost their lives, most of them buried inside their caved-in homes.

Periglacial loess deposits leave no doubt that wind erosion has played the major role in shaping the postglacial landscape. But loess and loess-like deposits also are found in areas far from glaciers, for example, near midlatitude deserts. There, winds have laid down accumulations of fine-grained material derived from the dry, dusty desert surface. Some other loess-like deposits may in fact be the work of water, not wind. The great loess deposits of Eurasia and North America, however, confirm the role of wind as an aggradational agent.

KEY TERMS

backslope (p. 503)

barchan (p. 503)

deflation (p. 501)

deflation hollow (p. 501)

desert pavement (p. 501)

dune (p. 502)

eolian (p. 499)

erg (p. 502)

loess (p. 504)

longitudinal dune (p. 504)

parabolic dune (p. 504)

slip face (p. 503)

surface creep (p. 502)

transverse dune (p. 504)

wind abrasion (p. 501)

yardang (p. 501)

REVIEW QUESTIONS

1. Describe the vertical limits of wind erosion.

2. Describe the typical sand-dune profile and how prevailing wind direction might be deduced from it.

3. Describe and differentiate among barchan, parabolic, longitudinal, and transverse dunes.

4. What is loess? Describe its major physical properties.

5. How have the thick loess deposits of the United States accumulated?

REFERENCES AND FURTHER READINGS

ARRITT, S. *The Living Earth Book of Deserts* (Pleasantville, N.Y.: Reader's Digest Assn., 1993).

BAGNOLD, R. A. *The Physics of Blown Sand and Desert Dunes* (London: Chapman & Hall, 2nd ed., 1954).

BROOKFIELD, M. E., and AHLBRANDT, T. S., eds. *Eolian Sediments and Processes* (Amsterdam: Elsevier, 1983).

COOKE, R. U., WARREN, A., and GOUDIE, A. S. *Desert Geomorphology* (Bristol, Pa.: Taylor & Francis, 1992).

EDEN, D. N., and FURKERT, R. J., eds. *Loess: Its Distribution, Geology and Soils* (Rotterdam, Netherlands: Balkema, 1988).

GLENNIE, K. W. *Desert Sedimentary Environments* (Amsterdam: Elsevier, 1970).

GOUDIE, A. S., and WATSON, A. *Desert Geomorphology* (London: Macmillan, 1990).

GREELEY, R., and IVERSEN, J. D. *Wind as a Geological Process: Earth, Mars, Venus, and Titan* (London/New York: Cambridge University Press, 1984).

LEIGHTON, M. M., and WILLMAN, H. B. "Loess Formations of the Mississippi Valley," *Journal of Geology*, 58 (1950), 599–623.

LUGN, A. L. *The Origin and Source of Loess* (Lincoln: University of Nebraska, Geological Studies N.S. 26, 1962).

MABBUTT, J. A. *Desert Landforms* (Cambridge, Mass.: MIT Press, 1977).

MARCUS, M. G., and BRAZEL, A. J. "Summer Dust Storms in the Arizona Desert," in D. G. Janelle, ed., *Geographical Snapshots of North America* (New York: Guilford Press, 1992), pp. 411–415.

PEWE, T. L. *Desert Dust: Origin, Characteristics, and Effect on Man* (Boulder, Colo.: Geological Society of America, Special Paper No. 186, 1981).

PYE, K., and TSOAR, H. *Aeolian Sand and Sand Dunes* (Winchester, Mass.: Unwin Hyman Academic, 1990).

SIEVER, R. *Sand* (New York: Scientific American Library, 1988).

TCHAKERIAN, V. P., ed. *Desert Aeolian Processes* (New York: Routledge, Chapman & Hall, 1994).

THOMAS, D. S. G. *Arid Zone Geomorphology* (London: Belhaven Press, 1989).

Coastal Processes

OBJECTIVES

◆ To establish the importance of coastal zones as areas of interaction between physical processes and human settlement.

◆ To examine the physical properties of waves and their significance in the operation of coastal processes.

◆ To discuss other sources of energy in the coastal zone and their erosional and depositional significance.

Some of the world's most spectacular scenery lies along the coasts of the continents. Sheer cliffs tower over surging waves. Curving beaches are cupped by steep-sided headlands. Magnificent bays lie flanked by lofty mountains. Great rivers disgorge into the open ocean. Glaciers slide and calve into the water. Coastlines are shaped by many forces: by waves from the sea, by rivers emptying from land, by ice, even by wind. Rising and falling sea levels, tides and currents, and tectonic forces all contribute to the development of coastal topography. The complexity and variety of coastal landforms and landscapes are the result, and these are the topics of Units 50 and 51.

Our interest in coastal processes stems from two considerations. On the one hand, we seek to understand how coastal landscapes are created and what processes are presently operating in these zones of interface between land and sea. The other motivation is more practical, because coastlines are probably the most intensively used landscapes for a variety of human activ-

UNIT OPENING PHOTO: Cape Hatteras on the Outer Banks of North Carolina is a tenuous outpost situated well off the mainland (*see* Fig. 51-6). It is also a splendid example of the low-relief coastlines that mark the U.S. Atlantic and Gulf seaboard south of New York City.

ities. In such heavily developed areas, coastal processes can have very significant consequences. Our ability to successfully manage these landscapes rests on our knowledge of the environmental processes operating within them.

COASTS AND SHORES

In this unit and Unit 51 we study the **littoral zone**, where land meets sea. In physical geography, the term **coast** refers in a general sense to the strip of land and sea where the various coastal processes combine to create characteristic landscapes, ranging from dunes and beaches to islands and lagoons. The term **shore** has a more specific meaning and denotes the narrower belt of land bordering a body of water (the most seaward portion of a coast). A *shoreline* is the actual contact border between water and land. Thus we often refer to a coastal landscape, of which the shoreline is but one part.

We have all heard about the special problems of coastal areas. Sometimes beaches must be closed because harmful waste materials we have dumped into the sea are washed onto them. Some popular beaches become ever narrower and must be protected by jetties or groins, or they will wash away. Parts of shores are threatened by urban pollution, their wildlife endangered. Overcrowding and expansion of waterfront towns and cities imperil local ecologies.

Most of the time, however, we seem to be unaware of these happenings because the changes tend to be slow, not dramatic. The beach we visit year after year looks pretty much the same. The strip of beachfront hotels and motels is lengthening, but gradually. Less obvious are the connections among the many processes at work in coastal zones. When we dredge the outlet of a port, build a series of jetties, construct a breakwater, or flatten a dune, the consequences may be far-reaching. Beach erosion may slow down in one place but speed up in another. The offshore turbidity (muddiness) of the water may change, affecting reef life. Coastal zones, therefore, are not only scenic and attractive: their landforms and landscapes result from the complex interaction of many processes.

WAVES AND THEIR PROPERTIES

Many forces help shape coastal landforms, but the key erosional agents are ocean *waves*. Most waves (not only in oceans, but also in seas and lakes) are generated by wind. Waves form when energy is transferred from moving air to water. Large waves form in water when wind *velocity* is high, wind *direction* is persistent, wind *duration* is protracted, and the *fetch* (the distance over which the wind blows) is long. When conditions are fa-

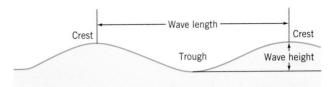

FIGURE 50-1 Wave height is the vertical distance between the wave crest and the wave trough. Wave length is the horizontal distance between two crests (or two troughs).

vorable, the ocean's upper layer is stirred into long rolling waves or swells that can travel thousands of kilometers before they break against a shore. When conditions are less favorable (changing wind directions, for example), waves are generally smaller but may have a more complex result.

Once a series of swells are well developed and moving across open water, we can observe their properties. The **wave height** is the vertical distance between the *crest*, or top, of a wave, and its *trough*, or bottom (Fig. 50-1). Wave height is important when it reaches the coast, because a high wave will do more erosional work than a low wave. The **wave length** (Fig. 50-1) is the horizontal distance from one crest to the next (or from trough to trough); a wave's *period* is the time interval between the passage of two successive crests past a fixed point.

In the open ocean, swells may not look large or high because they cannot be observed against a fixed point. But they often reach heights of up to 5 m (16.5 ft) and lengths ranging from 30 m (100 ft) to several hundred meters. Storm waves tend to be much higher, often exceeding 15 m (50 ft); the highest wave ever measured reached 34 m (112 ft) during a Pacific storm in 1933.

When swells travel across the open ocean, they seem to move the water itself in the direction of their movement. But this is not the case. In reality, the passing wave throws the water into an orbital motion. As shown in Fig. 50-2, a water parcel (or particle) affected by the passing wave takes a circular, vertical path. When the wave approaches, the parcel rises and reaches the height of the crest. Then it drops to the level of the trough, coming back to where it started when the wave arrived. Waves that move water particles in this circular up-and-down motion are called **waves of oscillation**. As Fig. 50-2 shows, the depth of a wave of oscillation is half its length: if a swell has a length of 100 m (330 ft), it will have a depth of 50 m (165 ft).

WAVES AGAINST THE SHORE

Once they have been formed by steady strong winds, swells (a series of long-period waves) may move across vast reaches of open ocean without losing their strength or energy. They retain their length, height, and period even far from their sources, and they can travel across

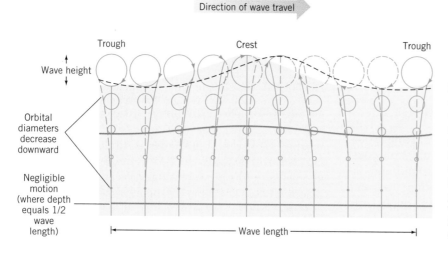

FIGURE 50-2 Orbital motion of water parcels within a wave of oscillation in deep water. To follow the successive positions of a water parcel at the surface, follow the arrows in the largest loops from right to left. This is the same as watching the wave crest travel from left to right. Parcels in smaller loops below have corresponding positions, marked by continuous, nearly vertical lines. Dashed lines represent wave form and water parcel positions one-eighth of a period later.

entire ocean basins. As they approach the coastline, they usually enter shallower water. Obviously, the free orbital motion of the water will be disrupted when the wave begins to "feel" (be affected by) the ocean bottom. At this point, the swell becomes a *wave of translation.* No longer do water particles in orbital motion return to their original positions. The wave has begun its erosional work.

Since we know that the depth of a wave is half its length, we can determine where it will "feel" bottom first. The wave in our example, with a length of 100 m (330 ft), will begin to interact with the ocean bottom when the water depth becomes less than 50 m (165 ft). There, the circular orbit of a water particle is compressed into an oval one (Fig. 50-3a, bottom panel). Contact with the bottom also slows the wave down, so that wave length

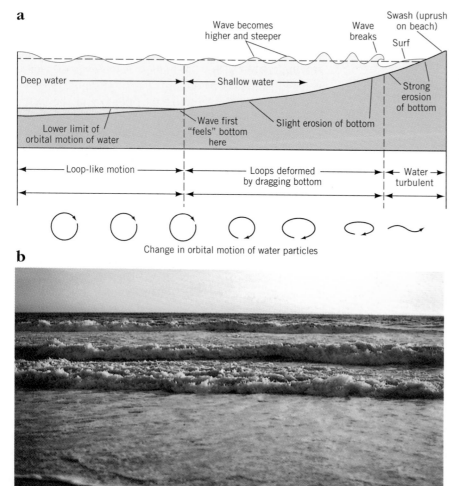

FIGURE 50-3 Waves are transformed as they travel from deep water through shallow water to shore (a), producing evenly spaced breakers as they approach the beach (b). In the diagram, circles, ovals, and wave lengths are not drawn to scale with the waves on the surface.

is forcibly decreased. As the water becomes still shallower, the coastward-moving wave pushes water upward, increasing wave height. Soon the wave becomes so steep that its crest collapses forward, creating a *breaker*. This happens along the breadth of the advancing wave, which is now capped by a foaming, turbulent mass.

From the beach we can see a series of approaching waves developing breakers as they advance toward the shore (Fig. 50-3b); such a sequence of breaking waves is referred to as the **surf**. When a wave reaches the shore, it finally loses its form and the water slides up the beach in a thinning sheet called **swash**. As anyone who has walked through the swash will remember, the dying wave still has some power, and the uprushing water carries sand and gravel landward. Then the last bit of wave energy is expended, and the water flows back toward the sea as **backwash**, again carrying sand with it. Along the shore, therefore, sand is unceasingly moved landward and seaward by wave energy.

Wave Refraction

When we look out across the surf from the top of a dune, it seems as though the surf consists of waves arriving parallel, or very nearly parallel, to the coastline. In actuality the parallel approach is quite rare, but the impression is produced during the **shoaling** process, the impact of shallow water on an advancing wave.

When a wave approaches a beach at an oblique angle, only part of it is slowed down at first—that part that first reaches shallow water. The rest of the wave contin-

ues to move at a higher velocity (Fig. 50-4). This process obviously bends the wave as the faster end "catches up" with the end already slowed by shoaling. This bending is known as **wave refraction**. By the time the whole wave is in shallower water, its angle to the shoreline is much smaller than it was during its approach through deep water. Notice, too, that the angled approach of the waves vis-à-vis the beach sets up a longshore current flowing parallel to the shoreline (Fig. 50-4).

When a coastline has prominent headlands (promontories) and deep bays, wave refraction takes place as shown in Fig. 50-5. The waves approach the indented coastline roughly parallel to its general orientation. They reach shallower water first in front of the headlands, so that they are slowed and their length reduced. The segment of the wave headed for the bay has yet to reach shallow water, so it continues at open-water velocity.

This has the effect of refracting the wave as shown in Fig. 50-5, concentrating its erosional energy on the point of the headland. Rock material loosened from the promontory is transported toward the concave bends of the bays, where a beach forms from it. Wave action, therefore, has the effect of straightening a coastline, wearing back the promontories and filling in the bays. Wave refraction is a crucial part of this process.

Longshore Drift

Let us return to the waves arriving at a slight angle on the beach. When a wave's swash rushes up the beach carrying sand and gravel, it does so at the angle of the arriving wave. But when the backwash carries sand back

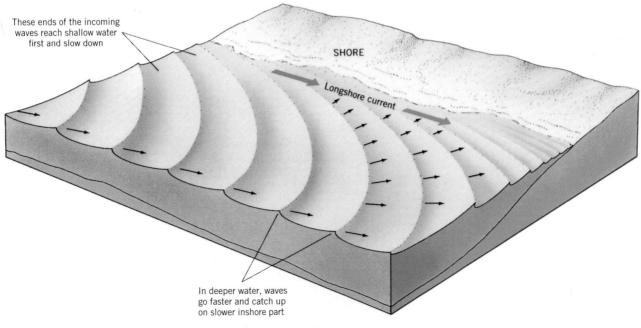

FIGURE 50-4 Refraction of incoming waves at the shoreline. Waves are bent so their angle to the coastline is much smaller than in the deep water at the beginning of their approach. However, the inshore angle of the waves is still sufficient to produce a longshore current that flows parallel to the shoreline.

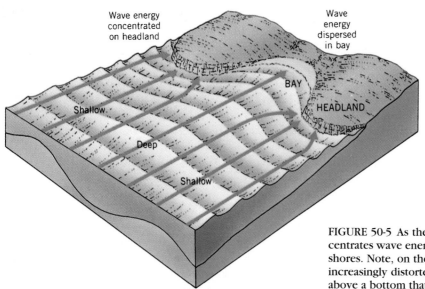

FIGURE 50-5 As the arrows indicate, refraction of waves concentrates wave energy on headlands and dissipates it along bay shores. Note, on the sea surface, how the incoming waves are increasingly distorted as they approach the irregular coastline above a bottom that is deepest in front of the bay.

seaward, it flows straight downward at right angles to the shoreline. The combined effect of this is to move sand along the beach, as shown in Fig. 50-6.

This process, called **longshore drift**, can easily be observed on a beach. Remember that the movement you see in one area of swash and backwash is continuously repeated along the entire length of beach. The larger process of **beach drift** moves huge amounts of sand along the shore and sometimes becomes so powerful that it threatens to move the whole beach downshore. When

that happens, engineers build groins and other structures at an angle to the beach and out into the surf, hoping to slow the drift of beach sand (Fig. 50-7).

While we can see beach drift occurring, longshore drift also affects materials in the surf zone where the waves of translation operate. We may not be able to observe it directly, but longshore drift in the breaker zone can create ridges of sand and gravel parallel to the shore. Those elongated ridges may interfere with the advance of the very waves that build them, growing longer and

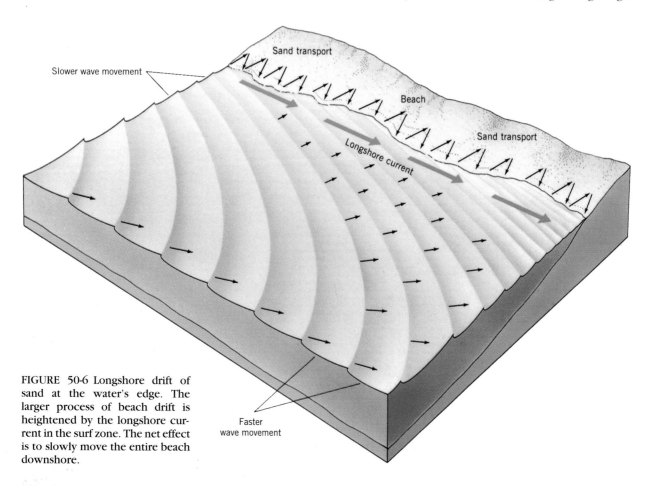

FIGURE 50-6 Longshore drift of sand at the water's edge. The larger process of beach drift is heightened by the longshore current in the surf zone. The net effect is to slowly move the entire beach downshore.

The speed and weight of the water smashing against the vertical bedrock contribute to erosion by *hydraulic action*. Hydraulic action is especially effective where rocks are strongly jointed or otherwise cracked (for instance, along bedding planes). Air enters the joints and cracks; when the water pounds the rock face, this trapped air is compressed. Next, the wave recedes, and the air expands almost explosively. This process, repeated over thousands of years, can fracture and erode coastal rocks quite rapidly.

Like rivers and glaciers, waves break pieces of rock from the surface being attacked, and these fragments enhance the waves' erosional effectiveness. This mechanical erosion process is known as **corrasion**. A wave loaded with rock fragments, large and small, erodes much more rapidly than water alone. Some coastal bedrock is also susceptible to chemical action by seawater. The breakdown of coastal bedrock by solution or other chemical means is referred to as *corrosion*.

Where offshore water is deep and coastal topography is steep, the onslaught of waves produces a set of degradational landforms that tell us immediately what processes are going on. No gently sloping beaches or sandy offshore islands grace these high-relief coastlines (Fig. 50-8). All the evidence points toward the hardness and resistance of the rocks, the erosional force of the waves, and the exposure of the coastline to storms.

Alternatively, where offshore water is shallow and waves break into surf, the coastal landforms also are characteristic. Aggradational landforms dominate, and

FIGURE 50-7 Groins, extending into the surf at a 90° angle, mark the beaches of the resort hotels that line the Atlantic shore of Miami Beach. These structures prevent the excessive loss of sand by beach drift, which piles up on the upcurrent side of each groin. The narrowness of the beach, however, indicates this method is not all that effective.

wider and eventually rising above the water surface as **barrier islands**. As explained in Unit 51, such landforms may even grow across the mouths of rivers and bays, often creating ecological as well as economic problems.

DEGRADATION AND AGGRADATION BY WAVES

Waves are powerful erosional agents and, as we have just noted, they also are capable of deposition. In the examples just given we described waves that, after developing in the open ocean, reach shallow water and lose their energy as they approach the shore. But the water does not always become so shallow near a coast. There are many places where the land descends steeply into the water, and where the water is hundreds of meters deep just a few meters offshore. In such places the waves advancing toward the land do not "feel" bottom, are not slowed down, and do not form breakers and surf. Where this happens, of course, the full force of the on-rushing wave strikes the coast.

FIGURE 50-8 A high-relief, submergent coastline north of Viña del Mar, Chile. All of the western coastline of the Americas displays evidence of tectonic plate motion; not only is coastal relief high, but also continental shelves comparable to the wide shelves off the east coasts do not exist. (Authors' photo)

Tides and Their Behavior

The earth's envelope of water—the hydrosphere—covers more than 70 percent of the planet. The surface of this layer of water unceasingly rises and falls in response to forces that affect its global distribution. This cyclical rise and fall of sea level is known as the **tide** recorded at any given place in the world ocean. The *tidal range* is the vertical difference between sea level at high tide and low tide.

Three principal forces control the earth's tides: (1) the rotation of the planet, (2) the gravitational pull of the moon, and (3) the gravitational pull of the sun. The earth's daily rotation has the effect of countering the gravitational pull of its own mass. Rotational velocity is greatest near the equator and lowest at the poles, so that the layer of water bulges slightly outward toward the equator. This is a permanent condition. But the earth orbits around the sun and is, in turn, orbited by the moon. This means that the gravitational pulls of moon or sun come from different directions at different times.

Tidal levels at a coastal location rise and fall rhythmically based on the earth's rotation and the 28-day lunar revolution, which produce two high tides and two low tides within a period slightly longer than 24 hours. When the earth, moon, and sun are aligned as shown in Fig. 50-9a, the effects of terrestrial rotation, lunar attraction, and the sun's attraction are combined, and the result is an unusually high tide, or *spring tide*. But when the moon's pull works at right angles against the sun's attraction and the rotational bulge, the result is a *neap tide*,

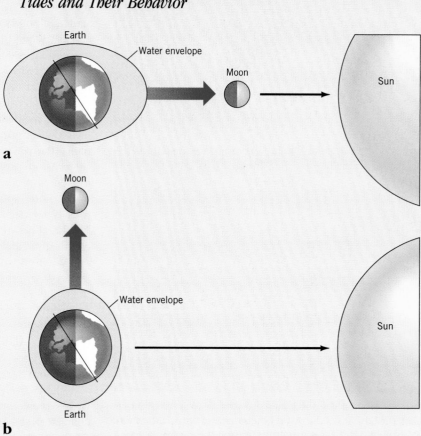

FIGURE 50-9 Schematic diagram of tides. Spring tide (a): the earth's rotational bulge and the gravitational pulls of both the sun and the moon combine to produce an unusually high tide. Neap tide (b): the moon's gravitational pull is at a right angle to the sun's pull and the terrestrial rotation bulge, resulting in the least extreme tides. The earth's water envelope and astronomical distances are strongly exaggerated.

which has the least extreme tides (Fig. 50-9b).

Tides play a major role in coastal erosion. They can generate strong tidal currents that rush into and out of river mouths. They carry waves to higher coastal elevations during spring-tide extremes. And when a severe storm attacks a coastline in conjunction with a spring tide, substantial erosion may occur. The contribution of tides to shoreline erosion also is influenced by coastal topography, both below the water (shallow and sloping or deep and steep) and above (long narrow estuaries or wide curving bays). Compared to the constantly pounding waves, tides are not prominent as coastal modifiers—but their impact is still significant.

coastal relief is usually low: beaches, dunes, and sandy islands reveal the dominant processes at work here (*see* unit opening photo). Wherever they erode or deposit, waves continually move loose material about. Gradually, larger fragments are reduced to smaller ones in a process called *attrition*. We associate a beach with sand, but much of the material along a shallow-water shoreline is even more finely textured.

a

b

FIGURE 50-10 High tide (a) and low tide (b) at Invergordon on Scotland's Moray Firth, demonstrating the sizeable tidal range that waterfront structures must accommodate. (Authors' photos)

TIDES AND SHORE ZONE CURRENTS

So far we have discussed the erosional and depositional action of waves that originate in the open ocean and that are formed by strong and persistent winds. Certainly such waves do the bulk of the degradational and aggradational work along shorelines, but other kinds of water movements also contribute to the shaping of the coastal landscape.

Effects of Tides

Sea level rises and falls twice each day (*see* Perspective: Tides and Their Behavior). Again, the beach tells the story: what has been washed ashore during the *high tide* lies along the upper limit of the most recent swash, ready for beachcombing during *low tide*. Thus the whole process of wave motion onto the beach, discussed previously, operates while the tides rise and fall. This has the effect of widening the sloping beach. During high tide, the uprushing swash reaches farther landward than during low tide; during low tide, the backwash reaches farther seaward than during high tide.

Along a straight or nearly straight shoreline, the *tidal range* (the average vertical distance between sea level at high tide and at low tide) may not be large, usually between about 2 and 4 m (6 and 12 ft). But in partially enclosed waters, such as estuaries, bays, and lagoons, the tidal range is much larger. The morphology (shape) of the inlet and its entrance affect the range of the tide. Probably the world's most famous tides occur in the Bay of Fundy on Canada's Atlantic coast; there, the tidal range is as much as 15 m (50 ft), creating unusual problems for people living on the waterfront. A similar situation occurs in the bays that surround Moray Firth on Scotland's northeastern coast, as Fig. 50-10 vividly demonstrates.

Tides have erosional and depositional functions. In rocky narrow bays, where the tidal range is great and tides enter and depart with much energy and power, tidal waters erode the bedrock by hydraulic action and cor-

rasion. The changing water level associated with tides sets in motion *tidal currents*, which rush through the sandy entrances of bays and lagoons, keeping those narrow thresholds clear of blockage. Where the tidal current slows down, the sediment it carries is deposited in a fan-shaped, delta-like formation (we examine such tidal features in Unit 51).

Tides occasionally take on the form of waves. A *tidal bore* is created when a rapidly rising high tide creates a wave front that runs up a river or bay. A dramatic example is sometimes seen in the lower course of Brazil's Amazon River, where the tide is known to rush in like a foamy breaker that never collapses forward. It is reported to start as a wall of water as much as 8 m (25 ft) high, moving upstream at a rate of 20 kph (13 mph). It loses its height as it advances upriver, but has been observed more than 400 km (250 mi) inland from the Atlantic coast.

Shore Zone Currents

We already have seen that tides generate currents. In this unit we have also learned how sand is moved by longshore drift arising from the refracted, oblique-angled arrival of waves onshore. However, not just the sand along the beach but also the water itself moves along the shore in the direction governed by the angle of the waves' approach. This water movement, parallel to the shore, is called a **longshore current**, and its generation and operation are shown in both Figs. 50-4 and 50-6. Longshore currents can also develop from tidal action and from storms along the coast. They are important agents because in total they move huge amounts of material. They are capable of erosion as well as deposition, creating lengthwise hollows and ridges in the surf zone. Moreover, they can erode beaches as well as build sand ridges across the mouths of rivers and bays.

Another type of current is the **rip current**. Even as the surf is surging landward, narrow stream-like currents cut across it, flowing from the shore seaward (Fig. 50-11a). These rip currents travel primarily in the surface layer of the water and can attain a high velocity, although

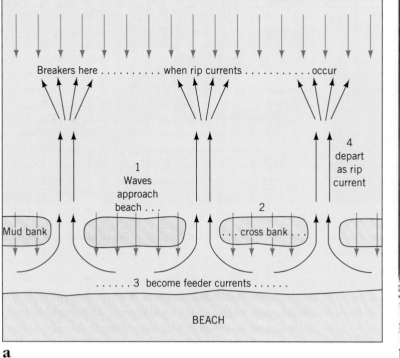

a   **b**

FIGURE 50-11 Rip currents are generated by small feeder currents in the shore zone, which at regular intervals rush seaward directly into the oncoming waves (a). Rip currents die out beyond the surf zone, and swimmers caught in them can easily exit to the side. Rip currents have an erosional function, moving the beach particles outward beyond the surf zone. This can be seen in their usually muddy composition (b).

they die out quickly beyond the surf zone. Rip currents have complex origins. They begin as small feeder currents in the shore zone, flowing parallel to the beach, sometimes behind low, surf-built sand ridges. At certain places along the beach, usually at fairly regular intervals, enough water gathers from these feeder currents to rush seaward, carrying a cloud of muddy sediment.

Standing on the beach, you may be able to see the effect of a rip current on the surf. Because the rip current advances against the incoming surf, the breakers are interrupted where the rip current encounters them. From a higher vantage point, you may even be able to see the patches of mud carried by the rip current contrasted against the less muddy surf (Fig. 50-11b). Rip currents vary in strength. When the tide is high and waves are strong, rip currents are especially powerful. When the waves are lower, the rip current is weaker.

The erosional work of rip currents is limited but is still of geographic interest. One spatial peculiarity just mentioned is that they tend to develop at rather regular intervals along the beach. This seems to be related to the feeder channels that supply the water from two directions (Fig. 50-11a). These longshore feeder currents hollow out their courses, so that the shore zone is marked by lengthwise depressions. Where the rip currents turn seaward, the beach surface is slightly lowered because the current carries much of its muddy sediment away from the beach. Thus rip currents do have an erosional function, moving fine beach particles outward beyond the surf zone.

THE ROLE OF STORMS

Most of the year, the world's coasts are slowly modified by the erosional and depositional processes described above. Except under special circumstances, these changes are slow to occur. But virtually all coastlines are vulnerable to unusual and even rare events that can greatly transform them in a very short period of time. These events are storms. In Unit 14, we discuss the various kinds of storms that can develop over water. In the lower latitudes, tropical cyclones or hurricanes can generate enormous energy in ocean waves. At higher latitudes, storms are most often associated with weather fronts and contrasting air masses.

Whatever the source, the powerful winds whipped up in these storm systems, in turn, spawn large waves. Propelled against the coastline, such waves produce a **storm surge**, a combination of rising water and forceful wave action. During a storm surge, waves attack coastal-zone areas normally untouched by this kind of erosion.

One severe storm surge can break through stable, vegetation-covered barrier islands, erode dunes lying well above the normal swash zone, and penetrate kilometers of coastal plain. When such a storm strikes the coast at the time of high tide, or during a spring tide, its impact is all the more devastating (see Fig. 51-8). These dramatic events notwithstanding, physical geographers disagree as to the long-term geomorphic effect of severe storms. It is certainly true that much of the coast and shore will return to pre-storm conditions after the surge.

But some effects of the storm may be long-lasting, if not permanent.

CRUSTAL MOVEMENT

When we study coastal processes, we must be mindful not only of the many and complex marine processes discussed in this unit, but also of the vertical mobility of the earth's crust. When we first observe a section of coastline, we classify it as a coast of erosion (degradation) or deposition (aggradation), depending on the dominant landforms we identify. That observation is based on the present appearance of the coast and the prevailing processes now at work.

Over the longer term, however, the coastal bedrock may be rising (relative to sea level) or sinking. For instance, we have learned that the Scandinavian Peninsula is undergoing isostatic rebound following the melting of the heavy icesheet that covered it until recently (*see* Fig. 46-8). This means that its coasts are rising as well. Along the coastlines of Norway and Sweden, therefore, we should expect to find evidence of marine erosion now elevated above the zone where wave processes are taking place. In other areas, the coast is sinking. The coastal zone of Louisiana and Texas (*see* Perspective in Unit 36) shows evidence of subsidence, in part because of the increasing weight of the sediments in the Mississippi Delta, but probably from other causes as well.

Add to this the rising and falling of sea level associated with the Late Cenozoic Ice Age, and you can see that it is impossible to generalize about coasts—even over short stretches. Ancient Greek port cities built on the waterfront just 2500 years ago are now submerged deep below Mediterranean waters as a result of local coastal subsidence. But in the same area there are places built on the waterfront that are now situated high above the highest waves. Unraveling the marine processes and earth movements that combine to create the landscapes of coastlines is one of the most daunting challenges of physical geography.

KEY TERMS

backwash (p. 511)

barrier island (p. 513)

beach drift (p. 512)

coast (p. 509)

corrasion (p. 513)

littoral zone (p. 509)

longshore current (p. 515)

longshore drift (p. 512)

rip current (p. 515)

shoaling (p. 511)

shore (p. 509)

storm surge (p. 516)

surf (p. 511)

swash (p. 511)

tide (p. 514)

wave height (p. 509)

wave length (p. 509)

wave refraction (p. 511)

waves of oscillation (p. 509)

REVIEW QUESTIONS

1. Why is an understanding of coastal processes an important part of physical geography?

2. Under what environmental conditions do large waves develop?

3. Describe what happens as incoming waves enter shallow water.

4. What is longshore drift, and how is it generated?

5. What are tides? What are their controlling forces?

REFERENCES AND FURTHER READINGS

BASCOM, W. *Waves and Beaches: The Dynamics of the Ocean Surface* (Garden City, N.Y.: Anchor/Doubleday, 2nd ed., 1980).

BIRD, E. C. F. *Coasts: An Introduction to Coastal Geomorphology* (New York: Blackwell, 3rd ed., 1984).

BURK, C. A., and DRAKE, C. L., eds. *The Geology of Continental Margins* (New York/Berlin: Springer-Verlag, 1974).

CARTER, R. W. G., ed. *Coastal Environments: An Introduction to the Physical, Ecological and Cultural Systems of Coastlines* (Orlando, Fla.: Academic Press, 1989).

CLAYTON, K. M. *Coastal Geomorphology* (London: Macmillan, 1972).

DAVIES, J. L. *Geographical Variation in Coastal Development* (London/New York: Longman, 2nd ed., 1980).

HANSOM, J. D. *Coasts* (London/New York: Cambridge University Press, 1988).

HARDISTY, J. *Beaches: Form and Process* (Winchester, Mass.: Unwin Hyman Academic, 1990).

JOHNSON, D. W. *Shore Processes and Shoreline Development* (New York: Hafner, reprint of 1919 original, 1956).

KETCHUM, B. H., ed. *The Water's Edge: Critical Problems of the Coastal Zone* (Cambridge, Mass.: MIT Press, 1972).

KOMAR, P. D., ed. *CRC Handbook of Coastal Processes and Erosion* (Boca Raton, Fla.: CRC Press, 1983).

PETHICK, J. S. *An Introduction to Coastal Geomorphology* (London: Edward Arnold, 1984).

SWARTZ, M. L., ed. *The Encyclopedia of Beaches and Coastal Environments* (Stroudsburg, Pa.: Dowden, Hutchinson & Ross, 1982).

SHEPARD, F. P., and WANLESS, H. R. *Our Changing Coastlines* (New York: McGraw-Hill, 1971).

SIEVER, R. *Sand* (New York: Scientific American Library, 1988).

VILES, H., and SPENCER, T. *Coastal Problems* (London/New York: Routledge, 1995).

Coastal Landforms and Landscapes

OBJECTIVES

♦ To examine the characteristics of a beach.
♦ To relate beaches to the coastline's topographic and tectonic setting.

♦ To discuss related coastal landforms of aggradation, such as sand dunes, offshore bars, and barrier islands.
♦ To discuss landforms typical of erosional coastlines.

♦ To relate erosional and depositional processes to a general classification of coastlines.

I n Unit 50, we review the numerous marine processes that contribute to the formation of coastal landforms and landscapes. In this unit, we examine those features themselves. The landforms of shorelines and the landscapes of coasts reveal the dominant processes at work. When we see cliffs and caves, we conclude that erosional activity is paramount. On the other hand, beaches and barrier islands indicate deposition. We therefore divide coastal landforms into two groups: aggradational and degradational features. However, we should keep in mind that most coastal landscapes display evidence of both erosion and deposition.

AGGRADATIONAL LANDFORMS

Undoubtedly, the most characteristic depositional landform along the coastline is a **beach**, defined as a coastal zone of sediment that is shaped by the action of waves. This means that a beach is wider than the part of it we can see: on the landward side it begins at the foot of a line of dunes or some other feature, but on the seaward side it continues beneath the surf. Beaches are constructed from sand and other material, derived from both local and distant sources. Most of the beach material in the coastal environment comes from rivers. When streams enter the ocean, the sediments they carry are deposited and transported along the shore by waves and currents. In addition, beach material may be produced locally: by the erosion of nearby sea cliffs and the physical breakdown of those particles as they are moved along the shore.

UNIT OPENING PHOTO: Scenic wonders off the coast of Capri, a southern Italian island in the Bay of Naples, including a stack (left) and a sea arch (center). (Authors' photo)

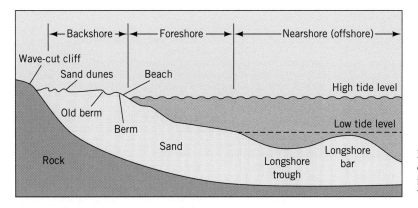

FIGURE 51-1 The parts of a beach shown in cross-sectional profile. The length of the profile is between 100 m and 200 m (330 and 660 ft). Vertical exaggeration is about twice.

Beach Dynamics

The character of a beach reflects the nature of the material composing it. Most beaches along the U.S. Atlantic coast are made of sand, their light color a result of the quartz fragments that make up the sand grains (*see* Unit 50 opening photo). In areas where dark-colored igneous rocks serve as the source for beach material, as they do in parts of Hawaii, beaches are dark-colored. Along the coast of northern California, Oregon, and Washington, high-energy conditions prohibit sand from being deposited on the beach, and larger particles make up the beach fabric; the resulting gravel and pebble beaches are often called *shingle* beaches.

A beach profile has several parts (Fig. 51-1). The **foreshore** is the zone that is alternately water-covered during high tide and exposed during low tide. This is the zone of beach drift and related processes. Seaward of the foreshore lies the **nearshore** (sometimes *offshore*), which is submerged even during an average low tide. One or more **longshore bars** (a ridge of sand parallel to the beach) and associated troughs often develop in this zone, where longshore drift, currents, and wave action combine to create a complex and ever-changing topography. Landward of the foreshore lies the **backshore**, which extends from the high-water line to the dune line. As Fig. 51-1 shows, the backshore consists of one or more sandy beaches called **berms**; these flat berms were laid down during storms and are beyond the reach of normal wave action.

Beach profiles show a considerable amount of variation. Where seasonal contrasts in wave energy are strong, a beach will have one profile during the winter and another during the summer (Fig. 51-2). The summer's long, steady, low waves carry sand from the nearshore zone onto the beach and create a wide summer

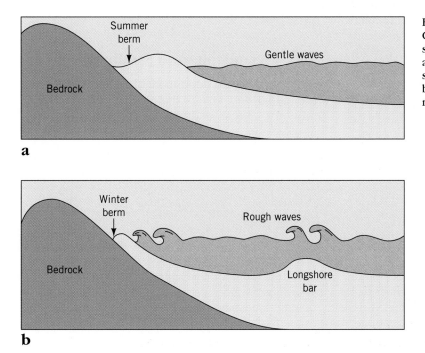

a

b

FIGURE 51-2 Seasonal variation in beach erosion. Gentle summer waves produce a wide berm that slopes gently landward (a). The rougher waves associated with winter storms produce a cold-season beach profile that shows the summer berm eroded to a narrower, steeper-sloped remnant (b).

berm that may develop a crest and slope gently landward. During the following winter, higher and more powerful storm waves erode much of the summer berm away, carrying the sand back to the nearshore zone and leaving a narrower winter berm. Thus a beach displays evidence of erosion as well as deposition.

Beaches are best viewed as open systems, characterized by inputs, outputs, and changes in storage. The size of the beach reflects the material in storage, and is therefore a measure of the balance between the availability of sediment and wave energy. Input of sediment is derived from local erosion, from offshore, and from upshore sections of the beach. Outputs of sediments can occur offshore or downshore, and the width of the beach reflects the magnitude of the inputs and outputs.

Two scenarios illustrate the behavior of a beach. Temporary increases in wave energy, say, associated with a storm, promote temporary erosion (increased output to offshore zones); when wave energy conditions return to normal, offshore material is redeposited on the beach and it rebuilds to its former configuration. Long-term changes in sediment supply, however, can disrupt this general balance and produce substantial, and rather permanent, changes in the beach. Most often, coastlines experience a decline in sediment supply from rivers because of upstream dams and reservoirs. Since wave energy is not affected, the decreased sediment supply results in heightened beach erosion. Accordingly, the beach gets smaller as the storage of sediment decreases to reflect the lower inputs from rivers.

The location and distribution of beaches is therefore related to both sediment availability and wave energy. Where coastal topography is being shaped by convergent lithospheric plate movement, coasts are steep, wave energy is generally high, and beaches are comparatively few. On the mainland of the United States, for example, beaches on the coast of the Pacific Northwest generally are discontinuous, short, and narrow (*see* photo, p. 524); but the East Coast, from the Mexican border to New York's Long Island, is beach-fringed almost continuously, and beaches tend to be wide. The global map of continental shelves (*see* Fig. 2-6) generally provides a good indication of the likelihood of beach development. Where shelves are wide, beaches usually are well developed. Along high-relief coastlines, where shelves are frequently narrow, beach forms are generally restricted to more sheltered locations, and the material composing the beach tends to be larger.

Coastal Dunes

Wind is an important geomorphic agent in coastal landscapes, and many beaches are fringed by sand dunes that are primarily the product of eolian deposition (Fig. 51-3). Lines of dunes are sometimes breached by a storm surge, but wind remains their prime aggradational agent.

FIGURE 51-3 The vegetated sand dunes lining the rear of Pismo State Beach on the central California coast.

Coastal zones are subject to strong sea breezes; sand on the winter berm is constantly moved landward by these winds and, less frequently, by storms. Even when an offshore storm does not generate waves high enough to affect the winter berm, its winds can move large amounts of berm sediment landward.

In this way, coastal dunes are nourished from the beach. Usually they appear simply as irregular mounds, rarely exhibiting the characteristic formations of desert dunes. Some, however, display a variation of the parabolic form (*see* Fig. 49-7c), reflecting the persistent wind direction during their development. An important point of difference between desert and coastal dunes is that coastal dunes, as a result of moist conditions, often are covered with vegetation and are relatively stable (Fig. 51-3).

Sandspits and Sandbars

We turn now to the depositional landforms of the surf zone and beyond. One of the most characteristic of aggradational coastal landforms is the **sandspit**. When longshore drift occurs and the shifting sediment reaches a bay or a bend in the shoreline, it may form an extension into open water as shown (twice) in the central portion of Fig. 51-4. In effect, the spit is an extension of the beach. It begins as a small tongue of sand, and grows larger over time. It may reach many kilometers in length and grow hundreds of meters wide, although most spits have more modest dimensions. Perhaps the most famous of all sandspits is New Jersey's Sandy Hook, which guards the southernmost entrance to New York City's harbor (Fig. 51-5).

Some sandspits continue to grow all the way across the mouth of a bay, and become **baymouth bars** (center of Fig. 51-4). The bay may simply be an indentation in

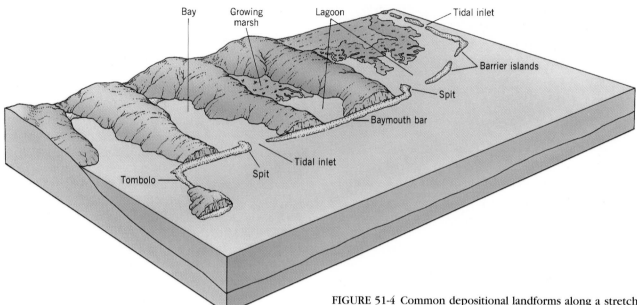

FIGURE 51-4 Common depositional landforms along a stretch of coastline.

the coastline, or it may be a river's estuary. At first, tidal currents may breach the growing bar and keep the bay open (where the tidal inlet is located to the left in Fig. 51-4); but if longshore drift is strong and sediment plentiful, the bay will soon be closed off.

This has important consequences, because if tidal action ceases the bay is no longer supplied with cleansing ocean water. If the mouth of an estuary is closed off, the river that enters it will drop its sediments in the bay instead of the ocean. Thus the bay becomes a lagoon, and its former saltwater environment changes as river water and sediment fill it. Behind the baymouth bar, the ecology of the new lagoon changes to that of a swamp or marshland (Fig. 51-4). A growing sandspit also may form a link between an offshore island and the mainland, creating a landform called a **tombolo** (Fig. 51-4, left front). Spits, bars, and tombolos can take on many different shapes as they bend, curve, and shift during their evolution.

Offshore Bars and Barrier Islands

A sure sign of offshore aggradation is the *sandbar* or **offshore bar** that lies some distance from the beach and is not connected to land. We referred to longshore bars in connection with the beach profile (Fig. 51-1), and such offshore bars can be observed to expand and contract depending on wave action and sediment supply. Once formed, offshore bars interfere with the very waves that built them; waves will break against the seaward side of a bar, then regenerate and break a second time against the shore itself (Fig. 51-2b). Some offshore bars become stable enough to attain permanence, rising above the water surface during low tide and being submerged only during high tide.

Along certain stretches of the earth's coasts lie large and permanent offshore bars, appropriately called **barrier islands**. These islands are made of sand, but they reach a height of 6 m (20 ft) above sea level, and average

FIGURE 51-5 Sandy Hook (top center), which protrudes for nearly 8 km (5 mi) into Lower New York Bay, marks the northern terminus of the New Jersey shore. In the hazy distance lie Staten Island and Brooklyn, two outer boroughs of New York City, and behind them (just left of center on the horizon) the skyscrapers of Lower Manhattan.

FIGURE 51-6 A satellite view of North Carolina's slender Outer Banks, separated from the mainland by Pamlico Sound (left of center). The easternmost point of land near the center is Cape Hatteras (*see* Unit 50 opening photo), a popular tourist destination highly vulnerable to the vagaries of winds and currents in its often turbulent marine environment.

from 2 to 5 km (1 to 3 mi) in width. They can lie up to 20 km (13 mi) from the coast, but more commonly are half that distance from shore, and they often stretch for dozens of kilometers, unbroken except by tidal inlets. The offshore barrier island strip (known locally as the Outer Banks) that forms North Carolina's Cape Hatteras is a classic example (Fig. 51-6).

Barrier islands may have had their origins as offshore bars during the last glaciation, when sea level was much lower than it is today. As sea level rose, these offshore bars migrated coastward, growing as they shifted. Since about 5000 years ago, when sea level stopped rising rapidly, the barrier islands have moved landward slowly. In the meantime, they have developed a distinctive profile. There is a gently sloping beach on the seaward side, a wind-built ridge of dunes in the middle, and a zone of natural vegetation (shrubs, grasses, mangroves) on the

landward side (Fig. 51-7). Finally, a lagoon almost always separates the barrier island from the mainland. Because barrier islands are breached by tidal inlets, however, tidal action keeps these lagoons from becoming swamps (Fig. 51-4, right side).

Because of their recreational and other opportunities, which often attract intensive development (*see* Fig. 50-7), barrier islands are of more than geomorphic interest. In fact, several large cities and many smaller towns have developed on these strips of sand, including Miami Beach, Galveston, and Atlantic City. Numerous long stretches of barrier island extend from South Texas to New York along the Gulf of Mexico and lower Atlantic coast of the United States. Between these islands and the mainland lies the Intracoastal Waterway, an important artery for coastwise shipping. Being low and exposed, the barrier islands are vulnerable to hurricanes, and severe storm waves can temporarily erase parts of them. In heavily developed areas, the hazard is particularly obvious (*see* Perspective: Hazards of Barrier Island Development).

DEGRADATIONAL LANDFORMS

Where wave erosion (rather than deposition) is the dominant coastal process, a very different set of landforms develops. Exposed bedrock, high relief, steep slopes, and deep water are key features of this topography. If there are islands, they are likely to be rocky remnants of the retreating coast, not sandy embankments being built in shallow water. In Unit 50, we discuss the processes of degradation by waves: hydraulic action, corrasion, corrosion, and the attrition of rock fragments. Just as a river seeks to produce a graded profile, so wave erosion works to straighten an indented, embayed coastline. Wave refraction concentrates erosional energy on the headlands that stick out into the water, while sedimentary material collects in the concave bends of bays, a process whose beginning we observed in Fig. 50-5.

The sequence of events that follows is depicted in Fig. 51-9. When headlands (a) are eroded by waves, steep **sea cliffs** develop (b). Waves cannot reach the upper parts of these cliffs, but they vigorously erode the bottom part of the cliff, seeking out joints, layers of softer strata, and other weaknesses. In so doing, the waves of-

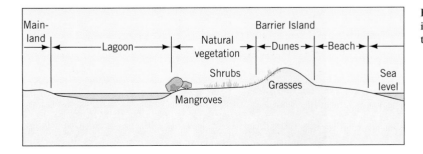

FIGURE 51-7 Cross-sectional profile of a barrier island and adjoining lagoon. Vertical exaggeration is about twice.

Hazards of Barrier Island Development

From Bar Harbor, Maine to the mouth of the Rio Grande at Brownsville, Texas more than 4000 km (2500 mi) away, 295 barrier islands lie along much of the U.S. Atlantic and Gulf of Mexico coastline. Given Americans' love of the seashore—and the fact that more than half of them reside within an hour's drive of a coast—it should come as no surprise that most of these islands have witnessed the development of lengthy oceanside strips of summer homes, resorts, high-rise condominiums, commercial and tourist facilities, fishing piers, and public beaches.

Barrier islands, as we know, are in constant motion, migrating slowly landward in response to increases in sea level (which have totaled about 30 cm [1 ft] over the past century along the eastern coast of the United States). The myriad structures built on loose sand atop these islands, therefore, are always threatened by erosional processes—which are strongest on the seaward side directly facing the most desirable beachfront development sites.

To protect themselves, barrier island communities construct seawalls parallel to the beach and groins or rock jetties perpendicularly outward into the surf (*see* Fig. 50-7); they also spend lavishly to replenish beaches by pumping in massive amounts of new sand from neighboring shorelines. These measures, unfortunately, buy only a few years' time and may actually worsen erosion in the longer run. Seawalls block the onrushing surf, but the deflected swash returns to the ocean so quickly that it carries with it most of the new sand that would otherwise have been deposited—and soon the beach has disappeared. Groins are more successful at trapping incoming sand, but studies have shown that saving one beach usually occurs at the expense of destroying another one nearby. As

for replenishing shrinking beaches by pumping in sand, all the evidence shows that this, too, is only a stopgap measure. For the past quarter-century, the U.S. Army Corps of Engineers has shifted massive amounts of beach sand along 500 miles of the eastern seaboard, with little more than a $10 billion expenditure to show for its herculean efforts.

The greatest hazard to these low-lying offshore islands are the 30 or so cyclonic storms that annually move over the East Coast. Even a moderate-strength cyclone can accentuate coastal erosion processes to the point where major property losses occur. The gravest of such threats to these vulnerable sand strips, of course, are the occasional bigger storms—which are long remembered by local residents.

A particularly nasty late-winter cyclone in 1962 smashed its way northward from Cape Hatteras to New England, leaving in its wake a reconfigured coastline (as the flooding ocean created new inlets across barrier islands) and over $750 million worth of storm damage in today's dollars (Fig. 51-8). (Another of these ferocious storms ravaged the New England coast in late 1991, causing the greatest damage since the landmark 1962 storm.) The worst devastation, however, is associated with tropical cyclones. After Hurricane Camille (one of this century's most powerful) attacked the Louisiana and Mississippi Gulf Coast in 1969, the United States for 20 years was spared this kind of awesome damage—until Hurricane Hugo roared across the South Carolina shoreline in 1989. Most importantly, this uncharacteristic lull of the 1970s and 1980s was accompanied by the largest coastal construction boom in history.

Only belatedly are the federal and state governments taking a hard look at all this development, and in many seaside

locales a new consensus is emerging that it was a mistake to build in such unstable environments. Thus the realization is dawning that nature is certain to win the battle of the barrier islands in the end—despite the best efforts of policymakers, planners, and coastal engineers to manage the precarious human presence that has cost so many billions of dollars to put into (temporary) place.

FIGURE 51-8 The March 1962 storm is still remembered as one of the most destructive in the recorded history of the U.S. Atlantic coast. This is the aftermath of that storm on Fire Island, the barrier island off the south-central shore of Long Island, about 80 km (50 mi) east of New York City.

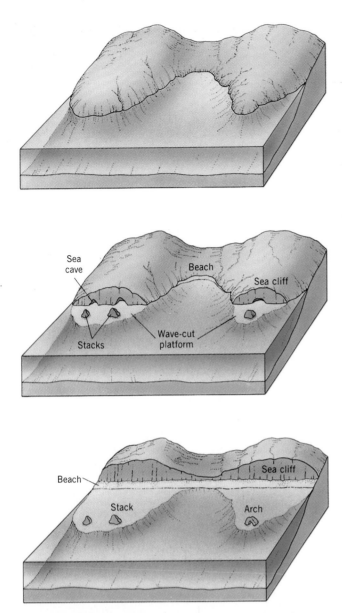

FIGURE 51-9 The straightening of an embayed, indented coastline by wave erosion. When degradation is the dominant coastal process, a very different set of landforms develops (compare to the aggradational landforms in Fig. 51-4).

FIGURE 51-10 Deep water, unobstructed fetch, and pounding waves drive back the shore along steep cliffs near the Oregon-California border. From the virtual absence of stacks offshore, we may conclude that the rocks being driven back here are not as resistant to erosion as harder crystallines are along other stretches of the Pacific coast of the United States. (Authors' photo)

ten create **sea caves** near the base of the cliff, undercutting it. Soon the overhanging part of the cliff collapses, so that wave action combines with mass movement to erode the coastal bedrock.

The cliff continues to retreat (Fig. 51-9b and c). A **wave-cut platform** (also called an *abrasion platform*) develops at the foot of the cliff (b), marking its recession. These platforms are nearly flat bedrock surfaces that slope seaward. At low tide we can see boulders and cobbles, broken from the cliff, lying on this platform. Soon the waves of a high tide (or a storm) will hurl these fragments back against the cliff face. As the headlands retreat, certain parts invariably prove to be more resistant than others (b and c).

Sections of the headlands survive as small islands, and as wave erosion continues, these islands are some-

times penetrated at their base and become **sea arches** (c). Other remnants of the headlands stand alone as columns called **stacks** (b and c). Arches and stacks (*see* unit opening photo) are typical of coastlines being actively eroded, but they are temporary features, and soon they, too, will be eroded down to the level of the wave-cut platform. Eventually, the headlands are completely removed, as are the beaches that lay at the heads of the bays. A nearly straight, retreating cliff now marks the entire coastal segment (Fig. 51-9c), with a portion of the wave-cut platform, covered by sediment, at its base.

Cliffs can form in any coastal rock, ranging from hard crystallines to soft and loose glacial deposits. England's famous White Cliffs of Dover are cut from chalk. Tall cliffs in the Hawaiian Islands are carved from volcanic rocks. Cliffs along coastlines of the Mediterranean Sea are cut from layers of sedimentary rocks (*see* Fig. 31-6). The speed of cliff retreat depends on a number of conditions, including the power of the waves and, importantly, the resistance of the coastal rocks. In parts of New England, particularly on Cape Cod, the coast is formed by glacial deposits, and this loose material retreats as much as 1 m (3.3 ft) per year. Along parts of the U.S. West Coast, deeper water and soft bedrock combine to produce an even faster rate of retreat; but in other segments, such as the coast of Oregon, the hard crystalline rocks wear away at a much slower rate (Fig. 51-10).

COASTAL LANDSCAPES

The events just described constitute a model for the evolution of a shoreline from an initially irregular shape to a straight, uniform beach. The reason so few beaches resemble this model is that many coastlines have experienced marked fluctuations in the conditions that shape shoreline evolution. Simply put, most beaches are too young to have achieved their equilibrium form. Recent sea-level fluctuations, as well as tectonic movements along plate margins, continually disrupt the operation of coastal processes. With this in mind, it is convenient to distinguish between two general types of coastlines: emergent and submergent coasts.

Emergent Coasts

The landscapes of uplifted or **emergent coasts** carry the imprints of elevation by tectonic forces. Some coastal zones have been uplifted faster than postglacial sea level rose. The net effect of this is that such features as cliffs and wave-cut platforms are raised above (sometimes tens of meters above) present sea level (Fig. 51-11). When elevated this way, a wave-cut platform is termed an **uplifted marine terrace**. Occasionally, such landforms as stacks and arches still stand on the uplifted marine terrace, indicating the stage at which wave erosion was interrupted by tectonic uplift.

Coasts where aggradational processes dominate also may be uplifted, but such depositional coastlines tend to lie in more stable lithospheric zones. There, the evidence is more rapidly erased by erosional processes, because uplifted landforms—dunes, berms, bars, spits—are far less resistant than bedrock cliffs and wave-cut platforms. Sometimes, cultural features reveal recent uplift. Stone structures of coastal settlements (including docks) may survive longer than soft sedimentary landforms. When such settlements lie well above the water, we can conclude that uplift has occurred. Some Maya buildings, constructed on the waterfront more than 1000 years ago, now lie elevated on uplifted segments of the Mexican and Central American coasts.

Submergent Coasts

More coastlines are **submergent coasts**: that is, drowned rather than uplifted. This submergence was caused in large part by the rise of sea level over the past 10,000 years. At the beginning of the Holocene, sea level stood perhaps 120 m (400 ft) below its present average mark. This exposed large parts of the continental shelves (*see* Fig. 2-6) that are now under water. Rivers flowed across these areas of dry land as they do today across the coastal plain, eroding their valleys to the edge of the ocean.

The courses of many such rivers, in fact, can be traced from their present mouths across the continental shelf to their former outlets. When Holocene glacial melting raised global sea levels by many meters, these marginal areas were submerged and the river valleys became submarine canyons. The water rose quite rapidly until

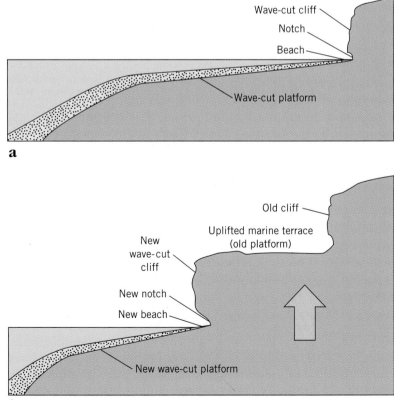

a

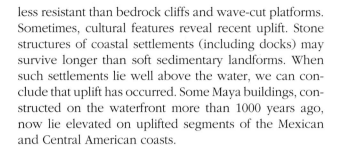

FIGURE 51-11 A wave-cut platform (a) is transformed into an uplifted marine terrace (b) when tectonic uplift elevates the coastal zone above the existing sea level.

b

about 5000 years ago. This was the time when the world's barrier islands formed and began to migrate landward. Over the past five millennia, sea level has continued to rise, but so slowly that many coastlines have stabilized.

In the drowned river mouths, however, we continue to witness the effects of submergence. Rivers often flow into these estuaries and leave no doubt regarding their origin. Sometimes the tops of nearly submerged hills rise as small islands above the water within the estuary. If sea level were still rising rapidly, this invasion of river valleys by advancing ocean water would continue today; but the rise of sea level has slowed sufficiently to permit long-shore drift to form spits and bars across many estuaries.

Evidence of submergence also can be seen on tectonically active, high-relief coasts and in areas affected by glacial erosion that were adjacent to coastlines. A combination of crustal subsidence and rising sea level produces very deep water immediately offshore (for instance, along the western coasts of North and South America). This exposes the coastal bedrock to the onslaught of powerful, deep-water waves unslowed by shoaling. Where fjords were carved by glaciers reaching the ocean, rising water has filled the U-shaped troughs above the level of the ice. Fjords, of course, are drowned glacial valleys, their waters deep, their valley sides sheer and often spectacular (*see* Fig. 47-6c). Submergent coastlines, therefore, display varied landscapes, all of them resulting from rising water, subsiding coastal crust, or both.

LIVING SHORELINES

Living organisms, such as corals, algae, and mangroves, can shape or affect the development of shores and coasts. A **coral reef** is built by tiny marine organisms that discharge calcium carbonate. New colonies build on the marine limestone deposits left by their predecessors, and this process can create an extensive network of coast-fringing ridges. To grow, corals need clear water, warm temperatures, and vigorous cleansing wave action. This wave action does erode the reef, but it also washes the growing coral. Ideally, the coral is covered with water during high tide and exposed to the air during low tide.

Where conditions are favorable, wide, flat-topped coral reefs develop. Sometimes they are attached to the shore; others lie offshore and create lagoons between shore and reef. Still other corals create atolls. **Atolls** are roughly circular reefs that surround a lagoon but without any land in the center. The round shape of atolls, many of which are found in the lower latitudes of the Pacific Ocean, was studied more than 150 years ago by Charles Darwin. He concluded that the corals probably grew on the rims of eroded volcanic cones. As the flattened vol-

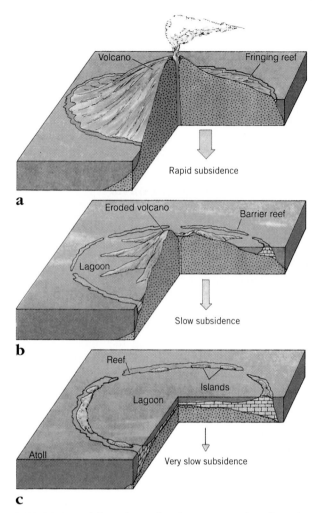

a

b

c

FIGURE 51-12 The relationship between coral atoll reefs and marine volcanoes. The coral reef originally develops around the rim of the subsiding volcanic cone (a). As the cone erodes the corals continue to build upward, leaving a lagoon surrounded by a ring-like reef (b). When the cone has disappeared below the water, the circular atoll is the only feature remaining on the ocean surface (c).

cano subsided (or sea level rose), the corals continued to build upward (Fig. 51-12).

Vegetation also influences the evolution of shorelines. In parts of West Africa, Southeast Asia, and the southeastern United States, mangroves and their elaborate root systems have become builders of shorelines. Once these unique plants have taken hold, the erosional power of waves and currents is harnessed by those root systems. As a result, there develops a zone of densely vegetated mud flats, creating a unique ecological niche.

As you can see, the landforms and landscapes of coastlines are formed and modified by many processes and conditions. No two stretches of shoreline are exactly alike because the history of a coastal landscape involves a unique combination of waves, tides, currents, wind, sea-level change, and crustal movement.

KEY TERMS

atoll (p. 526)	coral reef (p. 526)	offshore bar (p. 521)	stack (p. 524)
backshore (p. 519)	emergent coast (p. 525)	sandspit (p. 520)	submergent coast (p. 525)
barrier island (p. 521)	foreshore (p. 519)	sea arch (p. 524)	tombolo (p. 521)
baymouth bar (p. 520)	longshore bar (p. 519)	sea cave (p. 524)	uplifted marine terrace (p. 525)
beach (p. 518)	nearshore (p. 519)	sea cliff (p. 522)	wave-cut platform (p. 524)
berm (p. 519)			

REVIEW QUESTIONS

1. Describe a typical beach profile. Why would different beach profiles develop in summer and winter?

2. How do sandspits and baymouth bars form?

3. How do barrier islands form?

4. Describe the processes by which irregular, embayed coastlines become straightened.

5. What has been the fundamental cause of the many submergent coastlines encountered throughout the world?

REFERENCES AND FURTHER READINGS

BARNES, R. S. K., ed. *The Coastline* (New York: Wiley, 1977).

BEATLEY, T., et al. *An Introduction to Coastal Zone Management* (Covelo, Calif.: Island Press, 1994).

BIRD, E. C. F. *Coasts: An Introduction to Coastal Geomorphology* (New York: Blackwell, 3rd ed., 1984).

BIRD, E. C. F. *Submerging Coasts: The Effects of a Rising Sea Level on Coastal Environments* (New York: Wiley, 1993).

CARTER, R. W. G., ed. *Coastal Environments: An Introduction to the Physical, Ecological and Cultural Systems of Coastlines* (Orlando, Fla.: Academic Press, 1989).

CLAYTON, K. M. *Coastal Geomorphology* (London: Macmillan, 1972).

DOLAN, R., and LINS, H. "Beaches and Barrier Islands," *Scientific American*, July 1987, 68–77.

HANSOM, J. D. *Coasts* (London/New York: Cambridge University Press, 1988).

HARDISTY, J. *Beaches: Form and Process* (Winchester, Mass.: Unwin Hyman Academic, 1990).

KING, C. A. M. *Beaches and Coasts* (New York: St. Martin's Press, 2nd ed., 1972).

KOMAR, P. D. *Beach Processes and Sedimentation* (Englewood Cliffs, N.J.: Prentice-Hall, 1976).

PETHICK, J. S. *An Introduction to Coastal Geomorphology* (London: Edward Arnold, 1984).

SCHWARTZ, M. L., ed. *Barrier Islands* (Stroudsburg, Pa.: Dowden, Hutchinson & Ross, 1973).

SIEVER, R. *Sand* (New York: Scientific American Library, 1988).

SNEAD, R. A. *Coastal Landforms and Surface Features: A Photographic Atlas and Glossary* (Stroudsburg, Pa.: Dowden, Hutchinson & Ross, 1982).

STEERS, J. A. *Applied Coastal Geomorphology* (Cambridge, Mass.: MIT Press, 1971).

SUNAMURA, T. *Geomorphology of Rocky Coasts* (New York: Wiley, 1992).

TRENHAILE, A. S. *The Geomorphology of Rock Coasts* (London/New York: Oxford University Press, 1987).

VILES, H., and SPENCER, T. *Coastal Problems* (London/New York: Routledge, 1995).

"Where's the Beach? America's Vanishing Coastline," *Time*, August 10, 1987, 38–47.

UNIT 52

Physiographic Realms:
The Spatial Variation of Landscapes

OBJECTIVES

◆ To introduce physiographic realms and regions.

◆ To briefly discuss the six physiographic realms of North America.

W hen you fly home from college, or across the United States or Canada on vacation, you pass over a panorama of changing landscapes. Even from a height of 12 km (40,000 ft), the great canyons of the Colorado Plateau, the jagged relief of the Rocky Mountains, the undulating expanses of the Great Plains, and the echelon folds of the Appalachians leave clear impressions, and you would not mistake one for another. The landscapes of the continents are not jumbled and disorganized. On the contrary, each area has distinct and characteristic properties.

In our everyday language, we use words to identify those properties: the Rocky *Mountains*, the Great *Plains*,

the St. Lawrence *Valley*, the Columbia *Plateau*. In other parts of the world, you might in the same way have recognized the mountains of the Alps, the Amazon Basin, or the lowlands of western Siberia. In so doing, you prove to be a good geographer because physical geographers, like other geographers, use the **regional concept** to classify and categorize spatial information. And

UNIT OPENING PHOTO: In central Colorado, the sharp physiographic boundary between the Rocky Mountains and the western edge of the Great Plains is particularly prominent on the landscape. Looking down on the city of Boulder from the Front Range of the Rockies.

World Landscape Realms

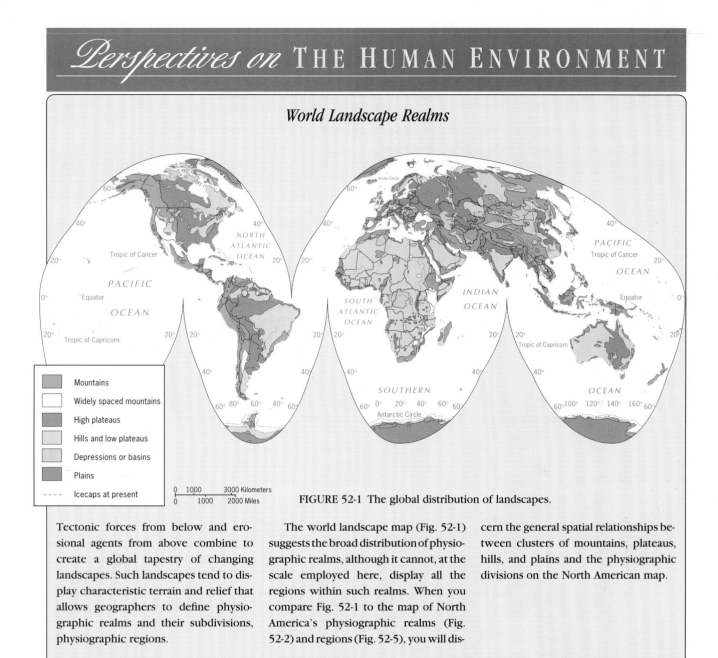

Mountains
Widely spaced mountains
High plateaus
Hills and low plateaus
Depressions or basins
Plains
- - - - Icecaps at present

0   1000   3000 Kilometers
0   1000   2000 Miles

FIGURE 52-1 The global distribution of landscapes.

Tectonic forces from below and erosional agents from above combine to create a global tapestry of changing landscapes. Such landscapes tend to display characteristic terrain and relief that allows geographers to define physiographic realms and their subdivisions, physiographic regions.

The world landscape map (Fig. 52-1) suggests the broad distribution of physiographic realms, although it cannot, at the scale employed here, display all the regions within such realms. When you compare Fig. 52-1 to the map of North America's physiographic realms (Fig. 52-2) and regions (Fig. 52-5), you will discern the general spatial relationships between clusters of mountains, plateaus, hills, and plains and the physiographic divisions on the North American map.

---

the Rocky Mountains and the Great Plains are just that: regions in the North American landscape.

Physical geographers, as we have seen throughout this book, are interested in the spatial arrangement of the natural phenomena on the earth's surface and the relationships among those phenomena. We investigated the major processes that operate in the atmosphere, and then studied the spatial distribution of climates across the earth. Similarly, following a discussion of some aspects of the biosphere, we examined the spatial distribution of soils and vegetation.

The geologic processes of the earth's crust and the forces of weathering, erosion, and deposition also have a spatial component. This component is expressed in what are known as **physiographic realms** and **physiographic regions**, which are characterized by a certain uniformity or homogeneity of landscapes, landforms, and other physiographic elements. When we think of the Rocky Mountains or the low-lying coastal plain of the southeastern United States, an image of a particular scenery comes to mind. It would be impractical to study physiographic realms and regions for the entire world (for a broad impression, however, *see* Perspective: World Landscape Realms). Therefore, in Units 52 and 53, we will concentrate on North America and the United States. Fortunately, virtually every landscape and landform found on earth can also be found on the North American continent.

DEFINING PHYSIOGRAPHIC REALMS AND REGIONS

Why do we attempt to regionalize our observations? Regionalization does for geographers what classification does for scholars in other academic disciplines. In regional human geography, the inhabited world is divided into approximately a dozen geographic realms, such as Europe, South Asia, and Subsaharan Africa. These realms, in turn, are subdivided into regions. Europe, for example, consists of five such regions: the British Isles, Western Europe, Northern Europe, Southern (Mediterranean) Europe, and Eastern Europe. These regions may be further subdivided: Northern Europe, for instance, contains the subregion of Scandinavia. In this hierarchical geographic classification, then, the *first-order* unit is the realm (Europe); the *second-order* unit is the region (Northern Europe); the *third-order* unit is the subregion (Scandinavia).

In physical geography, a similar hierarchical system can be used. When we view all of continental North America, six physiographic realms can be identified (Fig. 52-2). These are:

1. The Canadian Shield
2. The Interior Plains
3. The Appalachian Highlands
4. The Western Mountains
5. The Gulf-Atlantic Coastal Plain
6. The Central American Mountains

These are the first-order physiographic realms of the continent. Each contains second-order physiographic regions, many of which are familiar to us. The Western Mountains, for instance, include the Rocky Mountains (extending south from coastal northern Alaska through Canada into the conterminous United States) and the ranges paralleling the Pacific Coast. This realm also contains the Colorado Plateau and some other regions that are not mountainous at all. Thus the level of generalization in first-order realms is high.

Physiographic realms of the first order present a useful generalization for certain purposes, for example, in a comparison with equally general climatic distributions. But for more detailed analysis, second-order regions are more practical. Accordingly, the Western Mountains are subdivided into such regions as the Canadian Rocky Mountains, the (U.S.) Northern, Middle, and Southern Rockies, the Columbia and Colorado Plateaus, the Sierra Nevada, and so forth. Each of these regions, in turn, is subject to further subdivision at progressively lower orders. Thus the level of regionalization, like the level of classification, should always be linked to our objectives.

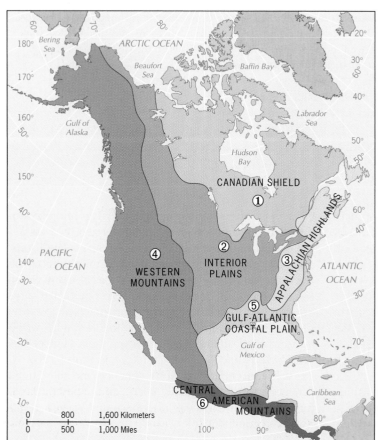

FIGURE 52-2 First-order physiographic realms of mainland North America. Each contains second-order physiographic regions (*see* Fig. 52-5).

In the case of physiographic realms, we do not want to amplify every drumlin cluster, every river valley, every complex of hills. What we want is an *overview* of the continent's grand panorama of landscapes.

This raises the question of the criteria we employ in defining physiographic realms or regions. (Physiographic regions are also sometimes called *physiographic provinces*, an older geographic term still in use in the regional vocabulary.) The term **physiography**, as we have noted, involves more than just the landscape and the landforms it is made of. It relates to all the natural features on the earth's surface, including not only the landforms but also the climate, soils, vegetation, hydrography, and whatever else may be relevant to changes in the overall natural landscape. For example, a realm such as the Interior Plains in Fig. 52-2 may be subdivided into several physiographic provinces because it extends through several climatic zones. The relief may remain generally the same, but the natural vegetation in one part of the realm may be quite different from that of another section. This, of course, is a response to climatic transition, soil differences, and other regional contrasts.

This leads us to the crucial question: How are the boundaries of physiographic regions established? Occasionally, there are no problems. Along much of the western edge of the Great Plains, the Rocky Mountains rise sharply (Fig. 52-3a; unit opening photo), terminating the rather level surface of the Great Plains and marking the beginning of a quite different region (and realm) in terms of relief, slopes, rock types, landforms, vegetation, and other aspects as well.

But as Fig. 52-3b indicates, the transition is far less clear and much less sharp in (most) other instances. The eastern boundary of the Great Plains is a good case in point. Somewhere in the tier of states from North and South Dakota through Nebraska, Kansas, and Oklahoma, the Great Plains terminate and the Interior Lowlands begin. Exactly where this happens depends on the criteria we use to define the two adjacent regions and the method we use to draw the dividing line. Sometimes a persistent linear landform, such as an escarpment, may serve effectively, overshadowing by its prominence the other transitions in the landscape (Fig. 52-3b). Where such a natural dividing line does not present itself, another solution must be found.

Nature rarely draws sharp dividing lines. Vegetation zones, climate regions, and soil belts merge into one another in transition zones, and the lines we draw on maps to delineate them are the products of our calculations, not those of nature. In such a situation, we can establish dividing lines for each individual criterion (vegetative change, climatic change, soil change, and so on), superimpose the maps, and draw an "average line" to delineate the physiographic realm, region, or subregion as a whole, as shown in Fig. 52-4. Still another solution is to

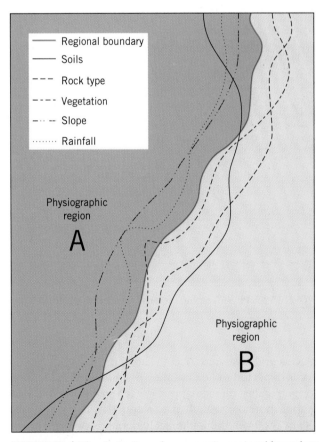

FIGURE 52-4 The derivation of a composite regional boundary (or "average line") from boundaries of several individual criteria.

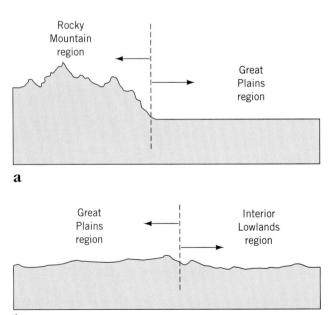

FIGURE 52-3 The topographic features that mark the western (a) and eastern (b) physiographic boundaries of the Great Plains region.

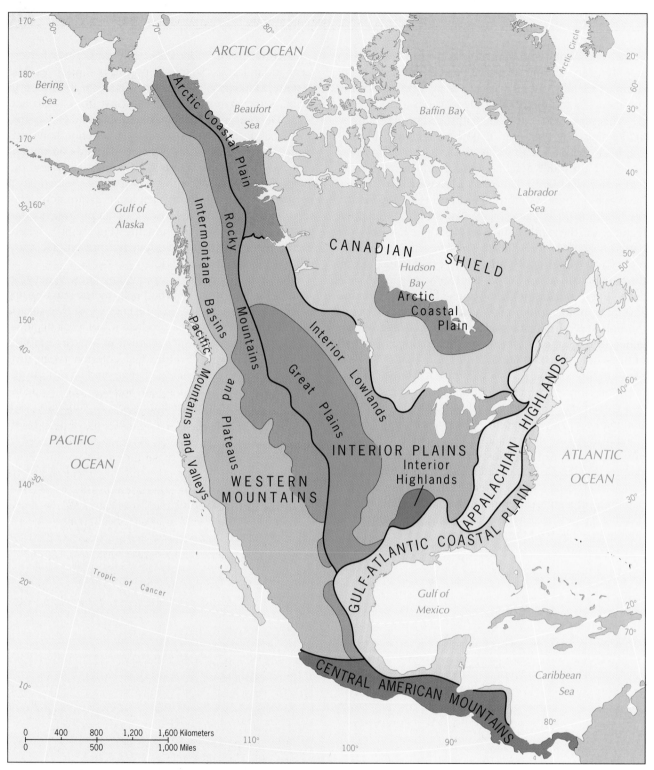

FIGURE 52-5 Physiographic regions (provinces) of North America.

place a grid of squares over the area in which the boundary is expected to lie, and then assign numerical values to each of the criteria mapped in each square; the physiographic boundary can then be mapped by computer. But as always, everything ultimately depends on our choice of criteria and our ability to map them or attach quantitative values to them.

PHYSIOGRAPHIC REALMS OF NORTH AMERICA

The map of North America in Fig. 52-5 is a more detailed version of Fig. 52-2. (As we noted earlier, Fig. 52-2 shows the continent divided into six physiographic realms.) In this unit, we examine North America at this highest level

FIGURE 52-6 The typical undulating terrain of the Canadian Shield in northern Ontario.

of spatial generalization in order to get an overall picture of its physiographic complexion. In Unit 53, we focus on several physiographic regions in the United States, regions that form the component parts of the broader continental-scale realms.

1. The Canadian Shield

The **Canadian Shield** is a vast area of mostly exposed igneous and metamorphic crystalline rocks, the original core of the North American landmass. It is a realm marked by low relief and generally thin, unproductive soils, but noted for its mineral deposits, especially iron, copper, nickel, and zinc. This realm was the source of the great continental icesheets of the Late Cenozoic Ice Age that scoured the rocks and hollowed out the basins now filled with the water of Canada's many lakes. Low hills and a few isolated mountains diversify a scenery that is characteristically monotonous and slightly rolling (Fig. 52-6). Where the meager glaciated soils are sufficiently developed to sustain vegetation, stands of spruce and fir trees clothe the countryside. Here stands much of the boreal forest described in Unit 27.

Along the southern coast of Hudson Bay, the Canadian Shield's crystalline *basement* rocks lie covered by a comparatively thin veneer of sedimentary rocks. The contact between these sediments and the crystallines, at the surface, is sometimes mapped as a physiographic boundary. Thus the Arctic Coastal Plain forms a distinct region within this physiographic realm (Fig. 52-5).

2. The Interior Plains

The physiographic realm known as the **Interior Plains**, extending eastward from the foot of the Rocky Moun-

tains to the inland edge of the Appalachian Highlands, constitutes another broad area of low relief. But this realm is mostly underlain by sedimentary rather than crystalline rocks. There also are contrasts in soils and vegetation between this physiographic realm and the Canadian Shield. The Interior Plains are in large part sustained by the glacial till that accumulated during the advance of the icesheets. When you fly over this realm north of the Ohio and Missouri Valleys, look for moraine belts that stretch across the surface, bearing witness to periods of glacial standstill. Other glacial landforms can also be easily identified.

The Great Plains region occupies the southwestern sector of this low-lying North American heartland, and is a region agriculturally known for its wheat and small-grain production as well as for livestock ranching. A look at the map of climate regions (Fig. 16-3) tells us that this province is drier than the adjacent Interior Lowlands to the east. The vegetation map (Fig. 27-2) confirms that assessment: the Great Plains is essentially a region of short, rather sparse grasses whereas the moister Interior Lowlands are covered by denser, taller grasses and trees. The soil map (Fig. 25-13) shows the Great Plains to be a region of semiarid mollisols, whereas the Interior Lowlands possess better-watered alfisols as well as pockets of mollisols. Of course, these differences occur across transition zones, and the relatively small-scale maps we are referring to are summaries based on artificial indexes. But the typical scenery of the Great Plains reflects a dryness that implies a number of significant contrasts vis-à-vis the Interior Lowlands, and there is ample justification for the identification of a discrete physiographic region here (Fig. 52-7).

To the south of the Interior Lowlands lies a comparatively small upland zone, identified in Fig. 52-5 as the

FIGURE 52-7 The semiaridity of the natural landscape of the Great Plains is apparent throughout this physiographic region, especially here in the ranchlands of western Montana.

Interior Highlands, which is usually separated into the Ozark Plateau in the north and the Ouachita Mountains in the south (*see* Fig. 53-1). The Ozark Plateau consists of slightly domed sedimentary layers, maturely eroded but still standing more than 450 m (1500 ft) above the average elevation of the Interior Lowlands to its immediate north (an area now occupied by the valley of the Mississippi River). The Ouachitas are reminiscent of a small-scale Appalachian ridge-and-valley zone. Like much of the Appalachians (the realm to be discussed next), the area is suited to forestry, cattle raising, and resort and vacation-home development. In fact, the Ouachitas are an extension of the Appalachians, with the intervening lowland somehow failing to undergo the same crustal readjustment that occurred in both the Appalachian Highlands to the east and the Ouachitas to the west. The Ouachitas generally are lower than the Appalachians, however, reaching no more than 780 m (2600 ft) in the central section and less than 300 m (1000 ft) along the region's periphery.

3. The Appalachian Highlands

The composite area made up of the low-relief realms of the North American interior we have just discussed is flanked on several sides by highlands and mountains. To the east lie the **Appalachian Highlands**, which extend northeastward from Alabama and Georgia to New England and beyond into Canada's Maritime Provinces (New Brunswick, Nova Scotia, Prince Edward Island, and

Newfoundland). As we observe in Unit 53, this realm contains several clear-cut subdivisions but, in general, it is characterized by elongated ridges. Many of these ridges exhibit a parallel or slightly zigzag pattern, rolling and sometimes rugged hills, some truly mountainous areas, and persistent intervening valleys.

Some of the contrasts within this realm stem from the fact that, although the northern sector was glaciated during Late Cenozoic times, the southern part was not. Appalachian rocks vary from folded sedimentary layers and maturely eroded horizontal strata to igneous intrusions. The familiar parallel, tree-clad ridges separated by populated valleys prevail, especially in the central zone of the region (*see* Fig. 36-15). Toward the west, plateau characteristics take over; rugged crystalline-supported topography dominates in many sectors to the east.

The Great Smoky Mountains (Fig. 52-8), a prominent part of this eastern zone of the Appalachian Highlands, are sustained by ancient igneous and metamorphic rocks, and are often referred to as the "Older" Appalachians. On these old rocks, the Appalachian Highlands reach their highest elevations—over 2000 m (6500 ft)—several hundred meters higher than the crestlines of the parallel ridges and valleys of the folded central zone.

4. The Western Mountains

Along its entire length, the Interior Plains realm of North America is flanked on the west by high mountains, which are much higher and far more rugged than those to the east and southeast. The **Western Mountains** undoubtedly constitute North America's most varied physiographic realm, its topography ranging from mountainous to flat, but dominated throughout by high relief.

The Rocky Mountains extend from northernmost Alaska to New Mexico and reappear in Mexico as the Sierra Madre Oriental (Eastern Sierra Madre). The Rockies, formed much later than the Appalachians, rise as much as 2700 m (9000 ft) above their surroundings to elevations that in places exceed 4200 m (14,000 ft) above sea level. None of the Appalachians' regularity exists in the Rocky Mountains, which consist of ranges extending in different directions for varying distances, valleys of diverse width and depth, and even large sediment-filled basins between the mountains.

The underlying rocks in this region include igneous batholiths (particularly in the northern sector), volcanics, metamorphosed strata, and sedimentary layers. There is intense folding and large-scale block faulting, all reflected in the high-relief topography (*see* Fig. 45-1b). The Rockies carry all the evidence of Late Cenozoic mountain glaciation, with cirques, horns, arêtes, **U**-shaped valleys, and aggradational landforms. As a result, this is one of the great tourist and ski-resort areas of the world. There is still some permanent snow on the Rockies in the conterminous United States; in Alaska, the glaciers still inhabit many valleys (*see* Fig. 47-2).

FIGURE 52-8 The haze-shrouded ridgelines in the heart of Great Smoky Mountains National Park, which lies astride the Tennessee-North Carolina border. The park's highest peak, as well as Tennessee's, is Clingman's Dome (elevation 2025 m [6642 ft]). The Smokies can be traversed, via U.S. Highway 441 through Newfound Gap, along one of the most scenic drives in the eastern United States.

As Fig. 52-5 shows, the Rocky Mountains physiographic province is a rather narrow but lengthy zone. To the west of it, along its entire length, lies the province of Intermontane Basins and Plateaus. This region, like the Rockies, begins in western Alaska, crosses Canada and the conterminous United States, and extends into Mexico. It is a complex region that can be subdivided, as we describe in the next unit. The western boundary of the Rocky Mountains region is not as sharp as its eastern boundary (Fig. 52-3a), because the Intermontane Basins and Plateaus lie at considerable elevations and do not usually provide the dramatic, wall-like effect that offsets the Rockies from the Great Plains (*see* unit opening photo).

The overall physiography of the Intermontane region is quite different, however. There are expanses of low relief, underlain by only slightly disturbed sedimentary strata or by volcanic beds; isolated fault-block mountains; areas of internal drainage; spectacular canyons and escarpments; and, in much of the region, landscapes of aridity. Because it includes such diverse areas as the Colorado Plateau, the Mexican Highlands, and the Columbia Plateau as well as the so-called Basin-and-Range province (Fig. 52-9), the Intermontane region could be considered not one physiographic province but several. Certainly, the strongest bond among these areas is their *intermontane* location—their position between the major north-south mountain corridors—with the Rockies on one side and the high ranges near the Pacific coast on the other. But internally there is a strong basis for dividing this region into several discrete provinces.

The Pacific Mountains and Valleys form the westernmost physiographic region in North America, again extending all the way from Alaska to Central America. This region incorporates three zones that lie approximately parallel to the long Pacific coastline. In the interior, farthest from the coast, lies a series of high continuous mountain ranges; in the conterminous U.S. portion of the region, these include the Cascade Mountains of Oregon and Washington and the Sierra Nevada of California. Bordering these ranges on the west are several large valleys, dominated by the fertile Central Valley of California and the Willamette-Puget Sound Lowland in the Pacific Northwest. Separating these valleys from the Pacific Ocean are the Coast Ranges, such as the Olympic Mountains of Washington and the Klamath Mountains of northern California.

Except for the intermittent plains and low hills in the valleys, the topography in this Pacific-adjoining province is rugged, the relief high, and the rocks and geologic structures extremely varied. Mountain-building is in progress, earthquakes occur frequently, and movement can be observed along such active faults as the famous San Andreas. The shoreline is marked by steep slopes and deep water, pounding waves, cliffs, and associated landforms of high-relief coasts. Elevations in the interior mountains rival those of the Rockies in places, reaching

FIGURE 52-9 A classic view of the alternating basin-and-range country that dominates the center of the Intermontane region. A satellite image of this terrain is shown in Fig. 53-10.

over 4200 m (14,000 ft) in the conterminous U.S. sector of the region (Fig. 52-10), and exceeding that level in the Canadian Rocky Mountains along the Alaskan coast, where some mountains are over 5400 m (18,000 ft) in height. This is the leading edge of the North American lithospheric plate (*see* Unit 32), and the landscapes of this coastal province decidedly reflect the tectonic forces at work here.

5. The Gulf-Atlantic Coastal Plain

Another major physiographic realm extends along much of the eastern seaboard coastal zone of North America, from the vicinity of New York City (including Long Island) southwestward to the Caribbean coast of Costa Rica in Central America. This is the **Gulf-Atlantic Coastal Plain**, which is bounded on the inland side by the Pied-

FIGURE 52-10 Mount Whitney (center) in California's southern Sierra Nevada, at 4418 m (14,495 ft), is the highest peak in the conterminous United States.

mont (the eastern foothills of the Appalachians), various components of the Interior Plains, the Sierra Madre Oriental, and the Central American mountain ranges. It incorporates all of Florida and attains its greatest width in the lower basin of the Mississippi River. The Coastal Plain extends well up the Mississippi Valley, and the river's delta forms part of the region (Fig. 52-11). In Texas, the boundary between the Coastal Plain and the Great Plains is in places rather indistinct, although a fault scarp does separate the two provinces in southern Texas.

In the eastern United States, the boundary between the Gulf-Atlantic Coastal Plain and the Appalachian Highlands is marked by a series of falls and rapids on rivers leaving the highlands and entering the plain. These falls have developed along the line of contact between the soft sedimentary layers of the Coastal Plain and the harder, older rocks of the Piedmont. Appropriately, this boundary is called the Fall Line, and in early American history the places where rivers crossed it marked not only the head of ocean navigation but also the source of local water power. Not surprisingly, these functions attracted people and economic activities, and major cities such as Philadelphia, Baltimore, and Richmond grew up on the Fall Line (Fig. 52-12).

The seaward boundary of the Coastal Plain is, of course, the continental coastline. This shoreline, however, has changed position in recent geologic times as sea level has risen and fallen. A case can therefore be made that the Gulf-Atlantic Coastal Plain really continues under the ocean to the edge of the continental shelf—the continental slope. The realm's coastline, as is pointed out in Unit 51, has characteristics of a low-relief coast, exhibiting offshore barrier islands, lagoons, beaches, and related landforms. The deepest indentations and irregularities occur north of the Carolinas, where Chesapeake Bay is the largest recess; otherwise, the shoreline is quite straight. Coral reefs appear off the coast in the warmer southern waters, and are especially well developed off southernmost Florida and Mexico's Yucatán Peninsula. The coastline is constantly changing in the area where the Mississippi Delta is building outward.

The whole of the Gulf-Atlantic Coastal Plain lies below about 300 m (1000 ft) in elevation, and comprises a realm of low relief and little topographic variety. Sedimentary rocks sloping gently seaward underlie the entire Coastal Plain. In the northernmost subregion, there is a cover of glacial material. Areas of low hills occur in the Carolinas and in northern Florida, and to the west of the Mississippi River a series of low ridges lies more or less parallel to the Gulf coastline. The lower Mississippi Valley and Delta form another distinct subregion, as does the area of karst topography in central Florida (*see* Fig. 44-6), where numerous sinkholes and small lakes pockmark the countryside. The Coastal Plain also lies substantially in near-tropical latitudes, and the soil map (Fig. 25-13) indicates the presence of characteristic tropical soils. The vegetation is quite varied (Fig. 27-2): there

FIGURE 52-11 New Orleans, with the downtown French Quarter in the foreground, is only one of the many major port cities that developed on the Gulf-Atlantic Coastal Plain. The view here is to the southeast down the Mississippi; the low-lying surface of the Mississippi Delta and the Gulf of Mexico can be seen in the distance.

are deciduous forests in the eastern and southeastern United States, grasslands in Texas and northern Mexico, and some savanna country in Yucatán and nearby Central America.

6. The Central American Mountains

The Gulf-Atlantic Coastal Plain reaches southward to encompass the eastern coastal strip of Mexico (including the wide Yucatán Peninsula) and the Caribbean-facing lowlands of eastern Nicaragua and northeastern Costa Rica. To the south and west of this extension of the

FIGURE 52-12 The Great Falls of the Potomac River, just upstream from the northwestern corner of Washington, D.C. (Maryland is to the right; the Virginia bank is on the left). Thus the nation's capital may also be regarded as a Fall Line city.

Coastal Plain rise the high volcanic **Central American Mountains**, a distinct physiographic realm (Fig. 52-2). This is the physiographic expression of the contact zone between the Cocos Plate on the Pacific side and the Caribbean Plate on the Atlantic side of Central America (*see* Fig. 32-3).

It is a landscape dominated by volcanic landforms and marked by much evidence of crustal instability (Fig. 52-13). Tall composite cones, wide craters, recent lava flows, and evidence of major mass movements prevail; destructive earthquakes also occur frequently in this geologically hazardous realm. In the prevailing warm and humid climate, however, the volcanic rocks rapidly yield fertile soils, and people often live clustered at the foot of active volcanoes. Where natural vegetation survives (deforestation is widespread here), it is dense and richly varied. This variety is related to the altitudinal zonation of Central America's natural environments, ranging from hot coastal lowlands to cool mountainous uplands (*see* Unit 19).

Physiographically, North America ranges from alpine to plainland, from Arctic to tropical, from verdant to barren. The first-order physiographic realms outlined in this unit provide the overall framework within which many

FIGURE 52-13 Popocatépetl (elevation: 5450 m [17,887 ft]) towers above the Valley of Mexico less than 65 km (40 mi) from the edge of Mexico City. Ominously, this huge volcano—with more than 25 million people living within 100 km (62 mi) of its crater—ended a long quiescent period in 1994 and may be entering an active stage. Yet Popocatépetl is but one of dozens of threatening volcanoes that line the backbone of southern Mexico and Central America.

diverse regions and subregions are set. In Unit 53, we explore the second-order physiographic regions of the conterminous United States.

KEY TERMS

Appalachian Highlands (p. 534)

Canadian Shield (p. 533)

Central American Mountains (p. 537)

Gulf-Atlantic Coastal Plain (p. 535)

Interior Plains (p. 533)

physiographic realm (p. 529)

physiographic region (p. 529)

physiography (p. 531)

regional concept (p. 528)

Western Mountains (p. 534)

REVIEW QUESTIONS

1. List the six physiographic realms of North America. On a sketch map, draw in their approximate boundaries.

2. Describe the general topography of each of the six physiographic realms of North America.

3. Which vegetation zones characterize each of these realms?

REFERENCES AND FURTHER READINGS

ATWOOD, W. W. *The Physiographic Provinces of North America* (New York: Ginn/Blaisdell, 1940).

BRIDGES, E. M. *World Geomorphology* (London/New York: Cambridge University Press, 1990).

FENNEMAN, N. M. *Physiography of the Eastern United States* (New York: McGraw-Hill, 1938).

FENNEMAN, N. M. *Physiography of the Western United States* (New York: McGraw-Hill, 1931).

GRAF, W. L., ed. *Geomorphic Systems of North America* (Boulder, Colo.: Geological Society of America, 1987).

HUNT, C. B. *Natural Regions of the United States and Canada* (San Francisco: Freeman, 2nd ed., 1974).

LOOMIS, F. B. *Physiography of the United States* (New York: Doubleday, 1938).

PIRKLE, E. C., and YOHO, W. H. *Natural Landscape of the United States* (Dubuque, Iowa: Kendall-Hunt, 3rd ed., 1982).

THORNBURY, W. D. *Regional Geomorphology of the United States* (New York: Wiley, 1965).

UNIT 53

Physiographic Regions of the United States

OBJECTIVE

◆ To discuss in greater detail the regions that constitute two of the physiographic realms of the conterminous United States.

In Unit 52, we survey the physiography of North America as a whole and regionalize it into six physiographic realms at a high level of spatial generalization. These are the first-order physiographic units of the North American continent, but when we study the physiography of the 48 conterminous United States we must be more specific and less general. In this unit we concentrate on second-order landscapes, and the six realms of North America are broken down into numerous physiographic regions or provinces, as shown for the United States in Fig. 53-1.

In the discussion that follows, you may find it useful to compare Fig. 53-1 to Fig. 52-5 (p. 532). Note, for example, that the larger-scale U.S. map shows no fewer than seven regions in what we earlier called the Appalachian Highlands physiographic realm. The Western Mountains are also seen to contain several regions, including three provinces in the Rocky Mountains alone. Moreover, it is possible to further divide each of the regions in Fig. 53-1 into subregions, based on even more

detailed and specific (third-order) criteria. In this unit we do not attempt to do so, but you may wish to try such a three-level hierarchical exercise for the regional physiography of the part of the country you know best. Neither do we have space to discuss every one of the 20-plus physiographic provinces mapped in Fig. 53-1. Our coverage here can only focus briefly on 12 prominent regions within two realms, one in the East and one in the West.

THE APPALACHIAN HIGHLANDS

The Appalachian Highlands realm actually consists of a group of several physiographic provinces (although Canada's St. Lawrence Valley lies almost entirely outside the

UNIT OPENING PHOTO: An all but cloud-free satellite image of the portion of North America occupied by the conterminous United States.

538

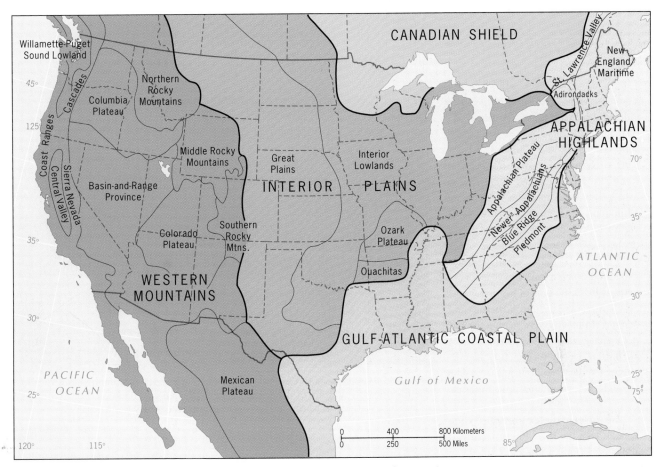

FIGURE 53-1 Physiographic regions (provinces) of the conterminous United States. At this national scale we gain the opportunity to refine our scheme in several places, and the regions are less generalized than on the continental map (Fig. 52-5).

United States). Most familiar to us, perhaps, is the ridge-and-valley topography of the younger or "Newer" Appalachians, a discrete region in Fig. 53-1; but this topography, as the map shows, exists only in the central corridor of the Appalachian realm. In the east-central and northeastern United States, there are six other different physiographic regions that constitute the remainder of the Appalachian Highlands realm delimited in Fig. 53-1. All seven are grouped together because they have a common geologic history and because they lie as highlands between lowland realms. We now examine five of these

regions in detail, proceeding from the interior toward the Atlantic coast.

Appalachian Plateau

There is much to differentiate the Appalachian Highlands internally, as you can see in Fig. 53-2. The **Appalachian Plateau**, noteworthy for its soft (bituminous) coal deposits, has an irregular topography carved by erosion out of horizontal and slightly dipping strata. Parts of this province might be described as low mountains, because

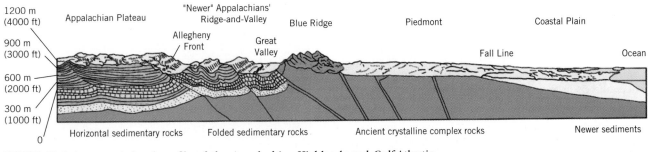

FIGURE 53-2 A cross-sectional profile of the Appalachian Highlands and Gulf-Atlantic Coastal Plain in southern Pennsylvania and New Jersey, showing the principal physiographic subdivisions.

Cumberland Gap

The westernmost region of the Appalachian realm is naturally divided into three plateaus. The most southerly of these subdivisions is the Cumberland Plateau, stretching southwest from eastern Kentucky across eastern Tennessee into northern Alabama. The Cumberland Mountains of southeastern Kentucky, which are the highest and most rugged of the Plateau's terrain, range in height from 600 to 1250 m (2000 to 4150 ft).

During the first hundred years of settlement by Europeans in the area that is now Virginia and the Carolinas, the Cumberland Mountains were an effective barrier against westward expansion. It was not until 1750, when Thomas Walker discovered a relatively low, clear passage through the mountains, that Kentucky was settled.

This passage came to be called Cumberland Gap (altitude 495 m [1650 ft]), and it ranks among the most significant mountain passes in American history (Fig. 53-3). Through it passed Abraham Lincoln's grandparents and Daniel Boone's Wilderness Trail to central Kentucky's fertile Bluegrass Basin. Indeed,

FIGURE 53-3 Cumberland Gap, one of the most important mountain pass routes in the historical geography of the United States. This passage enabled the beginnings of western settlement to proceed in mid-eighteenth-century Kentucky.

the Gap was so valuable in controlling the surrounding region that it was a strenuously contested site in the Civil War, changing hands between the armies of the North and South no fewer than three times. Today, Cumberland Gap Historical Park, located at the point where the three states of Virginia, Tennessee, and Kentucky meet, commemorates the pioneers who opened this vital western route.

---

the rivers have cut valleys over 450 m (1500 ft) deep and because summit elevations reach 1200 m (4000 ft) in such areas as central West Virginia. In fact, during colonial times these mountains were obstacles to the early settlement of the trans-Appalachian West (*see* Perspective: Cumberland Gap).

Despite forming a part of the Appalachian Highlands realm, the eastern interior boundary of the Appalachian Plateau is quite well defined by major escarpments—the Allegheny Front in the north (Fig. 53-2) and the Cumberland Escarpment in the south. It is the realm's western boundary facing the Interior Lowlands that is in places indistinct, because the Appalachian Plateau loses elevation and prominence toward the west.

"Newer" Appalachians

Compared to the Appalachian Plateau's irregular terrain, the **"Newer" Appalachians**' *ridge-and-valley* topog-

raphy forms a sharp enough contrast to delineate a discrete province. Parallel vegetation-clad ridges reach remarkably even summits between 900 and 1200 m (3000

FIGURE 53-4 The striking evenness of the summit level across the Tuscarora Mountains, a section of the ridge-and-valley terrain in central Pennsylvania (*see* Fig. 36-15 for a satellite view).

and 4000 ft) above sea level (Fig. 53-4). (A satellite image of the central Pennsylvania segment of this extraordinary topography can be seen on p. 384.) The eastern edge of this physiographic province is marked by a wider lowland corridor called the Great Valley of the Appalachians (Fig. 53-2). The Great Valley, containing fertile limestone- and slate-derived soils, has been tilled by the famous Pennsylvania Dutch farmers since before the American Revolution. It has also served as an important routeway: at the height of the Civil War, the Confederate Army attempted to encircle Washington, D.C. by moving up the Great Valley, but was repulsed in 1863 at the climactic Battle of Gettysburg as it fought to round the northern end of the Blue Ridge Mountains.

Blue Ridge Section

East of the Great Valley, extending from northern Georgia to Gettysburg, Pennsylvania, lies the **Blue Ridge Mountains** section of the Appalachians (Fig. 53-2). This province is underlain by crystalline rocks, mostly highly metamorphosed and ancient. In this respect, the Blue Ridge differs from the Appalachian Plateau and the "Newer" Appalachians in that the latter two are sustained by sedimentary rocks. The Blue Ridge section is indeed the "Older" Appalachians by virtue of the age of the rocks and the successive periods of mountain-building. In the mountainous Blue Ridge, where the Great Smoky Mountains dominate the province's southern end, the Appalachian Highlands reaches its highest elevations in western North Carolina (*see* Fig. 52-8).

The Piedmont

Toward the east, the mountainous topography of the Blue Ridge section yields to the much lower relief of the **Piedmont** (Fig. 53-2). This province—whose name is derived from the French term for foothills—is still an upland compared to the neighboring Gulf-Atlantic Coastal Plain beyond the Fall Line (*see* Fig. 52-12), but not nearly so rugged as the interior Appalachian ranges. In the Blue Ridge section, the Appalachians can rise to over 2000 m (6500 ft), but in the Piedmont the maximum elevation is about 900 m (3000 ft) along the region's inner boundary, declining from there toward the contact with the Coastal Plain. The underlying rocks of the Piedmont, however, are the same as those of the Blue Ridge: ancient, highly altered crystallines. Thus the difference is principally a matter of topography and relief. Were it not for the Fall Line, the Piedmont would merge almost imperceptibly into the Coastal Plain along its outer boundary (Fig. 53-2). The Piedmont has more relief than the Coastal Plain, with isolated higher hills rising above the general level of the countryside.

FIGURE 53-5 New Hampshire's Mount Washington crowns the Presidential Range of the White Mountains, at 1917 m (6288 ft) the highest elevation in the New England/Maritime province. The weather station atop the peak is kept busy tracking ferocious atmospheric conditions in every season; among its many records is the highest wind gust ever measured at the earth's surface (372 kph [231 mph]).

New England/Maritime Province

Figure 53-1 shows a northeastern region within the Appalachian Highlands realm—the **New England/Maritime province**. As the name of this physiographic province implies, this is a completely different environment. It is strongly affected by glaciation, dotted with lakes, penetrated by the waters of the Atlantic Ocean, and bounded by a higher-relief coastline. Major topographic features include Vermont's Green Mountains and New Hampshire's White Mountains (Fig. 53-5); like the Blue Ridge, they are made of ancient crystalline rocks. Elevations exceed 1800 m (6000 ft) in the mountains, but the overall topography has none of the regularity and parallelism that marks the ridge-and-valley terrain of the "Newer" Appalachians region. In fact, the New England/Maritime province is extremely varied and in places quite rugged. Moraines left by the glaciers help to give this region its special identity. If you were to fly over it, you would also be impressed by the considerable extent of the forest cover that still prevails compared to that of the central and southern portions of the Appalachian realm.

THE WESTERN MOUNTAINS

As shown in Fig. 52-2, the Western Mountains include all the highlands of western North America. In the conterminous United States, this incorporates the Rocky Mountains, the Pacific Mountains and Valleys, and the varied Intermontane Basin-and-Plateau country lying in be-

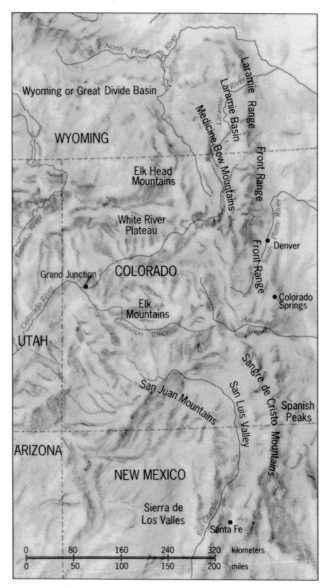

FIGURE 53-6 The topography of the Southern Rockies.

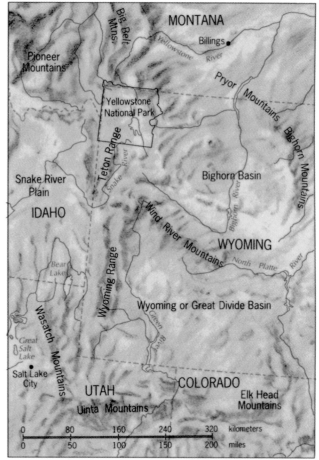

FIGURE 53-7 The topography of the Middle Rockies.

tween (*see* Fig. 52-5). The Rocky Mountains, as the map in Fig. 53-1 shows, are themselves divided into three major regions: the Southern, Middle, and Northern Rockies. The mountain ranges of the Pacific coast are, for the most part, longitudinally separated by lengthy valleys. And the Intermontane ("between the mountains") region consists of the Columbia Plateau, the Colorado Plateau, and the so-called Basin-and-Range province.

Southern Rocky Mountains

The **Southern Rocky Mountains** province lies across the heart of Colorado, as Fig. 53-6 indicates. The mountains constituting the eastern boundary rise sharply out of the adjacent Great Plains along a range called the Front Range in Colorado (*see* Unit 52 opening photo) but other names elsewhere. The Front Range and similar ranges in the Southern Rocky Mountains consist mainly of crystal-

line rocks with upturned sedimentary layers along their sides. The crystalline rocks near the Cripple Creek area of the Front Range (about 125 km [75 mi] southwest of Denver) yielded half a billion dollars of gold ore from 1890 until 1962, when the last mine closed. Elevations in many places exceed 4200 m (14,000 ft), and even the valleys between the ranges (here called *parks*) lie at elevations of 2100 m (7000 ft) or more. Some large rivers rise in this province, including the Rio Grande and the Colorado River, but it is the glacial period that gave this region its distinctive topography through scouring and deposition (*see* Fig. 45-1).

Middle Rocky Mountains

The continuity of the Rocky Mountains northward is broken by the Wyoming Basin (also called the Great Divide Basin), which on certain small-scale physiographic maps looks like a westward extension of the Great Plains. This basin constitutes a major east-west routeway, but it also signals a change in the character of the Rocky Mountains. The **Middle Rocky Mountains** province (Fig. 53-7) consists of ranges aligned in various directions, wide open valleys and basins, and a less congested topography than exists in the Southern Rockies.

Within the Middle Rockies, the Uinta Mountains of northeastern Utah lie east-west, but the next range to the west, the Wasatch Mountains, runs north-south. In the northern sector of the province, Wyoming's Bighorn Basin is almost encircled by a series of ranges. Most of these ranges are anticlinal uplifts similar to those of the Southern Rockies, but the Tetons to the west are a fault block. Folded sedimentary strata play a role in the northern part of the province; elsewhere, lava flows create some expanses of low relief, as in Yellowstone National Park, which occupies the northwestern corner of Wyoming.

Northern Rocky Mountains

Northwest of Yellowstone Park, the Rocky Mountain landscape again changes considerably. Gone are the persistent, conspicuous, high-crested ranges that provide such spectacular scenery in the Middle and Southern Rockies; and lost also are the high-elevation parks, basins, and valleys that separate the ranges. The **Northern Rocky Mountains** exhibit a confused topography, generally lower and sustained by an even more complex set of structures than exists in the Rockies to the south.

The boundary of this region can be seen quite clearly in Fig. 53-1. To the east lies the Great Plains province; to the south the Snake River Plain of southern Idaho forms a sharp topographic limit; and to the west the plain of the Columbia River borders the Northern Rockies. (The Snake and Columbia River plains are actually plateaus that abut the mountains.) Within this boundary, the Northern Rocky Mountains are cut from large batholiths, folded sedimentary layers, and extrusives (Fig. 53-8). Large-scale faulting and severe erosion, as well as glaciation, have combined to make this one of the country's most complex landscape regions.

Columbia Plateau

The **Columbia Plateau** region lies wedged between the Northern Rocky Mountains to the east and the Cascade Range of Washington and Oregon to the west. The Columbia Plateau itself is one of the largest lava surfaces in the world (*see* Fig. 34-3), and we call it a plateau only because it lies as high as 1800 m (6000 ft) above sea level. This intermontane province offers few obstructions to travel; the relief is rather low and the countryside rolls, except where volcanic and other mountains rise above the flood of lava that engulfed them. The southern boundary of the region is determined by the limits of this basaltic lava flow, and in many places it is not clear in the topography at all.

There is not much doubt that this region differs fundamentally from the mountainous provinces that flank it. The two major rivers that drain the Plateau, the Columbia and the Snake, form deep canyons in the lava. The Snake River Canyon is in the vicinity of the Seven Devils Moun-

FIGURE 53-8 In places the Northern Rockies evince some of the most rugged topography on this continent. This is the all but inaccessible heart of the Bitterroot Wilderness along the Idaho-Montana border.

tains in southwestern Idaho, and is well over a kilometer deep. Alluvial plains and loess accumulations (*see* Fig. 49-10) date from the glacial period, when the rivers were blocked and when the winds piled dust in deep layers.

Colorado Plateau

The **Colorado Plateau**, by contrast, is underlain chiefly by flat-lying or nearly horizontal sedimentary strata, weathered in this semiarid environment into vivid colors (*colorado* is a Spanish word for red) and carved by erosion into uniquely spectacular landscapes. The eastern boundary, of course, is shared with the Southern Rocky Mountains province (Fig. 53-1), a very different topography. On the western side, a lengthy fault scarp bounds the Colorado Plateau and separates it from the Basin-and-Range province (which also borders the region on the south). That southern boundary, by the way, is not marked by linear landscape features or by strong topographic contrasts; much of it is traditionally taken as the divide between the Plateau's major river, the Colorado, and the drainage basin of the Rio Grande.

The Colorado Plateau is traversed by several sets of steep-walled canyons with multicolored layers of sedimentary rocks. The Grand Canyon of northern Arizona, shown in Fig. 31-8 (p. 333), is of course the most famous. The dryness of this physiographic region keeps weathering processes to a minimum, preserving the valley walls and enhancing the coloration of the exposed strata (Fig. 53-9). Such rivers as the Colorado, San Juan, and Little Colorado have cut down vigorously into the plateau surface (*see* Fig. 43-11), which lies as high as 3500 m (11,600 ft) above sea level. The flat upper surface is broken by fault scarps, and erosion has created numerous mesas, buttes, badlands, and much more. No brief description could adequately summarize the distinct character of the Colorado Plateau, which is unlike any other physiographic province in North America.

Basin-and-Range Province

South of the Columbia Plateau and west and south of the Colorado Plateau lies the **Basin-and-Range province** (Fig. 53-1). Actually, this is not just one basin but an entire region of basins, many of them internally drained and not connected to other surface depressions by permanent streams. A large number of linear mountain ranges rise above these basins, and are generally from 80 to 130 km (50 to 80 mi) long and from 8 to 24 km (5 to 15 mi) wide. They rise 600 to 1525 m (2000 to 5000 ft) above the basin floors, reaching heights of 2100 to 3000 m (7000 to 10,000 ft) above sea level. These mountains are mainly oriented north-south and have a steep side and a more gently sloping side, which suggests that they are structurally related (*see* Fig. 42-4). Faulting is the most likely explanation, because the Basin-and-Range province looks like a sea of fault blocks thrust upward along parallel axes (Fig. 53-10).

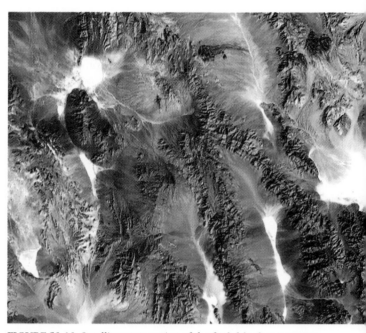

FIGURE 53-10 Satellite perspective of the fault-block mountains in east-central Nevada, also an exquisite example of alternating basin-and-range terrain.

Aridity is also a hallmark of this physiographic province. The basins are mostly dry, streams are intermittent, vegetation is sparse, and wind action plays a major role in sculpting the landscape. This was not always so, because old shorelines of now-extinct lakes have been found along the margins of some of the basins in the region. The most famous of all the existing bodies of water in this province is undoubtedly the Great Salt Lake in northern Utah.

This lake's volume has varied enormously in recorded times, but its high salt content suggests that it is a remnant of a much larger lake that has evaporated and shrunk to its present size (*see* Fig. 46-10)—a story of contraction and withering shared by all this region's bodies of water. Late in the nineteenth century, the Great Salt Lake's area was about 5000 km² (1900 mi²), but early in the twentieth century, during a drought, it almost disappeared. In recent years it has once again expanded (to about 6000 km² [2300 mi²]), flooding a lakeside resort and portions of a major interstate highway; but another serious drought could quickly reverse this trend.

Pacific Mountains and Valleys

West of the Columbia Plateau and the Basin-and-Range province lies the very complex region of **Pacific Mountains and Valleys** (*see* Fig. 52-5). The dominant orientation here is approximately north-south, parallel to the West Coast. This is the western edge of the North American Plate, and the orogenies are related to its westward movement.

In the north and south of this physiographic province, two elongated valleys separate lower coastal moun-

FIGURE 53-9 Brilliant colors mark the stunning sedimentary rock formations in Bryce Canyon National Park in southern Utah.

tains from higher interior mountains (Fig. 53-1). In the north, between the Cascades to the east and the Coast Ranges to the west, the Willamette-Puget Sound Lowland extends northward from south of Portland, Oregon to north of Seattle, Washington, where it becomes filled with ocean water. In the south, the Central Valley of California (containing the Sacramento and San Joaquin Rivers) separates the Sierra Nevada to the east from the Coast Ranges to the west (Fig. 53-11).

Although this region is quite clearly demarcated in the landscape, rising from the intermontane topography to the east and bounded by the Pacific Ocean on the west, it is a province with many different landscape subregions that are easy to identify. Unmistakably, the two great valleys just identified are such subregions, but even the mountainous areas vary in terms of age, origin, relief, and natural environment. The Sierra Nevada, for instance, consists mainly of great batholiths of granitic rocks uplifted by orogeny and its associated faulting. California's Coast Ranges, on the other hand, consist mostly of sedimentary strata caught in the deformations associated with the tectonically active contact zone where the North American and Pacific Plates adjoin. Thus a network of faults and related movements have crushed and tilted these rocks into the linear topography seen today. To the north, the Cascade Mountains are studded with great volcanoes (*see* Fig. 34-14); and west of Puget Sound lie the Olympic Mountains (Fig. 53-12), anchoring the northern

FIGURE 53-12 A panoramic view in the heart of Washington State's beautiful Olympic Mountains, which constitute the northernmost salient of the Coast Ranges that line the U.S. Pacific coast from Mexico to Canada.

end of the Coast Ranges, which line the entire length of the Pacific shore of the conterminous United States.

This westernmost U.S. physiographic province, with its considerable latitudinal extent, is also subdivided according to climatic and environmental conditions. As Fig. 16-3 reminds us, the north lies under marine west coast climatic conditions, with its associated forest vegetation. Southward, this gives way to a Mediterranean regime, and the natural vegetation becomes the characteristic chaparral. Still farther to the south, in the general vicinity of San Diego near the Mexican border, the Mediterranean environment in turn yields to the desert conditions of Baja California. Finally, it should also be noted that moisture declines sharply toward the east as the orographic (precipitation) effect produces the rain shadow that covers so much of the U.S. West (*see* Fig. 13-8).

THE PHYSIOGRAPHIC IMPRINT

These, then, are the prominent physiographic realms and regions of North America in general (Unit 52) and the United States in particular (Unit 53). What we see in the landscape today is the result of all the building forces and erosional systems we have discussed in this book. Geologic forces produced the structures in the upper layer of the crust that exert so much control over the formation of landscape. Crustal deformation from plate movement continues to modify what exists—slowly in most places, but quite rapidly in certain active zones. Solar radiation powers atmospheric circulation, and the hydrologic cycle rides on the movement of air, delivering water to the interiors of the continents and feeding drainage networks large and small. Water on the surface of

FIGURE 53-11 California's fertile Central Valley, one of the most productive agricultural regions in the United States. This is the Sacramento River (northern) portion of the Central Valley, looking east toward the mountain wall of the Sierra Nevada.

the earth comes under the force of gravity, and is subject to various laws of physics as it carries on its downward cutting and other erosional work. Weathering processes and mass movements contribute to the sculpting of the landscape, while deposition covers lower-lying areas. Soils develop and mature; vegetation anchors them.

The totality is summarized in one view of the landscape, which reflects and reveals the work of forces from below and above. The physiographic map records the total imprint made by all these environmental processes, and helps us understand the spatial variation of landscapes and landforms.

KEY TERMS

Appalachian Plateau (p. 539)

Basin-and-Range province (p. 544)

Blue Ridge Mountains (p. 541)

Colorado Plateau (p. 543)

Columbia Plateau (p. 543)

Middle Rocky Mountains (p. 542)

New England/Maritime province (p. 541)

"Newer" Appalachians (p. 540)

Northern Rocky Mountains (p. 543)

Pacific Mountains and Valleys (p. 544)

Piedmont (p. 541)

Southern Rocky Mountains (p. 542)

REVIEW QUESTIONS

1. What and where is the Fall Line?

2. How do the Northern Rocky Mountains differ from the Southern Rocky Mountains?

3. How does the Columbia Plateau differ from the Colorado Plateau?

4. In which U.S. physiographic region are volcanoes found?

REFERENCES AND FURTHER READINGS

ATWOOD, W. W. *The Physiographic Provinces of North America* (New York: Ginn/Blaisdell, 1940).

CONEY, P., and BECK, M. "The Growth of Western North America," *Scientific American*, November 1982, 70–84.

CURRAN, H. A., et al. *Atlas of Landforms* (New York: Wiley, 3rd ed., 1984).

FENNEMAN, N. M. *Physiography of the Eastern United States* (New York: McGraw-Hill, 1938).

FENNEMAN, N. M. *Physiography of the Western United States* (New York: McGraw-Hill, 1931).

GRAF, W. L., ed. *Geomorphic Systems of North America* (Boulder, Colo.: Geological Society of America, 1987).

HUNT, C. B. *Natural Regions of the United States and Canada* (San Francisco: Freeman, 2nd ed., 1974).

PIRKLE, E. C., and YOHO, W. H. *Natural Landscape of the United States* (Dubuque, Iowa: Kendall-Hunt, 3rd ed., 1982).

THORNBURY, W. D. *Regional Geomorphology of the United States* (New York: Wiley, 1965).

APPENDIX A

Units and Their Conversions

Appendix A provides a table of units and their conversion from older units to Standard International (SI) units.

LENGTH

Metric Measure

1 kilometer (km)	= 1000 meters (m)
1 meter (m)	= 100 centimeters (cm)
1 centimeter (cm)	= 10 millimeters (mm)

Nonmetric Measure

1 mile (mi)	= 5,280 feet (ft)
	= 1,760 yards (yd)
1 yard (yd)	= 3 feet (ft)
1 foot (ft)	= 12 inches (in)
1 fathom (fath)	= 6 feet (ft)

Conversions

1 kilometer (km)	= 0.6214 mile (mi)
1 meter (m)	= 3.281 feet (ft)
	= 1.094 yard (yd)
1 centimeter (cm)	= 0.3937 inch (in)
1 millimeter (mm)	= 0.0394 inch (in)
1 mile (mi)	= 1.609 kilometers (km)
1 foot (ft)	= 0.3048 meter (m)
1 inch (in)	= 2.54 centimeters (cm)
	= 25.4 millimeters (mm)

AREA

Metric Measure

1 square kilometer (km²)	= 1,000,000 square meters (m²)
	= 100 hectares (ha)
1 square meter (m²)	= 10,000 square centimeters (cm²)
1 hectare (ha)	= 10,000 square meters (m²)

Nonmetric Measure

1 square mile (mi²)	= 640 acres (ac)
1 acre (ac)	= 4840 square yards (yd²)
1 square foot (ft²)	= 144 square inches (in²)

Conversions

1 square kilometer (km²)	= 0.386 square mile (mi²)
1 hectare (ha)	= 2.471 acres (ac)
1 square meter (m²)	= 10.764 square feet (ft²)
	= 1.196 square yards (yd²)
1 square centimeter (cm²)	= 0.155 square inch (in²)
1 square mile (mi²)	= 2.59 square kilometers (km²)
1 acre (ac)	= 0.4047 hectare (ha)
1 square foot (ft²)	= 0.0929 square meter (m²)
1 square inch (in²)	= 6.4516 square centimeters (cm²)

VOLUME

Metric Measure

1 cubic meter (m³)	= 1,000,000 cubic centimeters (cm³)
1 liter (l)	= 1000 milliliters (ml)
	= 0.001 cubic meter (m³)
1 milliliter (ml)	= 1 cubic centimeter (cm³)

Nonmetric Measure

1 cubic foot (ft³)	= 1728 cubic inches (in³)
1 cubic yard (yd³)	= 27 cubic feet (ft³)

Conversions

1 cubic meter (m³)	= 264.2 gallons (U.S.) (gal)
	= 35.314 cubic feet (ft³)
1 liter (l)	= 1.057 quarts (U.S.) (qt)
	= 33.815 ounces (U.S. fluid) (fl. oz)
1 cubic centimeter (cm³)	= 0.0610 cubic inch (in³)
1 cubic mile (mi³)	= 4.168 cubic kilometers (km³)
1 cubic foot (ft³)	= 0.0283 cubic meter (m³)
1 cubic inch (in³)	= 16.39 cubic centimeters (cm³)
1 gallon (gal)	= 3.784 liters (l)

MASS

Metric Measure

1000 kilograms (kg)	= 1 metric ton (m.t)
1 kilogram (kg)	= 1000 grams (g)

Nonmetric Measure

1 short ton (sh.t)	= 2000 pounds (lb)
1 long ton (l.t)	= 2240 pounds (lb)
1 pound (lb)	= 16 ounces (oz)

Conversions

1 metric ton (m.t)	= 2205 pounds (lb)
1 kilogram (kg)	= 2.205 pounds (lb)
1 gram (g)	= 0.03527 ounce (oz)
1 pound (lb)	= 0.4536 kilogram (kg)
1 ounce (oz)	= 28.35 grams (g)

PRESSURE

standard sea-level air   = 1013.25 millibars (mb)
  pressure   = 14.7 lb per in²

TEMPERATURE

To change from Fahrenheit (F) to Celsius (C)

$$°C = \frac{(°F - 32)}{1.8}$$

To change from Celsius (C) to Fahrenheit (F)

$$°F = (°C \times 1.8) + 32$$

ENERGY AND POWER

1 calorie (cal)   = the amount of heat that will raise the temperature of 1 g of water 1°C (1.8°F)

1 joule (J)   = 0.239 calories (cal)

1 watt (W)   = 1 joule per second (J/s)
= 14.34 calories per minute (cal/min)

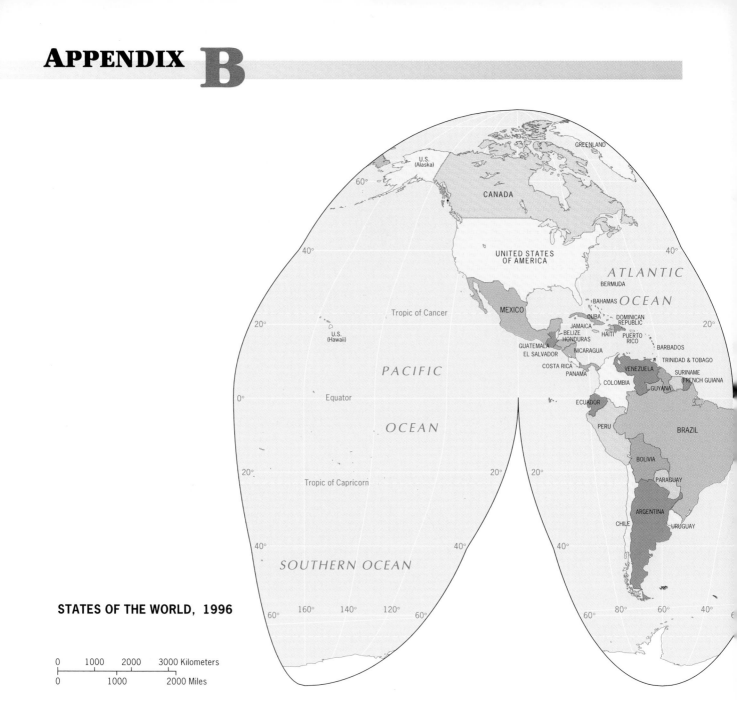

STATES OF THE WORLD, 1996

| 0 | 1000 | 2000 | 3000 Kilometers |
| 0 | | 1000 | 2000 Miles |

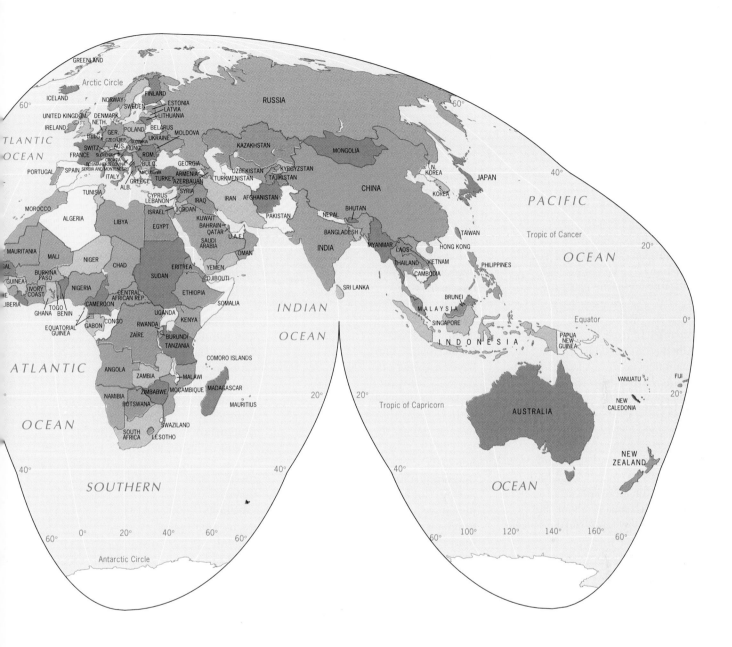

Pronunciation Guide

Aa (AH-AH)
Abilene (abba-LEEN)
Ablation (uh-BLAY-shunn)
Abyssal (uh-BISSLE)
Acacia (uh-KAY-shuh)
Aconcagua (ah-konn-KAH-gwah)
Adiabatic (addie-uh-BATTICK)
Adriatic (ay-dree-ATTIK)
Agassiz (AG-uh-see)
Akmola (AHK-moh-luh)
Akrotiri (ah-krah-TIRRI)
Albedo (al-BEE-doh)
Albuquerque (ALBA-ker-kee)
Aleutian (uh-LOOH-shun)
Algae (AL-jee)
Algeria (al-JEERY-uh)
Also Sprach Zarathustra (ull-ZOH shprahch zahra-TOOSE-strah)
Ameliorate (uh-MEEL-yer-ate)
Amensalism (uh-MEN-sull-ism)
Amino (uh-MEE-noh)
Anacostia (anna-KOSS-tee-uh)
Anemometer (anna-MOM-muh-der)
Anion (AN-eye-on)
Antarctica (ant-ARK-tick-uh)
Antimony (ANN-tuh-moh-nee)
Antipodally (an-TIPPUD-lee)
Aphelion (ap-HEELY-un)
Appalachia (appa-LAY-chee-uh)
Aqaba (AH-kuh-buh)
Aquiclude (AK-kwuh-cloode)
Aquifer (AK-kwuh-fer)
Archipelago (ark-uh-PELL-uh-goh)
Arequipa (ah-ruh-KEE-puh)
Arête (uh-RETT)
Aridisol (uh-RID-ih-sol)
Armero (ahr-MAIR-roh)
Artois (ahr-TWAH)
Asthenosphere (ass-THENNO-sfeer)
Aswan (as-SWAHN)
Atacama (ah-tah-KAH-mah)
Atoll (AT-tole)
Aurora borealis (aw-ROAR-ruh baw-ree-ALLIS)
Autotroph (AW-doh-troaf)
Azores (AY-zoars)
Azurite (AZH-zhuh-rite)

Bahamas (buh-HAH-muzz)
Bahía (buh-HEE-uh)
Baja (BAH-hah)
Bajada (buh-HAH-dah)
Bali (BAH-lee)
Bangladesh (bang-gluh-DESH)
Baobab (BAY-oh-bab)
Barchan (BAR-kan)
Barents (BARRENS)
Barnaul (bar-nuh-OOL)
Basal (BASE-ull)
Basalt (buh-SALT)
Bayeux (bye-YER ["r" silent])
Baykal (bye-KAHL)
Bedzin (bed-ZEEN)
Beijing (bay-ZHING)
Bellary (buh-LAHR-ree)
Bengal (BENG-gahl)
Benguela (ben-GWAY-luh)
Bergeron (BEAR-guh-roan)
Bering (BEH-ring)
Berkshire (BERK-sheer)
Bingham (BING-um)
Biome (BYE-ohm)
Bogor BOH-goar)
Boise (BOY-zee)
Bonneville (BON-uh-vill)
Borneo (BOAR-nee-oh)
Brahmaputra (brahm-uh-POOH-truh)
Breccia (BRETCH-ee-uh)
Buoy (BOO-ee)
Burkina Faso (ber-KEENA FASSO)
Butte (BYOOT)
Bylot (BYE-lot)

Cairo (KYE-roh)
Caldera (kal-DERRA)
Calve (KALV)
Calving (KAL-ving)
Camille (kuh-MEEL)
Capri (kuh-PREE)
Caribbean (kuh-RIB-ee-un/karra-BEE-un)
Caribou (KARRA-boo)
Castel Sant'Angelo (kuh-STELL sunt-UN-jeh-loh)
Catena (kuh-TEENA)

Cation (CAT-eye-on)
Caucasus (KAW-kuh-zuss)
Celebes (SELL-uh-beeze)
Celsius (SELL-see-us)
Cenozoic (senno-ZOH-ik)
Chang Jiang (chung-jee-AHNG)
Changnon (CHANG-nun)
Chaparral (SHAP-uh-RAL)
Chengdu (chung-DOO)
Cherrapunji (cherra-POON-jee)
Chesapeake (CHESSA-peek)
Chile (CHILLI/CHEE-lay)
Chimborazo (chim-buh-RAH-zoh)
Chinook (shin-NOOK)
Chornobyl (CHAIR-nuh-beel)
Chugach (CHEW-gash)
Chukotskiy (chuh-KAHT-skee)
Cineole (SIN-ee-ole)
Cirque (SERK)
Cirrus (SIRRUS)
Clayey (CLAY-ee)
Coachella (koh-CHELLA)
Cocos (KOH-kuss)
Conduit (KONN-doo-it)
Coniferous (kuh-NIFF-uh-russ)
Contentious (kon-TEN-shuss)
Coriolis, Gustave-Gaspard (kaw-ree-OH-liss, goo-STAHV gus-SPAH)
Corrasion (kaw-RAY-zhunn)
Costa Rica (koss-tuh REE-kuh)
Cretacious (kreh-TAY-shuss)
Crete (KREET)
Crevasse (kruh-VASS)
Croatia (kroh-AY-shuh)
Crystalline (KRISS-tuh-leen)
Cuesta (KWESTA)
Cumulonimbus (kyoo-myoo-loh-NIMBUS)
Cumulus (KYOO-myoo-luss)
Cvijic, Jovan (SVEE-itch, yoh-VAHN)

Dakar (duh-KAHR)
Davisian (duh-VISSY-un)
de Candolle, Alphonse (duh kawn-DOLE, ahl-FAWSS)
Debris (deh-BREE)
Deciduous (deh-SID-yoo-uss)

552

Deltaic (del-TAY-ik)

Denali (deh-NAH-lee)

Deoxyribonucleic (day-oxie-RYBO-noo-CLAY-ik)

Desertification (deh-ZERT-iff-uh-KAY-shun)

Diastrophism (dye-ASTRO-fizzum)

Diurnal (dye-ERR-nul)

Dokuchayev, Vasily (dock-koo-CHYE-eff, VAH-zilly)

Drakensberg (DRAHK-unz-berg)

du Toit (doo-TWAH)

Echidna (eh-KID-nuh)

Ecliptic (ee-KLIP-tick)

Ecuador (ECK-wah-dor)

Ecumene (ECK-yoo-mean)

Edaphic (ee-DAFFIK)

Eemian (EE-mee-un)

El Niño (el-NEEN-yoh)

Ellesmere (ELZ-meer)

Ely (EELY)

Emu (EE-myoo)

Eolian (ee-OH-lee-un)

Epeirogeny (eh-pye-ROJ-jenny)

Ephemeral (ee-FEMMA-rull)

Epiphyte (EPPY-fite)

Epoch (EH-pok)

Equinox (EE-kwuh-nox)

Eratosthenes (eh-ruh-TOS-thuh-neeze)

Estuary (ESS-tyoo-erry)

Eucalyptus (yoo-kuh-LIP-tuss)

Euphorbia (yoo-FOR-bee-uh)

Fahrenheit (FARREN-hite)

Falkland (FAWK-lund)

Felsenmeer (FEL-zen-mear)

Fenneman (FEN-uh-mun)

Fiji (FEE-jee)

Filchner-Ronne (FILK-ner ROH-nuh)

Findeisen (FIN-dyzen)

Firn (FERN)

Fissure (FISHER)

Fjord (FYORD)

Foehn (FERN ["r" silent])

Franz Josef (frunss YOH-zeff)

Fuji (FOO-jee)

Fungi (FUN-jye)

Galapagos (guh-LAH-pah-guss)

Galena (guh-LEENA)

Ganges (GAN-jeeze)

Gaseous (GASH-uss)

Geiger (GHYE-guh)

Geodesy (jee-ODD-uh-see)

Geostrophic (jee-oh-STROFFIK)

Geosynchronous (jee-oh-SIN-krunn-uss)

Geyser (GUY-zer)

Gibraltar (jih-BRAWL-tuh)

Gila (HEE-luh)

Gneiss (NICE)

Gnomonic (no-MONIC)

Gondwana (gond-WAHNA)

Goudie (GOWDY)

Graben (GRAH-ben)

Granule (GRAN-yule)

Grebe (GREEB)

Greenwich (GREN-itch)

Gros Piton (groh-peet-TAW)

Guatemala (gwut-uh-MAH-lah)

Guilin (gway-LIN)

Guinea (GHINNY)

Gulkana (gull-KAN-uh)

Gunz (GOONZ)

Guyana (guy-ANNA)

Guyot (GHEE-oh)

Gypsum (JIP-sum)

Gyre (JYER)

Halide (HAY-lyde)

Hatteras (HATTA-russ)

Heimaey (HAY-may)

Hematite (HEE-muh-tite)

Herbaceous (her-BAY-shuss)

Herodotus (heh-RODDA-tuss)

Hierarchical (hyer-ARK-ik-kull)

Hilo (HEE-loh)

Himalayas (him-AHL-yuzz/ himma-LAY-uzz)

Hindu Kush (HIN-doo KOOSH)

Holocene (HOLLO-seen)

Huang He (HWAHNG-HUH)

Huascarán (wahss-kuh-RAHN)

Hulwan (hil-WAHN)

Humus (HYOO-muss)

Hygroscopic (hye-gruh-SKOPPIK)

Icarus (ICK-uh-russ)

Igneous (IGG-nee-us)

Indigenous (in-DIDGE-uh-nuss)

Indonesia (indo-NEE-zhuh)

In situ (in-SYE-too)

Inuit (IN-yoo-it)

Ion (EYE-on)

Iquitos (ih-KEE-tohss)

Irrawaddy (ih-ruh-WODDY)

Isarithmic (eye-suh-RITH-mik)

Ischia (ISS-kee-uh)

Isohyet (EYE-so-hyatt)

Isostasy (eye-SOSS-tuh-see)

Iteshi (ih-TESHI)

Izu (EE-zoo)

Java (JAH-vuh)

Juan de Fuca (WAHN duh FYOO-kuh)

Kafue (kuh-FOO-ee)

Kalimantan (kalla-MAN-tan)

Kamchatka (komm-CHUT-kuh)

Kariba (kuh-REEBA)

Kaskawulsh (KASS-kuh-WULSH)

Katabatic (kat-uh-BATTIK)

Kauai (KOW-eye)

Kazakhstan (KUZZ-uck-stahn)

Kenya (KEN-yuh)

Khamsin (KAM-sin)

Khasi (KAH-see)

Kibo (KEE-boh)

Kilauea (kill-uh-WAY-uh)

Kilimanjaro (kil-uh-mun-JAH-roh)

Klamath (KLAM-uth)

Koala (koh-AH-luh)

Kobe (KOH-bay)

Kodiak (KOH-dee-ak)

Köppen, Wladimir (KER-pin ["r" silent], VLAH-duh-meer)

Kosciusko (kuh-SHOO-skoh)

Krakatau (krak-uh-TAU)

Küchler (KOO-kler)

Kurashio (koora-SHEE-oh)

Kuznetsk (kooz-NETSK)

Kyushu (kee-YOO-shoo)

La Niña (lah-NEEN-yah)

Laccolith (LACK-oh-lith)

Lahar (luh-HAHR)

Lancashire (LANKA-sheer)

Langarone (lahn-gah-ROH-neh)

Lanier (luh-NEAR)

Laredo (luh-RAY-doh)

Larvae (LAR-vee)

Lascaux (lass-SKOH)

Laurasia (law-RAY-zhuh)

Laurentian (law-REN-shun)

Lead [ice-surface channel] (LEED)

Leif Eriksson (LEEF ERIC-sun)

Lemur (LEE-mer)

Levee (LEH-vee)

Li (LEE)

Liana (lee-AHNA)

Lichen (LYE-ken)

Linnaean (LINNY-un)

Littoral (LITT-uh-rull)

Loess (LERSS)

Loihi (loh-EE-hee)

Loma Prieta (loh-muh-pree-AY-tuh)

Maasai (muh-SYE)

Machu Picchu (mah-CHOO PEEK-choo)

Madagascar (madda-GAS-kuh)
Mafic (MAFFIK)
Magellan (muh-JELLUN)
Malachite (MALLA-kite)
Malawi (muh-LAH-wee)
Mali (MAH-lee)
Manaós (muh-NAUSS)
Maori (MAU-ree/MAH-aw-ree)
Maquis (mah-KEE)
Marsupial (mar-SOOPY-ull)
Martinique (mahr-tih-NEEK)
Massif (mass-SEEF)
Mauna Kea (MAU-nuh KAY-uh)
Mauna Loa (MAU-nuh LOH-uh)
Mauritania (maw-ruh-TAY-nee-uh)
Maya (MYE-uh)
Mawsim (MAW-zim)
Meander (mee-AN-der)
Medieval (meddy-EE-vull)
Melbourne (MEL-bun)
Mercalli (mair-KAHL-lee)
Mercator, Gerhardus (mer-CATER, ghair-HAR-duss)
Mesa (MAY-suh)
Mesozoic (meh-zoh-ZOH-ik)
Mica (MYE-kuh)
Michoacán (mitcho-ah-KAHN)
Mindel (MINNDLE)
Minot (MYE-not)
Mirnyy (MEAR-nyee)
Mistral (miss-STRAHL)
Miyazaki (mee-YAH-zah-kee)
Moho (MOH-hoh)
Mohorovičić, Andrija (moh-hoh-ROH-vih-chik, ahn-DREE-uh)
Mohs, Friedrich (MOZE, FREED-rick)
Mojave (moh-HAH-vee)
Monadnock (muh-NAD-nok)
Mont Blanc (mawn-BLAHNK)
Montreal (mun-tree-AWL)
Moraine (more-RAIN)
Moray Firth (MAW-ray FERTH)
Morisawa (mawry-SAH-wah)
Muehrcke (MERR-kee)
Myanmar (mee-ahn-MAH)

Namib (nah-MEEB)
Namibia (nuh-MIBBY-uh)
Naples (NAYPLES)
Nazca (NAHSS-kuh)
Neap (NEEP)
Nefudh (neh-FOOD)
Negev (NEGGEV)
Nepal (nuh-PAHL)
Nevado del Ruiz (neh-VAH-doh del roo-EESE)
New Guinea (noo-GHINNY)

New Madrid (noo-MAD-rid)
Ngorongoro (eng-gore-ong-GORE-oh)
Niche (NITCH)
Niger [Country] (nee-ZHAIR)
Niger [River] (NYE-jer)
Nigeria (nye-JEERY-uh)
Nkhata (eng-KAH-tah)
Nouakchott (noo-AHK-shaht)
Novaya Zemlya (NOH-vuh-yuh zem-lee-AH)
Nuée ardente (noo-AY ahr-DAHNT)
Nunatak (NOON-uh-tak)

Oahu (uh-WAH-hoo)
Oblate spheroid (OH-blate SFEER-roid)
Obsidian (ob-SIDDY-un)
Okhotsk (oh-KAHTSK)
Orinoco (aw-rih-NOH-koh)
Orogenic (aw-ruh-JENNIK)
Orogeny (aw-RODGE-uh-nee)
Orographic (aw-roh-GRAFFIK)
Oscillation (oss-uh-LAY-shunn)
Osterbygd (AR-star-bade)
Ouachita (wah-CHEE-tuh)
Oxisol (OCK-see-sol)
Ouimet (WEE-met)
Oyashio (oy-yuh-SHEE-oh)

Pahoehoe (pah-HOH-hoh)
Paleontology (pay-lee-un-TOLLO-jee)
Paleozoic (pay-lee-oh-ZOH-ik)
Pamlico (PAM-lih-koh)
Pampas (PAHM-pahss)
Pangaea (pan-GAY-uh)
Paricutín (pah-ree-koo-TEEN)
Patagonia (patta-GOH-nee-ah)
Pedon (PED-on)
Pele (PAY-lay)
Pelée (peh-LAY)
Penck, Walther (PENK, VULL-tuh)
Peneplain (PEEN-uh-plane)
Periglacial (PERRY-GLAY-shull)
Petit Piton (peh-tee-peet-TAW)
Philippines (FILL-uh-peenz)
Phreatic (free-ATTIK)
Physiography (fizzy-OGG-ruh-fee)
Phytogeography (FYE-toh-jee-OGGRA-fee)
Piedmont (PEAD-mont)
Pierre (PEER)
Pinatubo (pin-uh-TOO-boh)
Planar (PLANE-ahr)
Planetesimal (planna-TEZZ-mull)
Platypus (PLATT-uh-puss)
Pleistocene (PLY-stoh-seen)
Pliny (PLY-nee)

Pliocene (PLY-oh-seen)
Podocarp (POD-oh-carp)
Podsol (POD-zol)
Polder (POHL-der)
Pompeii (pom-PAY)
Pontchartrain (PON-shar-train)
Popocatépetl (poh-puh-kah-TAY-petal)
Primeval (pry-MEE-vull)
Psychrometer (sye-KROM-muh-der)
Puget (PYOO-jet)
Pyrenees (PEER-unease)

Quartic (KWOAR-tik)
Quartzite (KWARTS-site)
Quasar (KWAY-zahr)
Quaternary (kwah-TER-nuh-ree)
Qinghai-Xizang (ching-HYE sheedz-AHNG)
Quito (KEE-toh)

Rainier (ruh-NEER)
Raisz (ROYCE)
Rawinsonde (RAW-in-sond)
Reykjavik (RAKE-yah-veek)
Rhine (RYNE)
Rhône (ROAN)
Rhumb (RUM)
Rhyolite (RYE-oh-lyte)
Richter (RICK-tuh)
Rio de Janeiro (REE-oh day zhah-NAIR-roh)
Rio de la Plata (REE-oh day lah PLAH-tah)
Roche Moutonnée (ROSH moot-tonn-NAY)
Rondônia (roh-DOAN-yuh)
Rotorua (roh-tuh-ROO-uh)
Ruhr (ROOR)
Ruwenzori (roo-wen-ZOARY)

Sahel (suh-HELL)
St. Lucia (saint LOO-shuh)
St. Pierre (sah PYAIR)
Saltation (sawl-TAY-shunn)
Salvador (SULL-vuh-dor)
San Andreas (san-an-DRAY-us)
San Gorgonio (san-gore-GOH-nee-oh)
San Joaquin (san-wah-KEEN)
San Juan (sahn-HWAHN)
Sangamon (SANG-guh-mun)
Santorini (santo-REE-nee)
São Gabriel de Cachoeira (sau GAH-bree-ell day kah-choh-AY-rah)
Saragosa (sarra-GOH-suh)

Schist (SHIST)

Scotia (SKOH-shuh)

Seine (SENN)

Seismic (SIZE-mik)

Seismograph (SIZE-moh-graff)

Serengeti (serren-GETTY)

Sesquioxide (SESS-kwee-OXIDE)

Seychelles (say-SHELLZ)

Sial (SYE-al)

Siberia (sye-BEERY-uh)

Sicily (SIH-suh-lee)

Sierra Madre Oriental (see-ERRA mah-dray orry-en-TAHL)

Sierra Nevada (see-ERRA neh-VAH-dah)

Silicic (sih-LISS-ik)

Silurian (sih-LOORY-un)

Sima (SYE-muh)

Slovenia (sloh-VEE-nee-uh)

Soledad (SOLE-uh-dad)

Solifluction (sol-ih-FLUK-shun)

Solstice (SOL-stiss)

Solum (SOH-lum)

Somali (suh-MAH-lee)

Soufrière (soo-free-AIR)

Spatial (SPAY-shull)

Spheroidal (sfeer-ROY-dull)

Spodosol (SPODDA-sol)

Stalactite (stuh-LAK-tite)

Stalagmite (stuh-LAG-mite)

Steppe (STEP)

Stratigraphy (struh-TIG-gruh-fee)

Strauss, Richard (SHTRAUSS, RIK-art)

Striation (strye-AY-shunn)

Suffosion (suh-FOH-zhunn)

Sulawesi (soo-luh-WAY-see)

Sulfur (SULL-fer)

Sumatra (suh-MAH-truh)

Sumbawa (soom-BAH-wah)

Sumbay (SOOM-bye)

Surficial (ser-FISH-ull)

Surtsey (SERT-see)

Susquehanna (suss-kwuh-HANNA)

Svalbard (SFALL-bard)

Swakopmund (SFAHK-awp-munt)

Swaziland (SWAH-zee-land)

Synoptic (sih-NOP-tik)

Syria (SEARY-uh)

Taiga (TYE-guh)

Taklimakan (tahk-luh-muh-KAHN)

Talus (TAY-luss)

Tana (TAHN-nuh)

Tanganyika (tan-gan-YEE-kuh)

Tangshan (tung-SHAHN)

Tanzania (tan-zuh-NEE-uh)

Tapir (TAY-per)

Terrane (teh-RAIN)

Tertiary (TER-shuh-ree)

Teton (TEE-tonn)

Thalweg (THAWL-weg)

Thames (TEMZ)

Thera (THEERA)

Theroux (thuh-ROO)

Tianjin (tyahn-JEEN)

Tiber (TYE-ber)

Tibetan (tuh-BETTEN)

Tierra del Fuego (tee-ERRA dale FWAY-goh)

Titanium (tye-TANEY-um)

Toba (TOH-buh)

Tombolo (TOM-boh-loh)

Torricelli (toar-ruh-CHELLY)

Travertine (TRAVVER-teen)

Trivandrum (truh-VAN-drum)

Tselinograd (seh-LINN-uh-grahd)

Tsunami (tsoo-NAH-mee)

Tucson (TOO-sonn)

Tunguska (toon-GOOSE-kuh)

Turkana (ter-KANNA)

Turukhansk (too-roo-KAHNSK)

Tuscarora (tuska-ROAR-ruh)

Uinta (yoo-IN-tuh)

Ukraine (yoo-CRANE)

Uranus (yoo-RAY-nuss)

Ustalf (YOO-stalf)

Uvala (oo-VAH-luh)

Vaiont (vye-YAW)

Valdez (val-DEEZE)

Venezuela (veh-neh-SWAY-luh)

Verkhoyansk (vair-koy-YUNSK)

Vesuvius (veh-ZOO-vee-us)

Viña del Mar (VEEN-yah-del-MAHR)

Vis-à-vis (vizz-uh-VEE)

Viscous (VISS-kuss)

Viti Levu (vee-tee-LEH-voo)

Waialeale (wye-ahl-ay-AHL-ay)

Wasatch (WAH-satch)

Weddell (weh-DELL)

Wegener, Alfred (VAY-ghenner, AHL-fret)

Willamette (wuh-LAMMET)

Wrangell (RANG-gull)

Wrangellia (rang-GHELLIA)

Wüste (VISS-tuh)

Xerophyte (ZERO-fite)

Yakutat (YAK-uh-tat)

Yenisey (yen-uh-SAY)

Yosemite (yoh-SEM-uh-tee)

Yucatán (yoo-kuh-TAHN)

Yugoslavia (yoo-goh-SLAH-vee-uh)

Yungay (YOONG-GYE)

Yunnan (yoon-NAHN)

Zagros (ZAH-gross)

Zaïre (zah-EAR)

Zambezi (zam-BEEZY)

Zimbabwe (zim-BAHB-way)

Zoogeography (ZOH-OH-jee-oggra-fee)

Glossary of Terms

A horizon The upper soil layer, which is often darkened by organic matter.

Aa Angular, jagged, blocky-shaped lava formed from the hardening of not especially fluid lavas.

Ablation (*see* **zone of ablation**)

Abrasion (glacial) A glacial erosion process of scraping, produced by the impact of rock debris carried in the ice upon the bedrock surface below.

Abrasion (stream) The erosive action of boulders, pebbles, and smaller grains of sediment as they are carried along a river valley; these fragments dislodge other particles along the stream bed and banks, thereby enhancing the deepening and widening process.

Absolute base level The elevational level lying a few meters below sea level.

Absolute zero The lowest possible temperature, where all molecules cease their motion; occurs at $-273°C$ ($-459.4°F$) or $0°K$ (Kelvin).

Abyssal plain Large zone of relatively low-relief seafloor constituting one of the deepest areas of an ocean basin.

Acid precipitation Abnormally acidic rain, snow, or fog resulting from high levels of the oxides of sulfur and nitrogen that exist as industrial pollutants in the air.

Active dune An unstable sand dune in a desert that is continually being shaped and moved by the wind; cannot support vegetation.

Active layer The soil above the permafrost table that is subject to annual thawing and freezing.

Active volcano A volcano that has erupted in recorded human history.

Actual evapotranspiration (AE) The amount of water that can be lost to the atmosphere from a land surface with any particular soil-moisture conditions.

Addition The soil-layer formation process involving the gains made by the soil through the adding of organic matter from plant growth, or sometimes when loose surface material moves downslope and comes to rest on the soil; expressed as a dark-colored upper layer, whose appearance is attributable to that added organic matter.

Adiabatic With air being a poor conductor of heat, a parcel of air at one temperature that is surrounded by air at another temperature will neither gain nor lose heat energy over a short period of time; when such nontransfer of heat occurs, the process is called *adiabatic*.

Adiabatic lapse rate When a given mass of air is forced to expand, its temperature decreases. If a parcel of air rises to a higher altitude, it expands and cools adiabatically; its lapse rate is therefore referred to as an *adiabatic lapse rate*.

Advection The horizontal movement of material in the atmosphere.

Aerosols Tiny solid or liquid particles suspended in the atmosphere.

Aggradation The combination of processes that builds up the surface through the deposition of material that was removed from elsewhere by degradation; contributes to the lowering of relief by reducing the height differences between the high and low places in an area.

Air mass A very large parcel of air (more than 1600 km [1000 mi] across) in the boundary layer of the troposphere that possesses relatively uniform qualities of density, temperature, and humidity in the horizontal dimension; it is also bound together as an organized whole, a vital cohesion because air masses routinely migrate as distinct entities for hundreds of kilometers.

Albedo The proportion of incoming solar radiation that is reflected by a surface; the whiter the color of the surface (albedo derives from the Latin word *albus*, meaning white), the higher its albedo.

Aleutian Low The local name for the northernmost Pacific Ocean's Upper-Midlatitude Low, which is centered approximately at 60°N near the archipelago constituted by Alaska's Aleutian Islands.

Alfisol One of the 11 Soil Orders of the Soil Taxonomy, found in moister, less continental climate zones than the mollisols; characterized by high mineral content, moistness, and sizeable clay accumulation in the **B** horizon.

Allogenic succession Plant succession in which vegetation change is brought about by some external environmental factor, such as disease.

Alluvial apron (*see* **bajada**)

Alluvial fan A fan-shaped deposit consisting of alluvial material located where a mountain stream emerges onto a plain; primarily a desert landform.

Alluvial soil (*see* **alluvium**)

Alluvium Sediment laid down by a stream on its valley floor; deposition occurs when the stream's velocity decreases and the valley fills with a veneer of unconsolidated material. Since soil particles washed from slopes in the drainage basin form a large part of these deposits, alluvial soils are usually fertile and productive.

Alpine glacier (*see* **mountain glacier**)

Alpine permafrost On the world map of permafrost, patches of high-altitude permafrost at lower latitudes associated with major highland zones such as the Tibetan Plateau and Canadian Rockies.

Amensalism Biological interaction in which one species is inhibited by another.

Andisol The newest of the 11 Soil Orders of the Soil Taxonomy; established to include certain weakly developed, parent-material-controlled soils, notably those developed on volcanic ash that are very finely distributed throughout the Pacific Ring of Fire, Hawaii, and the world's other volcanic zones.

Angle of incidence (*see* **solar elevation**)

Angle of repose The maximum angle at which a material remains at rest; all slopes have an angle of repose—the more solid a slope's constituent material the closer that angle is to 90°.

Anion A negatively charged ion, such as those of carbonates, nitrates, and phosphates.

Annual temperature cycle The pattern of temperature change during the course of a year.

Annular drainage A concentric stream pattern that drains the interior of an excavated geologic dome (*see* Fig. 42-10b).

Antarctic Circle The latitude (66½°S) marking the northern boundary of the Southern Hemisphere portion of the earth's surface that receives a 24-hour period of sunlight at least once each year.

Antarctic Icesheet The continental glacier that covers almost all of Antarctica, enabling the study of what conditions were like elsewhere in the world during much of the Late Cenozoic Ice Age.

Antecedent stream A river exhibiting transverse drainage across a structural feature that would normally impede its flow because the river predates the structure and kept cutting downward as the structure was uplifted around it.

Anticline An arch-like upfold with the limbs dipping away from its axis (*see* Fig. 36-12, left).

Anticyclone An atmospheric high-pressure cell involving the divergence of air, which subsides at and flows spirally out of the center; the isobars around an anticyclone are generally circular in shape, with their values increasing toward the center. In the Northern Hemisphere, winds flow clockwise around an anticyclone; in the Southern Hemisphere, winds flow counterclockwise around an anticyclone.

Antipode A location on the exact opposite point of the near-spherical earth; the North Pole is the antipode of the South Pole.

Apex (delta) The upstream end of a delta where the river begins to fan outward into several distributary channels.

Aphelion The point in the earth's orbit, which occurs every July 4th, where the distance to the sun is maximized (ca. 152.5 million km [94.8 million mi]).

Aquiclude Impermeable rock layer that resists the infiltration of groundwater; consists of tightly packed or interlocking particles, such as in shale.

Aquifer A porous and permeable rock layer that can at least be partially saturated with groundwater.

Aquifer, confined One that lies between two aquicludes; often obtains its groundwater from a distant area where it eventually reaches the surface.

Aquifer, unconfined One that obtains its groundwater from local infiltration.

Archipelago A group of islands, often elongated into a chain.

Arctic Circle The latitude (66½°N) marking the southern boundary of the Northern Hemisphere portion of the earth's surface that receives a 24-hour period of sunlight at least once each year.

Area symbols Portray two-dimensional spaces on a map, with colors or black-and-white areal patterns representing specific quantitative ranges (which are identified in the map legend).

Arête A knife-like, jagged ridge that separates two adjacent glaciers or glacial valleys.

Arid (B) climates The dry climates where potential evapotranspiration always exceeds the moisture supplied by precipitation; found in areas dominated by the subtropical high-pressure cells, in the interiors of continents far from oceanic moisture sources, and in conjunction with rain-shadow zones downwind from certain mountain ranges.

Aridisol One of the 11 Soil Orders of the Soil Taxonomy, and the most widespread on the world's landmasses; dry soil (unless irrigated) associated with arid climates, light in color, and often contains horizons rich in calcium, clay, or salt minerals.

Artesian well One that flows under its own natural pressure to the surface; usually associated with a confined aquifer that is recharged from a remote location where that aquifer reaches the surface (*see* Fig. 40-8).

Ash (volcanic) Solid, cinder-like lava fragments, smaller than volcanic bombs, that are exploded into the air during an eruption; most fall to the ground around the erupting volcano.

Aspect The directional orientation of a steep mountain slope; in the Northern Hemisphere, southerly aspects receive far more solar radiation than do northward-facing slopes, with all the environmental consequences such a differential implies.

Asthenosphere The soft plastic layer of the upper mantle that underlies the lithosphere, which is able to move over it.

Atmosphere The blanket of air that adheres to the earth's surface which contains the mixture of gases essential to the survival of all terrestrial life forms.

Atolls Ring-like coral reefs surrounding empty lagoons; grew on the rims of eroded volcanic cones.

Attrition A wave erosion process whereby larger rock fragments are reduced to smaller fragments as loose material is continuously moved about in churning water.

Aurora australis Name given to the *aurora borealis* phenomenon that occurs in the Southern Hemisphere's upper-middle and high latitudes.

Aurora borealis Vivid sheet-like displays of light in the night-time sky of the upper-middle and high latitudes in the Northern Hemisphere; caused by the intermittent penetration of the thermosphere by ionized particles.

Autotroph An organism that manufactures its own organic materials from inorganic chemicals; a good example is phytoplankton, food-producing plants that manufacture carbohydrates.

Axis (earth's) The imaginary line that extends from the North Pole to the South Pole through the center of the earth; the planet's rotation occurs with respect to this axis.

Azonal flow The *meridional* (north-south) flow of upper atmospheric winds (poleward of 15 degrees of latitude), particularly the subtropical and Polar Front jet streams; periodic departures from the zonal (west-to-east) flow of these air currents are important because they help to correct the heat imbalance between the polar and equatorial regions.

B **horizon** The middle soil layer, which often receives dissolved and suspended particles from above.

Backshore The beach zone that lies landward of the foreshore; extends from the high-water line to the dune line.

Backslope (dune) The windward slope of a sand dune.

Backwash The return flow to the sea of the thinning sheet of water that slid up the beach as swash.

Bajada A coalesced assemblage of alluvial fans that lines a highland front (also known as an *alluvial apron*); primarily a desert feature.

Barchan A crescent-shaped sand dune with its points lying downwind; convex side of this dune is the windward side.

Barometer Instrument that measures atmospheric pressure; invented by Torricelli in 1643.

Barrier island A permanent, offshore, elongated ridge of sand, positioned parallel to the shoreline and separated by a lagoon from it; may have originally formed as an offshore bar during the last glaciation, migrated coastward, and grew as it shifted.

Basal ice The bottom ice layer of a glacier.

Base level The elevational level below which a stream cannot erode its bed.

Batholith A massive, discordant body of intrusive igneous rock (pluton) that has destroyed and melted most of the existing geologic structures it has invaded.

Bay A broad indentation into a coastline.

Baymouth bar A sandspit that has grown all the way across the mouth of a bay.

Beach A coastal zone of sediment that is shaped by the action of waves; constructed of sand and other materials, derived from both local and distant sources.

Beach drift The larger process associated with longshore drift that operates to move huge amounts of beach sand downshore in the direction of the longshore current.

Bedding plane The distinct surface between two sedimentary rock strata.

Berm A flat sandy beach that lies in the backshore beach zone; deposited during storms and beyond the reach of normal daily wave action.

Bermuda High The local name for the North Atlantic Ocean's Subtropical High pressure cell, which is generally centered over latitude 30°N; also known as the Azores High.

Biodiversity Shorthand for *biological diversity*; the variety of the earth's life forms and the ecological roles they play.

Biogeography The geography of plants (*phytogeography*) and animals (*zoogeography*).

Biological weathering The disintegration of rock minerals via biological means; earthworms and plant roots are important in the development of soil, lichens contribute to the breakdown of rocks, and humans, of course, play various roles in the disintegration of rocks and the operation of soil-formation processes.

Biomass The total living organic matter, encompassing all plants and animals, produced in a particular geographic area.

Biome The broadest justifiable subdivision of the plant and animal world, an assemblage and association of plants and animals that forms a regional ecological unit of subcontinental dimensions.

Biosphere The zone of terrestrial life, the habitat of all living things; includes the earth's vegetation, animals, human beings, and the part of the soil layer below that hosts living organisms.

Birdfoot delta A delta whose distributaries are kept open, such as in the constantly dredged Mississippi Delta (*see* Fig. 43-12b).

Blocky (angular) soil structure Involves irregularly shaped peds with straight sides that fit against the flat surfaces of adjacent peds, thereby giving a soil considerable strength (*see* Fig. 24-5).

Bluffs The low cliffs that border the outer edges of an alluvium-filled floodplain.

Body wave A seismic wave that travels through the interior of the earth; consists of two kinds—**P** waves and **S** waves.

Boiling point (water) A key setting in the calibration of temperature scales; on the Celsius scale water boils at 100°C, while on the Fahrenheit scale water boils at 212°F.

Bomb (storm) A mild winter weather disturbance that drifts northward along the U.S. Middle Atlantic seaboard, and then suddenly "explodes" into a furious snowstorm that paralyzes the nation's most heavily populated region.

Bombs (volcanic) The just-solidified gobs of lava that rain down on a volcano during an explosive eruption; they originate as gas-rich magmas whose gases explode when they reach the surface, hurling projectiles of solid lava into the air (which soon fall back to the ground).

Bottomset beds The finest deltaic deposits, usually laid down ahead of the delta where the coastal water is quiet.

Boulder field (*see* **rock sea**)

Braided stream One carrying a high sediment load that subdivides into many intertwined channels that reunite some distance downstream; as these channels shift position, stretches of deposited alluvium become divided between them, giving the "braided" appearance.

Breaker The forward collapse of the wave crest as an incoming wave can no longer push water upward as it encounters the ever-shallower ocean bottom.

Breccia In clastic sedimentary rocks when pebble-sized fragments in a conglomerate are not rounded but angular and jagged.

Butte Small, steep-sided, caprock-protected hill, usually found in dry environments; an erosional remnant of a plateau.

***C* horizon** The bottom soil layer in which the weathering of parent material proceeds.

Calcite Calcium carbonate ($CaCO_3$), the common mineral that is the main constituent of limestone.

Caldera A steep-walled, circular volcanic basin usually formed by the collapse of a volcano whose magma chamber emptied out; can also result from a particularly powerful eruption that blows off the peak and crater of a volcano.

Calorie One calorie is the amount of heat energy required to raise the temperature of 1 g (0.04 oz) of water by 1°C (1.8°F); not the same unit as the calories used to measure the energy value of food (which are 1000 times larger).

Calving When an icesheet enters the sea, the repeated breaking away of the leading edge of that glacier into huge, flat-topped, tabular icebergs.

Canadian High The local name for northern North America's Polar High, centered over far northwestern Canada.

Canyon A steep-sided gorge.

Capillary action The rise of water upward through soil air gaps owing to the tension between water molecules.

Carbon dioxide (CO_2) cycle Dominated by exchanges occurring between the air and the sea; CO_2 is directly absorbed by the ocean from the atmosphere and is released during the photosynthesis of billions of small organisms known as plankton.

Carbonation The reaction of weak carbonic acid (formed from water and carbon dioxide) with minerals; carbonic acid, in turn, reacts with carbonate minerals such as limestone in a form of chemical weathering that can be quite vigorous in certain humid areas, where solution and decay lead to the formation of karst landscapes.

Carbonic acid A weak acid (H_2CO_3), formed from water and carbon dioxide, that is instrumental in the important chemical weathering process known as carbonation.

Caribbean karst The rarest karst topography, associated with nearly flat-lying limestones; underground erosion dominated by the collapse of roofs of subsurface conduits, producing the characteristic sinkhole terrain seen in Fig. 44-6.

Carnivore An animal that eats herbivores and other animals.

Cartography The science, art, and technology of mapmaking and map use.

Catena Derived from a Latin word meaning chain or series, refers to a sequence of soil profiles appearing in regular succession on landform features of uniform rock type; most frequently associated with hillsides where the same parent material has produced an arrangement of different soil types (*see* Fig. 24-7).

Cation A positively charged ion, such as those of calcium.

Cave Any substantial opening in bedrock, large enough for an adult person to enter, that leads to an interior open space.

Cave shaft A vertical cave entrance.

Celsius scale Metric temperature scale most commonly used throughout the world (the United States is an exception); the boiling point of water is set at 100°C and its freezing point at 0°C.

Cementation During the lithification process of compaction as the grains of sediments are tightly squeezed together, water in the intervening pore spaces, which contains dissolved minerals, is deposited on the grain surfaces and acts as a glue to further bond the grains together.

Cenozoic The era of recent life on the geologic time scale (*see* Fig. 37-7), extending from 66 million years ago to the present; subdivided into the Tertiary and Quaternary periods.

Chaparral The name given to the dominant Mediterranean scrub vegetation in Southern California.

Chemical weathering The disintegration of rock minerals via chemical means; in any rock made up of a combination of minerals, the chemical breakdown of one set of mineral grains leads to the decomposition of the whole mass.

Chinook Name given to the *foehn* winds that affect the leeward areas of mountain zones in the western plateaus of North America.

Chlorophyll The green pigment found in plant surfaces that ensures the correct wavelength of light is absorbed for the process of photosynthesis, and that enables this process to take place.

Cinder cone Volcanic landform consisting mainly of pyroclastics; often formed during brief periods of explosive activity, they normally remain quite small.

Cinders (volcanic) (*see* **ash [volcanic]**)

Circle of illumination At any given moment on our constantly rotating planet, the boundary between the halves of the earth that are in sunlight and darkness.

Circum-Pacific earthquake belt An aspect of the Pacific Ring of Fire, the lengthy belt of subduction zones that girdles the Pacific Basin; here occur the heaviest concentration of earthquake epicenters (as well as active volcanoes) on the world map (*see* Fig. 35-5).

Cirque An amphitheater-like basin, high up on a mountain, that is the source area of a mountain glacier.

Cirrus The cloud-type category that encompasses thin, wispy, streak-like clouds that consist of ice particles rather than water droplets; occur only at altitudes higher than 6000 m (20,000 ft).

Clastic sedimentary rocks Sedimentary rocks made from particles of other rocks.

Clay The smallest category of soil particles; clay particles are smaller than 0.002 mm (0.00008 in), with the smallest in the colloidal range possessing diameters of less than one hundred-thousandth of a millimeter.

Cleavage The tendency of minerals to break in certain directions along bright plane surfaces, revealing zones of weakness in the crystalline structure.

Climate The long-term conditions (over at least 30 years) of aggregate weather over a region, summarized by averages and measures of variability; a synthesis of the succession of weather events we have learned to expect at any given location.

Climatic controls Features of the earth's surface—such as the distribution of land and water bodies, ocean currents, and highlands—that shape the climate of a locale by influencing its temperature and moisture regimes.

Climatic normal The average value for one of many weather parameters during the period 1931–1960 for one of many worldwide locations, published in tables by the World Meteorological Organization.

Climatic state The average, together with the variability and other statistics, of the complete set of atmospheric, hydrospheric, and cryospheric variables over a specified period of time in a specified domain of the earth-atmosphere system.

Climatology The geographic study of climates. This includes not only climate classification and the analysis of their regional distribution, but also broader environmental questions that concern climate change, interrelationships with soil and vegetation, and human-climate interaction.

Climax community Achieved at the end of a plant succession; the vegetation and its ecosystem are in complete harmony (dynamic equilibrium) with the soil, the climate, and other parts of the environment.

Climograph A graph that simultaneously displays, for a given location, its key climatic variables of average temperature and precipitation, showing how they change monthly throughout the year.

Closed system A self-contained system exhibiting no exchange of energy or matter across its boundaries (interfaces).

Cloud A visible mass of suspended, minute water droplets or ice crystals.

Coalescence process The raindrop-producing process that dominates in the tropical latitudes, described on p. 132.

Coast General reference to the strip of land and sea where the various coastal processes combine to create characteristic landscapes; the term *shore* has a more specific meaning.

Cockpit karst In tropical karst areas, the sharply contrasted landscape of prominent karst towers and the irregular, steep-sided depressions lying between them; *cockpit* refers to the depressions.

Cold front Produced when an advancing cold-air mass hugs the surface and displaces all other air as it wedges itself beneath the preexisting warmer air mass; cold fronts have much steeper slopes than warm fronts and thus produce more abrupt cooling and condensation (and more intense precipitation).

Cold-air drainage A category of local-scale wind systems governed by the downward oozing of heavy, dense, cold air along steep slopes under the influence of gravity; produces katabatic winds (such as southeastern France's *mistral*) that are fed by massive pools of icy air that accumulate over such major upland regions as the Alps and the Rocky Mountains.

Collapse sink A collapsed sinkhole in which the rock ceiling of the sinkhole collapses into the underground solution cavity.

Collapse sinkhole In karst terrain, a surface hollow created by the collapse or failure of the roof or overlying material of a cave, cavern, or underground channel.

Colloids The tiny soil particles formed from the disintegration of soil minerals and humus; they can be observed in suspension in the soil solution, making it look turbid.

Colluvium Soil particles that have washed downslope and have come to rest on the slope; give rise to colluvial soils.

Color (mineral) One of the most easily observable properties of a mineral, used in its identification.

Column (cave) The coalescence of a stalactite and a stalagmite that forms a continuous column from the floor to the roof of a cave.

Compaction The lithification process whereby deposited sediments are compressed by the weight of newer, overlying sediments; this pressure will compact and consolidate lower strata, squeezing their grained sediments tightly together. Usually occurs in conjunction with cementation.

Composite volcano Volcano formed, usually above a subduction zone, by the eruption of a succession of lavas and pyroclastics that accumulate as a series of alternating layers; the larger and more durable composite volcanoes are called *stratovolcanoes*.

Comprehensive Soil Classification System (CSCS) The all-encompassing soil classification system, developed by the U.S. Department of Agriculture's Soil Conservation Service, based on the soils themselves; the CSCS, approved and adopted in 1960, is known as the *Seventh Approximation* because it was the seventh version of the original scheme that was refined throughout most of the 1950s. Since 1960, the CSCS has continued to be modified within a framework of ever more subdivisions, and by the late 1970s the name CSCS was phased out as soil scientists increasingly preferred the simpler term, Soil Taxonomy.

Compressional stress The stress associated with the convergence of lithospheric plates; the lithosphere is forced to occupy less space, and the rocks respond by breaking, bending, folding, sliding, squeezing upward and downward, and crushing tightly together.

Concave Refers to a surface that is rounded inward, like the inside surface of a sphere.

Concordant intrusion Intrusive magma that did not disrupt or destroy surrounding, existing geologic structures but conformed to them.

Condensation The process by which a substance is transformed from the gaseous to the liquid state.

Condensation nuclei Small airborne particles around which liquid droplets can form when water vapor condenses; almost always present in the atmosphere in the form of dust or salt particles.

Conduction The transport of heat energy from one molecule to the next.

Cone of depression The drop (drawdown) in a local water table that immediately surrounds a well when water is withdrawn faster than it can be replaced by water flowing through that tapped aquifer.

Confined aquifer (*see* **aquifer, confined**)

Conformal map projection A map projection that preserves the true shape of the area being mapped.

Conic map projection One in which the transfer of the earth grid is from a globe onto a cone, which is then cut and laid flat.

Coniferous Cone-bearing.

Conservation The careful management and use of natural resources, the achievement of significant social benefits from them, and the preservation of the natural environment.

Constant gases Atmospheric gases always found in the same proportions; two of them constitute 99 percent of the air, nitrogen (78 percent) and oxygen (21 percent).

Contact metamorphism Metamorphic change in rocks induced by their local contact with molten magma or lava.

Continental drift The notion hypothesized by Alfred Wegener concerning the fragmentation of Pangaea and the slow movement of the modern continents away from this core supercontinent.

Continental effect The lack of the moderating influence of an ocean on air temperature, which is characteristic of inland locations; this produces hotter summers and colder winters relative to coastal locations at similar latitudes.

Continental rise Transitional zone of gently sloping seafloor that begins at the foot of the continental slope and leads downward to the lowest (abyssal) zone of an ocean basin.

Continental (sheet) glaciers Huge masses of ice that bury whole countrysides beneath them.

Continental shelf The gently sloping, relatively shallow, submerged plain just off the coast of a continent, extending to a depth of ca. 180 m (600 ft/100 fathoms).

Continental shield A large, stable, relatively flat expanse of very old rocks that may constitute one of the earliest "slabs" of solidification of the primeval earth's molten crust into hard rocks; forms the geologic core of a continental landmass.

Continental slope The steeply plunging slope that begins at the outer edge of the continental shelf (ca. 180 m [600 ft] below the sea surface) and ends in the depths of the ocean floor at the head of the continental rise.

Continentality The variation of the continental effect on air temperatures in the interior portions of the world's landmasses; the further the distance from the moderating influence of an ocean (known as the maritime effect), the greater the extreme in summer and winter temperatures. Northeastern Eurasia is the classic example of such extreme annual temperature cycles.

Contouring The representation of surface relief using isolines of elevation above sea level; an important basis of topographic mapping.

Conurbation A coalescence of two or more metropolitan areas; the U.S. northeastern seaboard's Megalopolis consists of a continuous corridor of connected metropolises that stretches from south of Washington, D.C. to north of Boston.

Convection Spontaneous vertical air movement in the atmosphere.

Convection cell A rising column of air.

Convectional precipitation Convection is the spontaneous vertical movement of air in the atmosphere; convectional precipitation occurs after condensation of this rising air.

Convergent evolution Theory that holds that organisms in widely separated biogeographic realms, although descended from diverse ancestors, develop similar adaptations to measurably similar habitats.

Convergent-lifting precipitation Precipitation produced by the forced lifting of warm, moist air where low-level windflows converge; most pronounced in the equatorial latitudes where the Northeast and Southeast Trades come together in the Inter-Tropical Convergence Zone (ITCZ), especially over the oceans.

Convex Refers to a surface that is rounded outward, like the outside surface of a sphere.

Coral reef An aggradational reef formed from the skeletal remains of marine organisms.

Core (drilling) A tube-like sample of seafloor sediment that is captured and brought to the surface in a hollow drill-pipe; the drill-pipe is thrust perpendicularly into the ocean floor, and the sample it brings up shows the sequence of sedimentary accumulation.

Core (earth's) (*see* **inner core**; **outer core**)

Coriolis force The force that, owing to the rotation of the earth, tends to deflect all objects moving over the surface of the earth away from their original path. In the absence of any other forces, the deflection is to the right in the Northern Hemisphere and to the left in the Southern Hemisphere; the higher the latitude, the stronger the deflection.

Corrasion The mechanical coastal-erosion process whereby waves break off pieces of rock from the surface under attack; the rock-fragment-loaded water, now a much more powerful erosive agent, continues to be hurled against that surface.

Corrosion The process of stream erosion whereby certain rocks and minerals are dissolved by water; can also affect coastal bedrock that is susceptible to such chemical action.

Counter-radiation Longwave radiation emitted by the earth's surface that is absorbed by the atmosphere and reradiated (also as longwave radiation) back down to the surface.

Creep The slowest form of mass movement; involves the slow, imperceptible motion of a soil layer downslope, as revealed in the slight downhill tilt of trees and other stationary objects.

Crevasse One of the huge vertical cracks that frequently cut the rigid, brittle upper layer of a mountain glacier.

Cross-bedding Consists of successive rock strata deposited not horizontally but at varying inclines; like ripple marks on sand, this usually forms on beaches and in dunes.

Crustal spreading The geographic term for seafloor spreading; not all crustal spreading occurs on the ocean floor (*see* **rift valley**).

Crustal warping (*see* **diastrophism**)

Cryosphere The collective name for the ice system of the earth.

Cuesta A long ridge with a steep escarpment on one side and a gently dipping slope and rockbeds on the other.

Cumulonimbus Very tall cumulus clouds, extending from about 500 m (1600 ft) at the base to over 10 km (6 mi) at the top, often associated with violent weather involving thunder, lightning, and heavy winds and rains.

Cumulus The cloud-type category that encompasses thick, puffy, billowing masses that often develop to great heights; subclassified according to height (*see* p. 131).

Cycle of erosion The evolutionary cycle proposed by William Morris Davis that purportedly affects all landscapes (*see* Fig. 42-15).

Cyclic autogenic succession Plant succession in which one type of vegetation is replaced by another, which in turn is replaced by the first, with other series possibly intermixed.

Cyclogenesis The formation, evolution, and movement of midlatitude cyclones.

Cyclone An atmospheric low-pressure cell involving the convergence of air, which flows into and spirally rises at the center; the isobars around a cyclone are generally circular in shape, with their values decreasing toward the center. In the Northern Hemisphere, winds flow counterclockwise around a cyclone; in the Southern Hemisphere, winds flow clockwise around a cyclone.

Cyclonic precipitation Precipitation associated with the passage of the warm and cold fronts that are basic components of the cyclones that shape the weather patterns of the midlatitudes; often used as a synonym for *frontal precipitation*.

Cylindrical map projection One in which the transfer of the earth grid is from a globe onto a cylinder, which is then cut and laid flat.

Daily Weather Map Published daily by the U.S. National Weather Service's National Meteorological Center; includes a large synoptic surface weather map and smaller maps showing 500-mb height contours, highest and lowest temperatures of the preceding 24 hours, and precipitation areas and amounts of the preceding 24 hours.

DALR (*see* **dry adiabatic lapse rate**)

Daylight-saving time By law, all clocks in a time zone are set one hour forward from standard time for at least part of the year. In the United States, localities (e.g., the state of Arizona) can exempt themselves from such federal regulations.

Deciduous The term for trees and other plants that seasonally drop their leaves.

Deflation The process whereby wind sweeps along a surface and carries away the finest particles.

Deflation hollow Shallow desert basin created by the wind erosion process of deflation.

Deforestation (*see* **tropical deforestation**)

Deformation Any change in the form and/or structure of a body of crustal rock caused by an earth movement.

Deglaciation The melting and receding of glaciers that accompanies the climatic warmup after the peak of a glaciation has been reached.

Degradation The combination of processes that wear down the landmasses; implies the lowering, reducing, and smoothing of those surfaces.

Delta The often major sedimentary deposit surrounding and extending beyond the mouth of a river where it empties into the sea or a lake; frequently assumes a triangular configuration, hence its naming after the Greek letter of that shape.

Deltaic plain The flat, stable landward portion of a delta that is growing seaward.

Dendritic drainage A tree-limb-like stream pattern that is the most commonly observed (*see* Fig. 42-10e); indicates surface of relatively uniform hardness or one of flat-lying sedimentary rocks.

Dendrochronology The study of the width of annual growth rings in trees for dating purposes.

Density The amount of mass per volume in an object or a portion of the atmosphere.

Denudation The combined processes of weathering, mass movement, and erosion that over time strip a slope of its soil cover unless the local rate of soil formation is greater.

Depletion The soil-layer formation process involving the loss of soil components as they are carried downward by water, plus the loss of other material in suspension as the water percolates through the soil from upper to lower layers; while the upper layers are depleted accordingly, the dissolved and suspended materials are redeposited lower down in the soil.

Desert biome Characterized by sparse, xerophytic vegetation or even the complete absence of plant life.

Desert climate (BW) The most arid of climates, usually experiencing no more than 250 mm (10 in) of annual precipitation; characteristics are displayed in Fig. 17-8.

Desert pavement A smoothly weathered, varnish-like surface of closely packed pebbles that has developed on the upper part of an alluvial fan or bajada; no longer subject to stream braiding, such a surface is stable and may support desert vegetation.

Desertification The process of desert expansion into neighboring steppelands as a result of human degradation of fragile semiarid environments.

Dew The fine water droplets that condense on surfaces at and near the ground when saturated air is cooled; the source of this condensate is the excess water vapor beyond the saturation level of that air parcel, whose capacity to contain water decreases as its temperature drops.

Dew point The temperature at which air becomes saturated and below which condensation occurs.

Diastrophism The process whereby large continental areas undergo slight deformation without being faulted or folded; also known as *crustal warping*.

Diffuse radiation The proportion of incoming solar energy (22 percent) that reaches the earth's surface after first being scattered in the atmosphere by clouds, dust particles, and other airborne materials.

Dike A discordant intrusive igneous form in which magma has cut vertically across preexisting strata, forming a kind of barrier wall.

Dip The angle at which a rock layer tilts from the horizontal.

Direct radiation The proportion of incoming solar energy that travels directly to the earth's surface; globally, this averages 31 percent.

Discharge (stream) The volume of water passing a given cross-section of a river channel within a given amount of time; measured as average water velocity multiplied by the cross-sectional area.

Discordant intrusion Intrusive magma that did not conform to but cut across or otherwise disrupted surrounding, existing geologic structures.

Distributaries The several channels a river subdivides into when it reaches its delta; caused by the clogging of the river mouth by deposition of fine-grained sediment as the stream reaches base level and water velocity declines markedly.

Diurnal temperature cycle The pattern of temperature change during the course of a day.

Divide A topographic barrier, usually a mountain ridge, that separates two drainage basins.

Doldrums The equatorial zone of periodic calm seas and unpredictable breezes where the Northeast and Southeast Trades converge; the crews of sailing ships dreaded these waters because their vessels risked becoming stranded in this becalmed marine environment.

Dolomite Another soluble rock ($CaMg[CO_3]_2$), in addition to calcite-rich limestone, that can form karst topography.

Dormant volcano A volcano that has not been seen to erupt but shows evidence of recent activity.

Doubling time The number of years required for a given human population to double itself in size; for the world as a whole, in the mid-1990s that figure stood at 42 years.

Downthrown block The block that moves downward with respect to adjacent blocks when vertical movement occurs during faulting.

Drainage basin The region occupied by a complete stream system formed by the trunk river and all its tributaries (also known as a *watershed*).

Drainage density The total length of the stream channels that exist in a unit area of a drainage basin.

Drawdown The amount of the drop in a local water table that occurs when water is withdrawn through a well faster than it can be replaced by new groundwater in the aquifer; commonly forms a cone of depression around the well.

Drift (glacial) (*see* **glacial drift**)

Drift (ocean-surface) A term often used as a synonym for an ocean current, whose rate of movement usually lags well behind the average speeds of surface winds blowing in the same direction; currents are characterized by a slow and steady movement that very rarely exceeds 8 kph (5 mph).

Driftless Area An area in southwestern Wisconsin that was never covered by the continental glaciers that repeatedly buried adjacent areas of the U.S. Midwest.

Drought The below-average availability of water in a given area over a period lasting at least several months.

Drumlin A smooth elliptical mound created when an icesheet overrides and reshapes preexisting glacial till; long axis lies parallel to direction of ice movement.

Dry adiabatic lapse rate (DALR) The lapse rate of an air parcel not saturated with water vapor: 1°C/100 m (5.5°F/1000 ft).

Dry-bulb temperature On a psychrometer, the temperature reading of the thermometer whose bulb is not swaddled in a wet cloth.

Dune An accumulation of sand that is shaped by wind action.

Dune crest The top of a sand dune where the backslope and slip face meet.

Dust (volcanic) Fine solid lava fragments, smaller than particles of volcanic ash, that are exploded into the air during an eruption; these lighter materials fall to the ground near an erupting volcano, but also can be carried for substantial distances by prevailing windflows if they are exploded high enough into the troposphere.

Dust Bowl An environmental disaster that plagued the U.S. Great Plains for most of the 1930s. The continued heavy plowing of much of this region during a prolonged drought (in a futile attempt to keep raising wheat on a large scale) resulted in the loosened topsoil falling prey to the ever-present wind; huge dust clouds soon formed that carried the soil away and deposited it as dunes, forcing thousands of farm families to abandon their fields and migrate out of the Great Plains.

Dust dome The characteristic shape taken by the large quantities of dust and gaseous pollutants in a city's atmosphere (*see* Fig. 8-6).

Dynamic equilibrium State of a system in balance, when it is neither growing nor contracting but continues in full operation.

E **horizon** Soil layer located between the *A* and *B* horizons (*E* stands for eluviation); originally, the lower, lighter portions of the *A* horizon in certain soils were designated as *A₂* (with the darker, upper portion as *A₁*), but it eventually became clear that the *A₂* horizon had its own distinct qualities.

Earth flow A form of mass movement in which a section of soil or weathered bedrock, lying on a rather steep slope, becomes saturated by heavy rains until it is lubricated enough to flow (*see* Fig. 39-6).

Earth system The shells or layers that make up the total earth system range from those of the planet's deepest interior to those bordering outer space; this book focuses on the key earth layers of the atmosphere, lithosphere, hydrosphere, and biosphere.

Earthquake A shaking and trembling of the earth's surface; caused by sudden releases of stresses that have been building slowly within the planetary crust.

Easterly wave A wave-like perturbation in the constant easterly flow of the Northeast and Southeast Trade winds that produces this type of distinctive weather system; westward-moving air is forced to rise on the upwind side (producing often heavy rainfall) and descend on the fair-weather downwind side of the low-pressure wave trough.

Echelon faults A series of nearly parallel faults produced by compressional forces when the crust is horizontally shortened; sometimes the crustal block between two reverse faults is pushed upward to form a horst (*see* Fig. 36-3).

Ecological niche The way a group of organisms makes its living in nature, or the environmental space within which an organism operates most efficiently.

Ecosystem A linkage of plants or animals to their environment in an open system as far as energy is concerned.

Ecumene The portion of the world's land surface that is permanently settled by human beings.

Edaphic factors Factors concerned with the soil.

Eddies Localized loops of water circulation detached from the mainstream of a nearby ocean current, which move along with the general flow of that current.

El Niño A periodic, large-scale, abnormal warming of the sea surface in the low latitudes of the eastern Pacific Ocean that produces a (temporary) reversal of surface ocean currents and airflows throughout the equatorial Pacific; these regional events have global implications, disturbing normal weather patterns in many parts of the world.

Electromagnetic spectrum The continuum of electric and magnetic energy, as measured by wavelength, from the high-energy shortwave radiation of cosmic rays to the low-energy longwave radiation of radio and electrical power (*see* Fig. 3-15).

ELR (*see* **environmental lapse rate**)

Eluviation Means "washed out," and refers to the soil process that involves the removal from the *A* horizon and downward transportation of soluble minerals and microscopic, colloid-sized particles of organic matter, clay, and oxides of aluminum and iron.

Emergent coast A coastal zone whose landforms have recently emerged from the sea, either through tectonic uplift, a drop in sea level, or both.

Empirical Real-world, as opposed to an abstract theoretical model.

ENSO Acronym for *E*l *N*iño-*S*outhern *O*scillation; the reversal of the flow of ocean currents and prevailing winds in the equatorial Pacific Ocean that disturbs global weather patterns.

Entisol One of the 11 Soil Orders of the Soil Taxonomy, which contains all the soils that do not fit into the other 10 Soil Orders; is of recent origin, evinces little or modest development, and is found in many different environments.

Entrenched meanders Meanders of a rejuvenated stream incised into hard bedrock from overlying floodplain topography; caused by the uplifting of the land surface above the stream's base level.

Environmental lapse rate (ELR) The nonadiabatic lapse rate at any particular time or place; the troposphere's (nonadiabatic) normal lapse rate averages 0.65°C/100 m (3.5°F/1000 ft).

Eolian Pertaining to the action of the wind.

Epeirogeny The vertical movement of the earth's crust over very large areas that produces little or no bending or breaking of the uplifted (or downward-thrusted) rocks.

Ephemeral plant One that completes its life cycle within a single growing season.

Ephemeral stream An intermittently flowing stream in an arid environment; precipitation (and subsequent streamflow) is periodic, and when the rains end the stream soon runs dry again.

Epicenter The point on the earth's surface directly above the focus (place of origin) of an earthquake.

Epiphyte Tropical rainforest plant that uses trees for support, but is not parasitic.

Equal-area map projection One in which all the areas mapped are represented in correct proportion to one another.

Equator The parallel of latitude running around the exact middle of the globe, defined as 0° latitude.

Equatorial Low The Inter-Tropical Convergence Zone (ITCZ) or thermal low-pressure belt of rising air that straddles the equatorial latitudinal zone; fed by the windflows of the converging Northeast and Southeast Trades.

Equinox One of the two days (ca. March 21 and September 23) in the year when the sun's noontime rays strike the earth vertically at the equator; in Northern Hemisphere terminology, the March 21 event is called the spring (vernal) equinox and the September 23 event is called the fall (autumnal) equinox.

Erg A sand sea; a large expanse of sandy desert landscape.

Erosion The long-distance carrying away of weathered rock material, and the associated processes whereby the earth's surface is worn down.

Escarpment A very steep cliff; often marks the wall-like edge of a plateau or mountainous highland.

Esker A glacial outwash landform that appears as a long ribbon-like ridge in the landscape because it was formed by the clogging of a river course within the glacier, the debris from which remains after the ice melts.

Estuary The drowned (submerged) mouth of a river valley that has become a branch of the sea; in the United States, Chesapeake Bay is classic example.

Evaporation Also known as vaporization, the process by which water changes from the liquid to the gaseous (water vapor) state; it takes 597 cal of heat energy to change the state of 1 g (0.04 oz) of water at 0°C (32°F) from a liquid to a gas.

Evaporites Rock deposits resulting from the evaporation of the water these materials were once dissolved in; include halite (salt) and gypsum.

Evapotranspiration The combined processes by which water (1) evaporates from the land surface and (2) passes into the atmosphere through the leaf pores of plants (transpiration).

Evergreen The term for trees and other plants that keep their leaves year-round.

Exfoliation A special kind of jointing that produces a joint pattern resembling a series of concentric shells, much like the layers of an onion; caused by the release of confining pressure, the outer layers progressively peel away and expose the lower layers.

Extinct volcano A volcano that shows no sign of life and exhibits evidence of long-term weathering and erosion.

Extrusive igneous rocks Rocks formed from magma that cooled and solidified, as lava or ash, on the earth's surface.

Eye (hurricane) The open vertical tube that marks the center of a hurricane, often reaching an altitude of 16 km (10 mi).

Eye wall (hurricane) The rim of the eye or open vertical tube that marks the center of a well-developed hurricane; the tropical cyclone's strongest winds and heaviest rainfall occur here.

Fahrenheit scale Temperature scale presently used in the United States; water boils at 212°F (100°C) and freezes at 32°F (0°C).

Fall (Autumn) In Northern Hemisphere terminology, the season that begins at the fall (autumnal) equinox ca. September 23 and ends at the winter solstice on December 22.

Fall (Autumnal) Equinox In Northern Hemisphere terminology, the equinox that occurs when the sun's noontime rays strike the equator vertically on ca. September 23.

Fall (movement) The fastest form of mass movement that involves the free fall or downslope rolling of rock pieces loosened by weathering; these boulders form a talus cone or scree slope at the base of the cliff from which they broke away.

Fall Line In the northeastern United States, follows the boundary between the Piedmont and the Coastal Plain and connects the waterfalls that mark the inland limit of ocean navigation on the rivers that cross it; because these falls were also local sources of water power, they attracted people and activities and gave rise to such major cities as Richmond, Washington, Baltimore, and Philadelphia.

Fault A fracture in crustal rock involving the displacement of rock on one side of the fracture with respect to rock on the other side.

Fault-block mountains Horst-like highlands thrust upward along parallel axes by compressional forces; especially common in the Basin-and-Range Province of the western United States (*see* Fig. 42-4).

Fault breccia Crushed, jagged rock fragments that often lie along a fault trace.

Fault-line scarp A scarp that originated as a fault scarp but was modified, perhaps even displaced, by erosion (*see* Fig. 42-3b).

Fault plane The surface of contact along which blocks on either side of a fault move (*see* Fig. 35-7).

Fault scarp The exposed cliff-like face of a fault plane (*see*

Fig. 35-7) created by geologic action without significant erosional change.

Fault trace The lower edge of a fault scarp; the line on the surface where a fault scarp intersects the surface (*see* Fig. 35-7).

Fauna Animal life.

Feedback Occurs when a change in one part of a system causes a change in another part of the system.

Felsenmeer (*see* **rock sea**)

Fertigation Contraction of "fertilize" and "irrigation," an Israeli-perfected technique for the large-scale raising of crops under dry environmental conditions. Via subterranean pipes, the roots of genetically engineered plants are directly fed a mixture of brackish underground water and chemical fertilizers that supply just the right blend of moisture and nutrients for the crop to thrive.

Fetch The uninterrupted distance traveled by a wind or an ocean wave.

Field capacity The ability of a soil to hold water against the downward pull of gravity; also the maximum amount of water a soil can contain before becoming waterlogged.

Finger lake An elongated lake that fills much of an even longer, fairly narrow glacial trough (*see* Fig. 46-9).

Firn Granular, compacted snow.

Firn line (*see* **snow line**)

Fissure eruption A volcanic eruption that comes not from a pipe-shaped vent but from a lengthy crack in the lithosphere; the lava that erupts does not form a mountain but extends in horizontal sheets across the countryside, sometimes forming a plateau.

Fixed dune A stable sand dune that supports vegetation, which slows or even halts the dune's wind-generated movement.

Fjord A narrow, steep-sided, elongated estuary formed from a glacial trough inundated by seawater.

Flood An episode of abnormally high stream discharge; water overflows from the stream channel and temporarily covers its floodplain (which continues to build from the alluvium that is deposited by the floodwaters).

Floodplain The flat, low-lying ground adjacent to a stream channel built by successive floods as sediment is deposited as alluvium.

Flora Plant life; vegetation.

Flow regime A discrete region of outward ice flow in a continental glacier; possesses its own rates of snow accumulation, ice formation, and velocity.

Fluvial Denotes running water; derived from Latin word for river, *fluvius.*

Focus (earthquake) The place of origin of an earthquake, which can be near the surface or deep inside the crust or upper mantle.

Foehn The rapid movement of warm dry air (caused by a plunge in altitude), frequently experienced on the leeward or rain-shadow side of a mountain barrier, whose moisture has mostly been removed via the orographic precipitation process. The name is also used specifically to designate such local winds in the vicinity of central Europe's Alps; in the western plateaus of North America, these airflows are called *chinook* winds.

Fog A cloud layer in direct contact with the earth's surface.

Fold An individual bend or warp in layered rock.

Foliation The unmistakable banded appearance of certain metamorphic rocks, such as gneiss and schist; bands formed by minerals realigned into parallel strips during metamorphism.

Food chain The stages that energy in the form of food goes through within an ecosystem.

Foreset beds The sedimentary deposits that are built from the leading edge of the topset beds as a delta grows seaward; later, these will be covered by the extension of the topset beds.

Foreshore The beach zone that is alternatively water-covered during high tide and exposed during low tide; the zone of beach drift and related processes.

Fossil fuels Coal, oil, and natural gas, the dominant suppliers of energy in the world economy.

Fracture (mineral) When minerals do not break in a clean cleavage, they still break or fracture in a characteristic way; obsidian, for example, fractures in an unusual shell-like manner that is a useful identifying quality.

Freezing The process by which a substance is transformed from the liquid to the solid state.

Freezing nuclei Perform the same function for ice particles that condensation nuclei perform for water droplets; small airborne particles (of dust or salt) around which ice crystals can form when liquid water freezes or water vapor sublimates.

Freezing point (water) A key setting in the calibration of temperature scales; on the Celsius scale water freezes at 0°C, while on the Fahrenheit scale water freezes at 32°F.

Frictional force The drag that slows the movement of air molecules in contact with, or close to, the earth's surface. Varies with the "roughness" of the surface; there is less friction with movement across a smooth water surface than across the ragged skyline of a city center.

Front (weather) The surface that bounds an air mass, along which contact occurs with a neighboring air mass possessing different qualities; this narrow boundary zone usually marks an abrupt transition in air density, temperature, and humidity. A moving front is the leading edge of the air mass built up behind it.

Frontal precipitation Precipitation that results from the movement of fronts whereby warm air is lifted, cooled, and condensed; also frequently called *cyclonic precipitation*.

Frost action A form of mechanical weathering in which water penetrates the joints and cracks of rocks, expands and contracts through alternate freezing and thawing, and eventually shatters the rocks.

Frost creep The movement of particles within the active layer above the permafrost under the influence of gravity; on the surface, rocks will move downslope during the thawing phase.

Frost heaving The upward displacement of rocks and rock fragments within the active layer above the permafrost after they have been loosened by frost wedging; triggered by the formation of ice in the ground that expands the total mass of rock materials.

Frost shattering (*see* **frost wedging**)

Frost thrusting The horizontal movement of rocks and rock fragments within the active layer above the permafrost.

Frost wedging The forcing apart of a rock when the expansion stress created by the freezing of its internal water into ice exceeds the cohesive strength of that rock body.

Galaxy An organized, disk-like assemblage of billions of stars; our solar system belongs to the Milky Way galaxy, which measures about 120,000 light-years in diameter.

General circulation The global atmospheric circulation system of windbelts and semipermanent pressure cells. In each hemisphere, the windbelts include the Trades, Westerlies, and Polar Easterlies. The pressure cells include the Equatorial Low (ITCZ) and, in each hemisphere, the Subtropical High, Upper-Midlatitude Low, and Polar High.

Geodesy The precise study and measurement of the size and shape of the earth.

Geographic information system (GIS) A collection of computer hardware and software that permit spatial data to be collected, recorded, stored, retrieved, manipulated, analyzed, and displayed to the user.

Geography Literally means *earth description*. As a modern academic discipline, it is concerned with the explanation of the physical and human characteristics of the earth's surface. "Why are things located where they are?" is the central question that geographical scholarship seeks to answer.

Geologic structure Refers to landscape features originally formed by geologic processes, which are sculpted by streams and other erosional agents into characteristic landforms.

Geologic time scale The standard timetable or chronicle of earth history used by scientists; the sequential organization of geologic time units as displayed in Fig. 37-7, whose dates continue to be refined by ongoing research.

Geomorphology Literally means *earth shape* or *form*. The geography of landscape and its evolution, a major subfield of physical geography.

Geostrophic wind A wind that results when the Coriolis and pressure-gradient forces balance themselves out; follows a relatively straight path that minimizes deflection and lies parallel to the isobars (*see* Fig. 9-7).

Geosynchronous orbit Orbit in which a satellite's revolution of the earth is identical to the planet's rotational speed; therefore, the satellite is "fixed" in a stationary position above the same point on the earth's surface.

Geothermal energy Energy whose source is underground heat; where magma chambers lie close to the surface, groundwater is heated and emerges onto the surface as steam or hot water from a geyser or a hot spring.

Geyser A hot spring that periodically expels jets of heated water and steam.

Glacial creep One of the two mechanisms by which glaciers flow; involves the internal deformation of the ice, with crystals slipping over one another as a result of downslope movement.

Glacial drift Comprised of unsorted till and stratified drift.

Glacial erratic A boulder that was transported far from its source area by a glacier.

Glacial outwash Meltwater-deposited sand and gravel that are sorted into layers.

Glacial sliding One of the two mechanisms by which glaciers flow; involves the movement of the entire glacier over the rocks below it, lubricated by a thin film of water between the basal ice and the bedrock floor.

Glacial striations The scratches on the underlying rock surface made when boulders or pebbles (embedded in a moving glacier) were dragged across it.

Glacial surge Episode of rapid movement in a mountain glacier, as much as one meter per hour, that can last for a year or more.

Glacial trough A valley that has been eroded by a glacier; distinctively **U**-shaped in cross-sectional profile.

Glaciation A period of global cooling during which continental icesheets and mountain glaciers expand.

Glacier A body of ice, formed on land, that exhibits motion.

Global warming theory The notion, popular among scientists in the late 1980s and early 1990s, that human fossil-fuel consumption is causing atmospheric warming that will melt glaciers, raise sea levels, and inundate low-lying coastal areas. In the mid-1990s, however, competing theories of climate change are receiving considerable attention and gaining new proponents. *See* Perspective box, pp. 75-77.

Gneiss Metamorphic rock derived from granite that usually exhibits pronounced foliation.

Gondwana The southern portion of the primeval supercontinent, Pangaea.

Graben A crustal block that has sunk down between two fairly parallel normal faults in response to tensional forces (*see* Fig. 36-8).

Gradational process A process that works to wear down the geologic materials that are built up on the earth's landmasses.

Graded stream A stream in which slope and channel characteristics are adjusted over time to provide just the velocity required for the transportation of the load supplied from the drainage basin, given the available discharge.

Gradient (river) The slope of a river channel as measured by the difference in elevation between two points along the stream course.

Gravity The force of attraction that acts among all physical objects as a result of their *mass* (the quantity of material they are composed of).

Great circle The circle formed along the edge of the cut when a sphere is cut in half; on the surface of the sphere, that circle is the shortest distance between any two points located on it.

Greenhouse effect The widely used analogy describing the blanket-like effect of the atmosphere in the heating of the earth's surface; shortwave insolation passes through the "glass" of the atmospheric "greenhouse," heats the surface, is converted to longwave radiation that cannot penetrate the "glass," and thereby results in trapping of heat that raises the temperature inside the "greenhouse."

Ground heat flow The heat that is conducted into and out of the earth's surface; also known as *soil heat flow*.

Ground moraine Blanket of unsorted glacial till that was laid down at the base of a melting glacier.

Groundwater Water contained within the lithosphere; this water hidden below the ground accounts for about 25 percent of the world's fresh water.

Gyre The cell-like circulation of surface currents that often encompasses an entire ocean basin; for example, the subtropical gyre of the North Atlantic Ocean consists of the huge loop formed by four individual, continuous legs—the North Equatorial, Gulf Stream, North Atlantic Drift, and Canaries currents.

Habitat The environment a species normally occupies within its geographical range.

Hail Precipitation consisting of ice pellets, which form in a cloud that has more ice crystals than water droplets.

Hanging valley A valley formed by the intersection of a tributary glacier with a trunk glacier; when the ice melts away, the tributary valley floor usually is at a higher elevation and thus "hangs" above the main valley's floor.

Hardpan Colloquial term for an oxic horizon in the soil.

Hawaiian High The local name for the North Pacific Ocean's Subtropical High pressure cell, which is generally centered over latitude 30°N; also known as the Pacific High.

Headward erosion The upslope extension, over time, of the "head" or source of a river valley, which lengthens the entire stream network.

Heat-island intensity The maximum difference in temperature between neighboring urban and rural environments.

Hemisphere A half-sphere; used precisely, as in Northern Hemisphere (everything north of 0° latitude), or sometimes more generally, as in land hemisphere (the significant concentration of landmasses on roughly one side of the earth).

Herbivore An animal that lives on plants, or more generally the first consumer stage of a food chain.

Heterosphere The upper of the atmosphere's two vertical regions, which possesses a variable chemical composition and extends upward from an elevation of 80–100 km (50–63 mi) to the edge of outer space.

High mountains Terrain of less than 50 percent gentle slope whose local relief exhibits variations in excess of 900 m (3000 ft).

Highland (H) climates The climates of high-elevation areas that exhibit characteristics of climates located poleward of those found at the base of those highlands; the higher one climbs, the colder the climate becomes—even in the low latitudes. Thus **H** climate areas are marked by the vertical zonation of climates, as shown for South America's Andes in Fig. 19-10.

Hills Terrain of less than 50 percent gentle slope whose local relief exhibits variations of 0–300 m (0–1000 ft).

Histosol One of the 11 Soil Orders of the Soil Taxonomy; organic soil associated with poorly drained, flat-lying areas that, when drained, can become quite productive in root-crop agriculture.

Hogback A prominent steep-sided ridge whose rockbeds dip sharply.

Holocene The current interglaciation epoch, extending from 10,000 years ago to the present on the geologic time scale.

Homosphere The lower of the atmosphere's two vertical regions, which possesses a relatively uniform chemical composition and extends from the surface to an elevation of 80–100 km (50–63 mi).

Horizonation The differentiation of soils into layers called horizons.

Horn The sharp-pointed, Matterhorn-like mountain peak that remains when several cirques attain their maximum growth by headward erosion and intersect.

Horst A crustal block that has been raised between two reverse faults by compressional forces (*see* Fig. 36-3).

Hot spot A place of very high temperatures in the upper mantle that reaches the surface as a "plume" of extraordinarily high heat; a linear series of shield volcanoes can form on lithospheric plates moving over this plume, as happened in the case of the Hawaiian island chain.

Hot spring A spring whose water temperature averages above 10°C (50°F); often emanates from a crustal zone that contains magma chambers near the surface (*see* Fig. 40-10d).

Humid continental climate (Dfa/Dwa, Dfb/Dwb) Collective term for the milder subtypes of the humid microthermal climate, found on the equatorward side of the **D**-climate zone; characteristics are displayed in Fig. 19-5.

Humid microthermal climate (*see* **microthermal [D] climates**)

Humid subtropical climate (Cfa) The warm, perpetually moist mesothermal climate type that is found in the southeastern portion of each of the five major continents; characteristics are displayed in Fig. 18-1.

Humus Decomposed and partially decomposed organic matter that forms a dark layer at the top of the soil; contributes importantly to a soil's fertility.

Hurricane A tropical cyclone capable of inflicting great damage. A tightly organized, moving low-pressure system, normally originating at sea in the warm moist air of the tropical atmosphere, exhibiting wind speeds in excess of 33 m per second (74 mph); as with all cyclonic storms, it has a distinctly circular wind and pressure field.

Hydraulic action The erosional work of running water in a stream or in the form of waves along a coast. In a river, rock material is dislodged and dragged away from the valley floor and sides; where waves strike a shoreline, the speed and weight of the water, especially when air is compressed into rock cracks by the power of the waves, can fracture and erode coastal rocks quite rapidly.

Hydrograph A graph of a river's discharge over time (*see* Fig. 40-6).

Hydrography Refers to physical aspects of all the waters that occur on the earth's surface.

Hydrologic cycle The complex system of exchange involving water in its various forms as it continually circulates among the atmosphere, lithosphere, hydrosphere, and biosphere (*see* Fig. 12-2).

Hydrology Systematic study of the earth's water in all its states.

Hydrolysis A form of chemical weathering that involves moistening and the transformation of rock minerals into other mineral compounds; expansion in volume often occurs in the process, which contributes to the breakdown of rocks.

Hydrosphere The sphere of the earth system that contains all the water that exists on and within the solid surface of our planet and in the atmosphere above.

Hygrophyte Plant adapted to the moisture of wet environments.

Hygroscopic water Thin films of water that cling tenaciously to most soil particles but are unavailable to plant roots.

Hypothetical continent model The earth's landmasses generalized into a single, idealized, shield-shaped continent of uniform low elevation; elaborated in Fig. 16-2.

Ice age A stretch of geologic time during which the earth's average atmospheric temperature is lowered; causes the expansion of glacial ice in the high latitudes and the growth of mountain glaciers in lower latitudes.

Icecap A regional mass of ice smaller than a continent-sized icesheet (less than 50,000 sq km [20,000 sq mi] in size). While the Laurentide Icesheet covered much of North America east of the Rocky Mountains, an icecap covered the Rockies themselves.

Icecap climate (EF) The world's coldest climate, the harsher of the two subtypes of the **E**-climate zone; characteristics, insofar as incomplete data reveal, are displayed in Fig. 19-9.

Ice crystal process The most common process whereby precipitation forms and falls to earth, described on p. 132; rainfall in the tropical latitudes, however, is produced by the coalescence process.

Icelandic Low The local name for the northernmost North Atlantic Ocean's Upper-Midlatitude Low, centered approximately at 60°N near Iceland.

Icesheet (see continental [sheet] glacier)

Ice shelf Smaller icesheet that is a floating, seaward extension of a continental glacier, such as Antarctica's Ross Ice Shelf.

Ice tongues Relatively small outlet glaciers that extend into the sea at the margin of a continental glacier.

Ice-wedge polygons Polygonal features formed by the freezing and thawing of sediments that fill surface cracks caused by very cold winter temperatures in periglacial zones (*see* Fig. 48-6).

Igneous rocks The (primary) rocks that formed directly from the cooling of molten magma; igneous is Latin for "formed from fire."

Illuviation The soil process in which downward-percolating water carries soluble minerals and colloid-sized particles of organic matter and minerals into the **B** horizon, where these materials are deposited in pore spaces and against the surfaces of soil grains.

Impermeable A surface that does not permit water to pass through it.

Impurities (atmospheric) Solid particles floating in the atmosphere whose quantities vary in time and space; among other things, they play an active role in the formation of raindrops.

Inceptisol One of the 11 Soil Orders of the Soil Taxonomy; forms quickly, is relatively young (though older than an entisol), has the beginnings of a **B** horizon, and contains significant organic matter.

Incised stream (*see* **entrenched meanders**)

Inert gas A chemically inactive gas that does not combine with other compounds; argon, an inert atmospheric gas, makes up almost 1 percent of dry air.

Infiltration The flow of water into the earth's surface through the pores and larger openings in the soil mass.

Infiltration capacity The rate at which a soil is able to absorb water percolating downward from the surface.

Infrared radiation Low-energy, longwave radiation of the type emitted by the earth and the atmosphere.

Inner core The solid, most inner portion of the earth, consisting mainly of nickel and iron (*see* Fig. 29-5).

In situ In place.

Insolation **In**coming **sol**ar radi**ation**.

Intensity (earthquake) The size and damage of an earthquake as measured—on the Modified Mercalli Scale—by the impact on structures and human activities on the cultural landscape.

Interactive mapping In geographic information systems (GIS) methodology, the constant dialogue via computer demands and feedback to queries between the map user and the map.

Interception The blocking of rainwater from reaching the ground by vegetation; raindrops land on leaves and other plant parts and evaporate before they can penetrate the soil below.

Interface Surface-like boundary of a system or one of its component subsystems; the transfer or exchange of matter and energy takes place here.

Interfluve The ridge that separates two adjacent stream valleys.

Interglaciation A period of warmer global temperatures between the most recent deglaciation and the onset of the next glaciation.

International date line For the most part is antipodal to the prime meridian and follows the 180th meridian; crossing the line toward the west involves skipping a day, while crossing the line toward the east means repeating a day.

Inter-Tropical Convergence Zone (ITCZ) The thermal low-pressure belt of rising air that straddles the equatorial latitudinal zone, which is fed by the windflows of the converging Northeast and Southeast Trades.

Intrusive igneous rocks Rocks formed from magma that cooled and solidified below the earth's surface.

Inversion (*see* **temperature inversion**)

Ion An atom or cluster of electrically charged atoms formed by the breakdown of molecules.

Isarithmic mapping A commonly used cartographic device to represent three-dimensional volumetric data on a two-dimensional map; involves the use of isolines to show the surfaces that are mapped.

Island arc A volcanic island chain produced in a zone where two oceanic plates are converging; one plate will subduct the other, forming deep trenches as well as spawning volcanoes that may protrude above sea level in an island-arc formation (Alaska's Aleutian archipelago is a classic example.)

Isobar A line connecting all points having the identical atmospheric pressure.

Isohyet A line connecting all points receiving the identical amount of precipitation.

Isoline A line connecting all places possessing the same value of a given phenomenon, or "height" above the flat base of the surface being mapped.

Isostasy Derived from an ancient Greek term (*iso* the same; *stasy* to stand), the condition of vertical equilibrium between the floating landmasses and the asthenosphere beneath them; this situation of sustained adjustment is maintained despite the forces that constantly operate to change the landmasses.

Isotherm A line connecting all points experiencing the identical temperature.

Isotope Related form of a chemical element; has the same number of protons and electrons but different numbers of neutrons.

ITCZ (*see* **Inter-Tropical Convergence Zone**)

Jet stream The two concentrated, high-altitude, west-to-east flowing "rivers" of air that are major features of the upper atmospheric circulation system poleward of latitude 15° in both the Northern and Southern Hemispheres; because of their general occurrence above the subtropical and subpolar latitudes, they are respectively known as the *subtropical jet stream* and the *Polar Front jet stream*. A third such corridor of high-altitude, concentrated windflow is the *tropical easterly jet stream*, a major feature of the upper-air circulation equatorward of latitude 15°N. This third jet stream, however, flows in the opposite, east-to-west direction and occurs only above the tropics of the Northern Hemisphere.

Joint A fracture in a rock in which no displacement movement has taken place.

Joint plane Planes of weakness and separation in a rock; produced in an intrusive igneous rock by the contraction of cooling magma (sedimentary and metamorphic rocks also display forms of jointing).

Jointing The tendency of rocks to develop parallel sets of fractures without any obvious movement such as faulting.

Kame A ridge or mound of glacial debris or stratified drift at the edge of a glacier; often found as deltaic deposits where

meltwater streams flowed into temporary lakes near the receding glacier.

Karst The distinctive landscape associated with the underground chemical erosion of particularly soluble limestone bedrock.

Katabatic winds The winds that result from cold-air drainage; especially prominent under clear conditions where the edges of highlands plunge sharply toward lower-lying terrain.

Kelvin scale The *absolute* temperature scale used by scientists, based on the temperature of absolute zero (−273°C/−459.4°F); a Kelvin degree (°K) is identical to a Celsius degree (°C), so that water boils at 373°K (100°C) and freezes at 273°K (0°C).

Kettle Steep-sided depression formed in glacial till that is the result of the melting of a buried block of ice.

Kinetic energy The energy of movement.

L wave A surface seismic wave that travels along the earth's crust.

La Niña The lull or cool ebb in low-latitude Pacific Ocean surface temperatures that occurs between El Niño peaks of anomalous sea-surface warming.

Laccolith A concordant intrusive igneous form in which a magma pipe led to a subterranean chamber that grew, dome-like, pushing up the overlying strata into a gentle bulge without destroying them.

Lagoon A normally shallow body of water that lies between an offshore barrier island and the original shoreline.

Lahar A mudflow largely comprised of volcanic debris. Triggered high on a snowcapped volcano by an eruption, such a mudflow can advance downslope at high speed and destroy everything in its path; frequently solidifies where it comes to rest.

Land breeze An offshore airflow affecting a coastal zone, resulting from a nighttime pressure gradient that steers local winds from the cooler (higher-pressure) land surface to the warmer (lower-pressure) sea surface.

Landform A single and typical unit that forms parts of the overall shape of the earth's surface; also refers to a discrete product of a set of geomorphic processes.

Land Hemisphere The roughly one-half of the earth that contains most of the landmasses (*see* Fig. 2-4); the opposite of the water (oceanic) hemisphere.

Landscape An aggregation of landforms, often of the same type; also refers to the spatial expression of the processes that shaped those landforms.

Landslide A slide form of mass movement that travels downslope more rapidly than flow movements; in effect, it is a collapse of a slope and does not need water as a lubricant (can also be triggered by an earthquake as well as human activities).

Lapse rate The rate of decline in temperature as altitude increases; the average lapse rate of temperature with height in the troposphere is 0.65°C/100 m (3.5°F/1000 ft).

Late Cenozoic Ice Age The last great ice age that ended 10,000 years ago; spanned the entire Pleistocene Epoch (2,000,000 to 10,000 years ago) plus the latter portion of the preceding Pliocene Epoch, possibly beginning as far back as 3 million years ago.

Latent heat of fusion The heat energy involved in melting, the transformation of a solid into a liquid; a similar amount of heat is given off when a liquid freezes into a solid.

Latent heat of vaporization The heat energy involved in the transformation of a liquid into a gas or vice versa.

Lateral moraine Moraine situated along the edge of a mountain glacier, consisting of debris that fell from the adjacent valley wall.

Laterite Name sometimes given to a very hard iron oxide or aluminum hydroxide soil horizon that is reddish in color (*later* is Latin for brick); found in wet tropical areas.

Latitude The angular distance, measured in degrees north or south, of a point along a parallel from the equator.

Laurasia The northern portion of the primeval supercontinent, Pangaea.

Laurentide Icesheet The huge, Late Cenozoic continental glacier that covered all of Canada east of the Rocky Mountains and expanded repeatedly to bury northern U.S. areas as far south as the Ohio and Missouri Valleys.

Lava Magma that reaches the earth's surface.

Leaching The soil process in which downward-percolating water dissolves and washes away many of the soil's mineral substances and other ingredients.

Lead (ice surface) A channel of water that opens across a surface of sea ice during the warmer period of the year.

Leap year Occurs every fourth year, when a full day (February 29) is added to the calendar to allow for the quarter-day beyond the 365 days it takes the earth to complete one revolution of the sun.

Leeward The protected side of a topographic barrier with respect to the winds that flow across it; often refers to the area downwind from the barrier as well, which is said to be in the "shadow" of that highland zone.

Legend (map) The portion of a map where its point, line, area, and volume symbols are identified.

Levee, artificial An artificially constructed levee built to reinforce a natural levee, most often along the lowest course of a river.

Levee, natural A river-lining ridge of alluvium deposited when a stream overflows its banks during a periodic flood; when the river contracts after the flood, it stays within these self-generated "dikes" or levees.

Liana Tropical rainforest vine, rooted in the ground, with leaves and flowers in the tree canopy above.

Light-year The distance traveled by a pulse of light in one year. Light travels at a speed of 300,000 km (186,000 mi) per second; a light-year thus involves a distance of 9.46 trillion km (5.88 trillion mi).

Limestone A nonclastic sedimentary rock mainly formed from the respiration and photosynthesis of marine organisms in which calcium carbonate is distilled from seawater; finely textured and therefore resistant to weathering when exposed on the surface, it is susceptible to solution that can produce karst landscapes both above and below the ground.

Line symbols Represent linkages and/or flows that exist between places on a map.

Linear autogenic succession A plant succession that occurs when the plants themselves initiate changes in the environment that consequently cause vegetation changes.

Lithification Rock formation; the process of compression, compaction, and cementation whereby a sediment is transformed into a sedimentary rock.

Lithology The rock type of a local area, which greatly influences its landform and landscape development.

Lithosphere The outermost shell of the solid earth, lying immediately below the land surface and ocean floor (*lithos* means rock); composed of the earth's thin crust together with the solid uppermost portion of the upper mantle that lies just below.

Lithospheric plate One of the fragmented, rigid segments of the lithosphere (also called a *tectonic plate*, which denotes its active mobile character); these segments or plates move in response to the plastic flow in the hot asthenosphere that lies just below the lithosphere.

Little Ice Age The period of decidedly cooler global temperatures that prevailed from 1430 to 1850 (averaging ca. 1.5°C [3.7°F] lower than in the 1940s); during these four centuries, glaciers in most parts of the world expanded considerably.

Littoral zone Coastal zone.

Loam A soil containing grains of all three texture size categories—sand, silt, and clay—within the certain proportions that are indicated in Fig. 24-2; the term, however, refers not to a size category but to a certain combination of variously sized particles.

Local base level The base level for a river that flows into a lake at whatever altitude it may lie.

Loess A deposit of very fine silt or dust laid down after having been blown some distance (perhaps hundreds of kilometers) by the wind; characterized by its fertility and ability to stand in steep vertical walls.

Longitude The angular distance, measured in degrees east or west, of a point along a meridian from the prime meridian.

Longitudinal dune A long ridge-like sand dune that lies parallel to the prevailing wind.

Longshore bar A ridge of sand parallel to the shoreline that develops in the nearshore beach zone.

Longshore current A shore-paralleling water current, similar to the longshore drift of sand, that is generated by the refracted, oblique-angled arrival of waves onshore (*see* Figs. 50-4 and 50-6); can also develop from tidal action and from coastal storms.

Longshore drift The movement of sand along the shoreline in the flow of water (*longshore current*) generated by the refracted, oblique-angled arrival of waves onshore (*see* Fig. 50-6).

Longwave radiation Radiation emitted by the earth, which has much longer wavelengths—and involves much lower energy—than the solar (shortwave, higher-energy) radiation emitted by the sun.

Low mountains Terrain of less than 50 percent gentle slope whose local relief exhibits variations of 300–900 m (1000–3000 ft).

Lower mantle The solid interior shell of the earth that encloses the liquid outer core (*see* Fig. 29-5).

Luster (mineral) The surface sheen of a mineral that, along with color, can be a useful identifying quality.

Magma The liquid molten mass from which igneous rocks are formed.

Magnetosphere The upper portion of the thermosphere that constitutes the outermost of the atmosphere's layers; here the earth's magnetic field is often more influential in the movement of particles than its gravitational field is.

Magnitude (earthquake) The amount of shaking of the ground during an earthquake as measured by a seismograph.

Mangrove A type of tree, exhibiting stilt-like roots and a leafy crown, that grows in seawater on low-lying, muddy coasts in tropical and subtropical environments. In certain places, such as tidal flats and river mouths, these trees congregate thickly; in the United States, they are most evident along the southwesternmost shore of Florida where the Everglades meet the sea.

Map projection An orderly arrangement of meridians and parallels, produced by any systematic method, that can be used for drawing a map of the spherical earth on a flat surface.

Marble Metamorphosed limestone; the hardness and density of this rock is preferred by sculptors for statues that can withstand exposure to the agents of erosion for millennia.

Marine geography Physical side of this subfield treats coastlines and shores, beaches, and other landscape features associated with the oceanic margins of the continents.

Marine west coast climate (Cfb, Cfc) The perpetually moist mesothermal climate type associated with midlatitude coasts (and adjacent inland areas not blocked by mountain barriers) bathed by the prevailing Westerlies year-round, whose maritime influences produce temperature regimes devoid of extremes; characteristics are displayed in Figs. 18-6, 18-7, and 18-8.

Maritime effect The moderating influence of the ocean on air temperature, which produces cooler summers and milder winters relative to inland locations at similar latitudes.

Mass balance The gains and losses of glacier ice in the open glacial system.

Mass movement The spontaneous downslope movement of earth materials under the force of gravity; materials involved move *en masse*—in bulk.

Master horizons The soil horizons marked by capital letters: *O*, *A*, *E*, *B*, *C*, and *R* (*see* Fig. 23-5).

Meander belts On a wide floodplain, river meander development zones that themselves form giant meanders (*see* Fig. 43-7).

Meanders The smooth, rounded curves or bends of rivers that can become quite pronounced as floodplain development proceeds; also characteristic of many ocean currents, which, after the passage of storms, can produce such extreme loops that many detach and form localized eddies.

Meander scar A dried up, curved linear depression on a floodplain that is evidence of an old river channel that was once a meander or an oxbow lake.

Mechanical weathering Involves the destruction of rocks by physical means through the imposition of certain stresses, such as freezing and thawing or the expansion of salt crystals; also known as *physical weathering*.

Medial moraine A moraine—situated well away from a glacier's edges—formed by the intersection of two lateral moraines when a substantial tributary glacier meets and joins a trunk glacier.

Mediterranean climate (Csa, Csb) The west-coast mesothermal climate whose signature dry summers result from the temporary poleward shift of the subtropical high-pressure zone; characteristics are displayed in Figs. 18-9 and 18-12.

Mediterranean scrub biome Consists of widely spaced evergreen or deciduous trees and often dense, hard-leaf evergreen scrub; thick waxy leaves are well adapted to the long dry summers. Sometimes referred to as chaparral or maquis.

Megatherm Plant adapted to withstand considerable heat.

Melting A change from the solid state to the liquid state; at 0°C (32°F) it takes ca. 80 cal of heat energy to change 1 g (0.04 oz) of water from a solid into a liquid.

Mercalli Scale (Modified) Scale that measures earthquake intensity, the impact of a quake on the human landscape; ranges upward from intensity I to intensity XII.

Mercator map projection The most famous of the cylindrical projections, the only one on which any straight line is a line of true and constant compass bearing.

Meridian On the earth grid, a north-south line of longitude; these range from 0° (prime meridian) to 180° E and W (180° E and W, of course, are the same line—the international date line [written simply as 180°]).

Meridional circulation The north-south (azonal) movement of air.

Mesa Flat-topped, steep-sided upland capped by a resistant rock layer; normally found in dry environments.

Mesopause The upper boundary of the mesosphere, lying approximately 80 km (50 mi) above the surface.

Mesophyte Plant adapted to environments that are neither extremely moist nor extremely dry.

Mesosphere The third layer of the atmosphere, lying above the troposphere and stratosphere; here temperatures again decline with increasing elevation as they do in the troposphere.

Mesotherm Plant adapted to withstand intermediate temperatures that are not excessively high or low.

Mesothermal Climates exhibiting a moderate amount of heat; the **C** climates.

Mesothermal (C) climates The moderately heated climates that are found on the equatorward side of the middle latitudes, where they are generally aligned as interrupted east-west belts; transitional between the climates of the tropics and those of the upper midlatitudes where polar influences begin to produce harsh winters.

Mesozoic The era of medieval life on the geologic time scale (*see* Fig. 37-7), extending from 240 million years ago to 66 million years ago.

Metabolic heat The heat produced by human bodies (from the conversion of the chemical energy in the food we eat).

Metamorphic rocks The (secondary) rocks that were created from the transformation, by heat and/or pressure, of existing rocks.

Meteorology The systematic, interdisciplinary study of the short-term atmospheric phenomena that constitute weather.

Metropolis A metropolitan area consisting of a central city and its surrounding suburban ring; since U.S. suburbs in the 1990s are no longer "sub" to any "urb," that ring has now become an outer city in its own right, capturing an increasingly large share of the metropolis's population and economic activity.

Microclimate Climate region on a localized scale.

Microtherm Plant adapted to withstand low temperatures.

Microthermal Northern Hemisphere climates exhibiting a comparatively small amount of heat; the **D** climates.

Microthermal (D) climates The weakly heated continental climates of the Northern Hemisphere's upper midlatitudes, where the seasonal rhythms swing from short, decidedly warm summers to long, often harsh winters; mostly confined to the vast interior expanses of North America and Eurasia poleward of 45°N.

Mid-Atlantic Ridge The submarine volcanic mountain range that forms the midoceanic ridge that extends through the entire North and South Atlantic Oceans, from Iceland in the north to near-Antarctic latitudes in the south.

Midoceanic ridge High submarine volcanic mountain ranges, part of a global system of such ranges, most often found in the central areas of the ocean basins; here new crust is formed by upwelling magma, which continuously moves away toward the margins of the ocean basins.

Midstream bar A midchannel sandbar that is deposited where sediment-clogged water of a stream significantly slows in velocity.

Millennium A period of 1000 years.

Mineral Naturally occurring inorganic element or compound having a definite chemical composition, physical properties, and, usually, a crystalline structure.

Mineral spring A hot spring containing large quantities of minerals dissolved from surrounding rocks; its mineral water is sometimes used for medicinal purposes.

Mistral winds A katabatic windflow, affecting France's Rhône Valley each winter, that involves the cold-air drainage of the massive pool of icy air that accumulates over the nearby, high-lying French and Swiss Alps.

Mixing ratio The ratio of the mass of water vapor to the total mass of the dry air containing that water vapor.

Model The creation of an idealized representation of reality in order to demonstrate its most important properties.

Moho Abbreviation for the Mohorovičić discontinuity.

Mohorovičić discontinuity The contact plane between the earth's crust and the mantle that lies directly below it.

Mohs Hardness Scale The standard mineral-hardness measurement scale used in the earth sciences; ranges from 10 (the hardest substance, diamond) down to 1 (talc, the softest naturally occurring mineral).

Mollisol One of the 11 Soil Orders of the Soil Taxonomy, found in the world's semiarid climate zones; characterized by a thick, dark surface layer and high alkaline content.

Moment Magnitude Scale The most widely used measure of the severity of an earthquake's ground motion, which evolved from the Richter Scale developed in the 1930s; based on the size of the fault along which a quake occurs and the distance the rocks around it slip.

Monadnock A prominent, not-yet-eroded remnant of an upland on a peneplain.

Monsoon Derived from the Arabic word for "season," a regional windflow that streams onto and off certain landmasses on a seasonal basis; the moist onshore winds of summer bring the **wet monsoon**, whereas the offshore winds of winter are associated with the *dry monsoon*.

Monsoon rainforest climate (Am) The tropical monsoon climate, characterized by sharply distinct wet and dry seasons; characteristics are displayed in Fig. 17-4.

Moon Satellite that orbits a planet, probably originating from the clustering of planetesimals. All solar system planets except Mercury and Venus have such bodies; our Moon orbits the earth once every 27.3 days at an average distance of 385,000 km (240,000 mi).

Moraine A ridge or mound of glacial debris deposited during the melting phase of a glacier.

Morphology Shape or form.

Mountain (alpine) glaciers Rivers of ice that form in mountainous regions; confined in valleys that usually have steep slopes.

Mudflow A flow form of mass movement involving a stream of fluid, lubricated mud; most common where heavy rains strike an area that has long been dry and where weathering has loosened ample quantities of fine-grained material.

Multispectral systems The paraphernalia of remote-sensing platforms, which today increasingly employ sets of scanners simultaneously attuned to several different spectral bands, greatly enhancing observation quality and interpretive capabilities.

Muskeg In the northern coniferous forest biome, the particular assemblage of low-growing leathery bushes and stunted trees that concentrate in the waterlogged soil bogs and lake-filled depressions.

Mutation Variation in reproduction in which the message of heredity (DNA) contained in the genes is imperfectly passed on and from which new species may originate.

Mutualism Biological interaction in which there is a coexistence of two or more species because one or more is essential to the survival of the other(s); also called *symbiosis*.

Neap tide A least-extreme tide that occurs when the moon's pull works at right angles against the sun's attraction and the earth's rotational bulge (*see* Fig. 50-9b).

Nearshore The beach zone that is located seaward of the foreshore, submerged even during an average low tide; longshore bars and troughs develop here in this zone of complex, ever-changing topography.

Negative feedback mechanism A feedback mechanism that operates to keep a system in its original condition.

Net radiation The amount of radiation left over when all the incoming and outgoing radiation flows have been tallied (*see* Table 7-1, p. 78); totals about one-fourth of the short-wave radiation originally arriving at the top of the atmosphere.

Niche (*see* **ecological niche**)

Nonclastic sedimentary rocks Derived not from particles of other rocks, but from chemical solution by deposition and evaporation or from organic deposition.

Nonrenewable resource One that when used at a certain rate will ultimately be exhausted (metallic ores and petroleum being good examples).

Nonsilicates Minerals consisting of compounds that do not contain the silicon and oxygen of the silicates; include the carbonates, sulfates, sulfides, and halides.

Normal fault A tensional fault exhibiting a moderately inclined fault plane that separates a block that has remained fairly stationary from one that has been significantly downthrown (*see* Fig. 36-5).

North Pole The "top" of the earth, at latitude 90°N; one of two places (the other being the South Pole) where the planet's axis intersects the surface.

Northeast Trades The surface wind belt that generally lies between the equator and 30°N; the Coriolis force deflects equatorward-flowing winds to the right, thus recurving north winds into northeast winds.

Northern coniferous forest biome The upper-midlatitude boreal forest (known in Russia as the snowforest or *taiga*); dominated by dense stands of slender, cone-bearing, needleleaf trees.

Northern Hemisphere The half of the earth located north of the equator (0° latitude); northernmost point is the North Pole (90°N).

Nuclear winter The hypothesized global cooling, caused by massive smoke plumes, that would be generated by the widespread fires of a nuclear war.

Nuée ardente A cloud of high-temperature volcanic gas that races downslope following a spectacular explosion associated with unusually high pressures inside the erupting volcano; incinerates everything in its path.

Numerical weather prediction The computer weather forecasting method used by the National Weather Service, based on projections by small increments of time up to 48 hours into the future.

Nunatak A mountain peak of a buried landscape that protrudes through the overlying glacier.

O horizon In certain soils, the uppermost layer—lying above the *A* horizon—consisting entirely of organic material in various stages of decomposition.

Oblate spheroid The (minor) departure in the shape of Planet Earth from that of a perfect sphere; the earth, as an oblate spheroid, slightly bulges at the equator and is slightly flattened at the poles.

Occluded front The surface boundary between cold and cool air in a mature midlatitude cyclone; caused by the cold front undercutting and lifting the warm air entirely off the ground as in Fig. 14-9c.

Oceanic trench A prominent seafloor feature, often adjacent to an island arc, in a zone where two oceanic plates converge and one subducts the other; the greatest depths of the world ocean lie in such trenches in the Pacific east of the Philippines.

Offshore bar An offshore sandbar that lies some distance from the beach and is not connected to land; a longshore bar is an example.

Open system A system whose boundaries (interfaces) freely permit the transfer of energy and matter across them.

Open wave The early-maturity stage in the development of a midlatitude cyclone; surface cyclonic air motion transforms the original kink on the stationary front into an open wave, around which cold and warm air interact in the distinct ways shown in Fig. 14-9b.

Optimum (biogeographic) range For each plant and animal species, the geographic area where it can maintain a large, healthy population.

Orders of magnitude Sizes of geographic entities; Fig. 1-7 shows the entire range of magnitudes, including those that geographers usually operate within.

Organic matter One of the four soil components; the material that forms from living matter.

Orogenic belt A chain of linear mountain ranges.

Orographic precipitation The rainfall (and sometimes snowfall) produced by moist air parcels that are forced to rise over a mountain range or other highland zone; such air parcels move in this manner because they are propelled both by steering winds and the push of other air parcels piling up behind them.

Outer core The liquid shell that encloses the earth's inner core, whose composition involves similar materials (*see* Fig. 29-5).

Outwash plain Plain formed ahead of a receding icesheet by the removal of material carried in the glacier by meltwater; exhibits both erosional and depositional features.

Overthrust fault A compressional fault in which the angle of the fault plane is very low (*see* Fig. 36-4).

Overturned fold An extremely compressed fold that doubles back on itself with an axial plane that is oriented beyond the horizontal (*see* Fig. 36-17b).

Oxbow lake A lake formed when two adjacent meanders link up and one of the bends in the channel, shaped like a bow, is cut off (*see* Fig. 43-6)

Oxic horizon In moist equatorial and tropical areas, where humus is minimal, downward-percolating rainwater leaches the soil and leaves behind compounds rich in iron and aluminum; sometimes these aluminum and iron oxides are so concentrated they develop into a hard layer, red or orange in color, known as an oxic horizon (farmers call it *hardpan*).

Oxidation The chemical combination of oxygen and other materials to create new products (biologists call this process *respiration*).

Oxisol One of the 11 Soil Orders of the Soil Taxonomy, found in tropical areas with high rainfall; heavily leached and usually characterized by a pronounced oxic horizon, red or orange in color.

Oxygen cycle Oxygen is put back into the atmosphere as a by-product of photosynthesis, and is lost when it is inhaled by animals or chemically combined with other materials during oxidation.

Ozone hole The seasonal depletion of ozone over the Antarctic region, which very likely is caused by the discharge of artificial chemical compounds called chlorofluorocarbons (CFCs); evidence is mounting that a global thinning of the ozone layer is also taking place.

Ozone layer Also known as the ozonosphere, the ozone-rich layer of the atmosphere that extends between 15 and 50 km (9 and 31 mi) above the surface; the highest concentrations of ozone are usually found at the level between 20 and 25 km (12 and 15 mi).

Ozonosphere Synonym for the ozone layer.

P wave A seismic body wave that is also a "push" or compressional wave; moves material in its path parallel to the direction of wave movement, and can even travel through material in a liquid state.

Pacific Ring of Fire The Circum-Pacific belt of high volcanic and seismic activity, stretching around the entire Pacific Basin counterclockwise through western South America, western North America, and Asia's island archipelagoes (from Japan to Indonesia) as far as New Zealand.

Pack ice Floating sea ice that forms from the freezing of ocean water.

Pahoehoe Ropy-patterned lava (*see* Fig. 34-10); forms where very fluid lavas develop a smooth "skin" upon hardening that wrinkles as movement continues.

Paleozoic The era of ancient life on the geologic time scale (*see* Fig. 37-7) extending from 570 million years ago to 240 million years ago.

Pangaea The primeval supercontinent, hypothesized by Alfred Wegener, that broke apart and formed the continents and oceans as we know them today; consisted of two parts—a northern Laurasia and a southern Gondwana.

Parabolic dune A crescent-shaped sand dune with its points lying upwind; concave side of this dune is the windward side.

Parallel On the earth grid, an east-west line of latitude; parallels of latitude range from 0° (equator) to 90°N and S (the North and South Poles, respectively, where the east-west line shrinks to a point).

Parallelism The constant tilt of the earth's axis (at 66½° to the plane of the ecliptic), which remains parallel to itself at every position in its annual revolution of the sun.

Parent material The rocks of the earth, and the deposits formed from them, from which the overlying soil is formed.

Park A broad valley located among the high ranges of the Southern Rocky Mountains in Colorado.

Patterned ground Periglacial rock and soil debris shaped or sorted in such a manner that it forms designs on the surface resembling rings, polygons, lines, and the like (*see* Fig. 48-7).

Ped A naturally occurring aggregate or "clump" of soil and its properties.

Pediment The smooth, gently sloping bedrock surface that underlies the alluvial apron of a mountain front and extends outward from the foot of the highlands; primarily a desert feature.

Pediplane A surface formed by the coalescence of numerous pediments after a long period of erosion has led to parallel slope retreat

Pedology Soil science; the study of soils.

Pedon A column of soil drawn from a specific location, extending from the **O** horizon (if present) all the way down to the level where the bedrock shows signs of being transformed into **C**-horizon material.

Peneplain The concept of a "near plain" developed by William Morris Davis to describe the nearly flat landscape formed by extensive erosion over long periods of time.

Perched aquifer Pockets of groundwater situated above the level of the local water table; this type of confined aquifer is often associated with karst areas.

Perched water table A separate local water table that forms at a higher elevation than the nearby main water table; caused by the effects of a local aquiclude, as shown in Fig. 40-10a.

Percolation The downward movement of water through the pores and other spaces in the soil under the influence of gravity.

Perennial plant One that persists throughout the year.

Periglacial A high-latitude or high-altitude environment on the perimeter of a glaciated area.

Perihelion The point in the earth's orbit, which occurs every January 3, where the distance to the sun is minimized (ca. 147.5 million km [91.7 million mi]).

Permafrost The permanently frozen layer of subsoil that is characteristic of the colder portions of the **D**-climate zone as well as the entire **E**-climate zone; can exceed 300 m (1000 ft) in depth.

Permafrost table The upper surface of the permafrost (analogous to the water table); the soil above is subject to annual thawing and freezing.

Permeable A surface that permits water to pass through it; most natural surfaces are able to absorb, in this manner, at least a portion of the water that falls on them.

pH scale Used to measure acidity and alkalinity of substances on a scale ranging from 0 to 14, with 7 being neutral; below 7 increasing acidity is observed as 0 is approached, while above 7 increasing alkalinity is observed as 14 is approached.

Photochemical pollution A form of secondary pollution in which the effects of sunlight play a role.

Photosynthesis The process in which plants convert carbon dioxide and water into carbohydrates and oxygen through the addition of solar energy; carbohydrates are a significant component of the food and tissue of both plants and animals.

Phreatic eruption An extraordinarily explosive volcanic eruption involving the penetration of water into a superheated magma chamber; such explosions of composite volcanoes standing in water can reach far beyond a volcano's immediate area.

Phreatic zone (*see* **zone of saturation**)

Physical geography The geography of the physical world. Figure 1-4 diagrams the subfields of physical geography.

Physiographic province (*see* **physiographic region**)

Physiographic realm A first-order subdivision of the North American continent at the broadest scale; characterized by an appropriate uniformity of landscapes, landforms, and other physiographic elements.

Physiographic region A second-order subdivision of the North American continent at a more detailed scale than the physiographic realm; characterized by an appropriate uniformity of landscapes, landforms, and other physiographic elements. Also sometimes called *physiographic provinces*, an older geographic term still in use.

Physiography Literally means *landscape description*; refers to all of the natural features on the earth's surface, including landscapes, landforms, climate, soils, vegetation, and hydrography.

Phytogeography The geography of flora or plant life; where botany and physical geography overlap.

Phytomass The total living organic plant matter produced in a given geographic area; often used synonymously with biomass, because biomass is measured by weight (plants overwhelmingly dominate over animals in total weight per unit area).

Phytoplankton Microscopic green autotrophic plants at the beginning of the food chain.

Pingo A mound-like, elliptical hill in a periglacial zone whose core consists of ice rather than rock or soil (*see* Fig. 48-8).

Plain On a land surface of more than 50 percent gentle slope, a low-lying area that exhibits less than 90 m (300 ft) of local relief.

Planar map projection One in which the transfer of the earth grid is from a globe onto a plane, involving a single point of tangency.

Plane of the ecliptic The plane formed by the sun and the earth's orbital path.

Planet A dark solid body, much smaller in size than a star, whose movements in space are controlled by the gravitational effects of a nearby star.

Planetesimal Asteroid-sized body that, during the formation of the solar system, combined with swarms of similar bodies to eventually become compressed into forming one of the nine planets.

Plankton (*see* **phytoplankton; zooplankton**)

Plant nutrients In a soil, the plant-food substances that circulate through the humus-vegetation system (such as nitrogen compounds and phosphates).

Plant succession The process in which one type of vegetation is replaced by another.

Plate tectonics The study of those aspects of tectonics that treat the processes by which the lithospheric plates move over the asthenosphere.

Plateau On a land surface of more than 50 percent gentle slope, a tableland that exhibits more than 90 m (300 ft) of local relief; moreover, most of the gently sloping area occurs in the lower half of the plateau's elevational range, and plateaus are also bounded on at least one side by a sharp drop or rise in altitude.

Platy soil structure Involves layered peds that look like flakes stacked horizontally (*see* Fig. 24-3).

Pleistocene The epoch that extended from ca. 2 million to 10,000 years ago on the geologic time scale; includes the latter half of the last great (Late Cenozoic) ice age, which began ca. 3 million years ago, as well as the emergence of humankind.

Plucking (quarrying) A glacial erosion process in which fragments of bedrock beneath the glacier are extracted from the surface as the ice advances.

Plunging fold An anticline or syncline whose axis dips from the horizontal.

Pluton A body of intrusive igneous rock.

Pluvial lake A lake that developed in a presently dry area during times of heavier precipitation associated with glaciations; Glacial Lake Bonneville, the (much larger) forerunner of Utah's Great Salt Lake, is a classic example.

Point symbols Symbols that exhibit the location of each occurrence of the phenomenon being mapped, and frequently its quantity.

Polar (E) climates Those in which the mean temperature of the warmest month is less than 10°C (50°F); the tundra (**ET**) subtype exhibits warmest-month temperatures between 0°C (32°F) and 10°C (50°F), while in the (coldest) icecap (**EF**) subtype the average temperature of the warmest month does not reach 0°C (32°F).

Polar Easterlies The high-latitude wind belt in each hemisphere, lying between 60 and 90 degrees of latitude; the Coriolis force is strongest in these polar latitudes, and the equatorward-moving air that emanates from the Polar High is sharply deflected in each hemisphere to form the Polar Easterlies.

Polar Front The latitudinal zone, lying at approximately 60 degrees north and south, where the equatorward-flowing Polar Easterlies meet the poleward-flowing Westerlies; the warmer Westerlies are forced to rise above the colder Easterlies, producing a semipermanent surface low-pressure belt known as the *Upper-Midlatitude Low.*

Polar Front jet stream The upper atmosphere jet stream located above the subpolar latitudes, specifically the Polar Front; at its strongest during the half-year centered on winter.

Polar High Large semipermanent high-pressure cell centered approximately over the pole in the uppermost latitudes of each hemisphere.

Pollution (air) Air is said to be polluted when its composition departs significantly from its natural composition of such gases as nitrogen and oxygen; we are simultaneously concerned with factors that are detrimental to human comfort and, possibly, even life itself.

Pollution plume When prevailing winds exceed 13 kph (8 mph), dust domes begin to detach themselves from the cities they are centered over; as Fig. 21-9 shows, the polluted air streams out as a plume above the downwind countryside.

Porosity (rock) The water-holding capacity of a rock.

Positive feedback mechanism A feedback mechanism that induces progressively greater change from the original condition of a system.

Potential energy The energy an object has by virtue of its position relative to another object.

Potential evapotranspiration (PE) The maximum amount of water that can be lost to the atmosphere from a land surface with abundant available water.

Precambrian The era that precedes the Paleozoic era of ancient life on the geologic time scale (*see* Fig. 37-7), named after the oldest period of the Paleozoic, the Cambrian; extends backward from 570 million years ago to the origin of the earth, now estimated to be ca. 4.6 billion years ago.

Precipitation Any liquid water or ice that falls to the earth's surface through the atmosphere (rain, snow, sleet, and hail).

Predation Biological interaction, rare in occurrence, in which one species eats all members of another species.

Pressure (atmospheric) The weight of a column of air at a given location, determined by the force of gravity and the composition and properties of the atmosphere at that location. *Standard sea-level air pressure* produces a reading of 760 mm (29.92 in) on the mercury barometer; in terms of weight, it is also given as 1013.25 millibars (mb) or 14.7 lb per sq in.

Pressure gradient force The difference in surface pressure over a given distance between two locations is called the *pressure gradient*; when that pressure gradient exists, it acts as a force that causes air to move (as wind) from the place of higher pressure to that of lower pressure.

Prevailing Westerlies (*see* **Westerlies**)

Primary circulation (*see* **general circulation**)

Primary landform A structure created by tectonic activity.

Primary pollutant A gaseous or solid pollutant that comes from an industrial or domestic source or the internal combustion engine of a motor vehicle.

Primary rocks The igneous rocks, the only rock type created directly from the solidification of the earth's primeval molten crust.

Prime meridian The north-south line on the earth grid, passing through the Royal Observatory at Greenwich in London, defined as having a longitude of 0°.

Prismatic soil structure Involves peds arranged in columns, giving a soil vertical strength (*see* Fig. 24-4).

Proxy climatic data Indirect evidence of past climatic change; found all over the natural world, such as seafloor sediment deposits and the concentric annual growth rings of trees.

Psychrometer An instrument consisting of two thermometers; the bulb of one is swaddled in a wet cloth, while the other is not. It is used to measure relative humidity, specific humidity, and the mixing ratio.

Pyroclastics The collective name for the solid lava fragments that are erupted explosively from a volcano.

Quartzite A very hard metamorphic rock that resists weathering; formed by the metamorphosis of sandstone (made of quartz grains and a silica cement).

Quaternary The second of the two periods of the Cenozoic era of recent life on the geologic time scale (*see* Fig. 37-7), extending from ca. 2 million years ago to the present.

R horizon The layer at the base of the soil where the bedrock is breaking up and weathering into the particles from which soil is being formed; *R* stands for regolith.

Radial drainage A stream pattern that emanates outward in many directions from a central mountain (*see* Fig. 42-10a).

Radiant heat Heat flows in the form of shortwave or longwave electromagnetic radiation; the composite flows of radiant heat tallied on the surface and in the atmosphere make up net radiation.

Radiation The transmission of energy in the form of electromagnetic waves; a wide range of energy occurs within the electromagnetic spectrum (*see* Fig. 3-15).

Radiosonde Radio-equipped weather instrument packages that are carried aloft by balloon.

Rain Precipitation consisting of large liquid water droplets.

Rain shadow effect The dry conditions—often at a regional scale as in the U.S. interior West—that occur on the leeward side of a mountain barrier that experiences orographic precipitation; the passage of moist air across that barrier wrests most of the moisture from the air, whose adiabatic warming as it plunges downslope sharply lowers the dew point and precipitation possibilities.

Range (animal) The area of natural occurrence of a given animal species; often changes over time, and in some cases even seasonally (*see also* **optimum [biogeographic] range**).

Rawinsonde High-altitude radar tracking of radiosonde balloons, which provides information about wind speed and directions at various vertical levels in the atmosphere.

Recessional moraine A morainal ridge marking a place where glacial retreat was temporarily halted.

Rectangular drainage A stream pattern dominated by right-angle contacts between rivers and tributaries, but not as pronounced as in trellis drainage (*see* Fig. 42-10d).

Recumbent fold A highly compressed fold that doubles back on itself with an axial plane that is near horizontal (*see* Fig. 36-17a).

Regime (*see* **soil regime**)

Regional concept Used to classify and categorize spatial information; regions are artificial constructs developed by geographers to delineate portions of the earth's surface that are marked by an overriding sameness or homogeneity (in our case, physiography [*see* Units 52 and 53]).

Regional subsystem The particular interconnection, at any given place within the total earth system, of the four spheres or subsystems (atmosphere, hydrosphere, lithosphere, biosphere).

Regolith Weathered, broken, loose material overlying bedrock, usually derived from the rock below, but sometimes transported to the area. First stage in the conversion of bedrock to soil; located at the base of the soil in the *R* horizon (*R* for regolith).

Rejuvenated stream One that has newly increased its erosive power and cuts downward into its own floodplain and other alluvial deposits; caused by either a relative fall in the stream's base level or the tectonic uplift of the valley.

Relative humidity The proportion of water vapor present in a parcel of air relative to the maximum amount of water vapor that air could hold at the same temperature.

Relief The vertical distance between the highest and lowest elevations in a given area.

Remote sensing A technique for imaging objects without the sensor being in immediate contact with the local scene.

Renewable resource One that can regenerate as it is exploited.

Rescue effect When a successful species disperses into a less suitable environment, it often manages to survive because the species' numbers are constantly replenished by a continuing stream of migrants.

Residual soil The simplest kind of soil formation in which a soil forms directly from underlying rock; when this occurs, the dominant soil minerals bear a direct relationship to that original rock.

Respiration The biological name for the chemical combination of oxygen and other materials to create new products (chemists call this process *oxidation*). Heat increases plants' rates of respiration; respiration runs counter to photosynthesis because it breaks down available carbohydrates and combines them with oxygen.

Reverse fault The result of one crustal block overriding another along a steep fault plane between them (*see* Fig. 36-2); caused by compression of the crust into a smaller horizontal space.

Revolution One complete circling of the sun by a planet; it takes the earth precisely one year to complete such an orbit.

Rhumb line Any straight line drawn on a cylindrical Mercator map projection, which is automatically a line of true and constant compass bearing.

Richter Scale Open-ended numerical scale, first developed in the 1930s, that measures earthquake magnitude; ranges upward from 0 to 8+. Has evolved into the *Moment Magnitude Scale*.

Rift An opening of the crust, normally into a trough or trench, that occurs in a zone of plate divergence; *see also* **rift valley**.

Rift valley Develops in a continental zone of plate divergence where tensional forces pull the crustally thinning surface apart; the rift valley is the trough that forms when the land sinks between parallel faults in strips.

Rip current Narrow, short-distance, stream-like current that moves seaward from the shoreline cutting directly across the oncoming surf.

Roche moutonnée A landform created by glacial plucking; an asymmetrical mound that results from abrasion to one side (the side from which the ice advanced) and plucking on the leeward side.

Rock Any naturally formed, firm, and consolidated aggregate mass of mineral matter, of organic or inorganic origin, that constitutes part of the planetary crust; *see also* **igneous**, **sedimentary**, and **metamorphic rocks**.

Rock cycle Cycle of transformation that affects all rocks and involves all parts of the earth's crust (*see* Fig. 31-12): plutons form deep in the crust, uplift pushes them to the surface, erosion wears them down, and the sediments they produce become new mountains.

Rock flour The very finely ground-up debris carried downslope by a mountain glacier; when deposited, often blown away by the wind.

Rock glacier In high-relief periglacial areas, a tongue of boulder-like rock debris that has slowly moved downslope as a unit; may have been consolidated by a since-eroded matrix of finer sediments, or cemented together by long-melted ice.

Rock sea The area of blocky rock fragments formed when weathered rocks—particularly from frost wedging—remain near their original location; also known as a *blockfield* or *felsenmeer*, and a *boulder field* when the rock fragments are dominated by large boulders.

Rockslide A landslide-type of mass movement consisting mainly of rock materials.

Rock steps The step-like mountainside profile (in the postglacial landscape) often created as an eroding alpine glacier moved downslope.

Rock terrace Terrace of hard bedrock that results from the resumed downward cutting of a rejuvenated stream that removes the entire alluvial base of its floodplain.

Rotation The spinning of a planet on its axis, the imaginary line passing through its center and both poles; it takes the earth one calendar day to complete one full rotation.

Runoff The removal—as overland flow via the network of streams and rivers—of the surplus precipitation at the land surface that does not infiltrate the soil or accumulate on the ground through surface detention.

S wave A seismic body wave that is also a shear or "shake" wave; moves objects at right angles to its direction of movement, but (unlike **P** waves) cannot travel through material in a liquid state.

Sahel Derived from the Arabic word for "shore," the name given to the east-west semiarid belt that constitutes the "southern shore" of North Africa's Sahara; straddles latitude 15°N and stretches across the entire African continent (*see* Fig. 16-3). Suffered grievous famine in 1970s that killed hundreds of thousands, and since then has periodically been devastated by severe droughts.

SALR (*see* **saturated [wet] adiabatic lapse rate**)

Salt wedging A form of mechanical weathering in arid regions; salt crystals behave much like ice in the process of frost action, entering rock joints dissolved in water, staying behind after evaporation, growing and prying apart the surrounding rock, and weakening the internal structure of the host landform.

Saltation (stream) The transportation process that entails the downstream bouncing of sand- and gravel-sized fragments along the bed of a moving stream.

Saltation (wind) The transportation process whereby coarse particles, too large to be lifted and carried by the wind, are nonetheless still moved by the wind through bouncing along the ground, sometimes at high rates of speed.

Sand The coarsest grains in a soil; sand particles range in size from 2 to 0.05 mm (0.08 to 0.002 in).

Sandspit An elongated extension of a beach into open water where the shoreline reaches a bay or bend; built and maintained by longshore drift.

Sandstone A common sedimentary rock possessing sand-sized grains.

Santa Ana wind A hot, dry, *foehn*-type wind that occasionally affects Southern California; its unpleasantness is heightened by the downward funneling of this airflow from the high inland desert through narrow passes in the mountains that line the Pacific coast.

Saturated (wet) adiabatic lapse rate (SALR) The lapse rate of an air parcel saturated with water vapor in which condensation is occurring; unlike the *dry adiabatic lapse rate (DALR)* the value of the **SALR** is variable, depending on the amount of water condensed and latent heat released. A typical value for the **SALR** at 20°C (68°F) is 0.44°C/100 m (2.4°F/1000 ft).

Saturated air Air that is holding all the water vapor molecules it can possibly contain at a given temperature.

Savanna biome The transitional vegetation of the environment between the tropical rainforest and the subtropical desert; consists of tropical grasslands with widely spaced trees.

Savanna climate (Aw) Tropical wet-and-dry climate located in the transitional, still-tropical latitudes between the subtropical high-pressure and equatorial low-pressure belts; characteristics are displayed in Fig. 17-5.

Scale The ratio of the size of an object on a map to the actual size of the object it represents.

Scarp A small escarpment not necessarily bounding a plateau.

Schist A common metamorphic rock so altered that its previous form is impossible to determine; fine-grained, exhibits wavy bands, and breaks along parallel planes (but unevenly, unlike slate).

Scree slope A steep accumulation of weathered rock fragments and loose boulders that rolled downslope in free fall; particularly common at the bases of cliffs in the drier climates of the western United States (also called a *talus cone*).

Sea arch A small island penetrated by the sea at its base; island originated as an especially resistant portion of a headland that was eroded away by waves (*see* Fig. 51-9c)

Sea breeze An onshore airflow affecting a coastal zone, resulting from a daytime pressure gradient that steers local winds from the cooler (higher-pressure) sea surface onto the warmer (lower-pressure) land surface.

Sea cave Cave carved by undercutting waves that are eroding the base of a sea cliff.

Sea cliff An especially steep coastal escarpment that develops when headlands are eroded by waves (*see* Fig. 51-9).

Seafloor spreading The process wherein new crust is formed by upwelling magma at the midoceanic ridges, and then continuously moves away from its source toward the margins of the ocean basin; *see also* **crustal spreading**.

Seamount Undersea, abyssal-zone, volcanic mountains reaching over 1000 m (3300 ft) above the ocean floor.

Secondary circulation Regional air circulation systems; the monsoon phenomenon is a classic example.

Secondary landform A landform that is the product of weathering and erosion.

Secondary pollutant Produced in the air by the interaction of two or more primary pollutants or from reactions with normal atmospheric constituents.

Secondary rocks The rock types derived from preexisting rocks—sedimentary and metamorphic rocks.

Sediment yield The measurement of the total volume of sediment leaving a drainage basin.

Sedimentary rocks The (secondary) rocks that formed from the deposition and compression of rock and mineral fragments.

Seismic Pertaining to earthquakes or earth vibrations.

Seismic reflection When a seismic wave traveling through a less dense material reaches a place where the density becomes much greater, it can be bounced back or reflected.

Seismic refraction When a seismic wave traveling through a less dense material reaches a place where the density becomes greater, it can be bent or refracted.

Seismic waves The pulses of energy generated by earthquakes that can pass through the entire planet.

Seismograph A device that measures and records the seismic waves produced by earthquakes and earth vibrations.

Sensible heat flow The environmental heat we feel or sense on our skins.

Sesquioxide An oxide containing the ratio of 1½ oxygen atoms to every metallic atom.

Shaft (solution feature) In karst terrain, a pipe-like vertical conduit that leads from the bottom of a solution hollow;

surface water is funneled down the shaft to join a subsurface channel or the groundwater below the water table.

Shale The soft, finest-grained of the sedimentary rocks; formed from compacted mud.

Shearing stress Downslope pull.

Sheet erosion The erosion produced by sheet flow as it removes fine-grained surface materials.

Sheet flow The surface runoff of rainwater not absorbed by the soil (also called *sheet wash*); forms a thin layer of water that moves downslope without being confined to local stream channels.

Shield volcano Formed from fluid basaltic lavas that flow in sheets that are gradually built up by successive eruptions; in profile their long horizontal dimensions peak in a gently rounded manner that resembles a shield. The main island of Hawaii has some of the world's most active shield volcanoes.

Shifting agriculture A low-technology tropical farming system, usually associated with oxisols of low natural fertility, in which land is cultivated for a year or two and is then abandoned for many years to naturally renew its nutrients. Population pressures are today altering this system, and the rapid deterioration of the oxisols' already limited fertility is the result.

Shingle beach A beach consisting not of sand but of gravel and/or pebbles; associated with high-energy shorelines.

Shoaling The near-shore impact of ever shallower water on an advancing, incoming wave.

Shore Has a more specific meaning than *coast*; denotes the narrower belt of land bordering a body of water, the most seaward portion of a coast (the *shoreline* is the actual contact border).

Shoreline Within a *shore zone*, the actual contact border between land and water.

Shortwave radiation Radiation coming from the sun, which has much shorter wavelengths—and involves much higher energy—than the terrestrial (longwave, lower-energy) radiation emitted by the earth.

Sial Derived from the chemical symbols for the minerals **si**licon and **al**uminum; refers to the generally lighter-colored rocks of the continents, which are dominated by granite.

Siberian High The local name for northern Asia's Polar High, which is centered over the vast north-central/northeastern region of Russia known as Siberia.

Silicates Minerals consisting of compounds containing silicon and oxygen and, mostly, other elements as well.

Sill A concordant intrusive igneous form in which magma has inserted itself as a thin layer between strata of preexisting rocks without disturbing those layers to any great extent.

Silt The next smaller category of soil particles after sand, the coarsest variety; silt particles range in size from 0.5 to 0.002 mm (0.002 to 0.00008 in).

Sima Derived from the chemical symbols for the minerals **si**licon and **ma**gnesium; refers to the generally darker-colored rocks of the ocean floors, which are dominated by basalt.

Slate Metamorphosed shale; a popular building material, it retains shale's quality of breaking along parallel planes.

Sleet Precipitation consisting of pellets of ice produced by the freezing of rain before it reaches the surface; if it freezes after hitting the surface, it is called *freezing rain*.

Slickensides The smooth, mirror-like surfaces on a scarp face produced by the movement of rocks along the fault plane.

Slip face The leeward slope of a sand dune.

Slope (river) (*see* **gradient [river]**)

Slumping A flow type of mass movement in which a major section of regolith, soil, or weakened bedrock comes down a steep slope as a backward-rotating slump block (*see* Fig. 39-8).

Smog The poor-quality surface-level air lying beneath a temperature inversion layer in the lower atmosphere; the word is derived from the contraction of "smoke" and "fog."

Snow Precipitation consisting of large ice crystals called snowflakes; formed by the ice-crystal process whose crystals do not have time to melt before they reach the ground.

Snow line The high-altitude boundary above which snow remains on the ground throughout the year.

Soil A mixture of fragmented and weathered grains of minerals and rocks with variable proportions of air and water; the mixture has a fairly distinct layering, and its development is influenced by climate and living organisms.

Soil air Fills the spaces among the mineral particles, water, and organic matter in the soil; contains more carbon dioxide and less oxygen and nitrogen than atmospheric air does.

Soil body Geographic area within which soil properties remain relatively constant.

Soil components These are four in number: minerals, organic matter, water, and air (*see* Fig. 23-3).

Soil consistence The subjective measure of a moist or wet soil's stickiness, plasticity, cementation, and hardness; this test is done in the field by rolling some damp soil in the hand and observing its behavior.

Soil-formation factors These are five in number: parent material, climate, biological agents, topography, and time.

Soil geography Systematic study of the spatial patterns of soils, their distribution, and interrelationships with climate, vegetation, and humankind.

Soil heat flow The heat that is conducted into and out of the earth's surface; also known as *ground heat flow*.

Soil horizon Soil layer; the differentiation of soils into layers is called *horizonation*.

Soil moisture Rainwater that has infiltrated into the soil; any further movement of this water is by processes other than infiltration.

Soil Order In the Soil Taxonomy, the broadest possible classification of the earth's soils into one of 11 major categories; a very general grouping of soils with broadly similar composition, the presence or absence of certain diagnostic horizons, and similar degrees of horizon development, weathering, and leaching.

Soil profile The entire array of soil horizons (layers) from top to bottom.

Soil regime The variations in behavior of changeable elements in the soil-formation environment; the term implies there is some regularity in the (spatial) pattern, but recognizes change within it. Thus, even if their parent material remained constant all over the world, soils would differ because they would form under varying temperature, moisture, biogeographic, and other conditions.

Soil storage capacity For agricultural purposes, the product of the average depth in centimeters to which roots grow and the water storage per centimeter for that soil type.

Soil Taxonomy The soil classification scheme used by contemporary pedologists and soil geographers; evolved from the Comprehensive Soil Classification System (CSCS) that was derived during the 1950s.

Soil texture The size of the particles in a soil; more specifically, the proportion of sand, silt, and clay in each soil horizon.

Sol The Russian word for soil (Russians were pioneers in the modern science of pedology); used as a suffix for soil types in the Soil Taxonomy.

Solar constant The average solar energy received every minute at the top of the earth's atmosphere: 1.95 calories per sq cm (0.16 per sq in).

Solar elevation The number of degrees above the horizon of the noontime sun, the position at which the solar rays strike the surface at their highest daily angle. Also called the *angle of incidence.*

Solar system The sun and its nine orbiting planets (plus *their* orbiting satellites). In order of increasing distance from the sun, these planets are Mercury, Venus, Earth, Mars, Jupiter, Saturn, Uranus, Neptune, and Pluto.

Solifluction A special kind of soil creep in which soil and rock debris are saturated with water and flow in bulk as a single mass; most common in periglacial zones.

Solstice (*see* **summer solstice; winter solstice**)

Solum Consists of the *A* and *B* horizons of a soil, that part of the soil in which plant roots are active and play a role in the soil's development.

Solution (stream) The transportation process whereby rock material dissolved by corrosion is carried within a moving stream.

Solution sinkhole In karst terrain, a funnel-shaped surface hollow (with the shaft draining the center) created by solution; ranges in size from a bathtub to a stadium.

Source region An extensive geographic area, possessing relatively uniform characteristics of temperature and moisture, where large air masses can form.

South Pole The "bottom" of the earth, at latitude 90°S; one of two places (the other being the North Pole) where the planet's axis intersects the surface.

Southeast Trades The surface wind belt that generally lies between the equator and 30°S; the Coriolis force deflects equatorward-flowing winds to the left, thus recurving south winds into southeast winds.

Southern Hemisphere The half of the earth located south of the equator (0° latitude); southernmost point is the South Pole (90°S).

Southern Oscillation The periodic, anomalous reversal of the pressure zones in the atmosphere overlying the equatorial Pacific; associated with the occurrence of the El Niño phenomenon. As the sea-surface temperatures change and water currents reverse, corresponding shifts occur in the windflows above (*see* Fig. 11-8).

Spatial Pertaining to space on the earth's surface; synonym for geographic(al).

Species A population of physically and chemically similar organisms within which free gene flow takes place.

Species dominance The relationship between the most abundant and important species residing in a given geographic area and the lesser species that live in the same environment.

Species-richness gradient The phenomenon involving the general decline over distance in the number of species per unit area as one proceeds from the equatorial to the higher latitudes.

Specific humidity The ratio of the weight (mass) of water vapor in the air to the combined weight (mass) of the water vapor plus the air itself.

Spheroidal (granular) soil structure Involves peds that are usually very small and often nearly round in shape, so that the soil looks like a layer of bread crumbs (*see* Fig. 24-6).

Spheroidal weathering A product of the chemical weathering process of hydrolysis; in certain igneous rocks such as granite, hydrolysis combines with other processes to cause the outer shells of the rock to flake off in what looks like a small-scale version of *exfoliation* (*see* Fig. 38-6).

Splash erosion The process involving the dislodging of soil particles by large heavy raindrops (*see* Fig. 41-1).

Spodosol One of the 11 Soil Orders of the Soil Taxonomy, which develops where organic soil acids associated with pine needle decay cause the depletion of most *A* horizon minerals; that *A* horizon is characterized by an ash-gray color, the signature of silica that is resistant to dissolving by organic acids.

Spring (season) The Northern Hemisphere season that begins at the spring (vernal) equinox ca. March 21 and ends at the summer solstice on June 22. The Southern Hemisphere spring begins at the equinox that occurs ca. September 23 and ends at the solstice on December 22.

Spring (water) A surface stream of flowing water that emerges from the ground.

Spring tide An unusually high tide that occurs when the earth, moon, and sun are aligned as shown in Fig. 50-9a.

Spring (vernal) equinox In Northern Hemisphere terminology, the equinox that occurs when the sun's noontime rays strike the equator vertically on ca. March 21.

Spur A ridge that thrusts prominently from the crest or side of a mountain.

Stability (of air) A parcel of air whose vertical movement is such that it returns to its original position after receiving some upward force; however, if an air parcel continues moving upward after receiving such a force, it is said to be unstable.

Stack A column-like island that is a remnant of a headland eroded away by waves (*see* Fig. 51-9).

Stalactite An icicle-like rock formation hanging from the roof of a cave.

Stalagmite An upward-tapering, pillar-like rock formation standing on the floor of a cave.

Standard (sea-level) air pressure The average weight of the atmospheric column pressing down on the earth's surface. It produces a reading of 760 mm (29.92 in) on the mercury barometer; in terms of weight, it is also given as 1013.25 millibars (mb) or 14.7 lb per sq in.

Standard parallel The parallel of tangency between a globe and the surface onto which it is projected.

Stationary front The boundary between two stationary air masses.

Steady-state system A system in which inputs and outputs are constant and equal.

Steppe climate (BS) Semiarid climate, transitional between fully developed desert conditions (**BS**) and the subhumid margins of the bordering **A**, **C**, and **D** climate regions; characteristics are displayed in Fig. 17-9.

Stock A discordant pluton that is smaller than a batholith.

Stone net Hexagon-like arrangement of stones that is common in the periglacial soils of the Arctic and near-Arctic regions (*see* Fig. 48-7); created when ice forms on the underside of rocks in the soil, a process that wedges rocks upward and eventually results in their meeting to form lines and patterns.

Storm An organized, moving atmospheric disturbance.

Storm surge The wind-driven wall of water hurled ashore by the approaching center of a hurricane, which can surpass normal high tide levels by more than 5 m (16 ft); often associated with a hurricane's greatest destruction.

Strata Layers.

Stratification Layering.

Stratified drift One of the two types of glacial drift; the material transported by glaciers and later sorted and deposited by the action of running water, either within the glacier or as it melts.

Stratigraphy The order and arrangement of rock strata.

Stratopause The upper boundary of the stratosphere, lying approximately 52 km (32 mi) above the surface.

Stratosphere The atmospheric layer lying above the troposphere; here temperatures are either constant or start increasing with altitude.

Stratovolcano A large and durable composite volcano.

Stratus The cloud-type category that encompasses layered and fairly thin clouds that cover an extensive geographic area; subclassified according to height (*see* p. 131).

Streak (mineral) The color of a mineral in powdered form when rubbed against a porcelain plate; used in mineral identification.

Stream capacity The maximum load of sediment that a stream can carry within a given discharge (volume of water).

Stream competence The erosional effectiveness of a stream as measured by the velocity of water's movement: the faster the flow, the greater the ability to move large boulders.

Stream piracy The capture of a segment of a stream by another river.

Stream profile The longitudinal, downward curve of a stream from its head in an interior upland to its mouth at the coast.

Stream turbulence Irregularity in the water movement in a stream related to the interaction of sediment size and riverbed roughness.

Strike The compass direction of the line of intersection between a rock layer and a horizontal plane.

Strike-slip fault (*see* **transcurrent fault**)

Subduction The process that takes place when an oceanic plate converges head-on with a plate carrying a continental landmass at its leading edge: the lighter continental plate overrides the denser oceanic plate and pushes it downward.

Subduction zone An area in which the process of subduction is taking place.

Sublimation The process whereby a solid can change directly into a gas; the reverse process is also called sublimation (or deposition); the heat required to produce these transformations is the sum of the latent heats of fusion and vaporization.

Submarine canyon A submerged river valley on the continental shelf; before the rise in sea level (of ca. 120 m [400 ft]) over the past 10,000 years, the shelf was mostly dry land across which the river flowed to its former outlet.

Submergent coast A drowned coastal zone, more common than uplifted, emergent coasts; submergence caused in large part by the rise in sea level (ca. 120 m [400 ft]) of the past 10,000 years.

Subpolar gyre Oceanic circulation loop found only in the Northern Hemisphere; its southern limb is a warm current steered by prevailing westerly winds, but the complex, cold returning flows to the north are complicated by sea-ice blockages and the configuration of landmasses vis-à-vis outlets for the introduction of frigid Arctic waters.

Subsidence The vertical downflow of air toward the surface from higher in the troposphere.

Subsystem A component of a larger system; it can act independently, but operates within, and is linked to, the larger system.

Subtropical gyre Circulates around the Subtropical High that is located above the center of the ocean basin (*see* Fig. 10-3); dominates the oceanic circulation of both hemispheres, flowing clockwise in the Northern Hemisphere and counterclockwise in the Southern Hemisphere.

Subtropical High The semipermanent belt of high pressure that is found at approximately 30 degrees of latitude in both the Northern and Southern Hemispheres; the subsiding air at its center flows outward toward both the lower and higher latitudes.

Subtropical jet stream The jet stream of the upper atmosphere that is most commonly located above the subtropical latitudes; evident throughout the year.

Suffosion sinkhole A collapse sinkhole created when an overlying layer of unconsolidated material is left unsupported.

Summer In Northern Hemisphere terminology, the season that begins on the day of the summer solstice (June 22) and ends on the day of the fall (autumnal) equinox (ca. September 23).

Summer solstice The day each year of the poleward extreme in the latitude where the sun's noontime rays strike the earth's surface vertically. In the Northern Hemisphere, that latitude is 23½°N (the Tropic of Cancer) and the date is June 22; in the Southern Hemisphere, that latitude is 23½°S (the Tropic of Capricorn) and the date is December 22.

Superimposed stream A river exhibiting transverse drainage across a structural feature that would normally impede its flow because the feature was at some point buried beneath the surface on which the river developed; as the feature became exposed, the river kept cutting through it.

Surf The water zone just offshore dominated by the development and forward collapse of breaking waves.

Surface creep The movement of fairly large rock fragments by the wind, actually pushing them along the ground, especially during windstorms.

Surface detention The water that collects on the surface in pools and hollows, bordered by millions of tiny natural dams, during a rainstorm that deposits more precipitation than the soil can absorb.

Suspect terrane A subregion of rocks possessing properties that sharply distinguish it from surrounding regional rocks; a terrane consisting of a "foreign" rock mass that is mismatched to its large-scale geologic setting.

Suspension (stream) The transportation process whereby very fine clay- and silt-sized sediment is carried within a moving stream.

Suspension (wind) The transportation process whereby very fine clay- and silt-sized particles are carried high in the air by the wind.

Swallow hole In karst terrain, the place where a surface stream "disappears" to flow into an underground channel.

Swash The thinning sheet of water that slides up the beach after a wave reaches shore and has lost its form.

Swells Long rolling waves that can travel thousands of kilometers across the ocean surface until they break against a shore.

Symbiosis (*see* **mutualism**)

Syncline A trough-like downfold with its limbs dipping toward its axial plane (*see* Fig. 36-12, right).

Synoptic weather chart A map of weather conditions covering a wide geographic area at a given moment in time.

System Any set of related objects or events and their interactions.

Taiga The Russian word for "snowforest."

Taiga climate (Dfc/Dwc, Dfd/Dwd) Collective term for the harsher subtypes of the humid microthermal climate, found on the poleward side of the **D**-climate zone; characteristics are displayed in Fig. 19-1.

Talus cone A steep accumulation of weathered rock fragments and loose boulders that rolled downslope in free fall; particularly common at the bases of cliffs in the drier climates of the western United States (also called a *scree slope*).

Tarn Small circular lake on the floor of a cirque basin.

Tectonic plate (*see* **lithospheric plate**)

Tectonics The study of the movements and deformation of the earth's crust.

Teleconnections Relationships involving long-distance linkages between weather patterns that occur in widely separated parts of the world; El Niño is a classic example.

Temperate deciduous forest biome Dominated by broadleaf trees; herbaceous plants are also abundant, especially in spring before the trees grow new leaves.

Temperate evergreen forest biome Dominated by needleleaf trees; especially common along western midlatitude coasts where precipitation is abundant.

Temperate grassland biome Occurs over large midlatitude areas of continental interiors; perennial and sod-forming grasses are dominant.

Temperate karst Marked by disappearing streams, jagged rock masses, solution depressions, and extensive cave networks; forms more slowly than tropical karst.

Temperature The index used to measure the kinetic energy possessed by molecules; the more kinetic energy they have, the faster they move. Temperature, therefore, is an abstract term that describes the energy (speed of movement) of molecules.

Temperature gradient The horizontal rate of temperature change over distance.

Temperature inversion Condition in which temperature increases with altitude rather than decreases—a positive lapse rate; it inverts what we, on the surface, believe to be the "normal" behavior of temperature change with increasing height.

Tensional stress The stress associated with the divergence of lithospheric plates; as the lithosphere spreads, rocks are being pulled apart and they thin, break, and fault; sometimes blocks of crust sink down to form rift valleys.

Terminal moraine The rock debris, carried in and just ahead of the leading front of a glacier, that is deposited as an irregular ridge when the ice's forward progress stops; these ridges are important to earth scientists because they mark the farthest extent of an ice lobe.

Terraces, paired The higher-lying remnants of an old floodplain that stand above the bluffs lining the newer floodplain of a rejuvenated river; when they lie at the same elevation, these terraces are said to be paired (*see* Fig. 43-10).

Terrane A geological region of "consistent" rocks in terms of age, type, and structure; mismatched subregions can occur and are known as suspect terranes.

Tertiary The first of the two periods of the Cenozoic era of recent life on the geologic time scale (*see* Fig. 37-7), extending from 66 million years ago to ca. 2 million years ago.

Thalweg The deepest part of a stream channel; the line connecting the lowest points along a riverbed.

Thermohaline circulation Describes the deep-sea system of oceanic circulation, which is controlled by differences in the temperature and salinity of subsurface water masses.

Thermometer An instrument for measuring temperature; most commonly, these measurements are made by observing the expansion and contraction of mercury inside a glass tube.

Thermosphere The fourth layer of the atmosphere, lying respectively above the troposphere, stratosphere, and mesosphere; in this layer, temperatures increase as altitude increases.

Thrust fault A compressional fault in which the angle of the fault plane is very low (*see* Fig. 36-4); sometimes called an *overthrust fault.*

Thunderstorm A local storm dominated by thunder, lightning, heavy rain, and sometimes hail; exhibits a definite life cycle involving developing, mature, and dissipation stages (*see* Fig. 13-3).

Tidal bore A rapidly rising high tide that creates a fast-moving wave front as it arrives in certain bays or rivers.

Tidal current Tidal movement of seawater into and out of bays and lagoons.

Tidal range The average vertical difference between sea level at high tide and low tide.

Tide The cyclical rise and fall of sea level controlled by the earth's rotation and the gravitational pull of the moon and sun; daily, two high tides and two low tides occur within a period slightly longer than 24 hours.

Till One of the two types of glacial drift; the solid material (ranging in size from boulders to clay particles) carried at the base of a glacier that is deposited as an unsorted mass when the ice melts back.

Time zone An approximately 15-degree-wide longitudinal zone, extending from pole to pole, that shares the same local time. In the conterminous United States, from east to west, the time zones are known as Eastern, Central, Mountain, and Pacific. Figure 5-6 displays the global time zone pattern, which involves many liberties with the ideal 15-degree subdivision scheme.

Tombolo A sandspit that forms a link between the mainland and an offshore island.

Toponymy Place names.

Topset beds The horizontal layers of sedimentary deposits that underlie a deltaic plain.

Tornado A small vortex of air, averaging 100 to 500 m (330 to 1650 ft) in diameter, that descends to the ground from rotating clouds at the base of a severe thunderstorm, accompanied by winds whose speeds range from 50 to 130 m per second (110 to 300 mph); as tornadoes move across the land surface, they evince nature's most violent weather and can produce truly awesome destruction in the natural and cultural landscapes.

Tornado Alley A north-south corridor in the eastern Great Plains of the central United States that experiences tornadoes with great frequency; extends from central Texas northward through Oklahoma and Kansas to eastern Nebraska.

Tower (karst) In tropical karst landscapes, a cone-shaped, steep-sided hill that rises above a surface that may or may not be pocked with solution depressions.

Traction Transportation process that involves the sliding or rolling of particles along a riverbed.

Trans-Eurasian earthquake belt Second in the world only to the Circum-Pacific belt, a belt of high earthquake incidence that extends east-southeastward from the Mediterranean Sea across southwestern and southern Asia to join the Circum-Pacific belt off Southeast Asia (*see* Fig. 35-5).

Transcurrent fault A transverse fault in which crustal blocks move horizontally in the direction of the fault; also known as a *strike-slip fault* because movement at a transcurrent fault occurs along the strike of the fault.

Transform fault A special case of transcurrent faulting in which the transverse fault marks the boundary between two lithospheric plates that are sliding past each other; California's San Andreas Fault is a classic example.

Transformation The soil-layer formation process involving the weathering of rocks and minerals and the continuing decomposition of organic material in the soil; weathering is most advanced in the upper soil layers.

Translocation The soil-layer formation process involving the introduction of dissolved and suspended particles from the upper layers into the lower ones.

Transpiration The passage of water into the atmosphere through the leaf pores of plants.

Transported soil When a soil is totally independent of the 544underlying solid rock because the parent material has been transported and deposited by one or more of the gradational agents, often far from its source area.

Transverse dune A ridge-like sand dune that is positioned at a right angle to the prevailing wind; usually straight or slightly curved.

Transverse fault (*see* **transcurrent fault**)

Transverse stress The lateral stress produced when two lithospheric plates slide horizontally past each other.

Trellis drainage A stream pattern that resembles a garden trellis; flows only in two orientations, more or less at right angles to each other; often develops on parallel-folded sedimentary rocks (*see* Fig. 42-10c).

Tributaries In a stream system, the smaller branch streams that connect with and feed the main artery (trunk river); terminology also applies to glacier systems.

Tributary glacier A smaller glacier that feeds a trunk (main) glacier.

Trophic level Each of the stages along the food chain in which food energy is passed through the ecosystem.

Tropic of Cancer The most northerly latitude (23½°N) where the sun's noontime rays strike the earth's surface vertically (on June 22, the day of the Northern Hemisphere summer solstice).

Tropic of Capricorn The most southerly latitude (23½°S) where the sun's noontime rays strike the earth's surface vertically (on December 22, the day of the Northern Hemisphere winter solstice).

Tropical (A) climates Dominated by warmth (due to low-latitude location) and moisture (from the rains of the *Inter-Tropical Convergence Zone*); contained within a continuous east-west belt astride the equator, varying latitudinally from 30 to 50 degrees wide.

Tropical cyclone (*see* **hurricane**)

Tropical deforestation The clearing and destruction of trop-

ical rainforests to make way for expanding settlement frontiers and the exploitation of new economic opportunities.

Tropical depression An easterly wave of increased intensity, but exhibiting wind speeds of less than 18 m per second (34 mph); its low-pressure trough has deepened and begun to assume a rotating, cyclonic organization. Further intensification of the depression would next transform it into a tropical storm.

Tropical gyre Narrow, low-latitude oceanic circulation loop, in both the Northern and Southern Hemisphere, comprised of the equatorial currents and returning counter-currents; reinforced by the converging winds of the Northeast and Southeast Trades.

Tropical karst Dominated by steep-sided, vegetation-covered hill terrain; solution features are larger than in slower-forming temperate karst landscapes.

Tropical rainforest biome Vegetation dominated by tall, closely spaced evergreen trees; a teeming arena of life that is home to a greater number and diversity of plant and animal species than any other biome.

Tropical rainforest climate (Af) Due to equatorial proximity, exhibits the greatest effects of heat and moisture of any climate type; characteristics are displayed in Fig. 17-3.

Tropical savanna climate (*see* **savanna climate**)

Tropical storm An intensified tropical depression, exhibiting a deep central low-pressure cell, rotating cyclonic organization, and wind speeds between 18 and 33 m per second (34 and 74 mph); further intensification would transform the system into a tropical cyclone (hurricane).

Tropopause The upper boundary of the troposphere along which temperatures stop decreasing with height.

Troposphere The bottom layer of the atmosphere in which temperature usually decreases with altitude.

Truncated spurs Spurs of hillsides that have been cut off by a glacier, thereby straightening the glacially eroded valley.

Trunk glacier The main glacier that is fed by tributary glaciers.

Trunk river The main river in a drainage basin that is fed by all the tributary streams.

Tsunami A seismic sea wave, set off by a crustal disturbance, that can reach gigantic proportions.

Tundra biome Microtherm plant assemblage of the coldest environments; dominated by perennial mosses, lichens, and sedges.

Tundra climate (ET) The milder subtype of the **E**-climate zone, named after its distinct vegetation assemblage of mosses, lichens, and stunted trees; characteristics are displayed in Fig. 19-8.

Ultisol One of the 11 Soil Orders of the Soil Taxonomy; usually quite old, not especially fertile, and located in warm subtropical environments with pronounced wet seasons.

Ultraviolet radiation High-energy, shortwave radiation associated with incoming solar energy.

Unconfined aquifer (*see* **aquifer, unconfined**)

Unconformity A gap in the geologic history of an area as found in the rock record, owing to a hiatus in deposition, followed by erosion of the surface, with further deposition continuing later; more specifically, can also refer to the contact between the eroded strata and the strata of resumed deposition.

Underfit stream A small stream lying in a large river valley that seems incapable of having sculpted that valley; stream piracy is almost always the explanation.

Uplifted marine terrace In an emergent coastal zone, a wave-cut platform that has been exposed and elevated by tectonic uplift or a lowering of sea level (or both).

Upper mantle The viscous (syrup-like) interior shell of the earth that encloses the solid lower mantle; the uppermost part of the upper mantle, however, is solid, and this zone, together with the crust that lies directly above it, is called the lithosphere (*see* Fig. 29-5).

Upper-Midlatitude Low The semipermanent surface low-pressure belt, lying at approximately 60 degrees north and south, where the equatorward-flowing Polar Easterlies meet the poleward-flowing Westerlies; at this sharp atmospheric boundary, known as the Polar Front, the warmer Westerlies are forced to rise above the colder Easterlies.

Upthrown block The block that moves upward with respect to adjacent blocks when vertical movement occurs during faulting.

Upwelling The rising of cold water from the ocean depths to the surface; affects the local climatic environment because cold water lowers air temperatures and the rate of evaporation.

Urban heat island The form taken by an isotherm representation of the heat distribution within an urban region; the central city appears as a "highland" or "island" of higher temperatures on a surrounding "plain" of more uniform (lower) temperatures.

Uvala In karst terrain, a large surface depression created by the coalescence of two or more neighboring sinkholes.

Vadose zone (*see* **zone of aeration**)

Valley train Meltwater-deposited alluvium in a glacial trough; derived from the morainal material left behind by a receding mountain glacier.

Vapor pressure The pressure exerted by the molecules of water vapor in air.

Vapor-pressure gradient The difference in vapor pressure between two locations. A common situation involves air near a water surface, which contains a large number of water vapor molecules and thus exhibits a relatively high vapor pressure, and air at some distance from that surface, which contains fewer vapor molecules and therefore a lower vapor pressure.

Vaporization Synonym for evaporation.

Variable gases Atmospheric gases present in differing quantities at different times and places; three are essential to human well-being: carbon dioxide, water vapor, and ozone.

Varves Paired layers of alternating sediments caused by seasonal variations in deposition.

Velocity (stream) The rate of speed at which water moves in a river channel; this rate varies within the stream, as Fig. 40-4 shows.

Vent An opening through the earth's crust from which lava erupts; most eruptions occur through pipe-shaped vents that build volcanic mountains, but fissure eruptions also occur through lengthy cracks that exude horizontal sheets of lava.

Vertical zonation Characteristic of **H** climates, the distinct arrangement of climate zones according to altitudinal position; the higher one climbs, the colder and harsher the climate becomes—as shown for South America's Andes Mountains in Fig. 19-10.

Vertisol One of the 11 Soil Orders of the Soil Taxonomy, found in tropical as well as mesothermal wet-and-dry climates; this soil type is heavy in clay composition, cracking during the dry season and swelling with moisture when the rains return.

Viscous Syrup-like; capable of flowing slowly.

Volcanic ash (*see* **ash [volcanic]**)

Volcanic bombs (*see* **bombs, volcanic**)

Volcanic dome A small volcanic mound built by oozing lava without pyroclastic activity; often forms inside the crater following an explosive eruption.

Volcanism The eruption of molten rock at the earth's surface, often accompanied by rock fragments and explosive gases.

Volcano A vent in the earth's surface through which magma, solid rock, debris, and gases are erupted; this ejected material usually assumes the shape of a conical hill or mountain.

Volume symbols Describe surfaces on a map, which can be generalizations of real surfaces or representations of conceptual surfaces.

Wallace's Line Zoogeographer Alfred Russel Wallace's controversial boundary line that purportedly separates the unique faunal assemblage of Australia from the very different animal assemblage of neighboring Southeast Asia; Wallace's famous line (mapped in Fig. 22-5), introduced over a century ago, is still the subject of debate today.

Warm front Produced when an advancing warm air mass infringes on a preexisting cooler one; when they meet, the lighter, warmer air overrides the cooler air mass, forming the gently upward-sloping warm front (producing far more moderate precipitation than that associated with steeply sloped cold fronts).

Warm sector In an open-wave, midlatitude cyclone, the wedge of warm air enclosed by the cold and warm fronts.

Water balance Analogous to an accountant's record (which tallies income, expenditures, and the bottom-line balance), the measurement of the inflow (precipitation), outflow (evapotranspiration), and net annual surplus or deficit of water at a given location.

Water gap A pass in a ridge or mountain range through which a stream flows.

Water (Oceanic) Hemisphere The roughly one-half of the earth that contains most of the surface water; the opposite of the *land hemisphere* (*see* Fig. 2-4).

Water resources Subfield of physical geography involving its intersection with hydrology; systematic study of the surface and subsurface water supplies potentially available for human use.

Water table The top of the (phreatic) zone of saturation; does not lie horizontally but follows the general profile of the land surface above.

Water vapor The invisible gaseous form of water; the most widely distributed variable gas of the atmosphere.

Watershed (*see* **drainage basin**)

Waterspout A tornado that forms and moves over a water surface.

Wave crest The top of a wave; the wave's height is the vertical distance between the wave crest and the wave trough.

Wave-cut platform The abrasion platform that develops at the foot of a sea cliff, marking its recession; its nearly flat bedrock surface slopes seaward.

Wave height The vertical distance between the wave crest (top) and the wave trough (bottom).

Wave length The horizontal distance between one wave crest (or wave trough) and the next.

Wave of translation A swell nearing shore that has "felt" the rising ocean bottom and whose internal water motion (as a wave of oscillation) begins to be affected by it; the wave's erosional work has begun.

Wave period The time interval between the passage of two successive wave crests past a fixed point.

Wave refraction The near-shore bending of waves coming in at an oblique angle to the shoreline; shoaling slows part of the wave, which progessively bends as the faster end "catches up" (*see* Fig. 50-4).

Wave trough The bottom of a wave; the wave's height is the vertical distance between the wave crest and the wave trough.

Waves of oscillation Waves that move water particles in a circular up-and-down path; their depth is one-half their length.

Weather The immediate and short-term conditions of the atmosphere that impinge on daily human activities.

Weathering The chemical alteration and physical disintegration of earth materials by the action of air, water, and organisms; more specifically, the breakdown of rocks in situ, their disintegration and decomposition without distant removal of the products.

Westerlies The two broad midlatitude belts of prevailing westerly winds, lying between approximately 30 and 60 degrees in both hemispheres; fed by the Coriolis-force-deflected, poleward windflow emanating from the Subtropical High on the equatorward margin of the Westerlies wind belt.

Wet-bulb temperature On a psychrometer, the temperature reading of the thermometer whose bulb is swaddled in a wet cloth (this produces a lower temperature than that of the surrounding air due to cooling caused by the evaporation process).

Wilting point The practical lower limit to moisture contained within a soil; below this point, a crop dries out, suffering permanent injury.

Wind The movement of air relative to the earth's surface; winds are always named according to the direction from which they blow.

Wind abrasion The erosion of rock surfaces by windborne sand particles.

Wind-chill index An index that tells us subjectively how cold we would feel under given combinations of wind speed and air temperature.

Windward The exposed, upwind side of a topographic barrier that faces the winds that flow across it.

Winter In Northern Hemisphere terminology, the season that begins on the day of the winter solstice (December 22) and ends on the day of the spring (vernal) equinox (ca. March 21).

Winter solstice The day each year of the poleward extreme in latitude *in the opposite hemisphere* where the sun's noontime rays strike the earth's surface vertically. In the Northern Hemisphere, that date is December 22 when the sun is directly above latitude 23½°S (the Tropic of Capricorn); in the Southern Hemisphere, that date is June 22 when the sun is directly above latitude 23½°N (the Tropic of Cancer).

Wisconsin glaciation The most recent glaciation of the Late Cenozoic Ice Age, consisting of Early and Late stages.

Wrangellia A well-known assemblage of three suspect terranes in northwestern North America, collectively named after the Alaskan mountain range where the phenomenon was first identified (*see* Fig. 33-3).

Xerophyte Plant adapted to withstand the aridity of dry environments.

Yardang A desert landform shaped by wind abrasion in the form of a low ridge lying parallel to the prevailing wind direction; most common in dry sandy areas underlain by soft bedrock.

Zenith The point in the sky directly overhead, 90° above the horizon.

Zonal flow The westerly flow of winds that dominates the upper atmospheric circulation system poleward of 15 degrees latitude in each hemisphere.

Zone of ablation A glacier's lower zone of loss; *ablation* refers to all forms of loss at the lower end, including melting and evaporation.

Zone of accumulation A glacier's upper zone of growth, where new snow is added.

Zone of aeration The upper of the two subterranean zones that contains groundwater; lies above the water table and is normally unsaturated, except during heavy rainfall (also known as the *vadose zone*).

Zone of intolerance The geographic area beyond a plant or animal species' zone of physiological stress; conditions here are extreme for that species, and it cannot survive except possibly for short, intermittent periods.

Zone of physiological stress The marginal geographic area for a plant or animal species that surrounds its optimum range; here the species survives in smaller numbers, but increasingly encounters environmental stress as the outer spatial limit to its existence is neared.

Zone of saturation The lower of the two subterranean zones that contains groundwater; lies below the water table and is also known as the *phreatic zone*.

Zoogeographic realm The largest and most generalized regional unit for representing the earth's fauna; reflects evolutionary centers for animal life as well as the influence of barriers over time (*see* Fig. 28-4).

Zoogeography The geography of animal life or fauna; where zoology and physical geography overlap.

Zooplankton Microscopic animal life forms that float in the ocean and fresh water bodies; eaten by small fish, which in turn are eaten by larger fish.

Credits

CREDITS FOR LINE ART AND TABLES

UNIT 1 Fig. 1–7: From *Geography: A Modern Synthesis*, 3/ed. by Peter Haggett. Copyright © 1983 by Peter Haggett. Reprinted by permission of HarperCollins Publishers, Inc.

UNIT 3 Fig. 3–4: From Norman J. W. Thrower, *Maps and Man*, Prentice-Hall, 1972, p. 153, © Norman J. W. Thrower; Fig. 3–7: Reproduced by permission of John Wiley & Sons, from Arthur Robinson et al., *Elements of Cartography*, 5 rev ed., 1984, p. 99; Figs. 3–8, 3–11: Reproduced by permission of the Institute of Science and Public Affairs, Florida State University, from Edward A. Fernald and Donald J. Patton, *Water Resources Atlas of Florida*, 1984, pp. 26, 19; Figs. 3–9, 3–10: Reproduced by permission of the Florida State University Foundation, Inc., from Edward A. Fernald, ed., *Atlas of Florida*, 1982, pp. 238, 46; Figs. 3–12, 3–13: Adapted from Brian J. Skinner and Stephen C. Porter, *Physical Geology*, John Wiley & Sons, 1987, pp. 715, 711; Fig. 3–16: Adapted from "Spaceship Earth," *The Christian Science Monitor*, April 1, 1987, p. 19, drawn by Lisa Remillard and Heidi Mack, © 1987 TCSPS.

UNIT 4 Fig. 4–2: Adapted from *National Geographic Atlas of the World*, Sixth Edition Revised, 1992, plate 119. © National Geographic Society; Fig. 4–3: Reproduced by permission of The MIT Press, from I. Gass et al., *Understanding the Earth*, 1971, p. 65, © The MIT Press.

UNIT 5 Fig. 5–8: Adapted from Robert J. List, *The Smithsonian Meteorological Tables* (Washington, D. C.: Smithsonian Institution Press), by permission of the publisher. Copyright © 1971.

UNIT 6 Fig. 6–3: After C. D. Keeling et al., *Geophysical Monitoring for Climate Change*, National Oceanic and Atmospheric Administration (NOAA), 1988; Fig. 6–5: Adapted from *Understanding Our Atmospheric Environment*, by Morris Neiburger et al., Copyright © 1973 by W. H. Freeman and Company, reprinted by permission.

UNIT 7 Fig. 7–5: Reproduced by permission of Edward Arnold (Publishers) Limited, from J. G. Lockwood, *World Climatology: An Environmental Approach*, 1974, p. 43, © Edward Arnold (Publishers) Ltd.; Fig. 7–6: After M. I. Budyko (transl. N. A. Stepanova), *The Heat Balance of the Earth's Surface*, U. S. Department of Commerce, Office of Technical Services, 1958; Figs. 7–7, 7–8: Reproduced by permission of American Geographical Society, from M. I. Budyko, "The Heat Balance of the Earth," *Soviet Geography: Review and Translation*, Vol. 3, No. 5, May 1962, pp. 7, 9; Table 7–1: Reproduced by permission of The University of Chicago Press, from W. D. Sellers, *Physical Climatology*, 1965, pp. 32, 47, © The University of Chicago Press.

UNIT 8 Fig. 8–5: From *Geography: A Modern Synthesis*, 3/ed. by Peter Haggett. Copyright © 1983 by Peter Haggett. Reprinted by permission of HarperCollins Publishers, Inc.; Fig. 8–6: From *Urbanization and Environment* by Thomas R. Detwyler and Melvin G. Marcus, © 1972 by Wadsworth Publishing Company, Inc., reprinted by permission of the publisher; Fig. 8–8: From *Geography: A Modern Synthesis*, 3/ed. by Peter Haggett. Copyright © 1983 by Peter Haggett. Reprinted by permission of HarperCollins Publishers, Inc.; Fig. 8–11: From Roger G. Barry and Richard J. Chorley, *Atmosphere, Weather and Climate*, Methuen (Routledge), 5 rev. ed., pp. 24–25, © 1987 by Roger G. Barry and Richard J. Chorley; Table 8–1: From Frederick K. Lutgens and Edward J. Tarbuck, *The Atmosphere: An Introduction to Meteorology*, 2/E, © 1982, p. 58, adapted by permission of Prentice-Hall, Inc., Englewood Cliffs, N.J.

UNIT 9 Figs. 9–2, 9–3: From Roger G. Barry and Richard J. Chorley, *Atmosphere, Weather and Climate*, Methuen (Routledge), 2 rev. ed., pp. 8, 43, © 1987 by Roger G. Barry and Richard J. Chorley; Fig. 9–4: Adapted from *Life Science Library: Weather*. © 1965 Time-Life Books

Inc.; Fig. 9–6: From Howard J. Critchfield, *General Climatology*, 4/E, © 1983, p. 85, adapted by permission of Prentice-Hall, Inc., Englewood Cliffs, N.J.; Fig. 9–10: Adapted from Brian J. Skinner and Stephen C. Porter, *The Blue Planet: An Introduction to Earth System Science*, John Wiley & Sons, 1995, p. 351.

UNIT 10 Figs. 10–4, 10–5: Reproduced by permission of Edward Arnold (Publishers) Limited, from J. G. Lockwood, *World Climatology: An Environmental Approach*, 1974, pp. 146, 151, © Edward Arnold (Publishers) Ltd.

UNIT 11 Fig. 11–4: From *Oceanography: An Introduction to the Marine Sciences* by Jerome Williams. Copyright © 1962 by Little, Brown and Company. By permission, Little, Brown and Company. Fig. 11–7: Same source as Fig. 3–12, p. 382.

UNIT 12 Fig. 12–2: Same source as Fig. 3–12, p. 241; Fig. 12–6: Same source as Fig. 9–10, p. 329; Figs. 12–8a & b, 12–10: Reproduced by permission of Edward Arnold (Publishers) Limited, from J. G. Lockwood, *World Climatology: An Environmental Approach*, 1974, pp. 168, 62, © Edward Arnold (Publishers) Ltd.; Fig. 12–9: Reproduced by permission of The University of Chicago Press, from W. D. Sellers, *Physical Climatology*, 1965, p. 84, © The University of Chicago Press.

UNIT 13 Fig. 13–3: From S. Petterssen, *Introduction to Meteorology*, © 1958 McGraw-Hill, Inc. reproduced by permission of McGraw-Hill, Inc.; Fig. 13–5: Adapted by permission of *New Scientist*, from K. Hindley, "Learning to Live With Twisters," Vol. 70, 1977, p. 281; Fig. 13–8: Data from *Climate and Man: The 1941 Yearbook of Agriculture*, U. S. Department of Agriculture (USDA), 1941, p. 1085.

UNIT 14 Fig. 14–1: Reproduced by permission of Edward Arnold (Publishers) Limited, from J. G. Lockwood, *World Climatology: An Environmental Approach*, 1974, p. 96, © Edward Arnold (Publishers) Ltd.; Fig. 14–3: From *The New York Times*, September 20, 1989, copyright © 1989 by The New York Times Company, reprinted by permission; Fig. 14–6: Adapted from D. Riley and L. Spolton, *World Weather and Climate*, Cambridge University Press, 2 rev. ed., © 1981 by Cambridge University Press.

UNIT 15 Fig. 15–5: From Glenn Trewartha et al., *Elements of Geography*, 5 rev. ed., © 1967 McGraw-Hill, Inc., reproduced by permission of McGraw-Hill, Inc.; Fig. 15–9: After map of World Weather Extremes, Geographic Sciences Laboratory, U. S. Army Engineer Topographic Laboratories, Ft. Belvoir, Va., n.d.

UNIT 17 Fig. 17–13: Adapted from press release map, © 1985 National Geographic Society.

UNIT 18 Fig. 18–10: After NOAA/USDA Joint Agricultural Weather Facility, October 1988.

UNIT 19 Fig. 19–6: From Anne La Bastille, "Acid Rain: How Great a Menace?", *National Geographic*, November 1981, p. 667, © 1981 National Geographic Society.

UNIT 20 Fig. 20–3: Same source as Fig. 9–10, p. 364 (based on NCAR sea-level temperature records); Fig. 20–5: Adapted with permission from *Understanding Climate Change: A Program for Action*, National Academy Press, Washington, D.C., 1975; Fig. 20–7: From *The Weather Machine* by Nigel Calder. Copyright © 1975 by Nigel Calder. Used by permission of Nigel Calder.

UNIT 21 Figs. 21–2, 21–3: From L. P. Herrington, "Biophysical Adaptations of Man Under Climatic Stress," *Meteorological Monographs*, Vol. 2, No. 8, pp. 30–34, © 1954 by the American Meteorological Society; Fig. 21–5: From C. A. Woolum, "Notes From a Study of the Microclimatology of Washington, D. C.," *Weatherwise*, Vol. 17, No. 6, p. 6, 1964; published by Heldref Publications; Fig. 21–6: From *Urbaniza-*

tion and Environment by Thomas R. Detwyler and Melvin G. Marcus, © 1972 by Wadsworth Publishing Company, Inc., reprinted by permission of the publisher; Fig. 21–7: From Stanley A. Changnon, Jr., "Recent Studies of Urban Effects on Precipitation in the United States," in *Urban Climates*, Paper No. 254, pp. 325–341, © 1970 by the World Meteorological Organization; Fig. 21–9: From *Urbanization and Environment* by Thomas R. Detwyler and Melvin G. Marcus, © 1972 by Wadsworth Publishing Company, Inc., reprinted by permission of the publisher.

UNIT 22 Fig. 22–6: Adapted from *The Christian Science Monitor*, August 10, 1994, p. 14, drawn by Dave Herring © 1994 TCSPS; Fig. 22–10: Adapted from "Water," *National Geographic* Special Edition, November 1993, p. 61. © 1993 National Geographic Society.

UNIT 23 Fig. 23–3: From Léo F. Laporte, *Encounter With the Earth*, Canfield Press, p. 300, © 1975 by Léo F. Laporte.

UNIT 24 Fig. 24–1: From *Fundamentals of Soil Science*, eighth revised edition, by Henry D. Foth. Copyright © 1990 John Wiley & Sons, Inc. Reprinted by permission of John Wiley & Sons, Inc. Fig. 24–2: After USDA, Soil Conservation Service, n.d.; Fig. 24–7: From S. R. Eyre, *Vegetation and Soils: A World Picture*, Aldine, 1963, p. 259, © S. R. Eyre, reprinted by permission; Fig. 24–8: From Daniel H. Yaalon, *Transactions of the International Congress of Soil Science*, Vol. 4, V. 16, p. 120, © 1960 by the International Society of Soil Science.

UNIT 25 Fig. 25–1: After USDA, n.d.; Table 25–2: From *Fundamentals of Soil Science*, eighth revised edition, by Henry D. Foth. Copyright © 1990 John Wiley & Sons, Inc. Reprinted by permission of John Wiley & Sons, Inc.

UNIT 26 Fig. 26–1: From Léo F. Laporte, *Encounter With the Earth*, Canfield Press, p. 120, © 1975 by Léo F. Laporte; Fig. 26–3: After H. Lieth and E. Box, *Publications in Climatology*, C. W. Thornthwaite Laboratory of Climatology, Vol. 25, No. 3, 1972, p. 42; Figs. 26–5, 26–6: From *Geography: A Modern Synthesis*, 3/ed. by Peter Haggett. Copyright © 1983 by Peter Haggett. Reprinted by permission of HarperCollins Publishers, Inc.; Fig. 26–7: Figure adapted from *Fundamentals of Ecology*, Third Edition by Eugene P. Odum, copyright © 1971 by Saunders College Publishing, reprinted by permission of the publisher; Fig. 26–8: © *Biol. Bull.*, Vol. XXI, No. 3, pp. 127–151, and Vol. XXII, No. 1, pp. 1–38; Fig. 26–9: Reproduced by permission from C. B. Cox et al., *Biogeography*, 1973, p. 61, © Blackwell Scientific Publications Ltd.

UNIT 27 Fig. 27–2: Reproduced from *Plants and People*, Resource Publication of the Association of American Geographers, Thomas R. Vale, 1982; Figs. 27–3, 27–4: Figures adapted from *Environmental Science*, Third Edition by Jonathan Turk, Amos Turk, copyright © 1984 by Saunders College Publishing, reprinted by permission of the publisher.

UNIT 28 Fig. 28–6: Reprinted from *Guide to the Mammals of Pennsylvania*, by Joseph F. Merritt, by permission of the University of Pittsburgh Press, © 1987 by University of Pittsburgh Press.

UNIT 30 Fig. 30–4: Same source as Fig. 9–10, p. 126.

UNIT 31 Fig. 31–4: Adapted from Brian J. Skinner and Stephen C. Porter, *The Dynamic Earth: An Introduction to Physical Geology*, John Wiley & Sons, 3 rev. ed., 1995, p. 427.

UNIT 32 Figs. 32–3, 32–11: Same source as Fig. 3–12, pp. 23, 490; Fig. 32–4: Data from U. S. Coast and Geodetic Survey, and NOAA.

UNIT 33 Fig. 33–1: Same source as Fig. 9–10, p. 143; Fig. 33–3: Same source as Fig. 3–12, p. 509.

UNIT 34 Fig. 34–3a: Same source as Fig. 3–12, p. 90; Fig. 34–14: From Stephen L. Harris, *Fire Mountains of the West: The Cascade and Mono Lake Volcanoes*, © 1991 Mountain Press Publishing Company.

UNIT 35 Fig. 35–5: Data from U. S. Coast and Geodetic Survey; Fig. 35–6: After NOAA, n.d.; Tables 35–1, 35–2: From *The Way the Earth Works*, by P. J. Wyllie. Copyright © 1976 John Wiley & Sons, Inc. Reprinted by permission of John Wiley & Sons, Inc.

UNIT 36 Figs. 36–2, 36–5, 36–10, 36–12, & 36–13: Same source as Fig. 3–12, pp. 416, 416, 417, 424, & 426; Fig. 36–16: After U. S. Geological Survey, *1982 Annual Report*.

UNIT 37 Fig. 37–4: Same source as Fig. 3–12, p. 293; Fig. 37–7: After U. S. Geological Survey, n.d. (time scale updated to 1996).

UNIT 39 Fig. 39–5: Same source as Fig. 31–4, p. 234; Figs. 39–6, 39–8, 39–9, & 39–11: Same source as Fig. 3–12, p. 223; Fig. 39–10: After U. S. Geological Survey, Professional Paper 950, n.d.; Fig. 39–12: From Arthur Bloom, *The Surface of the Earth*, © 1969, p. 90, adapted by permission of Prentice-Hall, Inc., Englewood Cliffs, N.J.; Fig. 39–13: From Léo F. Laporte, *Encounter With the Earth*, Canfield Press, p. 97, © 1975 by Léo F. Laporte.

UNIT 40 Fig. 40–2: Reproduced by permission from John J. Hidore, *Physical Geography: Earth Systems*, Glenview, Illinois: Scott Foresman & Co., 1974, © John J. Hidore; Fig. 40–6: Reproduced by permission from J. P. Bruce and R. H. Clark, *Introduction to Hydrometeorology*, Elmsford, New York: Pergamon Press, 1966, © J. P. Bruce and R. H. Clark; Fig. 40–7: Adapted from Brian J. Skinner and Stephen C. Porter, *The Dynamic Earth: An Introduction to Physical Geology*, John Wiley & Sons, 1989, p. 199; Fig. 40–8: From Léo F. Laporte, *Encounter With the Earth*, Canfield Press, p. 249, © 1975 by Léo F. Laporte; Fig. 40–10: Figure from *Introduction to Physical Geography*, by Henry M. Kendall, Robert M. Glendenning, and Clifford H. MacFadden, copyright © 1974 by Harcourt Brace & Company, reproduced by permission of the publisher; Fig. 40–12: Same source as Fig. 9–10, p. 245.

UNIT 41 Fig. 41–7: Same source as Fig. 40–7, p. 221; Fig. 41–10: Same source as Fig. 3–12, p. 279; Fig. 41–13: From Marie Morisawa, *Streams: Their Dynamics and Morphology*, © 1968 McGraw-Hill, Inc., reproduced by permission of McGraw-Hill, Inc.; Table 41–1: From Arthur Bloom, *The Surface of the Earth*, © 1969, adapted by permission of Prentice-Hall, Inc., Englewood Cliffs, N.J.

UNIT 43 Figs. 43–2a, 43–10, & 43–14: Same source as Fig. 3–12, pp. 274, 287, & 290.

UNIT 44 Fig. 44–3: Same source as Fig. 3–12, p. 256; Fig. 44–4: From a portion of Glasgow North (Kentucky) Quadrangle, 7.5 minute series, U. S. Geological Survey, n.d.

UNIT 45 Figs. 45–2a & b: Same source as Fig. 3–12, both p. 345; Fig. 45–3: After NASA, n.d.; Fig. 45–6: Same source as Fig. 40–7, p. 270.

UNIT 46 Figs. 46–3, 46–8: Same source as Fig. 3–12, pp. 340, 365; Fig. 46–4: From J. F. Lovering and J. R. V. Prescott, *Last of Lands . . . Antarctica* (Melbourne University Press, 1979). Data from Gordon and Goldberg, "Circumpolar Characteristics of Antarctic Waters," American Geographical Society, Antarctic Map Folio Series 1970, and Kort, "The Antarctic Ocean," *Scientific American*, Vol. 207, 1962; Figs. 46–6, 46–10, & 46–13: Adapted from Richard Foster Flint, *Glacial and Pleistocene Geology*, John Wiley & Sons, 1957, pp. 227 ff. Copyright © Harrison L. Flint; Fig. 46–11: Adapted by permission of Doris Thornbury from William D. Thornbury, *Principles of Geomorphology*, John Wiley & Sons, 2 rev. ed., 1969, p. 399.

UNIT 47 Fig. 47–7: From Armin K. Lobeck, *Geomorphology*, McGraw-Hill Book Company, 1932; Fig. 47–9: Same source as Fig. 40–7, p. 268.

UNIT 48 Fig. 48–3: Same source as Fig. 3–12, p. 361.

UNIT 49 Figs. 49–4, 49–6, & 49–7: Same source as Fig. 3–12, pp. 316, 322, & 324; Fig. 49–10: After U. S. Bureau of Reclamation, 1960.

UNIT 50 Figs. 50–1, 50–4, & 50–6: Adapted by permission of John Wiley & Sons from Keith Stowe, *Essentials of Ocean Science*, 1987, pp. 84, 87, & 124. Copyright © 1987 John Wiley & Sons, Inc.; Figs. 50–2, 50–3a, & 50–5: Same source as Fig. 3–12, pp. 385, 386, & 387.

UNIT 51 Figs. 51–1, 51–4: Same source as Fig. 3–12, pp. 392, 393; Fig. 51–11: Same source as Fig. 40–7, p. 297; Fig. 51–12: Same source as Fig. 9–10, p. 302.

UNIT 53 Fig. 53–2: Adapted from *The Physiographic Provinces of North America* by Wallace W. Atwood, Copyright, 1940, by Ginn and Company. Used by permission of Silver Burdett Ginn Inc.; Figs. 53–6, 53–7: From William F. Powers, *Physical Geography*, Appleton-Century-Crofts, 1966, pp. 193, 195, © William F. & Marion Powers Trust.

CREDITS FOR PHOTOGRAPHS

TITLE PAGE Jose Fuste Raga/The Stock Market.
PART I OPENER Peter French/DRK Photo.

UNIT 1 *Opener.* Tony Stone Images/New York, Inc. Figures 1–2 & 1–3: National Geographic Society. Figure 1–5: Ardea London. Figure 1–6: Grant Heilman/Grant Heilman Photography. Figure 1–8: Jose Azel/Aurora & Quanta Productions.

UNIT 2 *Opener.* H. J. de Blij. Figure 2–6: Cartographic Division/National Geographic Society.

UNIT 3 *Opener.* NRSC Ltd./Science Photo Library/Photo Researchers. Figure 3–1a: The British Museum. Figure 3–17: Earth Satellite Corporation/Science Photo Library.

UNIT 4 *Opener.* © C.L.C./International Stock Photo. Figure 4–1: Dennis di Cicco/Peter Arnold, Inc. Figure 4–4: Francois Gohier/Ardea London. Figure 4–5: Courtesy NASA. Figure 4–6: © Stocktrek Photo Agency/Tom Stack & Associates. Figure 4–7: NASA/Mark Marten/Photo Researchers. Figure 4–8: © NASA/Peter Arnold, Inc. Figure 4–9: Courtesy NASA. Figure 4–10: NASA/Science Photo Library/Photo Researchers.

UNIT 5 *Opener.* Courtesy NASA. Figure 5–3: Courtesy John D. Stephens.

PART II OPENER © Richard Coomber/Planet Earth Pictures.

UNIT 6 *Opener.* Courtesy NASA. Figure 6–2: © Robert Landau/Westlight. Figure 6–4: © Dale Sanders/Masterfile. Figure 6–6: © JC Stevenson/Animals Animals/Earth Scenes. Figure 6–7: National Environmental Satellite, Data, & Information Service, and National Oceanic & Atmospheric Administration. Figure 6–8: © Johnny Johnson/DRK Photo.

UNIT 7 *Opener.* © Lowell Georgia/Science Source/Photo Researchers.

UNIT 8 *Opener.* NASA/Science Photo Library/Photo Researchers. Figure 8–1: The Mansell Collection. Figure 8–4: Courtesy NASA. Figure 8–7: Jim Mendenhall.

UNIT 9 *Opener.* Buff Corsi/© Focus on Nature.

UNIT 10 *Opener.* Courtesy NASA. Figure 10–6: Courtesy NASA.

UNIT 11 *Opener.* EROS Data Center, Courtesy Kalmbach Publishing Co. Figure 11–1: Courtesy NASA.

UNIT 12 *Opener.* EROS Data Center, Courtesy Kalmbach Publishing Co. Figure 12–4: Courtesy USGS. Figure 12–5: Clyde H. Smith/ f/STOP Pictures. Figure 12–7a: T.A. Wiewandt/DRK Photo. Figure 12–7b: Alan L. Graham/ f/STOP Pictures. Figure 12–7c: Robert Perron. Figure 12–7d: Warren Faidley/Weatherstock.

UNIT 13 *Opener.* Peter Menzel. Figure 13–2: Courtesy NASA. Figure 13–4: H. J. de Blij. Figure 13–6: W. Balzer/Weatherstock.

UNIT 14 *Opener.* Courtesy National Oceanic and Atmospheric Administration. Figure 14–2: National Hurricane Center. Figure 14–5: © John Lopinot/Palm Beach Post/Sygma Photo News.

UNIT 15 *Opener.* Frank Whitney/The Image Bank. Figure 15–2: Kavouras, Inc. Figure 15–3: Courtesy NOAA/NESDIS/Training Branch. Figure 15–8: Courtesy NOAA/NESDIS. Figure 15–10: Peter Menzel.

UNIT 16 *Opener.* Steve McCurry/Magnum Photos, Inc.

UNIT 17 *Opener.* Robert Caputo/Stock, Boston. Figure 17–1: M. Harvey/Panos Pictures. Figure 17–2: Frans Lanting/Minden Pictures, Inc. Figure 17–6: © Stephen J. Krasemann/DRK Photo. Figure 17–7: © Didier Givois/Agence Vandystadt/Photo Researchers. Figure 17–10: T.A. Wiewandt/DRK Photo. Figure 17–12: Steve McCurry/Magnum Photos, Inc.

UNIT 18 *Opener.* Joachim Messerschmidt/Leo de Wys, Inc. Figure 18–4a: Phil Schermeister/Tony Stone Images/New York, Inc. Figure 18–4b: Georg Gerster/COMSTOCK, Inc. Figure 18–5: Ardea London. Figure 18–11: John Livzey/Outline.

UNIT 19 *Opener.* © Bob Clemenz Photography. Figure 19–2: Robert McKenzie. Figure 19–3: Mark Kelley/Alaska Stock Images. Figure 19–6: Johnny Johnson/Tony Stone Images/New York, Inc. Figure 19–11: Robert Caputo/Aurora & Quanta Productions.

UNIT 20 *Opener.* © Larry Ulrich. Figure 20–1: Kenneth Garrett/Woodfin Camp & Associates. Figure 20–2: Courtesy H. N. Michael. Figure 20–6a&b: Courtesy National Optical Astronomy Observatory. Figure 20–8: Chip Hires/Gamma Liaison.

UNIT 21 *Opener.* David Woodfall/© Natural History Photographic Agency. Figure 21–4: Walter Hodges/Tony Stone Images/New York, Inc. Figure 21–8: The Hulton-Deutsch Collection. Figure 21–10: Eric Lars Bakk/Black Star.

PART III OPENER © Carr Clifton

UNIT 22 *Opener.* H. J. de Blij. Figure 22–1: G.I. Bernard/Oxford Scientific Films/Animals Animals/Earth Scenes. Figure 22–2: Randall Schaetzl. Figure 22–3: Frans Lanting/Minden Pictures, Inc. Figure 22–4: © Kevin Schafer/Kevin Schafer Photography. Figure 22–7: Hans Reinhard/Bruce Coleman, Inc. Figure 22–8: H. J. de Blij. Figure 22–9: ACME/Bettmann Archive.

UNIT 23 *Opener.* H. J. de Blij. Figure 23–1: Lindsay Hebberd/Woodfin Camp & Associates. Figure 23–2: Mickey Gibson/Animals Animals/Earth Scenes. Figure 23–4: William E. Ferguson. Figure 23–7: Dan Suzio/Photo Researchers. Figure 23–8: Randall Schaetzl.

UNIT 24 *Opener.* Grant Heilman/Grant Heilman Photography. Figure 24–3b: Randall Schaetzl. Figure 24–4b: Randall Schaetzl. Figure 24–5b: Randall Schaetzl. Figure 24–6b: Randall Schaetzl.

UNIT 25 *Opener.* F. Gohier/Photo Researchers. Figure 25–2: Bill Brooks/Masterfile. Figure 25–3: Randall Schaetzl. Figure 25–4: William E. Ferguson. Figures 25–5, 25–6, 25–8, 25–9, 25–10 & 25–11: Randall Schaetzl. Figure 25–12: H. J. de Blij.

UNIT 26 *Opener.* David Muench/AllStock, Inc. Figure 26–2: Frans Lanting/Minden Pictures, Inc. Figure 26–4: © Carr Clifton. Figure 26–10: Wendell Metzen/Bruce Coleman, Inc. Figure 26–11: Jen & Des Bartlett/Bruce Coleman Ltd.

UNIT 27 *Opener.* USGS/National Mapping Division/ EROS Data Center. Figure 27–5 & 27–6: H. J. de Blij. Figure 27–7: Leonard Lee Rue, Jr./Photo Researchers. Figure 27–8: David E. Rowley/Planet Earth Pictures. Figure 27–9: Tom & Pat Leeson/DRK Photo. Figure 27–10: © Frank S. Balthis. Figure 27–11: L. Veisman/Bruce Coleman, Inc.

UNIT 28 *Opener.* Clive Hicks/Bruce Coleman Ltd. Figure 28–1: © Renee Lynn/Davis/Lynn Photography. Figure 28–2: Tom McHugh/Photo Researchers. Figure 28–5 & 28–7: H. J. de Blij.

PART IV OPENER Earth Satellite Corp./Science Photo Library/Photo Researchers.

UNIT 29 *Opener.* Franco Salmoiragmi/The Stock Market. Figure 29–1: David J. Cross/Peter Arnold, Inc. Figure 29–7a: © John D. Cunningham/Visuals Unlimited. Figure 29–7b: © Steve McCutcheon/Visuals Unlimited. Figure 29–9: David Matherly/Visuals Unlimited. Figure 29–10: Cameron Davidson/Bruce Coleman, Inc. Figure 29–12: Thomas Kitchin/First Light, Toronto. Figure 29–13: Rob Crandall/Stock, Boston.

UNIT 30 *Opener.* James Mason/Black Star. Figure 30–1: Tom McHugh/Photo Researchers. Figure 30–2: © Jon Gnass. Figure 30–3a: Fred Ward/Black Star. Figure 30–3b: E.R. Degginger/Animals Animals/Earth Scenes. Figure 30–5 & 30–6: H. J. de Blij. Figure 30–7: © Carr Clifton.

UNIT 31 *Opener.* Tom Bean/DRK Photo. Figure 31–1 & 31–3: H. J. de Blij. Figure 31–5: William E. Ferguson. Figure 31–6: H. J. de Blij. Figure 31–7: Sinclair Stammers/Science Photo Library/ Photo Researchers. Figure 31–8: T. A. Wiewandt/DRK Photo. Figure 31–10 & 31–11: H. J. de Blij.

UNIT 32 *Opener.* H. J. de Blij. Figure 32–2 & 32–5: Cartographic Division/National Geographic Society. Figure 32–10: John S. Shelton. Figure 32–12: Fred Prouser/Reuters/Bettmann Archive.

UNIT 33 *Opener.* Loren McIntyre/Woodfin Camp & Associates. Figure 33–4: W.H. Robbins/Gamma Liaison. Figure 33–7: Earth Satellite Corporation. Figure 33–8: David Muench Photography.

UNIT 34 *Opener.* Steve Vidler/Leo de Wys, Inc. Figure 34–1: Olafur K. Magnussen/National Geographic Society. Figure 34–3b: Imagery. Figure 34–5: Steve Raymer/National Geographic Society. Figure 34–6: Tim Alipalo/Reuters/Bettman Archive. Figure 34–7: Patti Murray/Animals Animals/Earth Scenes. Figure 34–8: Peter Boonslar/Tony Stone Images/New York, Inc. Figure 34–10: H. J. de Blij. Figure 34–11a:

Cartographic Division/National Geographic Society. Figure 34–13: Jose Fusta Raga/The Stock Market.

UNIT 35 *Opener.* Richards/Sipa Press. Figure 35–3: Bettmann Archive. Figure 35–4: Asahi Shimbun/Sipa Press. Figure 35–8: UPI Telephoto/ Bettmann Archive.

UNIT 36 *Opener.* Craig Tuttle/The Stock Market. Figure 36–8: Georg Gerster/COMSTOCK, Inc. Figure 36–11: H. J. de Blij. Figure 36–15: Earth Satellite Corporation.

PART V OPENER Colin Prior/Tony Stone Images/New York, Inc.

UNIT 37 *Opener.* Michael J. Howell/International Stock Photo. Figure 37–1: Dave Lyons/Planet Earth Pictures. Figure 37–2: Mike Geissinger/ COMSTOCK, Inc. Figure 37–3: Meb Anderson/Picture Group. Figure 37–5: M. P. Kahl/Photo Researchers. Figure 37–6: H. J. de Blij.

UNIT 38 *Opener.* Jeff Gnass. Figure 38–1: H. J. de Blij. Figure 38–2: Kevin Schafer/Tom Stack & Associates. Figure 38–3: H. J. de Blij. Figure 38–4: John Shaw/© Natural History Photographic Agency. Figure 38–5: J. A. Wilkinson/Valan Photos. Figure 38–6: H. J. de Blij. Figure 38–7; Keith Gunnar/Bruce Coleman, Inc.

UNIT 39 *Opener.* Robert Holmes/Photo 20-20. Figure 39–1: Courtesy NASA. Figure 39–3: H. J. de Blij. Figure 39–4: © Richard Armstrong. Figure 39–7: San Francisco Chronicle. Figure 39–14: D. Gorton.

UNIT 40 *Opener.* Kjell Sandved/Bruce Coleman, Inc. Figure 40–1: Wolfgang Kaehler. Figure 40–3: © Willard Clay. Figure 40–11: Porterfield/Chickering/Photo Researchers.

UNIT 41 *Opener.* Courtesy NASA. Figure 41–1: USDA/Soil Conservation Service. Figure 41–3: Courtesy NASA. Figure 41–5: EROS Data Center. Figure 41–8: U.S. Army Corps of Engineers. Figure 41–9: © Dan Guravich.

UNIT 42 *Opener.* Henryk Kaiser/Leo de Wys, Inc. Figure 42–5: Shelly Grossman/Woodfin Camp & Associates. Figure 42–7: Aerials South East. Figure 42–8: Stockman/International Stock Photo. Figure 42–14: American Geographical Society. Figure 42–16: Grant Heilman Photography.

UNIT 43 *Opener.* Kal Muller/Woodfin & Camp & Associates. Figure 43–1: Tom Bean. Figure 43–2b: Courtesy NASA. Figure 43–3: S. C. Porter. Figure 43–9: Earth Satellite Corporation. Figure 43–11: Todd Powell/ ProFiles West, Inc. Figure 43–13: Earth Satellite Corporation.

UNIT 44 *Opener.* © Laurence Parent. Figure 44–1: John S. Shelton. Figure 44–2: H. J. de Blij. Figure 44–6: Courtesy NASA. Figure 44–7: Jim Tuten/Black Star. Figure 44–8: Robert Semeniuk/First Light, Toronto. Figure 44–9: © Larry Ulrich.

UNIT 45 *Opener.* Walter Frerck/Odyssey Productions. Figure 45–1a: H. J. de Blij. Figure 45–1b: Nicholas DeVore III/Bruce Coleman, Inc. Figure 45–4: Tom Bean/DRK Photo. Figure 45–7: S. C. Porter.

UNIT 46 *Opener.* Martin Rogers/Stock, Boston. Figure 46–1: Courtesy NASA. Figure 46–5: © Mark Newman. Figure 46–7: Jim Brandenburg/ Minden Pictures, Inc. Figure 46–9: Courtesy Advanced Satellite Productions, Inc. Figure 46–14: Lynda Dredge/Geological Survey of Canada.

UNIT 47 *Opener.* Sam Abell/National Geographic Society. Figure 47–1: Courtesy USGS. Figure 47–2 & 47–3: H. J. de Blij. Figure 47–4: Westermann Schulbuchverlag GmbH. Figure 47–5: Courtesy NASA. Figure 47–8: H. J. de Blij.

UNIT 48 *Opener.* Fletcher & Baylls/Photo Researchers. Figure 48–1: Natalie Fobes/AllStock, Inc. Figure 48–4: Imagery. Figure 48–6: Stephen J. Krasemann/DRK Photo. Figure 48–8: © Steve McCutcheon.

UNIT 49 *Opener.* Anthony Bannister/© Natural History Photographic Agency. Figure 49–1: Stephen J. Krasemann/AllStock, Inc. Figure 49–2: © Martin Miller. Figure 49–3: COMSTOCK, Inc. Figure 49–5: J. Hartley/ Panos Pictures. Figure 49–8, 49–11 & 49–12: H. J. de Blij.

UNIT 50 *Opener.* Tim Barnwell/Stock, Boston. Figure 50–3b: © Ric Ergenbright. Figure 50–7: Dick Davis/Photo Researchers. Figure 50–8 & 50–10: H. J. de Blij. Figure 50–11b: John S. Shelton.

UNIT 51 *Opener.* H. J. de Blij. Figure 51–3: © Robert Perron. Figure 51–5: Breck Kent. Figure 51–6: Courtesy NASA. Figure 51–8: UPI/Bettmann Archive. Figure 51–10: H. J. de Blij.

UNIT 52 *Opener.* David Muench Photography. Figure 52–6: Hank Andrews/Visuals Unlimited. Figure 52–7: Ned Gillette/The Stock Market. Figure 52–8: © Carr Clifton. Figure 52–9: Doug Greene/The Stock Market. Figure 52–10: © Jon Gnass. Figure 52-11: Nathan Benn/Stock, Boston, Figure 52–12: William E. Ferguson. Figure 52–13: Ken M. Johns/ Photo Researchers.

UNIT 53 *Opener.* Worldsat International/Science Photo Library/Photo Researchers. Figure 53–3: Mike Clemmer/Photo Researchers. Figure 53–4: Grant Heilman Photography. Figure 53–5: Todd Powell/ProFiles West, Inc. Figure 53–8: David Muench Photography. Figure 53–9: Fridmar Damm/Leo de Wys, Inc. Figure 53–10: Earth Satellite Corporation. Figure 53–11: Galen Rowell/Mountain Light Photography, Inc. Figure 53–12: Thomas Kitchin/Tom Stack & Associates.

Index

Earth *Magazine Hazards Case Studies*

EARTH MAGAZINE/WORLD WIDE WEB
HAZARDS CASE STUDIES

Introduction

Natural hazards are an integral part of the physical geography of the global environment, and none of our planet's six-billion-plus inhabitants is immune. Hardly a week goes by without a major news story from somewhere in the world about an extreme atmospheric or lithospheric event with disastrous consequences for some unlucky locale.

We have selected five articles from *Earth* Magazine that explore some of the most dramatic natural hazards: extreme weather events associated with El Niño, ranging from floods to droughts; tornadoes; hurricanes; volcanic eruptions; and earthquakes. Consistent with our approach in this book, each of the articles contains a blending of concepts and actual examples, and is supported by diagrams, maps, and photos. More importantly, each

article explores the topic in greater depth than was possible in the main body of this text.

Each article is preceded by a brief introduction that provides the context for users of this book, and includes a series of questions and issues to consider while reading. We also suggest that you explore these hazards even further by visiting our Hazards Website at

http://www.wiley.com/college/geography/hazards

Thus, when the next noteworthy hurricane or earthquake strikes, you will have the opportunity to monitor the event, learn first-hand about its human consequences, and keep using the interpretive skills you acquired in this course.

Contents

What's Wrong With the Weather?—El Niño Strikes Again, discusses the severe-weather hazards associated with ENSO (El Niño-Southern Oscillation). This topic was introduced in the Perspective box in Unit 11 (pp. 122-123) and then carried further in the Perspective box on "Teleconnections" in Unit 20 (p. 225).

ENSO's signature is the abnormal heating of the ocean surface in the eastern equatorial Pacific off northwestern South America and Central America to its north. This pooling of unusually warm water, which can last for more than a year, disrupts ocean currents and atmospheric pressure systems throughout the tropical latitudes of the Pacific—and in several mid-latitude zones beyond. Developed to its maximum extent, an ENSO event unleashes both flood and drought conditions that can have ruinous impacts on fishing, crop-raising cycles, and other human activities in locales as widespread as Peru, the western United States, Indonesia, Australia, India, and southeastern Africa.

This article discusses the global implications of the ENSO event of 1994-95, and highlights the suite of environmental hazards triggered by the temporary reversal of tropical Pacific water and air flows. It is particularly well focused on the wider geographic consequences of ENSO outbreaks. From your reading of the article, you should be able to respond to the following questions:

1. What are the negative human impacts of ENSO in coastal Peru and Ecuador as the upwelling of cold subsurface water is replaced by the warm waters of El Niño?

2. Why would an ENSO event be likely to disrupt seasonal weather patterns in California, Australia, and India?

3. What would be the likely impact in your home area if the local weather regime stayed much drier for several months? Much wetter? Much cooler? Much warmer?

4. In mid-1997, all signs pointed to the emergence of a significant El Niño event with the potential to equal or surpass the strongest ENSO outbreaks of recent times. Using the 1994-95 El Niño as a model, what similarities and differences can you note in a comparison to the ENSO event that took shape in 1997?

To answer Question 4, you will need access to the newest information on the ENSO event that began in early 1997. Newspaper and magazine articles will probably be helpful, but the best place for the latest and most comprehensive information is the website of the Climate Prediction Center/National Centers for Environmental Prediction, an agency of the federal government's National Oceanic and Atmospheric Administration (NOAA). You can easily get to this site and many others on the World Wide Web via our Hazards Website at:

http://www.wiley.com/college/geography/hazards

Once you arrive at the NOAA site, you will get the latest ENSO advisory and also be able to take advantage of the many other ENSO-related information services offered by this website.

As for the 1997 ENSO event, we regret that our deadline required us to sign off at mid-year, just at the exciting point when it seemed that the rapidly-strengthening El Niño was about to assume historic proportions. We are fully aware, of course, that ENSO's causal mechanisms are poorly understood, and that nature could be misleading us into making assumptions about an incipient pattern that could suddenly fall apart. But the last advisory from an always cautious government agency, just before going to press on June 30, leads us to believe that something extraordinary is in the making. NOAA bases its forecasts on a battery of computer models that have performed successfully in the past. In their June 26, 1997 advisory, the language does not equivocate: ". . . [ENSO] conditions have developed in the tropical Pacific. During the first half of June, the easterly winds collapsed over the entire equatorial Pacific, and actual westerlies developed in the region. The only other time that this has happened in the last 30 years was in mid-November 1982. . . . the recent sea surface temperature anomalies near the South American coast exceed those observed [in most previous ENSO episodes]. In fact, since 1950 only the 1982-83 ENSO episode [one of this century's strongest] featured temperature anomalies greater than those presently observed."

Caveat: In linking the *Earth* article to the material in this book, a comment is necessary. The term *thermocline*, used in the article's diagrams to indicate the boundary between the warm near-surface water layer and colder waters below, corresponds to the boundary between the light-blue and darker-blue zones on the front side of each block diagram in Fig. 11-8 on p. 123.

What's Wrong with the Weather? El Niño Strikes Again

by Keay Davidson

Noah might have been impressed by the deluge that hit California in

Keay Davidson is a science reporter for the San Francisco Examiner. His story on the origin of deep earthquakes appeared in the November 1994 issue of Earth. Reprinted with permission of Earth: The Science of Our Planet, *copyright 1994 by Kalmbach Publishing.*

January of 1995. For about two weeks, torrential rains marched in from the Pacific Ocean, killing at least nine people and causing more than $1.3 billion in damages. Roads and communities were inundated as rivers breached their banks, 30-foot waves slammed into the coastline, waterspouts whirred off the coast and mud slides buried cars and

closed roads in mountainous areas. In the Northern California town of Guerneville, nearly eight inches of rain fell in less than one day. Elsewhere in the state, skiers were delighted: By mid-January, the Lake Tahoe basin in the Sierra Nevada had received more than twice its normal snowfall for that point in the season.

At the same time, however, skiers in the northeastern United States were mourning. Because of the abnormally warm weather, even the snow machines couldn't keep enough powder on the slopes. And the weird weather wasn't confined to California and the Northeast. On average, temperatures in the United States were more than six degrees above normal, making it the warmest January on record. Meanwhile, Australia, the Caribbean and parts of Central American were beset by drought.

What's wrong with the weather? What's wrong is El Niño, a warming of the equatorial Pacific Ocean that disrupts weather patterns across vast stretches of the globe. Typically, El Niños occur every three to seven years when warm water in the upper reaches of the western Pacific drifts eastward and pools up off the coast of South America. This mysterious warming becomes noticeable around Christmas, which is why people on the coast of Peru and Ecuador dubbed the phenomenon El Niño, ''The Child,'' in reference to the infant Jesus.

The California rains dramatically demonstrated that El Niño's reach can extend far beyond the equatorial Pacific. As the warm water drifted toward South America, it spawned thunderstorms further east than usual. This shifted the path of the southern jet stream, a current of air that flows from west to east over the Pacific. Instead of coming ashore further north as usual, the jet stream slammed straight into California, carrying moisture and storms. Rain was not necessarily the only possible outcome. In some previous El Niños, changes in the jet stream brought drought to the same region.

Researchers emphasize, however, that El Niño did not *directly* cause the weird weather seen in the winter of 1995. Instead, the Pacific warming altered the circulation of the atmosphere and ocean in a way that made stormy weather more likely. El Niño ''stacks the deck a bit, changes the odds, so that certain [weather] patterns become more likely,'' explains meterologist Stephen Zebiak of Lamont-Doherty Earth Observatory in Palisades, New York.

Atmospheric scientists have developed sophisticated programs, or models, that can predict the onset and effects of El Niño with fair accuracy. They can point with pride, for instance, to the

In flooded Rio Linda, a homemade raft provided the only means of escape for this man's dogs.

successful forecast of the 1986–1987 El Niño, which arrived, peaked and then subsided as expected.

But forecasters' crystal balls have turned cloudy in recent years. In 1991, El Niño conditions began to develop in the Pacific, again as expected. The warming peaked in November had all but disappeared by summer 1992. But then, in January 1993, El Niño resurged, peaking again by late spring. Conditions in the Pacific were back to normal by the winter of 1993–94. But then in the spring of 1994, yet another El Niño began to develop. By the time California experienced its catastrophic rains, the warming was in full bloom—for the third time in four years. What's more, the latest satellite observations of the ocean indicate that this most recent El Niño was *twice* as strong as the last.

This repeated warming of the equatorial Pacific is unusual, but it is not the only new wrinkle in the science of El Niño. Recently, a team of researchers detected a moving region of abnormally warm water in the North Pacific they believe is the echo of the powerful 1982–1983 El Niño. Another team has linked the Pacific El Niño to a similar warming pattern in the Indian Ocean. This suggests that to forecast El Niño reliably, scientists may have to take into account the influence of past El Niños and climatic events thousands of miles from the Pacific Ocean.

Despite these complications, re-

searchers remain confident that forecasts of the onset, timing and effects of El Niños will continue to improve. And that's important because lives and economies are at stake. Communities dependent on skiing or agriculture can be hurt badly by unseasonable warmth or rain. Reliable, accurate forecasting of El Niño and its effects might help those communities to cut their losses. A forecast of drought, for example, might convince farmers to plant crop varieties adapted to dry conditions.

The historical roots of El Niño research lie in the early 20th century, when British meteorologist Sir Gilbert Walker noticed that the local barometric pressures on the eastern and western sides of the Pacific Ocean had a tendency over many years to see-saw back and forth. Atmospheric scientists call this the Southern Oscillation.

Walker attempted to link this barometric see-saw to changes in weather as far away as the United States and Canada. At least one of his contemporaries ridiculed him for suggesting that regional changes in the atmosphere could have global effects. We now know that Walker was onto something: The flip-flop of the Southern Oscillation is the hallmark of El Niño and the ultimate reason warm water drifts to the eastern Pacific.

The Pacific Ocean:
Normal Conditions

Strong trade winds blow west along the equator (white arrows), causing warm surface water (brown, orange and yellow) to pool in the western Pacific. Warm, moist air rises from the ocean surface, spawning thunderstorms. As the trade winds push surface water away from the South American coast, cold water rises to fill the void (black arrow) and the thermocline — the boundary between deep, cold water and the warmer layer above — dips sharply to the west.

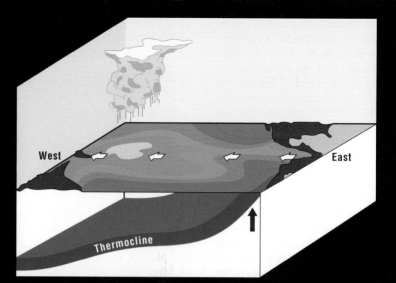

West   East

Thermocline

The Pacific Ocean:
El Niño Conditions

Trade winds slacken, allowing warm surface water to flow east (white arrows) from the western Pacific. Upwelling along the South American coast diminishes as the thermocline levels somewhat (black arrow). Thunderstorms, which normally remain in the western Pacific, follow the warm pool of water across the ocean. This alters the normal pattern of atmospheric circulation. One result of these changes is a redirection of jet streams, which upsets the weather in North America.

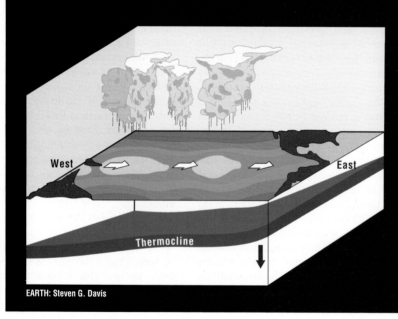

West   East

Thermocline

EARTH: Steven G. Davis

© EARTH: Steven G. Davis

Between El Niños, high pressure prevails in the east, causing the trade winds to flow briskly westward across the tropical ocean (see diagrams above). These winds drag warm surface water into the western Pacific, raising sea level there by as much as 24 inches. This westward drift robs warm water from the surface of the ocean off South America. In response, colder water wells up from below, loaded with nutrients and tiny creatures called phytoplankton. This is a favorite meal for fish, which is why the west coast of South America is normally such a good place to cast a net.

But when El Niño conditions develop, the catch of the day begins to diminish. For reasons as yet unknown, atmospheric pressure rises in the western Pacific and falls in the east, weakening the trade winds. This allows warm water in the western Pacific to drift back eastward. The result: A pool of unusually warm water extends across the equatorial Pacific.

As the upwelling of plankton-rich water decreases, the local fisheries begin to fall off drastically. This is why fishermen there could spot El Niños long before the advent of satellites and modern oceanography. As people in California well know, El Niño also has a dramatic effect on the atmosphere: Warm, moist air rushes upward from the surface of the ocean, spawning thunderstorms. As a result, the central Pacific becomes a rain-spattered corridor of storms.

El Niño is *supposed* to develop every three to seven years and confine itself to the equatorial Pacific. But the more scientists study El Niño, the more it defies previous assumptions. Scientists have some new problems to solve.

For example, why has El Niño returned three times in the past four years? Some blame a general warming of the atmosphere caused by the buildup of carbon dioxide and other "greenhouse gases." By raising global temperatures, the greenhouse effect may somehow disrupt the atmosphere in a way that makes El Niños more likely to develop. This explanation for more frequent El Niños is "a reasonable hypothesis that needs to be lookd at," says Kevin Trenberth, a scientist who studies El Niño at the National Center for Atmospheric Research (NCAR) in Boulder, Colorado. "There are definitely unprecedented things happening."

For instance, he says, the temperature of the atmosphere in the tropical Pacific has been rising since the late 1970s. Trenberth speculates that the ocean isn't cooling off as much as it once did between El Niños because it is absorbing more heat from the greenhouse-warmed atmosphere. This could conceivably diminish the length of the interval between ocean warmings and bring more frequent El Niños.

But some atmospheric researchers are skeptical, if not dismissive, of this explanation. "I don't believe it for a minute," says meteorologist James O'Brien, director of the Center for Ocean-Atmospheric Prediction Studies at Florida State University in Tallahassee. "This has happened before, so I'm not going to attribute it to some scary monster called greenhouse warming." Records of ocean temperatures, O'Brien says, show that El Niño returned repeatedly between 1939 and 1941.

Trenberth insists, however, that something has changed in the Pacific. In the last 20 years, there have been six El Niños. At the same time, he notes, there have been fewer periods in between of below-normal ocean cooling. These are known as La Niña, "the girl." To Trenberth, the scarcity of La Niñas means an overall warming trend in the Pacific Ocean. This could explain the repeated El Niños of the last four years.

The role of global warming is not the only topic of debate among El Niño researchers. Gregg Jacobs, an oceanographer at the Stennis Space Center in Mississippi, and colleagues at the University of Colorado say they've detected the relic—a kind of heat echo called a Rossby wave—of an El Niño that ended 12 years ago. They suspect the remnants of past El Niños may still roam the seas and perhaps even affect the climate.

Rossby waves aren't anything like ordinary, wind-driven surface waves, however. A Rossby wave is simply an abnormally warm region of the ocean, like an El Niño but not nearly as large. When El Niño's warmth encounters a continent, part of it bounces back, creating a Rossby wave.

According to Jacobs and his colleagues, the Rossby wave they've found was set in motion by the infamous El Niño of 1982–83. Often called "the El Niño of the century," the Pacific warming that struck in 1982 devasted marine life and brought disastrous droughts or torrential rains to various parts of the world. Some scientists consider it the greatest disturbance of the ocean and atmosphere in recorded history.

The researchers detected the Rossby wave using satellite observations of water movements in the Pacific Ocean. In their scenario, part of the warm pool that struck the Americas during the 1982–83 El Niño headed northwest after bouncing off the coast of North America. Right now, this elderly Rossby wave is headed toward Asia like an echo in an immense canyon. The researchers say it has already had a measurable effect on the ocean: The Kuroshio Extension, a current that normally flows east away from the coast of Japan, may have been deflected considerably north of its normal path in a collision with the Rossby wave.

If Rossby waves from long-gone El Niños are still rolling around the Pacific, then it's anyone's guess how they might affect the atmosphere and ocean. Some researchers have suggested that Rossby waves might introduce random "noise," like the background hiss of a radio, into the circulation of the ocean and atmosphere. This might explain why El Niño occurs at irregular instead of fixed intervals.

Even more intriguing, some scientists have raised the possibility that heat from this old Rossby wave could even influence the weather—that it may have actually played a role in the 1993 Mississippi floods. But NCAR's Trenberth has grave doubts about the whole basis of this speculation. In the first place, he doubts the wave is as old as Jacobs thinks. And if a Rossby wave from the 1982–83 El Niño can persist for so long, then why isn't the ocean cluttered up with the echoes of other El

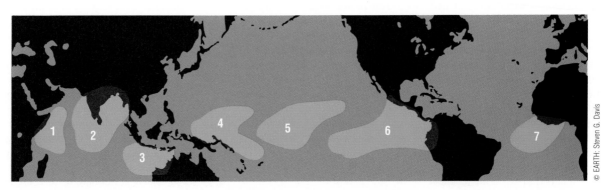

El Niño warming does not seem to be confined to the Pacific Ocean, as shown in this illustration. As an El Niño progresses, a pool of warm water moves east across the Indian Ocean (1 through 3) at about the same time a similar pool moves across the Pacific (4 through 6). Twelve to 18 months later, an additional warm pool (7) develops in the Atlantic Ocean.

Niños? "I don't want to say it's not possible," he says, "but it's not proven."

O'Brien is more enthusiastic about Jacobs' work. "It's a new way to think about the memory of the ocean," he says. He found the satellite observations particularly convincing. "If they hadn't had the satellite confirmation, nobody would have believed it, including me," he says.

On a broader front, scientists are also trying to understand El Niño's apparent long-distance connections outside the Pacific basin. For example, Yves Tourre of Lamont-Doherty Earth Observatory and Warren White of Scripps Institution of Oceanography in La Jolla, California, have recently completed the most detailed study yet of a warming in the Indian Ocean that appears to move in sync with the Pacific El Niño.

Scientists have known about this Indian Ocean El Niño for a while, but Tourre and White are the first to take its temperature below the surface—to a depth of 1,300 feet (400 meters), in fact. This measurement tells the researchers how much heat is stored beneath the surface of the ocean, and this information is useful in forecasting; the more heat stored, the longer and more severe an El Niño can be. Until this study, the Indian Ocean warming had been observed by satellites that could sense only the temperature of the ocean surface.

Tourre and White traced the movement of this warming based on ocean temperature data collected from 1979 to 1991 by cargo ships, fishing boats, research vessels and warships. These measurements showed that during the Pacific El Niños of 1982–83 and 1986–87, warm water off Africa's east coast drifted to the center of the Indian Ocean at the same time a warm pool reached the central Pacific. Then, as the warming in the Indian Ocean reached the Timor Sea, north of Australia, the Pacific El Niño reached its mature stage off South America.

Tourre and White also detected a warming and cooling cycle in the Atlantic Ocean that appears to be linked to the Indian and Pacific El Niños. At the peak of these warmings, winds blowing from the east over the Atlantic grew stronger. This cooled the topmost layer of the ocean off the coast of Africa. Twelve to eighteen months later, the winds died down, allowing the Atlantic waters to warm up.

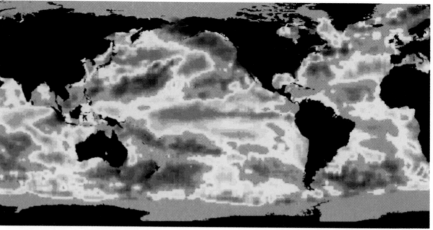

A map of sea surface temperatures measured November 1994 shows warm pools of water (red) building in the Pacific and Indian oceans—the beginnings of El Niño of 1995.

Source: Yves M. Tourre. Columbia University Lamont-Doherty Earth Observatory

The Pacific and Indian ocean El Niños behave as if connected by unseen processes in the atmosphere and ocean. Right now, Tourre and White admit, those links are far from clear. But the recognition of these warming and cooling patterns outside the Pacific might still prove useful in forecasting regional weather.

Like its counterpart in the Pacific, the Indian Ocean warming can drastically affect the atmosphere. It seems to suppress the summer monsoon rains farmers rely on for their livelihood. If dry summers in India and Australia could be forecast, says Tourre, it might give people time to prepare. Similarly, changes in ocean temperature and wind speed in the Atlantic are important because they affect the weather on adjacent continents. Again, establishing links between the Atlantic warming and cooling to weather patterns might provide better long-range forecasts.

Tourre and White's study only reiterates how much scientists' views of El Niño have changed. Once thought of as a regional warming that settled off the Pacific coast of South America every so often, El Niño is now regarded as only one aspect of a phenomenon that reaches virtually around the entire equatorial realm and has the ability to affect weather from San Francisco to Sydney.

Ironically, even if El Niño researchers achieve their ultimate goal—a complete model of the ocean and at-mosphere that can help them predict this chronic catastrophe—forecasters would still be wrong some significant fraction of the time. This is because the weather is driven by processes that don't progress in a highly predictable manner, like the falling of a line of dominoes. Instead, the weather is influenced by inherently unpredictable "nonlinear" processes.

That means that some degree of uncertainty may always plague the computer models used to forecast El Niño's effects on the weather. "The system is simply so complicated," says Nicholas Graham of Scripps Institution of Oceanography. "It's got some profound nonlinearities in it—billions of moving parts, connected by billions of springs and moving levers."

Despite the remaining questions and uncertainties about El Niño, researchers are pleased with the progress they've made since severe El Niños in the 1970s and 1980s demonstrated the phenomenon's great and far-reaching destructive potential. "If you go back seven or eight years ago, we couldn't predict El Niño very reliably," says Eric Barron of Pennsylvania State University, who chairs a National Research Council committee on climate research. "But the last several Niños have been [forecast] months in advance. I wouldn't say we've got the timing down cold, but we've made real progress."

Michael McPhaden, director of the Tropical Atmosphere Ocean Array project at the Pacific Marine Environmental Laboratory in Seattle, also sees a bright future for model-based forecasting of El

Niños. "Given the dramatic progress in the past 10 years, I'm very optimistic the models are going to improve with time—significantly."

Real success will come when scientists can anticipate El Niño's effects on regional weather as well as its time of arrival or departure. To that end, scientists such as Ants Leetmaa of the National Weather Service's Climate Analysis Center in Camp Springs, Maryland, is studying possible links between U.S. weather and the time of El Niño's peak warming.

Even though it may not be possible to flawlessly anticipate the behavior of an inherently unruly beast such as the atmosphere, "at the rate knowledge is advancing, we should be able to predict whether you're going to get excess drought *or* rain," says Florida State's O'Brien. That's because enough El Niños have been studied that meteorologists have identified some general trends in its effects on climate.

Consequently, once El Niño begins to show itself, the future becomes somewhat more predictable. Partly because of this increased understanding of El Niño, the National Weather Service has announced that it will begin issuing national forecasts of seasonal average temperatures and precipitation a full year in advance, four times the previous limit of three months.

Tornado Troopers discusses the Spring 1995 field experiences of atmospheric scientists in Oklahoma, who are searching the middle of "Tornado Alley" for nature's most vicious weather hazard in order to study these violent storms as closely as they can. Tornadoes were introduced in Unit 13, first in the general discussion of thunderstorms on pp. 140–141 and then as a separate topic in the Perspective box on pp. 142–143.

Tornadoes have always captured our imagination, and become prominent news stories when they inflict losses of life and property in the central and southern United States each year. Filmmakers seized on this fascination of the American public in the mid-1990s to produce a number of blockbuster movies based on natural disasters. None was more successful than the first of these, *Twister* (1996), whose plot revolved around the adventures of storm chasers in the southern Great Plains(!). The heightened awareness of tornadoes in this popular film's aftermath prompted its producers (Warner Bros.) to create a continuing Internet website that includes material on the latest scientific advances in the field. When you visit our website at

http://www.wiley.com/college/geography/hazards

you will find a hot link directly to this site.

An important part of this article is the large, five-stage diagram entitled "Anatomy of a Tornadic Thunderstorm." It nicely extends the text coverage of supercell formation discussed and diagrammed on pp. 140–141. By tracing the airflow patterns through *Earth*'s diagram, you can clearly observe the development of rotational mechanisms that intensify and produce the destructive vorticity circulation that is the hallmark of a tornado. This model will enhance your understanding of tornado formation, and enable you to answer questions about the local atmospheric processes involved. Although the rest of the article mainly treats applications rather than concepts of physical geography, you should be able to answer a pair of additional questions:

1. Why must scientific instruments be highly mobile in order to undertake meaningful observations of tornadic thunderstorms?

2. Why is the atmosphere above Tornado Alley one of the most active crucibles on Earth for the formation of tornadoes?

Of all the environmental hazards, North American television viewers most frequently encounter the aftermath of tornadoes on the news programs they watch. In many ways, we are almost too familiar with the news coverage: the obligatory ground and aerial views of the property destruction, the interviews with dazed survivors, and the announcers' exhortations as to how to prepare if you find yourself in the vicinity of a tornado. People living in tornado-prone areas quickly develop an abiding respect for the awesome destructive power of these storms. In fact, all of us who live east of the Rocky Mountains need to be aware of what to do if a local tornado warning is issued.

This information, plus everything else you've always wanted to know about tornadoes, is available on the Internet—and readily accessed via our website listed above. Tracking current tornado activity is most easily accomplished through the website of NOAA's National Severe Storms Laboratory, the same folks whose activities are described in the *Earth* article. From our website we can link you directly to the facility's Storm Prediction Center, so you can monitor their Hazardous Weather Update reports.

Caveat: The map showing the airflows that converge in the area of Tornado Alley indicates the southerly flow of cool dry air as originating in the prairies of Canada to the east of the Rocky Mountains. What you should also be aware of is the other major source of cool dry air: much of the western third of the United States between the Pacific coastal strip and the Rockies, thanks to the rain shadow effect and the relatively high elevation of the terrain (see Unit 13, pp. 141–144).

Tornado Troopers

by Daniel Pendick

Tornadoes are among the most dangerous and least understood forms of severe weather. This spring, a small army of scientists blitzed across Tornado Alley to catch up with a few.

The VORTEX armada pauses on a rural highway in southern Oklahoma, awaiting instructions on where to go to find a storm.

It's 9:15 A.M. on Monday, April 3, at the forecast center of the National Severe Storms Laboratory in Norman, Oklahoma, the storm-chasing capital of the United States. Eric Rasmussen and a handful of other meteorologists at the laboratory huddle around a cluster of computer screens, searching their silicon crystal balls for signs of impending action in the atmosphere.

What they're looking for, specifically, is any sign that "supercells" will form today within driving range of the lab. These towering, rotating thunderstorms often unleash damaging winds, hail bigger than golf balls and, about half the time, tornadoes. Hardly anyone in his right mind hopes that severe storms like these will strike anywhere nearby, but each of the storm-chasing troops now assembling at the lab's two-story, government-issue brick building is probably doing just that right now.

Made up of scientists and college students from around the United States, the troops are mustering in Norman for the largest series of coordinated storm chases ever. The computers are showing some faint indications that something is stirring in the atmosphere over Texas right now, raising hopes among the tornado troopers that today will bring the first chase of the season, which in this part of the country lasts from early April to late June. Heightening the sense of anticipation is the knowledge that this very day marks the 21-year anniversary of "Terrible Tuesday," a day that saw the worst

Daniel Pendick is an associate editor at Earth. *His story about geologist Stanley Williams and Colombia's Galeras Volcano appeared in the August 1995 issue. Reprinted with permission of* Earth: The Science of Our Planet, *copyright 1995 by Kalmbach Publishing.*

multiple tornado outbreak on record. Beginning on April 3, 1974, an incredible 148 tornadoes touched down over a period of 24 hours in 12 states from Michigan to Alabama. They killed 309 people, injured 5,300 and caused $600 million in damage.

But everyone also knows that April is just the beginning of the severe-weather season here. Statistically, the hottest time is May. So nature could easily confound the computers: The early atmospheric stirrings could die out, and the troops could be ordered to stand down until tomorrow. As in the real army, hurry up and wait may prevail.

The storm chasers eagerly await an order from Rasmussen to head out into the plains in their armada of cars, vans and aircraft. Every vessel in the fleet is bristling with equipment designed to probe the anatomy of severe storms and tornadoes. Their theater of operations is Tornado Alley, a region of the central United States that sees more twisters than anywhere else in the world. The chasers have dubbed their mission VORTEX—the Verification of the Origin of Rotation in Tornadoes Experiment. Although meteorologists have chased and observed storms in small groups since 1972, nothing of

VORTEX's scale and technical sophistication has ever been attempted.

Over the course of the next ten weeks, the chasers hope to use careful maneuvering—and a bit of luck—to position themselves near erupting supercells. Then, guided by Rasmussen in a mobile command post, the tornado troopers will deploy their array of sophisticated scientific instruments around the developing storms. The scientists believe that what they learn about tornado-spawning storms will lead to more accurate forecasts of severe weather. And that, they say, will lead directly to saved lives.

At full force, the VORTEX armada will consist of 21 vans and automobiles equipped with mobile weather stations and two airplanes carrying advanced radar equipment. Twenty lead scientists and about 100 students will be involved in the project over the course of the season. Last year was an unusually quiet one for severe weather, and the armada intercepted only a handful of severe storms. The project needs to succeed this year. "This may be the last time in ten years we get a chance to do this," Rasmussen says.

Rasmussen strolls over to a darkened room in the lab to give the first full-fledged morning briefing of the

© EARTH: Daniel Pendick

VORTEX forecasters Chuck Doswell (left), Jerry Straka (rear) and Erik Rasmussen (far right) confer in the "ops center" of the National Severe Storms Laboratory before deciding whether to send the tornado troops out on a storm chase.

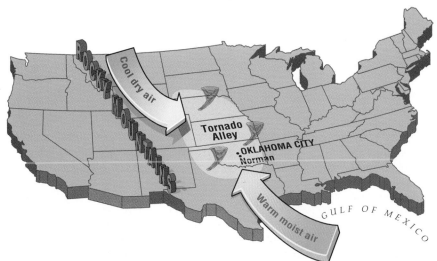

© EARTH: Steven G. Davis

Tornado Alley experiences the highest concentration of twisters in the world. It is located at the confluence of two rivers of air: a cool, dry one that is channeled south and east by the Rocky Mountains, and a warm, moist one that flows up from the Gulf of Mexico. When the currents converge, conditions in the atmosphere become ripe for severe storms, which often spawn tornadoes.

advice of the other forecasters on the project, Rasmussen finally makes up his mind. VORTEX is a "no-go" today, he says. Hurry up and wait prevails.

But not for long. By the next morning's briefing by Rasmussen, one of the computer models is forecasting conditions that could trigger a wild weekend of tornadic storms.

"If you believe the models," Rasmussen says, "Saturday looks like the end of the world."

Since at least the time of Aristotle, weather watchers have been awed by the inexplicable violence of tornadoes. Seeming to drop out of the sky without warning, these rapidly rotating columns of air can race through a town in minutes, smashing buildings into rubble. Monster tornadoes can pack winds in excess of 200 miles per hour into a space only a mile wide or less. Within such a twister, scientists believe, hot spots of 300-mile-per-hour wind can develop.

About 1,000 tornadoes are reported in the continental United States every year. Two percent of them—the true killer twisters spawned by supercells—cause about two thirds of all deaths. Nearly 18,000 people have been killed by twisters in the history of the United States. Major disasters have not been uncommon, especially before the advent of weather radars that can spot characteristic features of tornadic storms. On March 18, 1925, the infamous Tri-state tornado raced across Missouri, Illinois and Indiana at highway speed, killing 695 people and injuring 2,027, the worst such disaster to date.

Tornado Alley is so named because of a geographical coincidence. This region of the central United States, which includes northern Texas, Oklahoma, Kansas, Nebraska and southern South Dakota, is at the confluence of two rivers of air. One is a cold river from up toward Canada that's channeled south and then east by the Rocky Mountains. As the air descends from the higher altitude of the mountains to the Central Plains, it dries out. The other river, flowing north from the Gulf of Mexico, is moist and warm. When these rivers meet, they create a layering of cool, dry air aloft and warm, moist air near the ground—an atmospheric pressure cooker that routinely turns out severe thunderstorms and tornadoes.

In the heart of Tornado Alley lies Norman, the home base of some of the

season. He steps the troops through a series of weather maps generated by computer models that predict how the atmosphere over Tornado Alley will evolve in the next 12 hours. At the end of the forecast briefing, Rasmussen sums up the situation: "A lot of things have to come together today and I'm not too confident they will. I want it to look like a pretty good shot to justify going into Texas."

Over the next hour and a half, Rasmussen and his fellow meteorologists sift through data streaming in from balloon-launch stations, wind-measuring radars and weather stations throughout Tornado Alley. A complication has arisen: the various ingredients for storms are indeed present in the Alley, but they are not likely to be in the same place at the same time. Combining all this information and the

Earth   **9**

Anatomy of a Tornadic Thunderstorm

Overshooting

Stratosphere

Troposphere

Rotating updraft

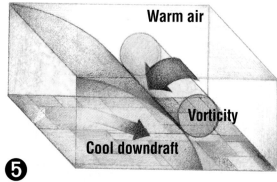

Warm air

Vorticity

Cool downdraft

❺

Just as vorticity developed in the middle altitudes of the
thunderstorm due to wind shear, it can develop within the
boundary regions between a cool downdraft and warmer
surrounding air — a region where wind speed and
directions vary. This creates vorticity, a pattern of flow in
the atmosphere that can create a rotating column of air
called a vortex.

   At first the vorticity develops in a horizontal orientation
depicted in this illustration as a rolling barrel. Then the
vorticity is bent into a vertical position by the downdraft
and drawn into the updraft. A whirling column of air, or
vortex, now develops beneath the base of the storm.
The vortex narrows and stretches, rotating faster
and faster. Eventually it extends completely
from the base of the storm to the ground,
becoming a tornado.

Downdraft

Matthew Groshek

Illustration not to scale.

Gust front

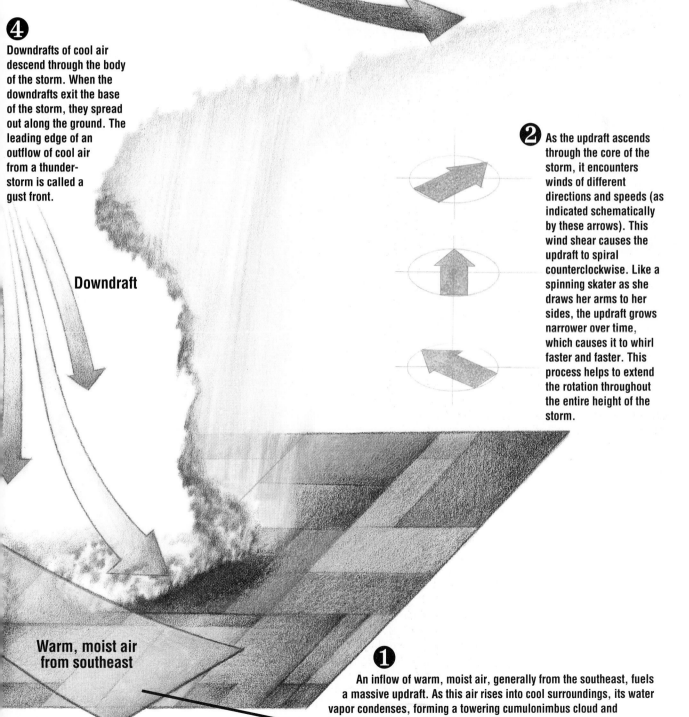

❸ Beginning at the bottom of the stratosphere, the atmosphere warms with increasing altitude. Consequently, after the updraft rises into the stratosphere a short distance — creating what scientists call an overshooting top — it is cooler than its surroundings and stalls. Jet stream winds blow the top of the storm across the bottom of the stratosphere, creating a cloud with a characteristic anvil shape.

50,000 to 60,000 ft.

❹

Downdrafts of cool air descend through the body of the storm. When the downdrafts exit the base of the storm, they spread out along the ground. The leading edge of an outflow of cool air from a thunder-storm is called a gust front.

Downdraft

❷ As the updraft ascends through the core of the storm, it encounters winds of different directions and speeds (as indicated schematically by these arrows). This wind shear causes the updraft to spiral counterclockwise. Like a spinning skater as she draws her arms to her sides, the updraft grows narrower over time, which causes it to whirl faster and faster. This process helps to extend the rotation throughout the entire height of the storm.

Warm, moist air from southeast

❶ An inflow of warm, moist air, generally from the southeast, fuels a massive updraft. As this air rises into cool surroundings, its water vapor condenses, forming a towering cumulonimbus cloud and releasing heat that sustains the updraft.

As a tornado sucks moist air and dust up into its low-pressure core, it darkens enough to be visible to the eye, as in this classic funnel-shaped twister photographed at dusk.

© Warren Faidley/International Stock Photo

most experienced scientist storm chasers in the world. Senior researchers at the severe storms lab like to point out that in 1972, when the first research-oriented chases began there, storm chasing was less than respectable. Perhaps because of the thrill-seeking element that has traditionally gone with chasing, some meteorologists considered it positively unscientific. But now careers in meteorology are made on field observing.

Although storm experts have a general grasp of the conditions that lead to severe weather, they aren't sure about the details. Mainly, they need to know more about the environment—the particular mix of pressures, temperatures and air circulation—in which tornadic thunderstorms evolve. That would help them answer a critical question for the millions of people who live in Tornado Alley: Why do some severe thunderstorms produce tornadoes and others don't? If the researchers can answer that question, they say they will be able

to cut down on false alarms, which encourage fatal complacency. More important, the scientists say there will be fewer instances of tornadoes spinning up unexpectedly, an improvement that would allow more people to get out of the way of tornadoes in time.

Despite the many remaining questions about how and why twisters form, the evolution of a tornadic supercell is understood in its essentials. In Tornado Alley, the process begins in the early spring, when warm, moist air from the Gulf builds up beneath the layer of cool, dry air aloft. When moisture fills a layer about a mile thick, severe weather becomes possible.

Over the course of a morning and afternoon, the Gulf air is juiced up by solar heating, gathering the energy it will need to rise in an updraft and punch into the upper atmosphere, forming a thunderstorm. But the air won't begin to rise by itself. Something has to lift it high enough until it's warmer than its surroundings, at which

point it will continue to rise on its own. Often the leading edge of a mass of cold air does just that. The cold front digs underneath the warm air like a plow, providing enough vertical "forcing" to send the warm air on its way upward.

Even with strong forcing from below, however, the moist air still has to eat through a thin layer of warm air that develops just beneath the cool, dry air aloft, what storm chasers call the "cap." Once it breaches the cap, the air can then rise freely to great altitudes. Because it often takes a long time for this to happen, severe thunderstorms often occur in the late afternoon or early evening.

Once the cap is broken, juiced-up air from ground level streams into the dry air aloft and condenses, forming a thick cumulus cloud. The condensing air releases heat and warms the rising air parcel, making it even more buoyant. The flow of moist air climbs higher and higher, building a cumulonimbus cloud tower. At the center of the tower is a

© National Center for Atmospheric Research/UCAR/NSF

An airplane stands outside a hanger at Denver's Stapleton Airport (now closed) while a tornado develops close overhead.

These shearing winds interact with the strong updraft of the storm, causing the rising air to spin in a counterclockwise direction. The rotating core of the storm then contracts and stretches vertically, extending the rotation through the entire height of the storm. Like a spinning skater drawing her arms to her sides, the updraft rotates faster and faster as it narrows.

The supercell is now a vast engine, drawing up moist air as fuel and converting part of that water vapor into rain and hail. It also has an exhaust: downdrafts of cool air that exit the base of the storm. One major flow, the rear-flank downdraft, exits the right rear corner of the moving storm; another, the front-flank downdraft, gushes from the leading edge of the storm along with rain and hail. These downdrafts create "gust fronts," the cool pulses of wind that arrive minutes before the main body of the storm.

Once the storm possesses a rotating updraft and cool downdrafts, it's primed to spawn tornadoes. The popular notion of how tornadoes form is that the storm itself drops a funnel to the ground. But in reality, most meteorologists now believe that supercells simply create an environment beneath themselves that favors the formation of tornadoes.

Computer models that simulate the complex plumbing of supercells say that tornadoes are born at boundaries between the cool downdrafts and the warm, rotating updraft of a storm. Across these boundaries, the speed, direction and temperature of the air changes dramatically—akin to the wind shear that creates rotation within the updraft of the thunderstorm. At these airflow boundaries beneath the storm, a distinctive pattern of air flow called *vorticity* develops.

Vorticity is hard to imagine because it's not a physical object, like a whirlpool. Instead, vorticity is a quality of the air itself that, when it interacts with the updraft of a supercell, can create a tornado.

The simplest way to picture the vorticity generated beneath a supercell is to imagine a barrel rolling on its side between the downdraft and updraft. This "horizontal vorticity" is then tipped into a vertical position by the downdraft and drawn into the updraft. The region of the updraft between the ground and the base of the storm now has weak, counterclockwise rotation—

strong updraft. The disturbance is now a thunderstorm "cell," named for the oval blob it makes on a radar screen.

Thunderstorms can grow and die in a matter of two hours as an intense updraft forms and then collapses in on itself, dropping a load of rain and hail in the process. But given the right conditions, an ordinary thunderstorm can keep developing, forming a persistent rotating core and complex internal plumbing. The storm is then considered a tornadic supercell. Towering to heights of 50,000 to 60,000 feet and oc-

cupying 1,000 cubic miles of space, large supercells can persist for as long as six hours and spawn multiple tornadoes as they barrel across the countryside.

Supercell growth is enhanced by what storm chasers call "upper air support": twists and turns in high-altitude jet-stream winds that sustain the growth of the storm tower. Another critical ingredient for supercells is wind shear: either changing wind direction or speed, or both, from the base of the cloud to an altitude of about 30,000 feet.

it is a vortex. Then, by a process similar to the one that created the core of rotation higher up in the storm, the vortex beneath the storm contracts and stretches into a tornado.

The makings of a twister can also form beyond the edges of supercell downdrafts. Vorticity can appear at any boundary between cool and warm air—the edge of a warm front or near a mass of cool air left behind by a passing storm. A supercell then comes along and grabs that vorticity into its updraft, spinning it up into a tornado.

At least that's what computer models fed by thirty years of spotty field observations have come up with. VORTEX is supposed to find out if the computer modelers have it right.

It's Sunday, April 9, and the end of the world hasn't arrived. At least not yet. But things have been heating up over southwestern Oklahoma since Rasmussen made his no-go decision on Monday. Even before his briefing today, it's clear that a storm chase, the first of the season, is a near certainty.

At the briefing, Rasmussen describes conditions he thinks may spawn severe storms over southwest Oklahoma late in the day. Balloon soundings of the atmosphere now indicate that there may be enough moisture under the cap to provide the necessary fuel. And it looks like a cold front moving down from the north will provide the match, plowing under the warm, moist air to ignite strong upward movement.

After the briefing, tornado troopers begin to muster on the laboratory grounds. Most of the armada will consist of "probes," cars topped with a yard-high tower of meteorological instruments to measure temperature, humidity and wind speed and direction. Two teams will deploy "turtles," heavily weighted, dome-shaped instrument packages the scientists place on the ground in the path of a storm to retrieve measurements from inside a twister. The armada also includes a camera team, a balloon-launching van that will send instruments up into the storm itself, and a P-3 Orion airplane on loan from government hurricane researchers based in Florida.

Each vehicle in the armada is equipped with a satellite tracking system that radios back its location to Rasmussen's mobile command post. There, his computer displays the location of each vehicle on an on-screen map of the

A P-3 Orion aircraft sits on the runway at Will Rogers International Airport in Oklahoma City shortly before a mission to observe severe storms. The plane is equipped with scanning radar systems that can probe the interiors of tornadic thunderstorms.

University of Oklahoma students Mike Magzig (left) and Mark Askelson prepare a "turtle," an instrument package that's designed to be left directly in a tornado's path and take readings from deep within the violently rotating winds of the vortex.

surrounding area that includes every known road, railroad crossing and bridge. From his weather-worn Ford van, Rasmussen—who is known as the FC, or field coordinator—will bring the armada to bear, casting it like a net over the storm and its immediate environment.

At 12:29 P.M, the armada heads for the turnpike toward southwest Oklahoma. In one of the instrumented cars,

Probe 7, University of Oklahoma student Charles Edwards handles the driving while the team leader, Dave Stensrud, fiddles with the laptop computer that controls the probe's electronics and instruments. In the back, Danny Mitchell, a computer specialist for a university meteorology lab, rides along as an observer.

Even if a tornado occurs today, many of the VORTEX teams actually will have only a slim chance of seeing one. To get a comprehensive set of observations, some of the probes will have to be many miles from the action. Other probes will be close, but in obstructed viewing spots. Close encounters with tornadoes will be few and far between. As in a real army, some troops get the glory, some clean the latrines.

By 1:10 P.M, the armada is heading southwest toward an area where warm and cold air masses are converging. These conditions must remain strong for moist air near the ground to break far enough into the dry air aloft to keep going.

3:10 P.M. Rasmussen has halted the armada along a rural road near Ratliff City. He begins to draw on current weather information radioed to him by "nowcasters" back at the lab and by scientists in the P-3 Orion plane circling the area. The sun is shining, juicing up the air.

"The most potentially unstable air is here," Probe 7 leader Dave Stensrud says. "If anything is going to happen, it will happen here."

Rasmussen wants to be sure, so he sends probe cars out in all directions, casting his net. Data from the vehicles and from distant weather stations indicate that a pool of moist air now blanketing the Ratliff City area is beginning to drift east. Rasmussen's voice crackles over the radio: "All teams: Let's roll. Northeast bound on Oklahoma seven."

3:45 P.M. The armada is heading for an area of thick cumulus clouds. As additional information scrolls across Rasumussen's computer screen, he becomes concerned that the piling up of air in southern Oklahoma has begun to weaken.

"We may have seen the best convergence we're going to see today," he radios to the troops, "but we can keep our fingers crossed."

5:45 P.M. The armada has turned north toward another region of building cumulus cloud. On the horizon, a 15,000-foot tower is trying to punch through the cap. It is the largest tower seen by the chasers all day. If it managers to get through, the storm will evolve quickly. The armada drives on, trying to get downwind of the tower in case if evolves into a storm.

6:00 P.M. Towers are starting to form all along the western horizon. Rasmussen stops the armada along the highway. A gigantic, frightening tower, backlit by the setting sun, is mushroom-

© EARTH: Daniel Pendick

Late in the day on VORTEX's first storm chase of 1995, a cumulus tower blooms on the horizon and mushrooms into the atmosphere. A VORTEX chase car is there to gather the data. It's equipped with instruments for measuring air temperature, humidity and wind speed and direction. The onboard computer stores the data and uses a satellite-tracking system to pinpoint the car's location.

© Joyce Franks/The Daily Ardmoreite

This home near Ardmore, Oklahoma, was destroyed by a May 7 tornado. The debris and tossed cars mark the path of the twister.

VORTEX's greatest quarry: This tornado, which touched down near Dimmitt, Texas, on June 2, tossed truck trailers, snapped utility poles and scoured a wide swath of asphalt right off a highway.

© Harold Richter

ing upward like an atomic cloud, climbing toward 20,000 feet.

If excitement could keep it going, the tower would rise into the stratosphere, 10 miles up. But energy in the atmosphere, not adrenaline, is what keeps these monsters going. And today there just isn't enough. In a few minutes, the top of the tower stalls, breaks off and drifts away.

6:40 P.M. After a half dozen hopeful scrambles in the direction of cloud towers that repeatedly bloomed and died, Rasmussen has halted the troops for the last time. In fading light at the end of the day, the Probe 7 team is standing around the car, shooting a few photos for the scrapbook. "Skip the legalities," Charles, the driver, says. "This thing has died." The towers had just enough energy to break through here and there, but not enough punch to keep going.

Forecaster Chuck Doswell of the severe storms lab ambles out of Rasmus-

sen's command-post van. He stares off to the west at the cloud towers rising and falling. Doswell, a veteran storm chaser, doesn't see today as a failure. It has been a typical chase in many ways—many hours on the road punctuated by moments of hope, all to see nothing. Yet he's pleased. "It was great," he says. "I had a blast. And we got a chance to work the kinks out of the system."

It's 7:07 P.M. Probe 7 pulls out onto the highway and heads west back to Norman. It wasn't the end of the world after all, just the end of the day.

The members of VORTEX '95 didn't have much to chase in April. But the situation soon changed. By June 8, when the project leaders called off the hunt, the armada had intercepted seven tornadic supercells and 11 tornadoes. On June 2 in Friona, Texas, one of these twisters picked up a railroad box car and sent it bouncing through a cemetery. Nearby, a twister in Dimmitt,

Texas, literally sucked a swath of asphalt right off Texas Route 86. As the tornado passed by, vehicles collected thousands of measurements from the ground, the P-3 Orion circled overhead, probing the insides of the storm from birth to demise. And from a distance of less than two miles, portable radar scanned the anatomy of the tornado. Rasmussen says the supercell that produced the Dimmit twister may become the "reference storm" for tornado studies for the next ten years.

The mountain of information collected during VORTEX may pose as many questions as it answers. Some storms refused to produce tornadoes when every sign said they should have; others spawned twisters in strange new ways. So VORTEX, the largest storm chase ever, will certainly not be the last. "We have a lot to learn," Doswell says, "and from time to time the atmosphere chooses to demonstrate that to us quite vigorously."

Hurricane Mean Season provides a state-of-the-art overview of hurricane tracking and prediction in the western North Atlantic, the source of these powerful tropical cyclones that seasonally threaten (between June 1 and December 1) the eastern seaboard of the United States. This topic was introduced in Unit 14 (pp. 149–153), which also included a Perspective box on the seasonal prediction of North Atlantic hurricane activity (p. 151).

A major theme of this article concerns the new knowledge that atmospheric scientists are continuing to acquire about hurricanes. Some is incorporated into the three-dimensional diagrams that accompany the article, which will enhance your understanding of the structure and behavior of hurricanes. Applications of that knowledge are already paying dividends: more sophisticated computer models have been developed, and the precision of storm-track forecasts has improved steadily since 1990. At the same time, TV coverage has been greatly upgraded, which you can see for yourself by tuning in *The Weather Channel* the next time a hurricane approaches the United States. Telecasters will now present a battery of the latest map images and charts, provide live reports from sites under hurricane warning, and offer the expert opinions of leading meteorologists with (or formerly with) Miami's National Hurricane Center.

The second leading theme of the article treats the predictability of annual hurricane activity, which may or may not follow 20-to-30-year cycles. The prognostications of William Gray in this area has already been noted in Unit 14, and the article brings his work into early 1996.

Unfortunately for Gray, he was wider from the mark than ever before by the time the 1996 hurricane season closed in December. His June forecast of ten tropical storms, with six becoming hurricanes and two of them exceeding Category 3 on the Saffir-Simpson scale, did not square with the actual year-end total of thirteen tropical storms, nine of which were hurricanes with six ranked Category 3 or higher. As for 1997, as of July 1 Gray had predicted eleven tropical storms, with seven to be hurricanes and three of them exceeding Category 3. This forecast, however, was based on a minimal El Niño influence (see first box in the article), a variable that tends to increase hurricane activity. But the rapid build-up of ENSO during mid-1997—as we discuss in the introduction to the El Niño article in this section—may force Gray to revise his numbers downward as the hurricane season progresses. For a current update on the accuracy of Gray's forecasts, check the Internet.

The World Wide Web contains several sites devoted to hurricane research, forecasts, and impacts, and when you visit our website at

http://www.wiley.com/college/geography/hazards

you will find direct links to a number of them. Many readers will be interested in following tropical storms and hurricanes in real time (because we teach physical geography in southern Florida, our students are particularly avid trackers). Our website will instruct you how to follow these storms by linking to the Tropical Prediction Center of NOAA's National Hurricane Center. In addition to a great deal of background material on a given storm and the hurricane season to that point, you will also have access to GOES satellite and radar imagery. At any time of the year you may also access archives of past hurricanes—and see for yourself the awesome destruction wrought by Andrew, Hugo, Camille, Agnes, and the other monster storms of the past three decades that struck the United States.

Caveat: The article points out that in 1996 scientists were expecting the arrival of a new Gulfstream jet plane to enable them to begin observations at altitudes up to 40,000 feet. Such close-up research into the atmospheric forces that steer hurricanes was expected to quickly yield new knowledge to further improve the accuracy of forecasting storm tracks. Technical difficulties delayed the arrival of this aircraft until 1997, but the plane was expected to be ready to go with the appearance of the first tropical storms in mid-summer.

Mean Season

by Daniel Pendick

Last year's wild Atlantic storm season may have been a preview of things to come. If so, hurricane researchers want to be ready.

T he atmosphere was in a very bad mood [during 1995] over the tropi-

Daniel Pendick is an associate editor of Earth. *His last feature, "Tornado Troopers," appeared in the October 1995 issue. Reprinted with permission of* Earth: The Science of Our Planet, *copyright 1996 by Kalmbach Publishing.*

cal Atlantic Ocean. Like some great hurricane factory, nature pumped out one storm after another. By the end of the hurricane season, 11 hurricanes had spun up in the tropics. Five of those storms were "major" hurricanes, with winds of 111 mph or more. It was one of the most active hurricane seasons on record. Perhaps most amazing of all, William Gray saw it coming.

Gray, a meteorologist at Colorado State University in Fort Collins, has issued forecasts of hurricane activity since 1984. Other hurricane researchers praise him for his uncanny ability to recognize patterns in the tangle of meteorological data he uses to make his forecasts. But even Gray cannot foresee what people in hurricane-prone regions most want to know: when and where the next season's hurricanes will make landfall and just how powerful the storms will be when they do. For this, people in the Caribbean and on the U.S. Gulf and East coasts rely on the meteorologists at the National Hurricane Center in Coral Gables, Florida.

Federal hurricane forecasters have made some notably excellent forecasts in recent years. In 1993, a powerful new computer model helped them predict that Hurricane Emily would veer out to sea after striking a glancing blow on the Outer Banks of North Carolina. And

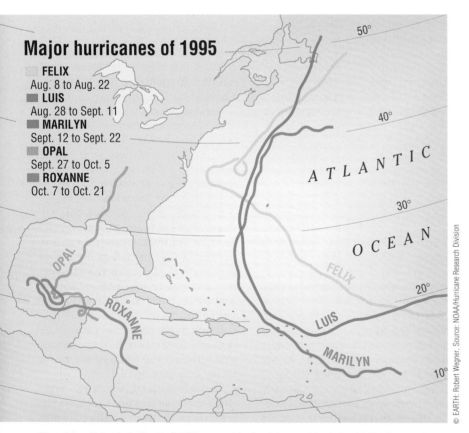

Courtesy Kimberly Baugh, Ethan David and Arno Granados/NOA/NGDC

During the busy 1995 hurricane season, four tropical storms—Iris, Humberto, Luis and Karen—cruised the North Atlantic at once on August 28th.

Major hurricanes of 1995

- **FELIX**
  Aug. 8 to Aug. 22
- **LUIS**
  Aug. 28 to Sept. 11
- **MARILYN**
  Sept. 12 to Sept. 22
- **OPAL**
  Sept. 27 to Oct. 5
- **ROXANNE**
  Oct. 7 to Oct. 21

ATLANTIC

OCEAN

© EARTH: Robert Wegner. Source: NOAA/Hurricane Research Division

Five of the 19 tropical storms of 1995 were major hurricanes, with winds of 111 miles per hour or more. The fierce winds and flooding caused by such storms account for 75 percent of hurricane destruction. The colored tracks indicate the paths taken by the hurricanes during their lifetimes, from the time they first intensified into tropical storms until they were no longer a threat and forecasters stopped tracking them.

they were right. As a result extensive—and expensive—storm preparations were avoided.

But what if Emily had hit land instead of turning out to sea? In that case, the forecasters might have been blamed for any deaths. So understandably, hurricane research today focuses on providing the most accurate warnings possible.

To improve forecasting, hurricane researchers study the internal working of tropical storms firsthand—by flying through them in airplanes outfitted with sophisticated radars and other instruments. After some 1,500 flights through dozens of storms, they've made progress. But a few tenacious problems remain, like the fact that forecasters can't always predict whether or how much a hurricane will intensify before it hits land. That's a problem for people in the path of a storm, who need to know if it's enough just to nail plywood across the windows or if they should leave town altogether.

The need for better hurricane forecasting may become more urgent in the near future. Gray's research suggests that the Atlantic is overdue for a dramatic upturn in major landfalling hurricanes. Since the end of the 1960s, hurricane activity has been relatively mild, he says, lulling coastal communities

William Gray's Hurricane Predictors

Tropospheric winds: Hurricanes and other storm systems form in the lowest layer of the atmosphere, the troposphere, which extends up to an altitude of about 40,000 feet over the Atlantic. William Gray, a meteorologist who issues annual forecasts of Atlantic hurricane activity, uses data from daily weather balloon launches in the Caribbean to measure wind shear—the change in wind speed or direction with increasing altitude. In the troposphere, wind shear works against the formation of hurricanes by interfering with vigorous upward circulation of air in storms. When shear is low, as in 1995, it's easier for hurricanes to form.

El Niño: This warming of the equatorial Pacific Ocean affects the strength of winds that blow from the west at 40,000 feet. Trade winds blow from due east beneath the high-level flow. Together, these opposing flows add up to high wind shear, the enemy of hurricane formation. In 1991–1994, repeated El Niños helped to minimize hurricane formation in the Atlantic . But in 1995 El Niño gave way to La Niña, a cooling of the Pacific that boosted hurricane formation.

Stratospheric winds: For reasons that remain unclear, hurricane activity is strongly affected by the direction of winds in the stratosphere, the layer of air above the troposphere. When stratospheric winds over the Atlantic blow consistently from the west, as they did in 1995, there is a 50 to 75 percent increase in hurricane activity. One proposed explanation for this is that westerly stratospheric winds create greater instability in the upper troposphere, enhancing the vertical circulation of air that helps hurricanes grow.

Climate in West Africa: In years when conditions are drier than average in West Africa, winds blow strongly from the west at 40,000 feet over the tropical North Atlantic. Below, the trade winds blow from due east. This creates wind shear in the troposphere, which stunts the growth of tropical storms into hurricanes. In wet years, winds blow from the east at 40,000 feet, greatly reducing wind shear.

When temperatures are higher and pressures lower on the West African coast than in central Africa, a flow of air from the south carries moist air northward. This clashes with dry air and enhances the growth of storm systems that drift out over the ocean and become tropical storms. This is a helper only: If this factor isn't present, hurricanes can still form.

Conditions in the Atlantic Ocean and Caribbean Sea: Below average pressures at sea level in the Caribbean and tropical Atlantic enhance hurricane formation. That's because these low pressures are the mark of high sea surface temperatures, which fuel storms. Also, lower pressures here lead to convergence, the coming together of air masses that causes vertical air motion, or convection. It is this convention of warm, moist air from the tropical ocean surface that sustains hurricanes.

Global warming? Scientists have speculated that global warming may raise the temperature of the oceans and cause more intense hurricanes, although not necessarily *more* hurricanes. But Gray says that global warming could also change circulation in ways that strongly inhibit hurricane growth, perhaps leading to weaker hurricanes overall. Right now, he says, nobody can predict which way it could go.

---

into a false sense of safety. As a result, development in hurricane-prone areas has soared. Gray and others say this is a recipe for disaster. It won't take more than a handful of major hurricanes striking land on the crowded and densely developed U.S. East Coast to cause damage in the tens of billions of dollars.

Last year [1995] was a mean season for hurricanes, but it was also an extraordinarily lucky one: Most of the storms veered northeast before actually reaching the coast of the continental United States. If Gray is right, that luck is about to run out.

Halfway through last year's hurricane season, forecasters watched the parade of storms with a mixture of dread and anticipation. Nature's hurricane assembly line was in full swing, and with each new storm Gray's forecast was looking better and better. By that time, he had six hurricanes in the bag out of the nine he'd forecast for the 1995 season. Two of them, Felix and Luis, were major hurricanes.

Then the season seemed to grind to a near halt. For most of September, only

one storm appeared, Marilyn. With maximum sustained winds of 115 mph, Marilyn killed eight people and either severely damaged or destroyed 75 percent of the homes on St. Thomas in the U.S. Virgin Islands. But Marilyn was just a taste of what was to come.

By November 30th, the end of the season, a total of 19 tropical storms and hurricanes had cruised the oceans for 121 days. Gray had been able to forecast the busy hurricane season as early as November 1994, and his success was typical: Since 1984, most of his forecasts have been either dead on or only slightly off.

To make his forecast, Gray relies on trends in the global climate that coincide with ups and downs in Atlantic hurricane activity. One such climate "predictor" is El Niño, the warming of the equatorial Pacific that disrupts weather across much of the globe. Shifts in air circulation caused by El Niño disrupt the vertical circulation in tropical storms which prevents them from growing into hurricanes. How other predictors affect hurricane activity isn't as well understood, even

though their influence is pretty firmly established.

What scientists *are* sure of is that the Atlantic hurricane assembly line begins over Africa. The collision of hot, dry air over the Sahara Desert with warm, moist air from the equatorial jungles gives birth to disturbances in the atmosphere called "hurricane seedlings." Each season, about 60 seedlings are blown westward into the tropical Atlantic by the trade winds. At first the seedlings are no more than clusters of thunderstorms. But in an average year, nine will evolve into named tropical storms and about six of those will become hurricanes.

On their way across the ocean, seedlings feed on the heat in warm surface water. The pressure in the center of a disturbance falls as the air inside it grows warmer and lighter. It also gains spin due to the effects of Earth's rotation, which in the Northern Hemisphere makes storms revolve counterclockwise. When the disturbance closes in on itself and becomes a discrete, rotating storm it's classified as a tropical depression. When its winds reach 40

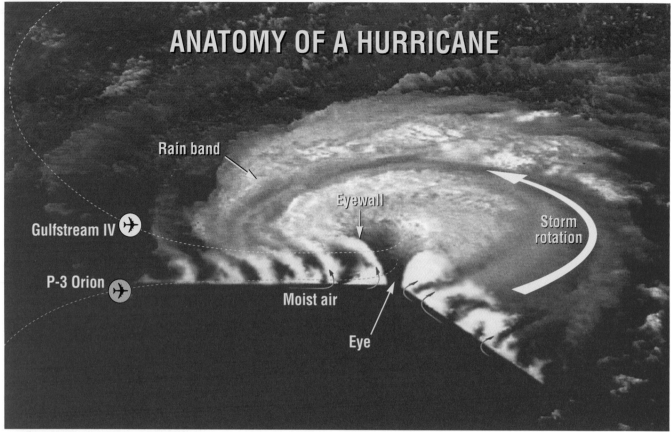

ANATOMY OF A HURRICANE

Rain band

Gulfstream IV

P-3 Orion

Eyewall

Storm rotation

Moist air

Eye

© Matthew Groshek. Source: Hurricane Research Division.

The most characteristic feature of a hurricane is its eye, a central area of low pressure. It's bordered by a ring of thunderstorms called the eyewall. Resembling the arms of a spiral galaxy, lines of clouds called rain bands swirl into the eye. Winds are fastest in the eyewall, the most dangerous region of the storm.

Hurricanes feed off warm surface water. Moist air near the sea surface rushes toward the eye. The air rises through the eyewall, where it condenses into rain and releases stored heat. This heat buoys the air upward, allowing additional moist air to enter the storm.

Researchers use P-3 Orion aircraft to study hurricanes. But they fly no higher than 20,000 feet through storms. So in 1996, the scientists will use a Gulfstream IV jet to reach the tops of hurricanes, at about 40,000 feet.

mph, it becomes a tropical storm and is given a name. When the winds reach 74 mph, the storm becomes a hurricane.

Using Gray's predictors to forecast how many seedlings will intensify into hurricanes is like adding weights to either side of a balance scale: The predictors can either help or hinder hurricane formation. In 1995, the scale shifted conspicuously in favor of storms. A combination of factors that occurs but once every 10 to 15 years added up to a kind of hurricane triple witching hour. Many strong seedlings formed in Africa, the winds were favorable and lots of moist tropical air was available for the storms to feed off. But the scales can tip suddenly: This year [1996], Gray forecasts, there will be only eight named storms, two of them major hurricanes.

Gray sticks by his statistical methods

of forecasting even though most of his colleagues put their faith in computer models that reduce the atmosphere's behavior to mathematical equations. He believes that the past is the best guide to what will happen in the future. It's not possible, Gray says, to reduce the behavior of the atmosphere to computer code. "You can't do it and it's pure arrogance to think that you can," he says.

But this is a minority opinion. Forecasters at the National Hurricane Center in Coral Gables, Florida, rely heavily on computer models that predict the behavior of storms based on current conditions. Not surprisingly, scientists at the federal Hurricane Research Division, housed at the Atlantic Oceanographic and Meteorological Laboratory in Miami, Florida, target their investigations on ways to improve forecasting

models. Gray's statistical methods have certainly led to accurate seasonal forecasts. But as Frank Markes, a scientist at the Hurricane Research Division, points out, "The one thing he can't tell you is whether a hurricane is going to hit your house."

Hugh Willoughby leans back in his chair, gazing out at some point in the stratosphere, utterly absorbed in the inner dynamics of hurricanes. Willoughby is the director of the Hurricane Research Division. At the moment he's trying to explain to a visitor what happens inside hurricanes when they undergo sudden changes in strength. He tries hard, too, periodically launching himself at a nearby blackboard to scribble diagrams.

If anybody knows what a hurricane looks like from the inside, it's Willoughby. He has racked up nearly 400

September 19, 1996, Somewhere Over the Atlantic Ocean . . .

Common sense dictates that deliberately flying an airplane into the heart of the most violent weather in the world is not such a good idea. "As a pilot," says Capt. Gerry McKinn, commander of the hurricane research plane Sky Hopper, "you fly into one of these things and in the back of your mind you're thinking, 'What the hell am I doing here?' As a pilot you're trained to avoid severe weather."

What he's doing is ferrying a team of scientists into Hurricane Marilyn. At the moment, the sky is icy blue and the air is smooth. But in two hours, the pilot will turn his wings into Marilyn's fierce winds and fly into the storm. The crew will cross the storm three times, laying down a network of special temperature probes as part of a scientific experiment. Marilyn is lumbering at about 18 mph toward the cold waters of the North Atlantic, the graveyard of tropical storms.

The hurricane hunters fly in P-3 Orions, sturdy aircraft originally designed to fly low and slow over the ocean, hunting submarines. McKinn spent 20 years in the Navy, tracking Soviet subs. But these days, he's engaged in a different kind of war; the P-3 is armed with scientific instruments instead of Harpoon missiles and nuclear depth charges. On a fully loaded research flight, the plane can carry nine crew, 10 scientists and observers, tons of radar gear and other scientific gizmos around for nine hours—*and* get the whole load home safely at the end of the work day.

So how do you fly an airplane through a hurricane and live to tell about it? It's not as difficult or dangerous as it may seem. The important thing is to keep the wings level and into the wind and maintain just the right air speed—not so slow that the wings lose lift and you plunge toward the sea, not so fast that the hurricane's winds start tearing pieces of the fuselage off.

The violent circulation inside hurricanes often causes the P-3s to lurch, shake, roll and dip. That's enough to give any pilot the willies. And it does: McKinn says it takes years of hurricane flights for pilots to achieve the required confidence. "After three or four years it becomes like driving through rush-hour traffic," he says.

The mission today is focused on the ocean surface, which is boiling with swells and whitecaps as the P-3 enters the storm. Scientist Pete Black is aboard to study the interactions between

the surface of the ocean and hurricanes. The ocean environment, it turns out, can have a great effect on the intensity of a storm. The smooth ride today is dramatic proof: With its heat supply cut off by a blanket of cold surface water, Marilyn is rapidly spinning down. As a result, turbulence is unusually light.

As the pilots carefully make their way through the storm, a member of the flight crew is dropping probes called drop sondes. These parachute to the surface, measure the temperature of the water down to 1,000 feet and radio the data back to the plane. Someday, Black says, the Air Force crews responsible for routine hurricane reconnaissance will deploy similar devices ahead of storms. That will tell them the temperature of the water in a storm's path. Forecasters will use this information to predict whether a storm is likely to intensify.

Today, the data from the drop sondes confirm that Marilyn in moving over Luis's 100-mile-wide cold wake, which is about 7 degrees Fahrenheit cooler than its surroundings. But Black finds something else, something entirely new. His probes are telling him that the multiple hurricanes that have recently passed through this area have left a cold pool of water on the surface, about 100 miles wide and 600 miles long. The pool, 5 to 6 degrees cooler than its surroundings, lingered for a third of the season. To Black and his colleagues, the cold pool calls to mind El Niño, the warming of the ocean surface in the Pacific that has far-flung effects on global climate. If a large pool of warm water in the Pacific can affect global climate,

Top: Looking out the window of a P-3 Orion aircraft that's taking researchers into the heart of Hurricane Marilyn. Bottom: A scientist drops a temperature probe from the plane to measure the interactions between the storm and the ocean.

Black speculates, then a pool of cold water in the Atlantic might, too.

As the Marilyn flight showed, sometimes these scientists are still surprised after decades of studying hurricanes. And no doubt all would agree that those surprises are worth risking a roller-coaster ride through a 10-mile-high storm.

—*Daniel Pendick*

passes into the eyes of hurricanes, more than any other scientist still flying. But he also understands the rules of evaporation, condensation and heat transport that govern storms. Both types of knowledge, practical and theoretical, are needed to understand hurricanes.

Because hurricanes are too powerful to be stopped, researchers focus on finding better ways to warn people ahead of time where a hurricane is likely to hit land and how strong it will be when it gets there. The accuracy of track forecasts—predictions of what area of coastline will be hit by a

Courtesy NOAA

In 1993, forecasters used an advanced computer model to predict that Hurricane Emily (above) would strike a glancing blow at North Carolina's Outer Banks and then head harmlessly out to sea. They were right. The image below is the model's three dimensional representation of Emily's close encounter with the continent. The central plume (in tan) indicates where moist surface air is rising around the storm's central eye; the red arrows show areas of strong upward air movement. At the base of the storm, red patches indicate the most intense rainfall, and dark green patches indicate lighter rain falling from the storm's swirling rain bands.

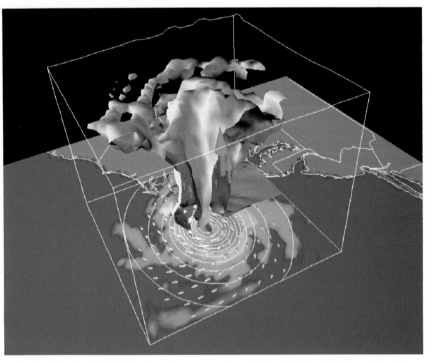

Courtesy NOAA/Geophysical Fluid Dynamics Laboratory, Princeton, N.J.

storm—has improved a great deal. Each year, forecasters have been able to narrow by 1 percent the length of coastline they have to warn in the paths of landfalling hurricanes. And that can save a lot of money. Adding up lost wages, retail sales, tourism dollars and productivity, it costs $300,000 to $1 million per mile to evacuate part of the coastline in South Florida.

Improvements in track forecasting are due chiefly to better computer models, particularly the one developed by federal researchers over the past 25 years at the Geophysical Fluid Dynamics Laboratory in Princeton, New Jersey. But even this sophisticated model can't always predict the sudden and potentially dangerous changes in intensity that hurricanes can undergo.

Hurricane strength is measured on the Saffir-Simpson intensity scale. This system takes into account central storm pressure, maximum winds and potential for storm surge—the wall of water hurricanes bulldoze ahead of themselves. Category 1 storms, with maximum sustained winds of 74 to 95 mph, usually do only minor damage. But major hurricanes—those of categories 3, 4, and 5—can have catastrophic effects. For example, the surge from a category 4 storm that struck Galveston, Texas, in 1900 drowned some 6,000 people.

Dramatic changes in hurricane intensity occur when storms undergo a process called rapid deepening. Suddenly the pressure in a storm plummets. This, in turn, speeds up the circulation in the eyewall, as if a slowly seeping bathtub drain were suddenly unplugged. Rapid deepening can transform a category 1 or 2 storm to a category 4 or 5 in just two days. But it's not entirely clear how this happens.

Hurricane researchers have studied the problem using mathematical models. According to current theory, rapid deepening happens when the top of a storm, which lies at about 40,000 feet, encounters a special kind of circulation pattern. Eddy currents loop in and out of the top of the storm, adding additional spin. This makes the top of the storm expand, creating a wider chimney for air to rise through from below. As a result, the storm sucks in warm, moist air from the sea surface—its fuel—at a faster rate. Upward circulation quickens, wind speeds in the eyewall rise and the central pressure falls: The storm intensifies.

© Al Diaz/The Miami Herald

In 1992, Hurricane Andrew cut a swath of destruction across South Florida, including the town of Homestead (above). Andrew was one of the costliest natural disasters in U.S. history, yet it was but one major landfalling storm. Some scientists believe the U.S. coast will see an upturn in major landfalling hurricanes in the near future.

If the eddies are real and scientists could document the conditions that give rise to them, forecasters might be able to anticipate rapid deepening based on atmospheric conditions ahead of a hurricane. The only way to settle the question, Willoughby says, is to fly through the tops of hurricanes and gather some hard data for the models to chew on.

There's just one problem, however: The P-3 Orion aircraft used to study hurricanes can't go to the altitudes where the eddies dwell. But [in 1997], the hurricane hunters will have at their disposal a small Gulfstream IV executive jet to reach those new heights. They'll be able to measure the direction and speed of winds by dropping parachuted probes equipped with satellite tracking devices. The data they beam back may reveal if the eddies exist.

Willoughby and his fellow researchers are excited about the coming season and their new plane. The spin-up eddies that may give birth to killer hurricanes lie high in the atmosphere, and that's where these scientists, desk-bound most of the year, are happiest.

"We've done a lot of philosophizing about this over the years and a lot of modeling," says Willoughby, who has generated his fair share of theoretical dissections of hurricanes. "But we're anxious to get into the upper troposphere to take a look for ourselves."

When Hurricane Andrew made landfall in South Florida on August 24, 1992, it was rapidly intensifying, snapping and snarling with sustained winds of more than 175 mph. The storm contributed to the deaths of 51 people in South Florida and Louisiana and left 180,000 homeless. In terms of insured property losses, Andrew was the most expensive natural disaster in American history, costing the insurance industry $16 billion in claims. The total economic cost of the storm is estimated at $30 billion. Unfortunately, if William Gray's long-term hurricane forecast is right, that storm was just a preview of things to come.

Gray says the last 25 years represents a lull in the great climate cycle that drives hurricane activity. From 1970 to 1994, four major hurricanes struck the U.S. East Coast. But in the previous 25 years, almost four times that number made landfall. Indeed, the repeated mean hurricane seasons of the 1950s led to the founding of the federal government's hurricane research program.

About 75 percent of hurricane damage is done by major storms—the Andrews, the Opals, the Marilyns. In the absence of repeated punishment, Gray says, people on the U.S. coast have become complacent. They have built not only on the edge of the coast and on shallow barrier islands, but above the water itself on wooden pilings, daring the next monster hurricane to come along and take a bite. If there is a significant upturn in the number of major hurricanes, as Gray forecasts, and if even a handful of them make landfall, the costs could be staggering.

Karen Clark, president and founder of Applied Insurance Research in Boston, Massachusetts, estimates the probable costs of future natural hazards for insurance companies. She has used a mathematical model to rerun the hurricane seasons of the 1940s, when 17 hurricanes of category 1 or greater hit the beach in the United States. Clark superimposed the effects of these storms on the insured property values of today's U.S. coastline and counted up the damages. If the hurricane seasons of the 1940s happened again, she found, the insured damage would be at least $60 billion, and perhaps twice that in overall economic losses.

Gray believes it's inevitable that the current lull in major hurricanes will break. He and his colleague Chris Landsea have looked at the record of Atlantic hurricanes and West African climate and found a pattern: Rainfall and hurricane activity wax and wane together on a 20 to 30 year cycle. In the 1950s and 1960s, when West Africa was relatively wet, nine major hurricanes struck the U.S. East Coast. But from 1970 to 1987, drought years in West Africa, only one major hurricane hit—Hurricane Gloria, in 1985. Gray believes that decade-scale changes in ocean circulation—a conveyer belt of currents that transport heat away from the equator—are behind these cycles. But other scientists are skeptical.

Wallace Broecker, a climatologist at the Lamont-Doherty Earth Observatory in Palisades, New York, is an expert on global ocean circulation and its role in climate change. He knows of no decade-scale cycles in the system that could be firmly linked to hurricane activity.

"I don't think there's any evidence that the conveyer has changed, nor is there any evidence that it hasn't changed," Broecker says. "Nor can you say [hurricane activity] has anything to do with the conveyer. It's just a very, very hard thing to assess."

Because of the kind of coastal development seen here in Miami Beach, Florida, an upturn in major landfalling hurricanes could bring unprecedented losses of life and property.

Lloyd Shapiro, a scientist at the Hurricane Research Division, is working to understand the physical underpinnings of Gray's forecasts—that is, how the climate predictors are physically linked to hurricane formation via "teleconnections" in the atmosphere and ocean. These forces, he concedes, are not completely understood. So right now, predicting what's going to happen in the future is "just guessing. Intelligent guessing, but still just guessing," Shapiro says. On the other hand, he adds,

Gray's speculations should not be discounted too quickly.

"Meteorology is not an exact science," he says. "Sometimes people forget that. There's so much intuition and gut feeling that goes into it, too."

Gray's educated gut is legendary, and it has served him well in decades of study and forecasting of hurricanes. But only time will tell for sure if 1995 was just another mean Atlantic storm season or a glimpse of the coming mean streak.

Fire and Water at Krakatau focuses on one of the most violent volcanic eruptions in modern history: the sudden explosion of an island volcano located just 40 km (25 mi) west of one of the world's most populous islands, Indonesia's Java. In Unit 34 (p. 365) we summarize this *phreatic* eruption and its major consequences, including the hail of pyroclastic rock fragments that fell over an area larger than Texas.

In their *Earth* article, authors Carey, Sigurdsson, and Mandeville point out that the deadliest hazard from Krakatau's eruption came from that part of this pyroclastic bombardment that cascaded downward closest to the slopes of the exploding mountain, fiery clouds that moved at speeds up to 100 km (63 mi) per hour, incinerating people, animals, and vegetation in their path. But it has long been known that the great majority of Krakatau's nearly 40,000 deaths were caused not by the eruption itself, but by *tsunamis* associated with it. These seismic sea waves radiated outward from Krakatau and swept onto the shores of Java and Sumatra as well as small islands, drowning entire towns and villages virtually without warning. The researchers theorize that those pyroclastic clouds, plunging into the water around the erupting mountain, were the main cause of the initial tsunamis, powerful enough to annihilate the entire populations of several nearby islands. Only after these pyroclastic-generated sea waves did the main (and largest) tsunami occur, resulting from the collapse of the entire structure and the formation of the *caldera* whose rim now forms several islands in the Sunda Strait.

The reconstruction of the events of August 26 and 27, 1883 required the coring of submerged pyroclastic deposits and the modelling of the mountain's pre-eruption morphology, so that the scientists could assess not only the volume of pyroclastic flows in all directions but also the severity and directional properties of the associated tsunamis. Their undersea research indicated that one long-held assumption, that the strongest tsunami (and the largest volume of pyroclastics) emanated from the northern flank of Krakatau, was in error. Other submerged evidence, coupled with information on the timing and severity of tsunamis from historical records, produced the frightening chronicle of Krakatau's destruction.

Of all the natural hazards that threaten humanity, volcanic eruptions may have the farthest reach. An earlier nineteenth-century eruption in Indonesia, Tambora in 1815, propelled so much dust and ash into the atmosphere that crops failed in Europe and North America as well as many other areas, making 1816—"the year without a summer"—a time of famine and disaster. And research on still another Indonesian eruption, this one before the dawn of civilization but well after the emergence of humanity, suggests that the gigantic explosion of a volcano named Toba, more than 73,000 years ago, affected global climate for such a long time that our very existence was threatened.

Death-dealing explosive eruptions can happen anywhere, and often they occur without warning. By combining your knowledge of the distribution of stratovolcanoes, the dominant patterns of atmospheric circulation, and the distribution of population, you are in a good position to assess the local and regional risk confronted by all of us who share this unpredictable planet.

There is a great deal of non-technical material about volcanism available on the Internet, much of it supported by fascinating photos, maps, and diagrams. When you visit our website at

http://www.wiley.com/college/geography/hazards

you will find hot links to connect you to many of the best sites. One of the most dramatic aspects of volcanism on the World Wide Web is the monitoring of ongoing volcanic events. For example, just as we went to press on June 30, 1997, a tense, two-year-long volcanic episode on the eastern Caribbean island of Montserrat was intensifying. Up to that point, more than a third of the island's 11,000 inhabitants had already abandoned the island-country; tragically, the first fatalities were recorded during those last days of June, villagers who had ignored advice to evacuate and were subsequently buried by streams of superheated rock and ash that arrived without warning. You may well want to investigate how this episode continued to the present time, and also examine other current examples of volcanic threats to human populations. Among the latter is Popocatépetl near Mexico City, shown and discussed in Fig. 52-13 on p. 537; its transformation from a dormant to an active stage has continued since 1994, and it may be several more years before this threatening episode comes to an end.

A number of websites monitor ongoing volcanic events across the globe, especially those affiliated with the U.S. Geological Survey. An especially interesting site we can link you to is *The Electronic Volcano* maintained by Dartmouth College, which bills itself as "a window into the world of information on active volcanoes" that can guide you to cartographic, photographic, and literary resources on other information servers or in libraries.

Caveat: Some scientists still use Krakatau's older spelling, "Krakatoa," which you may encounter when you visit certain websites.

Fire and Water at Krakatau

By Steven Carey, Haraldur Sigurdsson, and Charles Mandeville

Fiery ash and deadly waves brought death to thousands in Krakatau's 1883 eruption.

In 1883, the island of Krakatau in the Sunda Straits of Indonesia, west of Java, blew up in a catastrophic volcanic eruption, perhaps one of the most violent eruptions in recorded history. Enormous quantities of ash and aerosol gases ejected into the stratosphere and huge volumes of magma and seawater wrought destruction on the surrounding area. Islands in Krakatau's vicinity were wiped clean of vegetation and other life by violent burning winds from pyroclastic flows, then buried under thick volcanic ash. Krakatau island itself ceased to exist as it had been.

The *Simpati* at anchor, one of the vessels we hired in Indonesia to conduct our field work in the waters surrounding the remains of Krakatau.

Mont Pelée brought death to thousands in a similar eruption in the Caribbean in 1902. In this view, a pyroclastic flow enters the ocean on the southwest side of Mont Pelée.

Photo by A. Lacroix. The Department of Library Services, American Museum of Natural History

Drs. Steven Carey and Haraldur Sigurdsson, and graduate student Charles Mandeville from the University of Rhode Island Graduate School of Oceanography, Narragansett Bay Campus, hope to return to Krakatau in 1993 to continue their research in submarine pyroclastic flow behavior. Reprinted with permission of Earth: The Science of Our Planet, *copyright 1992 by Kalmbach Publishing.*

What was left was a fringe of island fragments, a huge submarine caldera and a death toll of over 36,000 lives in the region surrounding the volcano. To the north and northeast of Krakatau's remains, two new islands, Steers and Calmeyer, emerged steaming just above sea level.

Pyroclastic flows is a technical term for one of the most deadly phenomena of volcanic eruptions. These flows are fiery clouds of volcanic debris that cascade down the slope of a volcano at speeds often up to 100 kilometers (60 miles) per hour, engulfing whatever lies in their path. Historically, these clouds may have killed more people than any other aspect of volcanic eruptions. In 1902, 28,000 people died on the Caribbean island of Martinique in pyroclastic flows from the eruption of Mont Pelée. As of September 1991, Mount Unzen in Japan has claimed 39 victims, including three volcanologists caught in pyroclastic flows on its flanks during its latest period of unrest.

The flows from Krakatau's eruption have not been adequately investigated in the century or so following its eruption, despite the death and destruction they caused. So in 1990 we launched

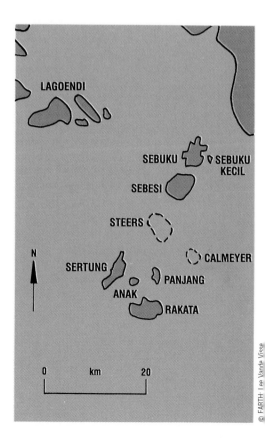

LAGOENDI

SEBUKU　SEBUKU KECIL

SEBESI

STEERS

CALMEYER

SERTUNG

ANAK　PANJANG

RAKATA

N

0　km　20

© EARTH / Lee Vande Visse

We collected samples of volcanic material on the islands as well as on the ocean floor. We needed to determine the sequence of events during Krakatau's eruptive phase and to construct a stratigraphy for the pyroclastic deposits.

an expedition from the University of Rhode Island Graduate School of Oceanography to Krakatau to investigate the 1883 pyroclastic flows. But we also went because many deaths associated with this particular explosive volcanic eruption were not from the clouds of fiery gas-laden debris and ash. Most fatalities resulted from immense *tsunamis*, walls of water up to 36 meters high (120 feet) that inundated coast lines of small islands in the Sunda Straits, the southern coast of Sumatra, and the west coast of Java. Just how these walls of ocean water were generated by the volcanic eruption was a question we hoped to answer.

We suspected the pyroclastic flows were the key. Indeed, 20 percent of all tsunamis associated with volcanic eruptions are generated by pyroclastic flows pouring into the sea adjacent to a

volcano. Many scientific observers, at the time of Krakatau's eruption and for many years since, have speculated concerning the nature of these deadly flows. We ourselves had unanswered questions about the interaction of these incredibly hot (up to 600°C or 1,100°F) flows with sea water of "normal" temperature. We also had unanswered questions about the direction that the majority of these flows traveled as they spread out from Krakatau. In previous studies researchers have proposed that a large amount of material from the exploding volcano entered the sea predominantly to the north. We questioned this assumption.

Krakatau presented a unique opportunity for us. We obtained cores of the submarine deposits from the 1883 pyroclastic flows and correlated them with dated deposits we obtained on the

© Steven Carey

An example of the type of pyroclastic material we found at the sea floor at our dive sites, identical to the 1883 deposits that we found exposed on many of the Krakatau islands.

Earth   27

© Charles Mandeville

Left: Paradise for fieldwork, but we never lost sight of the potential for other eruptions.
Above: We established our base camp on Sertung; the tents were to be our home for two months.
Below: Haraldur Sigurdsson, on board the *Simpati*, checks our diving gear and equipment.

surrounding islands to answer our questions about the flows and tsunamis that exacted such a violent toll in this eruption. One year, 59 scuba dives, 51 sediment cores, and innumerable bathymetric soundings later, we have a better picture of this violent event. Getting the picture wasn't easy, however.

In the mid-1980s, after working in the Lesser Antilles island arc of the Caribbean of the West Indies, we became convinced that we needed to investigate more thoroughly the nature of submarine pyroclastic flows. The flows we had studied were in deep water (6,000 feet or 1,800 meters) and thus difficult to sample. They also originated from eruptions that occurred tens of thousands of years ago, making it difficult to reconstruct the details of their origin.

So we decided to go to Krakatau. There we would find submarine volcanic flows in shallow water from a recent event (1883). We should be able to obtain the samples we needed to answer our questions concerning the behavior of such flows under water and how they were generated during the catastrophic eruption.

We spent almost two years preparing for our 1990 expedition. First we needed a source of funding for our research, which necessitated writing several grant proposals. Fortunately, both the National Geographic Society and the National Science Foundation were willing to finance our expedition.

Rakata—the island remnant of Krakatau—Panjang and Sertung are now part of a national park system, and all of these islands would be important to us. We would have to establish a base camp on one of them and investigate volcanic deposits on all of them, but we needed permission from the Indonesian government to conduct our work in the park. We also wanted to involve Indonesian scientists in a collaborative research effort. The Indonesian government kindly granted our request, and Sutikno Bronto of the Volcanological Survey of Indonesia in Bandung agreed to work with us.

In April 1990, we left the United States. Our scientific party consisted of Haraldur and Jean Sigurdsson, Charles Mandeville, and Steven Carey. The first leg of our journey took two days. Upon arrival, we spent several days in Jakarta, the capital city, arranging supplies, finalizing our permits, and hiring trucks and crew to transport the bulky equipment and other supplies we had air-freighted to Jakarta. We also needed boats.

We left with our trucks for the west coast of Java. There we hired two boats, the *Rapela* and the *Simpati*, and our dive master, Sundjoko Hardjosoekarto. We needed specialized equipment, particulary an air compressor to fill our scuba tanks.

At last, full of excitement, we left Java with our vessels and crew for the four-hour cruise to Krakatau. We landed on the island of Sertung, unloaded our equipment, and set up our base camp. From this site, we began two months work of investigations. We needed to correlate the submarine deposits with those simultaneously deposited on the islands. The size and shape of the collapse caldera had to be accurately established to determine the volume of material ejected and to determine where and how much was deposited as underwater flows. To accomplish this, we needed cores of the submarine deposits, samples of the volcanic layers on the islands, and a detailed depth profiling, obtained by bathymetric soundings, of the ocean surrounding the islands.

Several excellent accounts of the eruption exist, so we drew upon historical records. R. D. M. Verbeek, in his study of the eruption published in Batavia in 1885, stated that as much as 66 percent of the erupted material from the volcano could be found within a 15 kilometer radius (9 miles) of the fragmented island as submarine deposits. He also estimated that 12 cubic km (5 cubic miles) of material was ejected during the explosion. He based his figures on bathymetric soundings taken before and after the eruption.

We needed to make a series of scuba dives to obtain cores of the submarine pyroclastic flow materials. This phase of our expedition took a full month and was the most physically demanding portion of our entire operation. All in all, we made 59 dives and obtained 51 cores of the pyroclastic materials. We used a sediment-coring device that could obtain cores up to 1.3 meters in length (4 feet). These cores were collected in submarine deposits north, northeast, and west of the Krakatau island complex (see map on *Earth* p. 26). By comparing bathymetric data taken in 1886 with our data, we determined that our cores were taken from dive sites at depths 1 to 20 meters (3 to 70 feet) below the original surface of the deposits.

We wanted to ensure that our cores were taken from below the sequence surface to maintain the original stratification and avoid contamination with reworked material. We hoped the layering of the deposits would furnish much information concerning the sequence of events during the eruption, particularly about the sequence of the pyroclastic flow ejection and the way in which the fiery deposits reacted as they entered the much cooler ocean water.

The majority of the cores we obtained came from distinctive pyroclastic material. The primary deposit, massive and poorly sorted, was occasionally covered by a second deposit of volcanic sands and gravel. Pieces of pumice larger than the diameter of our sediment-coring device occasionally blocked our coring efforts. This required us to pull up the corer and try a new location. Rocky fragments ranging form 10 to 30 centimeters (4 to 14 inches) were found in addition to the larger pumice pieces, or clasts. These rocky fragments were embedded in a matrix of silty-sandy ash material that

we were to find identical to the exposed terrestrial deposits on the islands were surveyed. Based on the nature of the cored materials and the close correlation with the island deposits, we were able to identify these submarine materials as pyroclastic flows.

We also made dives in the northern region. Deeper ocean depths in this area, ranging from 50 to 90 meters (160 to 300 feet), define a moat between the Krakatau islands around the Steers and Calmeyer island platforms. It appears that little material was deposited in this deep channel during the eruption. In previous studies scientists have interpreted large hummocky features south of the platforms, rising to within 10 to 15 meters of the ocean surface (30 to 50 feet), as mounds on the surface of a debris avalanche, a cascade of exploded debris resulting from the collapse of Krakatau island during the eruption. After three dives to the crests of these features, we determined them to be instead composed of primary pyroclastic materials from the 1883 eruption.

We found evidence of dilute, denser-than-water turbidity currents on the now submerged, wave-cut platform of Calmeyer. Two dive sites had outcrops of a gray to light gray, well-sorted silty ash, with very find subhorizontal layering. Subsequent X-ray

analysis of these cores revealed extensive cross-bedding. We surmise that the turbidity currents may have resulted from the explosive activity on Calmeyer. Hot pyroclastic materials

A reconstruction of Krakatau island prior to the 1883 eruption, based on pre-eruption drawings and bathymetry data from Simke and Fiske, *Krakatau 1883*, Smithsonian Institution Press, 1983. Constructed by C. Mandeville. The patterned area of Rakata is what remains of Krakatau island today.

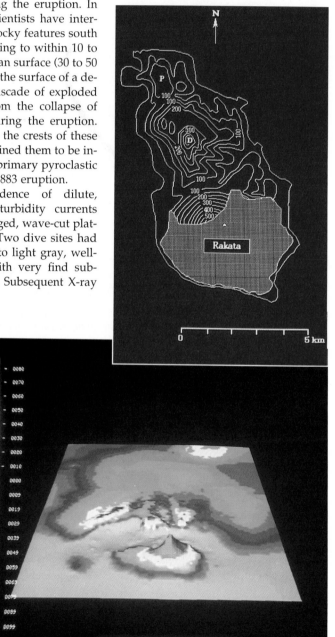

Current bathymetry and topograpny around the Krakatau island complex, looking north. Constructed by C. Mandeville (1990)

© Steven Carey

The sun sets on volcanic peaks in the waters surrounding Krakatau's island complex, uniquely beautiful amid the ever-present danger of another eruption. The next one may be Anak Krakatau. Continuing research will provide clues to the volcano's behavior.

caused steam explosions in the cooler seawater. The steam then mixed with particles of volcanic ash and flowed as a turbidity current.

The most impressive submarine deposit was found on the sea floor west of Sertung. This layer, some 90 meters in thickness (300 feet), was detected in water depths of 20 meters (70 feet) and consists of massive pyroclastic flow. At one dive site in this area, we found extremely large angular pumice fragments—fragments exceeding 1 meter in diameter (3 feet).

After obtaining the submarine cores, we spent another month surveying the islands on foot and collecting samples of the land-based volcanic material. We also took bathymetric soundings of the ocean. We surveyed Sebesi, Sebuku, Sebuku Kecil, Lagoendi, and the southern

coast of Sumatra, all of which were inundated by tsunamis in the 1883 eruption. The islands experienced a complete loss of life, with over 3,000 fatalities on Sebesi alone. On the southern coast of Sumatra, some 40 km from Krakatau (25 miles), thousands fled the deadly tsunamis on the coast, retreating up the sides of Raja Bassa volcano, only to succumb to burns from pyroclastic debris. Countless others never made it inland but were swept away by the walls of water.

We looked for evidence of the pyroclastic flows on the islands and found deposits correlating to the flows in all of these locations. A reworked pumice layer on Sebesi was proof that the main tsunami was preceded by at least one pyroclastic surge and followed by at least two others. Similar deposits were

found on the other islands and in five localities along the Sumatran coast. Typically, the lowest layer of the deposits is poorly sorted, 3 to 10 cm thick (1 to 5 inches) and massive silty-sandy ash, with large pumice pieces. This unit is overlain by an ash matrix approximately 12 to 24 cm thick (5 to 10 inches) and silty-sandy in texture, containing coral fragments and smaller pumice clasts. Often the deposits consist of three or four layers, some of them clearly reworked by water.

We interpreted the sequence of these deposits as an initial pyroclastic surge, followed by the tsunami deposits and then additional pyroclastic materials, which may have traveled up to 50 km (30 miles). The tsunami probably ripped the coral pieces from offshore reefs.

Based on the evidence collected during this expedition, it appears that pyroclastic flows dominated the land-based deposits from Krakatau's eruption. Data from the submarine cores suggests that the pyroclastic materials also dominated the extensive submarine deposits. Contrary to former assumptions about the emplacement of these flows, however, the submarine flows entered the oceans around Krakatau in all directions, not primarily to the north. Maximum accumulations are west of Sertung island, with a lesser volume north of Krakatau.

By digitizing our bathymetric data and constructing computer maps of the sea floor surrounding Krakatau's island system, we were able to compare the current sea floor elevations with a digitized map from pre-eruption data. We found that the caldera developed largely west of the main island. Great thicknesses of pyroclastic material, deposited underwater west of Sertung and spread out across the region to the coast of Sumatra, probably caused the immense tsunamis that hit the coastal areas.

We found little evidence of mixing between the fiery pyroclastic materials and seawater. The massive and poorly sorted nature of the submarine deposits indicates that they were transported by a thick, dense flow of particles. Had these flows been diluted by water, the deposits would appear well-sorted and stratified, and be vertically graded.

The deposition of large volumes of pumice, ash and other volcanic material would have displaced sufficient water to create the deadly walls of water that caused the deaths of so many of the region's inhabitants. Additional evidence supporting this are the field measurements indicating at least five pyroclastic flow units and the eyewitness accounts made by many observers, detailing the numerous tsunamis during the course of the eruption.

One of the tsunamis occurred at 10:02 A.M. on August 27, 1883, and was probably associated with the formation of the caldera during the climax of the eruption. This particular explosion and its accompanying tsunami appeared to be significantly more intense, according to the account written by Verbeek. The enormous explosion could have been the result of large-scale mixing of seawater and magma. A mud rain (a mixture of volcanic ash and water) began to fall over an extensive area east, northeast, and north of Krakatau island, possibly associated with this violent phase of the eruption.

As we sailed through the caldera we were repeatedly struck by the sheer size of the geologic feature that had been created in only two days. Its extent was a constant reminder of the power of this eruption and the fact that this area is still very active. We have learned much about the nature of this explosive event, but new questions arise as we analyze our data from 1990.

We hope to return to Krakatau in 1993 to continue our investigations. Krakatau spawned a new volcano in 1927, *Anak Krakatau*, meaning "child of Krakatau." It remains to be seen whether the child will follow in the footsteps of its parent.

Predicting Earthquakes: Can It Be Done? covers a crucial dimension of natural hazards. Although some volcanoes erupt or explode without warning, most give some sign that activity is imminent. The science of hurricane tracking and prediction has yielded huge dividends in the form of early warnings and timely evacuations. Even tornado watches and warnings have become substantially more accurate and effective.

But the science of earthquake prediction cannot produce comparable results. Year after year, earthquakes claim thousands of lives, striking in known high-risk zones where the danger is recognized and the impact expected—*and* in supposedly stable parts of tectonic plates where no quakes are anticipated at all.

Unit 35 (pp. 369 ff.) addresses the origins, severity (intensity), and distribution of quakes. Figure 35-5 (p. 373) maps the global distribution of recent earthquakes and thus reveals the world's high- and low-risk zones. The Pacific Ring of Fire stands out, as do plate-boundary zones from New Zealand to the Mediterranean. But note that there are epicenters in some of the Earth's most stable areas, including central Canada, central Australia, and central Brazil.

A comparison between Figs. 35-5 and 2-5 (p. 19) reveals how many hundreds of millions of people face the earthquake hazard every day. And what happened in China in 1975 and 1976 shows what earthquake prediction could mean. In 1975 Chinese seismologists, combining ancient methods (observing the behavior of domestic pets) and modern techniques (measuring local seismic activity) urged the evacuation of an area east of Beijing. More than 3 million people heeded the warning, and when a severe earthquake struck only a few hundred people died. But in the following year, the same scientists failed to forecast an even more severe earthquake not far away. The Tangshan quake may have killed as many as 750,000 (the official figure is 242,000), the twentieth century's highest toll in any natural disaster.

As author Davidson reports, the science of earthquake prediction remains relatively unproductive, its few successes so rare and random as to appear to be cases of chance. Precursors of earthquakes (increased seismic activity, cracking in rocks, slight terrain alterations such as "doming"), geoelectric potential (voltage differences in the affected area), and hydrologic factors (water flow in and on the zone under attention) stimulate predictions that do not pan out; quakes continue to occur unexpectedly. Only in Japan, where animal behavior continues to be a major ingredient in predictive science, have some noteworthy successes been recorded—*not* including the 1995 Kobe earthquake that killed more than 5,000 people (see table, p. 372). In general, Western scientists are reluctant to accept animal behavior as a precursor of an earthquake; they argue that the link has not been proven despite some East Asian successes.

The extremely complex, fault-riddled geology in which many earthquakes originate continues to defy prediction. Even the apparently logical deduction that "overdue" areas in high-risk zones are more likely to suffer from earthquakes compared to those recently affected is countered by reality: a major quake does not seem to "relieve" built-up tensions so as to defer the next assault of nature. In our attempts to cope with this omnipresent hazard, nature gives us no help.

The World Wide Web contains a number of sites dedicated to earthquake hazards, and when you visit our website at you will find direct links to many of them. The U.S. Geological Survey's National Earthquake Information Center provides an impressively current, "near real-time" list of the 21 most recent earthquakes that occurred during the approximately three days prior to the user's signing on. This information is presented in the form of a table that displays the following about these quakes: date and time, latitude and longitude of the epicenter, depth and magnitude of the tremor, and a brief description of the location. A major plus for physical geographers is the accompanying page-sized world map that locates the epicenters of each of these events. In fact, so much additional valuable material is now available via the Internet, that it is fast becoming inconceivable for any serious student of earthquakes to monitor a significant seismic event in any other way.

Predicting Earthquakes: Can It Be Done?

by Keay Davidson

Scientists once thought they'd forecast quakes like meteorologists forecast the weather. Now they're not so sure.

The most important lesson about January's [1994] killer earthquake in Los Angeles may be that no seismologist saw it coming.

The 6.6 magnitude temblor occurred at 4:31 A.M. on January 17 when a fault nine miles beneath Northridge in the San Fernando Valley suddenly ruptured. More than 50 people were killed and parts of the City of Angels were left in a shambles. Seismologists had known that the area is riddled with faults capable of producing strong earthquakes. But the most they had been able to say with any certainty was that the region is prone to earthquakes. None predicted that one would occur in Northridge.

It wasn't supposed to be this way. In the 1970s many earth scientists claimed that within a decade they'd be able to forecast earthquakes as meteorologists forecast the weather. Quake-casting, they said, would save lives by enabling government authorities to order evacuations when a big temblor threatened.

Now, however, scientists aren't so sure. Most suspected quake precursors such as odd animal behavior and subtle changes in the tilt of the landscape have not proven reliable. Time and again, quakes didn't strike when they

As a science reporter for the San Francisco Examiner, Keay Davidson frequently covers earthquakes. Reprinted with permission of Earth: The Science of Our Planet, *copyright 1994 by Kalmbach Publishing.*

"should" have, while others struck when they "shouldn't."

But would-be quake-casters still hope to achieve a modest goal: to estimate accurately the probability that a quake will strike a particular region sometime during a relatively long span of years. U.S. Geological Survey scientists claim their probabilities forecast anticipated the 1989 Loma Prieta earthquake, which rocked nearby San Francisco. They've also identified possible precursors to Loma Prieta, including waterfall surges, geyser eruptions and fluctuations in natural electromagnetic fields. If verified, these precursors might hold out a glimmer of hope that more precise quake predictions are possible.

Still, those are bright linings to a dark cloud: the failure of the world's most elaborate attempt at quake prediction—the Parkfield experiment. In the mid-1980s, seismologists forecast that this tiny, rural village in California would be hit by a magnitude 6 quake by January 1, 1993. It never happened. "We failed, no doubt about it," says USGS scientist Allan Lindh. He insists the failure only means forecasting efforts are "harder than we hoped," not that they are a waste. But after Parkfield, according to Jim Mori of the USGS in Pasadena, California, "in general, people in the United States are not very optimistic that they can predict an earthquake."

Back in the 1970s, theoretical developments made quake forecasting seem likely. Most geologists had by then accepted plate tectonics theory, according to which the continents travel atop moving crustal plates, like logs stuck in ice. Plate movement of a few centimeters per year exerts accumulating strain on active faults, strain that is eventually

released by quakes. According to seismic gap theory, the more time a fault has gone without a strain-releasing quake, the likelier the fault will soon shake again.

Another 1970s development was dilatency theory, which seemed to offer a mechanical explanation for quakes and a laundry list of precursors that should precede them. Dilatency theory held that as a fault accumulates strain, the surrounding rock expands and forms cracks and pores. Initially the cracks and pores are voids, full of air. Researchers can detect them by measuring their effect on seismic "P" (for pressure) waves generated by Earth's ceaseless seismic activity. Air-filled voids temporarily slow the P waves. But when groundwater seeps into the cracks, the P waves return to normal speeds. The water also lubricates the cracks, hastening a quake. This scenario suggested that one might forecast quakes by looking for P waves that first slowed and then later returned to normal speeds.

Dilatency theory also encouraged scientists to look for other precursors, including slight changes in surface terrain (caused by the expanding rock), which could be monitored with tiltmeters and surveying instruments, changes in groundwater levels and changes in natural emission of radon gas, a radium byproduct carried by groundwater.

On February 4, 1975, a 7.3-magnitude quake clobbered the Haicheng-Yingkow area east of Beijing in the People's Republic of China. Normally, such a temblor would have killed tens of thousands of people because much Chinese housing is not sturdy enough to withstand strong earthquakes. But Chinese seismologists had forecast the

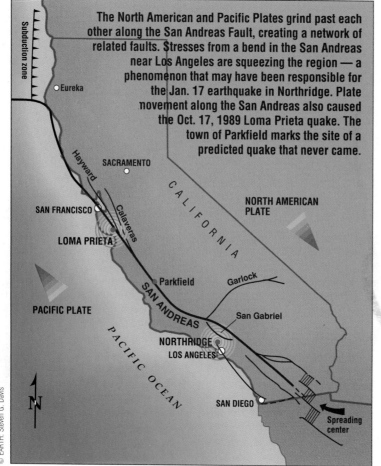

The North American and Pacific Plates grind past each other along the San Andreas Fault, creating a network of related faults. Stresses from a bend in the San Andreas near Los Angeles are squeezing the region — a phenomenon that may have been responsible for the Jan. 17 earthquake in Northridge. Plate movement along the San Andreas also caused the Oct. 17, 1989 Loma Prieta quake. The town of Parkfield marks the site of a predicted quake that never came.

© EARTH: Steven G. Davis

quake several hours in advance, so the government evacuated 3 million residents. Only 300 died. The Chinese seismologists had been monitoring seismic activity in the region and decided to recommend evacuation because of numerous small quakes that they thought were foreshocks to a much larger temblor.

Having triumphed at Haicheng-Yingkow, however, Chinese seismologists were humiliated in 1976 when they failed to foresee a huge quake in Tangshan. More than 200,000 people died.

In August 1979, a magnitude 5.7 quake hit Coyote Lake near San Francisco. Seismologists detected no clearcut precursors. This failure and the Chinese disaster convinced some experts that short-term quake-casts were likely to be unreliable.

But some seismologists believed they had an ace card up their sleeves: Parkfield. Nestled in the hills of central California, Parkfield has a population of 34, a one-room schoolhouse and a

water tower that proclaims in the "Earthquake Capital of the World." The first known quake here occurred in 1857. Thereafter, quakes averaging magnitude 6 came in 1881, 1901, 1922, 1934 and 1966. The resulting intervals—24, 20, 21, 12 and 32 years—weren't as regular as say, heartbeats. But the average interval of 22 years was closer to a cycle than anything else seismologists had seen. In addition, the 1934 and 1966 quakes were both preceded by foreshocks exactly 17 minutes in advance. And just before the 1966 quake, a water pipe broke—a possible precursor. In fact, the broken pipe is "one of the sacraments for the Parkfield 'believers,' " jokes USGS seismologist Andy Michael.

In 1984, three researchers from the USGS and the University of California at Berkeley predicted another quake would hit Parkfield around 1988—22 years after the 1966 quake, give or take a five-year margin of error ending January 1, 1993. This prediction made Parkfield world famous. The USGS

covered the hills with seismic instruments to record the progressive failure of the San Andreas Fault beneath the town.

When the deadline passed without a quake, some earth scientists began to think that the most fundamental assumption of the Parkfield experiment—that earthquakes occur with some regularity on faults—was seriously flawed. In the view of Paul Reasenberg of the USGS, for example, the Parkfield failure is "a significant and symbolic event in a changing paradigm"—a sign that Earth's inner workings are far more complex, perhaps even chaotic, than previously assumed.

Yet many experts, including geophysicist Jon Langbein, belive the Parkfield failure is more a technical slip-up than a major conceptual blunder. The quake might have occurred by now, they argue, if the region's state of tectonic stress hadn't been modified by a 6.5 quake in nearby Coalinga in 1983.

"The probability of getting a magnitude 6 quake at Parkfield is still higher than anyplace else," says Langbein's colleague, Allan Lindh.

Yet there's no denying that the Parkfield earthquake prediction *is* a failure, considering how confident seismologists were that the town would shake by 1993. The failure is especially troubling because the Parkfield segment of the San Andreas Fault is comparatively simple, geologically speaking—it's "very straight and vertical," says Andy Michael of the USGS. Movement of land on either side of a fault segment with such simple geometry should, it seems, be easy to model. So why hasn't the segment ruptured as the seismologists' model predicted? "If we can't figure out how this *simple* fault behaves, we know we're in real trouble," Michael says.

In contrast to the simplicity of the San Andreas near Parkfield, the fault's geometry is fairly complex as it passes through the heavily populated L.A. region, where it bends to the west. This bend is causing the area to be squeezed like an accordion. The resulting stresses have created a 3,600-square-mile network of faults, including the one that ruptured in the recent Los Angeles earthquake. Given the complexity of this fault system, which is not much more ordered than the cracks in a pane of fractured glass, seismologists hadn't a prayer of predicting that this quake would occur where and when it did.

The city of Tangshan, China, lay in ruins following a large earthquake on July 28, 1976, that killed more than 200,000 people. Chinese seismologists had been unable to repeat their earlier success, when they predicted that a huge earthquake would clobber the Haicheng-Yingkow area. On the scientists' advice, 3 million residents were evacuated and only 300 people died.

© China Stock Photo Library

Of the technologically advanced nations that have attempted quake forecasting, Japan remains the most enthusiastic. Over the last two decades, Japanese seismologists have spent close to $1 billion looking for precursors and other warning signs, such as weird animal behavior. One such sign that they've analyzed occurred on February 29, 1972, at 6:32 P.M., when an unusually large school of flying fish went into the nets of fishermen operating in the Pacific Ocean off Hachijo Island. The catch was about ten times more plentiful than normal. Shortly thereafter, the island was rocked by 7.4 magnitude earthquake.

The Japanese are especially fascinated by purported links between quakes and catfish behavior. For 16 years, researchers at the Tokyo Metropolitan Marine Experiment Staton monitored a tank of seven catfish 24 hours a day and compared their behavior with quake events. All seven became more active several days before one-third of the larger quakes, and five or six fish became perkier before 60 to 70 percent of the shakes, which is "bet-

ter than a random association," says researcher Yasuo Baba.

But American scientists say no controlled studies convincingly link weird animal behavior to quakes. And many believe that the links seen by the Japanese have more to do with the will to believe than with real cause-and-effect.

Even so, some Japanese precursor research is difficult to dismiss. From 1991 to 1993, three Japanese earthquakes with magnitudes higher than 5 were immediately preceded by unusual changes in geoelectric potential (voltage differences in the ground), Seiya Uyeda reported this past winter at the American Geophysical Union conference in San Francisco. Greek researchers have seen a similar link between geoelectric potential changes and quakes. But Uyeda, who has joint appointments at Tokai University in Japan and Texas A&M, cautions that further data is needed to establish a causal relationship.

Although American scientists are generally more skeptical than the Japanese about the prospects for finding reliable quake precursors, some still see

faint glimmers of hope. In 1993, the USGS issued an 85-page report, edited by Malcolm J. S. Johnston, that could be a watershed in the history of the search for quake precursors. On the one hand, it finds no evidence that Northern California's Loma Prieta quake was preceded by foreshocks, land deformation or subterranean geological strain. On the other hand, it cites potential evidence of "hydrologic" and electromagnetic precursors that might, just might, have been a warning that an earthquake was imminent.

Sometime after 4 P.M. on October 17, 1989, former park ranger Dan Friend was hiking through Big Basin Redwoods State Park in the Santa Cruz Mountains south of San Francisco. The state was suffering from a drought, and he noticed that some streams were almost dry. He was standing by Berry Creek Falls when the waterfall suddenly grew much louder. For the next four to five minutes, the water flow intensified up to five-fold. Puzzled, Friend continued hiking and decided to camp for the night by Berry Creek, surrounded by trees and amber-brown

The Earthquake Capital that didn't quake: The temblor that seismologists said would hit Parkfield, California, by January 1, 1993, never came. Cafe owner Jack Varian and cook Beverly stand outside the eatery that forms the social center of this town of 34 people. Varian holds a photo of the Parkfield Rodeo, a reminder of times when the town's population numbered several hundred.

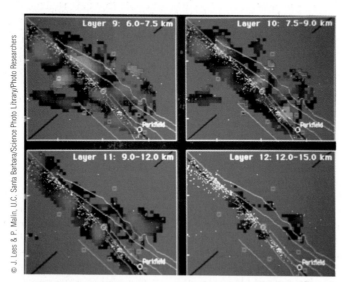

A seismic tomogram of the San Andreas Fault at Parkfield shows the location of microquakes. Each image shows a horizontal cross-section, 25 kilometers by 20 kilometers, at different depths. Green squares are seismometers. Blue areas represent brittle rock formations where the frequent microquakes were believed to relieve strain caused by movement of land on either side of the San Andreas. But the blue area beneath Middle Mountain (circle) appears unrelieved by microquakes. Seismologists expected this to be the site of the no-show Parkfield earthquake.

hills. At 5:04 P.M., the Loma Prieta quake struck. Boulders careened downhill. Unnerved, Friend moved on. On his way out of the park, he noticed that the once-dry streams were gurgling with water.

Had the waterfall surge been a precursor of Loma Prieta? Johnston suggests the surge was caused by fault movement "squeezing water out of the ground." Evenly Roeloffs of USGS points out that Berry Creek Falls sits at the northwest tip of the Zayante Fault, which may be connected to the San Andreas. Prior to the quake, she says, the Zayante Fault may have shifted and "compressed the Santa Margarita Sandstone, raising fluid pressure and, consequently, spring discharge." A speculative question: Could routine monitoring of hydrological phenomena help seismologists foresee quakes?

Which brings us to a geyser at the resort town of Calistoga, north of San Francisco. For years, its eruption times have been recorded by an infrared sensor. Researchers Paul G. Silver and Natalie J. Valette-Silver of the Carnegie Institution in Washington used a computer to analyze more than 150,000 eruptions. They found variations in the intervals between eruptions that occurred one to three days before the three largest quakes within a 150-mile (250-km) radius of the geyser, including Loma Prieta. Johnston is skeptical, though, because the geyser is far from the epicenter.

Scientists around the world have also speculated for decades about possible links between quakes and electromagnetic phenomena. The most impressive recent claim came in late 1989, when Stanford University electrical engineering professor Anthony Fraser-Smith and his colleagues were operating a detector of ultra-low-frequency natural electromagnetic radiation near Corralitos, California. After the Loma Prieta quake, Fraser-Smith and his team realized that their instrument had recorded an unusually surge of electromagnetic activity beginning a month prior on September 12. The signal markedly increased on October 5, especially at lower frequencies. Finally, on October 17, three hours before the earthquake, the lowest frequencies of the signal increased to very high levels.

Did these signals portend the Loma Prieta quake? A number of quake-related subterranean phenomena might have generated the electromagnetic

© David Parker/Science Photo Library/Photo Researchers

© J. Lees & P. Malin, U.C. Santa Barbara/Science Photo Library/Photo Researchers

surge, Fraser-Smith claims. These include "piezomagnetic" or "piezoelectric" effects in which rocks under pressure generate magnetic or electrical fields.

Alternately, the signal might have had a nonquake cause—say, magnetic storms. However, there's no evidence of magnetic storms at the time. Another possibility is that the signal came from the engines of tractors. The apple harvest was underway at the time, USGS scientists William Ellsworth and Randall White have pointed out. But Fraser-Smith doubts this because he has made measurements in other years and detected no similar signals.

For nearly three decades, the Chinese have recorded possible electromagnetic precursors to quakes. They are now beginning to release their raw data. According to Jean J. Chu, who has joint appointments at MIT and Academia Sinica, in Beijing, one data record indicates that the Tangshan quake of 1976 was preceded by an electromagnetic fluctuation that wasn't appreciated at the time. If it was a quake precursor and if it had been recognized as such, then a 200,000 people might not have died. "If we could get access to the huge databases in China—the door has opened a crack—then there's a good potential for progress being made in understanding of quake processes," Chu says. "And the potential for better quake forecasting would be a natural outcome."

Electromagnetic phenomena, however, are not the only promising leads seismologists are pursuing. About half of California's quakes are preceded by tremors, Lindh says. He compares these foreshocks to the cracking sounds made by a stick just before it breaks. (The Los Angeles earthquake was preceded by several tremors near Santa Monica, but seismologists say they were probably unrelated.)

In June 1988 and August 1989, several magnitude 5 quakes hit the northwest end of the Lake Elsman segment of the San Andreas Fault. Afterwards, the State of California issued a public warning that there was a chance of a magnitude 6.5 quake in the next few days. It didn't happen. But two months later, several miles away, the Loma Prieta quake struck. Had the Lake Elsman quakes been foreshocks? Possibly. They were "a full unit of magnitude larger than other events within a 15-kilometer radius for at least 74

Rich Liechti, a USGS technician, checks a creepmeter in Parkfield. These and other instruments placed around the central California town were supposed to give seismologists early warning of the expected earthquake.

© David Parker/Science Photo Library/Photo Researchers

years," Malcolm Johnston's USGS report says.

The trouble with foreshocks, of course, is that they look like any other quake. One doesn't *know* they're foreshocks until *after* the main shock. In the future, perhaps, computerized, real-time monitoring networks might watch for minor quake swarms preceeding devastating temblors.

Having put aside their unrealistic expectations that they would be capable by now of issuing reliable warning

a few days in advance of major earthquakes, American seismologists are trying a different, less ambitious approach: probabilities forecasting. And it may have already shown its mettle.

In 1983 Lindh estimated that there was a very strong probability—47 to 83 percent—that a magnitude 6.5 quake would strike the southern Santa Cruz Mountains over the ensuing 30 years. He based his prediction on an estimate of the time it would take for the San Andreas Fault segment in this region to

© Bonnie Kamin

The Loma Prieta earthquake on October 17, 1989, caused many buildings in San Francisco's Marina District to collapse, including this four-story house.

reaccumulate the strain that had triggered the great San Francisco earthquake of 1906. Five years later, in 1988, a USGS committee concluded that the southern Santa Cruz Mountains were more likely to be hit by a strong quake than any other place in Northern California over the next 30 years (although not as likely as Lindh thought). Sure enough, the 7.1 Loma Prieta quake struck. Its epicenter, Mount Loma Prieta, was in the southern Santa Cruz Mountains.

In 1990, a number of developments prompted the USGS to reassess the probability of a large earthquake occurring again in the Bay Area. Research suggested, for example, that landslippage along the Hayward and San Andreas faults amounted to several millimeters per year greater than previously expected. Models of crustal deformation showed the the Loma Prieta quake on the San Andreas had increased stress on adjacent faults. And a study by USGS scientist David Schwarz, in which he dug trenches to find evidence of past quakes, revealed that the Rodgers Creek Fault could rupture some time in the next 30 years.

Based on these and other factors, the USGS estimated that Bay Area residents stand a 67 percent chance of suffering another major earthquake in the next three decades. The worst risk, the report says, is posed by the Hayward Fault, which runs through densely populated Berkeley and Oakland on the east side of San Francisco Bay.

Are such probability forecasts really meaningful given the scantiness of the historical record on which they are partly based? The question is all the more urgent following a recent attack on a key paradigm of quake science: the seismic gap theory.

In 1979, William McCann of the Lamont-Doherty Earth Observatory and his colleagues used the seismic gap theory to estimate the quake risk around the Pacific Rim. They designated as high-risk areas those that hadn't sustained major quakes in the last century; as intermediate risk, those that had quaked between 30 and 100 years ago; and as low risk, those that had quaked within the last 30 years.

A decade later, UCLA geophysicist David Jackson and his colleague Yan Kagan reassessed McCann's risk esti-

mates. Their finding was unexpected: The "low-risk" zones had had five times as many quakes as the "high-risk" ones. In effect, an area that had just quaked seemed much more likely to quake again in the near future, whereas one would expect such an area to "rest" a long time before accumulating enough stress to shake again.

A caustic debate followed and continues to this day. Stuart Nishenko of the National Earthquake Information Center in Golden, Colorado, rejects Jackson and Kagan's rebuttal of seismic gap theory. The two scientists, he says, assessed an early, less sophisticated version of the theory than is now in use. Nishenko also says that since 1979 seismologists have learned that some magnitude 7 quakes, unlike magnitude 8 temblors, don't always relieve a significant amount of strain on a fault. Hence a second major quake may follow a first one sooner than it "should." Agreement comes from Lindh, one of the creators of the Parkfield experiment, who says that comparing magnitude 7 and 8 quakes is like comparing grapes and watermelons. "As admirable as Dave is, and as fine a scientist, he hasn't come anywhere close to removing the seismic gap hypothesis.''

But doubts about the theory persist, partly because of the work of geophysicist Steven Ward and graduate student Saskia Goes at the University of California, Santa Cruz. They say their computer analysis of 237 temblors shows that quakes occur more irregularly than generally thought. When they first presented their idea in 1992, ''people complained that our model was too simple and might not reflect reality,'' Goes says. So with the computer analysis ''we decided to look at as many earthquake data as we could find. The data seem to support the model: If you really look at it objectively, there's no regular pattern.''

Their data encompass quakes on 59 fault segments in Alaska, California, Chile, Italy, Japan and Mexico. The scientists found that the largest quakes on these faults—with magnitudes of 8 or greater—occurred with the greatest regularity. But even these temblors showed considerably more irregularity than implied by seismic-gap theory. Interestingly, Ward and Goes found that the more quake records that were available for a fault, the more the time intervals between the quakes varied. Thus, past assumptions of quake regularity may result partly from a statistical illusion: ''If you have [a quake record of] less than 10 events, things tend to look more periodic than they really are,'' Ward says.''You haven't looked long enough to get the oddball earthquakes.''

What's the bottom line? Goes acknowledges that ''the seismic-gap model may be valid for the largest events.'' But the USGS's 30-year probabilities forecasts for smaller quakes like Loma Prieta may rely too much on the seismic gap theory, he says. Perhaps, then, no one should be surprised that the Parkfield forecast flopped: Even that famous geological ''clock''

Water flow over Berry Creek Falls in Big Basin Redwoods State Park increased five-fold an hour before rock beneath a nearby mountain ruptured, unleashing the Loma Prieta quake.

hardly clicks like one, as its rupture intervals have varied from 12 to 32 years.

All in all, the prospects for quake forecasting are muddy. Seismologists may never be able to anticipate quakes as confidently as meteorologists foretell thundershowers. For the time being, it's best to heed experts' advice: If you live in a quake-prone area, bolt your house to its foundation; store food and water; keep fresh batteries for a flashlight and portable radio; learn how to use a wrench to turn off a gas main; etc. For the indefinite future, only common sense and civic precautions—not futuristic science—will save us from Earth's inevitable rumblings.